PRIMARY WOOD PROCESSING

Primary Wood Processing
Principles and Practice

2nd edition

by

JOHN C.F. WALKER
University of Canterbury, Christchurch, New Zealand

Springer

A C.I.P. Catalogue record for this book is available from the Library of Congress.

ISBN-10 1-4020-4393-7 (e-book)
ISBN-13 978-90-481-7120-0
ISBN-13 978-1-4020-4393-2 (e-book)

Published by Springer,
P.O. Box 17, 3300 AA Dordrecht, The Netherlands.

www.springer.com

Printed on acid-free paper

Printed in the Netherlands.

CONTENTS

PREFACE

Samuel Johnson (of Boswell fame) said about writing, '*whatever is written without effort is in general read without pleasure*' and '*the unexamined word is not worth reading.*' However, even the most thoughtfully written text is unlikely to be described as a good read.

This book is not an exhaustive review. Rather it is an uneven selection of examples whose interplay across disciplines hopefully illuminates what drives the practice of forest production, wood processing and consumer preferences. The choice of material is arbitrary reflecting personal biases. It is a summary of material presented to students at the NZ School of Forestry at the University of Canterbury.

This book takes liberties with other peoples' ideas, weaving them into fresh themes that are at best a work in progress. The philosopher Francis Bacon argued that '*science is a debate in progress not a body of knowledge*' and also '*truth emerges more readily from error than confusion*'. We have taken the position that it is preferable to venture a reasoned opinion and be proved wrong than to waffle; so don't swallow every sentence – '*doubt everything*'. The reader's task is to unravel the conceits of generalizing and the risks of particularizing.

Understanding grows from repetitively revisiting perspectives, like ascending a spiral stair. Unfortunately, in a text most topics are visited only once so sequencing of various themes may appear dyslectic, whilst the limited selection of material will be equally eclectic. Further, rather than burden the reader with the weight of evidence we have preferred to dwell at length on particular issues.

The selection of illustrative material raises a practical problem. The reader looks for familiar examples. To a European Norway and Sitka spruce and Scots pine might be an appropriate selection, to a North American reference to Douglas fir and the southern pines would be mandatory. In the Southern Hemisphere *Pinus caribaea* and *P. radiata* are clearly important. The more successful tropical hardwoods hardly get a mention. However, by emphasizing the principles governing processing and wood use, the hope is that reader will learn more than that offered by a narrow national perspective. We in New Zealand peep over the rim of the world and make our livelihood wherever we can, so it may be that a book written from that perspective has a better chance to capture the broad picture.

The undue emphasis on radiata pine demonstrates ignorance of better illustrative material elsewhere. A distinctive feature of radiata pine research is that it has been closely 'managed' by CSIRO in Australia and FRI in New Zealand. This has imposed a clarity that is less apparent when reviewing the literature on other species. It does not imply superior knowledge, just consistency of argument.

A myth of the forest profession is the belief that it alone engages in strategic planning over a long time horizon. Critical reassessments have to be ongoing and interactive: time does not excuse procrastination. The seeds of disappointment can be seen in the simplistic vision of fast-growing plantations if there is not a paranoid, concurrent will to improve wood quality. The lazy focus on density has been a notable disservice of wood science to the sector. This is not a matter of apportioning blame, merely recognition that industry suffered for the common unexamined assumptions of the time.

If there were a subliminal message in this second edition, it would be the interconnectivity of the subject matter, linking the fundamentals of cell chemistry and ultrastructure, the physics of instability *etc.* to the headbox, the log scanner *etc.* Consequently the intellectual infusion from elsewhere – the cast-offs from broader disciplines – together with meticulous attention to detail – the subtlety of thought – sustains every design concept. The capacity to innovate is quite astonishing.

If there is an explicit message it would be the necessity of seeing one's local plantation timbers in context, to understand the enormous variability within and between species, and to appreciate the opportunities offered by diversity. Short rotation forestry favours hardwoods, but the case has to be argued step by step.

I acknowledge the generosity of friends in sharing the challenge in writing this second edition – and in the infusion of new ideas.

I must thank Professor Rob Douglas for bringing clarity and elegance to a number of hideous diagrams and Jeanette Allen for so patiently and endlessly reformatting the text.

Finally, I greatly appreciate Aracruz Celulose S.A. of Brazil for the cover picture of the mosaic landscape with uniform clonal eucalypt plantations on the plateau tops and the Atlantic rain forest occupying the slopes and valley bottoms.

Christchurch, NZ, John Walker
January 2006

CHAPTER 1

THE STRUCTURE OF WOOD: FORM AND FUNCTION

BRIAN BUTTERFIELD

School of Biological Sciences, University of Canterbury, Christchurch, New Zealand

1. INTRODUCTION

Wood is a complex biological material. It is found in the stems and roots in most larger land plants. To the materials scientist and engineer it has measurable properties, to the economist it is a renewable natural resource, and to the consumer it is the material from which we obtain our timber (lumber), cardboard, paper, and a host of other reconstituted products. But whatever uses we have found for wood they are all secondary to its primary biological functions: namely to conduct water up to the leaves of living plants and to provide a structural framework for holding these leaves up to the sun.

While wood forms in the stems and roots of both the cone bearing plants (the gymnosperms popularly known as softwoods) and many dicotyledonous flowering plants (the angiosperms also called hardwoods), monocotyledonous flowering plants such as the palms have solid stems and roots but the material in them is not true wood and has very different properties.

Commercial timbers fall into these two categories – softwoods and hardwoods. Softwoods are the timbers of the narrow, needle-leaved trees such as the pines (*Pinus* sp), spruces (*Picea* sp), and firs (*Abies* sp). Most are tall evergreen trees that retain their foliage for most of the year. A few like European larch (*Larix decidua*) and swamp cypress (*Taxodium disticum*) are deciduous and lose their leaves completely in autumn and remain leafless throughout the winter. Hardwoods, on the other hand, are the timbers of the broadleaved trees such as oak (*Quercus* sp), ash (*Fraxinus* sp), and elm (*Ulmus* sp). Hardwood trees of the temperate regions are usually deciduous but some temperate hardwood species, such as the majority of the southern beeches (*Nothofagus* sp) and most tropical hardwood trees, retain their green foliage for much of the year and are said to be evergreen.

Softwoods and hardwoods have quite different cellular structures when viewed with a hand-lens or microscope. Softwoods (Figure 1.1) have a comparatively simple structure and are more uniform in appearance than hardwoods (Figure 1.2). They are made up of fewer cell types with the long, pointed fibrous cells termed tracheids providing both the structural support and the conducting pathways in the wood (Figure 1.3). Hardwoods, on the other hand, comprise several different cell types with highly specialized water conducting cells termed vessels (Figure 1.4) usually clearly visible with a hand lens in a cross-section of the wood. The function of structural support is carried out by another specialized cell, the fibre. These cellular differences between softwoods and hardwoods have a profound significance on the potential utilization of timber. Hardwoods, for example, have fewer fibre-like cells than softwoods and these are generally shorter in length. Hardwoods, therefore, are less suited for the production of strong papers but are well suited for smooth high quality writing paper. On the other hand, the range of cell types and the diversity of patterns of these cells in the wood, mean that hardwoods often have a more pleasing appearance and grain and are therefore more highly sought after for furniture or finishing timbers.

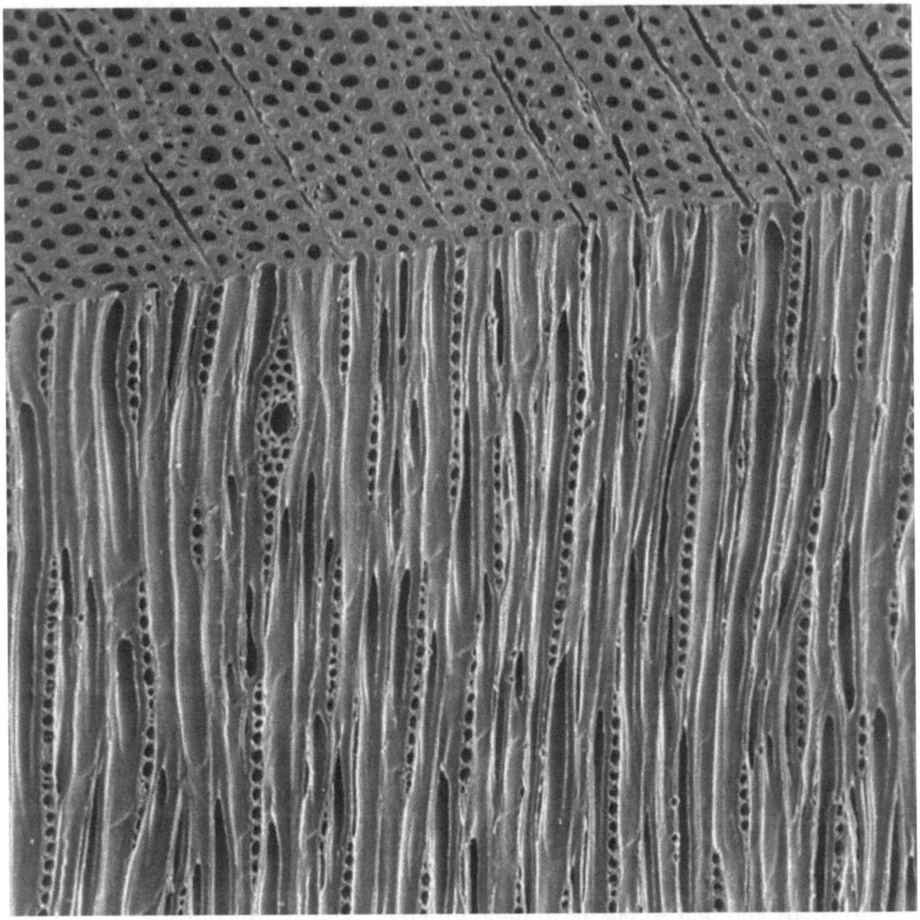

Figure 1.1. The transverse and tangential-longitudinal faces of the softwood European larch, *Larix decidua.* The wood comprises longitudinal tracheids forming the axial system of cells, and radial parenchyma mostly in uniseriate rays. Axial and ray canals are also present. Magnification x 125.

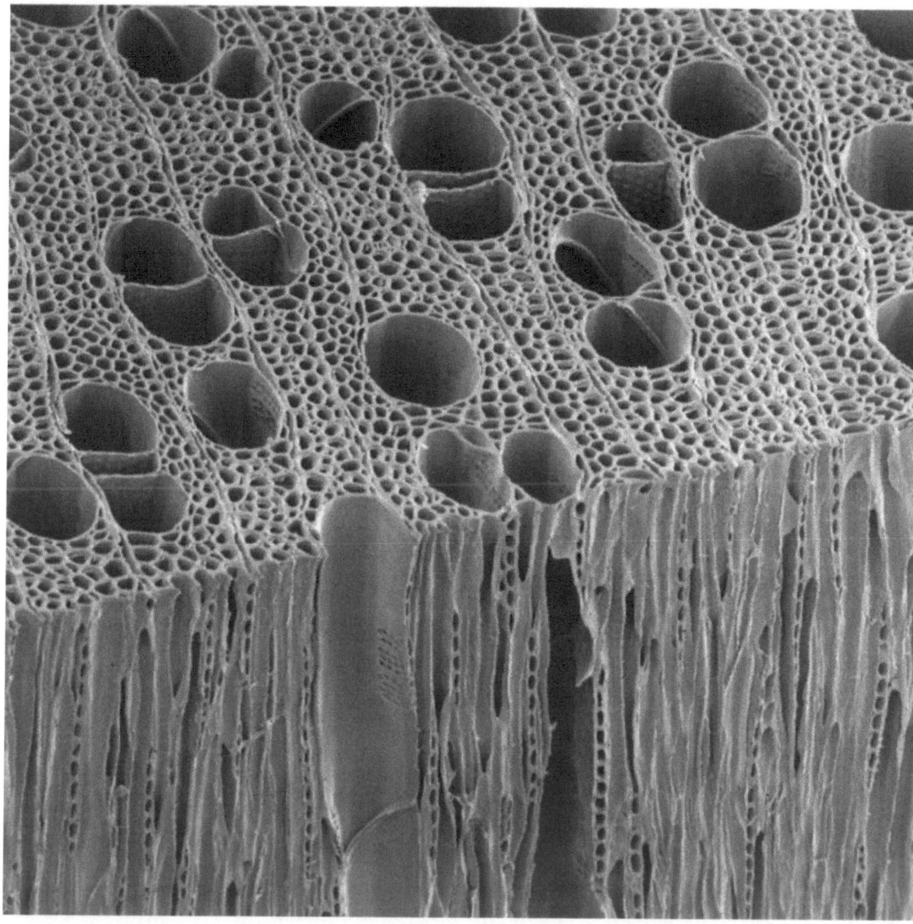

Figure 1.2. The transverse and tangential-longitudinal faces of the hardwood poplar, *Populus* sp. The wood comprises vessels arranged in radial groups and fibres both in the axial system of cells, as well as uniseriate parenchyma in the ray system. Magnification x 210.

Wood is produced by a thin zone of dividing cells near the outside of the trunk or branch just beneath the bark. Known as the vascular cambium, its cells are thin-walled and meristematic. The cambium is essential for the continued growth of the tree. As the crown of the tree gets larger with more leaves and branches, the stem must increase in diameter to support this extra load. More wood is added by the cambium to the trunk and the stem thickens. In temperate climates the cyclic production of new wood cells each spring and summer and the subsequent cessation of cambial divisions each autumn and winter leave the familiar pattern in the wood that we know as annual growth rings: however, drought may arrest growth to restart later in the season creating a 'false' ring. In variable tropical climates trees may produce new growth rings with each rainy period. In such cases the rings are

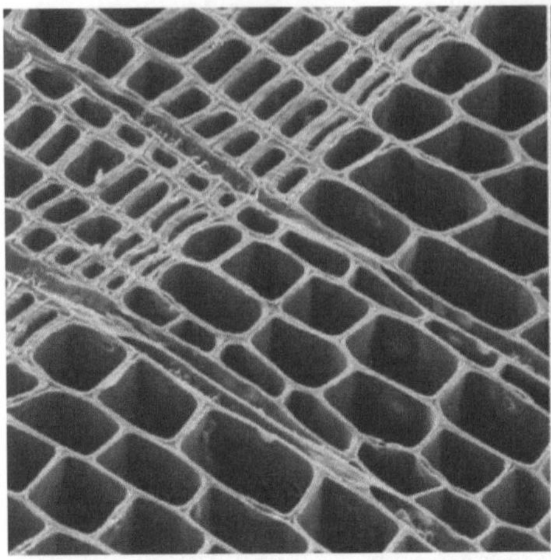

Figure 1.3. The softwood *Thuja plicata.* Tracheids function in both sap conduction and stem support in softwoods. A growth ring is clearly visible with the thicker-walled latewood tracheids to the top left, and the thinner-walled and larger lumened early-wood tracheids to the lower right. x 275.

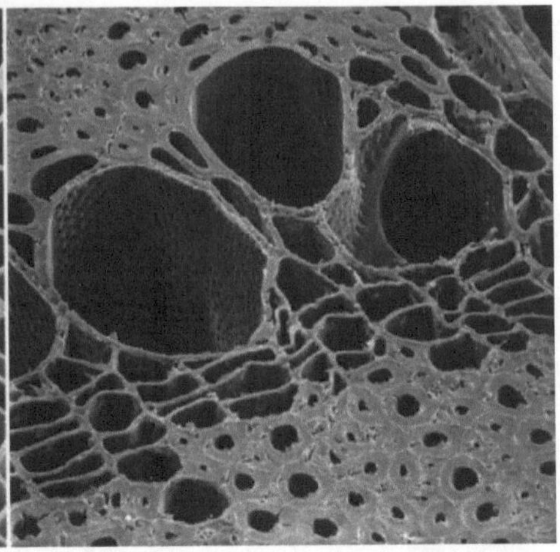

Figure 1.4. The hardwood *Knightia excelsa.* Vessels provide largely uninterrupted conduits for the conduction of sap, with thick-walled fibres functioning in support: earlywood with large vessels in the centre; latewood fibres to the lower right. x 235.

Figure 1.5. Indistinct growth ring boundaries in the wood of *Dacrydium cupressinum.* x 45.

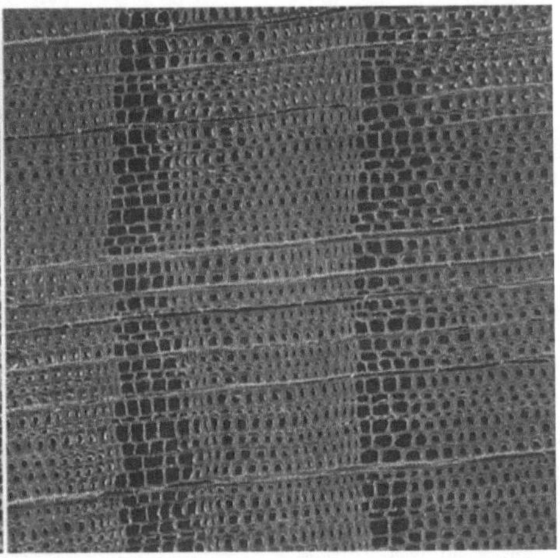

Figure 1.6. Very distinct growth ring boundaries in the wood of Douglas fir, *Pseudotsuga menziesii.* x 60.

obviously not annual. Growth rings are very distinct in some woods but are not well defined in others (Figure 1.5). This can be due to the differences in the cell types produced in the wood in the case of complex hardwoods, or it may be due to the severity of the winter season where the tree grows. Trees growing in milder climates with warm winters do not produce growth rings that are as sharp and distinct as trees growing in climates where the winters are more severe and extreme. In softwoods, growth rings show up as a result of differences in density across the year's growth. Early in the growing season the vascular cambium tends to produce tracheids that have large central cavities or lumens and thin walls (Figure 1.6). These cells function more for the conduction of water than in support of the trunk and the wood is called earlywood or springwood. Towards the end of the growing season the cambium produces cells that have smaller lumens and much thicker walls in keeping with the transition in function from conduction to primarily that of support. These cells form the latewood or summerwood. The tree thus concludes its year's growth with a cylinder of strong thick-walled cells suitable to help the tree overwinter. The transition from earlywood during the growing season can be quite gradual as in spruce or radiata pine, or it can be quite abrupt as in Douglas fir.

The wood at the centre of a trunk or stem is often harder and darker in colour than the wood nearer the bark. This darker central region is known as heartwood and its cells are dead and physiologically inactive. The outside of the trunk is known as sapwood and is active in water transport and other physiological activities. Sapwood is usually paler in colour than the heartwood though the heartwood in some species such as ash, fir, poplar and spruce is also quite pale. The cells of the heartwood are darker in colour on account of the enrichment of the cells by various extraneous chemicals known as extractives. These chemicals permeate both the cell wall and the cell lumen. Small quantities of the precursors of extractives may be found in the living cells near the sapwood-heartwood boundary. Extractives play an important role in slowing down the natural decay of wood by fungi. They also provide a measure of natural protection against the larvae of boring insects.

Heartwood is formed after several years of growth of the trunk or branch. It forms a cone within the trunk that spreads slowly outwards and upwards as the stem expands with age. Heartwood formation in *Pinus radiata* and the southern pines of North America begins when the tree is about 15 years old: young trees have little or no heartwood whereas older trees may have a substantive percentage of their trunk in heartwood. The precise cause of heartwood formation is not known but it is characterised by the accumulation of polyphenolic substances in the cells and a general reduction in the moisture content of the wood. Heartwood is of considerable interest because its colour renders it pleasing for furniture, panelling or craftware. In addition to its colour it may be more aromatic on account of the extractives. The durability of heartwood and its resistance to decay are quite variable. Heartwood is generally more difficult to penetrate with preservatives than sapwood and also more difficult to dry.

2. THE MICROSCOPIC STRUCTURE OF SOFTWOODS

Softwoods are built up primarily of axially-elongated cells termed tracheids. Tracheids have no living cell contents at functional maturity and comprise thick-walled conduits with their tips densely interlaced (Figure 1.1). Their length varies both with species and also with position in the stem. Tracheids tend to be longer at lower levels in the tree than higher up and also to be longer nearer the bark than at the stem centre. They also tend to be longer in the latewood than in the earlywood of each growth ring. In mature wood, tracheids range from between 2 to 5 mm in length but longer tracheids can occur in some species. With diameters of about 15-60 μm, tracheids are about 100 times longer than they are wide. Because they are cut off from dividing cambial cells, they tend to remain in radial files in the wood (Figure 1.3). As a result their tangential dimensions remain fairly uniform. The radial width of the tracheids is largest in the earlywood and smallest in the latewood where the cells appear radially flattened. The tracheids overlap one another along their thinner, wedge-shaped ends which appear sharply tipped in tangential view but are more rounded in outline in radial view. This cellular arrangement, which is a direct result of the pattern of the fusiform cambial cells in coniferous species, helps give softwoods a high 'along the grain' strength as well as allowing for maximum sidewall cell contact for the movement of water up the stem or branch.

Sap flow in softwood tracheids occurs from cell to cell through intertracheary bordered pits. Water is pulled up the trunk and branches of living trees by the transpiration pull created by the utilization and evaporation of water from the leaves. The transpiration pull lifts water to the crown of even the tallest trees, some over 100 metres above the ground. Under normal circumstances such negative pressures would cause the water column to break and a water vapour embolism to occur. The fact that such breakages are comparatively rare is due in part to the very small diameter of the conducting tracheids and to strong hydrogen bonding to the wall. The intertracheary pits connecting tracheids are specialized openings in the radial sidewalls of each tracheid (Figure 1.7) with specialized closing valves or pit membranes (Figures 1.8 and 1.9) which seal the cells that have been damaged or embolised by cavitation. Intertacheary pit membranes are comprised of the original pair of primary walls and intervening middle lamella. Subsequent digestion of the non-cellulosic components of the outer or margo region occurs permitting water to flow from cell to cell. The central zone that remains intact forms the torus. Where the water column is broken, the adjacent intertracheary bordered pit membranes are pulled towards their pit borders and the tori seal the pit apertures (the openings visible in Figure 1.7) so isolating a small number of wood cells that are affected by the air/vapour embolism. This irreversible process is referred to as pit aspiration (Figure 1.10). It occurs on drying of lumber.

Helical thickenings overlying the secondary wall are a regular feature of the tracheids of *Psuedotsuga menziesii, Taxus baccata, Torreya* and the woods of a few other species. In *Psuedotsuga* they are most prominent in the earlywood tracheids (Figure 1.11) and are sometimes absent in the latewood. Helical thickenings sometimes occur in the latewood tracheids of *Larix* and *Picea* species.

Some softwoods also have axially-elongated cells termed axial parenchyma cells present (Figure 1.12), sometimes referred to as longitudinal or wood parenchyma. These cells differ from tracheids in having thinner walls and a protoplast that may live for several years. The cell protoplasts die when the surrounding wood cells undergo the transition from sapwood to heartwood. Axial parenchyma cells often

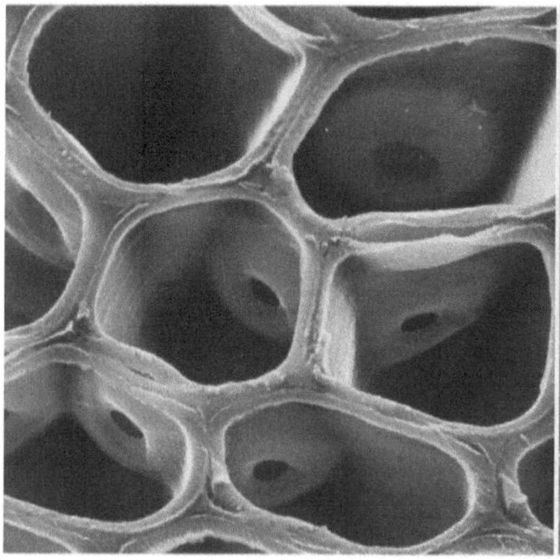

Figure 1.7. Intertracheid coniferous bordered pits in *Pinus nigra*. x 825.

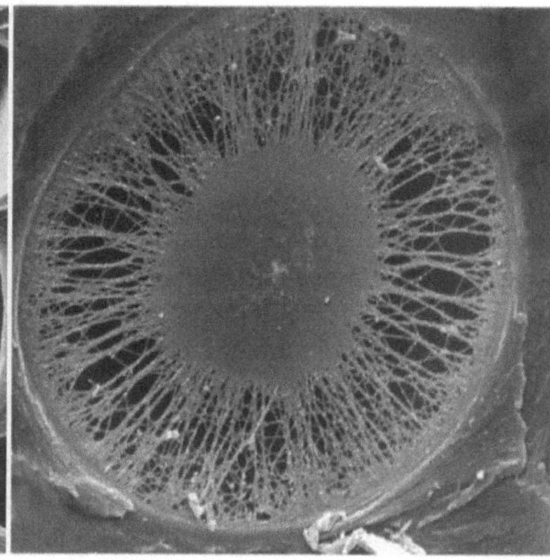

Figure 1.8. Pit membrane of an intertracheid bordered pit in *Phyllocladus glaucus* with the central torus and the microfibrils of the margo through which sap passes. x 3000.

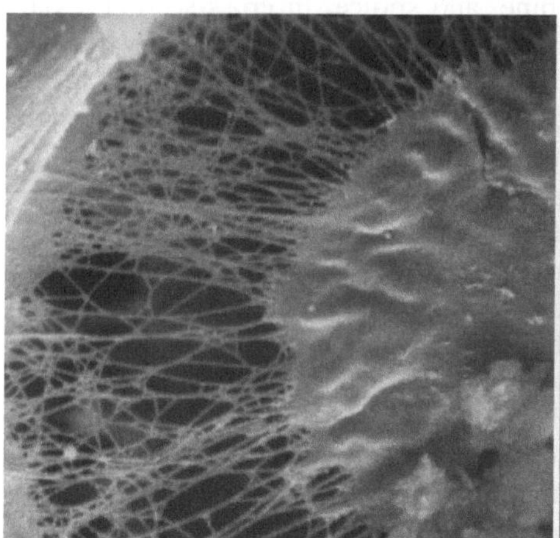

Figure 1.9. Detail of the margo and torus of a pit membrane in *Pinus radiata* showing the open texture of the microfibrils in the margo. x 12 500.

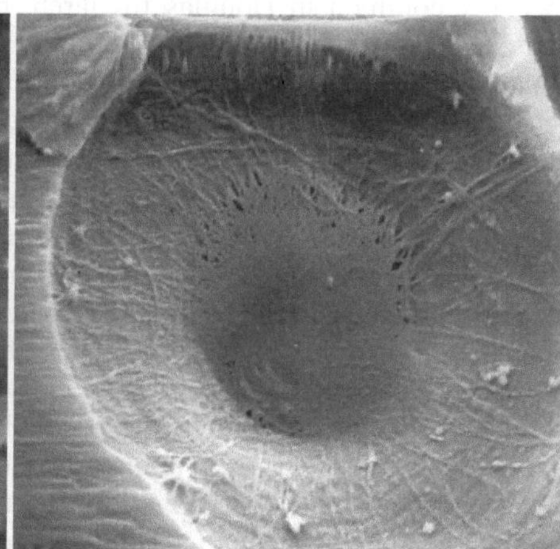

Figure 1.10. A surface view of an aspirated pit membrane in *Agathis australis*. The pit membrane has aspirated back against the pit border of the underlying tracheid. x 450.

contain starch grains, resins and other extraneous materials. Axial parenchyma are never common in softwoods though they can be found in most growth rings of *Chamaecyparis, Cupressus, Sequoia, Taxodium* and *Thuja*. They are sometimes present in *Tsuga, Larix, Psuedotsuga* and *Abies*. Individual parenchyma cells are axially-elongated, commonly with transverse endwalls, and are often arranged in axial files or strands. The cell walls may be thickened but are never as thick as the tracheid walls.

Cells that look superficially like axial parenchyma but in fact have bordered pits and assist with conduction are termed strand tracheids. Strand tracheids arise by the subdivision of axially-elongated cells that might otherwise have developed into normal undivided tracheids. Strand tracheids have no living contents at functional maturity. Some evolutionists believe that they represent an intermediate stage between the tracheid and the parenchyma cell. In some woods, such as larch and Douglas fir, they replace the parenchyma in the latewood.

Many softwoods contain a dark coloured, gummy substance termed resin. Resin is a mixture of complex organic substances. It can occur both within individual cells and, perhaps more commonly, in specialized canals between the tracheids (Figure 1.13). Resin canals or ducts are found in most softwoods. All the main North American commercial softwoods have them except the cedars and hemlock. Resin canals lie in both radial (Figure 1.14) and axial directions and interconnect to form complex networks. They develop during cellular differentiation by the breakdown of columns of thin-walled parenchyma cells in response to the hydrostatic pressure of the resin bleeding from neighbouring epithelial parenchyma cells. These epithelial cells surround each resin canal.

Resin canals are the source of the resin exudation in freshly felled trees, and are especially common in Douglas fir, larch, pine, and spruce. In cross-section, axial resin canals are more common in the outer portion of each growth ring. Over time, exudation pressures within the sapwood drive the resin into the heartwood further enriching the latter with organic substances. In the pines, axial resin canals are abundant and particularly large (100-200 μm in diameter). They are surrounded by thin-walled, unlignified epithelial cells that are prone to collapse. In Douglas fir, larch, and spruce, the epithelial cells become thick-walled, pitted, and lignified.

Traumatic resin canals can be caused by wind, frost, insect or fungal attack and mechanical damage to the cambium caused by thinning and pruning operations or by animals, e.g. possums or baboons. These traumatic resin canals tend to develop in the weaker earlywood and to form an arc within the growth ring. They can develop irrespective of whether or not the particular softwood forms normal resin canals. Internal shake or pitch pockets also develop in some softwoods. In *Pinus radiata* these may be as large as 40 mm wide, 100 mm long, and 5 mm thick. They are most prevalent in small to medium sized trees growing on windy sites where the cambium is subject to considerable flexure. In larger trees, resin pockets found some distance back in the wood behind the cambium probably formed when the tree was younger.

In addition to the axially-elongated tracheids, parenchyma, strand tracheids and resin canals, softwoods also have a system of cells radiating outward from the pith. These formations of cells, known as rays, contain radially orientated parenchyma cells (Figures 1.1 and 1.15), but in some woods, may also contain ray tracheids (Figure 1.16) and resin canals (Figure 1.14). The rays of most softwoods are only one cell wide, termed uniseriate. Part-biseriate rays occur in some woods but true biseriate rays (two cells wide) are uncommon. The shape and size of rays vary significantly between species and are often used as a diagnostic feature in the

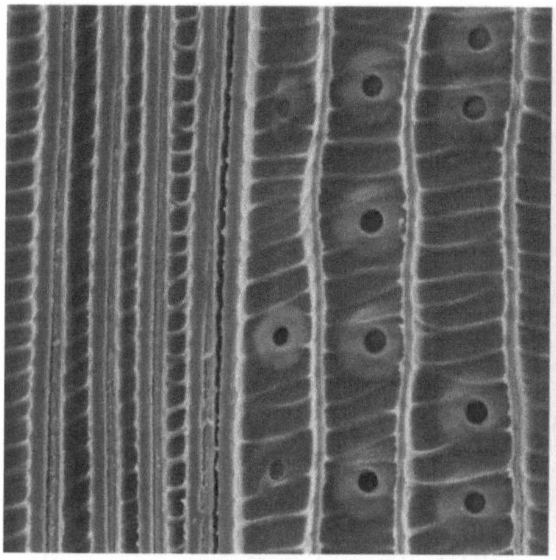

Figure 1.11. Helical thickenings overlying the S_3 layer in *Pseudotsuga menziesii*. A growth ring boundary divides this photo with latewood tracheids to the left. x 400.

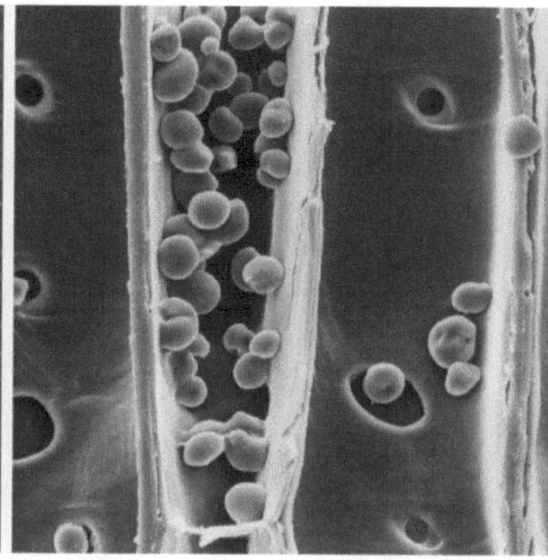

Figure 1.12. A strand of axial parenchyma cells containing starch grains in the wood of *Dacrydium cupressinum*. x 630.

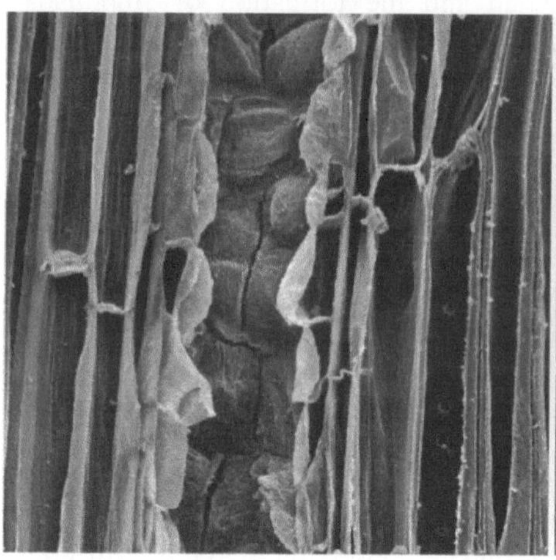

Figure. 1.13. An axial resin canal in *Pinus radiata* with associated epithelial parenchyma cells on either side. x 145

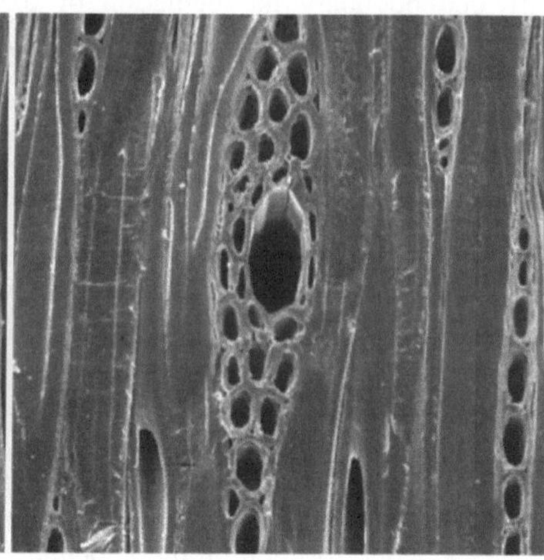

Figure 1.14. A radial resin canal in *Larix decidua*. x 200.

microscopic identification of wood. Rays vary from 2 to 40 or more cells in height but rays up to 15 cells high are found in most softwoods.

Radially-elongated parenchyma cells make up most rays. These cells resemble the axial parenchyma cells being rectangular in shape with more or less transverse endwalls. They have moderately thickened walls perforated by small simple pits.

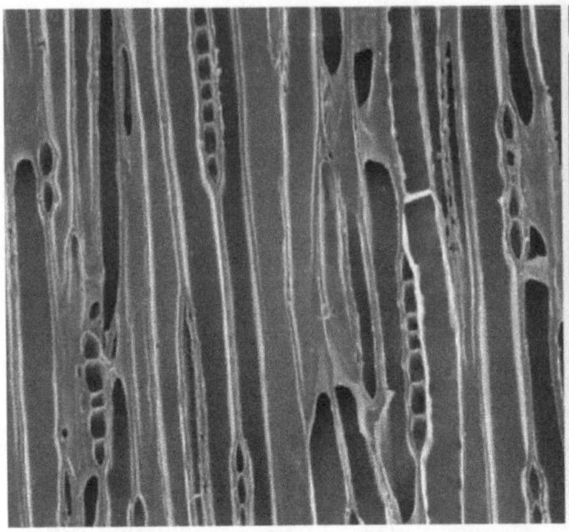

Figure 1.15. Uniseriate rays viewed from a tangential face in the wood of *Cedrus atlantica.* x 250.

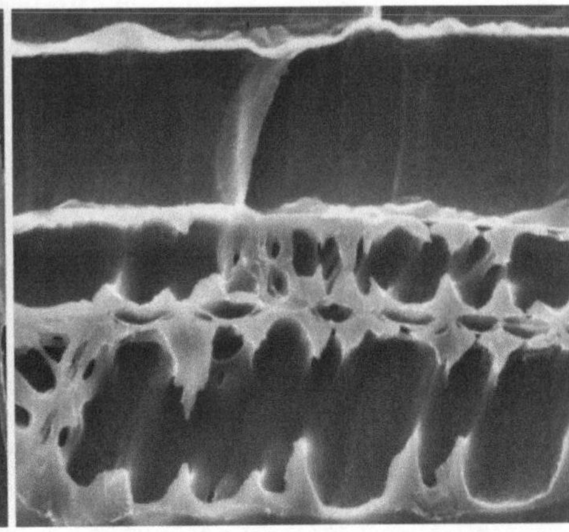

Figure 1.16. Ray axial parenchya cells (upper) and ray tracheids (lower) in *Pinus radiata.* x 550.

The cells are physiologically alive in the sapwood. However, near the sapwood-heartwood boundary the depletion of oxygen and the formation of embolisms accompany the hydrolysis of starch to sugars that in turn breakdown, oxidize and polymerize to yield polyphenolics. Extractives are formed and deposited. This is the onset of heartwood formation.

Ray tracheids are present in a number of softwoods including *Pinus* (Figure 1.16), *Picea, Larix, Psuedotsuga* and *Tsuga.* Sometimes the ray tracheids occur anywhere within the ray as in *Pinus* while in other genera they are usually confined to the upper and lower rows of cells. Ray tracheids can be distinguished from ray parenchyma cells by the complex bordered pits on their side and endwalls. Sometimes they have pronounced wall sculpturings termed dentate thickenings. Resin canals with accompanying epithelial cells occur in the rays of some woods. Where this happens the rays appear enlarged and multiseriate near their centres.

In softwoods leaning stems and branches are usually characterised by an eccentricity in cross-section, with the wood to the lower side having significantly wider growth rings than normal and consequently the pith is off-centre (Figure 1.17). The wood forming on the underside, termed compression wood, has different anatomical, chemical and physical characteristics. It develops in response to stress – gravity being the most common cause of this stress. Severe compression wood is dull reddish in colour. Compared to normal wood that is found in vertical stems and to the upper side of inclined stems, the individual tracheids are more rounded in cross-sectional outline, they have thicker walls, and intercellular spaces are common especially in the earlywood. Compression wood tracheids generally lack an S_3 wall layer, the microfibril winding angle in the S_2 wall layer is greater than in normal wood cells and the S_2 layer is commonly split by a series of close helical splits or checks (Figures 1.19 and 1.20). Compression wood has a lower cellulose and a higher lignin content than normal wood.

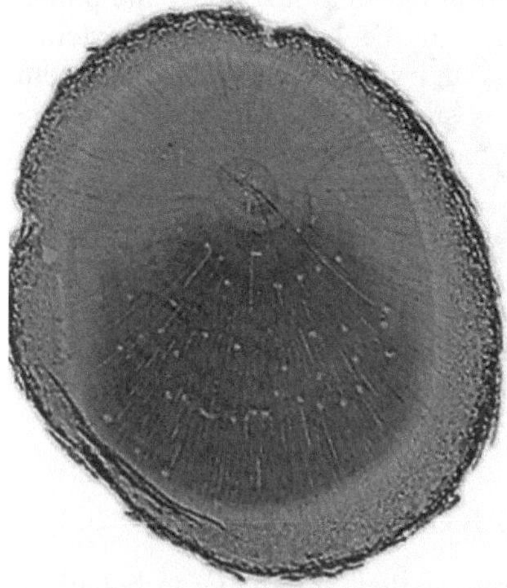

Figure 1.17. Cross-section of an inclined seedling stem of *Pinus radiata* showing the eccentric growth and modified wood to the lower side. x 8.

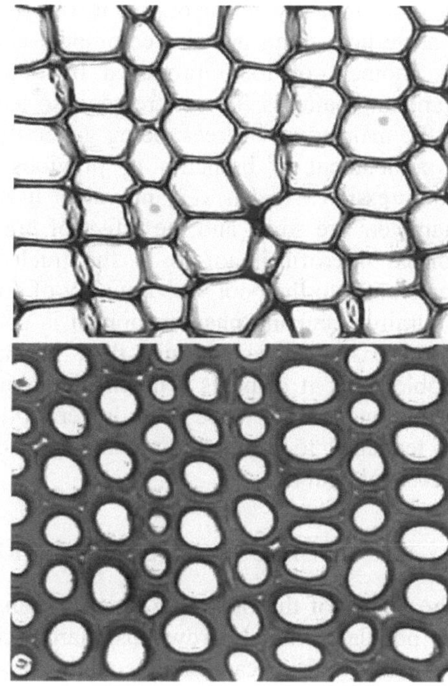

Figure 1.18. Same stem: normal tracheids to the upper side (upper micrograph) and compression wood tracheids to the lower side (lower micrograph). x 250.

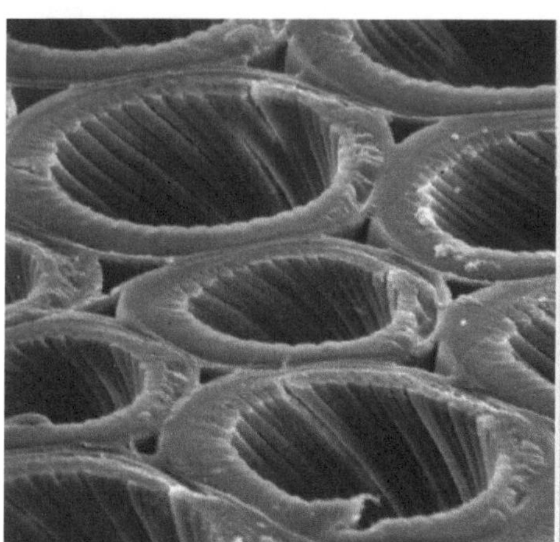

Figure 1.19. Compression wood tracheids in *Pinus radiata* showing helical splits in the S_2 wall layer and intercellular spaces. x 1100.

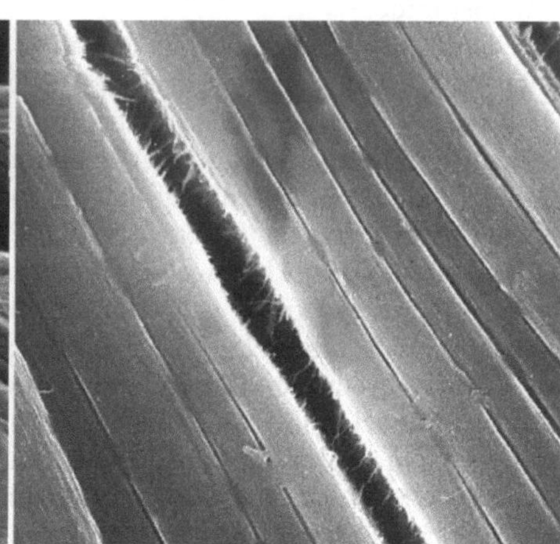

Figure 1.20. The S_2 wall layer of a compression wood tracheid in *Pinus radiata*. Note the large microfibril angle. x 3750.

Unfortunately few working in the timber industry appreciate that vertical and near vertical stems are also commonly asymmetrically stressed. This can be due to the moment of force produced by adjacent lateral branches and asymmetrical branching patterns. These stresses also lead to the production of compression wood in the main trunk of trees, being generally present in the stem just under the points of attachment of branches or previous branches in the case of pruned stems. Compression wood is very prevalent in seedlings and young trees where the stem diameters are small and the effect of branching is significant. This leads to above normal microfibril angles in the tracheids of the vertical stem, a feature that contributes to the poor wood quality of corewood. Although compression wood is a naturally occurring phenomenon, it is considered a defect in timber. Compression wood has a higher density than normal wood yet its presence in lumber is problematic: it displays a much higher longitudinal shrinkage than normal wood; often compression wood has a lower modulus of elasticity and impact strength; and its tendency to brittle failure under tensile load renders it very undesirable as a structural material.

3. THE MICROSCOPIC STRUCTURE OF HARDWOODS

The woods of the flowering plants, or angiosperms as they are known to botanists, are popularly called hardwoods. Hardwoods have a more complex overall structure than softwoods, because they contain more cell types arranged in a greater variety of patterns (Figure 1.2). The proportion, structure and distribution of the cell types combine to give the woods a more varied appearance and grain. As a consequence, they are in greater demand for their decorative appeal. Some hardwoods, such as willows and poplars, are quite bland in appearance and are of only utilitarian use to the timber trade.

Whereas in softwoods the functions of conduction and support are both carried out by the tracheids, in most hardwoods these functions are performed by different cell types. Vessels comprise many individual cells or vessel elements joined end-to-end to form long conducting conduits (Figure 1.2). These may extend the full height of the tree in the case of the ring porous woods, or more commonly extend for only short distances (often less than 200 mm) in most diffuse porous species. The structural load in hardwoods is carried by the fibres (Figures 1.27 and 1.28). These cells differ from softwood tracheids in a number of ways; they are comparatively short (0.25-1.5 mm long and generally less than 1.0 mm), more rounded in transverse outline and play virtually no role in the ascent of sap.

Vessels develop in characteristic patterns within each growth ring of a particular species. When they are grouped predominantly in the earlywood (Figure 1.22) the pattern is described as ring porous. This arrangement is common in many deciduous species such as elm (*Ulmus* sp) and oak (*Quercus* sp). Most evergreen trees have their vessels distributed throughout the growth ring and are described as diffuse porous (Figure 1.21). Within these two broad categories, vessels may be arranged in a solitary pattern or be arranged in multiples in various formations.

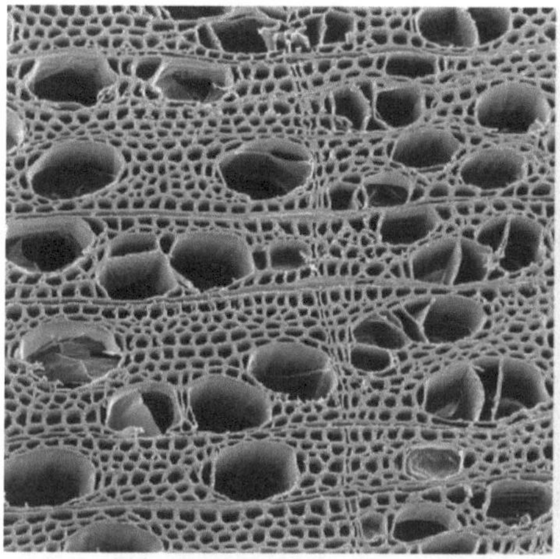

Figure 1.21. Vessels throughout the growth ring in the diffuse porous wood of *Populus robusta*. x 60.

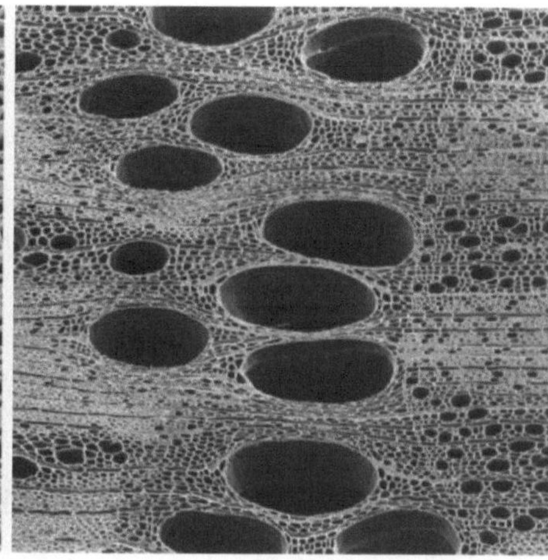

Figure 1.22. Large earlywood vessels in the ring porous wood of *Quercus robur*. x 40.

Vessels are difficult to visualize in their entirety due to their length and irregular pathways in wood. The complexity of these pathways has been investigated using video and cinematographic recordings of numerous individual transverse sections of wood so providing the clearest three-dimensional impression of vessel pathways.

Vessel length is also a parameter that has been intensively studied on account of its significance in timber preservation. The most common methods of measuring vessel length involve forcing a medium along a segment of stem which is progressively shortened until the vessels are cut at both ends. However, such techniques determine only the length of the longest vessel in a particular wood and give no indication of the lengths of shorter vessels nor what percentage of the total vessel population is represented by the longest vessels. Recent techniques involve infusing the wood with a dilute particle suspension, e.g. dilute latex paint, slicing the stem at 20 mm intervals and counting all the infused vessels. A simple calculation then enables the percentage of vessels in each length class to be determined. The most surprising feature of such analyses is the fact that most vessels in woods are very short – frequently 60-80% are less than 200 mm long. This has led to speculation that woods with many short vessels are in fact safer than woods with a few very long vessels. When cavitation occurs in a long vessel, a significant percentage of the wood's conducting potential has been lost, whereas the loss of a single short vessel, especially in woods with small-lumened vessels, results in little water conducting loss to the tree.

Sap ascending through vessels in a hardwood passes from one vessel element to another through pores or perforations in the endwalls of each cell. These pores are usually aggregated into what are termed perforation plates. Perforation plates develop during vessel element differentiation behind the cambium in hardwoods. By definition, a simple perforation plate (Figure 1.23) is a large single opening between two contiguous vessel elements, while a multiple perforation plate comprises a number of openings variously arranged. Primitive woods tend to have a series of more or less circular openings aggregated into a reticulate perforation plate, while slightly more advanced woods have a series of roughly parallel slit-shaped openings forming a scalariform perforation plate (Figure 1.24). The openings in a perforation

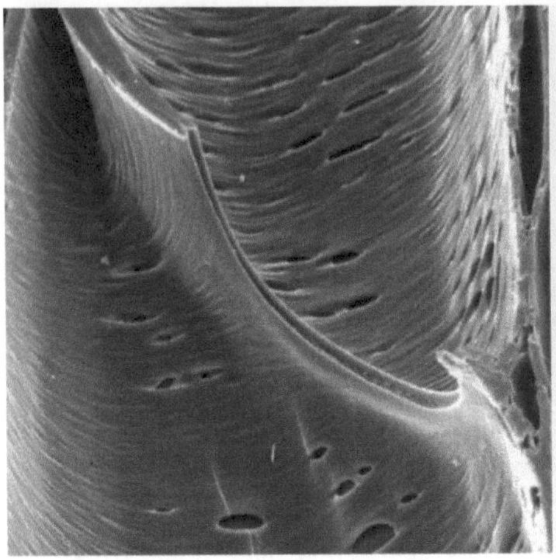

Figure 1.23. A simple perforation plate between vessel elements in *Beilschmiedia tawa*. x 450.

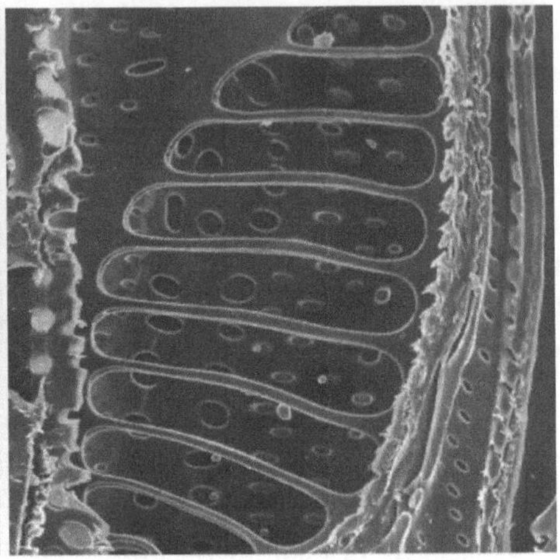

Figure 1.24. A scalariform perforation plate between two vessel elements in *Griselinia lucida*. x 240.

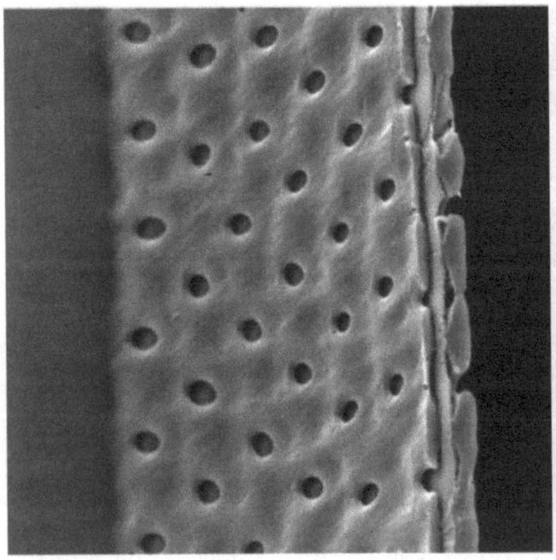

Figure 1.25. Inter-vessel bordered pits between two adjacent vessel elements in *Populus* sp. x 500.

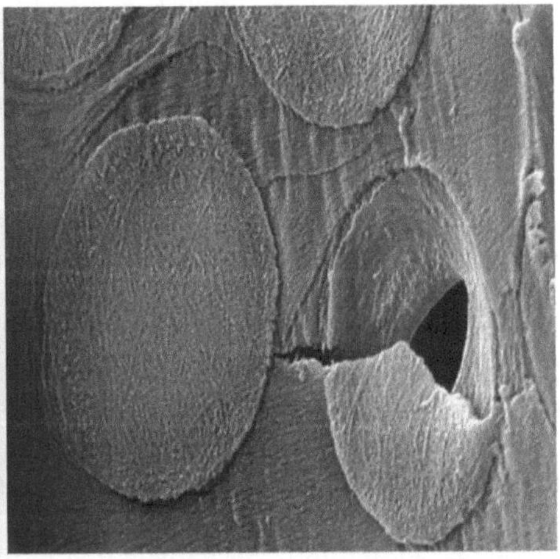

Figure 1.26. Inter-vessel pit membranes exposed by a split wall in *Populus* sp. x 2500.

plate are formed by the hydrolysis of the non-cellulosic components of the endwall of each differentiating vessel element and the subsequent loss of the remaining cellulosic web following the onset of the transpiration stream. The evolution of the vessel from the imperforate tracheid (a step that almost certainly occurred a number of times) required the development of perforation plates. The only difference between a pit and a perforation is the loss of the separating pit membrane in the per-foration. The aggregation of such pores into the perforation plate resulted in the formation of the vessel element. This evolution has been accompanied by a shortening of the mean vessel element length and an increase in width.

Sideways movement of sap between adjacent vessels occurs though the intervessel pits (Figure 1.25). These resemble softwood intertracheid bordered pits, but lack the separation of the pit membrane into a torus and margo. The small size of the pores in these membranes permits lateral water flow between adjacent vessels but probably prohibits air or vapour embolisms spreading to the adjacent, sap-conducting vessels.

Fibres make up a high proportion of the volume of most hardwoods. Fibres are imperforate, axially-elongated cells, with small lumens and ends that taper into pointed tips. The density of a hardwood is largely determined by the proportion of fibres to other cell types present in the wood. In a low density wood, the vessels occupy a major proportion of the wood volume, whereas denser woods have a larger proportion of thick-walled fibres. The secondary walls of fibres are usually sparsely pitted and the cells lack cell contents at functional maturity.

Fibres are usually classified into fibre tracheids, libriform fibres, and septate fibres. Libriform fibres (Figures 1.27 and 1.28) are longer then fibres tracheids and have moderate to very thick walls and simple pits. Their function is one of support. The shorter fibre tracheids have moderately thick walls and bordered pits. They function in both conduction and support although their occurrence in vesselled woods suggests that their function is primarily one of support. It is likely that they represent an intermediate evolutionary form between the softwood tracheid and the true libriform fibre. The fibres in some woods have their fibre lumens divided into chambers by septa. Such fibres are known as septate fibres. The septa only cross the fibre lumen and do not connect to the primary wall. They are produced by a late sequence of division in the fibre prior to death of the cytoplasm. Septate fibres resemble axial parenchyma in some woods and are most abundant in woods where the latter are poorly represented. This has led to the general belief that septate fibres have evolved as an alternative site for the storage of starches, oils and resins.

Vasicentric tracheids are found close to the vessels in some hardwoods, particularly in the earlywood of ring porous species. They are short tracheid-like cells with profuse sidewall pitting. They are often longitudinally bent and flattened transversely on account of the lateral expansion of the adjacent vessels.

Axial parenchyma cells (also called longitudinal parenchyma) are generally very abundant in hardwoods. Like vessel elements and fibres, axial parenchyma cells are derived from the axially-elongated fusiform initials of the vascular cambium but, whereas vessel elements and fibres (except septate fibres) remain unsegmented, axial parenchyma cells are formed by the transverse segmentation of the derivatives of fusiform initials. Axial parenchyma cells, therefore, tend to lie in vertical files with most cells having abrupt transverse endwalls. Axial parenchyma cells have relatively thin walls interrupted by small circular simple pits arranged irregularly. Instances of helical thickening have been recorded in a few woods. The protoplasts tend to be long-lived and to function in the development of tyloses in the heartwood and at sites of injury. Axial parenchyma cells may contain starch grains, crystals and other cell inclusions.

Hardwoods show a great diversity in the range and abundance of axial parenchyma. Three broad categories are defined: paratracheal axial parenchyma are always associated with the vessels; apotracheal axial parenchyma which have no relationship to the vessels but occur randomly amongst all the wood cells; and boundary axial parenchyma that always lie at the beginning or end of a growth increment. Within these three broad categories, wood anatomists identify a range of distribution types. These patterns are valuable in assisting with the microscopic identification of woods. Many woods have more than one type of parenchyma distribution present. Woods with large amounts of parenchyma will be light and of low hardness, although this may be offset by bands of thick-walled fibres.

The rays of hardwoods are generally larger and more variable than those found in softwoods. Rays are classified as uniseriate (Figure 1.29) or multiseriate (Figure 1.30) depending on whether they are one or more cells wide. Whereas the rays of softwoods are predominantly uniseriate, part-biseriate or rarely fully biseriate, hardwoods commonly have broad multiseriate rays (Figure 1.30). A few hardwoods such as *Salix* possess only uniseriate rays (Figure 1.29) and some genera have rays of two distinct sizes – small uniseriates and large, broad multiseriates. Many hardwoods have a continuum of ray sizes from small uniseriates through to large multiserisates.

The classification of various ray types present in hardwoods is complex. Rays are broadly defined as homogeneous or homocellular if they have only procumbent parenchyma cells, or heterogenous or heterocellular if they have axially-elongated parenchyma cells associated with their margins.

Ray parenchyma cells retain a living cytoplasm in the sapwood but lose their cell contents in the heartwood where they are often the storage sites of various extractives and crystals. The exact function of rays has been much debated. Since most radial sap flow occurs in the intercellular spaces and canals rather then through the cells, it is unlikely that rays contribute significantly to radial sap movement. It is also worthy of note that rays are completely absent in some woods, e.g. *Hebe*. The most likely function of rays is in the radial transport of material through the cambium to differentiating cells during radial growth. Once formed, rays then remain *in situ* in both the mature xylem and phloem. The observation that all tracheids have at least one ray contact along their length confirms that phloem to xylem transport along the ray is a likely pathway for carbohydrate movement during cell division, expansion, and wall deposition.

Woods with very large wide rays extending several millimetres, e.g. *Quercus, Knightia*, can have considerable decorative appeal especially if the timber or veneer is cut radially so as to expose the rays running in and out of the cut surface. In these woods the rays can comprise as much as 50% of the wood volume. More typically, the ray volume of hardwoods is around 15%.

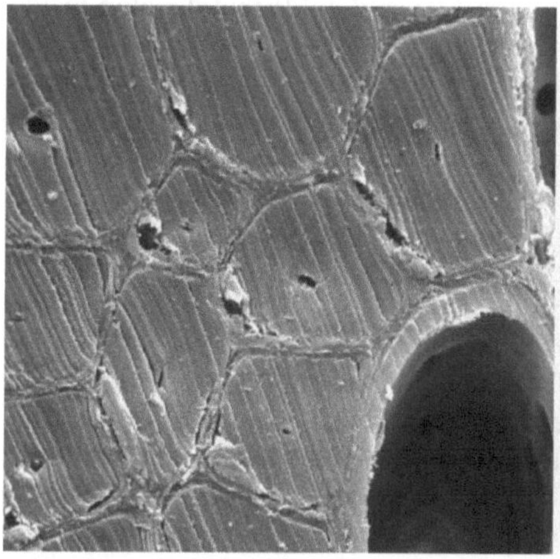

Figure 1.27. Libriform fibres seen in transverse face in the wood of *Hoheria angustifolia*. x 1800.

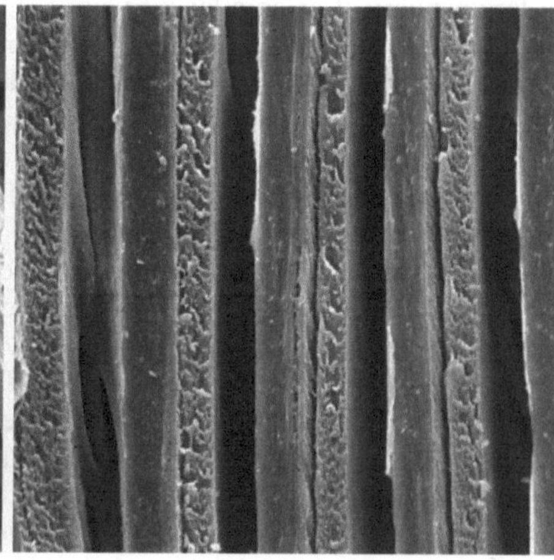

Figure 1.28. Libriform fibres seen in longitudinal view in *Beilschmiedia tawa*. x 1500.

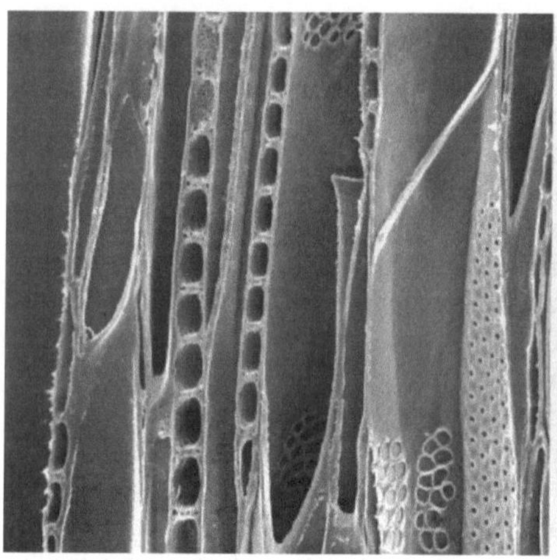

Figure 1.29. Uniseriate rays in *Salix alba* seen in tangential-longitudinal view. x 300.

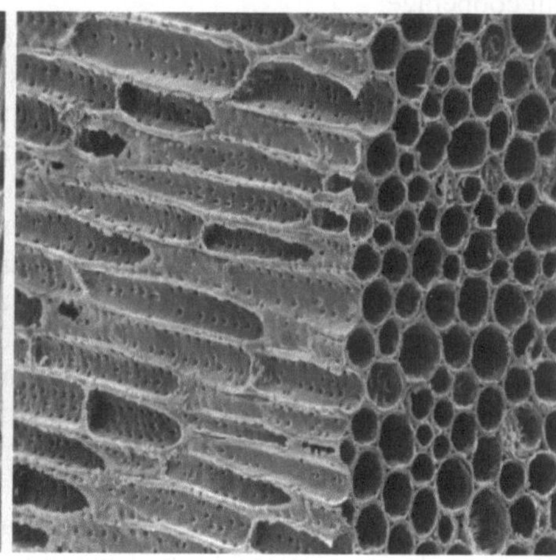

Figure 1.30. A multiseriate ray seen on radial-longitudinal (left) and transverse (right) faces in *Knightia excelsa*. x 175.

Hardwood vessel pit membranes are very homogenous structures comprising a tight web of cellulose microfibrils (Figure 1.26). These bordered pits do not aspirate to seal off cells that have cavitated as in softwoods. On the rare occasions where vessel pit membranes are seen to be aspirated, their microfibrils become stretched and the membrane is torn leaving an open pathway between vessels. Hardwood vessels, therefore, lack this mechanism for isolating sites of water cavitation and rely on tylosis and gum exudation to subsequently seal damaged vessel conduits. Tyloses are large spherical or angular intrusions (Figures 1.31 and 1.32) that develop inside the vessels of some species. The exact relationship between tylose formation and cavitation is not well understood, but tyloses are a common feature in the non-conducting heartwood and are also found in the sapwood at sites of injury. Tyloses develop from neighbouring paratracheal axial parenchyma cells by the intrusive growth of a special tylose-forming layer. This tylose-forming, or protective layer as it is sometimes called, is deposited over the inner wall of the parenchyma cells adjacent to vessels and covers the pit membranes as well as the wall areas. On cavitation, the pit membrane splits and the tylose-forming layer grows out into the adjacent vessel lumen. Tyloses are clearly an efficient way of sealing off already-damaged vessels but their generation is difficult to explain: water-filled vessels lying next to living parenchyma cells already exert a significant osmotic gradient, and together with the water tension in the conducting vessel set up forces that would be far more likely to induce tylose formation while the vessel is conducting than when it is empty. Also tyloses develop only in some hardwood species so they cannot be seen as a universal mechanism against cavitation in hardwoods. In other species, resins and gums are secreted at sites of injury to seal off damaged vessels. Tyloses present a significant nuisance value in obstructing the passage of preservatives into hardwood vessels, but their presence is a major reason for the use of some oaks for tight cooperage.

As with softwoods, the stems and branches in hardwoods that are subject to asymmetric loading produce wood that is characterized by different anatomical, chemical and physical properties. Thus tension wood is quite common in vertical stems, especially below the point of attachment of lateral branches. However, more usually it is associated with inclined stems and branches, the upper side shows a marked eccentricity with wider growth rings. This is the opposite situation from softwoods where the eccentricity and modified wood lies to the lower side of the inclined stem or branch. The modified wood in the eccentric growth in hardwood is referred to as tension wood on account of its position, and like the compression wood of softwoods, it develops in response to stress and acts to straighten the leaning stem. Tension wood is usually harder and denser than normal wood and is sometimes darker in colour. In sawn timber it shows up as having a woolly appearance.

Anatomically, tension wood shows greater variation than compression wood of softwoods. In its extreme form, tension wood contains fewer and smaller vessels than normal wood, and the fibres are modified by the deposition of an extra wall layer inside or replacing the normal S_3 wall layer (Figures 1.33 and 1.34). The extra wall layer is usually referred to as the gelatinous layer, and tension wood fibres with the extra G-layer are referred to as gelatinous fibres. The gelatinous layer is cellulose-rich with the microfibrils lying almost parallel to the long axis of the cell, sometimes forming a series of lamellae. The G-layer is clearly under tension because it usually retracts back into the cell when the fibre is cut transversely (Figure 1.33). Abnormally long pit apertures develop in tension wood fibres. In less extreme forms, the tension wood zone contains fibres that look apparently normal but have smaller microfibril angles than normal fibres. In a few rare cases, hardwoods may have the eccentric growth to the lower side of the stem as in softwoods.

Although the longitudinal shrinkage of tension wood is not as large as that of compression wood, it is still much higher than normal wood. Like compression wood, tension wood is a serious defect in timber owing to its uneven shrinkage and is now known to be quite common in vertical stems, especially below the point of attachment of lateral branches.

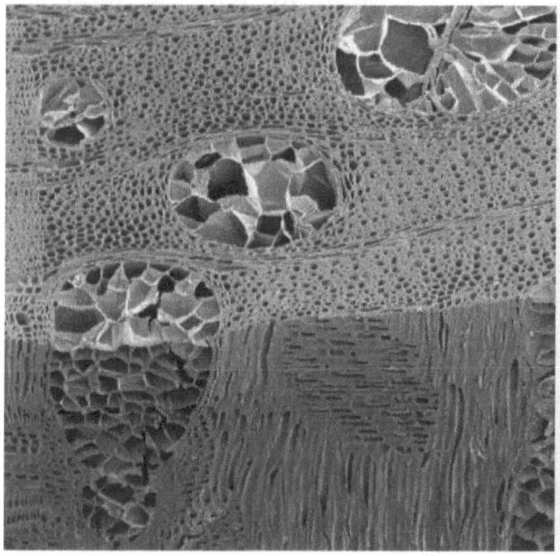

Figure 1.31. Tyloses completely filling the vessels in *Populus* sp. x 85.

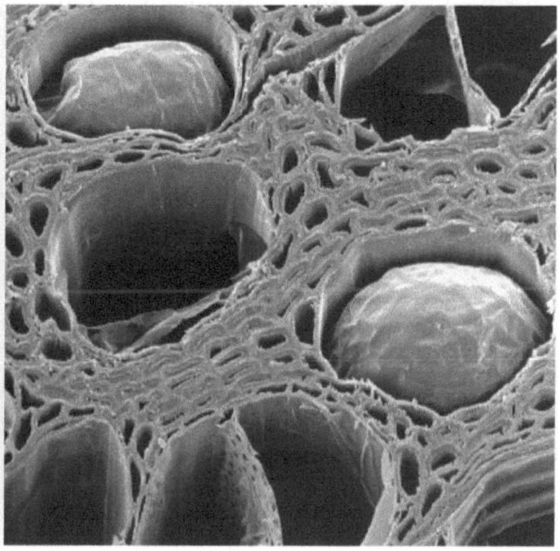

Figure 1.32. Tyloses filling two vessels in *Populus* sp. x 350.

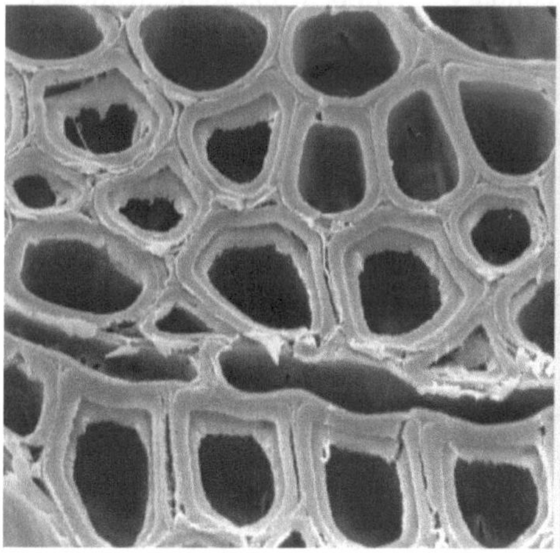

Figure 1.33. Tension wood fibres seen in transverse view of the wood of *Populus sp.* x 850.

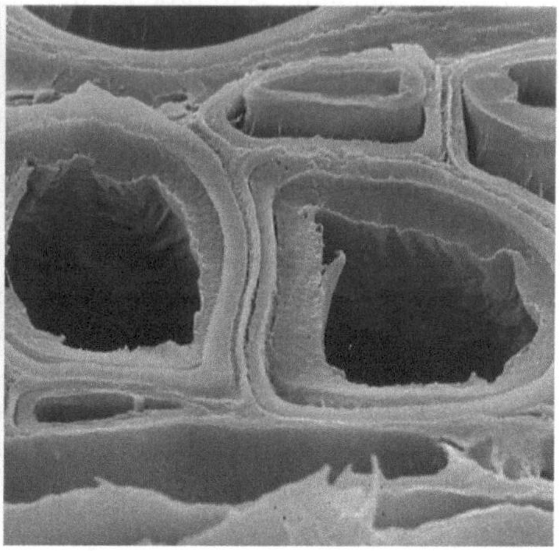

Figure 1.34. Tension wood fibres showing the detail of the extra G-layer on the inside of the normal secondary wall. x 1300.

4. THE MICROSCOPIC STRUCTURE OF BARK

Some 5-30% by volume of a log comprises bark. The presence of bark can present technical problems and increase the cost of log processing at the mill. When bark is stripped from the log, the failure surface occurs largely at the vascular cambium, but in practice the separated bark contains varying amounts of wood, while the stem retains some bark. Traditionally there have been few commercial uses for bark other than in horticulture and as a low-grade fuel, or for the limited manufacture of adhesives and other chemicals. Although the proportion of bark is high in smallwood (<100-150 mm diameter) and in branchwood chipped in whole-tree utilization, such material is in increasing demand for pulp, particleboard and fibreboard. Where used, often by necessity rather than design, its real disadvantages are offset by providing the manufacturer with a cheap, hitherto unutilized resource.

Bark is the term used to describe all the plant tissues outside the vascular cambium (Figure 1.35). The vascular cambium is the cylinder of soft dividing tissue that produces the wood to its inside and much of the bark to its outside. Ring barking a tree has the effect of breaking the continuity of the phloem, the tissue responsible for the bidirectional transport of the synthesized food products within the tree, and also of depriving the dividing cells of the cambium from any physical protection. As a consequence, ring barking normally leads to the death of any stem material above the cut.

The bark can be divided into two zones according to its development and function. The inner bark is derived from the vascular cambium and comprises the phloem containing the carbohydrate conducting cells and the youngest protective layers around the stem or root; while the outer bark comprises a series of largely dead zones of hard suberised cells functioning primarily as protection of the stem. Most woody plants have two cambial zones: the function of the inner zone, termed the vascular cambium, is to produce new secondary xylem or wood to its inside and new secondary phloem to its outside. The outer cambium termed the phellogen or cork cambium functions in producing phellem or cork cells to its outside and a few phelloderm or secondary cortex cells to its inside (Figure 1.36). The tissue complex of phelloderm, phellogen, and phellem forms what is known as the periderm. Each periderm functions for only a limited period of time. As the stem expands inside the cylinder of periderm, each periderm must be replaced by a new one deeper inside the bark. This sequence of successive periderms forms the so-called outer bark of the stem. New phellogens, or cork cambia, develop at varying depths within the inner bark, commonly cutting off older non-functioning phloem and phellem each time they develop. The pattern of formation of these new periderms gives the outside of the bark a characteristic pattern, common examples producing hammer bark (Figure 1.37) and strip bark (Figure 1.38). In many woody plants it is possible to strip off several layers of the outer bark without damaging the inner functioning phloem and periderm.

The inner bark comprises the functioning phloem tissues and the innermost functional periderm. These tissues are all functionally important to the life of the stem, the phloem for the bidirectional conduction of synthesized food products, and

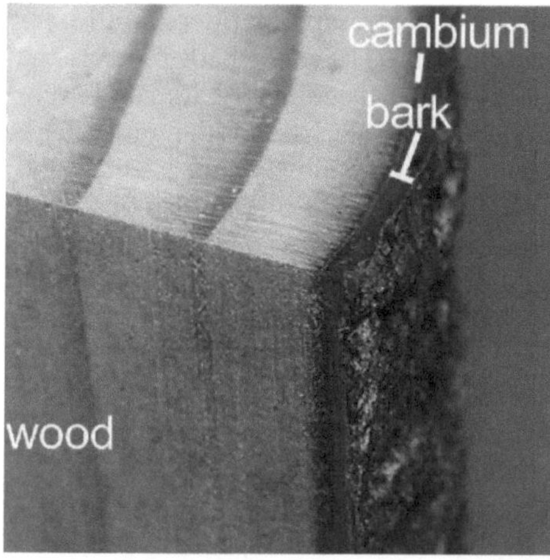

Figure 1.35. A block of *Pinus radiata* wood showing the position of the vascular cambium and bark.

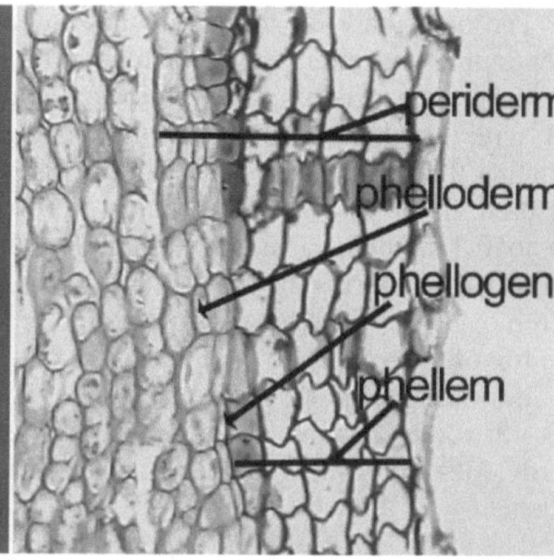

Figure 1.36. A light microscope section through the periderm of *Sambucus sp* showing the phelloderm, phellogen, and phellem of the outer bark.

Figure 1.37. The outside surface of the hammer bark of *Prumnopitys ferruginea.*

Figure 1.38. The outside surface of the strip bark of *Podocarpus totara.*

the periderm for the protection of the living tissues from physical and mechanical damage. Only a very narrow band of phloem immediately adjacent to the cambium is involved in conduction. Outside this narrow zone, the cells of the phloem have become distorted on account of the radial expansion of the stem. Non-functioning phloem cells show a number of modifications including the build up of callose and slime. The inner bark also contains parenchyma cells, rays, and in the case of the hardwoods, fibres as well.

The outer bark on the other hand is built up of successive periderms, all non-functioning and interspersed with arcs of dead phloem. The outer bark protects the tree from moisture loss, fire, disease and physical injury. Dead outer bark can fissure off with weathering. In some species this occurs regularly and the outer bark remains thin, e.g. *Picea* sp, while in other it erodes very slowly and is thick and fissured. In the forests of the Pacific Northwest, the outer bark of Douglas fir can be up to 0.3 m thick. The outer bark of the cork oak, *Quercus suber*, is regularly harvested and forms the basis of a flourishing industry. Provided only the outer bark is removed, new phellogens continue to replace the harvested outer bark throughout the life of the tree.

Bark has very little mechanical strength on account of its lack of fibres. The bark of softwoods has no thickened axially-elongated cells, but the bark of some hardwoods do sometimes contain small numbers of short fibres similar to wood fibres.

The ash content of bark is generally higher than that of the adjacent wood, with calcium and potassium accounting for a significant part. The high ash content, together with the dirt, sand and grit gathered during logging, reduces the usefulness of bark as a fuel in mill operations.

Waxes, fats and fatty acids, oils and resins are all readily extracted from bark using non-polar solvents, while tannins, gums, pectin, soluble carbohydrates and sugars can be removed by aqueous extraction. The amount of extractives that can be removed diminishes significantly after seasoning of the bark.

5. ACKNOWLEDGEMENTS

The SEM photos in this chapter were taken by the author, Dr Li Guizhen, Dr Brian. Meylan and Professor Ilona Peszlen.

CHAPTER 2

BASIC WOOD CHEMISTRY AND CELL WALL ULTRASTRUCTURE

JOHN WALKER

1. INTRODUCTION

All woods are composed of cellulose, hemicelluloses and lignin. Cellulose and hemicelluloses are polysaccharides while lignin is an oxygenated polymer of phenylpropane units. In addition there is a variable quantity of extraneous chemicals known collectively as extractives and small amounts of inorganic elements such as calcium, magnesium and potassium. The inorganic ash content is usually 0.1-0.3% by weight and rarely exceeds 0.5%, except in some tropical hardwoods where a high silica content (several percent) can cause rapid wear and blunting of machine tools.

In this chapter the structural components of wood – cellulose, the hemicelluloses and lignin – will be examined in turn while some features of extractives will be discussed briefly. When considering the individual constituents it is important to recognize that the cell wall is a complex structure in which the nanoscale architecture matters. Wood properties originate predominantly from an optimized hierarchical architecture, implying that the best approach to understanding is to consider the individual molecules and work up progressively to coarser and larger assemblies.

2. THE STRUCTURE OF CELLULOSE

Knowledge of the structure of cellulose is needed to interpret the properties of both lumber and paper. The long, thin crystalline filaments (microfibrils) of cellulose in the dominant S_2 layer are orientated roughly parallel to the fibre axis and are the principal cause of the anisotropic behaviour of most wood-based materials, e.g. wood and fibres are stronger in the fibre direction and weaker in the transverse directions. The same anisotropic behaviour is observed in the shrinkage of wood and in every other property one cares to consider. It is conceded that the geometry of the fibres themselves – long, thin and hollow – will contribute to this anisotropic behaviour, but this is less significant than the effect of the cellulose microfibrils. Further the crystalline nature of cellulose makes it resistant to chemical attack so that the majority of hemicelluloses and lignin can be removed during chemical pulping leaving behind a cellulose-rich fibre, which is the basis of many paper products. Only the very strongest acids and alkalis can penetrate and modify the crystalline lattice of cellulose.

Cellulose is a polymer derived from glucose: β-D-glucopyranose. Glucose is just one of a number of monosaccharides having the same chemical composition, $C_6H_{12}O_6$. Such six-carbon sugars are known as hexose sugars. All these sugars have five hydroxyl groups (-OH), which is why they are soluble in water. Their structure can be represented schematically by the general formula:

$$\underset{C_6}{CH_2OH.}\quad \underset{C_5}{CHOH.}\quad \underset{C_4}{CHOH.}\quad \underset{C_3}{CHOH.}\quad \underset{C_2}{CHOH.}\quad \underset{C_1}{CHO}$$

By convention the carbon atoms in the carbon skeleton are numbered as shown for ease of identification. The two terminal groups, ($-CH_2OH$) and (-CHO), differ functionally from the others. The four central carbons are bonded to four different groups. For example, in the case of the C_5 carbon these groups are a hydrogen atom, -H, a hydroxyl, -OH, the terminal $-CH_2OH$ and the rest of the molecule, -CHOH. CHOH.CHOH.CHO. An atom carrying four different groups constitutes a centre of asymmetry: there are two different ways of orientating these four groups spatially around the tetrahedral carbon atom which cannot be superimposed on one another (Figure 2.1a). These two enantiomorphs (from the Greek *enantia* meaning opposite) have similar chemical and physical properties but are mirror images of each other. By convention these pairs are distinguished by reference to the position of the hydroxyl on the penultimate carbon (the C_5 carbon in the case of hexose sugars). If this hydroxyl lies to the right in the Fisher projection (Figure 2.1b) it is a D-sugar; if it lies to the left it is a L-sugar (Guthrie, 1974). The other three central carbon atoms, C_4, C_3 and C_2, also display asymmetry. Thus there are 2^4 different ways of spatially representing the above hexose sugar formula (eight pairs of enantiomorphs). They have the same general chemical formula, $C_6H_{12}O_6$, but differ in the spatial positions of the hydrogen atom and the hydroxyl group about each of the four central carbon atoms. Three of these hexose sugars are abundant in wood: glucose, galactose and mannose.

Although these sugars react in ways common to aldehydes (chemicals with an aldehyde group, -CHO) there is much evidence that the sugars exist predominantly as a six-membered ring: both linear and ring structures occur but the equilibrium very heavily favours the ring structures (Figure 2.1c). Many chemical reactions involve first opening the ring, which in simple sugars is easily broken with an acid catalyst, followed by attack of the aldehyde form. In the six-membered ring structure the C_1 carbon is linked to the C_5 carbon through an oxygen atom. The C_1 carbon is bonded to four different groups so the C_1 now becomes a new point of asymmetry. In the case of D-glucose the six-membered ring has two forms known as α-D-glucose and β-D-glucose that differ only at the C_1 carbon position. Thus the structure of D-glucose can be represented in various ways (Figure 2.1c). Knowledge of carbon chemistry emphasizes that the carbon-to-carbon bonds within the ring must be approximately tetrahedral, as must be the bond angle at the oxygen in the ring. The six-member ring cannot be flat, as that would require bond angles of 120°

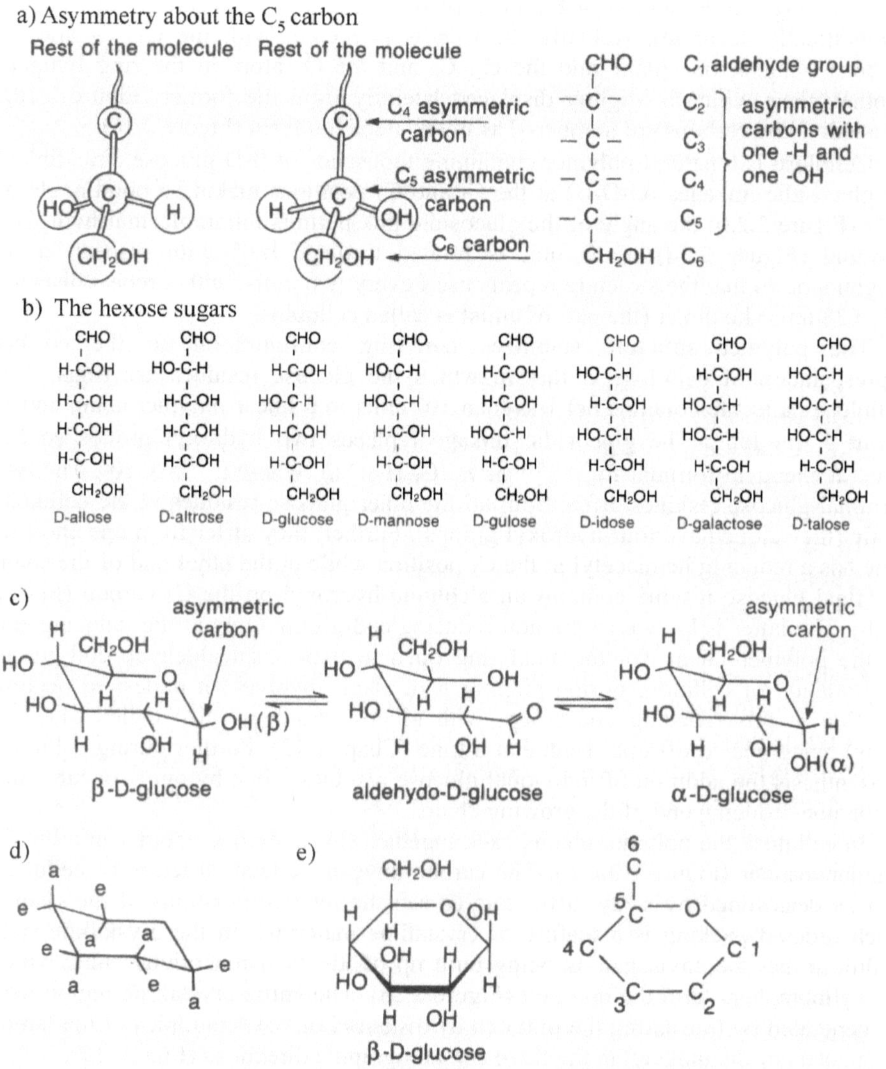

Figure. 2.1. Structure of glucose. (a) A carbon with four different groups constitutes a centre of asymmetry, with two ways of spatially orienting these groups about the carbon. (b) Fisher projections of the eight D-aldohexoses, drawn as open chain molecules, have an aldehyde end-group in the C_1 position. (c) In a six-membered ring a new centre of asymmetry, at the C_1 position, gives two further isomers. Thus D-glucose exists in the aldehydo form and in a ring conformation as either α-D-glucose or β-D-glucose. (d) Carbon bonds tetrahedrally so bond angles of 109.5° are required within the ring. For β-D-glucose all the hydrogen atoms are in axial (a)-positions whilst the larger hydroxyls and the C_6 group are splayed outwards, lying in the equatorial (e)-plane of the ring. (e) For convenience β-D-glucose can be drawn schematically as a flat ring with the carbons numbered as shown.

(the flat ring representation in Figure 2.1e is an accepted convention but it is not geometrically accurate). Actually the ring is puckered, with the C_1, C_3 and C_5 carbons lying in one plane and the C_2, C_4 and the O_5 atom in the ring lying in another plane which is slightly displaced laterally from the former (Figure 2.1d). The individual carbons are numbered as in the aldehydo form (Figure 2.1e).

Cellulose is a natural polymer containing thousands of β-D-glucose units linked by glucosidic linkages (C-O-C) at the C_1 and C_4 positions making a bond angle of 117° (Figure 2.2a): the angle of the glucosidic link permits intramolecular hydrogen bonding (Figure 2.3d). Each unit is rotated through 180° with respect to its neighbours, so that the structure repeats itself every two units, with a repeat distance of 1.038 nm. The dimer (the pair of units) is called cellobiose.

The polymer structure stabilizes the ring configuration so the correct representation of cellulose is that in which the glucose residues are rings. The cellulose molecule contains not less than 10^4 units in a linear polymer chain and is about 5 μm long. The glucosidic linkage replaces two hydroxyl groups so the general chemical formula for cellulose is $(C_{12}H_{20}O_{10})_n$ where $n \approx 5 \times 10^3$. The two terminal glucose residues differ from all the other glucose residues in the cellulose chain (they alone have four hydroxyl groups). Further, they differ from one another. One has a reducing hemiacetyl at the C_1 position while at the other end of the chain the final glucose residue contains an alcoholic hydroxyl on the C_4 carbon (Figure 2.2b). The latter is known as the non-reducing end-group. Only at the reducing end of the polymer chain can the final ring open to expose an aldehyde end-group. Degradation of cellulose during pulping with alkali involves an undesired peeling reaction at this reducing end-group, with glucose residues (also called glucosyl units) being peeled off one residue at a time (Chapter 13). Further during cellulose biosynthesis the addition of individual glucose residues – one by one – occurs only at the non-reducing end of the growing chain.

In cellulose the polymer chains pack together alongside one another in a highly regular manner (Figure 2.3a, b). The carbon-oxygen skeletal structure of cellulose can be determined by x-ray diffraction as can the relative positions of the chains. Such ordered packing is a feature of crystalline materials. In the crystalline state cellulose may be envisaged as being built up of identical repetitious units, which crystallographers term the unit cell (Figure 2.3a). The entire crystalline region may be generated by translating the unit cell by distances that are multiples of the lattice parameters of the unit cell in the three crystallographic directions (Figure 2.3a).

In this traditional representation one cellulose polymer chain lies at the centre of the unit cell (Figure 2.3b). There are four other polymer chains at the edges of the unit cell but these are 'shared, by the four unit cells that meet at that particular edge. Thus this unit cell contains two cellulose chains – a central chain and four others at the edges that are each shared with four unit cells.

The hemicelluloses, which are a significant component of wood, are also carbohydrates, often of very similar hexose sugars, e.g. galactose and mannose. Yet hemicelluloses are rarely if ever crystalline. One reason lies in the spatial distribution of the groups attached to the asymmetric carbons. Each of these carbons has one equatorial and one axial bond (Figure 2.1d). Only in cellulose (Figure 2.2a)

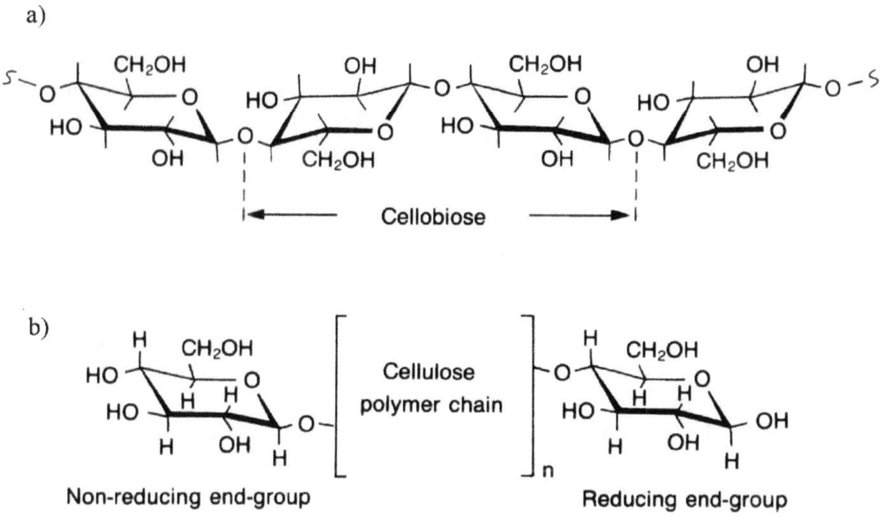

Figure. 2.2. Structure of cellulose. (a) The polymerization process or reaction eliminates a water molecule between the C_1 hydroxyl of a sugar and the C_4 hydroxyl of the adjacent sugar, with a single oxygen linking the two units. The C-O-C link is a glucosidic β-1,4' linkage. In cellulose synthesis repeated condensation reactions at one end of the cellulose chain results in very long chain polymers. The condensed glucose molecules are called glucose residues or glucosyl units. (b) The two terminal glucose residues differ from each other, one having a reducing hemiacetal end group, the other having an extra hydroxyl group. The reducing end group is linked to the rest of the polymer with a glucosidic bond on the C_4. Only this terminal residue can open between the ring oxygen and the C_1 to give an aldehyde (Figure 2.1c).

do the larger groups – the hydroxyls (-OH) attached to the C_2 and C_3 carbons and the CH_2OH attached to the C_5 carbon – all lie in the equatorial plane of the ring, pointing radially outwards from the ring as do the glucosidic linkages at the C_1 and C_4 positions (Figure 2.1d): and only the small hydrogen atoms lie axially (pointing alternately up or down out of the plane of the ring). This is significant. The large hydroxyl groups and the methylol group (the $-CH_2OH$ group) have more space and are further apart when lying in the equatorial plane than would be the case if any were occupying the axial positions. The equatorial distribution of the larger groups can be achieved without straining any bonds or distorting the molecule. Further this configuration permits two hydrogen bonds between adjacent residues in the same polymer chain (Figure 2.3d): between the hydroxyl group of the C_3 and the O_5' oxygen in the ring of the adjacent glucose residue; and between the oxygen of the hydroxyl on the C_6 and the hydroxyl on the C_2 of the adjacent residue. The 180° rotation between adjacent units in the same polymer chain places the C_2 and C_6 carbons on the same side of the polymer chain. Also the adjacent residues are inclined at a slight angle to one another (a mild zig-zag) so bringing the two hydroxyls sufficiently close to achieve effective intramolecular hydrogen bonding (bonding within the polymer molecule.) which stiffens the polymer. The immense tensile stiffness and strength of the cellulose molecule is due to covalent bonding and the axial alignment of the whole microfibril.

In the crystalline state cellulose chains pack in layers with all the hydrogen bonds lying in the plane of the layers. These parallel chains hydrogen bond with neighbouring polymer chains through the hydroxyl on the C_6 and the oxygen of the C_3 hydroxyl. This is an example of intermolecular hydrogen bonding between adjacent cellulose molecules (Figure 2.3d). In this way all available hydroxyl groups participate in hydrogen bonding. The distances between these hydroxyl groups and the oxygen in the ring are such that all favour hydrogen bonding, *viz.* in ice the length of the hydrogen bond O-H······O is about 0.276 nm, which is indicative of the optimal bond length for effective hydrogen bonding. Although hydrogen bonding is not nearly as strong as covalent bonding, strong lateral bonding between adjacent chains is provided by the large number of hydrogen bonds along the cellulose chain that collectively compensate for their relative weakness.

In summary, the cellulose chain lies in a flat-ribbon conformation in the a-c plane, with alternating glucosyl units locked into opposite orientations by intramolecular hydrogen bonds. Intermolecular hydrogen bonds hold adjacent cellulose molecules together edge-to-edge in flat sheets (Figure 2.3d).

The sheets of hydrogen-bonded cellulose chains (lying in the a-c plane of Figure 2.3d) stack on top of one another in the b-direction to form a three-dimensional crystalline structure (Figure 2.3c). In the b-direction of the unit cell the atoms that project out axially are hydrogen atoms. Fortunately, these atoms are very small so the cellulose layers can pack very close (*c.* 0.39 nm between layers): so close that the van der Waals forces stabilize this tight packing. Also, there is the further possibility of a multiplicity of very weak C-H······O bonds between layers. It is extraordinary that an abundant and ubiquitous material as cellulose relies on such weak hydrophobic forces to hold these relatively non-polar sheets together (Jarvis, 2003).

The cellulose chains in adjacent layers are in strictly ordered positions relative to one another despite the forces between the layers being comparatively weak. The cellulose chains in adjacent layers are staggered laterally by a distance of a/2 and are shifted axially by about a quarter of the unit cell dimension, 0.275c (Figure 2.3).

⟶

Figure 2.3. Structure of cellulose (Gardner and Blackwell, 1974). (a) A crystalline material is described by its unit cell. This is the smallest structural entity that, translated a unit distance in the three principal directions, generates the crystal structure. The cellulose molecules are aligned parallel to the c-axis of the unit cell. (b) In a two-chain model there is a central chain and four other chains at the cell corners that are shared by four adjacent unit cells. Thus a unit cell contains 1 + 4 x (1/4) = 2 cellulose chains. In the c-direction each unit cell contains two glucose residues rotated through 180° with respect to one another. For clarity only oxygen and carbon atoms in the cellulose chains at two corners and the centre of the unit cell are shown. (c) Plane view showing the position of the chains. There is hydrogen bonding between chains in the a-c plane, but only van der Waals forces in the b-direction. Although van der Waals forces are weak, close packing between sheets, 0.39 nm apart, means that the forces holding these sheets together are sufficient to ensure regular crystalline packing. (d) View of the a-c plane of the unit cell. Inter and intramolecular hydrogen bonding within and between cellulose chains occurs in this plane only.

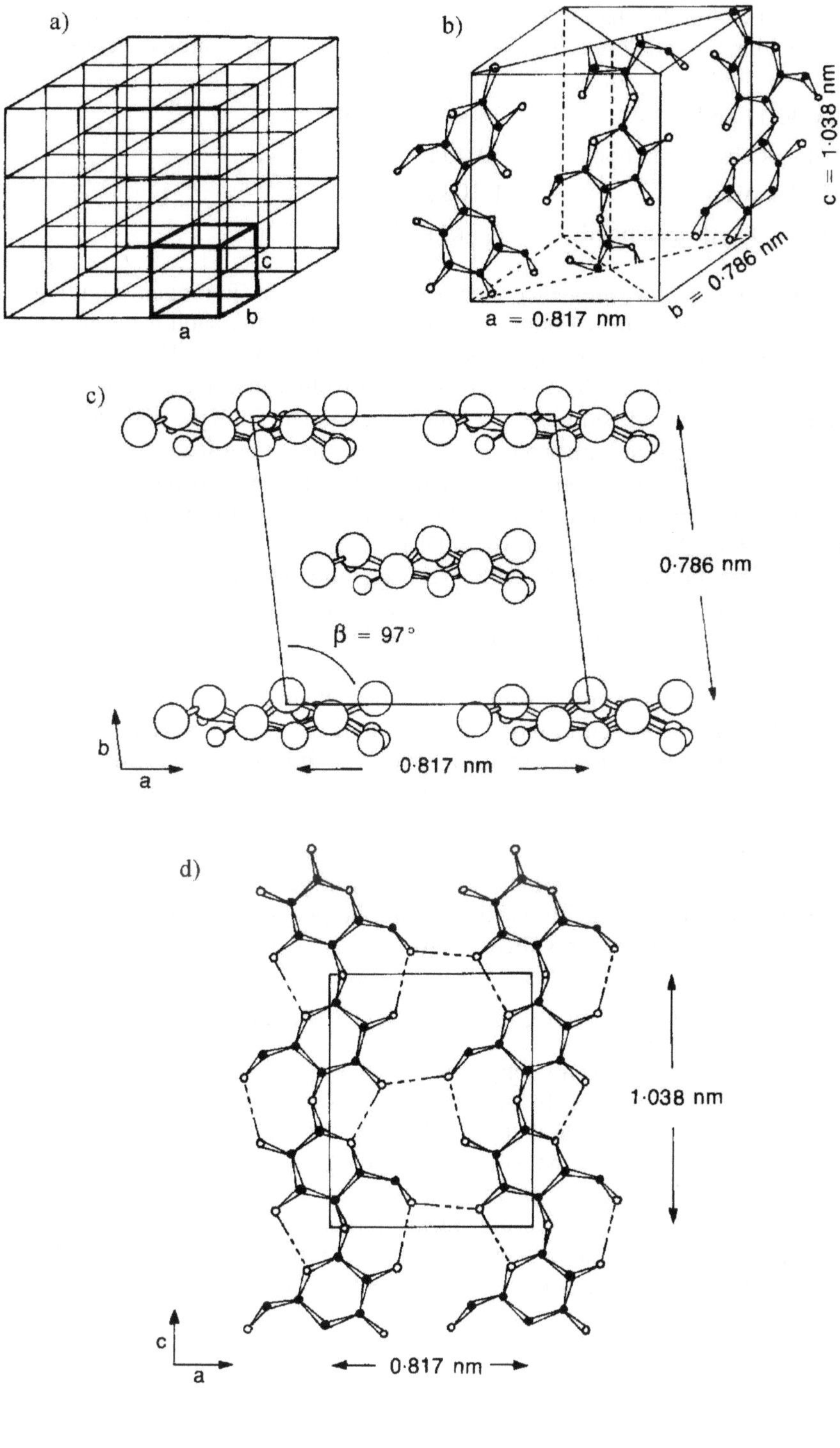
a)
c
b
a
b)
c = 1·038 nm
b = 0·786 nm
a = 0·817 nm
c)
0·786 nm
β = 97°
b
a
0·817 nm
d)
1·038 nm
c
a
0·817 nm

This maximizes the benefits of the weak secondary valence bonding that occurs between and within the chains. Such bonds are much weaker than the covalent bonds that exist within the cellulose chain. Relative bond strengths are approximately:

Covalent bond	200-800 kJ mol^{-1}
O-H······O hydrogen bond	10-40 kJ mol^{-1}
Van der Waals	1-10 kJ mol^{-1}
C-H······O hydrogen bond	2.5 kJ mol^{-1}

Recent studies have revealed subtle differences in hydrogen bonding and spacings to those shown in Figure 2.3, and these diagrams should be taken to be illustrative as they are not rigorously accurate (Sugiyama, 1999). Today we recognize that natural cellulose consists of one or other or a mixture of two forms (allomorphs) designated cellulose 1_α (which is dominant in algae and bacteria) and 1_β (which is dominant in higher plants). What is more extraordinary is that cellulose 1_α and 1_β have the same skeletal conformation of carbon and oxygen atoms in their molecular chains but they have two, mutually exclusive hydrogen bonding networks (Nishiyama *et al.*, 2002, 2003; Jarvis, 2005b).

X-ray diffraction can locate with precision only the positions of the heavier atoms of carbon and oxygen. The traditional structure of cellulose 1 (Figure 2.3) was deduced largely on the basis of x-ray diffraction. The positions of the hydrogen atoms are less determinate. Information on their locations (Atalla and VanderHart, 1999) relies on stereochemistry (indicating where there are spaces large enough to accommodate the hydrogen atoms within the unit cell), on expected bond lengths and angles (the strength of a chemical bond is a function of its proximity to its optimal length and on how well its spatial orientation avoids undue strain), on spectroscopy (NMR, infrared or Raman studies – the latter by observing bond vibration frequencies is helpful in establishing that the two allomorphs which have similar skeletal conformations have different hydrogen bonding patterns), and on electron and neutron diffraction (which locate the positions of the hydrogen atoms more accurately than does x-ray diffraction).

3. THE CELLULOSE MICROFIBRIL AND CELLULOSE BIOSYNTHESIS

Crystalline cellulose occurs in long thin filaments, called microfibrils. These are separated from one another by non-crystalline (amorphous) material. Both the structure of the microfibril and the mechanism for cellulose biosynthesis are of general interest. For fundamental studies cell biologists prefer to work with cellulose from algae, bacteria and fungi where isolation is easier – because they lack lignin – and because their microfibrils are much larger in cross-section with lateral dimensions up to 20 nm rather than only about 3 nm in higher plants. Delignified pulp fibres are favoured as a source material for specific studies of wood cellulose.

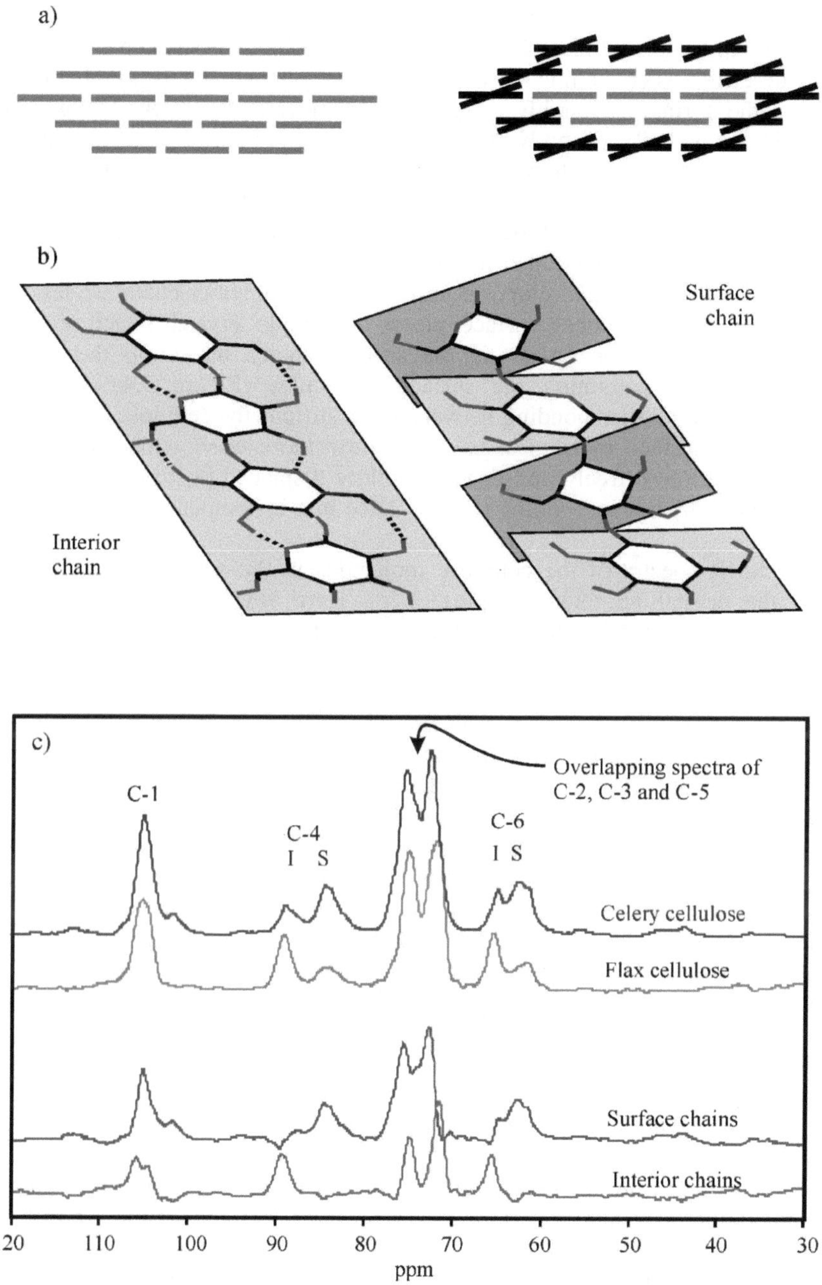

Figure 2.4. (a) Schematic model of a microfibril with metastable (LHS) and modified (RHS) surface. (b) Alternate twisting of surface glucosyl units by 30° results in less strain. (c) CP-MAS spectra for flax (30% surface chains) and celery cellulose (60% surface chains) can be split into spectra of surface (S) and interior (I) chains (Viëtor *et al.*, 2002).

In higher plants the microfibril is only about 3 x 3 nm in cross-section, containing up to 40 cellulose chains. Indeed it is problematic that so small an entity can be described as crystalline since the interior is so small and the number of chains lying on the surface is roughly equal to the number in the interior (Figure 2.4a). Molecules in the bulk of a body are subject to the attraction of molecules all around them, but at the surface they are subject only to the inward attraction of the molecules within the solid. This is true even where two crystalline surfaces impinge on one another. The mismatch between the two surfaces, however slight, means that atoms or molecules belong either to one crystalline region or to the other, not both. For cellulose the individual glucosyl units along the polymer chain are less free to vibrate than are individual surface atoms in simple crystal structures because intermolecular hydrogen bonds tie the surface chains back into the crystalline interior further constraining the surface molecules (Figure 2.4a) while some intramolecular hydrogen bonding between units stiffens the cellulose chain (Scallan, 1971). These surface molecules, having a little more energy than the molecules within, vibrate more freely and occupy slightly displaced positions. There will be less interchain hydrogen bonding (Figure 2.4b) that is compensated for by bonding to water.

The NMR spectra of the cellulose molecules in the surface differ from those within the crystalline region (Figure 2.4c). Here NMR spectroscopy picks up differences between the cellulose in the interior and those cellulose chains lying on the surface of the microfibril. Spectroscopic tools – NMR, infra-red, Raman – work in different parts of the electromagnetic spectrum and observe the way individual atoms or bonds vibrate. Different tools provide different information. With NMR the peaks are the 'signatures' of different atoms (generally carbon or hydrogen) in different locations (for example C_1 to C_6) and in different environments (for example in cellulose 1_α or 1_β; or on the surface or in the interior of the microfibril). For example, on some surface chains the C_5–C_6 bond can partially rotate and consequently the C_6 hydroxyl does not appear to hydrogen bond to interior chains in the same way – or as strongly – as interior chains bond to one another (Newman, 1997; Viëtor *et al.,* 2002). From the peak heights (Figure 2.4c) the proportion of cellulose chains on the surface to those in the interior can be estimated.

There is a parallelism of concept with earlier models. Preston (1974) considered the microfibril to have a crystalline interior surrounded by a partially crystalline cortex or sheath (Figure 2.5b). The cortex was conceived as a zone of gradually increasing lateral disorder several molecules thick. In the cortex the chains, while still parallel, are not as regularly packed. Instead of the surface of the microfibril being defined by a single layer of cellulose molecules the cortex concept suggested a gradual transition from the highly ordered interior to the disordered outer surface of the cortex (the term paracrystalline has been used to refer to this type of partial order). The model permitted the inclusion of structurally similar material that was not quite so particular about crystalline order, i.e. hemicelluloses such as xyloglucans in the primary wall and the glucomannans in the secondary wall of softwoods and xylans in hardwoods. The incorporation of some hemicelluloses – or hydrogen-bonded water molecules – would disrupt the local order and make the cortex increasingly vulnerable to the inclusion of further hemicelluloses and to increasing lateral disorder. Analysis of chemical pulps provided some support for this concept as it is not possible to obtain a high yield of cellulose free from hemicellulosic contamination, indicating that some hemicelluloses are bound to and cling tenaciously to the microfibrils.

All such models present a dilemma. X-ray diffraction patterns are analysed most easily in terms of two components (sharp diffraction images from the crystalline regions and diffuse scattering from non-crystalline regions), even though there

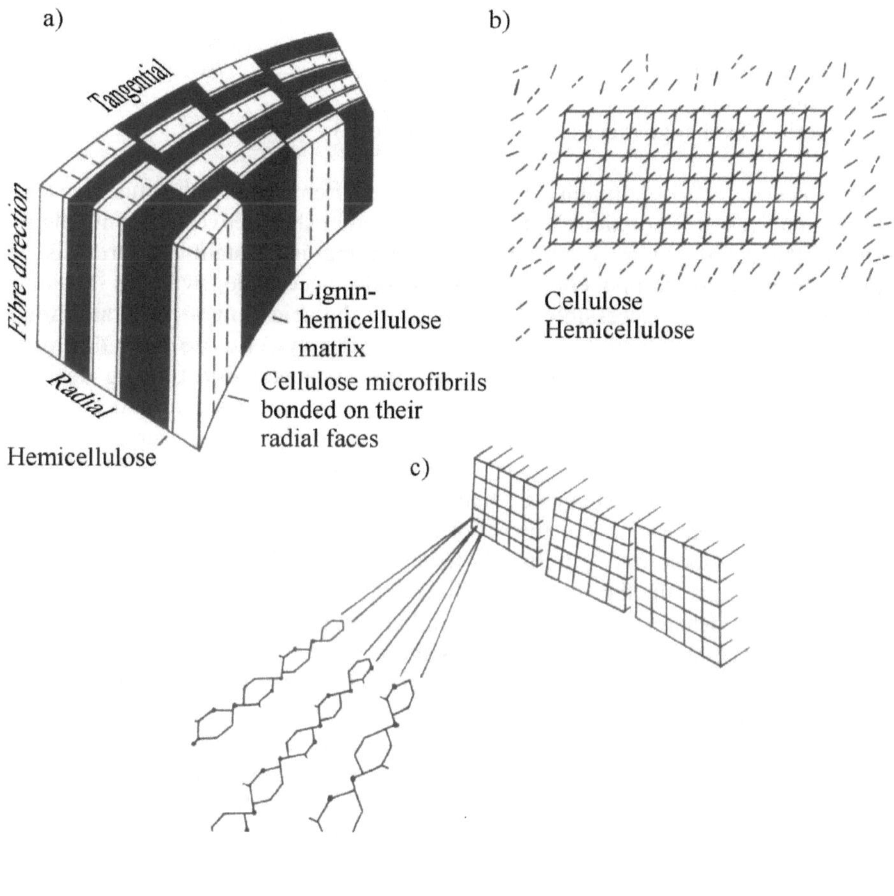

Figure 2.5. (a) The interrupted lamella model of Kerr and Goring (1975). (b) The microfibril with a crystalline interior of pure cellulose surrounded by a paracrystalline, partially ordered cortex of cellulose and hemicelluloses (Preston, 1974). (c) A slight misalignment between microfibrils is compatible with x-ray data (Marchessault and Sundararajan, 1983).

are regions that are neither perfectly organized nor completely disorganized, i.e. there are degrees of crystallinity dependent on the severity of local perturbations from the ideal unit cell dimensions. Further, x-ray diffraction tells us only about the distances between the heavier atoms in the molecule and the regularity with which they pack together, and nothing about accessibility. Other experimental techniques perceive this crystalline or inaccessible fraction differently. In the case of well-delignified wood fibres after chemical pulping, which approximate to pure cellulose, valid measurements of something called 'crystallinity' range from 45% to 90% (Marchessault and Sundararajan, 1983):

Hydrolysis	90%
X-ray diffraction	70%
Packing density	53%
Deuterium exchange (accessibility of the isotope)	45%

It is necessary to distinguish between a measurement and the model used to interpret that result. For example, hydrolysis distinguishes between a component of cellulose that rapidly hydrolyses and a component that is resistant to hydrolysis. The rapidly hydrolysed component must be readily accessible, whereas the more resistant cellulose is inaccessible simply because the acid cannot penetrate ordered crystalline regions. Substitution of hydroxyl groups in cellulose by -OD groups, using deuterated water (D_2O), is also a function of accessibility, being unable to penetrate to the interior chains of cellulose which are well-ordered and too tightly hydrogen-bonded to deuterate. The difference in the values for crystallinity by hydrolysis and deuterium exchange is therefore surprising. Scallan (1971) offered an elegant interpretation. He suggested that on the surface of the fibril the glucosidic bond is half hidden within the microfibril and so protected from rapid hydrolytic attack (Figure 2.4a) so only amorphous material is hydrolysed, whereas deuteration substitution occurs in the non-crystalline regions and on the surface of the microfibril only where each chain exposes half its hydroxyl groups (Figure 2.4a). The different crystallinity values given by these two techniques allowed Scallan to estimate the surface-to-volume ratio for the crystalline regions, which he concluded was compatible with the size of the microfibril.

In the case of algae such as *Valonia* sp. the microfibril is very large, 20 x 20 nm (with over a 1000 cellulose chains in the cross-section), and this is a single crystalline entity – at least there is no evidence of distinguishable subunits. Whereas in higher plants, e.g. wood, the *c.* 3 nm diameter microfibril (with 20-40 cellulose chains in the cross-section) appears to be the basic unit for biosynthesis. In consequence, higher plants display less sharp x-ray diffraction patterns, suggesting some disorder that might be attributed to an imperfect match between the microfibril arrays. The slightest misalignment between or a gradual twisting of microfibrils means that they cannot pack together to produce an enlarged single perfect crystalline microfibril (Figure 2.5c). The discontinuity of structure between adjacent microfibrils implies that defects, hemicelluloses or a partial monolayer of water molecules may intrude. According to Blackwell and Kolpak (1976) a gap of only 0.1 nm between the microfibrils is compatible with the x-ray data. For green alga, a consistent but imperceptible right-hand twist results in an 180° rotation of the microfibril every 700 nm (Hanley *et al.*, 1997). A lazy twist would make close-packing and registry between adjacent microfibrils ever more difficult.

As just noted, there is clear evidence that the size of the microfibril varies with the source of cellulose and this influences its crystallinity. In higher plants these are consistently some 3 nm thick, but they are some 10-20 nm wide in certain algae and in tunicates (sea squirts), the only animals that make cellulose (Jarvis, 2003). The crystallinity of natural celluloses is determined largely by the proportion of surface to interior cellulose chains (more crystalline in algae than in higher plants). Other factors (Larsson, 2003) that may contribute include conformational changes that may occur where microfibrils aggregate together, dislocations (very local severe misalignments where a single cellulose chain terminates within the microfibril) and subsequent structural changes when wood is processed (during chemical and mechanical pulping, and during drying when water is withdrawn from the cell wall). Lateral disorder due to imperfect packing is of considerable interest for the bonding between cellulose and the surrounding hemicelluloses, and to the lignin matrix – the implications of such bonding are discussed later.

The size and structure of the microfibril are under biological control rather than being determined by classical thermodynamics. In other words biosynthetically important processes determine the size of the microfibril rather than having a varying size distribution described by a probability function. There is much evidence that individual cellulose polymer chain and microfibril synthesis is one of almost simultaneous polymerization and then crystallization (Preston, 1974; Wilson and White, 1986; Delmer, 1999).

For bacteria and many algae a lineal multi-subunit array of enzyme complexes on the cytoplasmic face spins out (a loose figure of speech) cellulose molecules in the manner of a spider spinning its silk. After a finite lifetime, having produced a microfibril, these are deactivated and replaced. The act of producing the molecules forces the growing molecules through pores in the plasma membrane to secrete in the cell wall a microfibril containing about 1000 glucan chains (mostly cellulose 1_{α}).

For the higher plants the array is much smaller, whereby numerous mobile hexagonal rosettes each produces some 20-40 glucan chains (largely cellulose 1_{β}), with each of the six subunits containing a further catalytic subunits (Delmer, 1999): if as they emerge from their synthesizing complexes they were to achieve perfect registry (perfect matching of orientation of two or more crystalline entities) with other microfibrils from adjacent rosettes an enlarged crystalline entity would be achieved. This occurs in algae and bacteria within their lineal arrays but generally this fails to occur in higher plants (where limited aggregation and little registry occurs). In higher plants hemicelluloses appear to regulate – interfere with – cellulose aggregation. Hence the largest microfibrils are found in pure cotton, bacterial and some algal cellulose while smaller microfibrils occur in higher plants, whose cross-sectional dimensions appear to have been suppressed by partial adsorption of hemicelluloses on the microfibril surface (Tokoh *et al.,* 1998). Thousands of arrays of cellulose-synthesizing complexes in the plasma membrane are needed to produce the expected number of microfibrils.

Kinetic studies on cotton cellulose show that the degree of polymerization during the biosynthesis of secondary wall cellulose remains constant with a degree of polymerization (weight average) of about 13,000. The secondary wall cellulose has a high degree of polymerization and is essentially monodisperse. Since the microfibrils are far longer than that of the individual cellulose molecules (13,000 x 0.52 nm $\approx$ 7 μm) this implies that the cellulose molecules are terminated and reinitiated repeatedly during synthesis of a microfibril. These very localized disruptions to the crystalline interior result in localized disorder/strain at points long the length of the microfibril (Lewin and Roldan, 1975) and by analogy to metals such zones may be described in terms of 'dislocations' or 'pinned' point defects.

In contrast, the formation of the primary wall results in a lower degree of polymerization, around 6000, and a broad molecular weight distribution (Marx-Figini, 1969). The quantity of cellulose synthesized in the primary wall is small, it is polydisperse, and of low molecular weight.

A simple model for the cell wall (Figure 2.5a) is one in which the microfibrils extend along the length of the tracheid being embedded in a non-crystalline matrix of hemicelluloses and lignin (Kerr and Goring, 1975; Ruel *et al.*, 1978). The concept was based on work in which stained thin-sections of wood, cut radial to the cell wall, revealed regular striations with a repeat distance of about 7 nm for black spruce (*Picea mariana*) and about 8.5 nm for silver fir (*Abies pectinata*), i.e. for black spruce in the radial direction both the stained (lignin-rich) and unstained (cellulose-rich) bands were approximately 3.5 nm wide. When scanning tangentially (parallel to the middle lamella) the repeat distance was 15 nm, suggesting that microfibrils were associating laterally along their radial faces. The interrupted lamella model has a preferred tangential alignment of microfibrils forming interrupted concentric lamellae. According to Kerr and Goring (1975) with black spruce about a third of the hemicelluloses are associated intimately with the cellulose and two-thirds with the lignin, tentatively identifying the straight-chained glucuronoarabinoxylans with the cellulose.

It is not clear how best to describe the cellulose-matrix interface: whether in terms of an interphase boundary, paracrystalline cortex or a modified surface.

4. THE STRUCTURE OF HEMICELLULOSES

The use of the collective term, hemicelluloses, is convenient in spite of the fact that there are a number of hemicellulose macromolecules and that the individual hemicelluloses can react differently, for example during chemical pulping. The principal constituent sugars that are found in the hemicelluloses are the pentose sugars (five-carbon sugars), L-arabinose and D-xylose, and the hexose sugars, D-glucose, D-mannose and D-galactose (Figure 2.6).

β-D-galactose and β-D-mannose differ from β-D-glucose in the configuration at a single carbon, at the C_4 and C_2 positions respectively. The simple switch in the positions of the hydroxyl group and hydrogen atom at either C_4 or C_2 positions means that a hydroxyl group moves into an axial position. This makes hydrogen

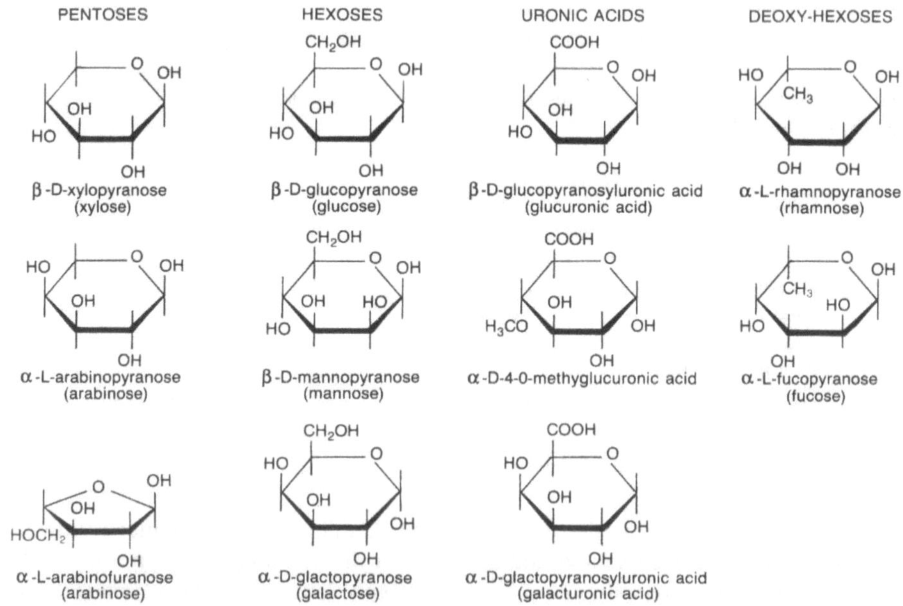

Figure 2.6. Common sugars found in the hemicelluloses.

bonding and close packing of polymer chains difficult and forces the layers further apart: chemists talk of non-bonding steric repulsion. The axial hydroxyl weakens the weak van der Waals forces between the layers. Large substitute groups often replace the hydroxyls on the polymer chain making a crystalline macromolecule even less likely. For example β-D-glucuronic acid and β-D-galacturonic acid both have a large carboxyl group, -COOH, in the C_6 position (such molecules are called polyuronides). The hydroxyls can be methoxylated (a methoxyl group, $-O-CH_3$, substituting for an hydroxyl group) or acetylated (an acetyl group, $-CO-CH_3$).

D-xylose is identical to D-glucose except it lacks the C_6 terminal methylol group ($-CH_2OH$). Therefore, one might expect xylans (polymers of xylose) to pack together as effectively as cellulose since all their hydroxyls lie in the equatorial plane. However the loss of the $-CH_2OH$ group means the loss of those hydrogen bonds involving the C_6 hydroxyl (Figure 2.3d). Xylan can be crystalline but in the native acetylated form it is amorphous.

Typical hemicellulose values are given in Table 2.1. The amount of hemicellulose and the structure and composition of the individual hemicelluloses vary between genera, species, cell type and location in the cell wall. Different hemicelluloses play similar and essential roles that have been conserved during evolution (Harris, 2005). An individual genus can have certain characteristic hemicelluloses: e.g. a high arabinogalactan content is peculiar to larch. Detailed discussion of the various hemicelluloses can be found in Sjöström (1981), Fengel and Wegener (1984), Gatenholm and Tenkanen (2002), and Entwistle and Walker (2005a). Here only the more significant hemicelluloses are examined briefly.

Table 2.1. The principal hemicelluloses in wood, expressed as a percentage of the oven-dry wood (Sjöström, 1981).

Hemicellulose	Occurrence	Percentage in wood (%)
Galactoglucomannan	Softwood	5-8
Glucomannan	Softwood	10-15
Arabinoglucuronoxylan	Softwood	7-10
Arabinogalactan^{+}	Larch	5-35
Glucoronoxylan	Hardwood	15-30
Glucomannan	Hardwood	2-5

$^{+}$normally only 1-5% in softwoods, larch is an exception.

In softwoods mannose is the most important hemicellulose monomer followed by xylose, glucose, galactose and arabinose. Most of the mannose is present as O-acetyl-galactoglucomannan (Figure 2.7a) of relatively low molecular weight (degree of polymerization of about 100). This is a mixed linear polymer of 1,4' linked glucose and mannose units linked through the C_1 and C_4 carbons of adjacent pyranose units. Some of these units are acetylated at the C_2 or C_3 positions with on average one substitute group for every three to four hexose units. The ratio of galactose to glucose to mannose in this polymer is about 0.1:1:4. In some species the polymer backbone may be branched at either one or two points. There is another galactose-richer fraction of galactoglucomannan with a galactose to glucose to mannose ratio of approximately about 1:1:3. The galactose is a single unit side-chain attached to the C_6 position by an α-1,6' linkage. This second galactose-rich fraction is generally referred to a galactoglucomannan whereas the fraction with little galactose is often described simply as glucomannan. Both 'glucomannans', taken together, account for 15-25% wt/wt of the cell wall in softwoods.

The other principal hemicellulose in softwoods is arabino-4-O-methylglucurono-xylan (degree of polymerization of 70-180). The backbone is composed of about 150 β-D 1,4' xylopyranose units which are partially substituted at the C_2 position by 4-O-methyl-α-D-glucuronic acid groups (approximately one group for every 5-6 xylose units). Also an α-L-arabinofuranose unit is linked by α-1,3' bond on approximately every 6 to 10 xylose units. Arabinofuranose is so called because it is a furanoside having a five-membered ring.

Xylans comprise 5-10% wt/wt of the cell wall mass in softwoods.

In hardwoods xylose is by far the most important hemicellulose monomer followed by mannose, glucose and galactose, with smaller amounts of arabinose and rhamnose. The xylose occurs predominantly as O-acetyl-4-O-methylglucuronoxylan (degree of polymerization of 100-200). The basic skeleton of all xylans is a linear

a) Partial chemical structure of O-acetyl-galactoglucomannan (a softwood hemicellulose).

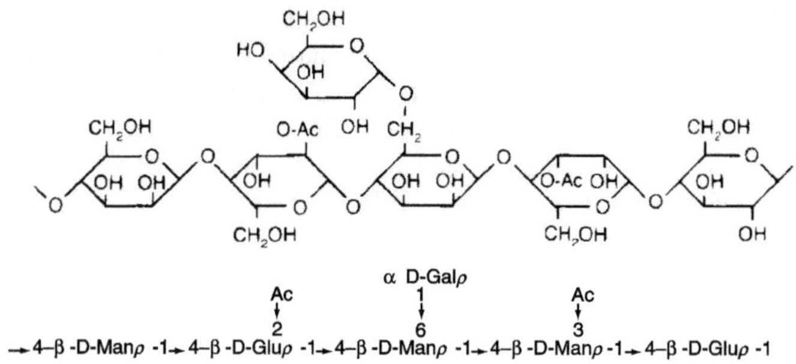

b) Partial chemical structure of arabino-4-0-methylglucuronoxylan (a softwood hemicellulose).

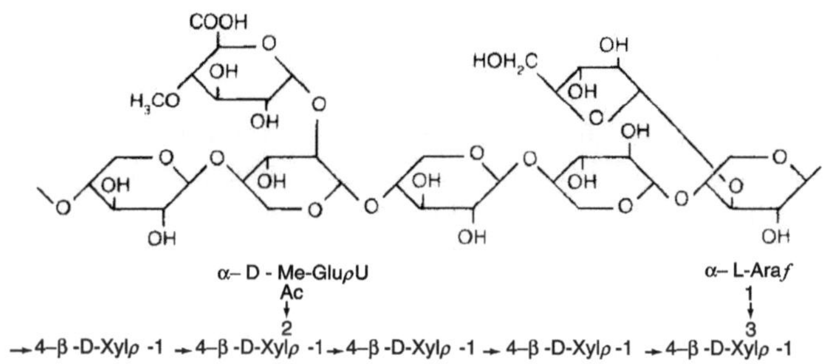

c) Partial chemical structure of 0-acetyl-4-0-methylglucuronoxylan (a hardwood hemicellulose).

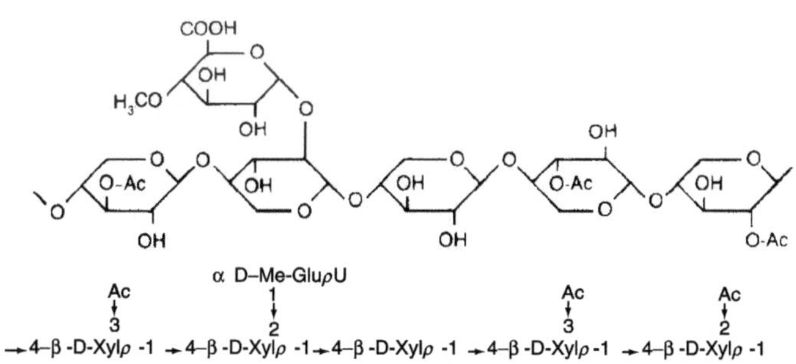

Ac = acetyl,-$COCH_3$
Me = methyl,-CH_3
U = uronic acid group,-COOH
Araf = arabinofuranose

Galp = galactopyranose
Glup = glucopyranose
Manp = mannopyranose
Xylp = xylopyranose

Figure 2.7. Some model hemicellulose structures (Fengel and Wegener, 1984). Softwood xylan is quite similar to hardwood xylan, the absence of acetyl groups on the softwood xylan being an obvious difference.

backbone of β-D 1,4' xylopyranose units, although this is always modified (40-70% of the xylose units are acetylated on the C_2 or C_3 position). Further, D-glucuronic acid or 4-O-methyl-D-glucuronic acid groups usually attach themselves to about one in ten of the xylose residues along the main chain, by an α-link to the C_2, or on occasions to the C_3 position. The ratio of xylose to glucuronic acid groups is about 10:1. With the 4-O-methylglucuronoxylans the chains is twisted like a ribbon, having a 3-fold screw axis (a 120° rotation per polymer residue), whereas cellulose has a 2-fold screw axis (180° rotation). Xylans comprise 15-30% wt/wt of the cell wall mass in hardwoods.

Glucomannan is present in hardwoods but is of minor significance compared to the more abundant xylans. It is a linear 1,4' copolymer with no substitution on the C_2 and C_3 positions (degree of polymerization of 60-70). The glucose to mannose ratio varies from 1:1 to 1:2.

In summary, the principal structural differences between cellulose and the hemicelluloses are:

- Hemicelluloses are mixed polymers, whereas cellulose is a pure polymer of glucose.
- Hemicelluloses have short side-chains, apart from arabinogalactan which is heavily branched. In contrast, cellulose is a long unbranched polymer.
- Hemicelluloses are low molecular weight polymers (forming short chains with a DP of 80-200), whereas cellulose has a very high degree of polymerization.
- Hemicelluloses may have large side groups substituting occasionally for the hydroxyls on the C_2, C_3 and C_6 positions.
- There will be variation in the abundance, type, distribution and degree of polymerization of side-chain substituents on the backbone.
- Their greater accessibility and susceptibility to hydrolysis compared to cellulose arise from their low molecular weights and non-crystalline structures. Their hydroxyl groups are free to hydrogen bond with water.

The hemicelluloses are structurally related to cellulose so one might expect their chemical reactions to be comparable, although the hemicelluloses are generally more reactive (Harris, 1975). The important differences in reactivity that are observed are strongly influenced by physical causes rather than just by differences in chemical reactivity. In general the more branched the polymer the more soluble it is and the more accessible it is to chemical attack.

The more heavily substituted hemicelluloses are confined to the non-crystalline regions of the lignin-rich matrix, while the least substituted parts of any polymer chain are more able to hydrogen bond to the cellulose microfibrils.

About one-third of all the hemicelluloses – those hemicellulose polymers with fewer large side groups and short side chains – appear to be adsorbed on the cellulose microfibrils, i.e. some glucomannans (softwoods) and xylans (hardwoods). One piece of evidence is the presence of small persistent amounts of residual hemicellulose after prolonged cooking of chemical pulps, when even the cellulosic material is being broken down. The hemicellulose chains align themselves parallel to the microfibrils, but they lack excellent lateral bonding as occasional bulky side-chains and substitution prevent regular close packing in all three dimensions.

The function of the hemicelluloses is uncertain. The intrinsic tensile strength of the cellulose molecule is considerable (134 GPa); lignin provides moderate compressive strength (6.9 GPa at 12% MC; 2.8 GPa where close to saturation) by preventing the slender microfibrils from buckling and impedes biodegradation (Yamamoto *et al.*, 2002). The obvious but simplistic role assigned to the hemicelluloses is to provide links between cellulose and lignin, permitting effective transfer of shear stresses: the hemicelluloses both hydrogen bond to cellulose and form occasional ester or ether bonds with lignin.

The roles for hemicelluloses are more complex – and tentative (Atalla, 2005).

First, as noted, certain hemicelluloses are adsorbed on the emerging microfibrils and appear to actively limit microfibril size in higher plants.

Further in the 'not-yet-lignified' primary wall xyloglucan (up to 20% wt/wt) tethers the microfibrils and forms a lightly-interlinked network to resist the swelling forces (osmotic pressure/turgor) that can reach several atmospheres. Possibly, the xyloglucans are incorporated on the microfibril surface for part of their length before stretching out to bind either with another microfibril or with another xyloglucan chain that in turn binds to another microfibril. Cell extension of the primary wall either involves gradual peeling of the xyloglucans from the microfibrils (Morris *et al.*, 2004) or unzipping (the Velcro effect) between hydrogen-bonded xyloglucans (Keckes *et al.*, 2003). With the cellulose microfibrils oriented close to 90° to the cell axis it is the hemicellulosic cross-links that control cell elongation by restraining lateral separation of the microfibrils. A second independent network of pectic polysaccharides governs the porosity of the wall.

In the secondary wall the hemicelluloses bridging between the microfibrils are expected to restrain them from both separating and sliding, and yet permit the very slow creep under load over long periods (Keckles *et al.*, 2003; Jarvis, 2005b; Newman, 2005). Apart from galactose, the hemicelluloses are linear polymers with occasional (random?) substitution and side branches. These disrupt local hydrogen bonding and may force the molecule to hinge or change direction. Studies of chemically extracted – and so modified – hemicelluloses allow one to calculate a statistical 'persistence length' of 7-10 nm (Jarvis, 2005b) which is greater than the separation distance between microfibrils. Thus hemicellulose chains are likely to curve gently between microfibrils or adjacent hemicellulose molecule – and not bridge straight across (the 'straight' segment between 'hinges' is too long). Such a trellis-like frame of hemicelluloses (Jarvis, 2005b) should be stiff in shear but weaker in tension between microfibrils (Figure 2.8a).

Trees need a mechanism to absorb energy – to dampen branches and stems swaying in the wind (Chapter 6). The shear and slip of the hemicelluloses is the probable means of achieving this.

Galactan is the one exception among the hemicelluloses. Its C-O-C interunit link at the C_4 is axial (Figure 2.6) and its natural shape is a coil with six monosaccharides per coil (Fukashima *et al.*, 1997). Rather than lying predominantly parallel to the microfibrils, one might speculate that the galactan may be able to coil between

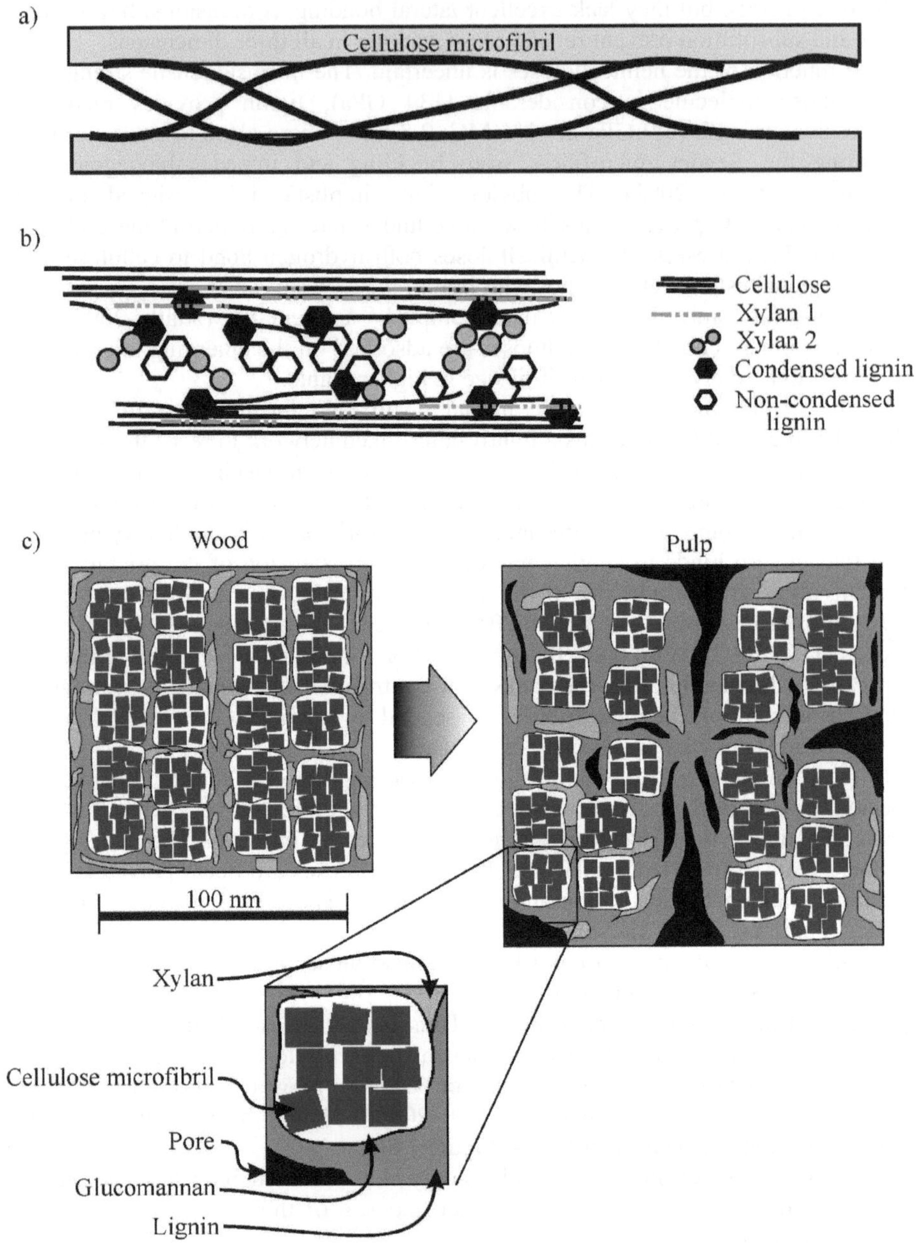

Figure 2.8. (a) Hemicelluloses with long persistence lengths are relatively inflexible and can only snake between microfibrils. (b) Schematic assembly of polydiverse xylans and lignins during secondary wall formation of a hardwood. (c) Microfibrils cluster in arrays surrounded by wider regions of matrix material (schematic from spruce wood).

microfibrils and so exert a transverse shrinkage and swelling pressure. This would explain its role in compression wood where a small percent of galactan is associated with disproportionate dimensional instability (Floyd, 2005). On hydration the galactan may then behave more like a wedge, creating disproportionate extra space between the microfibrils (unlike the other hemicelluloses) – and conversely for dehydration. Thus while the contribution of the other hemicelluloses to volumetric shrinkage is simply proportional to the number of water molecules hydrogen-bonding to the hemicelluloses, for galactan the volumetric shrinkage is disproportionately greater. A recent interpretation (Fahlén and Salmén, 2005), using atomic force micro-scopy, has the microfibrils grouping together to give two matrix zones: one intimate between the individual microfibrils – possibly with hydrated, non-acetylated glucomannans (softwoods) and xylans (hardwoods) – and the other a thicker region (10-15 nm) with lignin, the more acetylated hemicelluloses and large pore spaces for water (Figure 2.8c). The latter zone provides more room for the galactan to exert leverage.

A final point, the level of integration, of cellulose with the hemicelluloses and the hemicelluloses with lignin, is on such a fine scale approaching that of individual molecules that it becomes hard to conceive that they retain in full their individual identities. Indeed there appears to be some coordination and interaction in the parallel/sequential deposition of individual entities (Atalla, 2005). Certainly particular hemicelluloses – and proteins – appear to have some control of the structure and composition of lignin that is laid down subsequently with different lignin being found in particular regions of the cell wall (Atalla, 1996), i.e. the hemicelluloses are directing both where lignin is seeded within the cell wall and the type of lignin so deposited.

5. THE STRUCTURE OF LIGNIN

Lignin is an aromatic substance that is almost totally insoluble in most solvents. It cannot be broken down to monomeric units because, even when hydrolysed, it is very susceptible to oxidation and readily undergoes condensation reactions. For this reason studies of lignin structure and chemistry have been based on modified fragments extracted from very finely ground wood ('milled wood lignin'), on low molecular weight precursors, or on model chemical reactions examined in solution. Milled wood lignin from a softwood is estimated to have a molecular weight of about 11,000, which implies it is consists of some 60 phenylpropane units (Figure 2.9a), although in its unmodified form it is likely to be much larger (molecular weight of perhaps 90,000). Softwood lignins are based almost entirely on guaiacyl-propane units (G) that cross-link with one another to form an extensive polymer molecule (Figure 2.9c). Only minor amounts of *p*-hydroxyphenylpropane (H) and syringylpropane units (S) are present (Figure 2.9b, d), except in reaction wood where not only is the amount of lignin abnormally high but the proportion of *p*-hydroxyphenylpropane is considerable. The three principal monomeric units (H, G and S) differ according to whether or not there is a methoxyl ($-O-CH_3$) group at the C_3 and C_5 positions on the aromatic ring.

Hardwood lignins contain both guaiacylpropane (G) and syringylpropane (S) units, with the G/S ratio from 4:1 to 1:2, and smaller amounts of *p*-hydroxyphenylpropane (H). Hardwood lignins are of lower molecular weight, perhaps because the syringylpropane units are unable to cross link at the C_5 position as it is blocked by the additional methoxyl group.

No regular structure for lignin has been demonstrated, although an early presumption that the lignin molecule's random structure arose simply from non-selective free radical addition and condensation reactions has proved false. The broadest structural features of wood lignins are best described in terms of the variety of ways in which the basic monomeric units can be linked together (Fengel and Wegener, 1984). The large number of possible inter-unit linkages leads to a structure of great complexity (Figure 2.9e). The linkages can be:

- Head to tail. For example alkyl-aryl ether linkages between the C_4 and a side-chain carbon of another unit (most frequently β-O-4'), or a carbon-to-carbon C_β to C_5' linkage (β-5').
- Head to head. For example the C_5-C_5'linkages.
- Tail to tail. For example the α-dialkyl-ether linkages between the α carbon of one monomer and the α', β' or γ' positions on another: and also the α-α' and β-β'carbon linkages.

Lignin composition and bonding vary according to both species and tissue studied. It can be characterized by the relative abundance of the *p*-hydroxyphenyl (H), guaiacyl (G) and syringyl (S) propane units, and the distribution of the interunit linkages, such as those mentioned above. The inter-unit bonding is either a C-C bond or a C-O-C bond (Figure 2.9e).

The final building of the lignin polymer involves the enzymatic dehydration of the precursors, *p*-coumaryl alcohol, coniferyl alcohol and sinaply alcohol (Figure 2.10a). A number of resonance-stabilized phenoxyl radicals are possible (Figure 2.10b) and various polymer linkages are formed depending on the reactivity and relative abundance of each of the phenoxyl radicals (Figure 2.10c). Dimers formed *in vitro* by dehydrogenation polymerization (also described as oxidative coupling) form the same linkages as are also found in wood lignin. However in the developing cell wall the proportions of such linkages differ from those found *in vitro* (derived by mimicking such reactions in the laboratory where there is no explicit biological control over the way the monolignols polymerize – unlike all known biochemical processes). For example β-O-4' is the predominant interlink unit in natural lignins and β–β' is rare, whereas *in vitro* β–5' and β–β' are more abundant than β-O-4'. This implies that dilignols (dimers) and low molecular weights oligomers do not form initially in the cell wall and subsequently condense to produce the lignin →

Figure 2.9. The structure of lignin (Adler, 1977). (a) The phenylpropane skeleton with the carbon atoms labelled. (b) 4-hydroxyphenylpropane: found in bamboo and grasses but only a minor constituent of wood lignin. (c) Guaiacylpropane: the principal constituent of softwood lignin. It is also a significant component of hardwood lignin. (d) Syringylpropane: abundant in hardwood lignin, rare in softwoods. (e) Important structural linkages in softwood lignin.

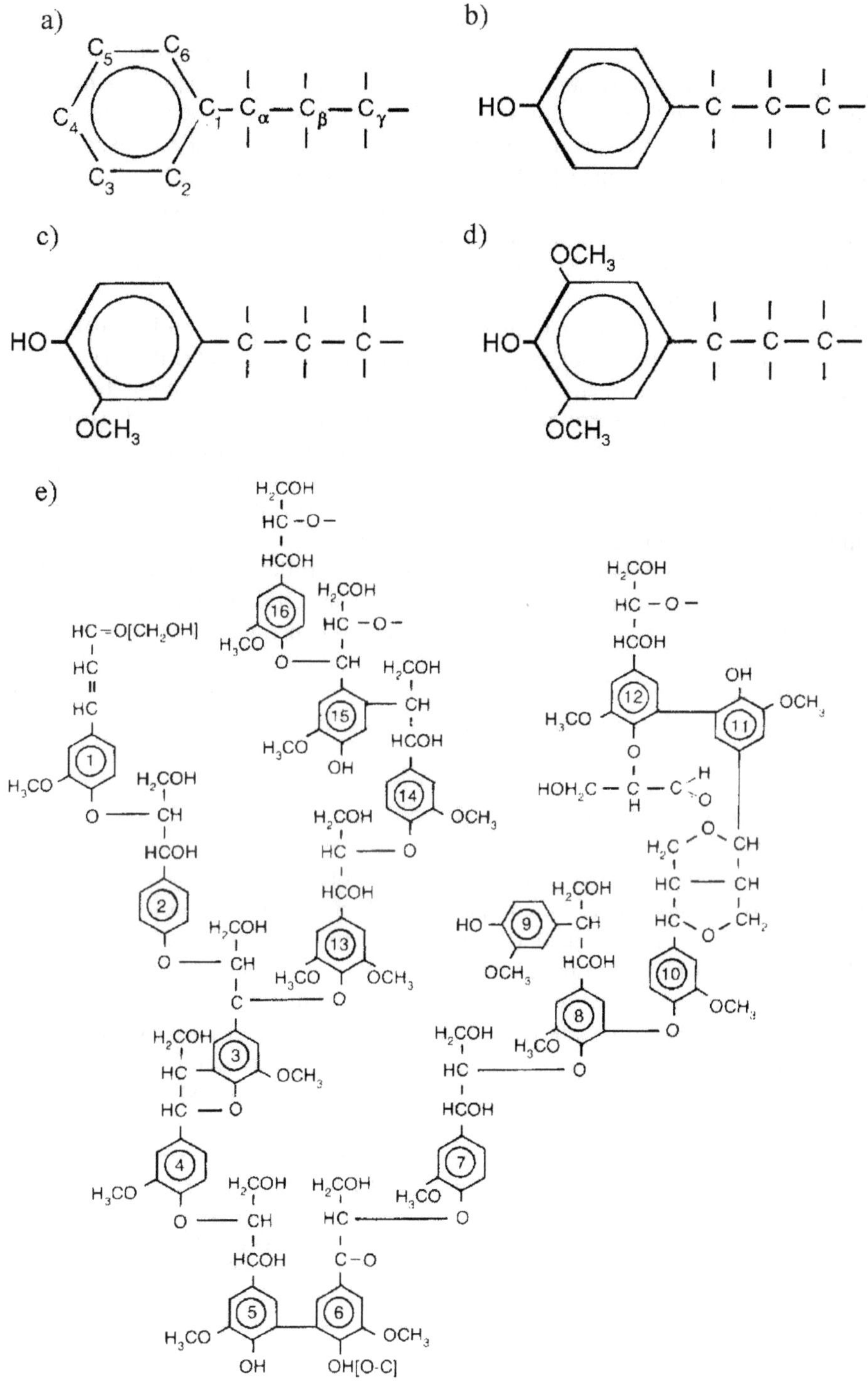
a)
b)
c)
d)
e)

macromolecule. Instead, current thinking is that after their synthesis individual monolignols are transported across the plasma membrane to some distant part of the developing wall where they are oxidized and polymerized by aligned 'end-wise' coupling to a growing polymer – one monomer at a time – according to some (uncertain as yet) self-replicating template (Boerjan *et al.*, 2003). The spatial arrangement presented by the monomer (H, G or S) is directed to ensure some predetermined coupling, such as β-O-4', to conform to and attach with the surface of the growing lignin molecule. Before polymerization can commence, the polymer appears to need an initiation ('seeding') site that is specific to some mono, di or trilignol (and whether H, G or S) with assembly directed by a dirigent protein (Latin *dirigo* = to guide/ align).

Hints of such control and specificity are revealed throughout cell lignification. The H-unit appears to be incorporated first in the cell corners and then H and G-units grow from the edge of the cell plate into the middle lamella; once the middle lamella is lignified, G-units polymerize within the cell wall of softwoods (from the outside in, i.e. from the primary wall to the S_3 layer). Last S-units are observed during the middle to late stages of lignification in hardwood fibre walls. All this is suggestive of a highly controlled process, i.e. some form of explicit control in the assembly of the final molecule (Atalla, 1996, 2005).

In summary, lignin polymerization is carefully controlled and occurs in different parts of the cell wall at different times utilizing specific monolignols with the assembly template for seeding polymerization being the hemicelluloses (or proteins) already adsorbed on the microfibrils – and the lignin molecule takes its structural form from that template.

The relative abundance of the principal bonds (Figure 2.10) in softwood lignin is:

$$\beta\text{-O-4}' > 5\text{-}5' \text{ and } \beta\text{-}5' > 4\text{-O-}5' > \beta\text{–}\beta' > \beta\text{-}1'$$

with a number of other linkages occurring in minor or trace amounts. The infrequent β-1' linkage is of some interest as on pulping this phenylpropane unit is cleaved to yield a linear three-carbon chain (the α, β, γ carbons) while the aryl fragment attaches to the other phenylpropane unit.

Every fourth or so guaiacylpropane unit (G) is trifunctional (bonding mainly at the O-4, 5 and β positions) allowing lignin to become branched. However with syringyl units (S) the C_5 linkage is absent, so for hardwoods rich in S-units the lignin tends to be of lower molecular weight and relatively unbranched – and easier to chemically pulp.

Figure 2.10. The structure of lignin (Adler, 1977). (a) Building units of lignin. (b) Enzymatic dehydrogenation of coniferyl alcohol results in a number of resonance-stabilized phenoxyl radicals. Subsequent polymerization creates the variety of linkages between phenylpropane units that is characteristic of lignin. (c) Common substructures and their proportions, as found in *Picea abies* and *Betula verrucosa* milled wood lignins. The proportion of individual bond types is noted beside each substructure for softwoods (S/w) and hardwoods (H/w).

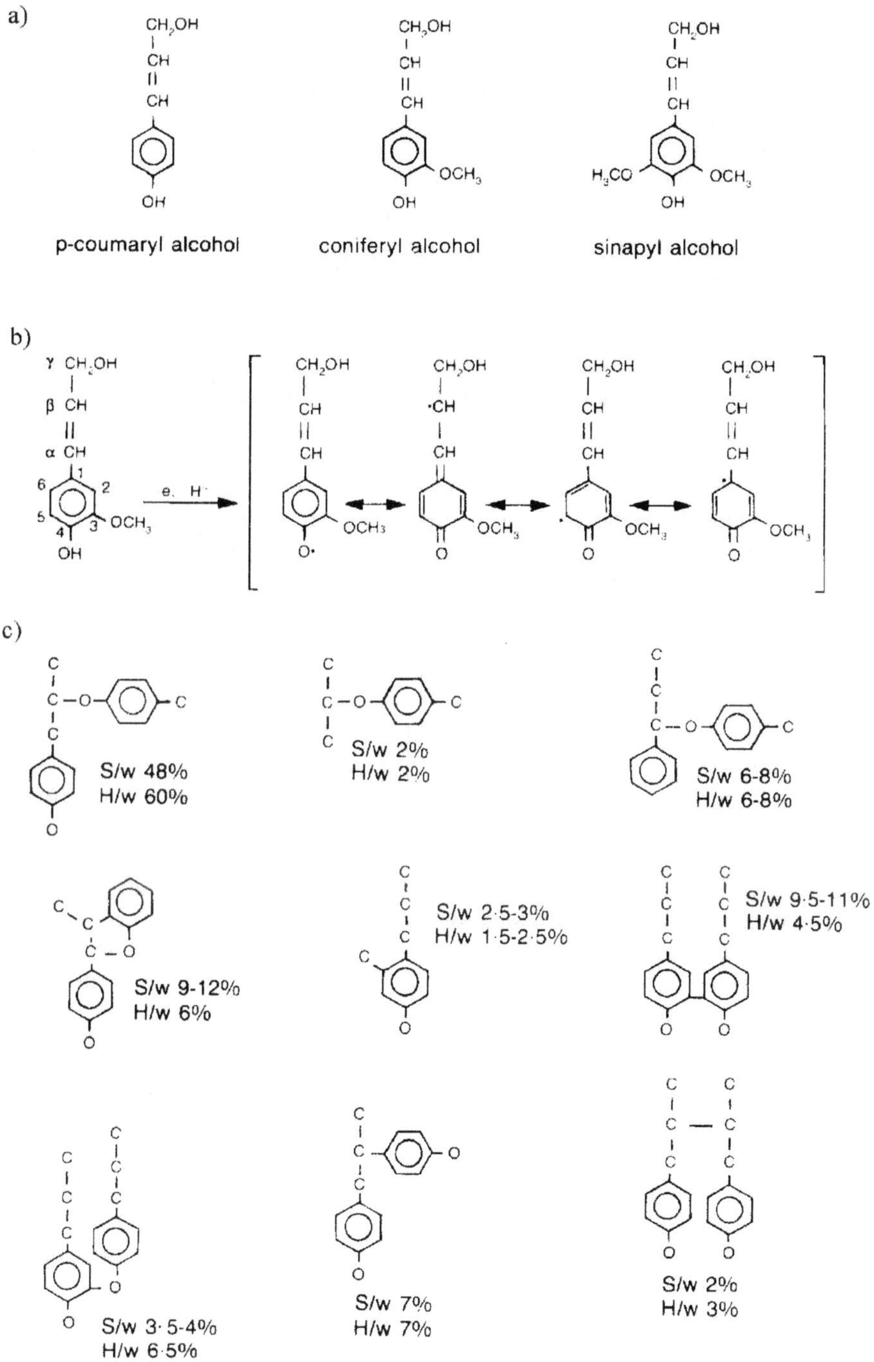
a)
p-coumaryl alcohol
coniferyl alcohol
sinapyl alcohol
b)
c)
S/w 48%
H/w 60%
S/w 2%
H/w 2%
S/w 6-8%
H/w 6-8%
S/w 9-12%
H/w 6%
S/w 2·5-3%
H/w 1·5-2·5%
S/w 9·5-11%
H/w 4·5%
S/w 3·5-4%
H/w 6·5%
S/w 7%
H/w 7%
S/w 2%
H/w 3%

The *p*-hydroxyphenyl (H) forms predominantly C-C linkages, on the C_3, C_5 and C_β carbons; whereas the syringylpropane (S) units is constructed predominately with β-0-4' linkages and portions of its molecule are quite linear. The bonding for the guaiacylpropane (G) is somewhat in-between. However, where H and G or S-units inter-link the bond is likely to be β-0-4'(Russell *et al.*, 2000). The β-0-4' bond permits planar stacking of the aromatic rings on top of one another while also providing a high degree of flexibility (rotational movement about the bond).

Lignin biosynthesis follows behind the laying down of structural carbohydrates and proteins that establish the overall architecture of the cell wall. Lignification initiation begins at distinct sites at different times and with differing proportions of H, G and S-units (Donaldson, 2001). According to Fujita *et al.* (1983) initial lignin deposition in *Cryptomeria japonica* occurs in the cell corners and then extends to the whole of the middle lamella. Lignification of the middle lamella commences at the same time as the formation of the S_1 layer by the carbohydrates and is completed before S_3 formation. A gradual lignification of the cell wall itself commences during the later stages of S_2 formation by the structural carbohydrates and lignification proceeds more rapidly after the entire microfibrillar network has been formed: lignin deposition coincides with the death of the cell. Once the cellulose and hemicelluloses have been laid down the lignin monomers diffuse into that part of the cell wall and polymerize *in muro:* probably requiring the presence of the polysaccharide or protein surfaces to assist in its initial seeding. This temporal sequence can have curious consequences: Donaldson (2001) observed that complete lignification of some latewood cells in radiata pine can be delayed until the following spring. Prior to lignification the cell wall is about 65% (vol/vol) water allowing space for diffusion of monolignols. After lignification the swollen cell wall is about 25% (vol/vol) water.

Hardwood lignin has a higher methoxyl content, $-OCH_3$, than does softwood lignin (18-22% vs. 12-16%). This is because hardwoods have a significant proportion of sinapyl alcohol in its synthesis, whereas softwood lignin is synthesized predominantly from coniferyl alcohol. The lignin of bamboo has an even lower methoxyl content which is attributed to the significant proportion of coumaryl alcohol monomer included in its synthesis (*c.* 30%). *p*-hydroxyphenylpropane is incorporated in the lignin of both hardwoods and softwoods, but its proportion is low, *c.* 10%, and it is selectively incorporated in the cell corners/middle lamella. In hardwoods further heterogeneity of lignin is revealed in the differing composition of lignin between the various cells and within these cells. In the case of *Betula papyrifera,* Saka and Goring (1985) observed that the corners of the middle lamella between all cells are rich in guaiacyl (>80%) as is the S_2 layer of vessels (*c.* 88%), whereas the S_2 layers of fibres are syringyl-rich (*c.* 88%).

As a general rule the water-conducting cells – tracheids in softwoods and vessels in hardwoods – have guaiacyl-lignin. On the other hand, hardwood fibres and parenchyma walls have both G and S-units (Wu *et al.*, 1992). Greater cross-linking from G-units and higher lignification of vessel walls (as compared to fibres), especially of ring-porous hardwoods or in earlywood vessels of diffuse-porous species, would allow the vessels to sustain higher capillary tension. The G-units and the greater amount of lignin contribute to the stiffness and collapse resistance of the vessel wall. Further one might expect a thicker S_3 layer with transverse microfibrils. Taken together, these factors support the conduction of water up tall stems.

Some hemicelluloses, predominantly arabinoxylans, are chemically linked to lignin. The bonding is through the C_2 and C_3 in the xylose residue and at the C_5' of the arabinose side-chain (Minor, 1983). Such linkages occur roughly once every 35 phenylpropane units. The presence of linkages of this kind, together with the fact that lignin interpenetrates the hemicelluloses, emphasize the impracticality of chemically isolating pure lignin, or isolating pure cellulose or hemicelluloses for that matter. Some contamination is inevitable. Further, as already noted even the mildest treatments to remove the cellulose and hemicelluloses also modify the lignin. Thus the isolated lignin differs from the original: the isolated lignin is fragmented to a degree and the bonds that are broken are susceptible to condensation reactions.

The formation of occasional ester, ether or ester-ether linkages (Koshijima and Watanabe, 2003) between the hemicelluloses and lignin creates a lignin-carbohydrate complex (LCC). For example, in the presence of hemicelluloses the coupling of phenoxy radicals can result in attachment through a benzyl ester bond to a carboxylic acid group on a hemicellulose (Koshijima and Watanabe, 2003). Thus a hemicellulose molecule can have a number of low molecular weight lignin entities attached as branches (a hemicellulose chain with multiple-pendants of lignin, generally through the α- or β-hydroxyl of the lignin). Where confined between microfibrils in the secondary wall the growing spherical entities are forced to elongate and align with microfibrils (Donaldson, 2001), and eventually to coalesce together creating an increasingly hydrophobic environment. The suggestion is that specific 1, 2 or 3-ring side-chains on the hemicelluloses can act as a template for lignin initiation by specific mono, di or trilignols (with assembly directed by a dirigent protein): hemicellulose side-chains are commonly of one or two units. These linkages make the removal of lignin – the aim of chemical pulping – more tricky. The residual lignin surviving towards the end of a pulp cook is bound tenaciously such that subsequent alternative strategies are necessary to effect their removal (in the bleach plant).

In materials science, with industrial composites the amount of material at the interfaces is small compared to that within the bodies of the individual components: there are two clearly distinct components or phases. Consequently the properties of the composite are determined by the bulk properties of the two components. Following this approach, wood has been viewed as a two-phase system of crystalline microfibrils embedded in a non-crystalline matrix of hemicelluloses and lignin. From this perspective the hemicelluloses and lignin are so intermixed that at times one needs to consider the two as a single-phase continuum and at other times as distinct, individual molecules. The intimate mixing of the hemicelluloses with lignin at the nanoscale level may not always admit an explicit definition of an interface between the hemicelluloses and lignin despite the hemicelluloses being hydrophilic and the lignin being hydrophobic. Yet the hemicelluloses and lignin can behave independently with each displaying a moisture-dependent glass-to-rubber (brittle-to-ductile) transition at distinctly different temperatures in green wood (Kelly *et al.*,

1987). Further, in the cellulose phase the surface chains of the microfibrils are subtlely different in their hydrogen bonding to those in the crystalline interior.

Jurasek (1996, 1998) provides a conceptual 'model' simulating the assembly of lignin. For the middle lamella representative macromolecules were 'grown' from initiation sites subject to known constraints such as the natural frequencies of bond types (β-O-4', 5-5' *etc*) and the required stereo-selective alignment of the monomer to match the docking surface of the growing polymer (not all bond types will be feasible), i.e. bond angles (bond energy minimization) predetermines the preferred bond type(s) and whether H, G or S can be incorporated. Jurasek's 'virtual' lignin grew by end-wise extension as a random coil with occasional branching (tri-functional units). Where side branches of the same polymer grew sufficiently close to bond they were allowed to cross-link, forming closed loops/rings of phenylpropane units (*c.* 13 such loops per 100 units). The majority of these loops were short with 3-4-5 chain loops, so they would not be very effective in resisting polymer fragmentation and solublization during pulping. The middle lamella lignin grew as compact balls until about 20 nm in diameter at which point lignin from other initiation sites cross-linked and merged. The resulting polymer reached a density of 1350 kg/m^3 – a value typical of natural lignin (Ramiah and Goring, 1965). On sectioning the virtual lignin revealed a degree of porosity sufficient to allow diffusion of small molecules, *c.* 1 nm in diameter, which agrees with porosity values from laboratory studies. The virtual lignin was highly cross-linked.

Within the secondary wall, Jurasek's virtual lignin was grown in the 7 nm space between microfibril surfaces and was forced to share this space equally with the pre-existing hemicelluloses, with half the hemicelluloses lining some of the microfibril surface (glucomannans) and the other half (xylans) in the remaining space (an effective film thickness of 3-5 nm) to be ultimately shared with the lignin. In contrast to the middle lamella, the lignin evolved a looser stringy structure with cavities of various sizes, some already occupied by the hemicelluloses but others, arising from the restricted bonding possibilities for the lignin, were available for water. Here the heterogeneity of bonds proved essential for the snug packing within these confined spaces. The lignin polymers were shorter and were less able to form closed loops as only half the bonding space was accessible, because of the pervading presence of the bounding microfibrils – and so fewer choices for bonding. Such lignin would be expected to break down faster during pulping than that in the middle lamella. Further, the aromatic rings were constrained by the narrow spaces to align roughly parallel to the cell-wall lumen surface (in agreement with Raman spectroscopy). Indeed the deposition of lignin in the secondary wall approximates to thin films of lignin squeezed into complex spaces intermixing with hemicellulose molecules and sandwiched between the faces of the microfibrils (Figure 2.8b).

In chemical pulping the dissolution of lignin is as much a function of accessibility as of lignin reactivity. Earlier, Goring (1983) had observed that the middle lamella and cell wall lignins of spruce wood differ in structure and chemical reactivity. The middle lamella lignin was less reactive than the cell wall lignin. Goring suggested that the lower methoxyl content of the middle lamella lignin would result in a greater number of C-C bonds at the C_3 and C_5 positions. Such bonds are hard to break during pulping. Dissolution of lignin in the middle lamella is needed to achieve fibre separation; but the lignin distributed throughout the wall must be attacked first and pulping is not specific (hemicelluloses are degraded too). The chemicals have to diffuse/move into the wood chips and the lignin fragments have to diffuse out across the cell walls. Lignin fragment remain in place until small enough, e.g. a DP of *c.* 10. However as pulping progresses and the intracellular bonds weaken, the fibre wall swells allowing larger lignin fragments to diffuse. The dissolution of lignin is achieved through breaking the ether linkages, primarily of the abundant β-aryl ether type (β-O-4' linkage) that account for more than half of the total inter-unit linkages, and the solublizing of lignin fragments. More than two-thirds of the phenylpropane units in softwood lignin are linked by ether bonds (C-O-C), the rest being carbon-carbon bonds (Sjöström, 1981). Carbon-carbon bonds (e.g. 5-5' and β-5') are much more difficult to cleave. The faster chemical pulping of hardwoods can be accounted for by the presence of syringyl lignin, the lower molecular weight of hardwood lignins and the reduced frequency of carbon-carbon bonds. On the other hand, hardwoods with a high S/G ratio are more susceptible to fungal delignification.

In the secondary wall it has been proposed (Goring *et al.*, 1979) that the lignin is sandwiched between lamellae of polysaccharides, with the thickness of the lignin lamella/film being about 2 nm (Figure 2.5a). The evidence for this is the uniform thickness of solublized lignin fragments when spread on a liquid surface to form a monolayer, regardless of the molecular weight of the lignin fraction used, i.e. larger macromolecular fractions have the same thickness but spread further. The size and disc-like shape of macromolecular fragments after chemical delignification can be measured under the electron microscope using various techniques. Again, the estimated thickness of the lignin fragments was about 2 nm.

Ruel *et al.* (2005) used immuno-labelling to examine the cell wall disposition of polydiverse linear xylans in hardwoods, specifically distinguishing between largely unsubstituted xylans (that bind to the microfibrils) and those substituted by 4-O-methyl groups (that is intermixed with the lignin). At the same time they distinguished between the three monolignols, distinguishing particularly between the condensed guaiacyl polymers and the non-condensed mixed G-S polymers. The spatiotemporal consequences for lignin deposition is that early lignification at any point in the cell wall involves condensed G-lignin interacting with the linear xylans deposited on the microfibrils in the water-swollen unconsolidated wall. Subsequently as the space becomes more confined the G-S lignin forms by endwise polymerization around the substituted xylans in an increasingly hydrophobic environment during secondary wall thickening. The latter may interact with the already present condensed type of lignin and act as cross-linking molecules (see schematic representation in Figure 2.8b). Such associations between the two types of lignins can be established via hydrophobic bonds between the aromatic rings, with the consequence that as secondary thickening and lignification proceed, water is expelled and the hydrophobicity of the wall is increased.

Clearly there are commercial interests in developing clonal lines of trees that have modified or reduced amounts of lignin for fast-growing, short-rotation crops grown exclusively for pulpwood (a practice largely confined to the subtropics). Such clonal lines would be of particular interest for pulping softwoods, where extensive cross-linking in the H and G-units makes for slower delignification. However, one needs to be cautious as lignin contributes to the compressive and bending strength of structural lumber. Also the pulp industry generally survives by purchasing the unwanted residues from sawlog forest crops at a huge discount, as high as 90%, to that paid by sawmillers: the tail should not wag the dog.

Low-yield chemical pulp fibres – largely free of lignin and of a large part of the hemicelluloses – are generally beaten or mechanically refined to improve subsequent paper properties. The process takes advantage of the weakened bonding between microfibrilar arrays (Figure 2.8c) and lamellae and as a result the fibres can unravel or separate into numerous concentric layers (Figure 2.11).

6. THE CELL WALL STRUCTURE OF A SOFTWOOD TRACHEID

Bailey and his co-workers worked out the basic structure of the softwood tracheid (Figure 2.12) in the 1930s, e.g. Bailey and Kerr (1935); the orientation of the microfibrils within the cell wall was determined by polarizing microscopy and by

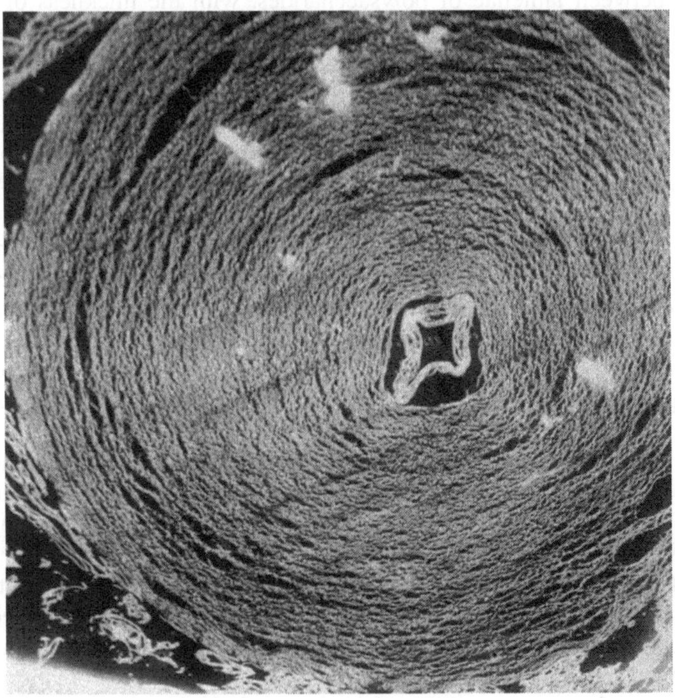

Figure 2.11. Cross-section of delignified swollen lamellae in a spruce fibre after removal of lignin by sulphite pulping, solvent exchange (to swell the fibre) and polymer impregnation (to replace the solvent by a hard polymer) prior to sectioning (Young, 1986).

x-ray techniques (Preston, 1974; Wilson and White, 1986); while the electron microscope revealed the cell wall ultrastructure in more detail and with greater clarity (Butterfield and Meylan, 1980).

The principal features are the middle lamella (ML), a primary wall (P), and a compound secondary wall (S). The orientation of the microfibrils in the various layers is shown schematically. By convention, the direction in which the microfibrils wind around the cell is defined by reference to their slope on the outer surface of the cell layer when viewed externally. The helix can be either Z or S. The microfibrils form a Z helix if they wind up from the lower left to the upper right (as in the S_2 layer in Figure 2.12), and a S helix if they wind up from the lower right to the upper left. The direction of the helix is determined by comparing the direction of the microfibrils with the direction indicated by the middle stroke of the letters Z or S.

The middle lamella is the intercellular region. It contains material that holds adjacent cells together. Structurally, it is not part of the cell wall. While the cells are enlarging it is largely pectic and only later does the middle lamella become highly lignified. There are no cellulose microfibrils in the middle lamella.

Cells that have been formed recently at the vascular cambium have only a very thin primary cell wall. Even in the fully lignified cell the primary wall is very thin (0.1 μm) and can be hard to distinguish from or isolate from the middle lamella: many studies analyse the two together (ML+P) and relate results to the compound middle lamella (CML), a term which embraces both middle lamella and primary wall. The primary wall displays both elasticity and plasticity (permanent extension) during early cell growth and extension – at this stage in tracheid cell development

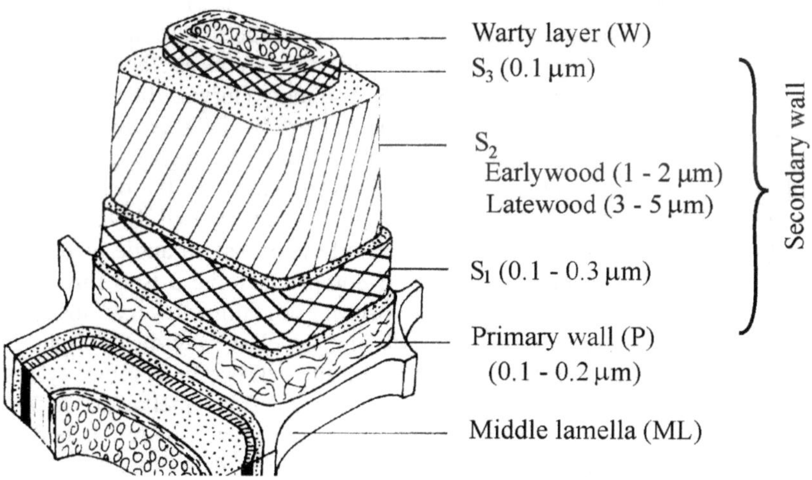

Figure 2.12. Schematic representation of the microfibrillar structure within the cell wall of a softwood tracheid (Côté, 1967).

the secondary wall has yet to be laid down. In the primary wall the microfibrillar network is unstructured and the microfibril orientation is random except near the corners of the cell where they run axially along the length of the cell. The microfibrils are embedded in a gel-like matrix of hemicelluloses and pectic compounds. Pectinaceous material is colloidal and made up of uronic acids derived from the same kind of sugars as the hemicelluloses. The primary wall only becomes lignified after the deposition of the secondary wall has commenced. In its development the primary wall is maintained in tension by the turgor pressure within the living cell while the microfibrils winding round the developing wall at a large angle to the cell axis readily admit considerable longitudinal extension. There is also limited radial expansion while tangential expansion is prevented by adjacent cells in the cambium. The microfibrils are interlinked by xyloglucan chains. In theory the optimal shape for such an elongating cell under high turgor pressure is a cylinder with rounded ends, i.e. with a circular cross-section rather than square, rectangle or polygonal. Consequently turgor pressure generates forces that seek to separate the cells at the cell corners. These forces scale with size, so large diameter cells are more likely to separate at the middle lamella or delaminate within the primary wall (Jarvis, 1998, 2005a). Thus large earlywood cells are more vulnerable than latewood cells. Intercellular failure is inhibited by pectic compounds that provide cross-links between adjacent primary walls, e.g. the galacturonans form stable salt bridges involving Ca^{++} and Mg^{++} ions. These pectic polysaccharides and structural proteins may be forming a three-dimensional network that later interpenetrates/intermixes with the microfibril-xyloglucan network in the primary wall so preventing cell separation while permitting cell elongation. Within-ring internal checking, evident largely after kiln drying of lumber, may originate in poor intercellular adhesion prior to lignification. It is unclear whether the initiation of the failure is in the ML, at its interface with the primary wall or at the interface between the P and S_1 wall, although the failure zone migrates to the P/S_1 boundary (Jarvis, 2005a).

The secondary wall is laid down after the primary wall. Three distinct layers, the S_1, S_2, and S_3, are recognized. The cellulose microfibrils are highly organized and lie broadly parallel to one another within these layers. However the orientation of the microfibrils differs within the three layers of the secondary wall. In the thin S_1 layer (0.1-0.3 μm) the microfibrils lazily wind round the cell at an angle of between 50° and 75° to the cell axis. The S_1 wall itself comprises some 3 to 6 thin concentric lamellae. Within each lamellae the microfibrils are aligned very closely to one another, while the orientation between adjacent lamellae may differ slightly or the direction in which the microfibrils wind round the cell can be completely reversed, switching from a S to Z helix or *vice versa*. Both S and Z lamellae are found in the S_1 layer. Generally the S lamellae are more developed. After the few lamellae of the S_1 have been laid down the orientation of the microfibrils rapidly changes to that found in the S_2 layer. The microfibrils in the S_2 layer are densely packed and steeply inclined, making an angle of only 10° to 30° with the tracheid axis. There are a large number of lamellae in this layer ranging from 30 in earlywood to 150 in dense latewood. All are similarly orientated (but with some variation about the mean angle, *c.* ± 5-10°) and wind around the cell in the Z direction. The S_2 layer is 1-4 μm thick in earlywood and 3-8 μm in latewood. Finally in the thin S_3 layer (0.1 μm) the orientation of the microfibrils changes again to a predominantly S helix with the microfibrils steeply inclined relative to the cell axis (60°-90°). Thus the predominant orientation of microfibrils across the secondary wall is S-Z-S in the S_1, S_2 and S_3 layers respectively. The S_2 layer is by far the thickest and is the dominant feature of thick-walled latewood tracheids. Earlywood tracheids are thinner walled and this is principally due to a thinner S_2. The distribution of the microfibrils about the mean microfibril angle in the in the secondary wall layers is represented schematically in Figure 2.13.

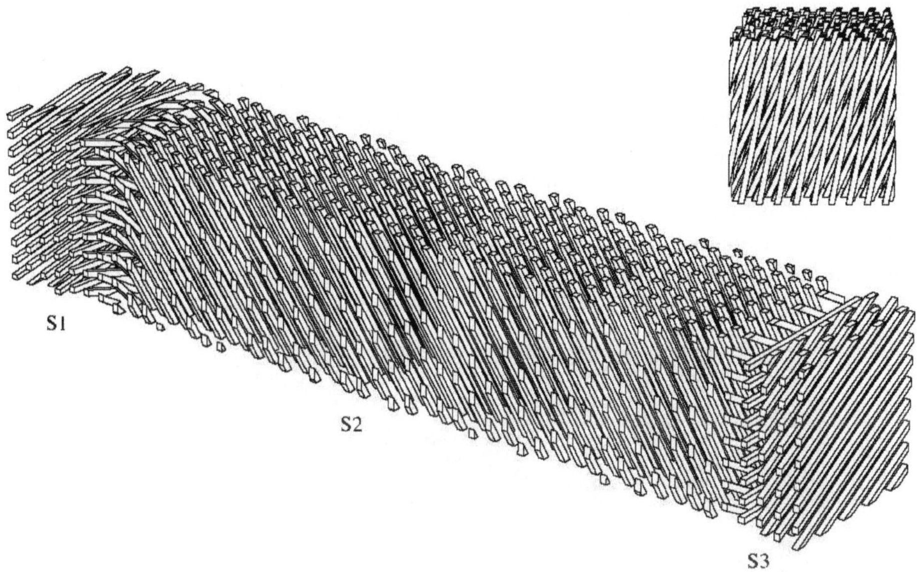

Figure 2.13. Microfibrils in the secondary wall, visualized using a 'trellis-like' concept; with inset of the tangential-longitudinal face in the S_2 layer (Harrington, 2002). There is a broad distribution of microfibril angles about the mean. Most literature only quote the mean value.

Finally a thin warty layer may be deposited on the cell wall, which appears to be composed of proteinaceous or lignin-like material. Where present the warty layer completely lines the cell lumen and pit cavities. It is widely found in softwood tracheid elements.

The wall structure determines largely the chemical, mechanical and physical properties of wood. The cell wall has many features characteristic of, and sometimes mimicked by, modern fibre composites of which the familiar resin treated fibre-glass and the more esoteric carbon-fibre composites used in aircraft jet engines are but two examples. The mechanical characteristics of wood can be analysed using a reinforced matrix model. The cellulosic microfibrils are the very strong (they have a high failure stress) and stiff (highly inextensible) reinforcing material. The non-crystalline material between the microfibrils is the bulking matrix, which can redistribute stresses between microfibrils, deform (change shape) and so accommodate any shear stresses within the wall, and support the microfibrils in compression (prevent them from buckling) and carry compressive loads in its own right. What is already a complex model is complicated further by the fact that adsorbed water within the cell wall has a profound effect on the cell wall structure and properties. The relative proportions of the various layers are beautifully revealed (Figure 2.14): the delignified tracheids have been made to balloon by using chemicals that swell the cell wall.

7. DISTRIBUTION OF CELL CONSTITUENTS

The distribution of cellulose, hemicelluloses and lignin can be estimated using various techniques. First, by separating postcambial cells at various stages of development so that the cells have only the primary wall tissue, or successively $P+S_1$, $P+S_1+S_2$, and $P+S_1+S_2+S_3$ wall material. This provides information on the relative proportions of cellulose and hemicelluloses within these layers. However at this stage of postcambial development the cells may not be lignified. The lignin distribution in mature cells can be determined by UV spectroscopy. Finally, chemical analysis of the mature cell wall allows one to allocate various amounts of

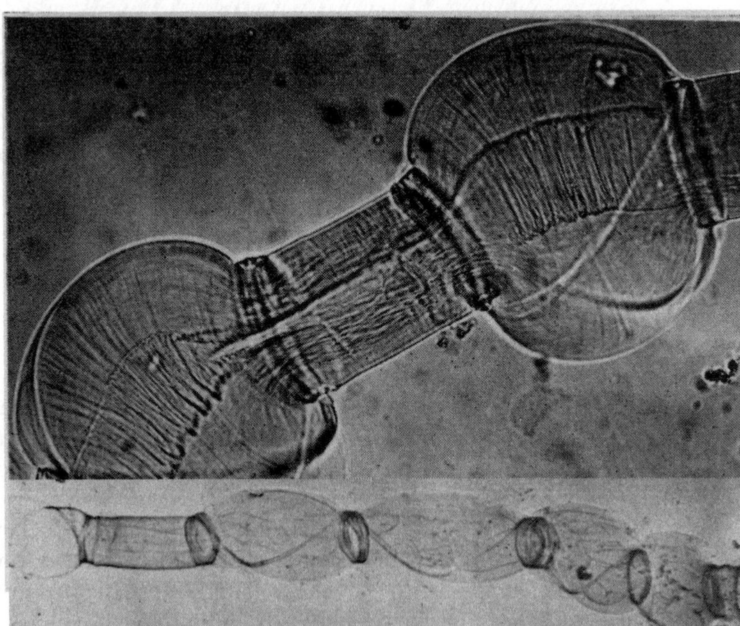

Figure 2.14. Ballooning of delignified softwood tracheids after treatment with very strong swelling agents (Wardrop and Dadswell, 1950).

the cell constituents to the cell wall layers. A schematic interpretation is shown in Figure 2.15. The cell corners of the middle lamella are lignin-rich (*c.* 70% wt/wt). The compound middle lamella, ML+P, contains more than 50% (wt/wt) lignin. The proportion of lignin in the S_2 region is about 20% (wt/wt). However, because the S_2 layer is dominant, approximately three-quarters of the total lignin is in the secondary wall with only a quarter in the lignin-rich middle lamella and cell corners (Donaldson, 2001; Fukazawa and Imagawa, 1983).

The cellulose content of hardwoods is marginally higher than for softwoods (Table 2.2). These figures are derived by first determining the total carbohydrate component and then subtracting the analysed hemicellulose values. A rough estimate of the cellulose content is given by the α-cellulose content, which corresponds to that portion of the wood cellulose which is insoluble in strong, cold alkali (17.5% NaOH). The latter figure is somewhat high due to contamination by various hemicelluloses. For example, for *Pinus radiata* the α-cellulose is 42-45% while the true value for cellulose is around 41-42%.

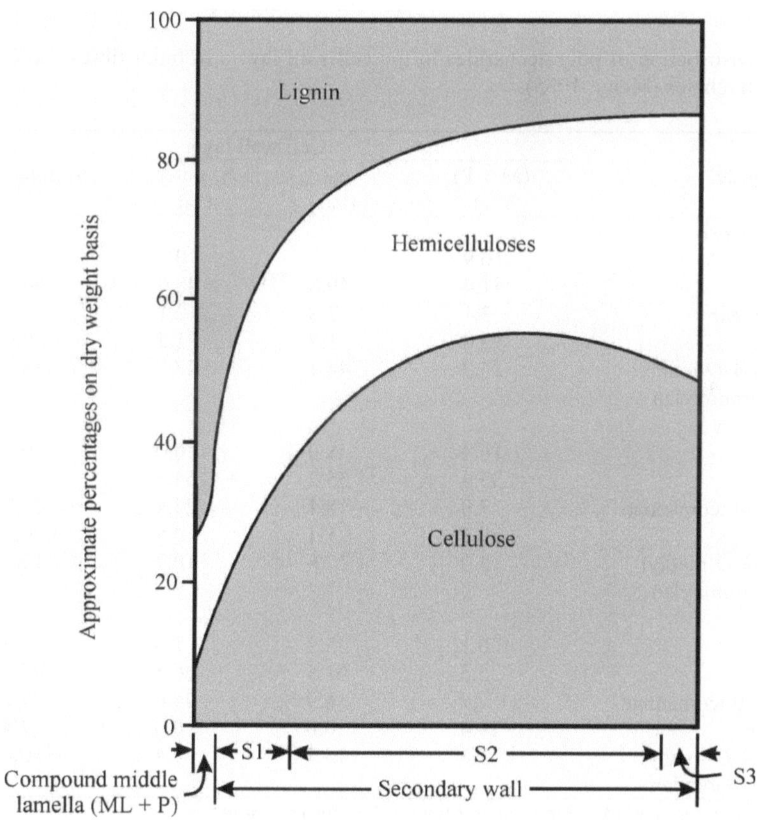

Figure 2.15. Schematic distribution of the principal cell wall constituents within a softwood tracheid (Panshin and De Zeeuw, 1980).

There is a large difference in the mean lignin content of hardwoods and softwoods (Table 2.2). The load bearing fibres of hardwoods are thicker-walled than are the tracheids of softwoods. One could speculate that the greater rigidity of a thick wall implies less demand for an intrinsically high within-wall rigidity that is provided by lignin.

A fuller chemical analysis reveals the principal polysaccharides (Table 2.3).

Table 2.2. Typical composition for the softwoods and hardwoods.

Polymers	Softwoods (%)	Hardwoods (%)
Cellulose	42 ± 2	44 ± 2
Hemicelluloses	27 ± 2	28 ± 5
Lignin	28 ± 3	24 ± 4
Extractives	3 ± 2	4 ± 3

Table 2.3. Distribution of polysaccharides in the cell wall layers of birch fibres and of pine and spruce tracheids (Meier, 1985).

	Cell wall layer			
Polysaccharide	(M + P) (%)	S_1 (%)	S_2 outer (%)	S_2 inner + S_3 (%)
Betula verrucosa				
Galactan	16.9	1.2	0.7	0.0
Cellulose	41.4	49.8	48.0	60.0
Glucomannan	3.1	2.8	2.1	5.1
Arabinan	13.4	1.9	1.5	0.0
O-Acetyl-4-*O*-methyl-glucuronoxylan	25.2	44.1	47.7	35.1
Pinus silvestris				
Galactan[a]	16.4	8.0	0.0	0.0
Cellulose	33.4	55.2	64.3	63.6
O-Acetylglucomannan	7.9	18.1	24.4	23.7
Arabinan	29.3	1.1	0.8	0.0
Arabino-4-*O*-methyl glucuronoxylan	13.0	17.6	10.7	12.7
Picea abies				
Galactan[b]	20.1	5.2	1.6	3.2
Cellulose	35.5	61.5	66.5	47.5
O-Acetylglucomannan	7.7	16.9	24.6	27.2
Abarinan	29.4	0.6	0.0	2.4
Arabino-4-*O*-methyl glucuronoxylan	7.8	15.7	7.4	19.4

[a] Contains also pectic acid, which has not been taken into account.

[b] The galactan content indicated is somewhat too high, since some galactose is part of the glucomannan (double counting).

Similarly a more detailed distribution of lignin is shown in Table 2.4. The lignin concentration (wt/wt) is lower in the S_2 layer than in either the S_1 or S_3: presumably the highly regular packing of the microfibrils in the S_2 layer leaves less room for the deposition of lignin. However the thickness of the S_2 means that much of the lignin is located in this layer.

Table 2.4. The distribution of the lignin in the tracheid wall of loblolly pine (Saka and Goring, 1985).

Wood	Morphological region	Tissue volume (%)	Lignin % total	Lignin Conc (wt/wt)
Earlywood	S_1	13	12	0.25
	S_2	60	44	0.20
	S_3	9	9	0.28
	ML	12	21	0.49
	ML_{cc}	6	14	0.64
Latewood	S_1	6	6	0.23
	S_2	80	63	0.18
	S_3	5	6	0.25
	ML	6	14	0.51
	ML_{cc}	3	11	0.78

ML_{cc} refers to the cell corners.

8. WOOD EXTRACTIVES

There are detailed accounts of wood extractives and their chemistry by Sjöström (1981), Hillis (1987), Fengel and Wegener (1989), Gang *et al.* (1996) and Hon and Shiraishi (2001) so only broad aspects of their isolation and chemistry are described here. The term wood extractives is used to describe the numerous compounds that can be extracted from wood using polar and non-polar solvents. Both the nature and amount of extractives in wood are important in wood utilization. The amount of extractives in wood can range from 1 to 20% depending upon species and position within the tree. Sometimes the effects of extractives on wood utilization are large relative to the small amounts often present in the tree. Extractive content, and the related term resin content, generally refer to compounds extractable with organic solvents. However extractives technically include resins that are soluble in organic solvents and carbohydrates that are water-soluble. Where the total extractive content is required, it is usual to use solvents such as methanol followed by aqueous extraction (Uprichard, 1963, 1971; Uprichard and Lloyd, 1980). The types of compound isolated depend to a large extent upon the polarity of the solvents used for extraction. Sometimes the solvent used relates to a particular aspect of wood utilization. Thus in studies related to pitch formation in papermaking, softwoods are extracted with solvents such as dichloromethane that are known to remove compounds such as resin acids, fatty acid triglycerides and other esters, and neutral compounds including sterols, which as a group mutually inhibit crystallization.

Sometime in the tree's life, heartwood formation begins adjacent to the pith, and gradually extends outward and upward resulting in a tapered cone of heartwood surrounded by sapwood (*c.* 15-30 mm wide). The extractives that characterize heartwood are stored in specialized cells – living ray parenchyma – to be secreted subsequently into the cells that become heartwood. Extractives follow completely different biosynthetic pathways to the synthesis of cell wall constituents even though they utilize as entry points similar precursors, e.g. guaiacylpropane, to those for lignification. For example the lignans in their coupling as dimers, e.g. pinoresinol a dimer of guaiacylpropane (Figure 2. 14), and subsequently as oligomers differ from lignin and perform distinctively different non-structural functions. Reflecting these differences, they are formed after lignification is complete, and in the (somewhat atypical) example of western red cedar, *Thuja plicata*, these lignans (plicatic acid and its derivatives) together with tropolones (7-membered cyclic compounds) are deposited in massive quantities (20% wt/wt) in the heartwood. Some of these compounds exist as highly insoluble oligomers that can only be extracted by finely grinding the wood before using the appropriate solvents.

Extractives function as antifeedants, antioxidants, antivirals, bacteriacides, cytotoxins, fungicides *etc.* and their presence accounts for the colour of heartwood, whether black in ebony, a red-brown in cedars, Douglas fir and hemlock, yellow-orange in the southern pine or indistinguishable from pale sapwood with poplar.

In response to some biological hazard, comparable extractives can be released from ray cells into the adjacent sapwood to isolate and localize the event.

8.1. Extractives determination and separation procedures

Extractives are generally removed from finely-ground wood samples by Soxhlet extraction with solvent for periods ranging from 4-18 hours. In the quantitative determination of extractives the air-dry ground wood sample (normally about 5 g oven-dry equivalent) is held in a paperboard thimble and continuously percolated and leached with solvent in a Soxhlet extraction unit equipped with an extraction flask and condenser; the number of extraction cycles generally being about four cycles per hour. The extract so obtained is evaporated and dried to constant weight and the extractive content determined. In the detailed study of individual extractives, for example those present in species about which little is known, extraction is undertaken on a larger scale. Techniques such as gas chromatography, mass spectrometry, and H^1 and C^{13} nuclear magnetic resonance spectroscopy are often used in identification or structural elucidation of extractives by wood chemists.

8.2. Wood extractives and their location in the tree

Wood extractives vary in nature and amount within and between species, and within trees there is generally a decrease in extractive content with tree height. With both softwoods and hardwoods, extractives are more abundant in heartwood and generally these differ in chemical composition from those in sapwood, although in *Pinus* sp. some extractives are common to both (Hillis, 1962). Wood extractives range from low molecular weight, volatile monoterpenes to higher molecular weight substances such as triterpenes and sterols, and from hydrocarbons to complex polyphenolic structures.

8.3. Softwoods

Studies on *Pinus* species have shown that the nature and amount of extractives depend upon the percentage of heartwood present and thus on tree age. In *Pinus radiata*, heartwood starts forming once the trees are about 12 to 15 years old. Heartwood extractives occur in greatest amount in inner growth rings near the pith (Uprichard, 1971; Lloyd, 1978) especially in the butt log of mature trees (Table 2.5). The high level of resin in the inner zone appears due to a process of enrichment with sapwood extractives via the transverse resin canals (Harris, 1965). Resin acids predominate in heartwood and comprise from 70-80% of total extractives, however in sapwood there are approximately equal amounts of resin acids and fatty acids (Table 2.6). An important feature of the resin constituents of pines is that a mixture of resin acids in turpentine occur in the resin canals, and the fatty acid esters and unsaponifiable materials occur in the ray parenchyma resin. In some processes, for example refiner mechanical pulping some separation of these chemical components can occur.

Table 2.5. Variation in percentage of acetone extractives (oven-dry wood basis) with position in a 40 yr-old radiata pine tree (Uprichard and Lloyd, 1980), i.e. there is a higher extractive content near the pith at the base of the tree (the samples with more growth rings).

Number of growth rings in disc	Number of growth rings from the pith, and extractive content (%)							
	1-5	6-10	11-15	16-20	21-25	26-30	31-35	36-40
15	3.5 H/S	1.8	1.5					
25	5.4	1.5 H/S	0.8	0.7	1.0			
35	7.4	2.0 H/S	1.1	1.0	0.9	1.1	1.2	
40	9.7	2.8 H/S	1.0	1.0	0.9	0.9	0.8	0.9

H/S denotes the approximate heartwood/sapwood boundary (shaded).

Table 2.6. The components of extractives, as a percentage of total extractives in heartwood and sapwood of radiata pine (Uprichard and Lloyd, 1980).

Compounds	Heartwood	Sapwood
Fatty acids (free)	2 %	1 %
Fatty acid esters	11	41
Resin acids	71	41
Phenols	6	3
Unsaponifiables (neutrals)	10	14

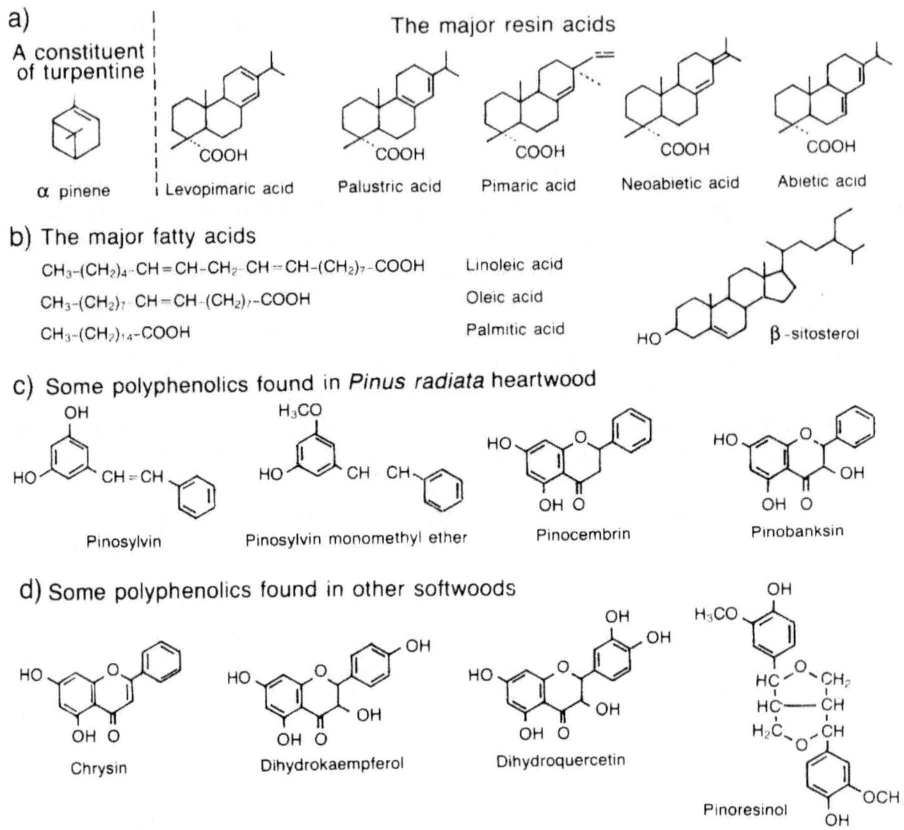

Figure 2.16. Some important extractives of softwoods. (a) A mixture of resin acids in turpentine is present in resin canals of *Pinus radiata.* (b) Ray parenchyma resin of *Pinus radiata* consists mainly of fatty acid esters and unsaponifiable materials (a major component of which is β-sitosterol). (c) Polyphenolics found in *Pinus radiata* heartwood. (d) Some polyphenols that are significant in certain other softwoods.

The turpentine fraction in *Pinus radiata* contains volatile monoterpenes: the two main constituents are α- and β-pinene. Levopimaric, palustric, pimaric, neoabietic and abietic are some of the principal resin acids. The fatty acids occur mainly as esters and the unsaponifiables are sterols, alcohols and hydrocarbons. Figure 2.16 shows some of the compounds present in softwoods.

The early chemotaxonomic studies of Erdtman (1952) showed that the heartwood of pines contained flavonoid (C_6-C_3-C_6) and stilbene (C_6-C_2-C_6) compounds, which could aid species identification. Although the two sections of the pine genus, *Haploxylon* and *Diploxylon,* both contain the stilbenes pinosylvin and pinosylvin monomethyl ether, the groups can be distinguished since the *Diploxylon* group (pines with two or three needles per cluster) contains only flavanones whereas the *Haploxylon* group (five needles per cluster) contains not only flavanones but also flavones, for example chrysin.

Flavonoid compounds are common in other softwoods. Both Douglas fir and western larch contain the dihydroquercetin (Gardner and Barton, 1960), the distribution of this constituent being rather variable within the tree, but is highest at the heartwood-sapwood boundary. In studies on New Zealand grown *Larix decidua* and *Larix leptolepis* (Uprichard, 1963) it was shown that, in the 45 yr-old trees examined, the flavonoid polyphenols (dihydroquercetin and dihydrokaempferol) increased steadily from the centre of the tree to the heartwood-sapwood boundary after which polyphenols dropped to neglible amounts.

Lignans, compounds formed by the condensation of two lignin monomer units, occur frequently in softwoods, an example of these is pinoresinol which occurs in spruce and other species (Fengel and Wegener, 1984).

8.4. Hardwoods

Hardwoods contain monoterpenes, of which camphor from *Cinnamonum camphora* is probably the most notable. They also contain fatty acids and alcohols similar to those in softwoods, the fatty acids being present as triglycerides (Fengel and Wegener, 1984). Diterpenoid compounds occasionally occur in hardwoods. Triterpenoid compounds (C30 compounds) occur frequently in hardwoods, either as alcohols or acids: the alcohol betulin and the methyl ester of acetylated betulininic acid from birch wood are shown in Figure 2.17. Triterpene acids are often present in *Quercus* and *Terminalia* species while sterols such as β-sitosterol are present in many hardwoods. Lignans also occur frequently in hardwoods, some of them containing guaiacyl units and some made up of syringyl units, for example syringaresinol.

Compounds such as ellagic acid (Figure 2.17) occur in *Eucalyptus* and *Terminalia* species, often in the form of 'ellagitannins' or glucosides. Because ellagic acid is very insoluble it can give rise to difficulties in the chemical pulping of eucalypts. Hydroxylated stilbenes are present in certain *Eucalyptus* sp. (Hathway, 1962), for example 3,5,4'-trihydroxy stilbene (Figure 2.17).

A wide variety of flavonoid compounds have been isolated from hardwoods, including flavones, flavanols, flavanones, isoflavones, and chalcones, most of which are highly hydroxylated, and which vary in hydroxylation pattern. Two of the many compounds present are shown in Figure 2.17; robinetin is the 3-hydroxy-flavone extracted from *Robinia* and *Intsia* sp. while the chalcone, okanin, is found in *Cyclodiscus* sp.

Heartwoods of many tropical species are rich in tannins. The basic unit in some of these structure is considered to be derived from flavan-3,4-diols and related compounds, for example melacacidin from *Acacia melanoxylon* (Hathway, 1962), the structure of which is shown in Figure 2.17. The stereochemistry of the flavan-diols has been the subject of much study by Roux and his collaborators (Fengel and Wegener, 1984).

a) Typical triterpenoids

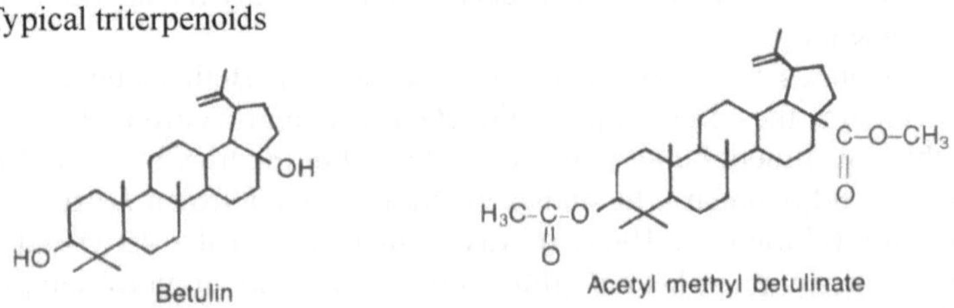

b) Polyphenols and related compounds in heartwood

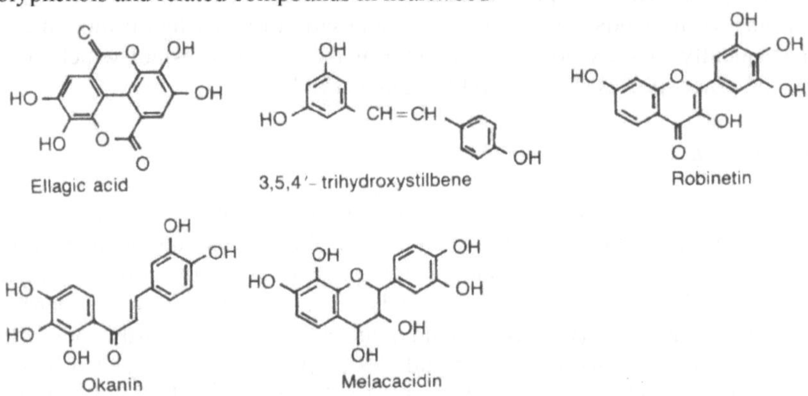

Figure 2.17. Some hardwood extractives. (a) In hardwoods the resin is almost entirely located in the ray parenchyma. The extract consists mainly of fatty acids and esters (not shown), triterpenoids being present in some species. (b) Polyphenolic extractives: Ellagic acid; 3,5,4'-trihydroxystilbene; Robinetin, 3,7,3',4',5'-pentahydroxyflavone; Okanin, 3,4,2',3',4'-pentahydroxy chalcone; Melacacidin, 7,8,3'4'-tetrahydroxyflavan 3,4-diol.

8.5. Effects of extractives on aspects of wood utilisation

This final section on extractives explains why extractives are important in utilization, and why processing studies on wood ignore them at their peril. They can be important to solid wood processing as well as to pulp and paper production, and the problems to which they give rise are still being examined and sometimes cured. In some instances wood extractives are beneficial! The effects of extractives listed below are only indicative and are not arranged in order of importance.

8.5.1. Durability

The polyphenols, i.e. stilbenes such as pinosylvin, in *Pinus* sp. are considered to be responsible for the better durability of pine heartwood compared to sapwood. The flavonoid dihydroquercetin is considered to be responsible for the durability of Douglas fir heartwood; while tectoquinone, 2-methyl anthraquinone, is responsible for the durability of teak; and cedrol contributes to the durability of Chinese fir,

Cunninghamia lanceolata (Lu *et al.,* 1987; Huang *et al.,* 2004). The diterpenoid phenol ferruginol contributes to the durability of *Cryptomeria japonica* and *Dacrydium colensoi* (Kai, 1991).

Historically, heartwood of certain timbers would be used where decay was foreseen, a practice that declined with the rise in wood preservation. Today environmentally friendly solutions to decay relying on naturally durable wood are increasingly sought.

8.5.2. Colour

The polyphenols and tannins in hardwoods largely contribute to wood colour, particularly heartwood colour, and in earlier times some hardwoods were sought after for dyestuffs. Colour change is undesirable with darker timbers that are rich in extractives, where they are to be used for joinery or furniture. With light-coloured woods such as pines the effect of ultraviolet light is complex (Hon, 1991). Some pines can have very pale sapwood. Unfortunately the attractive white surface can yellow where directly exposed to sunlight (Sinclair and Vincent, 1964).

8.5.3. Tanning

The tannins of some woods, acacias, chestnuts and oaks – and also their barks – were once of importance for the tanning of leather. Yet they have adverse influence on gluing (inhibiting polymerization of adhesives) and in pulping. Today, in the state of Rio Grande do Sul, Brazil, 100,000 ha of *Acacia mearnsii* provide short fibre for chip export while the 27-28% tannin in its bark supports another industry.

8.5.4. Woods injurious to health

Some timbers are known to induce allergic reactions with people processing them but the effects, which are presumably due to extractives, are generally difficult to relate to specific compounds. The topic is well described by Woods and Calnan (1976) and by Kai (1991).

8.5.5. Contribution to wood density

In studies of plantation forest species it is common practice to carry out surveys of wood density (and other properties) by means of 5 mm diameter increment core samples at breast height. Often the unextracted wood density (oven-dry wood substances plus the extractives) is quoted in the literature.

In studies of wood chemistry it is important to remove extractives with a multipurpose solvent such as methanol before wood density determination so that the correct extractive-free density data are gathered. 'Wood density anomalies' led to extractive studies on larch species in the 1960s (Uprichard, 1963), with both polyphenols and arabinogalactan making large contributions to the apparent (unextracted) basic wood density.

8.5.6. Staining and corrosion

The presence of polyphenols and tannins in woods such as oaks, western red cedar, eucalypts and *Nothofagus* sp. leads to blue-black tannin stains during sawing of moist timber, or in the grain around metallic wood fastenings when the wood remains moist.

In kraft pulping the substitution of *Thuja plicata* for western hemlock, *Tsuga heterophylla*, led to increased digester corrosion (Gardner and Hillis, 1962). It was shown that steam-volatile tropolones known as thujaplicins were responsible for corrosion at the top of the digesters, and that polyphenols with a catechol group, which have a tendency to complex with iron, were responsible for the remainder of the corrosion.

8.5.7. Extractive exudation

Extractive exudation is a common feature of larches, pines, and spruces, and is particularly marked in heartwood samples. It can be controlled to some extent by sealing with an aluminium primer. During high temperature drying of resinous samples of radiata pine heartwood some migration of extractives occurs so that resin accumulates either on or near the surface in a way that is deleterious to wood finishing.

Hemingway (1969) showed that in the high temperature drying of yellow birch veneer, fatty acids and other extractives influenced the wetting of veneers in a complex manner and showed that the observed effects were due to acetone extractives.

Volatile emissions from kilns are subject to environmental audits.

8.5.8. Cement curing

Sugars, carbohydrates and tannic acids in wood retard or prevent hydration of the inorganic mix – inhibiting the curing of cement – when manufacturing wood-cement products. There are various strategies to counter such effects (Moslemi, 1991).

8.5.9. Chemical pulping

Studies by Erdtman in the 1950s showed that the heartwood of *Pinus* sp. resists sulphite pulping because it contains the stilbene pinosylvin, which condenses with lignin during the cook. Sulphite pulping is influenced also by the presence of resins (unsaturated fatty acids) in wood chips and to overcome pitch formation in the papermill it has been recommended that the chips are well seasoned before pulping (Fengel and Wegener, 1984).

In the kraft pulping of softwoods the resin acids and fatty acid esters present consume alkali required for delignification and are therefore undesirable. These can be recovered as tall oil by concentration of the spent liquor, separating off the acid soap (which also contains neutral components) and acidifying the sodium salts with sulphuric acid from which a blend of resin acids, fatty acids and neutrals is obtained.

The resin acids and fatty acids are subsequently purified by distillation. Some isomerization of levopimaric acid to abietic acid occurs during kraft cooking.

When pulping eucalypts (Gardner and Hillis, 1962) the ellagic acid present in heartwood can form salts that adversely influence the viscosity of concentrated black liquor: black liquor is the spent cooking liquor that is concentrated and burnt in order to produce heat and to recover mill chemicals. Also with hardwoods rich in polyphenols much of the alkali added for pulping is consumed by polyphenols rather than being used for delignification.

8.5.10. Mechanical pulping and papermaking

Corson and Lloyd (1987) observed selective removal of resin acids from refiner mechanical pulp, the ratio of resin acids to fatty acids in the aqueous refiner effluent being 80:20 compared to a ratio of 55:45 in the incoming chips. In modern refiner mechanical pulping the chips and pulps commonly pass through plug screws of 3:1 compression ratio. The pressates from these units are enriched in resin acids, because of the ready exudation of resin acids from resin canals (Suckling *et al.*, 1993). Such findings are important since it means that the undesirable resin acid rich effluent concentrates, which are of relatively small volume, may be selectively treated in future.

Traditionally mechanical pulps have given rise to pitch problems in papermaking. Some of the problems are of a seasonal nature. It is thought that this may relate to the change of resin viscosity with seasonal change in resin composition. Papermill chemists have developed various methods for dealing with pitch formation. Earlier the use of papermakers alum, aluminium sulphate, was common since this reagent can form resin acid complexes which ideally can bond to the papermaking fibres of the formed sheet. If instead pitch precipitates on mill equipment then there are pitch problems. Today special chemicals such as talc or polyethylene oxide are used to control and even remove pitch from the papermill.

8.6. Conclusions

The above examples are a few of the many instances where the presence of extractives and their influence are to a large extent out of all proportion to the comparatively small amounts present. On the other hand in tropical hardwoods the amounts of tannins and related compounds may be as much as 20% of oven-dry wood and here the potential for problems is more apparent. Overall the important role that extractives may play in wood utilization must not be underestimated.

CHAPTER 3

WATER IN WOOD

JOHN WALKER

1. INTRODUCTION

It would be hard to exaggerate the crucial role that water plays at every stage in the processing and use of wood regardless of the end product. The problems that the presence of water creates can rarely be avoided and often the solutions are partial at best. Wood tissue is formed in a saturated environment and the non-crystalline cell wall constituents, the hemicelluloses and lignin, are both affected to some degree by it's presence. One of the principal problems encountered is that wood shrinks as it loses moisture and swells again as it regains moisture. Wood is not dimensionally stable. Here the wood-water system is examined at some length.

2. SOME DEFINITIONS

2.1. Moisture content

Wood is a porous material that contains air and water as well as wood substances. As a result the weight of a piece of wood is not constant. Wood loses or gains moisture depending on the environmental conditions to which it is exposed. Further, the volume of a piece of wood is not constant. Wood shrinks and swells as it loses and gains moisture. It is therefore essential to know how much water a piece of wood contains before attempting to determine any other property.

Foresters and wood technologists define the moisture content of wood in terms of the initial weight of the piece of wood and the final weight of the wood after oven-drying to constant weight at 103 ± 2°C The difference in the two values is assumed to be due to loss of water by evaporation during drying.

$$\textit{Moisture content} = \frac{\textit{Original weight - oven - dry weight}}{\textit{Oven - dry weight of wood}} \times 100\% \qquad (1)$$

It should be noted that the moisture content is expressed as a percentage of the oven-dry weight rather than as a percentage of the original weight. There are a number of reasons for adopting what at first sight appears to be a curious definition:

- Industry is concerned primarily with the amount of wood in a log. The moisture within the log is of no value.
- The oven-dry weight provides a stable reference point.

- In consequence the chemical composition of the dry matter in wood is expressed as a percentage of this oven-dry weight.

It follows from this definition that the original moisture content of a piece of wood, which weighed 0.6 kg initially and 0.4 kg after oven-drying at 103°C, is:

$$\textit{Original moisture content} = \frac{0.6\ -\ 0.4}{0.4} \times 100\% = 50\% \tag{2}$$

If the oven-dry weight had been 0.2 kg the original moisture content would have been 200%: the moisture content can exceed 100%. This arises simply because of the way moisture content has been defined. A piece of wood having a moisture content of 200% consists of one-third dry woody matter and two-thirds of water.

It might seem more logical to define moisture content as follows:

$$\textit{Moisture content} = \frac{\textit{Original weight - oven-dry weight}}{\textit{Original weight of wood}} \times 100\% \tag{3}$$

This definition is used in the pulp and paper industry and in those log sales where the logging trucks are weighed on a weighbridge. The weight of dry matter and the moisture content of the logs or chips are determined by oven-drying samples or by applying an agreed formula. In some situations this definition can be less satisfactory because the initial weight changes with time as the material dries.

It is a simple matter to express the results either way. Referring to the example already quoted, of a piece of wood weighing 0.6 kg initially and 0.4 kg after oven-drying, the second definition would yield a moisture content of 33.3% rather than 50%. If the oven-dry weight had been 0.2 kg rather than 0.4 kg the moisture content would be 66.7% rather than 200%. Purely as a matter of convention the first definition is used in this text.

The moisture content of green, freshly felled timber is very variable. With softwoods the moisture content of sapwood is generally much greater than that of the heartwood: a narrow dry-wood zone can extend beyond the coloured heartwood by about half an annual ring and its presence in young trees is the first indication that heartwood formation is about to develop. With most hardwoods the moisture content of sap and heartwood are roughly comparable. The values in Table 3.1 are averages and the range of values found when sampling individual logs can be considerable (± 15-25% of the mean value). The moisture content of sapwood in particular can vary between spring and autumn, the moisture content can vary within the stem, and there are also differences due to geographic location and site.

2.2. Measurement of moisture content (James, 1988)

The oven-dry method of measuring the moisture content is accurate but it involves a delay of a day or so in obtaining results, and there is a practical limit to the size of pieces to be dried and the number of boards that can be cut up for testing. Where wood contains volatile extractives the loss of this material can be confused with the loss of water. In that situation drying under reduced vacuum or distillation in a closed system would be preferable with the water collected and measured directly.

Table 3.1. Green moisture content values for heartwood and sapwood of certain species (data averaged where applicable and rounded to the nearest 5%).

Species		Common name	Heartwood (%)	Sapwood (%)
Hardwoods				
Betula lutea	1	Yellow birch	75	70
	2		65	70
Fagus grandifolia	1	American beech	55	70
	2		60	80
Ulmus americana	1	American elm	95	90
	2		90	85
Eucalyptus nitens	3	Shining gum	115	125
Softwoods				
Picea sitchensis	1	Sitka spruce	40	140
	2		50	130
Pinus elliottii	3	Slash pine	40-45	180
Pinus ponderosa	3	Ponderosa pine	40-45	160
Pinus radiata	3	Radiata pine	40-45	150
Pinus taeda	1	Loblolly pine	35	110
	3		40-45	170
Pseudotsuga menziesii	1	Douglas fir	40	115
	2		40	115
	3		45	145
Sequoia sempervirens	1	California redwood	85	210
	3		180	220
Tsuga heterophylla	1	Western hemlock	85	170
	2		95	170

[1] USDA, 1987; [2] Dinwoodie, 1981; [3] NZ Forest Research Institute data.

Electrical moisture meters provide a quick and reasonably accurate non-destructive alternative. The direct-current resistance of the timber is measured; or either the alternating-current capacitance or power loss can be measured. Direct-current resistance moisture meters are more common: a pair of needles, a fixed distance apart, is driven into the wood across or along the grain (depending on the manufacturer's instructions) and the electrical resistance measured. The procedure is reasonably accurate between the fibre saturation point (defined later) at 30% and about 6% moisture content (at which point the resistance becomes too great to measure with reasonable accuracy). In this moisture content range, the relationship between electrical resistance and moisture content is represented by a log-log plot.

In the case of Douglas fir the resistance drops a million-fold from about 100 GΩ at 6% moisture content to 50 kΩ at 30% moisture content. Above the fibre saturation point the meter is less reliable as the resistance reduces from 50 kΩ at 30% moisture content to 10 Ω at 60% moisture content (Forrer and Vermass, 1987). If a moisture gradient exists within the wood, as occurs during drying, the electrical current follows the path of least resistance and the indicated reading will be higher than the average moisture content. If the probes are insulated with only the tips exposed the moisture profile through the wood can be estimated as the needles are pushed further into the lumber. The conductivity of any wood is dependent on the presence of ions: the reading will vary with species, whether heart or sapwood, temperature and the presence of preservative. Resistance meters are calibrated for a particular species and manufacturers advise on the appropriate correction factors to apply for other woods.

By placing two metal plates on either side of a sample a capacitor is formed with the lumber acting as the dielectric. The moisture content of the wood has a major influence on the properties of the dielectric and hence affects capacitance and power loss. The water molecule is electrically neutral, although it has a minute negative charge on the oxygen atom and positive charges of half that magnitude on each of the two hydrogen atoms. This separation of charge means that the water molecule possesses a small permanent dipole moment. When there is no applied electric field the water molecules are randomly orientated, but when placed between the plates of the capacitance meter the negatively charged oxygen atoms point towards the positively charged plate and the positively charged hydrogen atoms point towards the negatively charged plate. The water dipoles are being orientated parallel to the applied electric field. Moisture content is determined by measuring either the capacitance or power loss. In effect capacitance meters measure the energy stored within the wood, which is mainly achieved by the alignment of the water dipoles to the applied electric field. The capacity of the wood to store energy increases with moisture content. If an alternating field is applied the water molecules have to jostle against one another in their efforts to align themselves with the polarity of the electric field, and this dissipates energy and results in power loss.

Capacitance meters are sensitive to wood density and are harder to calibrate. They have a major advantage in ease of use. The meter need only be laid on the wood and the capacitance measured between two plates that lie side-by-side in a hand-held device. Alternatively meters can be installed with plates on either side of the boards in a production line.

Incidentally, the energy loss appears as frictional heat and so can be employed to dry wood and to cure resins that are used to glue pieces of wood together (Pound, 1973). The energy dissipated as heat within wood is related to the applied voltage, the frequency of the electric field and the power factor of the material (the capacity of the moist wood to dissipate electrical energy). With radio-frequency heating a frequency of 2-30 MHz is usually employed. At these frequencies the water dipoles have just about managed to align themselves with the electric field before the field is reversed again and they have to realign themselves some 10^6 times a second and this generates a lot of heat. Radio-frequency heating has the advantage of generating heat within the wood rather than waiting for the heat to diffuse in as occurs where hot presses are used. It has applications in the curing of thick plywoods (>25 mm), and in preheating partially consolidated mattresses of particleboard prior to hot pressing. It is used in curing glue lines in laminated members and in finger joints. Pressure is essential to prevent the resin boiling where intermolecular friction raises the temperature above 100°C: if the resin foams it forms a very weak joint.

2.3. The density of wood

The density of wood can be considered once its moisture content has been defined. In physics the density of a material is defined as the mass per unit volume (kg m^{-3}). The situation is not so simple with wood because changes in moisture content affect both its mass and volume. Therefore it is necessary to specify the moisture content of the wood as well as its density. Thus:

$$\text{Density at x\% moisture content} = \frac{\text{Mass of wood at x \% moisture content}}{\text{Volume of wood at x \% moisture content}} \qquad (4)$$

Three separate measurements are needed to define unambiguously the density of wood: the mass and volume of the wood at x % moisture content and the oven-dry mass (drying to constant weight at 103°C). The latter value is needed to specify the moisture content at which the density was determined. Three specific definitions follow:

$$\text{Oven - dry density} = \frac{\text{Oven - dry mass of the wood}}{\text{Oven - dry volume of the wood}} \qquad (5)$$

$$\text{Air - dry density} = \frac{\text{Mass of wood in equilibrium with atmospheric conditions}}{\text{Volume of wood in equilibrium with atmospheric conditions}} \qquad (6)$$

$$\text{Green density} = \frac{\text{Mass of the wood when green}}{\text{Volume of the wood when green}} \qquad (7)$$

Unfortunately the air-dry and green densities are not very reproducible. The air-dry density is measured after wood has been left to dry in the open and it has attained a moisture content that is governed by the changing atmospheric conditions, e.g. temperature and humidity. The moisture content of commercial air-dried lumber can differ by 3-6% according to the local climate, the season of the year and the species of wood. Further the equilibrium value, around which the moisture content fluctuates, depends on the climatic zone in which the sawmill is located. For example, the average moisture content in the drier parts of Australia and the United States can be as low as 8%, whereas in moister climates it approaches 20%. Clearly the air-dry density of lumber is quite variable. There are similar difficulties in meaningfully characterizing the green density. The term 'green' is applied loosely to freshly felled logs or to sawn timbers that still contain most of the moisture present at the time of felling. The green density of wood in a living tree is not constant. It will vary between summer and winter. Windblown trees or logs left unduly long in the forest or mill can dry and their green densities may be atypically low. Green density is species and age dependent. The latter is sensitive to the proportion of heart and sapwood as there are usually wide differences in the moisture content of sapwood and heartwood in softwoods (Table 3.1). Green density of timber is of practical interest, as some forest owners may sell timber on a green weight basis. However the purchaser is interested in the amount of wood fibre in the logs and not in the quantity of water that he is obliged to take as well. An equitable price for the sale of logs can only be reached when the amount of woody material in the logs is known and this has to be determined by sampling the logs to obtain amongst other things an average green density.

As industry is more concerned with the amount of woody tissue in a given volume of timber an alternative definition of density, the nominal density, is often preferred:

$$\textit{Nominal density at x\% moisture content} = \frac{\textit{Oven-dry mass of wood}}{\textit{Volume of wood at x\% moisture content}} \tag{8}$$

Thus, if the green density of a piece of wood at 50% moisture content is 600 kg m^{-3} its nominal density will be 400 kg m^{-3}. The wood will contain 400 kg m^{-3} of oven-dry material per cubic metre of green wood.

Basic density relates to a specific condition, derivable from the previous definition:

$$\textit{Basic density} = \frac{\textit{Oven-dry mass of wood}}{\textit{Volume of the wood when green}} \tag{9}$$

The term 'basic' emphasizes that both the parameters measured, the oven-dry mass and the swollen volume, have constant and reproducible values. Basic density is the most useful descriptor of wood density.

Green density, the weight of undried wood per unit volume, provides an unreliable indication of basic density. Within a species and within a tree there is usually a negative correlation between basic density and moisture content, indicating that the greater the green density of the wood the lower the basic density. A material having a low basic density has a high moisture content and *vice versa*.

3. THE DENSITY OF WOOD TISSUE

The density of oven-dry cell tissue of all woody plants is roughly 1500 kg m^{-3}. It can be accurately determined with a pycnometer, a small vessel (*c.* 10 ml) with a ground glass stopper having a concentric capillary bore to allow excess liquid to escape when filling the vessel. The pycnometer can be filled with a highly reproducible volume of liquid and allows the density to be determined to 1 part in 100,000 or more. To determine the density of a solid the pycnometer is weighed empty; with a liquid of known density; and with the liquid plus a known weight of the solid. Three points need to be considered (Weatherwax and Tarkow, 1968):

- There must be full penetration of all accessible checks and fissures in the wood. For example, thin microtomed sections are used to minimize the risk of incomplete penetration of the fluid into some wood tissue such as ray cells. There are advantages in using a fluid having a small molecular size and low viscosity.
- Some sub-microscopic pore space may exist within the cell walls which are inaccessible to some fluids.
- There are varying degrees of interaction between fluids and the wood substances. For example, water would be expected to form hydrogen bonds with available hydroxyl groups of the wood tissue.

When taking pycnometric measurements of cell wall density using a non-swelling organic liquid like silicone it is assumed that the fluid penetrates all accessible crevices. Such solvents measure the bulk density of the oven-dry cell wall. But since such organic liquids do not swell and penetrate the cell wall there is the possibility that some sub-microscopic pore space exists within the cell wall that remains inaccessible. The density determination therefore refers to the oven-dry density of the cell wall, including any inaccessible void spaces within the wall. A value of 1465 kg m^{-3} was obtained for *Picea sitchensis* using silicone as the displacement fluid.

However, when water was used the pycnometric density of the cell wall was calculated to be 1545 kg m^{-3}. This is some 5% greater than the previous value (Table 3.2). Two factors are responsible for this difference:

- Submicroscopic void spaces within the cell wall are only accessible to fluids like water that swell and penetrate the cell wall. Non-swelling fluids such as silicone therefore over-estimate the volume occupied by the cell wall material and so underestimate its density.
- Water (and any other swelling fluid) which forms hydrogen bonds becomes adsorbed within the cell wall. It is postulated that localized orientation and alignment of the water molecules on the water/wood-substance interfaces means that the volume occupied by these water molecules adjacent to the interfaces is very slightly reduced. Therefore the density of this interfacial water is slightly greater than that of bulk ('normal') water. If the density of this adsorbed interfacial water is assumed to be the same as that for bulk water then the estimated volume of this water will be slightly greater than the true volume, and the calculated volume for the wood substance as measured by volume displacement in water will be slightly less than its true value. The estimated density for the cell wall matter will then be slightly greater than its true density.

A further experiment is needed to determine the relative importance of these effects. When a water-swollen specimen is placed in successive water-alcohol solutions of increasing alcohol concentration the water in the swollen cell wall can be replaced by the alcohol (a process known as solvent exchange). Similarly it is possible to replace the alcohol by an inert solvent like hexane by solvent exchange (two stages are needed since hexane is insoluble in water). The swollen cell wall is now completely penetrated by an inert non-swelling solvent. Hexane is inert in the sense that the only attraction between the hexane and the wood substances is the weak van der Waals force, whereas with water stronger and more selective hydrogen bonds occur. Consequently the density of the hexane at the hexane/wood substance interfaces is the same as that of bulk hexane. The density of the cell wall material, measured with hexane after solvent exchange, is 1533 kg m^{-3} (Table 3.2).

Yes, it is possible that a little of the more strongly adsorbed water or alcohol is not displaced during solvent exchange by the hexane which would introduce an error here and in subsequent calculations.

The difference between the oven-dry cell wall density determined using silicone and after solvent exchange using an inert liquid like hexane must be due to the presence of residual sub-microscopic pores within the oven-dry cell wall, that are only penetrated when the cell wall is swollen. Accordingly, the inaccessible void volume per kg of cell wall material can be determined from the difference between the apparent volume of the cell wall material in silicone and the true volume as measured in the solvent-exchanged hexane. This is 0.0303 x 10^{-3} m^3 per kg, or about 4% on a volume basis, i.e. (0.0303/0.6825) x 100% (Table 3.2).

The degree of adsorption compression of the water at the wood substance interfaces can be estimated by comparing the apparent volume of the cell wall where using water and solvent-exchanged hexane as displacing fluids in the pycnometer. It is assumed that all the sub-microscopic pore space (0.0303 x 10^{-3} m^3 per kg of oven-dry cell wall material) is accessible to both displacing fluids, so that the difference in the apparent volume of the cell wall (0.0052 x 10^{-3} m^3 per kg) in hexane and water must be due to the orientation and densification of the adsorbed water at the wood substance/water interfaces. The cell wall of most species is fully saturated when the moisture content is 30% (evidence for this is discussed later) so 1 kg of oven-dry cell wall material will adsorb 0.3 kg of water that normally might be expected to occupy 0.3 x 10^{-3} m^3. From Table 3.2, this water appears to experience an average adsorption compression of 0.0052 x 10^{-3}/0.300 x 10^{-3}, i.e. about 1.7%, and the mean density of the adsorbed water is estimated to be 1018 kg m^{-3}, rather than 1,000 kg m^{-3} as for bulk water. Densification of the adsorbed water does not imply better ordering of the molecules, merely closer packing. For example, water is denser than ice even though a crude estimate, based on the difference between the latent heat of sublimation (2838 kJ kg^{-1}) and fusion (334 kJ kg^{-1}) at 0°C, implies that 12% of the hydrogen bonds are broken when ice melts. Consequently some choose to describe this strongly adsorbed water as being ‘destructured’.

This analysis is a little simplistic (Alince, 1989) in that it treats the water and the cell wall as two distinct phases rather than as a solid solution and, besides, the adsorbed water might encourage a realignment of the cell wall material itself which may then pack more closely, i.e. the cell wall material could equally well be densified rather than just the water.

Three general observations can be drawn from the study of Weatherwax and Tarkow (1968). First, 1500 kg m^{-3} is a reasonable approximation for the density of the oven-dry wood tissue. One would expect the actual value to vary slightly from species to species and between earlywood and latewood as the proportion of the cell wall constituents varies somewhat: the densities of cellulose, the hemicelluloses and lignin are roughly 1600, 1500 and 1400 kg m^{-3} respectively. Secondly, the amount of void space within the oven-dry cell wall is small, *c.* 4%. Again some variation is to be expected: the value will be influenced by how well the microfibrillar lamellae pack together. Thirdly, the destructuring and increased density of the adsorbed water implies that this water may not have the same characteristics as those of bulk water: this will be discussed later.

Table 3.2. Apparent cell wall density of sitka spruce in various fluids (Weatherwax and Tarkow, 1968).

Displacement fluid	Sitka spruce: physical state	Apparent cell wall density (kg m^{-3})	Apparent volume of cell wall material (m^3 kg^{-1})	Loss in apparent volume of cell wall material, based on the volume of unswollen cell wall (m^3 kg^{-1})
Silicone	Unswollen	1465.0	0.6825×10^{-3}	-
Water	Swollen	1545.7	0.6470×10^{-3}	0.0355×10^{-3}
Solvent – exchanged hexane	Swollen	1533.3	0.6522×10^{-3}	0.0303×10^{-3}

4. THE AMOUNT OF AIR IN OVEN-DRY WOOD

Wood contains air and water as well as wood substances. Wood, when oven-dry, contains only air and woody tissue. The oven-dry density of any species is a direct reflection of the amount of space occupied by wood tissue. Wood densities can vary from 50 kg m^{-3} with balsa (*Ochroma lagopus*) to 1400 kg m^{-3} with lignum vitae (*Guaiacum officinale*), although most commercial species have densities between 350 and 800 kg m^{-3}. The proportion of air space to total wood volume can be estimated from the oven-dry density of the timber. Consider a timber having an oven-dry density of 500 kg m^{-3}. A cubic metre of such a timber would contain 500 kg of woody material whose density is roughly 1500 kg m^{-3}. This means that the wood substances occupy some 0.33 m^{-3} leaving 0.67 m^{-3} as air space, i.e. two-thirds of the volume is air space. This calculation emphasizes the porosity of most dry wood.

4.1. Maximum moisture content and estimation of basic density

The maximum moisture content for a wood can be estimated easily. In a timber having a basic density of 300 kg m^{-3} the oven-dry cell tissue (density 1500 kg m^{-3})

will occupy 0.2 m^3 per m^3 of swollen wood. Consequently 0.8 m^3 per m^3 is available for water (800 kg of water per m^3) giving a maximum moisture content of 267% (800/300 x 100%). More usually this calculation is reversed, to estimate the basic density of an irregularly shaped piece of wood. All that is needed is the saturated green weight of the wood (y) and the weight (x) after oven-drying:

$$Basic\ density = \frac{x}{(x/1500) + (y - x)/1000}\ kg\ m^{-3} \tag{10}$$

In the case just examined the basic density is, as expected:

$$Basic\ density = \frac{300}{(300/1500) + (1100 - 300)/1000} = 300\ kg\ m^{-3} \tag{11}$$

5. THE FIBRE SATURATION POINT

Water in wood can exist as either absorbed (also called free water) in the cell lumens and intercellular spaces or as adsorbed (also called bound water) within the cell walls. When wood dries water first evaporates from the lumens and intercellular spaces. The fibre saturation point is defined as the moisture content at which all the absorbed water has been removed but at which the cell walls are still fully saturated. This occurs at a moisture content of 25 to 35%. In most instances it is adequate to presume the fibre saturation point to be 30% moisture content.

The concept of the fibre saturation point cannot be over emphasized. Its importance lies in the fact that the manner in which water is held is different in the adsorbed and absorbed states. The fact that the absorbed water can be removed before stripping off the adsorbed water indicates that a distinction can be made between these two states of sorption. For example, when green wood is dried there is no appreciable change in its mechanical properties until the fibre saturation point is reached. Below this moisture content they increase almost linearly with any further decrease in moisture content. Furthermore, wood only shrinks when the adsorbed water is removed from the cell walls.

Absorption refers to the take up of liquid by capillary condensation within a porous solid as a result of surface tension forces. It is accompanied by a limited reduction of the vapour pressure over the liquid as a result of the concave liquid-vapour mensicus. The energy required to evaporate the liquid is thus only slightly greater than that required to evaporate from a flat surface. The sap filling the lumens of wood is absorbed water. This bulk water can persist only at very high humidities, evaporating from successively smaller capillaries as the humidity falls. Table 3.3 states that lumens with radii exceeding 1 μm will drain and be free of all absorbed water when the relative humidity falls below 99.9% and so forth.

By contrast adsorption is different. Adsorbed water within the cell wall remains even at very low vapour pressures indicating that the attractive force between the adsorbent (in this case wood) and the adsorbate (water) is much greater than the attractive force of the adsorbate for itself. The process is accompanied by the evolution of heat. Physical adsorption occurs where the adsorbed molecules are held by secondary valence forces, e.g. H-bonding. The adsorption of water within the wood cell wall is an example of such physical adsorption. The initial binding energy is *c.* 20 kJ mol^{-1}, which is high and indicative of strong hydrogen bonding. This kind of adsorption is reversible. When wood is dried the adsorbed water is desorbed and evaporates off. The intermolecular forces holding the adsorbed water are greater than those holding the absorbed water, so the absorbed is the first to be removed.

Table 3.3. The maximum size of capillary that will fill with water varies with the relative vapour pressure, as does the minimum size of capillary from which water will evaporate when the relative vapour pressure falls to a particular value. Values are approximate and assume the surface tension is that of water at room temperature.

Relative vapour pressure	Capillary radius, μm
0.999	1.06
0.995	0.210
0.99	0.106
0.98	0.053
0.97	0.035
0.95	0.020
0.90	0.010

6. HYSTERESIS AND ADSORBED WATER IN THE CELL WALL

It is important to appreciate that water molecules do not penetrate a porous cell wall. Rather the dry cell wall expands as water is adsorbed within the non-crystalline regions of the wall and the accompanying volumetric swelling of the cell wall roughly corresponds to the volume of water adsorbed. Conversely when green wood is dried the volumetric shrinkage of the cell wall corresponds to the volume of adsorbed water that is removed.

The cell wall is hygroscopic: that is it adsorbs/desorbs moisture and so maintains equilibrium with the water vapour in the atmosphere. The hydroscopicity of the cell wall is due to the presence of accessible hydroxyl groups that exist throughout its structure. In the non-crystalline regions of the wall the molecules are not arranged in a coherent and ordered fashion and their hydroxyl groups are not always able to form hydrogen bonds with one another. Consequently when the cell wall adsorbs water, the water penetrates these non-crystalline regions and forms hydrogen bonds with accessible hydroxyl groups, mainly on the hemicelluloses. However, the water cannot penetrate the crystalline microfibrils where all the hydroxyl groups are already mutually bonded.

The amount of moisture adsorbed by the cell walls is directly related to the humidity of the surrounding air. Relative humidity or relative vapour pressure is defined as the ratio of the amount of water vapour in the atmosphere to the amount of water vapour present when the atmosphere is saturated with water vapour at the same temperature. When a thin section of freshly felled wood is dried very slowly, always allowing it to attain equilibrium with successively drier air, then the moisture content of the wood follows a moisture content – relative humidity path shown in the first desorption curve in Figure 3.1. If the same oven-dried wood is then exposed to progressively higher relative humidities its moisture content increases as shown by the adsorption curve. At 100% relative humidity the cell walls have readsorbed as much water as they are capable and the wood has reached the fibre saturation point. On redrying the wood follows a desorption curve that lies below the original green wood desorption curve until the curves merge at around 65% relative humidity. Any repetition of the adsorption-desorption cycle follows the established adsorption and desorption curves that delineate the hysteresis loop. These two curves define the maximum and minimum equilibrium moisture contents that can be expected for that wood at a given relative humidity. The hysteresis coefficient, defined as the ratio of the adsorption to the desorption equilibrium moisture content at the same relative humidity, is 0.8-0.85 for most timbers over the relative vapour pressure range of 0.1 to 0.9. For example in attaining a relative humidity of 45%, a piece of wood could have dried to an equilibrium moisture content of 10% or it could have adsorbed moisture from the oven-dry state and attained an equilibrium moisture content of 8%. Any point within the hysteresis loop bounded by the two curves can be achieved: if the wood is partially dried and then exposed to increasing humidity the new adsorption curve departs from the desorption curve at A and moves across the centre of the hysteresis loop to rejoin the adsorption curve at B (Figure 3.1).

Below 99% relative humidity, the moisture content is always less than or equal to the fibre saturation point. In other words hysteresis is associated with the sorption of moisture within the cell walls and is not affected by condensation of water in the void spaces. The hysteresis loop can be interpreted in terms of the accessibility of hydroxyls in the wood to adsorbed water. Consider the desorption/adsorption sequence illustrated in Figure 3.1. In the living tree the cell wall is laid down in the fully swollen state. Consequently the hydroxyl groups in the non-crystalline regions of the wall had the opportunity to form hydrogen bonds with the adsorbed water as well as with other cell wall constituents. On drying the adsorbed water is removed from the cell walls; the individual cell wall constituents are drawn close together: and so are able to form some new strong hydrogen bonds among themselves. Not all of these newly formed hydrogen bonds can be broken again during subsequent rehydration and fewer are available for bonding with water. Consequently the initial desorption curve lies above all subsequent desorption curves which define the upper boundary of the closed hysteresis loop. The adsorption curve lies below the desorption curve because the adsorbed water molecules have to penetrate and push apart the consolidated cell wall constituents and replace some of the newly formed intermolecular hydrogen bonds with fresh bonds with the readsorbed water.

The adsorption-desorption isotherms shown in Figure 3.1 relate to an experiment at 32°C. A rise in temperature causes a decrease in the equilibrium moisture content of wood at all relative humidities (Figure 3.2). There is also a corresponding reduction in the moisture content of the wood at the fibre saturation point. Both

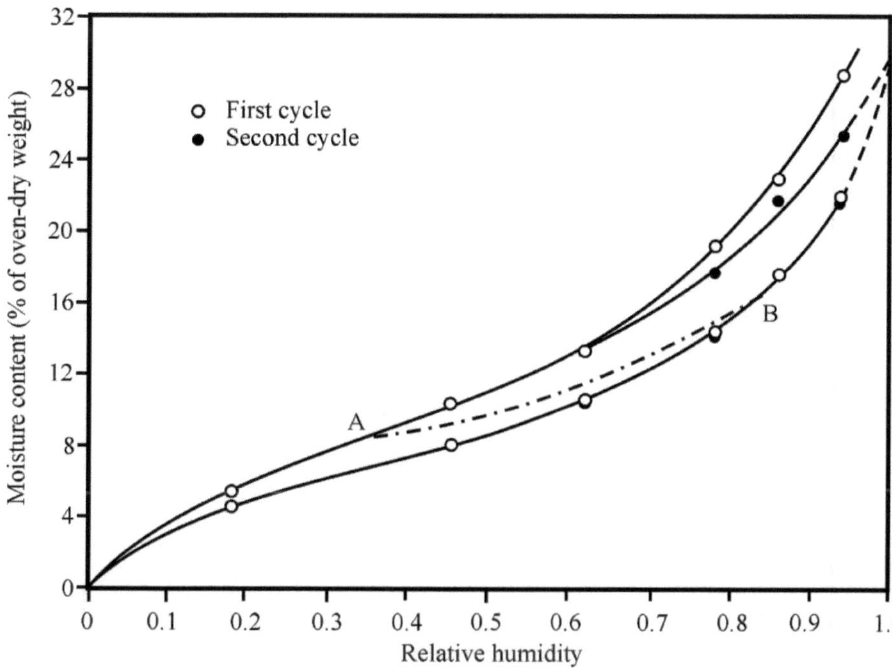

Figure 3.1. Sorption hysteresis curves for Douglas fir at 32.2°C (Spalt, 1958).

trends are due to the natural tendency for adsorbed water to evaporate more readily at high temperatures where the molecules possess greater thermal or vibrational energy. This is an example of the Le Chatelier's principle:

> Any change in one of the variables (temperature in this case) that determines the state of a system in equilibrium (adsorbed water in wood in equilibrium with water vapour in the surrounding environment) causes a shift in the position of equilibrium in a direction that tends to counteract the change in the variable under consideration.

Water vapour + wood $\rightleftharpoons$ *adsorbed water in the wood + release of heat.* (12)

Raising the temperature tends to displace the equilibrium towards the left, i.e. less adsorbed water and a lower equilibrium moisture content. In this particular example the fibre saturation point drops from around 31% at 25°C to 23% moisture content at 100°C (Figure 3.2).

If fresh green wood is heated in water one would expect the amount of adsorbed water in the cell walls to decrease, the volume of the swollen cell walls to contract and the wood to shrink, even though it is still green. In practice the matter is more complicated due to the presence of growth stresses in wood that are partially relieved at high temperatures (Yokota and Tarkow, 1962).

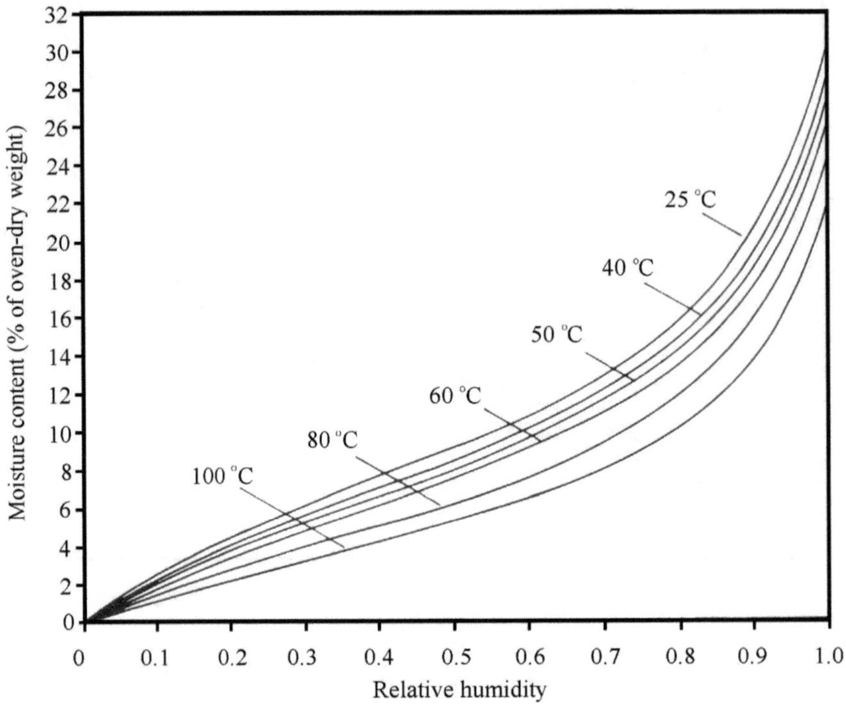

Figure 3.2. Desorption isotherms for green, never-dried sitka spruce as a function of temperature (Stamm, 1964).

There is a difficulty with the definition of the fibre saturation point. Strictly it corresponds to the moisture content when the cell walls are fully saturated (which would occur where the relative humidity is 1.0) and all the lumens are free of water. However at high relative humidities, above about 0.98, the lumen tips and pits begin to fill by capillary condensation, causing a sharp upward break in the sorption curve. In practice it is extremely hard to measure moisture content at these humidities and extrapolation from lower relative humidities is not particularly accurate.

7. MEASURING THE FIBRE SATURATION POINT

A method that distinguishes between the water in the cell walls and that in the lumens and pits is needed. Feist and Tarkow (1967) used a water-soluble polymer, polyethylene glycol (PEG) having a molecular weight of 9000, which is a sufficiently large molecule to preclude it penetrating the cell walls but which is still small enough to diffuse into the lumens and pits. Then, assuming that the concentration of PEG polymer solution in the lumens and pits is the same as that of the PEG solution in which the wood is immersed, the fibre saturation point can be calculated by:

- Determining the concentration of PEG (by interferometry) in the external solution once equilibrium has been established;
- Removing the wood from the PEG solution and weighing;
- Extracting all the PEG from the wood with excess water, and analysing this solution for PEG content, again by interferometry;
- Oven-drying the PEG-free wood and weighing.

The amount of water in the lumens can be calculated knowing the amount of PEG extracted from the wood and assuming the concentration there is the same as in the external solution. The amount of water in the cell walls is the difference between the amount of water in the wood and that calculated to reside in the lumens.

Feist and Tarkow (1967) found that the fibre saturation point obtained with virgin green wood was about 35-40% moisture content: at least for softwoods having a basic density of 300 kg m^{-3} or more. The fibre saturation point was a little lower for wood that had been dried and rewetted. The difference between the green and rewetted samples could arise from the formation of effective hydrogen bonds between the non-crystalline cell wall material when the wood was first dried. With balsa (*Ochroma lagopus*) they obtained a much higher fibre saturation point, 52%. They suggest that with decreasing basic density the thickness of the cell walls decreases and hence the walls display less resistance to swelling. The equilibrium moisture content that corresponds to the fibre saturation point represents a balance between the solution pressure within the swollen cell wall and resistance to this arising from the mechanical rigidity of the wall. Balsa had the lowest oven-dry density of the species examined, 250 kg m^{-3}.

It is worth making a simple calculation. Assuming that the fibre saturation point corresponds to 30% moisture content, then at the fibre saturation point there are 300 kg of water (density 1000 kg m^{-3}) for every 1000 kg of oven-dry cell wall (density 1500 kg m^{-3}). At fibre saturation point the cell wall is fully swollen and the proportion of the swollen wall that is occupied by water will be:

$$\frac{\textit{Volume occupied by the water}}{\textit{Total volume of the swollen wall}} = \frac{300/1000}{(1000/1500) + (300/1000)} = 0.3 \qquad (13)$$

30% of the swollen cell wall is occupied by water and 70% is occupied by the cell wall constituents. The properties and distribution of this adsorbed water located within the cell wall is the subject of attention in the rest of this chapter.

8. THEORIES OF ADSORPTION

8.1. The Langmuir equation for monomolecular adsorption

Langmuir (1918) derived a general equation for monomolecular adsorption. Under equilibrium conditions the rate of adsorption of a gas onto a surface must be equal to the rate of evaporation from the surface. If the fraction of the surface covered is x then the rate of evaporation is proportional to that fraction and equals to k_1x, where

k_1 is a constant at a given temperature. At the same time, the rate of adsorption depends on the fraction of the surface not covered, (1-x), on the rate at which the gas molecules strike the surface (which is proportional to the gas pressure, p), and on the probability that molecules striking the surface will be adsorbed (the adsorption constant). Thus the rate of adsorption is $k_2p(1-x)$, where k_2 is a temperature dependent constant. Under equilibrium conditions the rate of adsorption must equal the rate of evaporation:

$$k_1x = k_2 p(1-x) \tag{14}$$

and the proportion of the surface covered by a monolayer is given by:

$$x = k_2 p/(k_1+k_2p) \tag{15}$$

This is the Langmuir equation. The quantity of adsorbed moisture approaches asymptotically a maximum moisture content of 5% with increasing relative vapour pressure. The general shape of this equation is shown in Figure 3.3.

Adsorption is more complicated than the Langmuir equation suggests and some implicit assumptions should be examined critically, namely that:

- The surface is homogeneous and all surface sites for adsorption are equally favoured.

  Real surfaces are heterogeneous and the affinity of the surface for the adsorbate (the gas) will depend on local structure. Heterogeneity leads to a decrease in binding energy (energy of adsorption) with increasing surface coverage as the most favoured sites are occupied first.
- There is no lateral interaction between adsorbed molecules. This implies that individual molecules will evaporate from a well-covered surface as readily as they will from an almost bare surface.

  In practice lateral attraction between adsorbed molecules on a well-covered surface increases the heat of adsorption since the adsorbed molecules are not only bonded to the surface but also to other adjacent adsorbed molecules.

These two effects counterbalance one another. A further assumption is that:

- Only unimolecular adsorption occurs, where the amount of material adsorbed approaches asymptotically a limiting value with increasing vapour pressure. Obviously the quantity of material adsorbed is proportional to the fraction of the surface covered. A plot of surface covered *versus* vapour pressure is equivalent to a plot of the quantity of material adsorbed *versus* vapour pressure.

The Langmuir equation is able to explain the initial parabolic wood-water adsorption curve up to a relative humidity of about 0.2, equivalent to a moisture content of about 5%. However the Langmuir equation does not explain the convex portion of the sigmoidal curve at higher relative humidities (Figure 3.1).

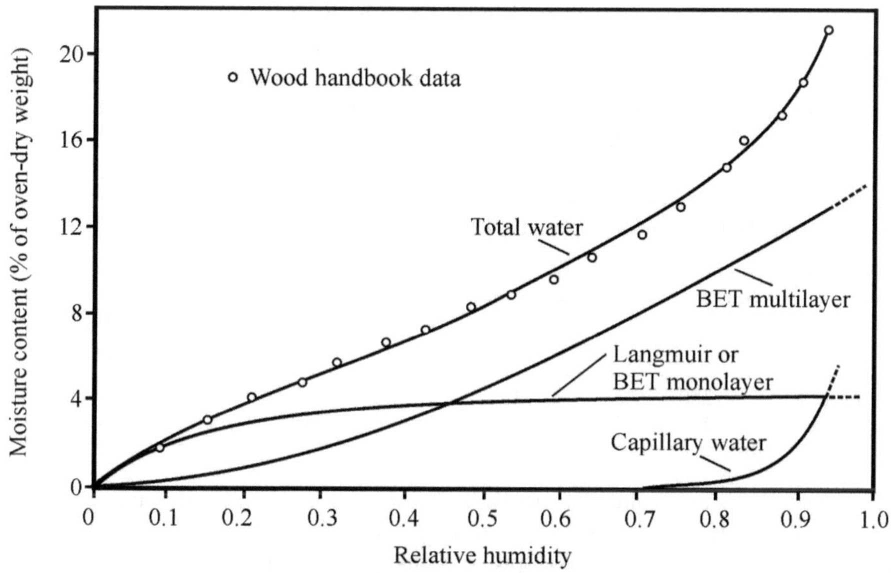

Figure 3.3. Sorption equations fitted to experimental data for the sorption of water by wood at 40°C (Simpson, 1980). The Langmuir isotherm is parabolic, corresponding to the formation of a BET monolayer. Multilayer adsorption describes sorption behaviour better but sorption at the highest moisture contents, where capillary condensation occurs, is underestimated.

8.2. BET theory for polymolecular adsorption

Brunauer, Emmett and Teller (1938) derived a sigmoidal adsorption curve (the BET equation named after these authors) assuming that multilayer adsorption was possible rather than just monomolecular adsorption (Figure 3.3). They assumed that the forces which produce condensation are chiefly responsible for multilayer adsorption. They argued that the formation of further layers of adsorbed gas/liquid on the surface at pressures below 100% relative humidity is analogous to the formation of multimolecular clusters in a non-ideal gas with the difference that in the gas phase the formation of a cluster requires the creation of a vapour-liquid surface whereas in multilayer adsorption a liquid surface has already been formed once a monolayer has been laid down on the surface, i.e. a molecule approaching this surface no longer sees the original surface but the 'liquid-like' monolayer of adsorbed gas. In consequence little surface tension has to be overcome during the formation of subsequent layers.

The analysis of multilayer adsorption follows that of Langmuir for monomolecular adsorption. Adsorbate-adsorbent bonding is involved in the first layer while adsorbate-adsorbate bonding is involved in forming all subsequent layers. The heat of adsorption for the first layer, E_{ads}, is very different from that for subsequent layers where the heat of adsorption is assumed to be the same as the heat of liquefaction, E_{liq} (only approximately true). The mathematical derivation of the

BET need not concern us here. What is significant is that when the BET equation is applied to the adsorption of water by wood (Figure 3.3) the BET allows one to estimate both the surface area on which water is being adsorbed within the cell wall and the average number of layers of water adsorbed on these surfaces. The results are quite startling (Table 3.4).

Table 3.4. Analysis of water vapour adsorption by wood (*Acer saccharum* and *Picea sitchensis*) using the BET equation (Stamm, 1964).

Species	Surface area ($m^2 g^{-1}$)	Relative humidity at which a monolayer is formed	Heat of adsorption ($J g^{-1}$)	Average number of molecular layers at saturation
Sitka spruce	250	0.22	360	6.5
Sugar maple	210	0.21	350	7.5

The surface area calculated from BET theory of roughly 250 m^{-2} per gram of oven-dry sitka spruce is far greater than the total surface area of the cell lumens, which is about 0.2 $m^2 g^{-1}$ for sitka spruce (density 400 kg m^{-3}). This demonstrates that the water penetrates the cell walls and creates enormous internal surfaces. The area of these 'internal surfaces' is a thousand times greater than the surface area of the lumens. The BET equation estimates that the water film on these surfaces to be about 6-7 molecules thick at the fibre saturation point. The more active and accessible sorption sites adsorb more and the less active will adsorb fewer. The thickness of such a film between two adjacent surfaces can be calculated from the total number of layers (*c.* 13) and the size of the water molecule. The film thickness is around 4-5 nm. The water-wood tissue interfacial area is assumed to remain constant as water is adsorbed (which is not strictly true) while the distance between such surfaces increases as water is adsorbed. The surfaces are 'conceptualized' as concentric lamellae with films of water sandwiched in between. Figure 2.11 provides a partial appreciation of the extent of such internal surfaces. Cross-linking in the wall and the restraint provided by the layered structure of the wall itself will inhibit adsorption, implying that the number of polymolecular layers at saturation is less than in would be the case were there no residual swelling pressure. The BET equation is only one of a number of theoretical approaches that seek to describe the adsorption behaviour of wood (Skaar, 1988; Hartley and Avramidis, 1993). All, including the BET model, have limitations and are at best approximations of what is happening within the cell wall. These equations fit the observed adsorption curve better by including a small component of absorbed capillary water at high humidities (>80% relative humidity). Such absorbed water would occupy minute capillaries (>5 nm) and might account for 4% of the total uptake to fibre saturation, i.e. 4% out of the 30% notionally adsorbed at the fibre saturation point (Figure 3.3).

The actual adsorption process is not fully understood. In particular there is a dearth of experimental detail on the actual distribution of the adsorbed water within the cell wall at different moisture contents. At low relative humidities the water molecules penetrate the non-crystalline regions of the cell wall in the process of which they break some weak and distorted hydrogen bonds between the polymer molecules (cellulose, hemicelluloses and lignin) and displace adjacent polymer molecules relative to one another. The adsorbed water molecules then form their own hydrogen bonds with the newly accessible polymer hydroxyl groups, creating a polymer hydrate. The moisture needed to form a monomolecular surface layer within the amorphous regions is calculated to correspond to a moisture content of about 5%. This is estimated by dividing the moisture content at the fibre saturation point by the average number of polymolecular layers. At high relative humidities the adsorbed water forms polymolecular layers between the wood polymer molecules. These water molecules are only indirectly bonded to the wood polymers by water-water hydrogen bonds. An adsorbed film needs to be only 1-2 nm thick to mask almost completely the influence of the underlying substrate (Tabor, 1969), but this would correspond to a significant proportion of the half-layer between two adsorbing surfaces within the cell wall.

The water in the cell walls can be replaced gradually and successively by miscible solvents (water→methanol→pentane), before evaporating off the non-wetting, low surface tension, non-polar liquid. Unlike the evaporation of water the capillary forces in pentane are insufficient to consolidate the cell wall: there is little shrinkage and the wall remains essentially unshrunk and porous. The internal surface area can be estimated by measuring the Langmuir (monolayer) adsorption of nitrogen. The values obtained are about half to two-thirds of the internal surface area estimated from BET adsorption of water (Table 3.4). Merchant (1957) used sulphite spruce pulp rather than solid wood. This does not affect the general conclusions. The small surface tension of pentane and incomplete replacement of the water and methanol during solvent exchange would result in the collapse of some of the pore space on evaporation so the small reduction in surface area is to be expected. The surface measured by adsorption of nitrogen gas in the porous delignified cell wall is still enormous and it is a true internal surface that physically exists.

One might anticipate that if the water in the cell wall were to be frozen and then the ice were to be subliminated off there should be no liquid capillary tension, no cell wall shrinkage and it should be possible to create a porous cell wall. However, sublimating the water molecules from the cell wall at –20°C does not prevent collapse of the internal pore structure (Merchant, 1957). This implies that the cell wall water is not actually frozen at this temperature: the cell wall still shrinks and very little internal surface is created. Indeed there is evidence (Tarkow, 1971) that at least some adsorbed water does not lose the mobility characteristic of the liquid phase until very low temperatures (<-80°C). Of course water in the lumens behaves like bulk water and freezes at a temperature between -0.1°C and -2.0°C, depending on the concentration of dissolved sugars in the sap.

9. DISTRIBUTION OF WATER WITHIN THE CELL WALL

In order to measure the fibre saturation point Feist and Tarkow (1967) used a sufficiently large water soluble polymer to preclude it penetrating the cell wall. Stone and Scallan (1968) reversed this approach and used a series of much smaller polymer probes, calculated to have 'equivalent spherical diameters' in the range 0.8-56 nm, to investigate the pore size and pore volume within the swollen cell wall. Provided the openings in the wall are larger than the size of the molecular probe the polymer can enter the cell wall, diluting the concentration of the polymer in the external solution. Thus the solution is diluted, but not by as much as it would have been if the polymer were to have complete access to all the adsorbed water in the cell wall. Quantitative relationships between accessible pore volume and probe diameter can be obtained. Working with fresh wood meal from black spruce, *Picea mariana*, Stone and Scallan found that about a third of all the adsorbed water was in pores inaccessible to their smallest molecular probe (0.8 nm in diameter) and that the maximum pore width was only 3.6 nm, with a medium pore size of only 1.6 nm (Figure 3.4). The 'geometry' of the pore space is ill-defined. Its size relates to its accessibility to the polymer probes. The smallest dimension of the pore space must exceed that of the probe molecule if that molecule is to diffuse into the pore opening. The actual pore size may be somewhat larger as the initial monolayer of water may not be displaced to the polymer probes.

Li *et al.* (1995) applying nuclear magnetic resonance techniques (see later) to delignified and beaten kraft pulp fibres derived a cell wall water fraction equivalent to 160% MC that was confined to elongated pores or 'channels' extending at least for tens of microns (μm) in the microfibril direction. The slightly higher moisture content of the beaten kraft fibres compared to the unbeaten, low-yield kraft pulp in Figure 3.4 would be a consequence of the beating that encourages further intra-wall delamination and internal swelling (Figures 2.11 and 2.14).

These polymer probe studies help interpret some aspects of biodegradability and chemical pulping of wood (Lindström, 1986). For instance fungal decay of wood ceases at low moisture contents (<21-22%), presumably because enzymatic degradation requires access to the cell wall through pores that are large enough to admit small enzyme molecules. Other factors should be considered: for instance the peroxide molecule plays a role in initial decay. Peroxide is a small molecule and would be expected to penetrate the cell wall more easily than the larger enzyme molecules. Further, the presence of ionic groups on the 'capillary' wall can exclude even non-ionic solutes. Finally in anticipation of discussion on chemical pulping, the lignin has to be broken into fragments small enough to diffuse out of the wall.

10. WHERE IS THE ADSORBED WATER WITHIN THE CELL WALL?

BET theory considered the cell wall water as a surface phenomenon, with a strongly adsorbed monolayer and less strongly bound polymolecular water. It is worth making a few naïve calculations.

First, the proportion of the swollen cell wall occupied by water at fibre saturation point (30% MC) is:

$$\frac{\textit{Volume occupied by the water}}{\textit{Total volume of the swollen wall}} = \frac{300/1000}{(1000/1500) + (300/1000)} = 0.3 \qquad (16)$$

Further, hemicelluloses account for 27% of the oven-dry mass (Table 2.2), having an average oven-dry density of about 1500 kg m^{-3}. Hence, the proportion of the swollen cell wall occupied by the hemicelluloses is:

$$\frac{\textit{Volume occupied by the hemicelluloses}}{\textit{Total volume of the swollen wall}} = \frac{270/1500}{(1000/1500)+(300/1000)} = 0.19 \quad (17)$$

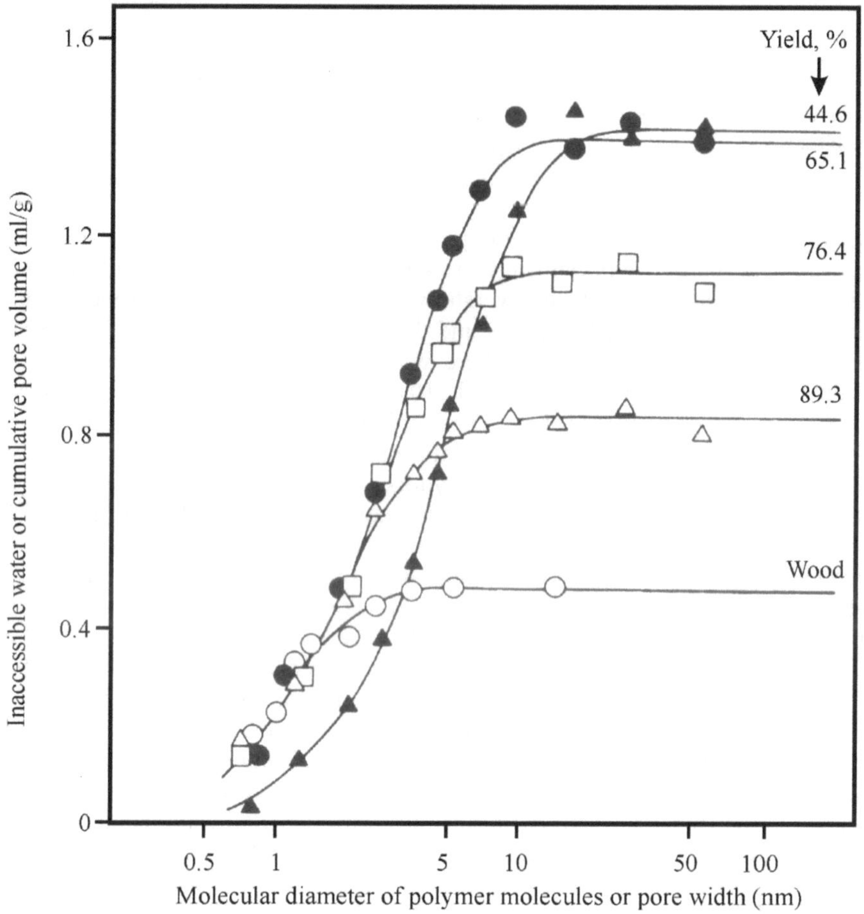

Figure 3.4. Water adsorbed in the cell wall of *Picea mariana* (Stone and Scallan, 1968). The amount of adsorbed water that is accessible to a polymer molecule increases as the size of the polymer molecule decreases. During pulping the cell wall is opened up with both pore volume and pore size distribution increasing with the degree of delignification. The yield is the ratio of the weight of oven-dry fibre remaining after pulping to the initial weight of oven-dry wood.

Now assume (very approximately) that all the water in the cell wall is associated with the hemicelluloses then roughly half the swollen cell wall is the hydrated hemicellulose interphase between the cellulose and lignin (Table 3.5).

Table 3.5. Mass and volume proportions of cell wall material assuming densities of cellulose, lignin, extractives, hemicelluloses and water to be 1600, 1400, 1400, 1500 and 1000 kg m^{-3}.

	% of swollen wall, wt/wt	% of swollen wall, vol/vol	Molecular weight	Molecular weight ratio	Ratio of water to hydroxyls for each hemicellulose residue
Cellulose	42	27%			
Lignin	28	23%			
Extractives	3				
Hemicelluloses	27	50%	162	9.0:1.0	1.0:3.3
Water	30		18		

Second, at fibre saturation point the proportion of water (300 kg of water/1000 kg of oven-dry wood) is approximately the same as that for the hemicelluloses (270 kg/1000 kg of oven-dry wood) on a weight/weight basis. The molecular weight of a hemicellulose residue, $C_6H_{10}O_5$, is 162 and that of water is 18, a ratio of 9:1. So if there were the same mass of each there would be 9 water molecules for each hemicellulose residue, but in fact there is 10% more water (30 vs. 27 on a wt/wt basis) so the ratio is 10:1 (Table 3.5). Further, assuming that there are three unsubstituted hydroxyls on each hemicellulose residue this calculation implies that there will 3.3 water molecules for each hydroxyl on the hemicellulose residue. In reality this is an underestimate. There will be more water molecules per accessible hydroxyl group. This is because the number of accessible hydroxyls on the hemicelluloses will be reduced to the extent that some hemicelluloses are adsorbed onto the microfibril surfaces, that some hydroxyls are substituted, and that there will be inter-hemicellulosic hydrogen bonding. However, the purpose of this tentative calculation is to indicate that relatively few water molecules – whether 3, 6, or 9 – need cluster around accessible hydroxyls to account for the adsorbed water.

Another approach treats the cell wall water as a solution phenomenon, with the water of hydration accounting for the first 4-6% of moisture adsorbed or the last moisture lost during desorption (essentially the same 4-6% that accounts for the monolayer in the BET analysis). All further adsorbed cell wall water is the water of solution. In a similar vein Harley and Avramidis (1993) calculate that where moisture contents exceed 20-25% the water of solution forms clusters of 10 or more water molecules that corresponds to a hydraulic spheroid about 0.6 nm in diameter.

Finally, absorbed 'lumen' water has been detected by NMR (contributing a minute 0.3% wt/wt) when the moisture content was only 17% (Araujo *et al.*, 1992). This blurs the definition of the fibre saturation point in that small amounts of absorbed water are present at moisture contents well below the fibre saturation point.

Despite the uncertainties of where the adsorbed water is located, the naïve/conceptual definition of fibre saturation point remains of considerable practical utility.

11. CHARACTERISTICS OF ADSORBED WATER IN THE CELL WALL

Adsorbed water in biological tissue behaves differently to absorbed water. This is thought to be due to destructuring of the water in proximity to cell wall hydroxyls and to the finely dispersed state of adsorbed water within the cell wall. Thin films of water have the same surface energy as do two surfaces separated by bulk water. This holds true even when the film is only 2 nm thick, emphasizing that surface forces are very short ranged. Such a thin film is able to completely mask the influence of the underlying substrate (Tabor, 1969). However, the BET model concludes that much of the adsorbed water lies in films no thicker than 2 nm. Early wide-line nuclear magnetic resonance (NMR) studies revealed that the adsorbed water exists in two states (Nanassy, 1974; Riggin *et al.*, 1979). The NMR signal arising from the hydrogen nuclei can be resolved into two components (Figure 3.5). One is the broad spectral output produced by the hydrogen nuclei in dry wood (giving a broad spectra due to differing dipolar-interactions between specific neighbouring protons replicated thoughout the wood tissue) and in the localized portion of the adsorbed water. The other, which is superimposed on this, is a narrow spectral output produced by the mobile, less strongly adsorbed water. The water molecules that are hydrogen bonded directly to the polymeric materials of the cell wall appear to be held in relatively fixed positions as elements of inter-chain links and the macromolecules restrict the motion and mobility of these water molecules but do not completely immobilize them. The NMR spectrum arising from these water molecules is broadened because the magnetic field that each individual hydrogen nuclei experiences is very slightly different due to the relative positions of surrounding molecules and molecular groups on the polymer chains in their immediate vicinity, and due to their inability to align themselves with the magnetic field. Beyond the monolayer, this localized structure grades into a mobile phase. The narrow spectral output corresponds to this mobile component. It is a little broader than that for pure water implying that this adsorbed water is less mobile than pure water although it appears to have the solvent properties of ordinary water and to be only weakly bonded to the polymer system. According to Froix and Nelson (1975) adsorbed water retains this mobile component even at very low moisture contents (<5%). As the moisture content increases the proportion of adsorbed molecules that are mobile increases rapidly. At the same time the quantity of the less mobile portion (the broad spectrum component) increases somewhat, evidence that more internal surfaces are being opened up as further water is adsorbed.

The adsorbed water is held quite tenaciously. The amount of heat needed to remove the adsorbed water from wood is greater than that needed to remove the absorbed water. This can be measured quite simply by immersing oven-dry ground-wood particles in a known quantity of water and measuring the temperature rise (Stamm, 1964). The heat released on completely wetting a gram of oven-dry ground-wood is about 80 Joules. The amount of heat released decreases as the ground-wood is preconditioned at progressively higher moisture contents, i.e. a gram of water is absorbed on an infinite amount of preconditioned ground-wood. The differential heat of wetting can be calculated from the slope of the curve at any given moisture content (Figure 3.6).

For most woods the initial differential heat of sorption (at 0% moisture content) is approximately 1250 J g^{-1}. This equates to 22.5 kJ mol^{-1} (18 g of water in one mol) which is typical for hydrogen bonding. By the time a complete monolayer has formed, by about 4-5% moisture content, the differential heat of wetting is around half the initial value (9-10 kJ mol^{-1}). This value would be amongst the weakest hydrogen bond values reported. The first water molecules to be adsorbed on the oven-dry ground-wood have a complete choice of accessible cell wall hydroxyls and

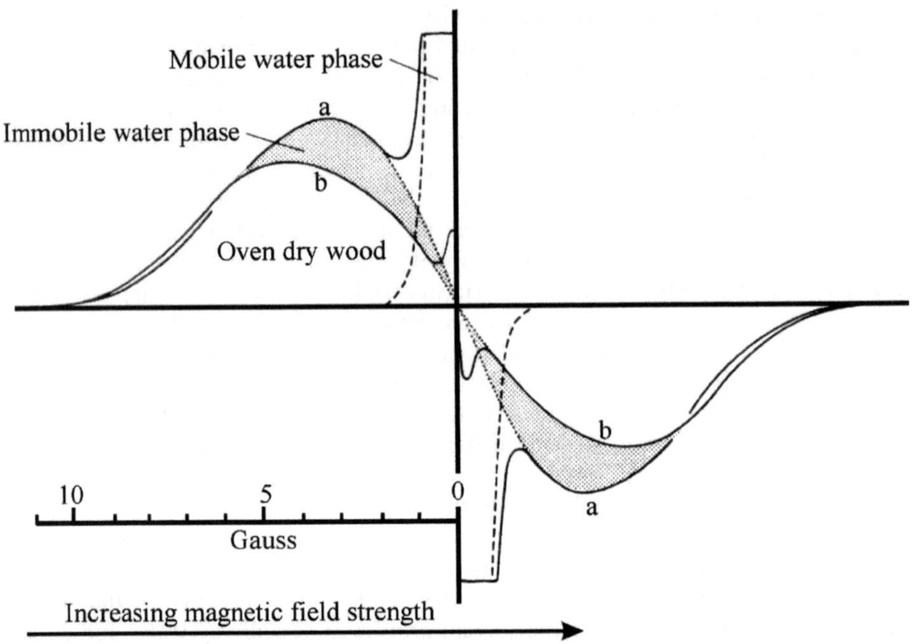

Figure 3.5. NMR spectrum of wood as a function of moisture content (Nanassy, 1974). The spectral intensity of the broad component is due to hydrogen nuclei in dry wood tissue (6% of oven-dry mass) and to a monolayer of strongly adsorbed water molecules. The intensity of this component increases somewhat with moisture content, at least to the fibre saturation point. This implies that new internal surfaces are being created as the moisture content increases. The narrow component corresponds to the more mobile multilayers of adsorbed water. The narrow component is truncated because the idea was to record the broad spectrum which required expanding the vertical scale, so that the peak of the narrow spectrum is well off scale. With quantitative NMR techniques the areas under the broad and narrow components of the spectrum provide a measure of the number of hydrogen atoms in these two states.

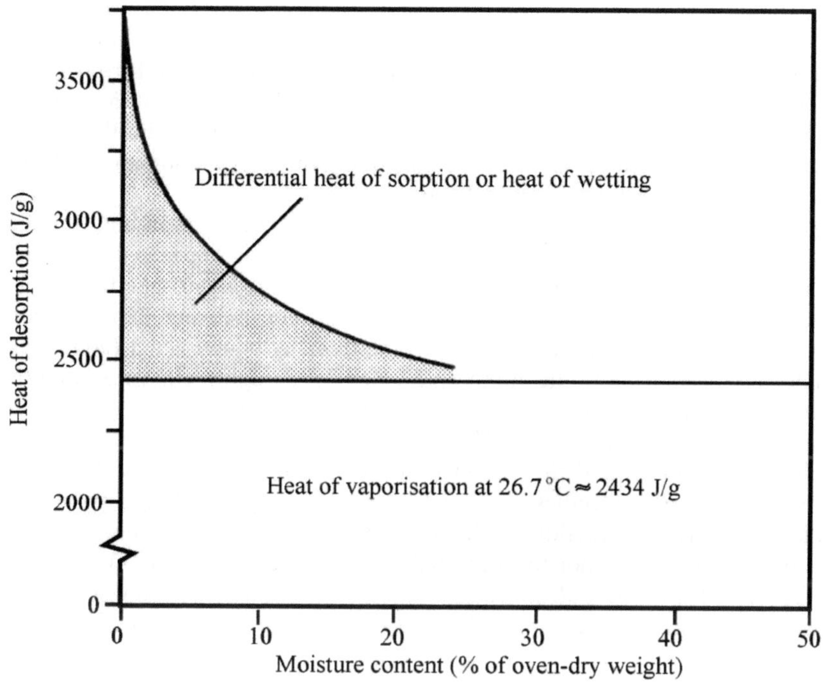

Figure 3.6. Energy required to evaporate water from *Araucaria klinkii* at 26.7°C (Kelsey and Clarke, 1956).

are able to form the strongest bonds. Later when the monolayer is almost complete the water molecules may not be able to get appropriately positioned to form really strong bonds due to the presence of previously adsorbed molecules (a feature known as steric hindrance) and the water molecules may have to do work to open up the cell wall structure sufficiently to be adsorbed at the bonding position. Further adsorption will be less favoured as multilayer adsorption is involved and the differential heat of wetting gradually diminishes to zero at around the fibre saturation, at which point the characteristics of the last few molecules of adsorbed water will correspond to those of bulk water.

NMR techniques distinguish between nuclei (usually hydrogen nuclei when studying wood, i.e. ^{1}H-NMR) in terms of their mobility by revealing the conditions under which they resonate within a very strong magnetic field. This provides information about their immediate environment. Each hydrogen atom will have its own distinctive 'signature' depending on whether it is a –H or –OH attached to any of the six carbons of a cellulose residue in the crystalline microfibril or on a non-crystalline hemicellulose residue, or part of a water molecule (or incorporated in lignin). Using pulsed NMR techniques, various relaxation times are obtained. This technique measures how quickly a resonance spectra decays – a different procedure to that shown in Figure 3.5. The pulsed NMR decay signal for protons in woody tissue falls away in 1-30 μs, in bound water in 1-10 ms, and in lumen water in 10-100 ms (reflecting different scales of molecular motion). Using this technique with redwood, Araujo *et al.* (1993) showed that the fibre saturation point fell from 35.5% to 28.0% to 21.7% as the temperature was raised from 4° to 26.5° to 55°C. These FSP values are somewhat different to those for spruce in Figure 3.2. Further, NMR techniques can be extended to estimate the size distribution of the lumens – clearly distinguishing between earlywood and latewood.

Finally, Salmén and his colleagues (Berthold *et al.*, 1996; Olsson and Salmén, 2004) offer a more specific variant of the solution model (as opposed to the BET layer model) in which cell wall water uptake is seen as occurring at specific sites by cluster adsorption. The adsorbed water, both the non-freezing bound water (equivalent to monomolecular adsorption) and the freezing bound water (equivalent to polymolecular adsorption in the alternative school of thought), cluster around the polar groups, –OH, $-COO^-$, and SO_3^- (in paper products), and their counterions/cations, H^+, Na^+, Ca^{2+}. The dissociation of the ion pair results in considerable hydration around the cation. They estimate that there are 1.35 non-freezing bound water molecules for each accessible hydroxyl group on the macromolecular polymers (cellulose, hemicelluloses and lignin). The cell wall 'freezing' bound water does not form normal ice structures due to the influence of the polar groups of the cell wall materials, and this water has non bulk-water-like thermodynamic parameters, i.e. unlike the free water in the lumen it doesn't freeze at 0°C.

CHAPTER 4

DIMENSIONAL INSTABILITY IN TIMBER

MORWENNA SPEAR[1] AND JOHN WALKER

[1]*School of Agriculture and Forest Sciences, University of Wales, Bangor, UK*

1. INTRODUCTION

Dimensional instability is one of the major impediments in the processing and use of lumber. Three separate facets need to be distinguished and considered: shrinkage on drying, movement in service, and the responsiveness of lumber to a fluctuating environment. The issues demand the attention of workers and management at every stage of manufacture. There are more complaints about the instability of lumber than any other matter and rectifying problems is expensive. Figure 4.1 provides unhappy examples of the consequences of movement in service.

Some typical moisture content values for green wood are noted in Table 3.1. These values are considerably greater than the fibre saturation point. Absorbed water at the surface will evaporate and the lumber will dry provided the surrounding atmosphere is not totally humid. Indeed the absorbed water in the lumens cannot remain there in equilibrium with the atmosphere unless the relative humidity of the air is in excess of 99% (Table 3.3). If the wood is left under cover – keeping the rain off – it will eventually dry to a moisture content that will vary according to the temperature and humidity of the air (Figure 3.2). This moisture content will be below the fibre saturation point so all the absorbed water and some of the adsorbed water will have evaporated. If an even lower moisture content is required it is necessary to use a kiln to lower the relative humidity and raise the temperature (Figure 3.2).

2. SHRINKAGE AND SWELLING OF WOOD

Wood only shrinks when water is lost from the cell walls and it shrinks by an amount that is proportional to the moisture lost below fibre saturation point. To a first approximation the volumetric shrinkage is proportional to the number of water molecules that are adsorbed within the cell wall, and that in turn is related to the number of accessible hydroxyls on the cellulose, hemicelluloses and lignin, and to the amount of cell wall material, i.e. the basic density of the wood (Figure 4.2).

The axial, radial and tangential shrinkages, which together account for the volumetric shrinkage, are directed by features of wood structure that resist shrinkage, e.g. the quantity of ray tissue in the radial direction, or by features of the

Figure 4.1. (a) The parquet floor has lifted because the wood was dried to too low a moisture content (CSIRO, Australia). (b) Old paint is liable to crack as the wood moves.

ultrastructure, e.g. the tendency for microfibrils to align roughly in the axial direction.

Early in the drying process it is inevitable that the moisture content at the centre of the piece will be above the fibre saturation point while the fibres at and near the surface will be well below the fibre saturation point. There will be a moisture gradient within the wood and the system will not be in equilibrium. In this situation the surface fibres will have started to shrink and the overall volume of the piece will be reduced even though the average moisture content is above fibre saturation. This accounts for the shrinkage of the wood at mean moisture contents a little above the fibre saturation point (Figure 4.2).

The volumetric shrinkage to the oven-dry state is determined by measuring the green and oven-dry volume:

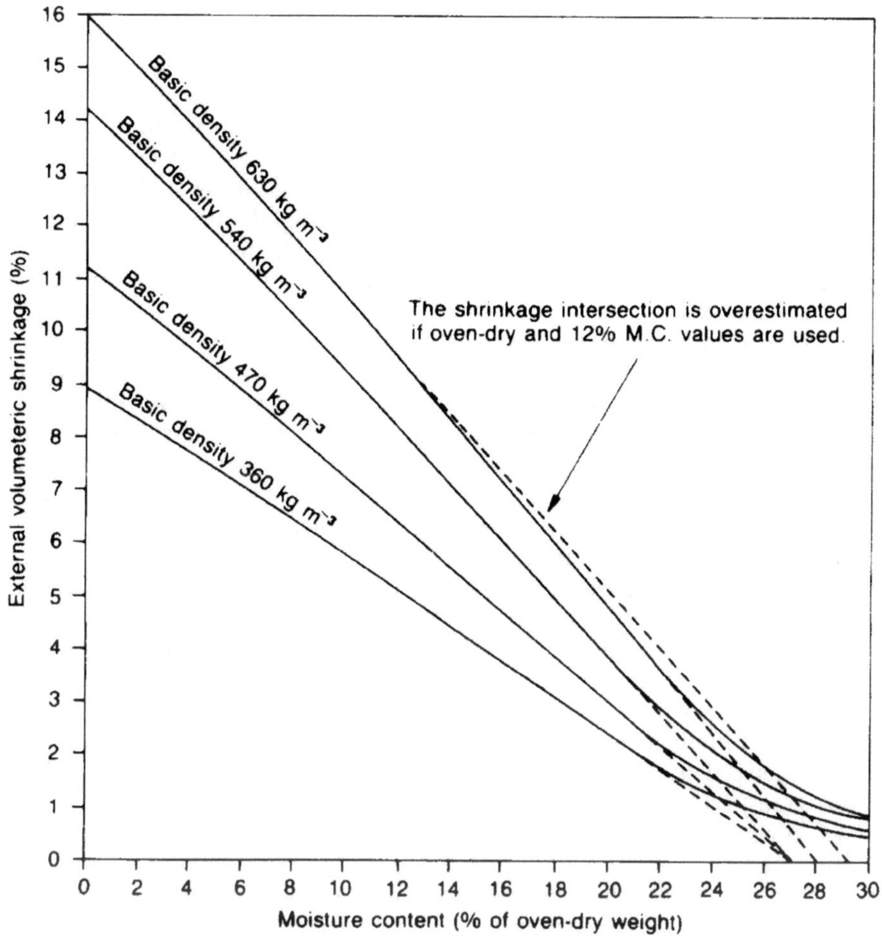

Figure 4.2. External volumetric shrinkage of 7/8 in. (22 mm) boards of loblolly pine (Stamm, 1964). The dotted lines extrapolate to the shrinkage intersection point, which is an estimate of the fibre saturation point.

$$Volumetric\ shrinkage\ (\%) = \frac{green\ swollen\ volume - oven\text{-}dry\ volume}{green\ swollen\ volume} \times 100\% \quad (1)$$

This can be estimated if the basic density is known:

$$Volumetric\ shrinkage\ (\%) = MC\ at\ fibre\ saturation\ (\%) \times basic\ density \times 10^{-3} \quad (2)$$

where the numeric constant, 10^{-3}, has units of $m^{-3}\ kg^{-1}$, i.e. the volumetric shrinkage corresponds to the volume of adsorbed water per unit volume of green wood. High density woods have proportionately more cell wall and less lumen volume, and so shrink and swell more. The cell wall water occupies space that exists only when the adsorbed water is present, i.e. as water is removed from the cell wall during drying the volume of the cell wall decreases by an equivalent amount and the external volume of the wood shrinks accordingly: further there is more adsorbed water per cubic metre of green wood in those species that have a high basic density. From this equation the estimated volumetric shrinkage to the oven-dry state for lumber having a basic density of 630 kg m^{-3} is 18.9%, slightly higher than the 16% recorded (Figure 4.2). This equation provides a useful, quick estimate of the volumetric shrinkage. In this calculation it is sufficient to assume that the fibre saturation point is 30%.

The plot (Figure 4.2) of the external volumetric shrinkage *versus* moisture content is not linear at low moisture contents (<8% moisture content). Shrinkage is slightly less than one would expect from extrapolation. This would follow if the last bit of water retained during drying is adsorbed more strongly and is somewhat denser than bulk water. This applies particularly to those water molecules bonded directly to the cell constituents, which are the last to be removed when drying. Also this non-linear shrinkage would be observed if the residual micropore void spaces were formed within the drying cell wall as the wood approached the oven-dry state.

Figure 4.2 suggests a quick, simple method for estimating the fibre saturation point. The shrinkage intersection point method involves measuring the original green volume of the wood and both volume and moisture content at two positions along the line in Figure 4.2. Ideally the volumetric shrinkage is measured at around 20-25% and again around 8-12% moisture content. By extrapolating to zero shrinkage the moisture content at the shrinkage intersection point is estimated. The shrinkage intersection point method provides a surrogate estimate of the fibre saturation point, accurate perhaps to $\pm 2\%$. The shrinkage intersection point itself varies according to whether it is the green virgin value or the rewetted value that is being measured (Figure 3.1): the original green shrinkage intersection point is always a little greater than that obtained when air or oven-dry wood is re-saturated. One might be tempted to use the oven-dry shrinkage value (it is easy and convenient to determine) together with a shrinkage value at around 12% moisture content (which approximates to the equilibrium moisture content inside unheated buildings in temperate regions), but this extrapolation will over-estimate the shrinkage intersection point because of the departure from linearity at very low moisture contents in Figure 4.2.

The volumetric shrinkage intersection point can be estimated from the volumetric shrinkage plot in Figure 4.2. The radial and tangential shrinkage intersection points can be estimated similarly. The tangential shrinkage intersection point is generally somewhat greater than the volumetric shrinkage intersection point and that in turn is greater than the radial shrinkage intersection point. A difference of 2-4% between the tangential and radial shrinkage intersection values is not unusual. This emphasizes that the volumetric shrinkage intersection point is not a particularly precise term, nor is the fibre saturation point. However both are of practical utility.

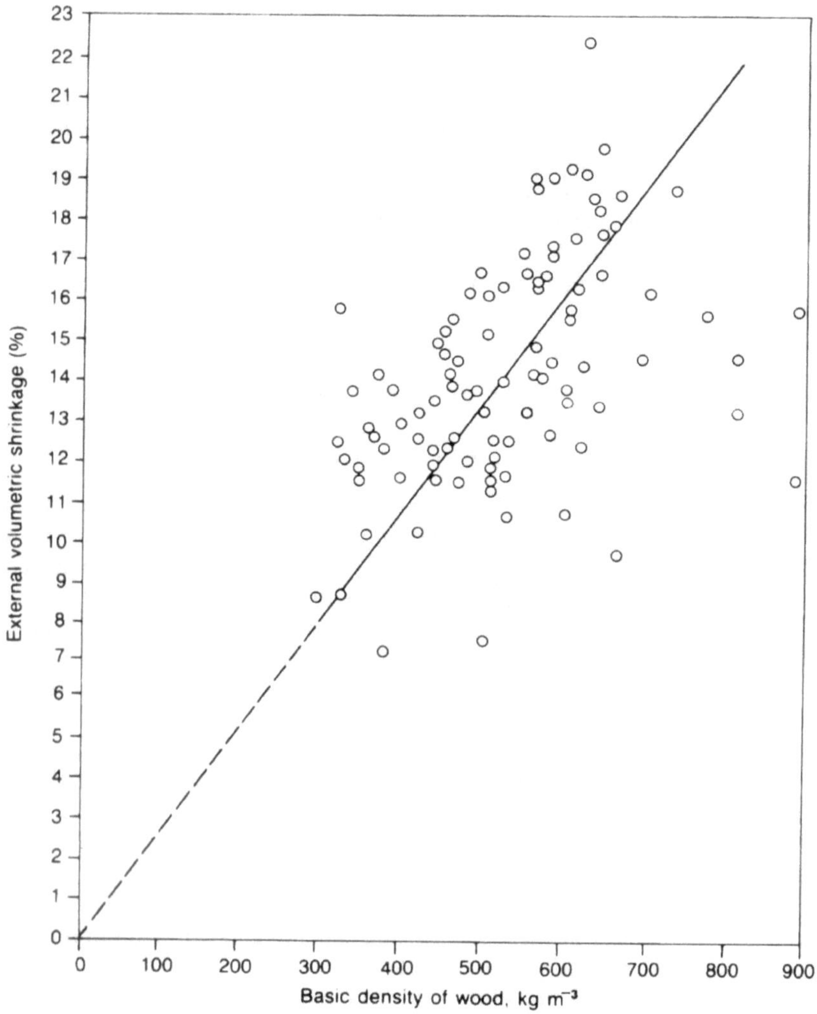

Figure 4.3. External volumetric shrinkage of 106 different North American hardwoods from the green to oven-dry state plotted against their respective basic densities (Stamm, 1964).

Figure 4.3 is a plot of the external volumetric shrinkage for a large number of hardwoods. The regression line drawn through the data corresponds to the general Eq. (2) that has already been considered.

In this case the moisture content at the fibre saturation point, or more properly the shrinkage intersection point, is 27%. The points lying below the regression line imply that the shrinkage intersection point for these timbers is less than 27%, while those lying above the line have a higher shrinkage intersection point and shrink more than would be predicted by the regression equation. Half the specimens deviate less than 11% from their predicted shrinkage value (Stamm, 1964). This equation explicitly links the external volumetric shrinkage to the loss of adsorbed water.

Implicit is the further assumption that the diameter and volume of the lumens do not change on drying. Optical microscopy confirms this to be a reasonable approximation (Beiser, 1933). Where there is a slight increase in the diameter of the lumen on drying the external volumetric shrinkage will be a little less than predicted, where the lumen decreases in size the external volumetric shrinkage will be somewhat more than predicted. Some variability is to be expected.

In the same way, the volumetric swelling from the oven-dry state can be estimated using the following relationship:

$$\begin{aligned} \textit{Volumetric swelling, \%} &= \textit{(moisture content (\%) at fibre saturation)} \\ &\quad \times \textit{oven - dry density (kg m}^{-3}\textit{)} \times 10^{-3} \end{aligned} \tag{3}$$

3. EXTRACTIVE BULKING

The amount of extractives in the heartwood of temperate species is generally only a few percent, whereas some tropical and sub-tropical species have much larger amounts. There is little information on the distribution of extractives between the cell wall and the lumen (where they mostly line the cell wall). However, Kuo and Argenbright (1980) have observed that roughly three-quarters of all the extractives in the inner heartwood of mature coast redwood, *Sequoia sempervirens*, and incense cedar, *Libocedrus decurrens*, was located in the cell wall, but this figure dropped to 50 and 60% respectively at the sap/heartwood boundary. In the case of coast redwood the cell wall extractives could be removed with hot water whereas with incense cedar organic solvents were necessary. The good dimensional stability and low shrinkage on drying of these two species are attributed to the presence of extractives bulking the cell wall. In the case of coast redwood the shrinkage intersection point is 24%.

Consider the following example of a timber having an unextracted basic density of 530 kg m^{-3}, comprising 500 kg m^{-3} of wood and 30 kg m^{-3} of hydrophilic (water-soluble) extractives deposited within the cell wall.

If it had been assumed that the wood was extractive-free then the estimated oven-dry volumetric shrinkage would be:

$$\begin{aligned} \textit{Volumetric shrinkage, \%} &= \textit{(moisture content (\%) at fibre saturation)} \\ &\quad \times \textit{basic density (kg m}^{-3}\textit{)} \times 10^{-3} \\ &= 30 \times 530 \times 10^{-3} = 15.9\% \end{aligned} \tag{4}$$

and its volumetric shrinkage on drying to 15% moisture content would be approximately half this figure, i.e. 7.95%.

However, the extractive-free basic density of this wood is actually 500 kg m^{-3}, so if it were not bulked by extractives it should only shrink by 15% on oven-drying and by 7.5% on drying to 15% moisture content.

This wood has 30 kg of extractives in a cubic metre of green wood and, assuming a density of 1400 kg m^3 for these extractives (Tarkow and Krueger, 1961), they will occupy 0.021 m^3 of the swollen cell wall per m^3 of swollen wood. If the extractives were not present, the extractive-free basic density would be 500 kg m^{-3} and the wood would shrink by 15%, i.e. the cell wall would shrink by 0.15 m^3 per m^3 of swollen wood. However the water soluble extractives occupy 0.021 m^3 of the cell wall for every m^3 of swollen wood, so the wood and the cell wall can only shrink by 0.150-0.021 m^3, i.e. 0.129 m^3 per m^3 of swollen wood rather than by 0.150 m^3 per m^3 which would be expected of an extractive-free wood having a basic density of 500 kg m^{-3}. The oven-dry volumetric shrinkage will be only 12.9% rather than 15.0%.

Further, extractive bulking of the cell walls reduces the hygroscopicity of wood at high humidities (Spalt, 1958), so the shrinkage from green to 15% moisture content is only 5.4% (12.9-7.5%) whereas shrinkage below 15% moisture content is 7.5%, being unaffected by the presence of extractives. Stone and Scallan's data (Figure 3.4) showed that the molecular probes only penetrate the largest openings in the cell wall, and the larger the molecular probe the larger the openings need to be. Thus it is logical for the water-soluble extractives to be confined to the larger pore spaces in the swollen cell wall, and consequently the bulking effect should be most evident at high humidities. The estimated shrinkage to 15% moisture content is only 68% (5.4/7.95) of that expected of a wood having a density of 530 kg m^{-3}. Using the earlier equation, the fibre saturation point is estimated to be a little over 25% moisture content.

If extractives were located in the lumens rather than in the cell walls this timber would shrink by 15% on oven-drying. However, the expected shrinkage for extractive-free wood of this density (530 kg m^{-3}) would be 15.9%. The estimated shrinkage on drying to 15% moisture content is 94% (7.5/7.95) of that predicted (7.95%) and the estimated fibre saturation point would be 28%.

Clearly shrinkage is dependent on the extractive-free basic density, the amount of extractives in the wood, and the distribution of the extractives between the cell wall and the lumen. The bulking of the cell wall with extractives explains part of the shrinkage variation observed in Figure 4.3.

In field-based research measurements of basic density are based frequently on unextracted wood and the mass of extractives is lumped in with the wood tissue.

4. ANISOTROPIC SHRINKAGE AND SWELLING OF WOOD

The shrinkage of wood is different in the three principal directions: longitudinal, radial and tangential (Table 4.1). Typical oven-dry shrinkage values for medium density woods would be:

Longitudinal shrinkage, α_{long}:	0.1–0.3%
Radial shrinkage, α_{rad}:	2–6%
Tangential shrinkage, α_{tang}:	5–10%

Table 4.1. Shrinkage and movement in heartwood of certain timbers (PRL, 1972; BRE, 1977).

Species	Density[1] (kg m^{-3})	Shrinkage[2] (%)		Movement[3] (%)		Moisture content at 60 and 90% humidity	
		Radial	Tang	Radial	Tang		
Oak, white, *Quercus alba*	750	3.0	5.5	1.3	2.6	12.5	21
Oak, European, *Quercus robur*	720	4.0	7.5	1.5	2.5	12	20
Teak, *Tectona grandis*	640	1.5	2.5	0.7	1.2	10	15
Mahogany, *Swietenia* sp.	540	2.0	3.0	1.0	1.3	12.5	19
Douglas fir, *Pseudotsuga menziesii*, UK	530	2.5	4.0	1.2	1.5	12.5	19
Radiata pine, *Pinus radiata*	480	2.5	4.0	1.2	2.0	12.5	20
Spruce European, *Picea abies,* Scandinavia	470	2.0	4.0	1.0	2.1	12	20
Sitka spruce, *Picea sitchensis*, UK	400	3.0	5.0	0.9	1.3	12.5	19
Western red cedar, *Thuja plicata,* Canada	370	1.5	2.5	0.4	0.9	9.5	14
Western red cedar, UK	–	–	–	0.8	1.9	13	21

[1] Mean air-dry density at 12% moisture content, kg m^{-3}.
[2] Drying from the green condition to 12% moisture content.
[3] Movement of wood when relative humidity decreases from 90% to 60% at 25°C.

In normal wood the longitudinal shrinkage from the green to the oven-dry condition is an order of magnitude less than transverse shrinkage. Tangential shrinkage is usually 1.5 to 2.5 times that of the radial shrinkage. Thus:

$$\alpha_{tang} > \alpha_{rad} >> \alpha_{long}$$

To a first approximation the longitudinal shrinkage can be ignored. The volumetric shrinkage then becomes:

$$\alpha_{vol} = \alpha_{tang} + \alpha_{rad} - \alpha_{tang} \cdot \alpha_{rad},$$

and if the cross product can be neglected, then

$$\alpha_{vol} \approx \alpha_{tang} + \alpha_{rad}$$

Lumber is not used in the oven-dry state. For many uses the desired in-service moisture content falls within the range of 8-15% moisture content. The shrinkage will then be only a half to three-quarters of the oven-dry shrinkage value. Therefore the timber trade is more interested in shrinkage values from the green to 12% moisture content (Table 4.1). These values indicate to the sawmiller the green tolerances required to produce wood of specified dry dimensions. Again, in plywood manufacture green veneer must be clipped over-size in order to be of the desired width after drying: e.g., when clipping green veneer for 1.2 x 2.4 m (4 x 8 ft) panels the sheets would have to be clipped 60 mm oversize to allow for 5% tangential shrinkage, i.e. 1.26 m plus tolerances to allow for lay-up and trim.

5. THEORIES FOR ANISOTROPIC SHRINKAGE

5.1. Microfibril angle (Barber and Meylan, 1964)

The cell wall of wood can be considered to consist of a non-crystalline matrix of lignin and hemicelluloses in which strong, stiff cellulosic microfibrils are embedded. The crystalline microfibrils exhibit no tendency either to adsorb moisture or to change in length or cross-section. On the other hand the non-crystalline isotropic matrix can lose and gain water and shows a considerable tendency to shrink and swell. In isolation one would expect the matrix to shrink or swell equally in all directions, that is $\alpha_x = \alpha_y = \alpha_z = \alpha_0$ and $\alpha_{vol} \approx 3\alpha_0$ i.e. α_0 is the isotropic shrinkage in any direction. However, the microfibrils in the cell wall restrain the matrix from shrinking in the direction parallel to the microfibril axis, and so the cell wall is forced to shrink excessively in the plane transverse to the microfibrils.

Barber and Meylan (1964) developed a model based on matrix-microfibril interaction. In its simplest form their model ignores the different characteristics of the various layers in the cell wall. Instead it assumes that the behaviour of the wood is determined by the thick, dominant S_2 layer, where the microfibrils can be inclined to the fibre axis at some angle between 5° and 50° (Figure 4.4).

Barber and Meylan (1964) considered the shrinkage in an element of a tracheid wall (Figure 4.4) assuming the microfibrils lie in turn parallel to the fibre axis, at 30° to the axis, and at 45° to the axis (Figure 4.5a). Where $\theta = 0°$, there is very little longitudinal contraction (in the x direction) and the longitudinal strain ratio, α_x/α_0, is also small, i.e. $0 < \alpha_x/\alpha_0 << 1$. Here strain ratios, α_x/α_0, α_y/α_0, and α_z/α_0, express the dimensional change in a particular direction relative to that expected of an isotropic body. Thus where $\theta = 0°$, the volume shrinkage is observed only in the width and thickness of the wall element (in the y and z directions). Assuming the volumetric shrinkage is approximately $3\alpha_0$ and that shrinkage in the x direction is strongly restrained by the microfibrils, then where $\theta = 0°$ the strain ratios, α_y/α_0 and α_z/α_0, must be approximately equal to 1.5 (Figure 4.5b).

Where the microfibrils lie at some angle between 0° and 45° to the fibre axis the change in width still exceeds the change in length ($\alpha_x/\alpha_0 < \alpha_y/\alpha_0$). However the microfibrils are no longer aligned parallel to the x axis but lie at some angle, θ, to it (in the xy plane) so any shrinkage in the y direction seeks to shorten the microfibrils and put them in compression even if the longitudinal shrinkage were zero. If the microfibrils are incompressible (implying a very high modulus of elasticity) the decrease in the microfibril angle, θ, that results from shrinkage in the y direction

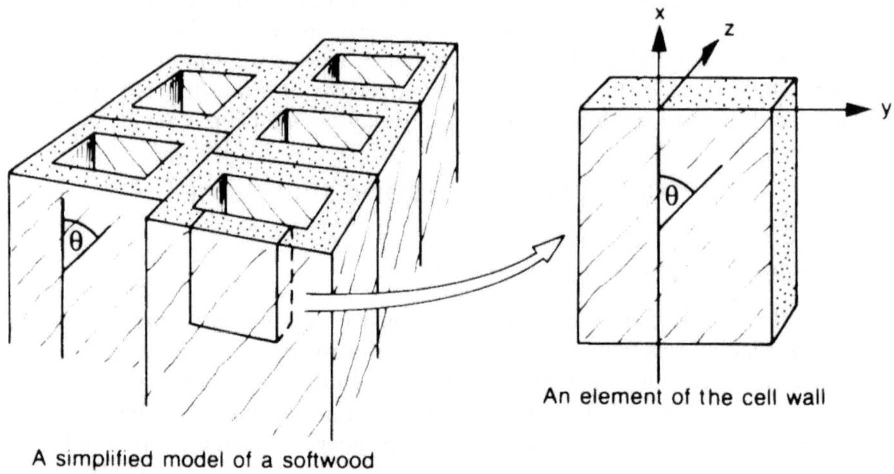

Figure 4.4. A cell wall element. The cell is aligned in the x direction, the y direction lies parallel to the wall (width), z is transverse to the wall (thickness) and θ is the microfibril angle within the S_2 layer (Barber and Meylan, 1964). The y and z directions are not defined relative to the radial and tangential directions: they approximate to polar coordinates.

must be counterbalanced by a slight longitudinal expansion of the wall element in the x direction: this allows the microfibril length to remain unchanged. Thus where $\theta = 30°$ the axial shrinkage, α_x, becomes slightly negative, i.e. the wood swells slightly in the x direction on drying, $\alpha_x/\alpha_0 \leq 0$. The microfibrils offer some restraint to shrinkage in the y direction, $1 < \alpha_y/\alpha_0 < 1.5$, and by inference $\alpha_z/\alpha_0 > 1.5$.

For $\theta = 45°$ shrinkage in the x and y directions must be the same (by symmetry). Thus: $0 < \alpha_x/\alpha_0 = \alpha_y/\alpha_0 < 1$, and α_z/α_0 attains its maximum value (> 1.5). The shrinkage ratios in the three principal directions are summarized in Figure 4.5b. The longitudinal shrinkage ratio, α_x/α_0, is small but positive where $\theta = 0°$, becomes zero or negative around $\theta = 30°$ and then rises again rather sharply. The transverse or width shrinkage ratio, α_y/α_0, starts large but diminishes with increasing angle so that where $\theta = 45°$ it is equal to the longitudinal shrinkage ratio, α_x/α_0. The contraction in thickness of the cell wall, α_z/α_0, is large where $\theta = 0°$ and increases slowly with increasing angle, θ, until it reaches a maximum where $\theta = 45°$. These curves show

Figure 4.5. Anisotropic shrinkage (Barber and Meylan, 1964). (a) Shrinkage in the xy plane of an isolated element of the cell wall as a function of microfibril angle. Heavy lines represent the green dimensions, normal lines represent the dimensions after drying, and dotted lines the isotropic dimensions after drying (were the microfibrils to offer no restraint to shrinkage). The microfibril angle decreases slightly on drying if the microfibrils do not change length (AA = BB). (b) Resultant strain ratios in the x (length), y (width) and z (thickness) directions as a function of θ. (c) Actual shrinkage depends on the stiffness of the microfibrils (E) and the shear modulus (S) of the matrix. The latter varies with moisture content.

a) Effect of microfibril angle on shrinkage of the cell wall element.

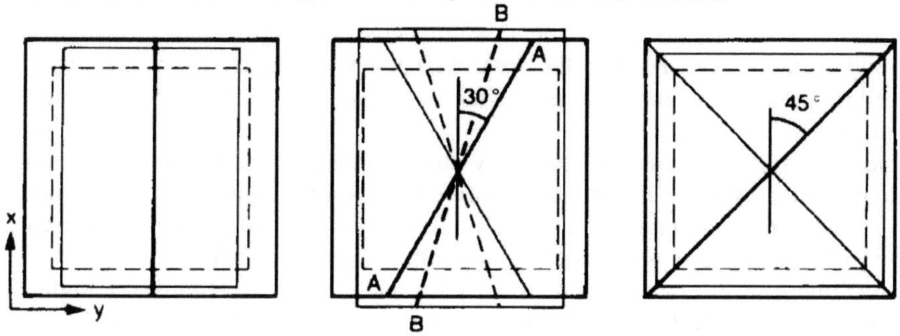

b)

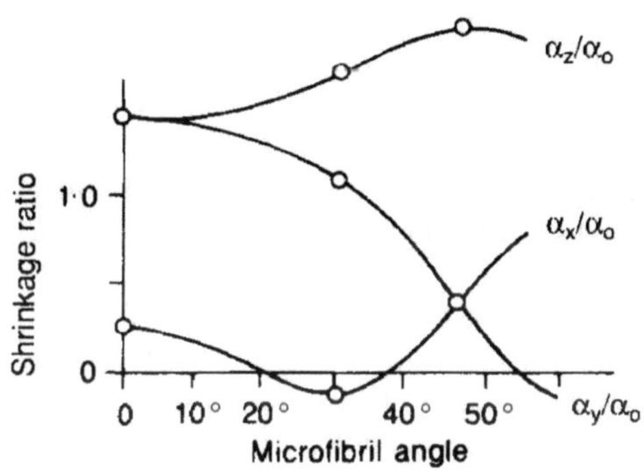

Shrinkage ratios in the three principal directions

c)

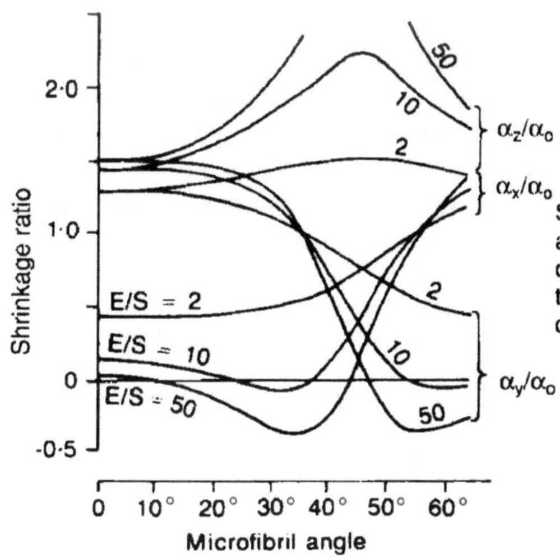

Shrinkage ratios will vary according to the stiffness (E) of the microfibrils and the shear modulus (S) of the matrix.

symmetry in that α_x/α_0 between $\theta = 45^\circ$ and $\theta = 90^\circ$ is the mirror image of α_y/α_o between $\theta = 0^\circ$ and $\theta = 45^\circ$ and *vice versa*. There is symmetry in α_z/α_0 about $\theta = 45^\circ$ (Figure 4.5b). Clearly although the relation between longitudinal shrinkage and microfibril angle is non-linear the longitudinal shrinkage is determined by the microfibril angle and is generally negligible for θ less than 35° but increases rapidly thereafter.

The magnitude of the shrinkage in any direction depends on the stiffness, i.e. the elastic modulus (MOE or E), of the microfibrils and on the rigidity (S) of the matrix (also called the shear modulus). Various E/S ratios generate a series of curves (Figure 4.5c). E/S varies with species and with density. In principle a material with very stiff microfibrils embedded in a matrix having a low shear modulus (a high E/S ratio) will show noticeable longitudinal expansion on drying if its microfibril angle is lies between 20° and 40°. Furthermore, on drying the E/S ratio decreases gradually as the shear modulus of the matrix is sensitive to the moisture content.

The theory of Barber and Meylan (1964) is supported by experimental observations. Corewood and compression wood, with their high microfibril angles, are two obvious examples that display large longitudinal shrinkage. Meylan (1968) examined the longitudinal and tangential shrinkage in the corewood of *Pinus jeffreyi* where the microfibril angle can exceed 40° and observed longitudinal shrinkage as high as 7% which was well in excess of the tangential shrinkage in 'normal' samples (Figure 4.6).

The same theory can explain the transverse anisotropy of wood. Barber and Meylan (1964) noted that the microfibrils in the radial walls are less uniformly arranged and on average lie at larger angles than those in the tangential walls, the difference being as much as 15°. This is because bordered pits in the radial walls force the microfibrils to deviate around them. In order to simplify their analysis of radial and tangential shrinkage Barber and Meylan (1964) assumed that the microfibril angle in the radial walls is always 1.5 times that in the tangential walls. As a consequence of the microfibril angle being greater in the radial walls, both terms which together determine tangential shrinkage – the shrinkage in the thickness of the radial walls and in the width of the tangential walls – are both greater than the two terms which determine radial shrinkage – the shrinkage in the thickness of the tangential walls and in the width of the radial walls, i.e. $(\alpha_z/\alpha_0)_{\text{rad wall}} > (\alpha_z/\alpha_0)_{\text{tang wall}}$ and $(\alpha_y/\alpha_0)_{\text{tang wall}} > (\alpha_y/\alpha_0)_{\text{rad wall}}$ (Figure 4.5b). Thus where $\theta_{\text{tang wall}}$ lies between 10 and 30° and $\theta_{\text{rad wall}}$ lies between 15 and 45° tangential shrinkage will be greater than the radial shrinkage. Their full analysis was not quite that simple since it is not possible for the radial and tangential walls of a tracheid to behave independently. Subsequently their modelling was generalized to include the effects of all cell wall layers, but this did not affect the general conclusions.

5.2. Earlywood-latewood interactions (Pentoney, 1953)

In this theory the difference in radial and tangential shrinkage of many species grown in the temperate zone is attributed to differences in density of early and latewood. The shrinkage of dense latewood cells is greater than that of earlywood.

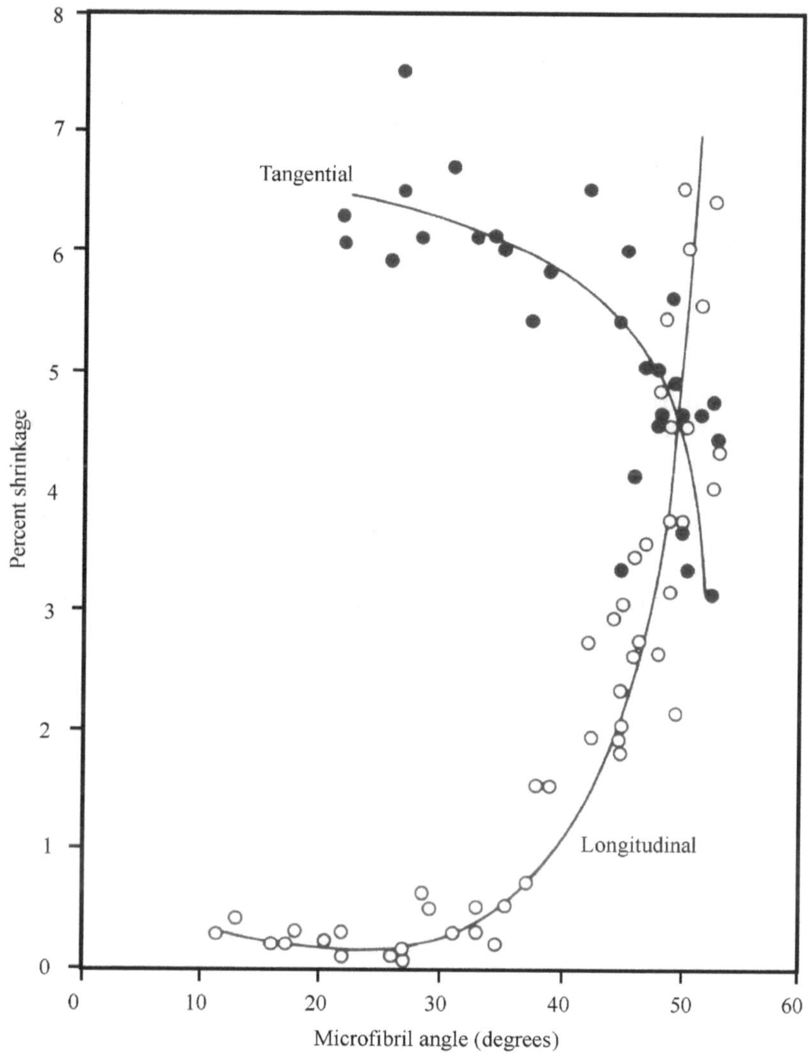

Figure 4.6. Relationship between microfibril angle and longitudinal and tangential shrinkage in *Pinus jeffreyi* (Meylan, 1968).

In the tangential direction the earlywood and latewood must move together (they act in parallel). The earlywood is forced to shrink more than it would wish by the latewood, while the latewood is slightly restrained by the weaker earlywood and does not shrink quite as much as it would wish. Where latewood is considerably denser and stronger than earlywood the bands of latewood force the weak bands of earlywood to shrink tangentially by almost as much as an isolated band of latewood would shrink (they act in parallel). In the radial direction both the earlywood and the latewood shrink independently (they act in series) and the total shrinkage corresponds roughly to the weighted mean shrinkage of the two components. Pentoney estimated that a tangential to radial shrinkage ratio as high as two would be possible where the density of the latewood is 2.5 times that of earlywood and the earlywood occupies two-thirds of the annual ring.

5.3. Ray tissue

This theory assumes that ray tissue shrinks less in the radial direction than does the axial tissue and therefore ray tissue restrains radial shrinkage. This is to be expected since the rays are lying in the direction of maximum strength, along the ray cell axis, with their microfibrils in the various wall layers having a strong bias parallel to the axis of the ray cell (Harada and Côté, 1985), whereas the axial cells are operating in the direction of minimum strength, in the radial direction, i.e. perpendicular to the cell axis. Shrinkage data support the view that ray restraint is at least one factor responsible for transverse shrinkage anisotropy, particularly of the broad-rayed timbers such as oak and beech where the ray volume is some 17-22% of the wood tissue (Table 4.2).

Table 4.2. Influence of ray tissue on the radial shrinkage of wood (Skaar 1988).

Species	Radial shrinkage: green to oven-dry			
	Ray-free wood (%)	Wood with fine rays only (%)	Wood with all ray tissue (%)	Isolated ray tissue (%)
		→ increasing restraint to radial shrinkage →		
Acer pseudoplatanus	4.9	–	4.3	3.8
Arctocarpus integra	4.8	–	3.8	1.3
Canarium zeylanicaum	4.0	–	3.4	0.2
Cardwellia sublimis	–	3.5	–	0.9
Casuarina iukmanni	–	–	3.3	1.2
Fagus grandifolia	12.7	6.7	–	2.3
Grevillea robusta	–	–	3.7	1.2
Helicia terminalis	–	1.9	–	0.8
Quercus borealis	12.0	–	5.1	2.5
Quercus borealis	6.8	4.8	–	2.6
Quercus ilex	–	6.0	–	3.1
Quercus spp.	–	–	4.9	3.2
Quercus kelloggii	–	5.8	3.0	2.1
Xykomelum pyriforme	–	–	2.0	0.7

5.4. Anisotropic elasticity (Boutelje, 1973)

For softwoods the elastic moduli E_{long}, E_{rad} and E_{tang} are in the ratio of approximately 50, 2, 1. Wood is very stiff in the axial direction. In the transverse plane wood is twice as stiff in the radial direction than it is in the tangential direction. This means that a given stress will result in an extension or compression in the tangential direction which will be twice that in the radial direction. A major reason for this is the regular ordering of tracheid cells in the radial direction. This can be seen in Figure 1.1. In the radial direction the load is carried by the radial walls. In the tangential direction because the cells do not lie in ordered files the load has to be transferred from tangential wall to tangential wall via the radial cross walls, which carry the load in bending. If wood is twice as stiff in the radial direction it follows that the resistance to radial shrinkage will be twice that for tangential shrinkage. The consequent anisotropic shrinkage is not related to details of cell wall ultrastructure but arises simply from the arrangement of the tracheid walls in the transverse plane. Such radial ordering is not found with hardwoods where the presence of large vessels disrupts the alignment of cells.

Clearly shrinkage anisotropy is a complex issue. A number of factors can contribute and the relative importance of each will vary between timbers. In some cases a large microfibril angle might be significant, as in corewood and in compression wood. Ray tissue will be important in species such as beech and oak. Contrasting earlywood and latewood densities is a likely cause in Douglas fir, but would be irrelevant for a tropical hardwood. The effects of elastic anisotropy would be more apparent in low density softwoods.

6. MOVEMENT AND RESPONSIVENESS OF LUMBER

Ideally lumber should be dried to the moisture content approximating to its average in-service moisture content. This ensures that shrinkage is complete prior to going into service. The desired moisture content depends on whether the lumber is for interior or exterior use. The moisture content will be sensitive to the local climate. Figure 4.7 illustrates the recommended values within the United States. The variation in recommended values between states is not unexpected (7.6% in lower Arizona and 15.2% in coastal Oregon and Washington). Climatic conditions in Australasia, Europe and Japan dictate the need for 10-12% kiln-dried furniture stock rather than the customary 6-8% needed for most of the United States (Araman, 1987). Such differences should be taken into account when exporting otherwise the material would move in service.

If wood is conditioned at 20°C and 65% relative humidity the sapwood of most timbers will equilibrate to about 12% moisture content. This corresponds to the typical moisture content of lumber in an unheated building in many temperate regions. There are exceptions. *Nothofagus fusca*, one of the Southern Beeches, would dry to 9.5% or so under these conditions, so drying to 12% would be inadequate (Harris, 1961). Medium density fibreboard and particleboard also equilibrate at a lower moisture content of about 8-9%. For this reason Hoadley (1979) observed that it is more logical that wood products be dried to a specified relative humidity than to a specified moisture content (Table 4.1).

Even supposing wood has been dried to the desired moisture content, and the sawmiller aims to dry to within 2% of the specified value, lumber will move in

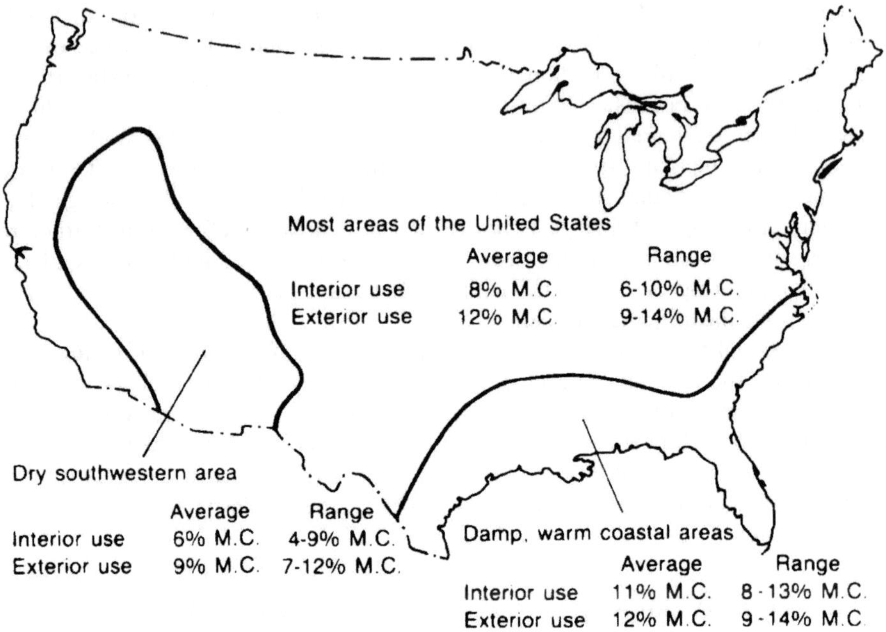

Figure 4.7. Recommended moisture content (MC) for lumber at the time of installation (USDA, 1987). The range relates to the moisture content values of individual pieces of timber found within a packet.

service. For example, exterior decks, patios and cladding are subject to fluctuating environmental conditions, especially where they are exposed to full sunlight. They will shrink during a dry hot summer and swell again during the winter months as moisture is re-adsorbed within the cell walls. Movement is defined arbitrarily as that occurring when the relative humidity is reduced from 90% to 60% (Table 4.1).

Movement does not directly equate with shrinkage. Shrinkage relates to the drying of lumber to the desired equilibrium moisture content while movement relates to the performance in service. For example, a species having a low shrinkage intersection point will not shrink much on drying but may move a lot in service and *vice versa*. For example European oak and Sitka spruce move less than one might expect, knowing their shrinkage values (Table 4.1). The exceptionally low shrinkage and movement of redwood and western red cedar are related to both the resinous nature of these timbers and their low extractive-free basic densities.

The large variation in equilibrium moisture content within buildings in different parts of the United States (Figure 4.7) may be a surprise. An extreme example relates to heated buildings in the Northeastern States. There the humidity drops below 20% in the cold winter months which equates with a moisture content of about 4% while in the summer it can rise to 80% which equates to a moisture content of about 16% (Hoadley, 1979). This enormous change in moisture content gives rise to a movement of about 3% during the year, i.e. the transverse dimensions of small (responsive) boards will change by 3% over the course of a year. Air conditioning is far less severe than central heating as it provides humidity as well as heat during the winter.

Of perhaps equal importance is the responsiveness of lumber to a fluctuating environment. Harris (1961) observed that some species take weeks or even months to regain equilibrium on being transferred to another environment, and concluded that less responsive woods are unlikely to display the full range of movement when exposed to such changes of humidity during the course of the year. Heartwood of all species tends to be more impermeable and picks up moisture more slowly than sapwood. *Nothofagus fusca* is an outstanding example of this. Notoriously difficult to dry, this strong, durable hardwood has the excellent property of being equally difficult to wet and consequently is an exceptionally stable timber in service.

Radial shrinkage is significantly less than tangential shrinkage, so an obvious way to reduce both shrinkage and movement would be to cut the timber so that the radial direction corresponds to the most critical dimension. For example in flooring the width of the board should correspond to the radial direction so unsightly gaps are less likely to appear if the boards have not been dried sufficiently. Also the risk of buckling and lifting of the floor is less likely if the boards have been over-dried or if the floor is accidentally flooded and the boards swell again (Figure 4.1). However cutting timber to expose a true radial face is uncommon. It involves more turning of the logs in the sawmill and production is slowed down. Further, less timber is recovered from the logs. What may appear to be a logical solution is not easily achieved or sought in the sawmill. In choosing material to cover a large surface it is preferable to select a species which shows little movement, is not particularly responsive to fluctuation environmental conditions (heartwood), and to ensure that the design allows for any movement that will inevitably occur, for example by incorporating expansion gaps at regular intervals.

7. COATINGS (TURNER, 1967; CASSENS AND FEIST, 1991)

The traditional way of limiting the movement of timber has been to apply 2-3 coats of paint (*c.* 0.1 mm film). Paint coatings prevent rapid surface absorption of water and the development of steep, superficial moisture gradients in exposed wood. For example, the average seasonal variation in moisture content of exterior timbers can be reduced by a factor of three or more where using three coats of a traditional linseed oil house paint (Orman, 1955).

Painting is also important to give wood resistance to weathering and photochemical degradation. Weathering involves the roughening of the surface with raised grain in latewood alongside the sunken more heavily eroded earlywood, together with checking and an attractive, silver-greying of the surface. Weathering is not simply the consequence of physical cycles of wetting and drying. Wood undergoes photochemical degradation, where lignin in particular breaks down into water-soluble fragments as a result of degradation by ultraviolet light. The chromophoric groups, such as phenolic hydroxyl groups, carbonyls (>C = O), quinones and hydroperoxy radicals, are particularly susceptible to absorbing the energy of sunlight in the visible and UV and so initiate degradation and give rise to the change in colour. In humid regions the desired patina fails to develop; instead one is left with an unsightly dark, blotchy discolourations due to mildew. Thus, one objective of all paints and stains is to protect the surface by incorporating photo-stabilizers as additives: typically these adsorb the solar radiation and dissipate the energy at frequencies that are less harmful to wood.

Wood of high density, wood with strongly contrasting early and latewood, and flatsawn boards with their exaggeratedly-wide latewood bands will display more movement in service. This stresses the wood-finish interface encouraging premature failure. Hence the favoured woods for exterior use are western red cedar or redwood rather than Douglas fir or southern yellow pine; and cottonwood or poplar as against oak amongst hardwoods.

The durability of a paint film (7-10 years) relies on a complex interaction involving the wood surface, the coating, the application technique and service conditions. Paints contain substantial quantities of inorganic pigments (*c.* 50% solids) that provide colour and protect the wood by adsorbing UV-radiation. The fine pigments are suspended in a resin carrier that subsequently cures and binds the pigment to the wood surface. Oil paints, and oleoresinous paints, rely on the cross-linking of polyunsaturated fatty acids in linseed or tung oil. The oleoresinous paints also include rosin, petroleum resins, terpene resins, coumarone-indene resins or phenolic resins to increase cross-linking and provide a greater sheen to the paint film. Alkyd resins are formed by the esterification of multifunctional alcohols, such as glycerol, with acids or cyclic anhydrides: phthallic anhydride is often used as the anhydride reacts forming an ester, and leaves a carboxylic acid that can also undergo esterification, allowing polymer development. Where oils are used as the base for alkyd paints they are heated with glycerol to form monoglycerides with which the phthallic anhydride may then react to give a better polymer network. Finally for oil-based or alkyd-based systems, a low-viscosity organic solvent is added to thin the paint and aid in the spreading and superficial penetration of the wood surface. With water-based systems such as acrylic paints, the pigment-resin is dispersed as an emulsion in water, only coalescing as a continuous film once the water evaporates.

Traditional oil-based paint formulations suffer from photo-oxidation, that results in a chalky appearance as the surface resin is eroded. The end result is the tendency for the film to become less elastic over time, to embrittle and then to fail so allowing water ingress. The 0.1 mm paint film fails by fatigue from the movement of the wood induced by endless fluctuations in the superficial moisture content. The trend has been away from oil-based paints to water-based systems such as acrylic emulsions. These are more permeable to water vapour and so the timber moves more, but the acrylic film has a greater degree of elasticity and, provided it is properly applied to clean surfaces, performs extremely well. The technical characteristics sought in such coatings are good adhesion, extensibility, impermeability to water (but not to water vapour) and a slow rate of chalk (erosion).

Aesthetics, cultural preferences and tradition play a strong part in coating selection. For example, in Scandinavia full-bodied stains containing iron oxide pigment have been favoured (3-6 year maintenance cycle). Stains have a higher pigment loading than traditional paint, resulting in a non-gloss film that emphasizes the underlying wood texture – often rough-sawn – while still obscuring and protecting the wood grain.

Conceptually, the ultimate progression is to simple, superficial water-repellent, preservative treatments using wax, a fungicide/mildewcide, a resin or drying oil and a solvent. These are surprisingly efficient in reducing water absorption and in reducing the movement of wood (Feist and Mraz, 1978). As the surface weathers, they require reapplication. However, surface preparation is far easier than with paint systems, simply scrubbing down biennially before applying the new coat. Moderately pigmented (tinted) penetrating stains are available.

A final point, good design is not just a matter aesthetics. People expect a paint finish to remain fresh and to require little maintenance for the first 5-10 years. Dark colours absorb direct sunlight and such painted surfaces can achieve temperatures of 50°C or more, resulting in superficial drying and exaggerated shrinkage of the substrate. Thus good design requires attention to functionality. Another example, narrow-dimensional timber members will move less reducing the likelihood of doors and windows jamming. Wide eves (soffits or roof overhangs) protect exterior cladding from the intense, overhead summer sun.

8. DIMENSIONAL STABILIZATION (ROWELL AND BANKS, 1985)

Research in dimensional stabilization over the last 50 years has achieved limited but specific goals: it has yet to fulfil its broader objectives. Niche markets, such as the use of polyethylene glycol (PEG) in the treatment of marine and waterlogged artefacts are well established. New technologies for commercial scale treatment of timber by acetylation, and for various heat treatments, have reached the market in the last decade, and may increase in importance in future.

An endless variety of treatments have been contemplated by researchers. These involve filling accessible spaces (either the lumens or pores in the cell wall) with inert material, or some form of chemical modification of the wood tissue.

To interpret their effectiveness swelling coefficients are first calculated using the method reported by Stamm (1964):

$$S\ (\%) = ((V_w - V_d) / V_d) \times 100\ (\%) \qquad (5)$$

where V_w and V_d are the volumes of water-saturated and oven-dry wood respectively. Then the anti-shrink efficiency (ASE) can be calculated according to:

$$ASE\ (\%) = ((S_c - S_m) / S_c) \times 100\ (\%) \qquad (6)$$

where S_c is the volumetric swelling coefficient of control (unmodified) samples and S_m is the volumetric swelling coefficient of modified samples. For example, with anhydride modification methods ASEs of up to 90% have been reported. The degree of dimensional stabilisation imparted to the wood is closely related to the weight percent gain (Hill and Jones, 1996).

Treatments may involve slow diffusion when the wood is green, or impregnation and curing after the wood has been dried. Laboratory scale chemical reactions often use swelling agents and catalysts such as pyridine or dimethyl formamide to increase the reaction rate. These would require rigorous clean up after treatment, or alternative slower treatment conditions when scaled up for an industrial operation. To date the cost of additional preparation and processing has precluded wide uptake of many of these treatments. This slow progress has left many in favour of alternative strategies, such as selecting naturally durable plantation species, whose heartwood is more stable, with the cachet of being environmentally friendly.

8.1. Bulking with leachable chemicals (Stamm, 1964; Rowell and Konkol, 1987)

The particular case of bulking by extractives has been discussed. Bulking refers to the condition where chemicals are deposited within the swollen cell wall, replacing some of the water. As a result when the wood is dried the presence of bulking chemical within the cell wall prevents some of the shrinkage that would normally occur. The bulking agent does not strengthen wood.

An effective bulking agent should be highly soluble in water. Salts have been tried. The cheapest such as sodium chloride are not particularly soluble (1 part salt to 5 parts water in a saturated solution) and so cannot reduce shrinkage very much. Furthermore such treated wood feels damp and the salts corrode fittings.

Polythylene glycol (PEG) is the best-known bulking chemical and at low molecular

$$HO\text{-}CH_2\text{-}(CH_2OH\text{-}CH_2)_n\text{-}CH_2OH$$

weights of 200-600 PEG is a liquid that is miscible with water while at higher molecular weights PEG is a waxy solid whose solubility in water is limited unless maintained in solution by heating (30-70°C). The choice of molecular weight is dictated by two considerations. Use too low a molecular weight and the polymer would be volatile, the wood will feel damp, and the polymer can readily leach out again. Use too high a molecular weight and the polymer will be much less soluble in water, less able to penetrate the cell wall and so be less effective as a bulking agent. Furthermore, a high molecular weight polymer diffuses into the wood very slowly (indeed once it exceeds a certain size it cannot penetrate the swollen cell wall) which increases treatment time and reduces economic viability. A molecular weight of about 1000 is a compromise. PEG is used with green timber. The swollen wood is soaked in increasingly concentrated, heated solutions of PEG (< 50% by weight) for a few days to many months, depending on the size of the timber, before drying gently. The cross-sectional shrinkage and swelling of the treated wood is reduced to less than 1% when the PEG content exceeds about 35%. The treated material feels damp and needs a finish to seal the wood if it is to be handled. The technique has some popularity amongst wood turners and hobbyists.

The most spectacular uses of PEG have been in marine archaeology, in the treatment and protection of waterlogged ships after they have been raised to the

Figure 4.8. The 1200 tonne man-of-war, *Vasa*, sank in Stockholm harbour in 1628 and was raised in 1961 to undergo massive conservation work over more than two decades (Statens Sjöhistoiska Museum, Vasavarvet, Stockholm).

surface (Figure 4.8). There are problems in the conservation of such timbers, that centre around the fact that the timbers is partially decayed, and the polysaccharide components can be extensively depolymerized, which means that the timber becomes much weaker. If simply dried such timbers are susceptible to excessive fracturing and collapse. Further, with the depolymerization and destruction of cellulose there is less restraint to longitudinal shrinkage and the anisotropic behaviour of wood is much reduced: on drying the wood shrinks excessively in the longitudinal direction. However, the integrity of the structure can be maintained by bulking the wood and, as important, by minimizing any stresses generated during subsequent dehydration. The preservation of large structures is an immense task.

The treatment of the 'Vasa' extended over two decades before all members were treated and the ship completely reconstructed (Håfors, 1990). The treatment involved initially spraying the timbers with dilute solutions of PEG and gradually increasing the concentration over a number of years. Two fractions of PEG were used, with molecular weights of 400-600 and 1500-4000. Boric acid and borax were added to prevent further biodeterioration during the extensive treatment programme in which the largest members had to be treated continuously for up to 17 years. With this combination of molecular weights it was possible to achieve good penetration and reasonable bulking of cell walls throughout the cross-section and minimal shrinkage when finally drying the timbers for the public display. Even after such an extensive treatment, with the exception of the outer 10 mm or so, the PEG loading within the structural timbers is quite low (particularly so of oak). At low concentrations the PEG only bulks the cell wall while at higher concentrations it also fills the lumens, so providing some mechanical support to the outer fibres.

8.2. Bulking with thermosetting resins (Meyer, 1984)

Water-soluble thermosetting resins offer a non-leachable, strength-enhancing means of bulking timber. Partially polymerized phenol-formaldehyde (PF) diffuses into the swollen cell wall and after drying the resin is polymerized using heat and a catalyst. Shrinkage is inevitable as cell wall bulking is not total and water molecules are eliminated from between monomers during condensation to a high molecular weight polymer. A resin content of about 35% is sufficient to produce a reduction in swelling of 70-75%. The material has good decay and acid resistance.

An ideal process would allow the treatment of substantial cross-sections of timber with inexpensive, low viscosity resins that could be polymerized without the application of heat and with minimal volume change. Unfortunately low-viscosity monomers liberate a large amount of heat during polymerization, are highly volatile, and are unsuited to any thermally-initiated process.

Recently Furuno *et al.* (2004) used a pH neutral, low molecular weight PF resin to achieve 60% ASE, with all the resin located in the cell-wall. The pH neutral system gives a colourless resin retaining the natural wood appearance, whereas the longer established alkaline treatments cause the wood to become red-brown. These procedures are usually limited to veneers because the diffusion-soak time increases rapidly with thickness and there are curing problems. While the polymerization reaction is initiated by heating, the reaction is then exothermic, releasing heat, so the subsequent build up of heat and pressure can result in vaporization and expulsion of the resin, resulting in poor utilization and a non-uniform distribution. Some pre-polymerization of the resin reduces the amount of heat released, but the viscosity of the resin is increased so it is harder to treat thick pieces (> 10 mm). Where a thick final product is wanted this can be achieved most efficiently by laminating.

Furfurylation is a closely related process that has been commercialised in Norway. Initially polyfurfuryl alcohol (PFA) was investigated as having potential to replace PF resins in board manufacture. Recent research has led to polymerization of furfuryl alcohol within wood during heating to around 100°C. The furfuryl alcohol polymerizes under acidic conditions, and has also been shown to graft onto the wood substrate (Lande *et al.,* 2004a). Furfurylated wood is a dark coloured due to the PFA; however this has led to successful niche marketing of the hard, stable, dark timber as an alternative to tropical hardwoods such as Wengé (*Millettia laurentii*). An ASE of 50% has been reported for furfurylated timber with a weight percent gain (WPG) of 32%; and an ASE of 70% for a WPG of 47% (Lande *et al.*, 2004b).

In one sense full impregnation of timber is a false goal. If the surface layer of the timber can be impregnated with a water resistant resin, the centre can remain untreated, and yet be protected by the surface envelope layer. So today the emphasis is on superficial, shell treatments (Schneider and Witt, 2004).

8.3. Polymer loading of the lumens (Meyer, 1984)

Bulking is not necessary to ensure dimensional stability: an alternative is to reduce the permeability of the wood by filling the lumens with resin. The solution used to impregnate the timber need not penetrate the cell wall or even be water-soluble. It may be sufficient to physically fill the lumen spaces. The timber needs to be dried to about 10% moisture content before pulling a vacuum to remove the air and then forcing the monomer into the lumens under pressure.

The polymerization of such monomers or low molecular weight polymers can be achieved with ionizing radiation, e.g. gamma rays. Ionizing radiation offers one significant advantage. The reaction is initiated without the application of heat and so can be terminated instantaneously by removal of the radiation source whereas, as already noted, with the traditional catalyst the chemical reaction once started cannot be terminated. The radiation penetrates the impregnated material so that the reaction takes place uniformly throughout the timber, and the rate of evolution of heat during polymerization can be controlled by the intensity of the radiation. A number of vinyl monomers have been used, including methyl methacrylate, and a styrene/acrylonitrite mixture.

Resin impregnated wood products have never been able to capture a large market despite having many admirable characteristics. At best these treatments have found small niche markets, of which parquet flooring is by far the largest. Some years ago, one estimate puts their market share in the U.S. at 5000 m^3 a year. Most properties of the wood are improved. Greatly improved hardness and wear resistance, resistance to chemical staining and a good finish are all as important characteristics as material stability.

Several recent wood modification systems are based upon the principle of polymer loading, for example, maleoylation (using an alkyd reaction between maleic acid, or its anhydride, and glycerol); impregnation with low molecular weight cellulose; or impregnation with reactive linseed oil resin. In some of these cases grafting to the wood substrate has also been shown to occur. However the intention is not to deliberately swell and hold the wood cell wall in a bulked state, but to bind the resin to (but not into) the cell lumen by chemical bonding as well as mechanical keying. Consequently these treatments often result in slightly lower ASE values than the bulking or substitution mechanisms (Epmeier *et al.*, 2004).

Water logged archaeological artifacts can be radiation cured. Here the specific niche may be in conserving the most severely degraded wood, consisting mostly of lignin, whose strength is minimal and which would otherwise crumble on drying. Both water-soluble monomers and solvent-soluble resins introduced by solvent exchange have been tried. Minimal dimensional change on polymerization is an important requirement in treating such fragile material. The disadvantage is that best conservation practice prefers treatments that can be reversed if desired as any treatment may reveal subsequent undesirable side affects.

8.4. Cross-linking (Stamm, 1964; Rowell and Konkol, 1987)

Where wood is stabilized by cross-linking, the dry dimensions are not significantly altered by the treatment. The wood is prevented from swelling rather than being prevented from shrinking. This is a non-bulking process in which neighbouring polysaccharide chains are cross-linked, e.g. with methylene bridges ($-CH_2-$). Good dimensional stabilization (50-70% ASE) can be obtained with as little as a 4% weight increase due to cross-linking.

Formaldehyde in the presence of an acid catalyst acts as a cross-linking agent in wood. Since the reaction only takes place where the wood is dry, it is usual to treat the wood in the vapour phase at an elevated temperature over paraformaldehyde in the presence of a strong acid. Reductions in swelling of about 85% can be obtained. However not all of this can be attributed to cross-linking as a certain amount of the paraformaldehyde condenses and bulks the cell wall. This unbound chemical is both volatile and leachable.

Cross-linking induces dimensional stability in the presence of all solvents. The wood is resistant to decay fungi. Unfortunately, mechanical properties are poorer than those of the untreated wood. This can be attributed to the acid catalyst that will hydrolyse the carbohydrate polymers. The material is particularly brittle, possibly due to the short inflexible cross-linking.

8.5. Substitution: acetylation (Stamm, 1964; Rowell and Konkol, 1987; Hill, 2006)

Swelling of wood is a consequence of the large number of hydroxyl groups in the wood substance. Acetylation aims to reduce the degree of swelling by substituting hydrophobic acetyl groups for the accessible hydroxyls. The principal interest is in vapour phase acetylation at 100°C:

$$\text{Wood-OH} + (CH_3CO)_2O \rightarrow \text{Wood-OCOCH}_3 + CH_3COOH \qquad (8)$$

although higher molecular weight carboxylic acid anhydrides can also react in a similar manner. Cyclic anhydrides have also been investigated as these react with the wood to leave a free carboxylic acid to which other reagents may subsequently be grafted.

Surprisingly, the high dimensional stabilization achieved (a 75-80% reduction in swelling for a 15-20% increase in weight) is primarily due to bulking of the walls by the large acetyl groups, leaving the cell wall in a permanently swollen state. The reduction in the hygroscopicity of the wood, by the reduction of the number of hydroxyl moieties per unit volume, appears to have a secondary role in the stabilisation achieved.

The mechanical properties of wood are hardly affected by acetylation. There is no embrittlement since there is no three-dimensional resin network: the hydroxyls are replaced by acetyl groups. There is no loss of toughness, and the wood is very resistant to attack by fungi, termites and marine organisms.

Various attempts have been made to produce acetylated timber, initially for the Japanese market in the 1980s and 90s. More recently a successful commercial acetylation process has been established in the Netherlands. In addition to a 70-80% increase in ASE, it is reported that durability of both hardwoods and softwoods are improved. For example, beech with a 10% weight gain has a durability class 2, while beech with gains of 15% or more achieve a durability class 1. It has also been shown that acetylated wood has improved stability to photodegradation by ultraviolet light.

8.6. Inorganic deposition (Saka, 2001)

Wood-inorganic materials potentially offer dimensional stability, fire-retardant properties, decay and termite resistance while retaining the physical character of wood. Of interest are the metal alkoxides, $M(OR)_n$ in which M is typically B, Al, Si or Ti; the alkoxide R is $-OCH_3$ or $-OC_2H_5$ and n = 3-5. These can be hydrolysed with water and then poly-condensed by dehydration to form sols. The sol forms within the cell wall if the wood has been dried (the cell wall water initiates the reaction), or in both wall and lumen if the wood is green. The sol is then solidified into a wet gel, heated to around 120°C and dried, resulting in the crystallization of B_2O_3, Al_2O_5, SiO_2 or TiO_2. The result is a protective barrier of amorphous 'glass' – a barrier that is durable, non-carcinogenic, non-toxic and fire-resistant.

8.7. Thermal treatments (Evans, 2003; Hill, 2006)

The process of heat treatment of timber has been used in various forms since the middle ages, where Central European houses were designed to recognise that heat and smoke from the fireplace provided a preservative effect on structural timbers. Over-heating of timber however was known to weaken the beams as pyrolytic degradation of the holocellulose occurred. Stamm (1964) recognised that heat treatments of timber which excluded oxygen reduced the rate of pyrolysis, allowing a greater level of control in improving dimensional stability without excessively weakening the timber. This is the principle which underlies the modern interest in thermally treated timbers, and while several different treatment methods have reached the market during the 1990s all rely on the exclusion of oxygen, or the replacement of air with steam, to limit the pyrolytic effect. The controlled heating releases acetic acid and aldehydes from the hemicelluloses. Lignin cleavage reactions also occur, followed by autocondensation reactions forming methylene bridges between aromatic rings; thus an increase in crosslinking in the lignin may also contribute to dimensional stabilisation (Tjeerdsma *et al.*, 1998). Most modern treatment methods heat timber to between 180 and 250°C, although many aim to remain below 200°C to limit the extent of degradation.

The ThermoWood® process has been developed in Finland (Finnish ThermoWood Association, 2003) and both hardwoods and softwoods are available in two classes – ThermoWood D (durability) and ThermoWood S (stability), where the ThermoWood D grade is treated at a higher temperature to allow use in exterior applications out of ground contact (hazard class 2). The process uses steam to limit the extent of pyrolysis, and control crack formation. A similar system called retification has been developed in France. This uses a nitrogen atmosphere to reduce the oxygen content of the kiln to around 2%, and thus limit pyrolytic degrade of the wood at the elevated temperatures. Both systems are successfully supplying an increasing market.

Heat treatment by immersion of the timber in hot oil, thus excluding air during treatment, has also been developed for the market. Rapeseed oil, or in some cases linseed oil and oil-based resins, have been used to impregnate the timber and provide a rapid thermal transfer to the wood, while excluding oxygen. The thermally treated timber market has benefited from interest not only in improved dimensional stability, but also the improved resistance to biodegradation seen in these timbers (Welzbacher and Rapp, 2002). This area is set to expand as traditional CCA preservatives are phased out due to recent legislation.

9. PANEL PRODUCTS

All of the above treatments, unless deliberately superficial, are not easy to apply to lumber because of the lengthy time for diffusion and reaction, or because of the difficulties in achieving effective penetration. In recent years, therefore, interest has also centred on improving the dimensional stability and durability of reconstituted panels: hardboard, particleboard, medium density fibreboard and new derivatives.

Residual stresses locked into the boards during pressing are often released when boards become damp in service. These stresses arise from the compression, crushing and distortion of the particles during manufacture. The resultant springback on wetting is additional to the normal transverse swelling of wood fibres. Of particular concern is the high thickness swelling of boards. Porous board edges are especially vulnerable to moisture ingress and can swell excessively. Dimensional stabilisation of fibres or particles by substitution methods, such as acetylation, would primarily improve particle stability, but it is hoped that the reduced hygroscopicity would also reduce the magnitude of springback.

Panel products are especially susceptible to movement, and any gains in performance might more readily offset the cost of treatment than in solid wood. Fibres and chips have high surface area to volume ratios, which means that they are more rapidly impregnated with stabilizing agents. Further fibres and chips need to be dried (< 5%) and then hot pressed meaning that it is possible to integrate reagent fixation within the production process. In one trial Rowell *et al.* (1989) simply dipped the comminuted material in acetic anhydride, and with a 15% uptake of chemical reduced the thickness swelling of boards made from treated material to at least a sixth of the control value. The potential benefit is very great, but manufacturing costs will be high. Unfortunately acetylation makes the chip surface hydrophobic and this interferes with the spread of adhesive on the chip surfaces.

CHAPTER 5

WOOD QUALITY: IN CONTEXT

SHAKTI CHAUHAN[1], ROBERT DONNELLY[2], CHIH-LIN HUANG[3], RYOGO NAKADA[4], YIN YAFANG[5] AND JOHN WALKER

[1]Institute of Wood Science and Technology, Malleswaram, Bangalore, India; [2]Donnelly and Associates, Curitiba, Brazil; [3]Corporate Research, Weyerhaeuser Inc, Federal Way, Washington, USA; [4]Tohoku Regional Breeding Office, Forest Tree Breeding Centre, Takizawa, Iwate, Japan; [5]Research Institute of Wood Industry, Chinese Academy of Forestry, Beijing, China

1. INTRODUCTION

Context is everything. It informs, clothes and gives shape to the simplest idea. In isolation wood quality has little meaning; only where set against a particular set of end use requirements is it possible to categorize the desired wood characteristics and properties that a particular product needs.

The market imposes acute constraints on wood products. First a tenuous thread – linking wood characteristics to wood properties, to product specifications, and finally to consumer desires and needs – generates weak or confusing signals as to what the market requires. Second, this is complicated by the long time, from 5 to 50 years, between establishment and commercialisation. However, it is an exaggeration to see a long time horizon as being unique or relevant to current forest practice. For example, it can be 10 to 30 years between first discovering a mineral resource and bringing that mine to production. Further, forestry is resolutely, if incrementally, moving to shorter rotations. Pulpwood crops on seven year production cycles are routine in a number of countries in the southern hemisphere, while one should set the bar at no more than 20 years for a short rotation sawlog regime.

Discussion of the influence of time for forest practices is deferred until later. At this stage it is useful to consider further this tenuous thread linking the forest resource to consumer desires and needs, as examination brings clarity – or at least systematic order – to the concept of wood quality.

2. MARKET PULL OR PRODUCT PUSH

2.1. Buying tomatoes and selling wood (Dickson and Walker, 1997)

In supermarkets the fruit and vegetable section remains the last bastion of commodity trading. No one has succeeded in growing and supplying branded, fresh,

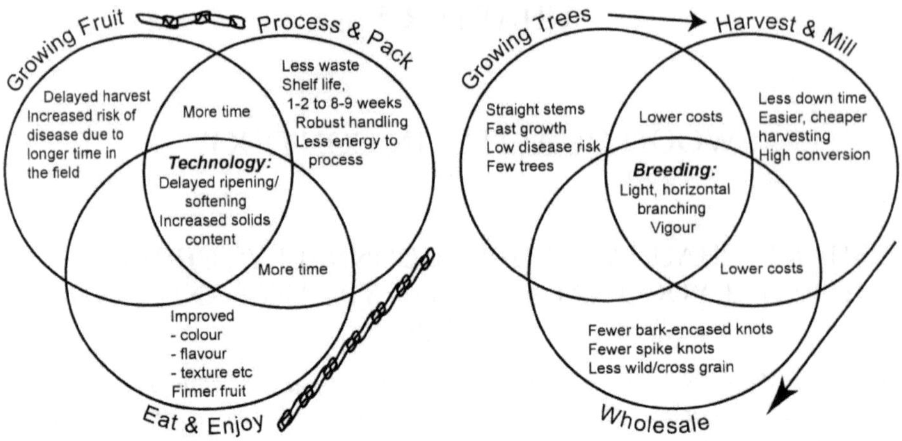

Figure 5.1. Selection strategies for fresh tomatoes and wood: genetic modified fruit (market pull) contrasted with classical breeding for tree form, vigour and health (product push).

flavoursome produce all year round. Consequently it is interesting to juxtapose the best efforts of the suppliers of fruit and of wood in order to understand those features in common and contrasting strategies (Figure 5.1).

To supply field-grown tomatoes, the essential input requirements are disease and herbicide resistance, as well as lower production and harvesting costs; passed on through the intermediary where shelf-life, firmness and solids-content (for ketchup) are critical: and so to the consumer – who seeks good-looking fruit, together with retention of freshness, taste and texture. It does not end there: the buyer is thinking of eating the fruit or delighting a companion with its succulence. In responding to this market pull the breeder has sought primarily to neutralize the genes responsible for softening. This allows harvesting to be deferred while the tomatoes develop full flavour and texture before passing ripe but firm fruit down the supply chain with minimal risk of spoilage. Juxtapose this with the current commodity practice of premature picking of hard, tasteless, green tomatoes, trucking in cooled containers before finally ripening with ethylene gas. No wonder consumers are willing to pay a huge premium of 50-250% for fully flavoured fruit over the price for the generic product. In this context disease and herbicide resistance, as well as lower production and harvesting costs, are 'merely' essential prerequisites for a successful enterprise. The real goal is to meet the consumer's desire for flavour, freshness and texture.

Contrast the tomato on a dinner-plate with a stick of lumber in the DIY store (which itself is a recent global phenomenon). Of course this comparison involves an element of exaggeration. Modern plantation forestry is just 25-150 years old, depending on which species, industry and country one considers. Historically, the tree breeder focussed on the interests of the forest grower and the complaints of the sawmiller. It was reasonable that stem straightness, vigour and health should be prime candidates in the first round of genetic improvement. However, although highly appropriate for the forest manager – and essential prerequisites to a successful forest enterprise – these attributes are irrelevant to the desires and needs of the consumer, who is more likely to be interested in colour, odour and attractive grain. Market demand is about aesthetics reflecting local culture, beliefs and preferences as well as the ephemeral fashions of the time.

Maybe Figure 5.1 is a slight exaggeration, but it is only in recent years that tree breeders have even contemplated increasing density or modifying wood chemistry and structure. Commercial deployment of trees with improved intrinsic wood qualities is in its infancy.

2.2. The wood quality chain (Walker and Nakada, 1999)

Table 5.1. Traditional perspectives of forest and wood quality do not identify with the pull of the market. Industry must switch from being production-driven to being consumer-led in which individual preferences matter. This table ignores non-technical issues such as cost, sustainability and eco-labelling.

Consumer desires	*Market concept*	
Housing	Selling an investment: status, style and substance	
Furniture	Selling a dream: furniture is a statement of your personality	
Newsprint	Selling advertising, sport, sex and politics	
Wood characteristics	*Wood/paper properties*	*Desired attributes and product specifications*
Density, i.e. cell diameter and wall thickness	Stiffness Longitudinal shrinkage Stability/warp-free	Construction – machine stress grades: requiring stiffness, straightness and stability
Microfibril angle Reaction wood Spiral grain	Machinability/finish Figure/grain/texture Hardness	Furniture – quality finish: calling for hard surfaces; tight joints and little movement in service; ability to stain
Permeability Heartwood Extractives	Ease of drying Colour Odour	Newsprint – high speed printing, resilience, adequate brightness, opacity, low-cost
Tracheid length Coarseness More cellulose Less lignin	Tear strength Brightness (less bleaching)	Paperboard and packaging – requiring good handling and serviceability (burst, wet-strength), clear print

A hierarchical succession of steps marks out the wood quality chain, beginning with the intrinsic characteristics and features of the wood cell (Table 5.1). The spatial arrangement hints at the relationships between characteristics, properties and product specifications, but they are less obvious than is usually acknowledged. For furniture one would expect adequate density (revealed in properties such as stiffness, strength and hardness), straight grain, absence of reaction wood and tight knots (easier to dress and finish) or conversely cross-grain (harder to machine but with more figure), extractives or heartwood (distinctive colour) *etc.* The desired product specifications for furniture are exemplified by surfaces that don't mark easily (requiring hardness), in tight joints and little movement (low shrinkage), and an ability to stain uniformly (porosity of the surface) *etc.* Even when one selects a material that meets such criteria time and markets may decide otherwise – silvicultural strategies are prey to future uncertainties. Wood (solid, veneered or panel) is only one possible platform competing to satisfy the aspirations of the consumer – glass or metal may be in fashion. Who specifically needs… prefers… wants… warp-free wooden mouldings, stiff timber studs or even real solid wood for that matter? The consumer presumes an acceptable and functional level of wood quality and focuses instead, and rightly, on what is wanted – style, status and (financial) substance; for colour, odour, a fresh feel and… creativity. Inescapably, most woods are commodities.

The paradox is that while industry may seek improved product specifications, the delivery requires attention at a more fundamental level in the selection for improved wood characteristics and properties. Selection implies choice and a focus on only a few characteristics. But first, one needs to understand how these key characteristics and properties relate to one another.

3. INDUSTRY REQUIREMENTS

Each industry has its own distinctive set of requirements for the type and quality of wood and each has to contend with a very variable resource. However, a quality resource for chemical pulp is not the same as that sought for particleboard or even for mechanical pulp. Fortunately this allows each industry to compete for that segment of the wood supply that it can use best. The prices offered, as a consequence of their differing assessments of the quality of a particular resource, determine who purchases that material. A sawmill will pay more for a large log than for an equivalent volume of smaller wood, because lumber can be cut more economically from large logs and generally a better grade is obtained. The fibres of the discarded slabwood from a large log differ little from those in the adjacent sawn timber and are ideal for making strong paper by kraft pulping. Fortuitously, the pulp mill buys these slabwood chips at a fifth-to-tenth of the price paid for the sawlog from which they are derived. Small top logs are very satisfactory for mechanical pulp and particleboard manufacture and can also produce adequate kraft pulp. Each industry applies its particular selection criteria, e.g. log diameter and length, branch distribution and size, density *etc.* for sawlogs; cell wall thickness, fibre length and fibre content for pulp logs. Price is synonymous with quality only if identical selection criteria apply across a range of end uses. On the contrary, even softwood log grading rules for lumber, which classify material according to its appropriateness for that purpose, distinguish between board grades with desired visual features and structural grades which emphasize stability, stiffness and strength. Consequently sawmills cutting board and structural grades evaluate logs differently.

The wood quality of a fast grown plantation species differs markedly from the same species growing, often to a much greater age, in its natural environment (Bao *et al.,* 2001). Plantation timber is a 'new' timber, in terms of its changed characteristics and properties, for example the trade in New Zealand has always distinguished between old-growth 'Oregon' imported from North America and the local, plantation-grown 'Douglas fir'.

In Brazil the principal eucalypts are *E. grandis, E. saligna, E. urophylla* and several others. However, more than half the plantations owned by the paper companies are hybrids, most notably *E. grandis x E. urophylla* (*E. urograndis*). Since there has been so much breeding to develop specific characteristics and properties and so much deliberate hybridisation, at times it may be more appropriate to discuss 'density groups' of eucalypts rather than species. An example of the huge variability can be seen at Aracruz, who developed clones of *E. grandis* as well as of the hybrid *E. urograndis* with a wide range of basic density. Aracruz Wood Products advertises 'Lyptus' lumber (eucalyptus) available in densities ranging from 450 to 750 kg m^{-3}.

4. SPATIAL DISTRIBUTION WITHIN TREES

Traditionally, around the world the terms juvenile wood and mature wood have been taken to relate to cambial age, i.e. juvenile wood is the wood surrounding the pith that is formed by the young (juvenile) cambium. Confusingly, in some Southern Hemisphere countries the terms corewood and outerwood refer to the same radial gradient in wood quality. To avoid – or add to (?) – any potential confusion, this text follows the new convention proposed by Burdon *et al.* (2004) that has yet to achieve broad consensus.

4.1. Corewood and outerwood

Wood quality varies within trees both in the radial and axial directions. Burdon *et al.* (2004) propose a two-dimensional framework with the radial variations described in terms of corewood and outerwood and the axial variations described in terms of juvenile and mature wood. Arbitrarily, corewood has been described as a cylindrical zone enclosing the first few growth rings around the pith. Typically, for fast grown pines this zone around the pith is considered to be of poor quality, having a number of undesirable features (Zobel, 1975):

- A low basic density in the corewood, primarily a consequence of thin cell walls and the formation of relatively little latewood, means that the timber is less strong.
- A high moisture content (before heartwood formation) and a low basic density in the corewood means the green density of young thinnings or top logs exceeds that of mature butt logs and harvesting costs are high per tonne of oven-dry fibre.
- Longitudinal shrinkage is greater (> 1%) making sawn timber and plywood less stable products. This is a consequence of both a larger microfibril angle (30-50°) in the S_2 layer of the wall and spiral grain.

• There is a tendency for fast grown corewood to contain above average amounts of compression wood.
• Production from a chemical digester is reduced as the lower basic density means that the amount of oven-dry wood fibre within the digester is reduced. Furthermore, corewood has a lower percentage of cellulose.
• Fibres are shorter than in outerwood, giving chemical pulps that have lower tear strength.

The properties that most prejudice the use of corewood as lumber are its low stiffness and strength, together with poor stability. In turn these are a consequence of the low density, large microfibril angle and spiral grain, all of which are quite strongly heritable. Thus there are prospects for improving the quality of corewood in fast grown softwoods by selection and tree breeding.

Any definition of the corewood zone is arbitrary. Jane (1956) observed that 'the period during which corewood is produced varies amongst trees and even in individuals of the same species, but, in general, it is safe to assume that wood after about the 50th growth ring will possess the structure of outerwood'. Jane wrote at a time when wood production was largely from abundantly available, old-growth timber. More recently Harris and Cown (1991) discussing plantation grown radiata pine noted that 'for sawn timber, its [corewood] most damaging features will be confined to the first 3-5 annual growth layers from the pith – yet outerwood in which all wood properties including density have stabilized may not be developed until after the 25-30 growth layers.' For a 25 yr-old radiata pine it has been convenient to describe corewood as occupying a cylindrical zone enclosing the first 10 rings; this means that the proportion of corewood in the log increases from a significant 35% in the butt log to 50%, 60%, 75% and 90% in other logs further up the stem (Cown, 1992), i.e. 50% of the merchantable timber is corewood.

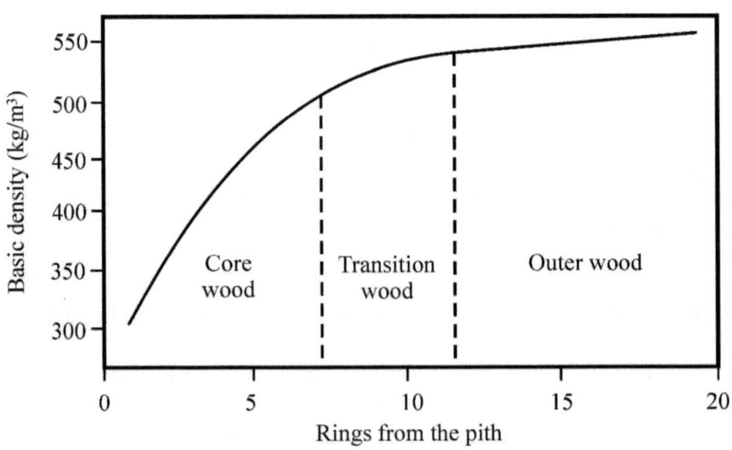

Figure 5.2. Typical density profile for pine. The introduction of a transition zone is not particularly useful.

The defining feature of corewood is that wood characteristics and properties are changing rapidly, whereas in outerwood the changes are gradual at most: but for Douglas fir the corewood zone is quite prolonged and changes more gradually when compared with that for *Pinus radiata* in New Zealand.

As the rate of change diminishes it is questionable whether the wood is still corewood, resulting in the curious proposal of introducing an indefinite transition zone with equally arbitrary boundaries of its own (Zobel and van Buijtenen, 1989). In practical terms the defining feature of corewood is the steep initial gradient in wood quality over the first few growth rings (Figure 5.2). Based on graphic solution methods, the corewood zone of loblolly pine ranges from 4-8 years in the coastal plain to 10-12 years in the piedmont areas of the southeastern US (Clark and Saucier, 1989).

4.2. Juvenile and mature wood

'Biological logic argues for a two-dimensional characterization of wood properties: juvenile *vs*. mature for the progression up the stem, and corewood *vs*. outerwood for the radial progression' (Burdon *et al.,* 2004): wood quality improves from pith-to-cambium (from corewood to outerwood); and it improves from ground level up the stem (from juvenile to mature wood) but most obviously over the lowest 3-5 metres. The variation in wood quality up the stem, while less prominent than the pith-to-cambium variation, still has important implications, most acutely for the butt logs of fast grown pines. Changes from juvenile to mature wood appear largely to be expressed in progressively more moderate microfibril angles on moving up a few metres from the base of the tree. This generates a series of conic sections tapering upwards over the first 3-5 metres, above which each conic section becomes a cylindrical section extending further up the stem: this 3-5 metre juvenile zone is reflected in the butt-swell and the propensity to form compression wood (Burdon, 1975) in young trees. Such juvenile-mature wood changes match the classical concept of maturation (foliage, stem morphology and onset of reproduction). There is limited evidence that where cuttings or grafts from mature trees are planted the young trees express mature wood features from the outset.

While the juvenile zone only extends upwards for a few metres, this zone is encompassed by the butt log that, by tradition, is valued because of its size. As with the corewood-outerwood boundary, there are tree-to-tree and species-to-species variations in the vertical extent of juvenile wood and in the rate of change in wood quality toward the tree top. Cuttings from older branches lack juvenile responses to silvicultural practices and the benefits of physiologically aged cuttings are improved form and stiffness, i.e. minimizing the juvenile core but at the cost of some loss of vigour.

4.3. Plantations and natural forests

The exhaustion of natural forest resources is forcing industry to come to terms with the corewood of plantation timber. As its properties are better understood industry is finding appropriate end uses that reflect the intrinsic properties of the individual piece of lumber. There is enormous variability in wood quality between trees so that corewood of one tree can be of better quality, e.g. of higher stiffness, than outerwood of another tree.

The psychological significance of corewood should be appreciated. Historically the softwood industry of the Pacific Northwest of North America relied on old-growth, and more recently second-growth, Douglas fir. In the former case the basic density is about 15-25% greater than that found in fast grown plantations (< 30 years old), which contain a very high proportion of corewood. Sawmillers have had to accept the fact that high density material, > 500 kg m^{-3}, coming from stands 75 years or older is being replaced by lower density material, < 450 kg m^{-3}, from fast grown intensively managed stands that are only 40-50 years old.

Indeed in some parts of the world plantation-grown trees harvested for commercial uses are composed entirely of juvenile wood, examples being 10-15 yr-old *Paulowania* and *Populus* sp. in parts of China (Bao *et al.,* 2001). Fortunately the differences between corewood and outerwood, and between the juvenile wood and mature are not nearly as obvious in hardwoods (or maybe they are less well categorized).

5. DENSITY

Even today, to many people density is synonymous with wood quality: this is a truth that misses the point. Certainly improving wood quality through selection of higher density material was justified in the 1950s to 1970s and is justified still when screening a new plantation species. For both hardwoods and softwoods density is strongly heritable which favoured its inclusion in early tree improvement programmes. Further, the genotype *x* environment interaction is often low so that any improvement should be sustained across a variety of sites.

However, plantation forestry today no longer resembles the one that gave rise to that original insight. For example, discussion on wood quality often revolves around distortion, instability and stiffness, mirroring the preoccupations of the sawmilling sector. Yet none of these properties is affected by density in the manner so popularly presumed. Density does not predispose lumber to behave in a particularly way, it merely magnifies the effects of other intrinsic wood quality characteristics, whether good or bad. Density describes the quantity of matter, not its intrinsic qualities. Today density has become a concave/convex fairground mirror that, while revealing, also distorts the reflection.

For almost all softwoods and low-to-medium density hardwoods increasing wood density has been ranked above all other desirable objectives in traditional wood quality improvement programmes. Basic density provides an index of wood quality to which all end users are able to relate. To the sawmiller a high density indicates the timber will be stiff and strong; to the pulpmill it indicates that a given volume of wood will yield more pulp than would a low-density timber. But too high a basic density (> 600 kg m^{-3}) – more a problem with some hardwoods – is as undesirable as too low a basic density (< 400 kg m^{-3}) – more a problem with softwoods (also *Paulownia* and *Populus* sp.). For woods above 600 kg m^{-3}, furniture and building timbers are uncomfortable heavy, often unnecessarily stiff and strong, and joints can open if under-dried. For pulp it means that the dense fibres will require prolonged beating before the lumens collapse to give a dense, well-bonded sheet of paper.

Obviously basic density of wood relates to the amount of dry matter per unit of green volume. The compound characteristic of density (Figure 5.3) means that two samples can have the same density but very different wood and fibre properties. Differences are due to interactions between wall thickness and cell diameter, the proportion of the cell wall that is occupied by the S_2 layer, the amount of cellulose in the wood (age related) and the microfibril angle in the S_2 layer (Cave, 1969). Compression wood has a large microfibril angle and inflated density where compared with those of normal wood of the same age and the same height. For a given density, samples with compression wood have poorer mechanical properties, and its stiffness is not related to density (Dohr, 1953). For severe compression wood, there is a slightly negative relationship between density and stiffness.

Further basic density is complicated by the presence of extractives, which vary from less than 1% of the oven-dry mass in sapwood to well over 10% in the heartwood of some species. Extractives increase the weight of a wooden member without contributing to its strength, and consume chemical without contributing to the pulp yield. In rigorous studies the extracted (extractive-free) basic density may be needed in order to compare pulp yields or mechanical properties between samples or species.

Density has proved to be a useful, surrogate indicator for many properties of wood and paper. Its other virtues are that it can be measured reasonably quickly and cheaply. In essence an increment core taken at breast height is used to predict the mean density of the tree.

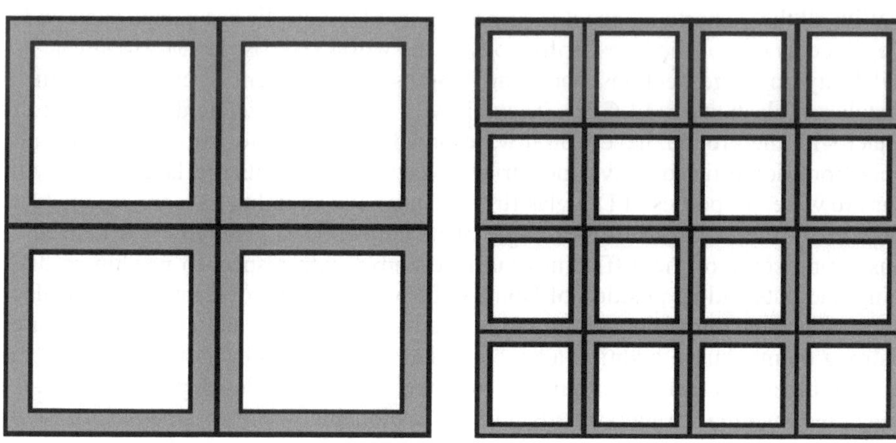

Figure 5.3. Both examples have the same basic density, but one has four-times fewer fibres whose walls are 50% thicker.

The principal sources of variation in wood density relate to:

- Within-ring variations;
- Within-tree variations;
- Between trees on the same site;
- Between populations of the same genotype growing in different regions.

5.1. Within-ring variations

Most species, apart from *Araucaria* sp. and diffuse-porous hardwoods, show contrasting differences in wood density across the growth ring. This is a response to seasonal climatic variations and the formation of latewood. The density variation across a growth ring far exceeds the density variation between trees. As an extreme case, Harris (1969) cites the contrast between latewood (870 kg m^{-3}) and earlywood (170 kg m^{-3}) in adjacent growth rings in the outerwood of a sample of Douglas fir, *Pseudotsuga menziesii.* A more typical within-ring and between-ring variation for Douglas fir is shown in Figure 5.4. Douglas fir is used as a furnish for fibreboard so any mixture of species having widely different densities should be as acceptable.

The coarse texture of Douglas fir creates problems in wood use. It is only moderately easy to work. Care is needed in planing because the soft earlywood may be compressed/crushed rather than cut, recovering slowly to create a corrugated surface. The strong contrast in hardness between earlywood and latewood makes for uneven wear. The surface does not paint well, and early failure on the broad latewood bands of flat-sawn material is sometimes experienced. Nailing can cause splits. Differential glue absorption between earlywood and latewood can cause starved joints. Differential shrinkage between earlywood and latewood can cause latewood bands to shell out on weathered flat-sawn surfaces.

With all these shortcomings it is surprising that Douglas fir has gained such wide acceptance as one of the finest softwoods in the world (Harris, 1993). However, its availability in large sections and long lengths, its excellent strength properties (modulus of elasticity *c.* 13 GPa), its stability in use, and moderate durability out of contact with the ground, have won it widespread acceptance for structural use at all levels from domestic to heavy industrial. Except where knots are large or growth rings are wide, properties of Douglas fir – stiffness and straightness – are acceptable: unlike many pines the microfibril angle of its corewood is generally less than 35°. Considering some of the difficulties outlined above with respect to machining and gluing, the outstanding position of Douglas fir plywood on world markets may also appear somewhat surprising. Here, too, it seems that strength and axial stability are its strong points, but in addition technology has played a major part, particularly in defining the log characteristics suitable for veneering, and in developing appropriate gluing and manufacturing systems. It is questionable whether Douglas fir will continue to be perceived so favourably when supply is largely from plantations.

Pinus caribaea, P. merkusii and *P. oocarpa* in Malaysia produce little latewood during the first two to four growth layers, but latewood develops strongly thereafter. Even with subsequent latewood formation the contrast between earlywood and

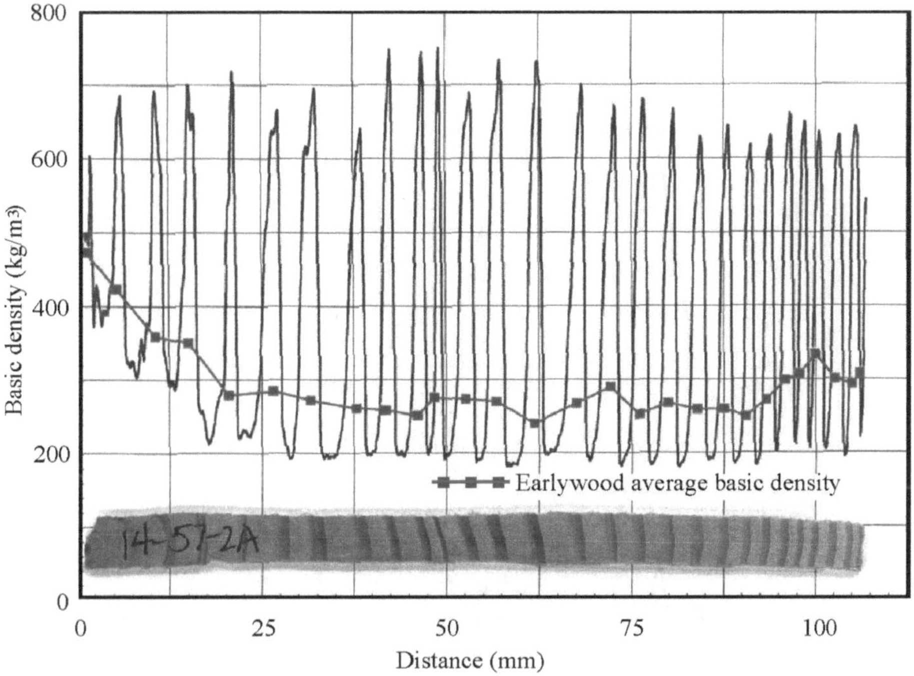

Figure 5.4. Within and between-ring variations in basic density with distance from the pith for Douglas fir (Weyerhaeuser, unpublished).

latewood density is low, approximately 1:1.5, whereas the corresponding ratios for *Pinus radiata*, *P. taeda* and Douglas fir are 1:1.8, 1:2.3 and 1:5.0 respectively (Harris, 1973). The modest differences in density between earlywood and latewood means that these tropical pines and *P. radiata* can be described as even textured, while their wide growth rings would classify them as coarse grained. Northern species such as spruce and hemlock are much sought after for certain pulps on account of their uniformity of density: with these species the difference in density between earlywood and latewood is comparatively small and the transition from earlywood and latewood is gradual.

5.2. Within-tree variations

For the hard pines, Douglas fir and some other, but by no means all softwoods, the wood adjacent to the pith is of lower density and is of poorer quality than the wood in the rest of the tree. Many important plantation species are found in these groups. In contrast, for *Cryptomeria* sp., true fir, hemlock and spruce the basic density decreases for the first few annual rings from the pith before levelling off or increasing moderately toward the cambium.

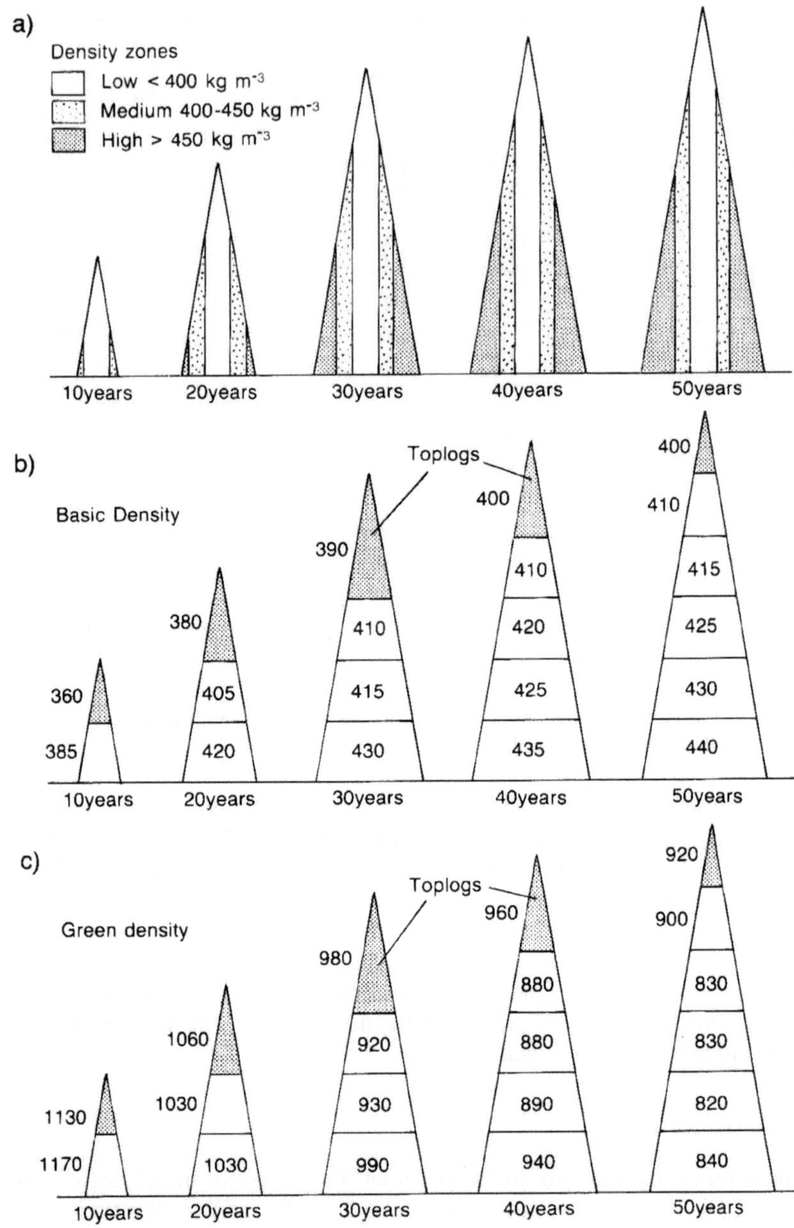

Figure 5.5. Within-tree variations in density for New Zealand radiata pine (Cown, 1992). (a) Basic density in the tops of old trees is similar to that from 10 yr-old trees. (b) Basic density increases with physiological age: the butt log of an old tree has proportionately more outerwood and has a higher basic density. (c) Green density is highest in top logs where the basic density is lowest and the moisture content corresponding to full saturation is very high.

or hardwoods, the situation is complex.

> All possible patterns of wood density variation appear in hardwoods. The middle to high density diffuse-porous hardwoods generally follow a pattern of low basic density near the pith and then an increase, followed by a slower increase or levelling off toward the bark. The low density, diffuse-porous woods, such as *Populus*, seem to have a somewhat higher density at the pith, although some have a uniform density from pith to bark [while *Populus deltoids* increases in density from pith to bark (Shukla *et al.*, 1994)]. The ring-porous hardwoods tend to have a high density at the centre, which decreases and then increases to some extent toward the bark (Zobel and Buijtenen, 1989).

As just noted, for many plantation species the corewood is of lower density than is the outerwood. Further Figure 5.5 indicates that there is little difference in basic density between the corewood-mature wood zone in the topmost part of the stem and the corewood-juvenile wood in the butt log that had formed years earlier when the green crown of the younger tree was much lower.

On progressing up the tree there is proportionately more corewood, and the basic density of the stem-section decreases while both green density and moisture content increase. Thus young thinnings, top logs and whole trees when grown on short rotations contain less biomass and more water, such that young radiata pine logs weigh over 1000 kg m^{-3} but contain only 400 kg m^{-3} of oven-dry fibre (Figure 5.5).

This (arbitrary) transition from corewood to outerwood occurs between the fifth and thirtieth growth ring from the pith depending on the species and the characteristic or property being examined (Zobel and Buijtenen, 1989). They suggest that the corewood zone so far as basic density is concerned coincides with the first 5-6 rings for *Pinus elliottii, P. caribaea* and *P. radiata*, the first ten rings for *P. taeda* and twenty rings or more for *P. ponderosa*. With New Zealand *P. radiata* basic density increases quite markedly for the first 10-15 rings changing only slowly thereafter (Figure 5.6) but by convention corewood is taken to be the first 10 rings. But equally there is logic in defining the outerwood as beginning when the basic density exceeds 400 kg m^{-3}, in which case corewood would be restricted to only 5 rings in low latitudes but would extend out to 15 rings in the south of New Zealand.

These corewood-outerwood density trends, when compounded by the high proportion of corewood in fast grown, short rotation plantations, have given rise to the mistaken perception that fast growth *per se* is detrimental, whereas it is the preponderance of poor quality corewood in the fast grown tree that is the critical feature. Corewood can account for 50% of the stemwood of 25 yr-old well thinned, fast grown *Pinus radiata*.

Growth rate has little effect on the wood properties of diffuse-porous hardwoods. These have approximately the same proportion of vessels across the annual ring, regardless of the growth rate. However growth rate has a noticeable influence on the density of ring-porous hardwoods, which usually produce denser wood when fast grown. The volume of low density vessel tissue produced early each year in a ring-porous hardwood is constant regardless of the total radial growth during each growth period: therefore the wider the growth ring the smaller the proportion of earlywood with its vessel tissue and the greater the proportion of denser, vessel-free latewood.

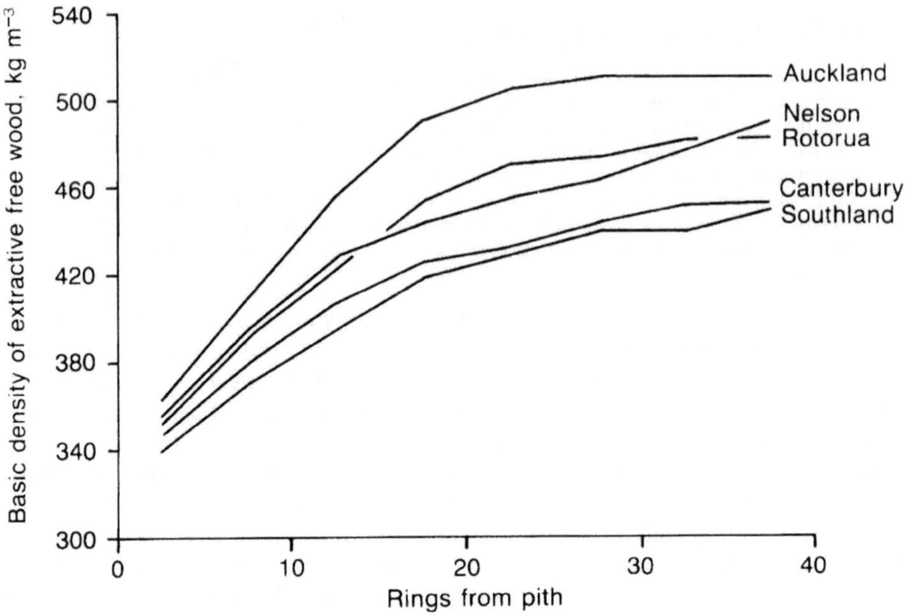

Figure 5.6. Within-tree variations in extractive-free basic density of radiata pine from various regions in New Zealand (Cown, 1992). Temperature and rainfall are major influences on wood density, outerwood being more sensitive than corewood: outerwood density decreases by 7 kg m^{-3} per degree increase in latitude and for each 100 m rise in altitude.

Bhat and Indira (1997) and Bhat (1999) observed that faster growth (6.00 mm/yr *vs.* 2.8 mm/yr in the control) in 5 yr-old teak trees with the application of fertilizer resulted in wood with a lower vessel percentage and an 8% increase in density.

5.3. Within-stand variations

Regardless of species or where the forests are established, the variation in wood properties between trees is considerable. For a typical stand of *Pinus radiata* the range of basic density is shown in Figure 5.7. Here the unextracted basic density in the corewood rings 5-10 is 348 ± 22 kg m^{-3}, whereas in rings 18-22 it is 430 ± 30 kg m^{-3} (both with a coefficient of variation of *c.* 7%). Where tree improvement programmes emphasise high basic density the between-tree variations will be reduced and the distribution will centre on the medium-to-high density range shown (Figure 5.7). Unfortunately the absolute variation in corewood density is less than that of outerwood so an equivalent increase in corewood density is harder to achieve, in which case the within-tree variability may actually increase.

Another example of within-stand variability relates to Douglas fir (Figure 5.8). The basic density of the cross-section decreases on ascending the stem, but, because

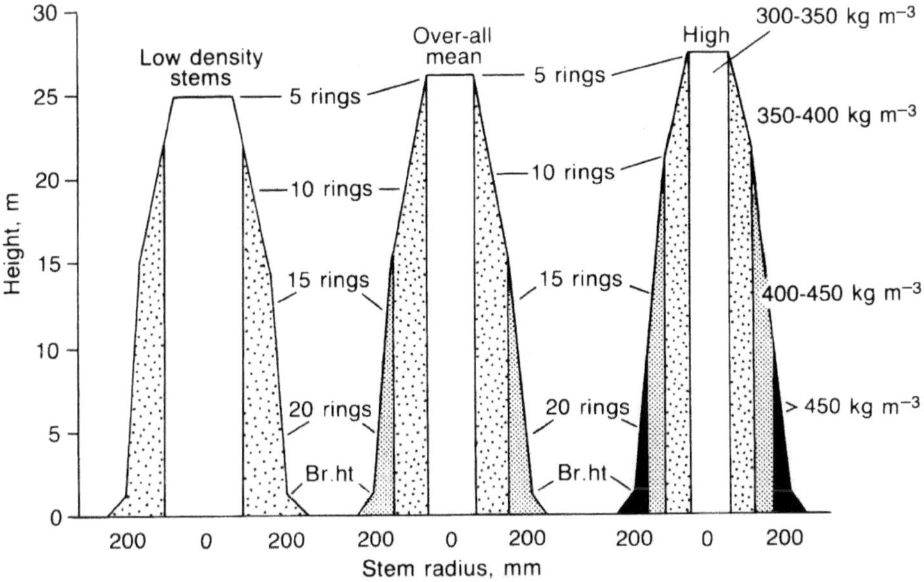

Figure 5.7. Variations in extracted basic density within selected stems from a typical 24 yr-old stand of radiata pine in the central North Island of New Zealand (Cown and McConchie, 1983). Ten trees were chosen after assessing outerwood density at breast height in 193 stems using increment cores (unextracted densities of 430 ± 30 kg m^{-3} with a range from 357 to 512 kg m^{-3}). For the low, mean and high density stems the mean whole-tree basic densities (unextracted) were 354, 380 and 395 kg m^{-3}, while the corresponding outerwood basic densities (unextracted) were 375, 433 and 494 kg m^{-3}.

the between-tree distributions are so broad, the section density at the top of one tree may exceed the section density at the base of another tree. For a comparable sample (same age, same site, *etc.*) a between-tree variation in average cross-sectional basic density of 15 to 25 kg m^{-3} can be expected. High density trees tend to have both more latewood and higher average earlywood and latewood densities. Unfortunately in the short term only a 5% overall increase in wood density of Douglas fir is likely from genetic improvement which will do little to offset the effect of lower density due to the shorter rotations envisaged in the Pacific Northwest (McKimmy, 1986).

5.4. Inter-regional variations

The environment exerts strong control over average basic density. Figure 5.6 illustrates the variation for radiata pine plantations across New Zealand. For industries merely seeking biomass that is a compelling reason for locating to lower latitudes (Auckland) where, for example, the wood of ring 20 is some 20% greater than that of the southerly forests (Southland). The general trend is one of higher basic density with lower altitude or latitude of the site. The effect is less obvious in the first 5-10 growth rings.

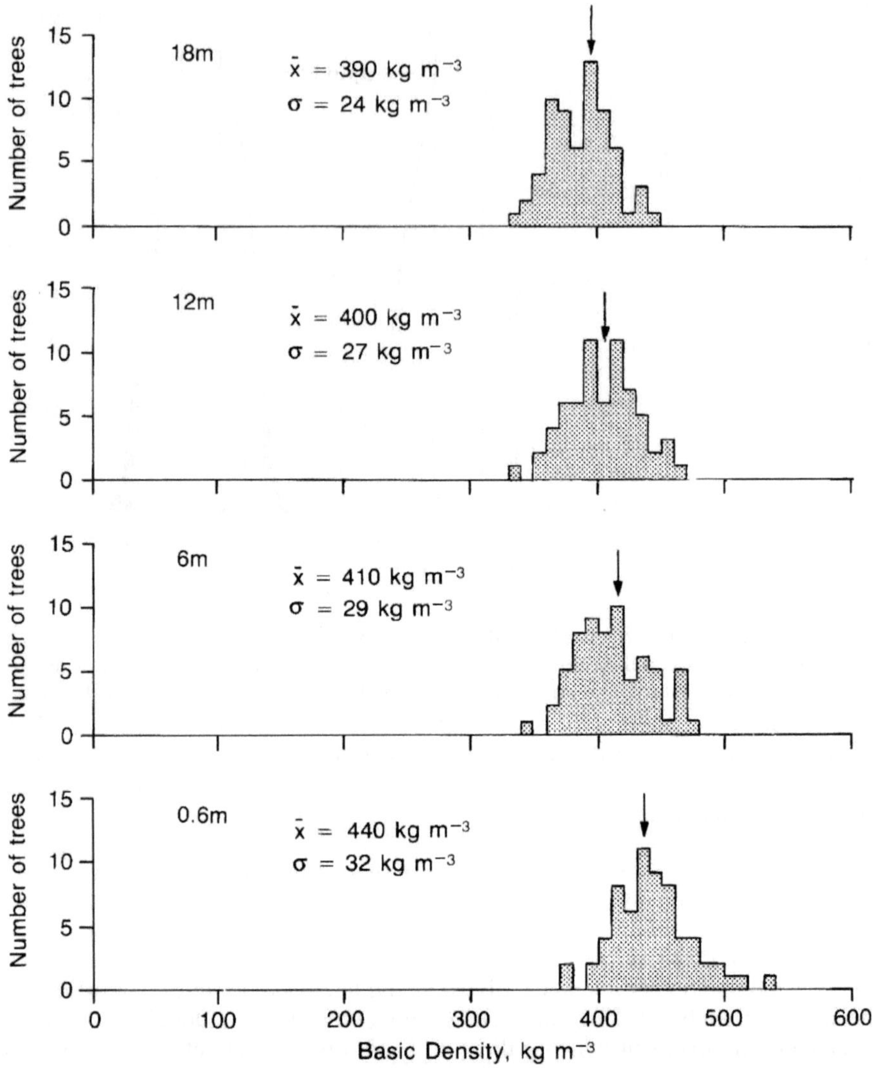

Figure 5.8. Frequency distributions at several points up the stem illustrating the variations in cross-sectional basic density amongst sixty-four Douglas fir trees in a 55 yr-old stand. Arrows indicate the distribution means (Megraw, 1986).

Figure 5.6 provides the broad conceptual approach. Like all big pictures it oversimplifies. The line for each region is constructed by collecting data from trees in 5-ring increments (the cross-sectional averaged increment core density for all trees, taking rings 1-5, 6-10, 11-15, 16-20, 21-25, 26-30, 31-35 and 36-40) and averaging again for all trees. This introduces uncertainty. Downes *et al.* (2002) have

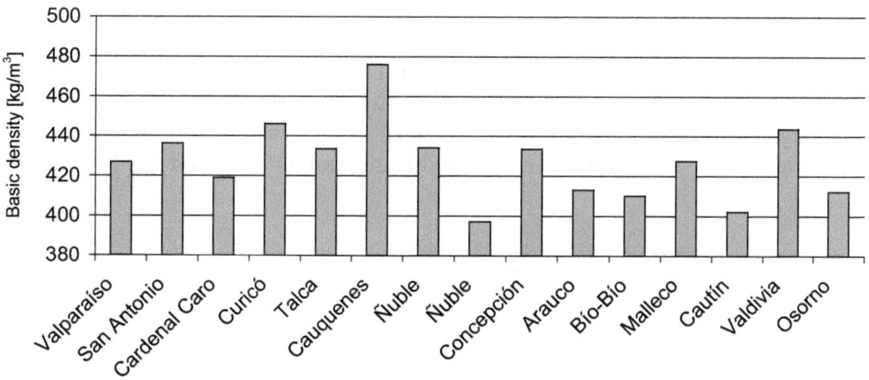

Figure 5.9. Density of radiata pine shows no latitudinal gradient in Chile; the average age of the stands is 22 (Delmastro *et al.*, 1982).

noted that true underlying trends can be obscured by such clumping of data. Thus a predetermined decision to define corewood as the first 10 rings results in sampling in 5-ring increments – or more correctly the methodology predetermined the definition of corewood. Raymond and Anderson (2005) observed that for *Pinus radiata* in New South Wales basic density is constant over the first 8 growth rings before abruptly increasing. This contradicts the traditional picture of pine corewood.

While the north-south trend of declining density appears so logical, it is worth noting that same north-south trend for the same species in Chile (Figure 5.9). The provinces/forests are spread along a latitudinal distance of over 1000 km, yet there is no discernable density gradient. One explanation is that the rainfall gradient (drier in the north) counters the temperature gradient; equally Chile has two mountain ranges parallel to the Pacific Ocean with great east-west influences over temperature and rainfall; and locally there are different conditions of water availability during the period of latewood formation. The desire to generalise should not deny the need to particularise. Decarte's axiom '*de omnibus dubitandum*' (doubt everything) is useful advice.

The wide distribution of Douglas fir may be of more general interest. Natural populations of coastal Douglas fir include the lower elevations of British Colombia through to the higher elevations in northern California; whereas east of the Cascade range, inland Douglas fir grows in a warmer, drier environment. Across these sites there are significant differences in the density of Douglas fir (Figure 5.10).

5.5. Density: a pragmatic convenience masquerading as insight

Many have claimed that density is the most important characteristic in determining wood properties. For example, Zobel and van Buijtenen (1989) argue:

> ...[basic density] is of key importance in forest products manufacture because it has a major effect on both yield and quality of fibrous and solid wood products and because it can be changed by silvicultural and genetic manipulation. Therefore, [basic density]

largely determines the value and utility of wood and overshadows the importance of other wood properties.

The basis for such claims lies in the general observation that denser timbers are stiffer and stronger. In Figure 5.11 mean 'whole-tree' stiffness and density are plotted against one another species-by-species using North American and European data from Table 4-11a in the USDA Wood Handbook (1999); and Table 2 in Lavers

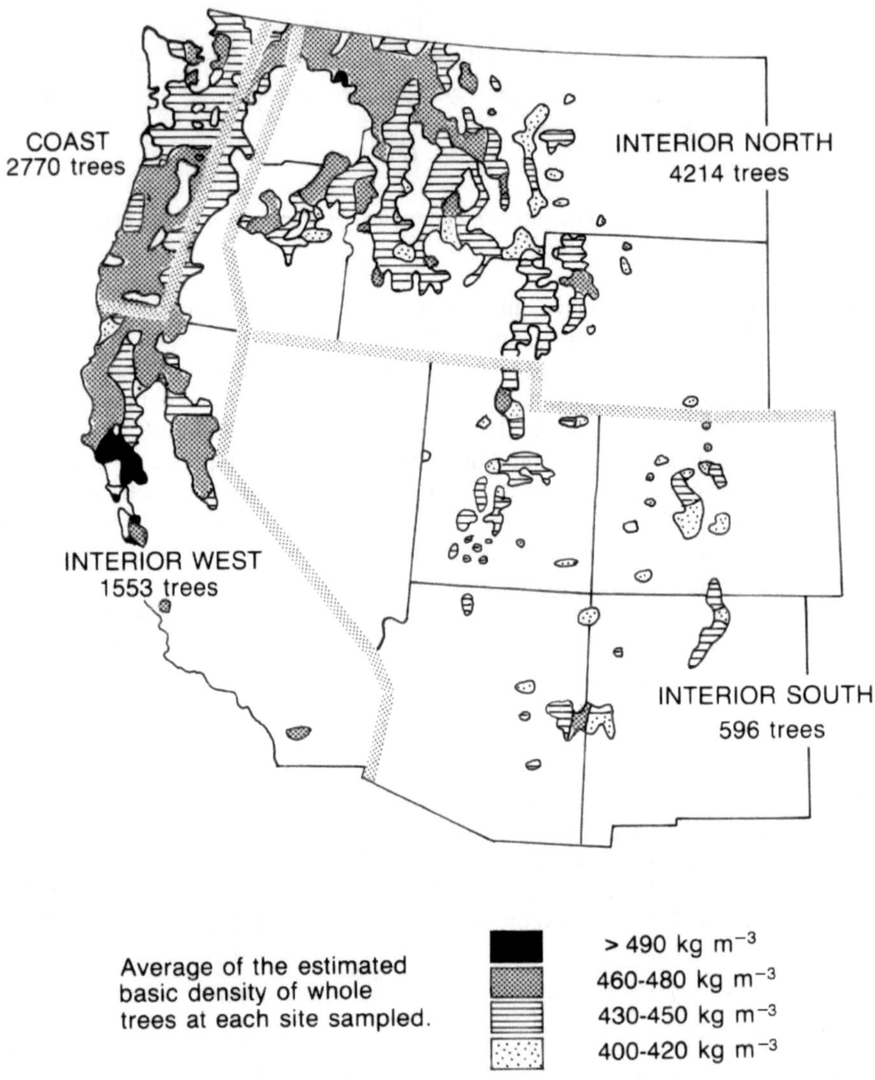

Figure 5.10. Variations in the mean basic density of Douglas fir growing in various parts of the United States (USDA, 1965).

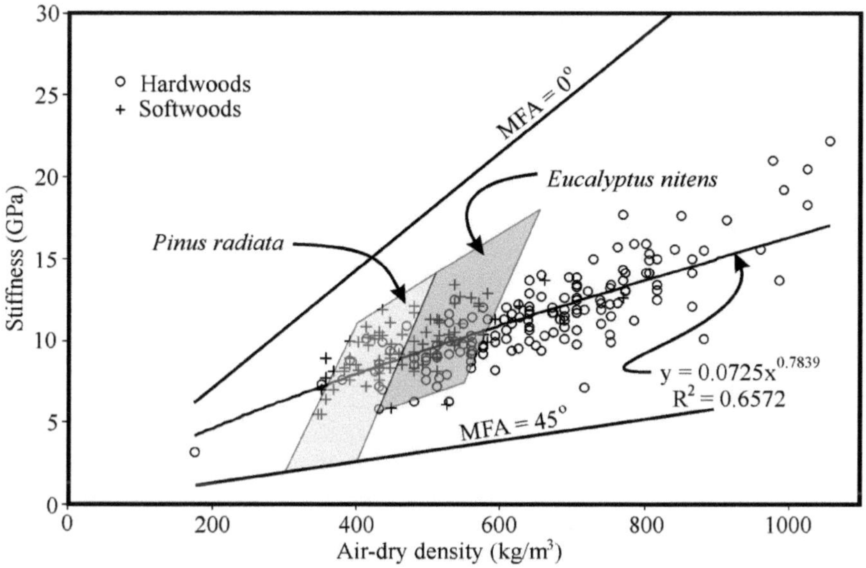

Figure 5.11. Mean 'whole-tree' values of stiffness and density for hardwoods and softwoods. Superimposed is the range in values expected of lumber from *Pinus radiata* and *Eucalyptus nitens.* Variations within species are greater than differences between mean values for species.

(1974). Expressing the relationship in the form MOE = k (density)^n, then n is either 0.7 or 1.01 for hardwoods and 0.85 or 0.82 for softwoods according to the US and European data sets. Using the mean values for species, the data indicates that, at best, if you double the density you don't even double the stiffness: there is no leverage.

Whole-tree data ignore variations in density and stiffness within trees. Stiffer and stronger outerwood happens to be denser than corewood. However, it is necessary to distinguish between the density of a wood, which is mere mass (the quantity of matter), and the intrinsic characteristics of the cell wall (the quality of matter). Thus on going from pith to cambium in pine the mean density might increase by 50% (Figure 5.6) and so one would conclude on the basis of mere mass that the outerwood would be 50% stiffer, yet the increase in stiffness is typically three-fold.

Fujisawa *et al.* (1993) compare density trends in a number of sugi clones. Sugi, *Cryptomeria japonica,* is one of those softwoods where density decreases with distance from the pith (Figure 5.12). Again for sugi, Hirakawa and Fujisawa (1995) compare density and stiffness in corewood and outerwood. They observe that the corewood is denser and less stiff than the outerwood for fast and slow growing clones (Table 5.2). There is a negative correlation between density and stiffness.

Finally, a simple study by Simperingham (1997) demonstrates that the assumption that low density corewood produces wood of low elastic modulus is not necessarily valid. Here 100 x 100 mm members were cut enclosing the pith, i.e. all

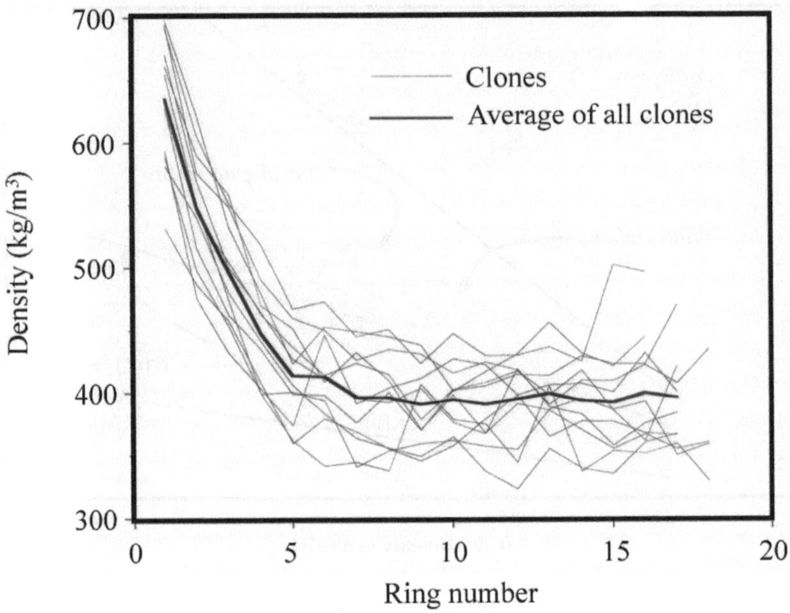

Figure 5.12. Clonal variation in air-dry density of sugi, *Cryptomeria japonica,* with ring position (Fujisawa *et al.,* 1993).

Table 5.2. Diameter at breast height, density and clearwood stiffness (MOE) for select 30 yr-old clones of sugi, with high and low stiffness clones being compared in three growth categories (Hirakawa and Fujisawa, 1995).

Clone	DBH, cm	Air dry density, kg m^{-3}		Young's modulus, GPa	
		Corewood	Outerwood	Corewood	Outerwood
Fast growing					
Takahagi 16	36	351	293	4.9	5.8
Kooriyama 1	35	317	272	2.9	3.2
Medium growth					
Taga 6	26	421	354	6.8	7.0
Usui 3	27	380	350	3.3	3.6
Slow growing					
Takahagi 15	19	413	375	7.3	7.5
Nakoso 1	16	322	314	3.5	4.8

the pieces were corewood with no more than five growth rings. The members were ripped into two pieces, kiln-dried, dressed and tested in bending (both on edge and on face) and their air-dry densities measured. The interesting feature was that the with-pith lumber varied in stiffness by a factor of three (from 3 to 10 GPa) whereas the air-dry density ranged from 320 to 500 kg/m^{-3}. In this study there was no correlation between stiffness and density (Figure 5.13).

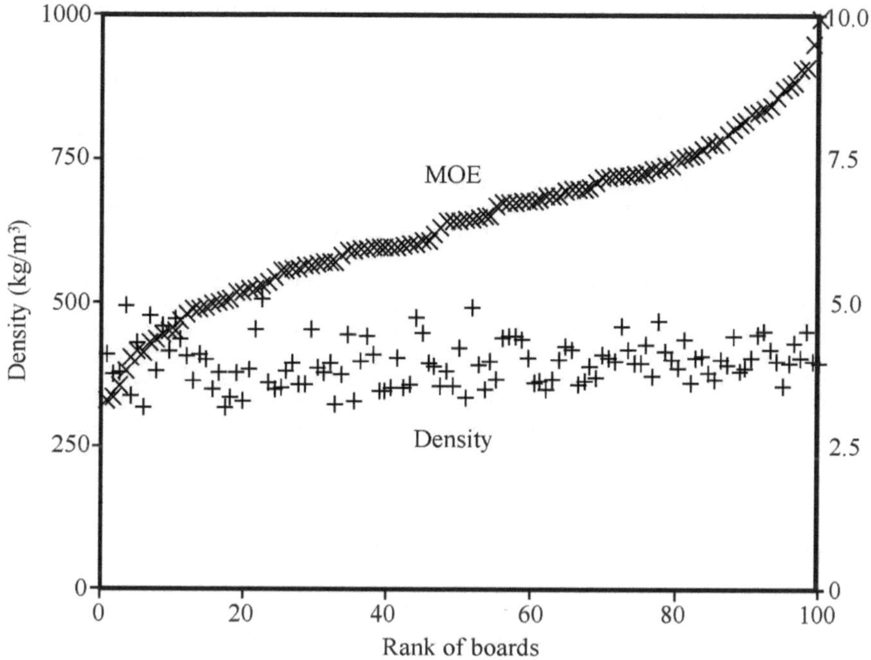

Figure 5.13. Modulus of elasticity (MOE) and unextracted air-dry density for with-pith radiata pine (Simperingham, 1997).

Perhaps the last word should be that of Uprichard (2002):

> There appears to be a general assumption, based on little evidence, that a tree suitable for pulping and papermaking will be unsuitable for timber, since softwood trees of very high density are unsuitable for papermaking. In the context of radiata pine, *with its limited density range* [italics added for emphasis], such arguments appear irrelevant.

The purpose of this section has been to emphasise that density describes the quantity of matter and not the intrinsic qualities of matter. Both vary within and between trees. Generally one can find some statistical relationship between the two. Thus there is an indirect utility for line managers who can be exact in their requirements for density, e.g. sawmills may divide their wood supply into density sorts to meet different processing issues (drying) and market needs. However, density does not predispose lumber to behave in a particularly way, it merely magnifies the effects of intrinsic wood quality behaviour, whether good or bad. For example outerwood is stiffer and more stable, not because the wood is denser but because the microfibril angle is lower. Indeed as will be discussed in Chapter 6, it is more efficient – biologically – for the tree to change the microfibril angle in the cell wall to achieve superior stiffness than to achieve the same end result merely by having more cell wall material.

Product specifications can set desirable minimum and maximum threshold values for density. This was anticipated in a more nuanced interpretation by Nicholls and Dadswell (1965) of the consequences in selecting for density.

> The criterion associated with wood density is subject to qualification. Though an increase in density is associated with better strength properties in sawn timber, and, in the case of pulping processes, produces greater yield of pulp for a given volume of wood, high basic density may not always be desirable. Because a close correlation exists between percentage latewood and basic density in coniferous timbers, and as latewood fibres are observed to have much thicker cell walls than those of the earlywood of the same growth ring, high basic density will be associated with a preponderance of thick-walled cells. Not only may a high latewood content be unsatisfactory in a utility softwood from the point of view of uniformity of texture, and nailing and working quality, but it can also adversely affect wood properties. Pulps made only from latewood fibres have different paper-making properties to those produced from earlywood alone, and it has been suggested that for *P. radiata* less than 20% of latewood would be satisfactory for general purposes (Watson and Dadswell, 1962). The criterion for basic density selection, therefore, is not straight forward, and it may well be that improvement in uniformity is more important than high or low basic density. Zobel (1963) is of the opinion, in fact, that not only could a much more uniform wood be produced for breeding, but that improved uniformity alone could repay the breeding costs many times over. Fielding and Dadswell (1961), speaking specifically of *P. radiata*, thought that little purpose would be served by attempting to breed for greater strength in this wood, but that greater tree-to-tree uniformity of strength is desirable.

6. TREE FORM: SIZE, COMPRESSION WOOD AND KNOTS

Rightly, initial tree improvement programmes placed particular emphasis on fast growth, forest health and good tree form (straightness of stem; and the lightness and frequency of branching).

Fast growth reduces the time needed to produce a commercial log of the desired size. Fast growth *per se* does not greatly influence wood quality (except for ring-porous hardwoods). However, in a fast grown stem of particular diameter the proportion of corewood is greater than in a slower grown tree and it is this feature of plantations that has the greatest downside influence on wood properties.

Improved tree form increases the value of the log and reduces harvesting and processing costs. Stem straightness and light branching improve wood quality in that there is less reaction wood in the stem (especially near the pith), small flat-angled branches are less likely to have ingrown or encased bark trapped at their upper junction with the stem, small branches are less resinous, and the volume of reaction wood associated with the knot is reduced. Fortunately severe reaction wood is negatively correlated with stem straightness that is under strong genetic control, so selection for straight stems reduces the severity of reaction wood.

With softwoods the average knot volume in a stem is generally between 0.5-2.0%, although the volume of wood affected by knots is much greater. The alignment of the axial tracheids is disturbed as they sweep around the knot and the volume of disturbed wood tissue is 1-3 times that of the knot itself. Knot volume is proportionately greater in young trees and in open grown stands of *Pinus sylvestris*

Table 5.3. Percentage of knotwood within the stem of 44 yr-old Scots pine, *Pinus sylvestris,* on various sites in Sweden (Nylinder, 1958; see Timell, 1986, p. 924).

Position as a percentage of stem height	Spacing of trees in the stand [metres]		
	0.75 x 0.75	1.5 x 1.5	3.0 x 3.0
	Percentage of knot wood		
90-100	1.55	1.92	1.98
80-90	1.77	1.94	2.26
70-80	1.63	1.69	1.97
60-70	1.14	1.09	1.51
50-60	0.74	0.88	1.23
40-50	0.52	0.53	0.98
30-40	0.34	0.44	0.97
20-30	0.18	0.30	0.75
10-20	0.11	0.25	0.71
0-10	0.11	0.20	0.55
Average	0.44	0.52	0.92

(Nylinder, 1958) where the trees are more vigorous, resulting in larger branches and less branch mortality or suppression (Table 5.3). Wide initial spacing and thinning favour fast growth, a large corewood zone, stem taper and large branches that are slow to self-prune. Branches grow until they experience canopy closure and the time to closure is one factor determining branch and knot size. Knot volume increases up the stem as the taller dominant trees acquire growing space at the expense of their suppressed neighbours. Similarly branch size increases with the quality of the site, which promotes vigorous growth. The deeper green crown in thinned stands promotes the incidence of larger branches.

In Nylinder's study the average diameter of knots at breast height was only 9.7 mm where the spacing was 0.75 x 0.75 m as against 19.6 mm at 3 x 3 m spacing. Knots are small compared with those in fast grown softwoods in warmer climates. For example, the average knot size in 30 yr-old New Zealand grown *Pinus radiata* ranges from around 25 mm to over 80 mm, mainly as a result of spacing and site. Close spacings between trees are needed to prevent branches, and so knot size, from becoming excessive. In addition the proportion of knot wood in the butt log is quite limited because at these spacings conifer branches only increase rapidly in diameter for the first two or three years after which they suffer suppression and growth virtually ceases, whereas radial growth of the stem continues year after year (Von Wedel *et al.*, 1968). At the same time close spacings offer the benefits of a smaller corewood zone, and less stem taper. However growing trees too close together produces small stems and requires a long rotation if large diameter logs are required.

Increasing spacing not only results in larger knots, it leads to more stem malformation. Malformed and less vigorous stems can be removed in a production thinning or thinning-to-waste operation. The loss of merchantable volume due to thinning is compensated for by concentrating timber production on fewer stems. Clearfelling costs are reduced as there are fewer logs to harvest and these are larger. Further, the loss in wood volume may be illusory (Bunn, 1981) as mortality and stem breakage on clearfelling in heavily stocked stands must be set against the earlier loss of volume when thinning-to-waste.

Large knots drastically reduce strength and are a major cause of downgrade in lumber. They are undesirable in fibre products. Knots are very dense, some 2-3 times as dense as stem wood and are frequently above 1000 kg m^{-3}. This is a consequence of compression wood formation and heavy resinification after mortality or pruning: resin can account for 30% or more by weight of knot wood. Knots are hard to penetrate with chemicals and are inadequately pulped, while in mechanical pulping knots are resistant to defibration.

The significance of branches – and knots in lumber – is out of proportion to the percentage of stem volume that they occupy. First, trees experience asymmetric loading from heavy branches such that compression wood is observed often in the immediate vicinity of knots (usually in streaks extending below the branch). The second effect arises from the stresses in the immediate vicinity of the branch or knot.

Intuitively, one recognizes that the swelling of the stem around branch whorls is the means of accommodating the bending forces introduced by heavy branches. It is also a defence mechanism against severe wind, although where trees break instead of being uprooted, failure occurs in the immediate vicinity of a large branch.

Lumber has no such defence mechanism as it is straight edged. The knot and the cross-grain in its immediate vicinity is a far more serious defect in the lumber than the branch is in the tree, and grading rules place limits on knot size. For this reason forest growers have to be mindful in selecting intensively for and keeping branch diameters small. The implications for lumber strength are discussed in Chapter 10.

7. SOFTWOOD PLANTATION SILVICULTURE

There is an inevitable tendency to assume that the silvicultural systems with which one is familiar have some general validity. More often they are determined as much by cultural, fiscal and political forces as by silvicultural or physiological insight. The arguments presented in this study of New Zealand forestry may have a general applicability and validity – and yet they provide also a warning of presumption and single-mindedness. Ideas should never be writ on stone. New Zealand forestry was driven by production and 'growth and form'; and not by wood quality. Where prospective markets had been identified the ability to supply those markets with quality wood was presumed. The most significant aspect of plantation silviculture is the effect on wood properties of the reduction in the age of clearfelling. The younger the trees are the greater the proportion of corewood and the poorer the overall wood quality. All other factors are secondary.

Natural mortality of untended stands places boundaries on practical options in the management of forests. Competition within heavily stocked stands is intense. First the canopy must close, restricting branch growth at the base of the green crown, to be followed by branch mortality. Eventually the smaller suppressed trees die. The first major plantings of *Pinus radiata* in the central North Island of New Zealand (1925-35) were not thinned. Galbraith and Sewell (1979) observed that

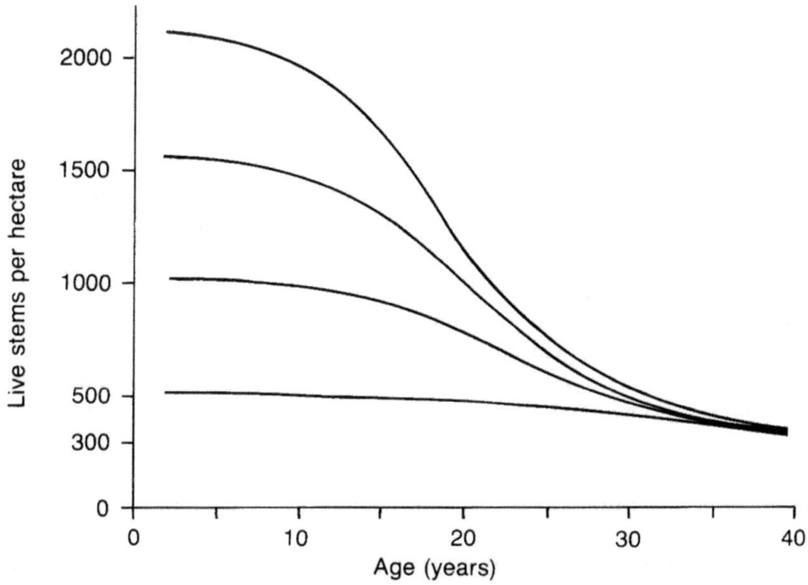

Figure 5.14. Natural mortality in untended radiata pine as a function of stocking and age (Galbraith and Sewell, 1979). In tended stands it is desirable to thin with sufficient intensity to anticipate natural mortality so the final crop trees do not experience competition.

irrespective of the initial stocking the final stocking, arising from natural mortality, came down to the same figure of around 300 stems ha^{-1} by age 45 (Figure 5.14). There is nothing inevitable about this figure. Stockings for *Pinus radiata* as high as 1500 stems ha^{-1} in 30 yr-old stands have been observed in Chile (Delmastro *et al.*, 1982). This may be due to differences in climate and site, and the presence of fewer pathogens. However, in these early New Zealand plantings, the loss of stemwood volume due to mortality was considerable, being up to one-third of the total increment by the age of 35 (Sutton, 1984). These stands had to be maintained on extended rotations, typically 45-50 years – to smooth out a major discontinuity in plantings – to provide a constant wood supply over a prolonged period.

The end result was a heavily stocked final crop whose stem characteristics included live knots near the pith and dead, bark-encased knots in the outerwood in the lower half of the stem (Walker, 1984). The trick in processing such material was to minimise the influence of defects, first by zoning and then by sawing each zone to best advantage. Thus acceptable framing timber was cut from the outside of the butt and second log: despite the presence of dead, bark-encased knots, this is a zone of stiff, high density wood and not overly large knots. The same loose knots prevented the sawing of board grades from the outerwood of these two logs, but some moderate quality boards was cut from near the pith where the branches would have been alive when the wood was laid down, and the knots were sound and intergrown. Short, clear lengths cut from shop or factory grade were best taken from the internodal regions between branch whorls of the second log. At this height in the stem the internodes are long since the tree would have been growing most vigorously at that stage. Further up the tree stem-cone holes are too numerous to yield the better board grades, although some low-grade boards could be cut from the second and third logs provided the tree was of reasonable diameter and the branches were still alive. Unlike most softwoods, radiata pine is prone to stem cone formation in the upper parts of the stem: cone formation occurs both as separate cone whorls and at existing branch whorls. The best hope for recovery from the fourth log was to cut framing from the pithy corewood by enclosing the pith within the centre of the sawn section. Top logs and thinnings provided low density corewood fibres having thin-walled cells. This chipwood was acceptable for particleboard, fibreboard and mechanical pulp. High density, long fibres from the slabwood was well suited for strong high-tear kraft pulps used for linerboard and kraft sack paper.

Such trees provided a versatile mix of lumber and fibre, although the top board grades could be met only by finger jointing and the proportion of No. 1 framing was disappointingly low (30%), because of the difficulty of keeping knot diameters below the specified 33 mm limit for that grade. High stockings, no thinning and long rotations to produce framing, boards with sound knots and pulpwood was never going to be an economic proposition.

Since the 1950s most New Zealand plantations have been under some form of active management. Initially high stockings (2500 stems per hectare) of *Pinus radiata* were advocated for the following reasons:

- To control branch size in unthinned stands. However, if stocking is to be the tool to control branch size the initial spacings between trees needs to be around 1.8 x 1.8 m on some sites (Table 5.4). High stockings ensure early canopy closure and branch mortality as the green crown moves up the tree (pine and larch are less shade tolerant than are spruce and fir). The same problem of branch control is found with *Picea sitchensis* growing in Britain, where an initial stocking of 2500 stems per hectare (2 x 2 m spacing) and no subsequent thinning has been advocated to ensure that a significant proportion (> 75%) of the sawn timber is of the better framing grades (Brazier *et al.,* 1985). However the consequences are stark. Expect either small logs or long rotations.
- To allow for selection of the best trees. Selection criteria include vigour, straightness and uniform local spacing after thinning. The first major plantings in New Zealand (1925-35) were established with unimproved seed. In these young unmanaged stands 37-42% of stems had multiple-leaders (Macarthur, 1952) and an acceptable final crop was achieved only because the high initial stocking resulted in heavy natural mortality (generally of the less vigorous and malformed stems) or the deliberate thinning of stands. Successive tree improvement programmes mean it is no longer necessary to remove four out of every five stems and initial planting (550-1100 stems per hectare) can be as low as three or two times the final stocking (300-400 sph) while still achieving a final crop with all trees of good form and vigour. The contrast between the original stands and those planted with select seed 50 years later is dramatic.

While it is technically possible to reduce knot size and to minimise the volume of the corewood by restricting growth through high initial stockings, productive thinning must be delayed until the green crown has receded up the tree, at a stand height of 14-18 m. Further, any control of branch growth due to high stocking is lost once the stand is thinned and the branches in the live crown will be greater than would have been the case had the stand not been thinned. Production thinning of such stands yields expensive wood since the piece size and the volume per tree extracted is small, and the lumber is of low grade. Delaying production thinning to increase the piece size and the volume of extracted wood brings with it the risk of windblow. Indeed there are considerable costs in growing such a forest, the most significant being that the rotation age has to be extended a further 5-10 years before the piece size in the surviving stems makes final harvesting economic (age 35+).

Log size has a major impact on forest economics. The percentage of the log that can be recovered as lumber declines quite noticeably once the log size drops below about 400 mm while harvesting, transport and handling costs escalate. Lower initial stockings and subsequent thinning shorten the rotation length but it is doubtful that the control of knot size would be sufficient to produce a significant proportion of framing timber. Here experience in Australia, Chile and the Cape Province of South Africa on the one hand and New Zealand and Kenya on the other diverge. Adequate control of branch size is achievable in many Australian radiata pine regions by judicious manipulation of stocking levels through commercial thinnings. More especially, coarse branching is not a feature of slower grown *Pinus radiata* stands on lower rainfall sites of lower fertility. Some slash *x* caribaea hybrid clones have very small branches and narrow crowns yet they can be good volume producers.

As already discussed wide initial spacings result in large knots, but the effect in practice depends on the grading rules that apply. The knot volumes and sizes noted by Nylinder (1958) in Sweden are small (Table 5.3) and the sawn outturn yields quality board and structural timbers. This contrasts with the faster growing, heavier

Table 5.4. Effect of initial spacing on branch size at various sites for *Pinus radiata* in New Zealand (Sutton, 1970): branch size is the mean of the 16 largest knots from four quadrants and at four heights. Historically for No. 1 framing, knots should not exceed one-third of the cross-section, which in a piece of 100 x 50 mm corresponded to a knot of 33 mm. This was achieved only with the stockings shown in the shaded zone.

Site (Forest)	Initial spacing (m x m)				
	1.8 x 1.8	2.4 x 2.4	3.0 x 3.0	3.6 x 3.6	4.8 x 4.8
Rotoehu	40	-	47	43	-
Kaingaroa	33	47	50	-	-
Gwavas	-	-	-	38	-
Ashley	-	31	34	39	-
Woodhill	25	25	33	37	46
Golden Downs	25	28	33	35	-
Eyrewell	24	27	33	36	39

limbed softwoods in warmer climates. With these faster growing species small knots sizes cannot be as easily achieved through close stocking and subsequent thinnings, and the knotty timber is invariably of poor quality (Table 5.4). A partial solution to the larger knots endemic to fast grown pine in New Zealand was to use larger framing members (100 x 50 mm green and 90 x 45 mm dry-dressed compared to 90 x 35 mm dry-dressed pine framing in Australia) and to cut framing from the stronger outerwood; but now New Zealand has adopted the smaller section size. The inability to reduce knot size through initial spacing provided the impetus to use early pruning and thinning as a tool to improve wood quality, especially in the large butt log of fast growing softwoods. In New Zealand pruning has increased the cost of the log delivered to the mill by only 10-15%, principally because harvesting and transport costs loomed so large (Fenton, 1972). The quandary is whether to plant at wider spacing and to both thin and prune the butt log (yielding very low grade corewood); or to adopt a close initial spacing (*c.* 2500 stems per hectare) and to control knot size in the bottom two logs through early canopy closure, only thinning (to about 350-400 stems per hectare) much later (at 14-16 m) at the cost of a prolonged rotation.

Where quality lumber is sought there is a clear incentive for having low final stockings. The basic premise must be to plant as few trees as possible (550-1100 stems per hectare) while still allowing for some selection when thinning-to-waste – to remove those trees that lack vigour or are of poor form. This concentrates the merchantable wood on the final 300+ sawlog trees. The wider initial spacing is possible with reliable, improved planting stock. However the forester must now prune the butt log as there is no control over branch growth at these wide initial spacings. Further it is essential to prune as early as possible to minimise the size of the knotty occluded core and to maximise the valuable clearwood in the butt log (Figure 5.16b). The aim is to confine the occluded core to a cylinder of no more than 150 mm in diameter. Most of the value in the tree resides in the pruned butt log (*c.* 60%) while the industrial pulpwood contributes very little (*c.* 2%). Silvicultural treatments need to be timed precisely as a delay of 12 months from the prescribed pruning date results in an enlarged knotty core that would delay clearfelling by 3 years if the same proportion of clearwood is to be obtained. More likely the rotation length will be kept short (age 25-30) and the proportion of clearwood will be much reduced (Bunn, 1981).

Thinning is an integral part of the pruning operation as it gives the final trees the growing space to put on greater diameter growth and produce more clearwood. An unfortunate consequence of wider spacings is the larger branch size (> 50 mm)

Figure 5.15. Brazil: in a class of its own – but not for long. (a) On the left is a 3 yr-old stand of *Pinus taeda* with a target of 34 m^3/ha/yr over bark at age 20 for sawlogs: on the right is a 3 yr-old stand of *Eucalyptus grandis* with a target of 84 m^3/ha/yr over bark at age 7 for pulpwood; 54 m^3/ha/yr is the commercial average (International Paper do Brasil Ltda). (b) A stand of 10 yr-old *Eucalyptus urophylla x grandis* hybrid, planted at 5 x 2.4 m, and now 35-40 m tall. It is atypical in that pulpwood is grown for 7 yrs but this stand was planted early in anticipation of the 900 000 tonne bleached eucalypt kraft pulp mill commencing production in 2005. Height/dbh can reach 150:1 and in a gentle breeze entire stands will sway in unison (forest operations of Veracel Celulose S.A., a joint-venture between Aracruz and Stora Enso).

171

above the pruned log, although some of the deleterious effects of these large intergrown knots are offset by the larger diameter of the second log (compared to that in a more heavily stocked stand of similar age). In managed plantations the costs of silvicultural operations compound over time. This more than anything drives down the age of clearfelling and thereby increases the proportion of corewood. The same argument applies to efforts to improve the site, to reduce competition and to the use of fertilisers and trace elements to boost growth. There may be a slight drop in basic density during the first year or two after fertilising, but this is offset completely by faster growth. Indeed the principal effect of the use of fertilisers is to reduce the time to harvesting and it is the shorter rotation rather than any drop in basic density associated with fertilization that is of major significance.

Since clearfelling the first plantations from the 1920s-1940s there is evidence that intrinsic wood quality of subsequent crops has deteriorated, in particular the wood is less stiff and has shorter tracheids. Until now this has been attributed to the younger age of clearfelling, down from 45-50 years to 25-30 years. However, Lasserre *et al.* (2004) observe that for 11 yr-old pines corewood stiffness was reduced from 6.7 GPa at high stockings (2500 sph as in early plantings) to 5.0 GPa at low stockings (833 sph as practiced today). If the negative relationship between DBH and stiffness – big trees are less stiff – is included as a covariate the adjusted stiffnesses are 6.4 GPa (2500 sph) and 5.3 GPa (833 sph) respectively, so the effect of stocking is reduced to 1.1 GPa. One is then comparing the effect of stocking between trees of equal DBH. The study included ten clones. With clones (genotype) the difference between the best clone and the worst clone was 1.3 GPa for both stockings so the difference between the mean and the best of the clones was only 0.65 GPa. However these clones had not been bred for improved wood properties, so the future benefits are greatly understated. This work implies that both initial stocking and genotype should be used as complementary approaches for improving corewood stiffness. Wind and tree sway are likely factors in these findings.

However, only hardwoods have the potential to produce the wood quality needed for sawlogs when grown on very short rotations, that is 10-12 yrs, a fact that has little to do with the growth rate (Figure 5.15).

Perhaps New Zealand pine forestry was not as smart at it thought. There is a current trend to switch back to higher initial and final stockings and to longer rotations: maybe with a commercial thinning. These self-defeating changes seek to increase the proportion of outerwood ahead of any future genetic gain to be delivered by the breeders. Extended rotations do not offer a long term solution and are, at best, an acknowledgment of past failures to attend to wood quality. To date there has been no will to diversify, to contemplate planting fast growing hardwoods with better corewood and superior wood qualities.

There is a tendency for *Pinus radiata* in the central North Island to form one to five branch clusters in a growing season, with *c.* 70% having either three or four branch clusters. The number of whorls decreases on moving from the north to the south of New Zealand, with a corresponding increase in the length of the internodes: only 12% of the sawlog length yields clear lengths of 0.6 m or more in the north but almost 50% in the south (Bannister, 1962; Cown, 1992). The prospect of cutting short clear lengths for componentry and finger-jointing is much greater in the south.

Figure 5.16. Some management options. (a) Selecting the best species and provenance is a key to successful plantation forestry. The *Pinus ponderosa* in the foreground is the same age, 45 yr-old, as the *P. radiata* in the background. Seed for this ponderosa pine stand came from British Columbia. Poor growth was partly due to selecting too northerly a provenance that did not correspond to the southern latitude of New Zealand. Even with better selection it is doubtful whether its growth would be comparable to that of radiata pine. (b) Early thinning and pruning increase the diameter of final crop trees and allow the production of clearwood from pruned logs. No more than half of the live crown is removed on pruning at age 5-7. (c) and (d) In unpruned stands clearwood is limited to the internodes. Only a limited amount of cuttings is greater than 0.6 m long. By breeding select long-internode trees it is possible to obtain long-internodal lengths (> 0.6 m) from over half the internodes on some sites (Kininmonth and Whiteside, 1991).

Table 5.5. Silvicultural practices that influence wood properties.

Silvicultural practice	Influence on wood quality
Planting stock	• Species: clones or select/control-pollinated seed; rooted cuttings
Initial spacing	• The onset of mortality and log size of final crop trees
	• The control of branch and hence knot size
Thinning	• Growth rate and hence rotation length
	• Affects indirectly whole tree density by reducing the rotation length
	• Increases knot size in the live crown
Pruning	• Early pruning eliminates large knots otherwise present in butt logs
Rotation length	• Affects mean age of wood and hence basic density

The branching habit of radiata pine has suggested two separate tree improvement programmes: that of breeding a more multinodal tree with lighter, more frequent branching for most purposes (easily-pruned clearwood, framing and pulpwood), and a long internode type which would allow for the cutting of short (0.6-1.8 m) lengths of clearwood from between the large knot whorls of the unpruned trees (Figure 5.16c,d). Because multinodality is much more frequent it has proved easier to find and select genotypes that combine the multinodal form with vigorous growth and straight stems than it is to find long-internode genotypes with desirable characteristics (Carson, 1988).

Perhaps the surprising feature of the New Zealand breeding programme is that the next generation of trees will be of lower density than the first generation. The overall drop of 20-30 kg m^{-3} arises from the reduced rotation length (from 45-50 years to 25-30 years) and the weak negative correlation between growth rate and basic density (Cown, 1992). Although basic density is considered a highly heritable trait, selection of the appropriate high density families will rectify the situation only slowly. A more immediate alternative is to vegetatively propagate superior clonal stock displaying high density (or, better, displaying high stiffness).

Sutton (1984) emphasises two general silvicultural principles: 'that no aspect of silviculture can be considered in isolation', and 'that for any silvicultural practice both yield and tree quality are largely predictable'. To which he added 'that tree quality is very largely determined by the early silvicultural treatments', e.g. late pruning is a waste of time as the pruned envelope will be too narrow to provide much clearwood. The principal practices controlling wood properties are summarised in Table 5.5.

Figure 5.17 is an example from the southeastern US of contrasting silvicultural prescriptions in two locations. In the piedmont early slow growth is accelerated by aggressive thinning, e.g. from 1000 to 225 sph at age 15. On the coastal plain where early growth is better – and where thinning ought to be earlier at age 12 – thinning too late, e.g. from 1000 to 275 sph at age 21, resulted in forgone outerwood growth.

The viability of any intensive silvicultural programme is tied closely to the rotation length and the premium to be paid for quality material, whether that be for

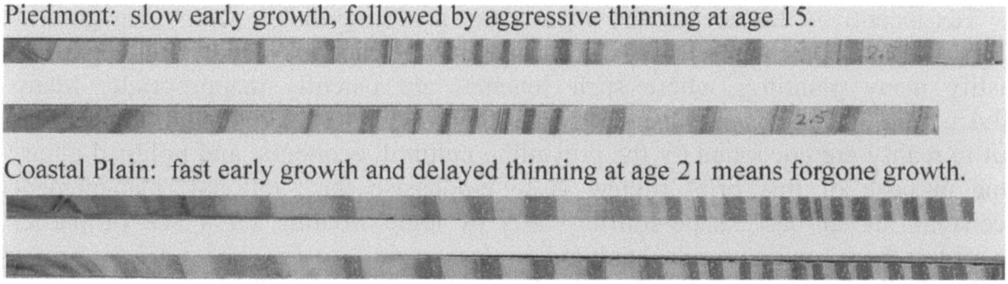

Figure 5.17. Pith-to-bark increment cores for *Pinus taeda* growing in the southeastern US.

clearwood or structural lumber. As rotation lengths are extended the compounding effect on the costs of managing the stands in their early years becomes so great that these operations cannot be justified unless enormous premiums are to be paid for the higher grades of lumber that are produced as a consequence of that early silviculture. For this reason the management of temperate forests established in less benign climates than that of countries such as New Zealand has tended to emphasise high initial stockings to achieve effective branch control. However, high initial establishment costs associated with high initial stockings and the extended rotation length suggest that the issue has been only partially addressed.

A review of silvicultural systems for *Picea sitchensis* in the UK (Macdonald and Hubert, 2002) provides further examples of 'product push' with low valued pallets, packaging and fencing absorbing two-thirds of production – and Sitka spruce is classified as non-durable and resistant to treatment.

Ideally the market for lumber and its interaction with species, silviculture and site determine the appropriate management policies in any local situation, rather than the tax regime or the various subsidies that all too often distort management decision making. New Zealand has considered itself to be a good example of plantation forestry, offering an internal rate of return of about 6-8% above that of inflation. However the sale of the State Forests in the 1980s and 1990s suggests that for the people of New Zealand the profit from exotic forestry has been largely illusory. Forests were not established solely to maximise the return on investment, but for a plethora of worthy reasons such as soil protection, regional employment, and in the private sector, favourable tax treatment. Some State Forests were too small to offer economies of scale, were unnecessarily dispersed and located at a distance from both mills and export ports. Further there was a lack of consistency: many stands were thinned and pruned too late, negating the benefits of such silviculture. Ironically the State Forest Service, having advocated enthusiastically a most intensive programme of thinning and pruning to produce clearwood, when transformed to a self-funding State Owned Enterprise drastically reduced the number of stands being pruned (down 46% in its first year of operation) and being thinned (down 20%) in order to increase its cash flow and provide a skinny dividend to its shareholder (the Government). Even that failed to satisfy the philosophical and political objectives of the Government. These commercial forests have been sold to private enterprise. Bilek and Horgan (1992) provide an excellent commentary.

Tax incentives, tax breaks and government meddling distort forest management decision-making throughout the world, and by an Alice-in-Wonderland logic can justify many plantings where such schemes are patently inappropriate. Many silvicultural and management systems have their own internal logic and consistency, but in reality are cocooned by the prevailing cultural, economic and political ethos. One purpose of this brief review is to emphasise that 'rational' management decisions are at best shape-shifting rafts of logic floating on a sea of partial variables: there can be no rules, only principles.

Leadership in short rotation forestry lies in the sub-tropics. Table 5.6 represents a typical regime for *Pinus taeda* in Brazil, i.e. all trees are pruned in the first lift and potential final crop trees only in the second lift. Improved genetics and management have raised average productivity of pine plantations from 18 το 23 to 33 m^3/ha/yr in the successive decades, in the 1970s, 80s and 90s respectively. Whether corewood quality issues have been addressed remains to be seen.

Table 5.6. Typical silvicultural system for loblolly pine, *Pinus taeda,* in Brazil on a high productivity site, with an initial stocking of 1111 stems/hectare.

Age	Pruned height [m]	Pruned stems	Remaining trees
3	2.5	1111	1111
4	4.5	750	1111
5	5.0	750	1111
6		Thinning-to-waste	750
7	6.0	400	750
10		Commercial thinning	400
20			Clearfell

This New Zealand study makes little mention of wood quality as it was largely off the agenda. Little effort was spent on improving important properties like stability, stiffness and strength – all received virtually no practical attention. No wonder the principal markets for UK spruce remain low valued commodities. The failure to address this issue is examined in Chapter 6. However, this failure should be placed in context. Current understanding of wood behaviour, properties and structure is reminiscent of that facing metals in the 1940s: the time lapse reflects the vastly more complex ultrastructure of the wood cell wall and the difficulty in securing molecular structural information compared to the simple crystalline structure of metals (Entwistle and Walker, 2005b). However, that means that the opportunities are enormous, once one sheds ridiculous, naïve suppositions such as '… increasing wood density has been ranked above all other desirable objectives of any wood quality improvement programme' (Walker, 1993 p. 155). The basic sciences are available (Chapter 2), future prospects are visible at least in outline (Chapter 6), and opportunities for genetic improvement of wood quality are there to be grasped. What so often is lacking is an industry willing to embrace substantial change – change that may foreshadow greater emphasis on hardwoods for markets that have been traditionally the preserve of softwoods.

8. EUCALYPTS FOR WOOD PRODUCTION (DONNELLY ET AL., *2003*)

With hardwoods opportunities for creative thinking are even greater. By 2010 eucalypts are projected to account for about 25% of global non-coniferous roundwood production. Yet in 2000, eucalypt accounted for only a very small part, *c.* 2%, of sawnwood production, i.e. the major market for eucalypt is pulp, with the market for lumber being relatively insignificant. Further, Donnelly *et al.* (2003) observe that although Australia is the home of almost all eucalyptus species, the 3.1 million m^3 of sawlog production comes almost entirely from native forests. This roughly matches the estimated 2.9 million m^3 coming from plantations worldwide (with 37% from Brazil alone). Indeed in Australia the plantation area is only about 15% of that in Brazil, which has *c.* 3 million hectares.

It would be hard to overstate the importance of South America and Brazil in particular. By 2010 South America is projected to account for 55% of the world's supply of eucalypt plantation roundwood, followed by Asia with 20%. Donnelly *et al.* (2003) estimate that eucalypt sawlogs from plantations will increase to 10.6 million m^3 by 2015, with a total pruned log supply of 1.4 million m^3. Intensively managed silvicultural systems for eucalypts are in their infancy.

The considerable demand for eucalypt as domestic fuelwood is not examined. However, the charcoal market supplying Brazil's iron and steel industry is of more interest: 42% of Brazil's 50-60 million m^3/yr of eucalypt production is for charcoal, with another 46% for pulp. Prior to the industrial revolution, smelting of iron ore used charcoal and was responsible for the deforestation of considerable areas in northern Europe. Notably in China, cast iron and coal firing were known as early at the 5th century BC, while they may have discovered how to smelt ore with coke by the 12th century. However in Brazil the use of charcoal continues to be favoured over the use of coke in part because of the absence of sulphur in wood charcoal and the very low net carbon emissions. Cheap, poor quality imported coal, low in calorific value and high in ash (30%), provides a (questionable) basis for substitution against charcoal – and even then only when the exchange rate is favourable. Under the Kyoto protocol charcoal is carbon neutral, an advantage compared to coke.

The reduction of the ore is carried out in a blast furnace where the coke/charcoal is both a reducing agent and a source of energy. It must supply a porous structural support for the iron ore so that the hot air and released CO/CO_2 percolate up through the ore while the molten iron flows down to the hearth of the furnace The reaction at the surface of the charcoal/coke is:

$$2C + O_2 = 2CO + \text{energy}, \tag{1}$$

and with the iron ore

$$3CO + Fe_2O_3 = 2Fe + 3CO_2 \tag{2}$$

Coal cannot be used directly because of its structural properties, the volatiles and its impurities – hence the use of metallurgical coke or charcoal. After smelting the

Bessemer converter refines the iron to steel by oxidizing out impurities such as sulphur and surplus carbon arising from the reduction of the iron oxide.

In Brazil, the other principal industrial market for eucalypt has been pulp. Optimal pulpwood crops follow a 7-yr cycle for pulp production incorporating two coppice rotations before finally replanting after 21 years. However continuously improving genetic gain is making earlier replanting more attractive.

Coppicing can be attractive. In San Paulo State, Brazil, there are stands of *E. globulus* that have been continuously coppiced for five rotations (50-60 yrs). The choice is one of cash flow. Small companies often cannot afford the much higher costs of fresh planting ($800/ha) *versus* coppicing ($100/ha for thinning coppice shoots and weed control). They accept the loss of vigour (*c.* 15% less volume) and that the stand becomes progressively less uniform as openings develop within the stand. Equally, they cannot afford the high costs of clonal breeding that can deploy new clones that out-perform the older coppicing clones (*c.* 5-10% more volume). On the other hand the inherent maturity of coppice implies improved corewood wood quality from aged material that must be set against uncertain (confidential) improvement in the intrinsic wood quality of newer clones.

The largest eucalypt pulpwood operations in Brazil are gradually reducing their reliance on coppicing (down from 50% to 20% or less). Currently retaining some coppice remains attractive for those companies that are having to expand rapidly their forest lands ahead of a sharp increase in pulp production – increases of up to a million tonnes of pulp per annum – with new land having first call on available clonal stocks.

While routine (conservative) operational growth rates in the region of 30-50 m^3/ha/yr are taken as normal, it is worth remembering that 30-40 years ago growth would have been around 15 m^3/ha/yr.

Modest prospects for world hardwood pulpwood, both in terms of abundant volume and declining price, are powerful incentives to consider other silvicultural and process options. However the characteristics and properties that have been sought by the paper industry somewhat limit the utility of these select, high productivity clones were they to be redirected to lumber. Despite this caveat, electing to extend rotations (from 7 to 16 years), to undertake thinning and pruning, and so to market a range of products from a multi-product regime is at the forefront of eucalypt management. In this way pulpwood arises from commercial thinnings and from forest and mill residues – pulpwood becomes a by-product. However, capturing a commercial thinning for pulpwood carries an opportunity cost by postponing the time of clearfelling, until the sawlogs reach the desired diameter.

8.1 Management and silviculture of eucalypt

Their diversity of habitat is such that generally one can find a species or hybrid that offers superior growth rates to most other species under almost any condition and, being diffuse-porous hardwoods, the density of their wood is not greatly influenced by their growth rate. A further plus is their genetic diversity.

Critical wood quality issues for a pruned eucalypt crop differ significantly from those for a pine crop. Desirable features of eucalypts include good colour of the heartwood (although *E. dunnii* and *E. nitens* are light coloured) and generally a moderate-to-high density with correspondingly superior hardness, stiffness and strength. The most vexatious issues are brittleheart, tension wood, splitting, high shrinkage, slow-drying and collapse. This formidable but not intractable list is not altogether too different to the red and white oaks of North America. It is important to recognise that all species have undesirable attributes – and these can be addressed. It is salutary to remember that in New Zealand in the 1940s, a major plantation softwood like *Pinus radiata* was considered fit only for packaging and even its potential for pulp was uncertain: this was some 20 years after huge, initial plantings.

These attributes of eucalypts, both positive and negative, are somewhat at variance with desirable pulpwood characteristics of density, fibre length, cellulose content *etc.* that pulpwood breeders concentrate on. Sawlogs call for different selection strategies in breeding, but the large natural variation within-species implies that considerable gains are achievable.

A classic problem for all fast-growing plantations targeting the solid wood industries is that the original untended and genetically unimproved logs have too many and too large knots, even for species that are supposedly self-pruning. Fast growth necessitates minimizing competition and keeping the stand open which results in large, deep crowns. This means pruning, frequently and early, to produce one or two log lengths within which the high frequency of knots are confined to a low-value defect core that should be keep as small as possible. It is imperative that pruning be of live (green) branches to avoid stain and fungal infection.

All eucalypts are intolerant of competition such that in an unmanaged stand crown dominance and suppression begin at an early age, resulting in a wide diameter variation and some mortality. Further at high stockings and under competition less dominant trees bend away in an attempt to out-flank competition for solar radiant energy (so forming tension wood). This is a reflection of early, fast growth such that the site is occupied rapidly with a rapid rise and fall in both basal area increment and current annual increment. While native eucalypt is self-pruning producing long knot-free boles, this develops slowly over many decades and self-pruning is poorly expressed in plantations. It appears that the faster diameter growth prevents normal branch shedding, stub ejection and occlusion processes. For high quality, pruned sawlog production early thinning is essential: delay to attract a low volume commercial thinning after current annual increment has peaked is counter-productive. The first live prune should co-inside as closely as possible with canopy closure removing about a third – and certainly no more than half – of the green crown. Further, there is a suggestion that the height-to-diameter over bark at breast-height should be no more that 100:1 otherwise the over-slender trees are too susceptible to wind-sway and in response have high growth stresses and more tension wood: height growth in early years can be *c.* 25 mm/day. This is another argument in favour of low initial stockings.

Radical regimes are being proved in Uruguay. These were developed for the most productive sites in regions where markets do not exist for pulpwood. The schedules involve annual lifts in pruned height (up to 10.5 m) beginning after only

18 months to produce a small, occluded defect core (< 120-150 mm diameter) together with thinning-to-waste when very young. By concentrating on fewer stems individual trees have sufficient unimpeded growth to be clearfelled as young as age 16. Donnelly *et al.* (2003) emphasize that intensive management of eucalypts for high quality sawlogs is relatively untried and it is difficult to ascribe hard values for such timber. However, there is an established market for clear sawn eucalypt in the United States with prices that are 30% less than cherry and 10% less than Philippine mahogany.

Methol (2001, 2002) has modelled such regimes using data from COFUSA (Uruguay). His work provides justification for pursuing such a strategy (Table 5.7). The modelling is tentative, as the first stands will not be harvested for 3-4 years. This particular regime (refer Donnelly *et al.*, 2003), using appropriate but hypothetical costs (Uruguay) and current revenues (Brazil), generated a pre-tax internal rate of return of 22.7%, which meets the most demanding criterion for international investment. Uruguay has only begun to deploy clonal technology in the last few years. This series of simulations by Methol indicates that in all cases shorter rotations and more open stands create value, with lower returns occurring where pulpwood or only low pruning are sought.

Neighbouring Brazil has better land for eucalypts, where in some states the hybrid *E. urophylla x grandis* is preferred being more drought tolerant than *E. grandis*. Cubbage *et al.* (2005) examined investment returns for over 20 plantations and native forests across the Americas, providing an index/rank of performance for various species. They obtained similarly high internal rates of returns for *E. grandis* and *E. dunnii,* in this case in Brazil. No wonder the forest sector is looking to South America for future industry leadership.

Table 5.7. Eucalyptus grandis growth and yield projections for a highly productive site, initial stocking of 900 stems/hectare.

Age	Thinning to stems/hectare	Pruned stems	Extracted volume
2	400	400	Thin to waste
2-6		Prune in annual lifts to 10.5 m	
6	300	300	18 m^3 pulpwood
9	150	150	60 m^3 pulwood 47 m^3 sawlog
16	Clearfell		40 m^3 pulpwood 212 m^3 sawlog 231 m^3 pruned logs

CHAPTER 6

WOOD QUALITY: MULTIFACETED OPPORTUNITIES

SHAKTI CHAUHAN[1], ROBERT DONNELLY[2], CHIH-LIN HUANG[3], RYOGO NAKADA[4], YIN YAFANG[5] AND JOHN WALKER

[1]Institute of Wood Science and Technology, Malleswaram, Bangalore, India; [2]Donnelly and Associates, Curitiba, Brazil; [3]Corporate Research, Weyerhaeuser Inc, Federal Way, Washington, USA; [4]Tohoku Regional Breeding Office, Forest Tree Breeding Centre, Takizawa, Iwate, Japan; [5]Research Institute of Wood Industry, Chinese Academy of Forestry, Beijing, China

1. INTRODUCTION

Chapter 5 examined the difficulty in recognising the pull of the market; it considered the impact of wood quality gradients (mainly of density) that vary within and between trees from the grower's and end user's perspective; and explored how forest management and silviculture can be tailored to specific wood products. This chapter revisits the same themes but here these are addressed by examining the way wood properties impinge on the practical work of foresters and mill managers. The key topics are stiffness and stability.

2. MOVING CLOSER TO MARKETS

Nicholls and Dadswell (1965) argued that the breeding of pine should be based on density, tracheid length and spiral grain. This strategy addressed the needs of that time as these characteristics could be measured without undue difficulty and so offered a pragmatic means of delivering improved wood and pulp properties.

That task would be formulated differently today. The market for solid wood wants stiffer wood, more stable wood, harder wood *etc*. This is handled in two ways. For the future resource, it is logical to focus directly on these properties or to improve those underlying characteristics that most directly determine such properties. Thus for softwoods a reduction in microfibril angle takes precedence over an increase in basic density. A higher density implies increasing the rate of photosynthate production or increasing the allocation to stemwood in preference to roots and branches if the volume increment is to be sustained. On the other hand, changing the microfibril angle is a more efficient way of improving stiffness – requiring no change to synthate production or allocation.

For the existing resource, it is logical to focus directly on the same properties of stiffness, stability… and to find tools that recognise these properties: to determine those regions, those stands and those trees and logs that have above – or below – average stiffness. There is no merit in cutting a low stiffness log for structural lumber only to get a lamentable grade outturn. Currently the market is looking to acoustics and infrared spectroscopy to sort the existing resource.

For the existing resource the focus is on sorting so each log can be sent to where it can be best processed. In addition to the changing demand for products and the increasing mix of young plantation logs, the merger of timberlands and the closure of mills have also changed the regional log supply conditions.

Mills producing structural products must know how to avoid getting too much low stiffness material from this new wood basket. Also there can be seasonal fluctuations in log quality supply and product requirements. Harvesting may be restricted during the winter when timber from high elevations or wet sites becomes unavailable, and alternatives sources have very different intrinsic wood qualities. Thus inventory and harvesting schedules need to be well coordinated. Mills anticipate these trends based on experience but sometimes will need to pay extra to source the right material to meet the demands of important customers.

3. STIFFNESS

The terms stiffness and strength are often confused (Figure 6.1). In practical terms low stiffness is visualised in the excessive deflection of a joist or the sag (or springiness) of a scaffold plank, while strength is concerned with how strong the beam is – what load it can carry before it fails. Stiffness relates to the ease with which one can stretch a material in tension, compress it under load or shear it. Strength relates to the ability of a material to sustain a load. In a stress-strain diagram (Figure 6.1) the slope of the initial straight line determines the stiffness of the material, while the failure point defines the tensile strength.

The (horizontal) x-axis in Figure 6.1 records the strain. Strain is the change in length per unit length, $\delta l/(l+\delta l)$, so if the body is stretched by 1% the strain is 1% or 0.01. Most materials fail in tension on being stretched by less than 1-3%. Rubber is an exception, easily stretching by 20-100% before breaking. In Figure 6.1, the (vertical) y-axis records the applied stress. Stress is the force per unit area, typically expressed in Newtons per square metre (N m^{-2}) or Pascals (1 Pa = 1 N m^{-2}): the Pascal is a unit of pressure while the Newton is the unit of force. Stiffness is a measure of the force per unit area required to stretch a sample elastically a given amount of strain. Thus for the specimen with a microfibril angle, θ, of 5° the strain at the elastic limit is about 1.7% or 0.017 and the normalized maximum stress is 190 MPa. The stiffness (the modulus of elasticity or MOE) is stress/strain or 190/0.017 MPa or 11.2 GPa. Note that the strength of the material, 190 MPa, is numerically about 1 to 2% of that of the MOE. This 100 to 50-fold difference arises because the elastic limit occurs at about 1-2% strain (and stiffness is stress divided by strain). This estimate assumes wood behaves as a linear elastic material up to the point of failure, which is approximately true for θ is 5° – the stress-strain plot appears approximately straight until failure. Where θ is 26° the elastic limit, the end of the linear elastic region, occurs at about 1.5% strain while the specimen fails at around

8% strain. The initial slope of the stress-strain plot (from which the elastic modulus is calculated) becomes progressively less steep, for specimens with microfibril angles of 20°, 26° or 50°, i.e. the stiffness or modulus of elasticity decreases progressively. Rubber is not only stretchy but requires little force to stretch and so has a low elastic modulus.

Reiterer *et al.* (1999) observed that their 200 μm thick specimens stretched considerably once the elastic limit had been exceeded, but only where the MFA is quite large (> 15-20°). Then it stretches even further for each increment of strain – in this region, beyond the linear elastic limit, the material deforms irrecoverably by viscoelastic or plastic flow. Finally the sample breaks in tension. The strength of the material, i.e. the failure stress, is read from the y-axis. The stiffness of all woods ranges from 0.5-20 GPa and strength ranges from 1-40 MPa, from the corewood of low density species to the outerwood of very dense species.

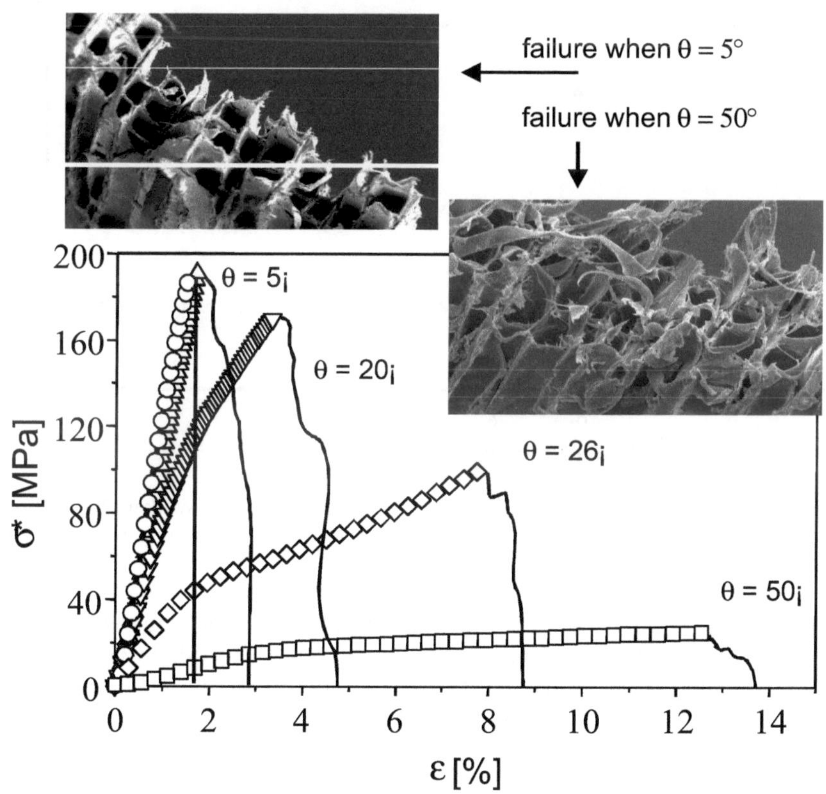

Figure 6.1. Stiffness and strength of Norway spruce, *Picea abies,* in tension (Reiterer *et al.*, 1999). The data has been normalized, i.e. all values have been adjusted by multiplying by the cell wall density (1500 kg m^{-3}) and dividing by the actual density of the sample.

3.1. Stiffness of wood

As one might expect, the within-tree stiffness distribution reflects the corewood-outerwood and juvenile-mature wood patterns observed for many characteristics and properties: furthermore between-tree variations are very noticeable.

The within-tree pattern of stiffness can be seen in Figure 6.2. This study by Xu and Walker (2004) examined some sixty trees from a single 27 yr-old stand. First the logs were sawn to give a 100 mm thick central cant and 40 mm thick flitches on either side of the cant. These were recut to 100 x 40 mm boards that were kiln-dried, dressed and machine stress graded at 150 mm intervals along their lengths. Finally boards were reassembled back into the original logs so that board positions, relative to the pith, could be noted. Figure 6.2 shows the average stiffness along boards both by log type and by distance from the pith with P_1 boards containing pith while P_5 are the outermost boards adjacent to the cambium in butt logs.

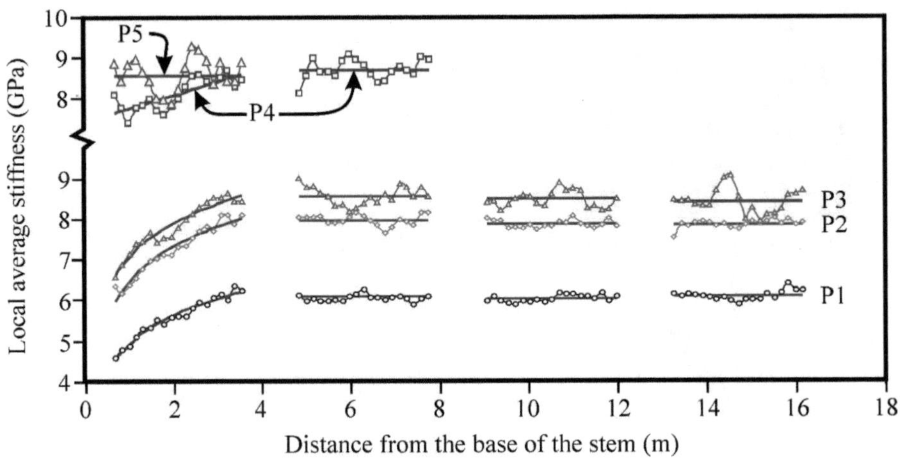

Figure 6.2. Mean within-tree stiffness profiles for *Pinus radiata* (Xu and Walker, 2004). For clarity, boards from positions P_4 and P_5 have been offset to avoid overlapping boards from position P_3.

The corewood-outerwood radial stiffness gradient in the upper logs is steepest between the with-pith P_1 boards (6 GPa) and the adjacent P_2 boards (8 GPa), with little change thereafter, from P_2 to P_5 (9 GPa). The poorest wood, in the juvenile-corewood zone, displays a dramatic loss of stiffness in the lowest 3 metres of the stem (4.5 GPa at the bottom end of the P_1 boards). As will be argued later, wood stiffness is only adequate where it exceeds about 7 GPa or 1 million psi. Thus on average the with-pith boards (P_1) are unsuited for both structural and non-structural purposes. The same applies to the lowest part (bottom 3 metres) of P_2 boards from the butt log. The mean stiffness of all other boards – from P_3 to P_5 (for butt logs) and from the P_2 outwards (for the 2nd 3rd and 4th logs) – exceeds this minimum threshold.

To show the between-tree stiffness distribution the same butt logs were grouped into five stiffness classes (Figure 6.3). For trees in the lowest quintile, entire boards in the P_1 to P_3 positions fail to reach the minimum threshold of 7 GPa. Even for butt logs of median stiffness (the 3rd quintile) something like half the boards in P_2 position appear marginal (at least toward the lower end of each board). Contrast that with boards from the uppermost quintile group of butt logs: the entire log exceeds the minimum threshold value.

For the upper logs the between-tree stiffness distribution is captured by considering the stiffness values at the upper end of the butt logs: there are no further changes up the tree as the juvenile zone only extends about three metres above ground level (Figure 6.2). Thus for the lowest quintile group the boards from positions P_1 to P_3 do not make 7 GPa. Boards from the P_1 position are marginal from logs in the next three quintiles, but boards from P_2 and beyond exceed the 7 GPa threshold. A threshold of 8 GPa provides a higher standard for times of oversupply.

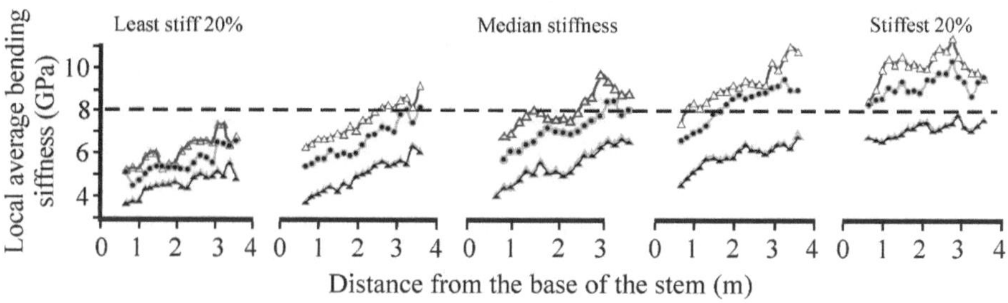

Figure 6.3. Stiffness within butt logs for *Pinus radiata*, grouped in quintiles (Xu, 2000).

There are three lessons from this study.

- First, there is obviously a significant potential for improving stem stiffness. It is perfectly feasible for tree breeders to produce planting stock having those properties that are seen in the topmost quintile. This would be a transformational improvement.
- Second, the existing forest resource is hugely variable. Currently sawmillers buy the natural distribution of log stiffnesses, accepting that 20% of the logs are pathetic (in the lowest quintile). Hence the interest in technologies such as acoustics and near infrared spectroscopy that are capable of presorting logs into stiffness classes that can be priced and marketed accordingly.
- Third, the very poor properties in the lowest 3 metres of the butt log – in the corewood-juvenile wood zone – has never been recognised. Logically, it makes sense to sell a short 2.5 to 3.0 m butt log, possibly to get excellent LVL/veneer from the outerwood and targeting low value products from the enlarged low stiffness peeler core, or to full-taper saw to maximize lumber from the outerwood. The poor corewood-juvenile wood zone at the base of the tree – like a spreading pedestal – is beautifully captured in Figure 6.4. Knowing this, to

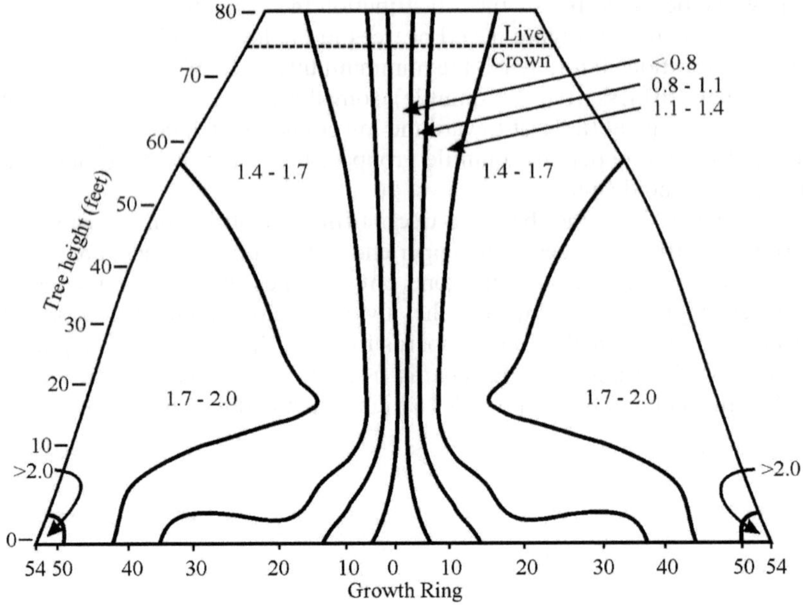

Figure 6.4. Variation in stiffness in a single loblolly pine, *Pinus taeda*. The 0.8 to 2.0 contour values are in millions of psi, where a million psi is approximately 7 GPa (So *et al.*, 2004).

continue cutting a 4.2 to 6.0 m butt log defies logic – even accepting that some of this is already being left as a high stump, where there is oversize butt swell or a severely swept/distorted butt end.

Slower-growing Douglas fir in the Pacific Northwest has an advantage compared to most pines in that it displays a more rapid improvement in MFA up the stem, i.e. the juvenile-corewood zone appears to be no more than two metres tall and the radial extent of very low stiffness is reduced. Further up the stem the corewood-outerwood gradient in MFA is similar to that found in pines.

3.2. Stiffness thresholds: achieving mediocrity or magnificence?

Forest owners have made two mistakes. They have paid too much attention to outerwood properties (often arguing for prolonging the rotation in order to grow additional superior wood) and they have failed to understand the implications of minimum threshold values. In contrast, engineers require quantitative data on many properties, of which stiffness and strength would be among the most important. The 'number' that they need is not the average value of some property but a safe minimum value, usually the lower 5th percentile value. They are only interested in knowing how poor the stiffness, or how weak the piece might be; they have no interest in its potential stiffness or strength above these minimum threshold values.

The stiffness of individual boards can be measured by passing through a machine stress grader. From the deflection the stiffness is calculated and its strength is estimated (Figure 10.13). This puts the spotlight on the poor wood that fails to make the grade especially in the corewood-juvenile zone. The strength-stiffness ellipse in Figure 6.5 is derived from countless destructive measurements of these properties (Figure 10.14). There are two essential points to understand. First, there are stiffness thresholds that trigger substantial price steps. Second, the critical point is that the biggest step occurs on moving from the lowest valued box or dunnage grade to a framing grade at around one million psi or 7 GPa. The premium paid for significantly stiffer lumber is relatively small. The conclusion is that there is only a small demand for super-stiff lumber (for engineered wood products), but a huge penalty in failing to meet the minimum threshold.

Owners of softwood plantations have made a third mistake: they have attributed stiffness variation to density and so missed the plot. Of course a co-varying density trend exists, and it works/has worked for procurement managers. Until recently, forest owners knew little of the microfibril angle (MFA), while most scientists involved in wood quality and geneticists did not appreciate the direct link between MFA and stability and stiffness.

Forest owners need only be curious about MFA since acoustics (or near-infrared spectroscopy) provide rapid, surrogate methods for determining stiffness.

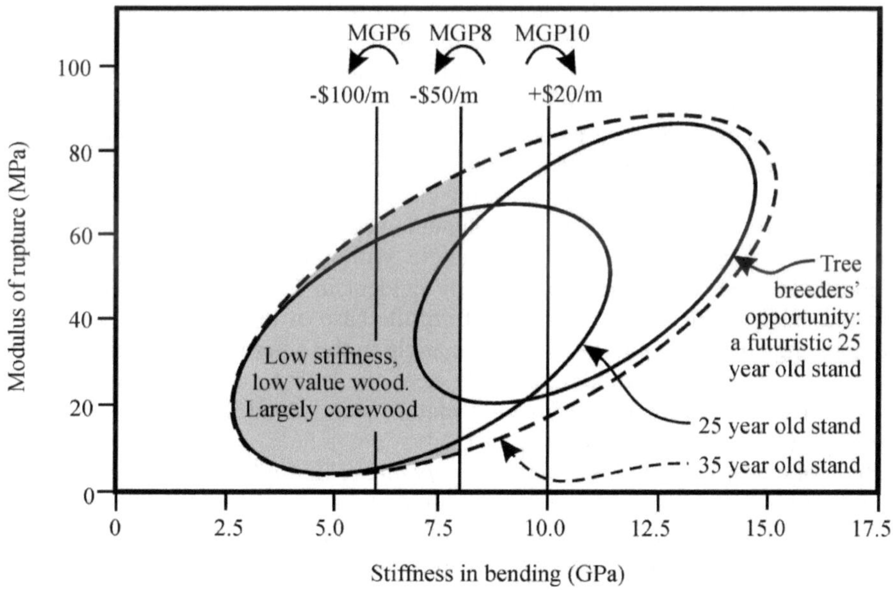

Figure 6.5. Wood quality envelope for *Pinus radiata.* Unimproved pine has a surfeit of low quality wood, but note the threshold has been set at 8 GPa. Failure to meet threshold values devastates profits. The area within the envelope is not equally populated: the right hand side (or top-right quadrant) is less strongly represented. Tree breeding has the realistic potential to transform profitability, as would serious investment in superior species (Figure 5.11).

4. MICROFIBRIL ANGLE

Until recently, few have recognized the enormous impact that the microfibril angle has on wood properties, especially in the corewood of plantation softwoods. In particular the microfibril angle strongly determines the stability and stiffnesss of wood within the first 10-15 growth rings of the pith (Barber and Meylan, 1964; Cave, 1968; Preston, 1974).

The microfibril angle (MFA) refers to the mean helical angle that the cellulose microfibrils in the S_2 layer make with the longitudinal axis of the cell (Figure 2.12). There are a variety of methods for measuring MFA each giving consistent results but which can differ systematically from one another by as much as 5°. To interpret the effects of MFA on wood quality it is important to use consistent methods and to follow the trends, within and between trees. Further, while this review only considers the MFA, a complete analysis recognizes that the microfibrils, even within the S_2 layer, are not uniformly aligned (Figure 2.13) and that knowledge of the dispersion or distribution of the microfibrils about the mean MFA value is needed (Cave, 1997a,b; Evans, 1999; Reiterer *et al.*, 1999).

The microfibril angle was first measured in the 1930s and by the mid-1960s the critical role of MFA in determining various wood properties such as stiffness, shrinkage and stability was well documented (Cave, 1968; Meylan 1968; Preston, 1974). While the commercial relevance of this work was recognized, in practical terms application was held up by under-developed technology – measurement of MFA using an x-ray diffractometer with photographic film took several hours in the 1960s compared to a minute or less with a diffractometer coupled to a CCD-device (akin to a very sensitive digital camera) today. Further, the forest sector thought it had a silver bullet with the use of density. Hence even today density is the most studied wood characteristic (Walker, 2004). Emphatically, research in the 1990s established MFA to be the dominant wood characteristic underlying poor wood quality in many fast-grown, short-rotation plantation softwoods (Butterfield, 1998).

There are two trends for microfibril angle within trees. First, there is the corewood-outerwood trend that is particularly evident in fast-grown pines (Figure 6.6). In this instance on moving outwards from the base of the tree the MFA falls from a high angle near the pith to a moderate angle in the outerwood; further up the stem the MFA falls from a moderate angle near the pith to a low angle in the outerwood. The second trend, the juvenile-mature wood trend, is only obvious in the lowest 3 metres of the stem (Figure 6.6).

A study of cultivars of sugi, *Cryptomeria japonica*, (Figure 6.7) demonstrates the practicality of selecting for low MFA at an early age, since MFA invariably decreases with age, i.e. on 99 occasions out of a hundred. Having selected a clone with a low initial MFA there are only two possibilities, either the MFA declines noticeable with further growth rings or it declines only gradually. The cultivar 'Kumotoshi' is preferred.

Microfibril angle is a characteristic that has a profound influence on the wood properties of stiffness and stability. This is especially true where considering the

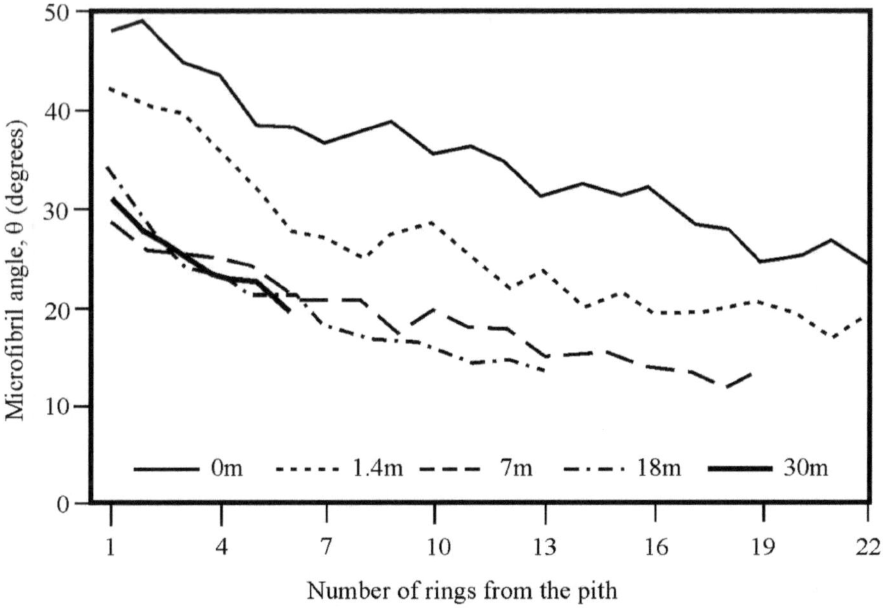

Figure 6.6. Variation in microfibril angle within *Pinus radiata* (Donaldson, 1992).

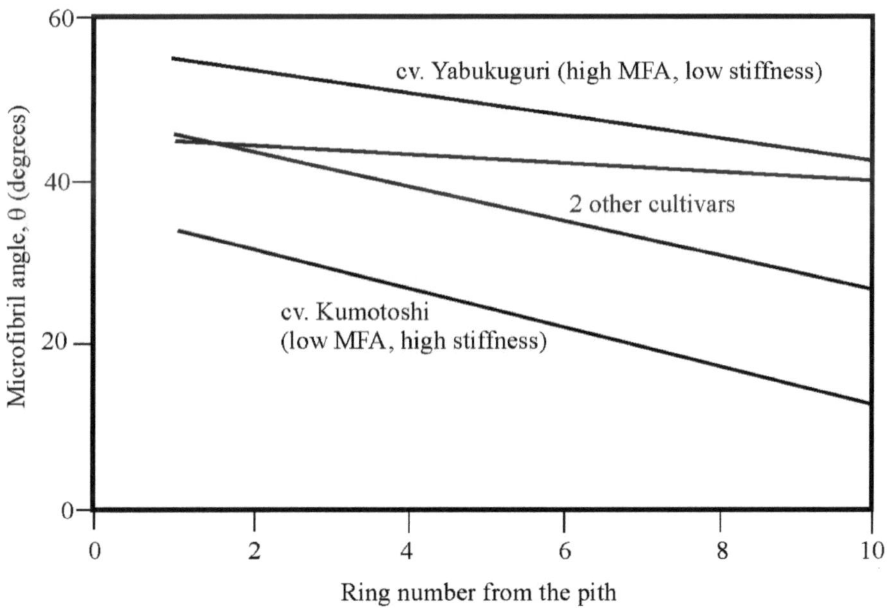

Figure 6.7. Variation in microfibril angle in cultivars of sugi, *Cryptomeria japonica* (Fujisaki, 1985).

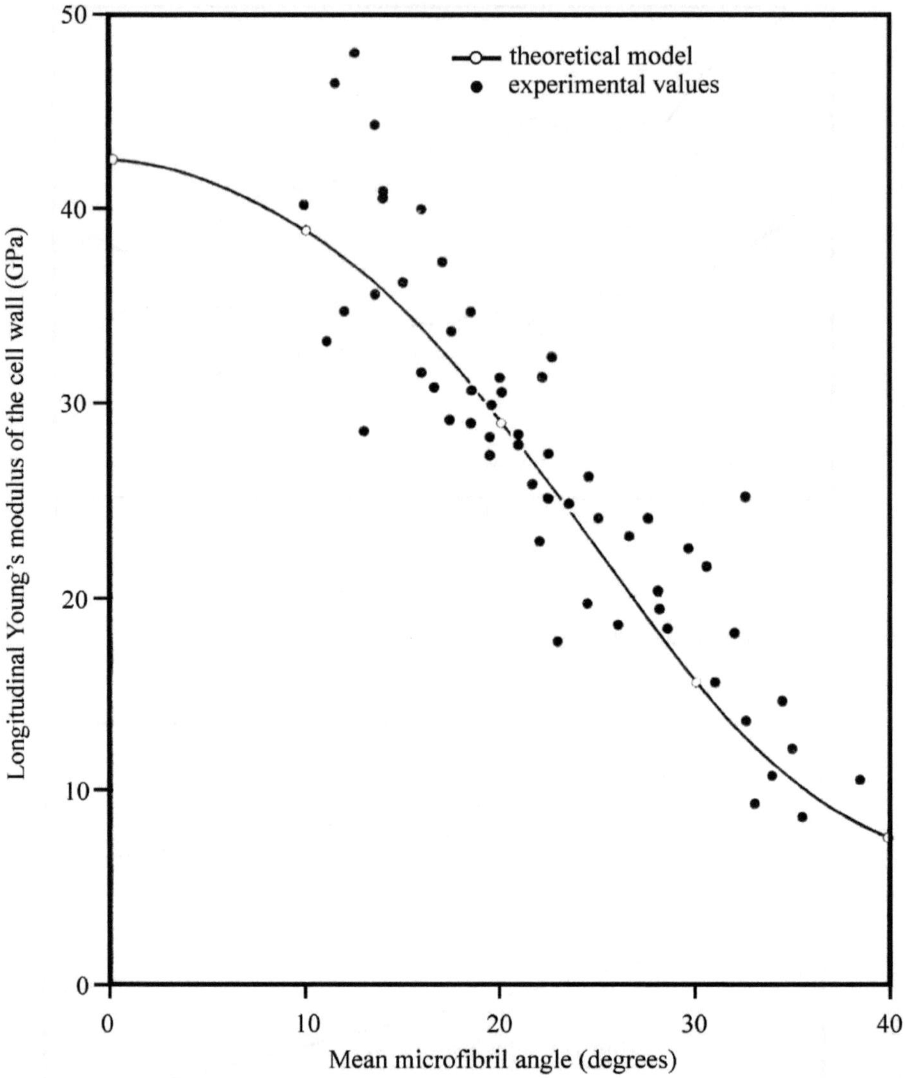

Figure 6.8. Influence of MFA on cell wall stiffness of *Pinus radiata* (Cave, 1968). The data has been normalized to that for cell wall stiffness, by dividing the sample stiffness by the sample density and then multiplying by the density of the cell wall itself (1500 kg m^{-3}).

juvenile-corewood zone of the butt log. Cave (1968) measured the MFA of 2 x 2 x 60 mm specimens and then proceeded to pull then in tension to observe the corresponding stiffness (Figure 6.8).

There are some important observations. The observed change in microfibril angle, decreasing from 40 to 10°, has an enormous influence on longitudinal stiffness (MOE), from 8 to 39 GPa. Further, the range of MFAs found amongst fast grown softwoods (Figures 6.6 and 6.7) means that there will be a correspondingly large stiffness variation between trees. The decline in MFA with ring number accounts for the increased stiffness of the outerwood of plantation softwoods. Attributing the same to co-varying density trends is an imperfect diagnosis masquerading as insight.

Numerical finite element models of Astley *et al.* (1998) take account of chemical composition, microfibril angles and their dispersion within all the cell wall layers (not simply the S_2) and cell geometry (shape, size, wall thickness). Where adapted to realistic models of earlywood/latewood entities, these simulations reduce the disparity between stiffness as the MFA changes from 45 to 10° to a factor of about 3 as opposed to about 5 in the original model (although that work was supported by experimental data) by Cave (1968).

Thus changes in microfibril angle provide the explanation for the wide range of stiffnesses found in with-pith corewood (Figure 5.12) and for the general observation that in moving from corewood to outerwood stiffness can increase three-fold while basic density increases by only 50%. The contribution of density is underwhelming.

However, hardwoods are different. The MFA is never that large, rarely exceeding 30°, even in the corewood-juvenile wood zone (Bao *et al.,* 1998; Lima e*t al.,* 2004; Yang and Evans, 2003). In this context, a study of *Eucalyptus delegatensis* by Evans and Ilic (2001) recorded MFAs ranging from 20 to 8.5° and densities ranging from 452 to 990 kg m^{-3}: samples included isolated earlywood, latewood and tension wood tissue so accounting for the large range of density and stiffness. Obviously both characteristics contribute to stiffness: the lowest stiffness of 10.2 GPa was found for a specimen of density 537 kg m^{-3} and MFA of 20°, while the specimen with greatest stiffness of 28.3 GPa had a density of 900 kg m^{-3} and MFA of 9°. There are problems associated with high density (too heavy, excessive shrinkage) and low microfibril angle (growth stresses), so one can argue that a stiffness in excess of 15 GPa may not be so obviously desirable.

Thus corewood of teak with its modest microfibril angle (15-25°), reasonable density (550-600 kg m^{-3}), stiffness (12.5-14.5 GPa), shrinkage (8-11% for heartwood) and relatively uniform radial distribution of properties has the potential to meet the most critical end-use requirements usually expected of older material (Bhat, 1998). Fast growth in young trees is positively correlated with percentage heartwood and reduced MFAs. This, together with scope for genetic selection of individual trees, means that 20-30 yr-old teak plantations justify their role as an important new tropical resource for sustainable forest management (Bhat *et al.*, 2001, 2005).

Equally interesting, a fast growing, low density, plantation hardwood such as *Paulownia lanko*, oven-dry density > 250 kg m^{-3}, has a very low MFA in both corewood and outerwood of about 10°, and poor stiffness, 3.9 GPa (Bao *et al.*, 1998). *Paulonia* sp. is biomechanically efficient, on the basis of the high stiffness/density ratio (as is balsa). Consequently the potential to improve the stiffness of such a timber is limited, which is irrelevant as it is not a structural wood. Despite the very low basic density of *Paulonia tomentosa* of 234 ± 24 kg m^{-3} (USDA, 1999), it is highly desired for chests and wardrobes in traditional Japanese bedroom furniture because it is so dimensionally stable which is important where the humidity can be extremely high at certain times of the year. Further its light-weight suits the construction of large drawers to hold the bulky kimono. Logically it is too light and marks too readily, but that is acceptable in a society which is modern while retaining and prizing traditional values.

Although hardwoods are different in that they tend to have lower microfibril angles, Evans (2000) has shown that for both hardwoods and softwoods stiffness (MOE) can be estimated with a standard error of 1 GPa knowing microfibril orientation, wood density and chemical composition. Further, MFA and density appear to vary independently between trees and may therefore be manipulated separately in tree improvement programs. These MFA studies ought to be seen in the broader context of plant adaptation (Givnish, 1995). The axiom that 'form follows function' needs to be extended beyond external appearance to consider how factors such as density and MFA are employed by individual trees, species and genera in their particular strategies for survival – whether light-demanding or shade-tolerant; pioneering or late successional; timberline or rainforest; short-lived or long-lived.

A major misconception is that softwood lumber whose stiffness is inadequate for structural purposes (< 7 or 8 GPa) can be diverted to quality non-structural purposes. The problem is that wood of low stiffness is likely to have a large microfibril angle (> 35°). If some part of the piece has such a high MFA it will show excessive longitudinal shrinkage (Figure 4.6). Uniform longitudinal shrinkage would not be of great concern. However, invariably, there is a MFA gradient across the piece such that there will be differential shrinkage on drying: a board containing wood from, say, rings 1 to 6 is likely to have different MFAs associated with different ring numbers (Figure 6.6). Such boards are liable to either cup or crook on drying (Figure 8.14) and move in service due to humidity changes. Gaby (1972) observed that boards containing pith and boards containing compression wood are most vulnerable. In practice, the danger zone for plantation softwoods lies in the first 5-15 rings, especially in the butt logs (Figure 6.9). Such behaviour has been modelled (Ormarsson, 1999) and used to validate various sawing strategies that minimize these problems.

A Weyerhaeuser patent illustrates the dilemma of low quality wood. Low stiffness boards (< 7 or 8 GPa) also shrink longitudinally (Figure 6.10). Further, these boards do not just shrink longitudinally but are liable to distort since such boards will incorporate a number of growth rings having different MFAs (Figure 4.6). The dominance of very low stiffness wood in the juvenile-corewood zone of many genetically-unimproved softwoods means that the wood in this zone lacks both stiffness and stability. The most valuable log – the butt log – has both the stiffest outerwood and the poorest corewood (Figures 6.2 and 6.4).

There are three mechanical properties that are particularly important for trees: stiffness, strength and work to fracture – the ability to absorb and dissipate energy under dynamic conditions such as strong wind gusts. Gordon and Jeronimidis (1980) observe that the specific work of fracture for *Picea sitchensis* is superior to steel (on a weight-for-weight basis). More recently Reiterer *et al.* (1999) observe, for *Picea abies,* a very great increase in tensile strain (at maximum stress) with increasing MFA, from 1.8% at 5° MFA to 13% at 50°. Subsequently Reiterer *et al.* (2001) measured the work to failure, which is the area under the curves in Figure 6.1. They observed that the maximum work to failure occurs at about 25-30°. Perhaps the explanation for the diversity of microfibril angles that is observed in the corewood-juvenile wood zone should be sought in stem physiology and functional morphology (Givnish, 1995). In turn the ability to absorb energy must be linked to the abundance of hemicelluloses surrounding the microfibrils as perfectly elastic bodies (crystalline microfibrils) find it hard to dissipate energy, yet trees cannot continue to sway indefinitely.

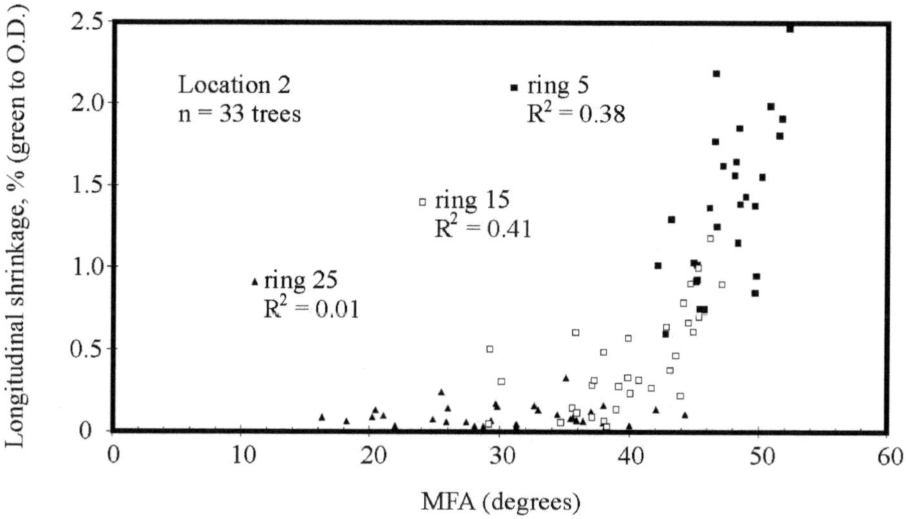

Figure 6.9. Longitudinal shrinkage in loblolly pine, *Pinus taeda*, as a function of MFA for rings 5, 15 and 25 in butt wood (Megraw *et al.*, 1998).

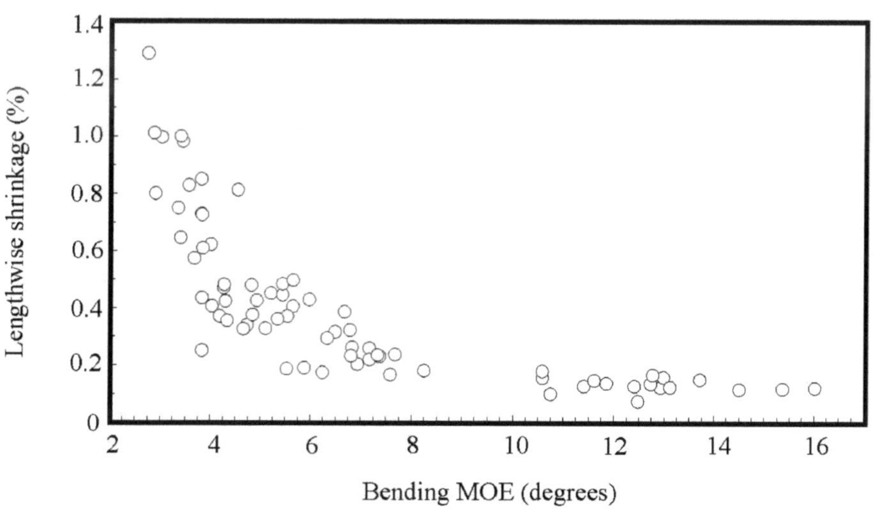

Figure 6.10. Longitudinal shrinkage shrinkage in specimens of loblolly pine, *Pinus taeda*, as a function of stiffness (International Patent, 2000 filed by Weyerhaeuser).

MFAs around 30-40° seem to be the critical zone for certain physical responses of a wood fibre, for example the growing strain of a maturing xylem cell switches between contraction and expansion when MFA is around 30° (Guitard *et al.*, 1999).

5. BREEDING FOR INCREASED STIFFNESS

> When we say that a wood property is under environmental control we mean that it varies considerably with a change in the environment. When a wood property is under strong genetic control, it may vary without regard to the environment under which the tree is grown or its properties may stay constant despite the trees having been grown in differing environments (Zobel and Buijtenen, 1989).

For these reasons the consequences of introducing a species to a new region are generally difficult to predict and it is desirable to grow that species in a limited way before becoming committed to a major plantation programme. The end results are often unforseen and there have been numerous disappointments.

Further, there is a genetic *x* environment interaction which if large would mean that the ranked performance of families or clones will differ across sites or regions. In that case it might warrant going to the expense of developing different breeding populations for different regions. Hopefully this is can be avoided.

> Predictions of relative performance of seed-orchard parents appear to be valid for all parts of the country [*New Zealand*] where radiata pine has been tested. Genotype-by-environment interaction does not appear to be large enough to warrant selection for adaptation to local site and climate conditions. The predicted increase in genetic gain calculated for regionalised *vs* non-regionalised seed orchard strategies was small (Carson, 1991) and would require substantial extra cost for progeny testing to capture. It appears instead that designer seedlots produced by varying the mix of selection traits for end-product improvement will increase profitability more than specialisation aimed at increasing adaptation to local climate and site (Carson, 1996).

A genetic *x* environment interaction is frequently present but can be managed in trials within the same breeding zone. For example, geneticists usually focus on those families that show consistent response across various sites within a breeding zone to avoid complicated deployment schemes. Ecologists may like to study G *x* E interactions. Geneticists prefer to circumvent the issue.

For unstable, low-stiffness softwoods – and one is tempted to brand all fast grown pines and spruces in this category – one challenge is to achieve sufficient stiffness within the first few growth rings that the pejorative inference of corewood ceases to have any sting. Indeed, the combination of high MFA, compression wood and large knots poses problems in using the corewood of fast grown plantation softwoods. This is why there is so much to gain from a breeding programme focussed on improving wood quality in corewood.

Most tree breeding programmes are working from populations that are only one or two generations from the 'wild'. These breeding populations were selected for a handful of visible traits – volume growth, stem straightness, branching habit, disease resistance and, possibly, basic density. The selection intensity would have been, perhaps, one tree in a thousand having the best mix of such traits. Collecting seed from such trees, cross-pollinating their progeny (controlled or wind-pollinated), reselecting new breeding families *etc.* provided the seed for new forest plantings. This initial screening minimized undesirable features that destroy value. Diseased, crooked, heavy-limbed runts are obvious 'losers'. This is the initial entry requirement to the global supply challenge that places ever more stringent demands on forest commodities. The end result of these early selections is impressive in terms of increased performance. For example, the volume growth rate of radiata pine is some 30-35% greater than 50 years ago.

The problem is that there are so many possibilities. If one selects with an intensity of 1 in 100, then selecting for multiple traits rapidly increases the intensity of selection, e.g. 1 in 10^{2n}. This dilemma was a factor favouring the historical preference for density. Fortunately, trees are amongst the most variable of living organisms. More fortuitous still, wood quality characteristics are highly variable and strongly heritable (Shelbourne, 1997) so selection can be very effective from a broad-based population, e.g. stiffness is highly heritability (as high as for density). The tricky bit is to maximize the genetic gains while minimizing the potential adverse effects on the operation. For example high stiffness elite families must display good growth and form as well as ensuring that they are good seed producers and easy to propagate. A high selection intensity is achievable since acoustics provides a fast and efficient screen for stiffness.

Family forestry is expressed in the forest as the mean properties of all individual trees within each family. Even so, there still remains a wide variation in wood characteristics amongst individual trees from the same family with a few individuals performing much better than the average for the family. Therefore, much more impressive gains are achievable by intensively selecting a few outstanding individuals. Clonal propagation has been practiced on a significant scale, most notably with *Eucalyptus* sp. in South America. Here, this entailed the selection of half a dozen clones from a breeding population of several thousand (Brandão, 1984): a new improved clone may be deployed and an older one retired every 2-5 years. These elite clones, which are replicated by the million, have been chosen for desired fibre characteristics for specific pulping processes or paper products, e.g. coarseness, density, fibre length, percent cellulose, wall thickness *etc.* Concurrently, operational production from these pulpwood crops has increased from about 15 to 45 m^3/ha/yr in the last 40 years. However, clonal trees selected for pulp and paper, if grown on for sawlogs may not be ideal for solid wood products. Other selection criteria may apply. There is the suspicion that those criteria – longer fibre length, low lignin content, increased crystallinity – may correlate with an excessively low MFA that may in turn be implicated in high growth stresses.

In searching for the best material for solid wood, the challenge is to set the threshold values for the corewood-juvenile wood zone as high as practical. For radiata pine in New Zealand, as a minimum, by ring 3 one might seek an air-dry density of 400 kg m^{-3}, a MFA of 35°, spiral grain of less than 3°, volumetric shrinkage to 12% MC of less than 4% and an MOE of 6.5 GPa. These thresholds are hugely controversial and significant as they predetermine the marketability of the future resource. Tree breeders are not seeking elite athletes. Rather, breeding involves a paranoid culling of the awful. A nation's health is not judged by its Olympic athletes but by the draining cost of its depressed and obese.

For solid wood, the potential is demonstrated by considering stiffness. The mediocrity of radiata pine is exciting to breeders because the wood can be both so poor and so good. Sorensson *et al.* (2002) observed that the age at which the stem (at 3-metres height) achieved a stiffness of 8 GPa (a conservative, but challenging value) can vary between age 3 and 13 when considering the best and worst single tree in 100, or between age 5 and 9 when considering best and worst 1 tree in 5. In effect, they defined an acceptability threshold of 8 GPa and proceeded to demonstrate that the undesirable, low-stiffness corewood zone (< 8 GPa) would correspond to the first 3 to 13 rings or the first 5 to 9 rings depending on the chosen percentile within the population. Table 6.1 and Figure 6.3 are two sides of the same coin – looking from different perspectives. Other owners with different species ought to strive to set the highest practical thresholds to ensure relative commercial advantage.

Table 6.1. Modelling of clearwood stiffness of *Pinus radiata* clones having average growth (Sorensson *et al.*, 2002).

Selected percentile class	Average MOE (GPa) ± Std	Age to 8 GPa ± Std
Top 1%	8.6 ± 0.4	3.3 ± 0.4
Bottom 1%	4.0 ± 0.4	12.4 ± 2.1
Top 20%	7.6 ± 0.1	4.5 ± 0.2
Bottom 20%	5.1 ± 0.2	9.3 ± 0.4
Average for all 1000 clones	6.4 ± 0.1	6.7 ± 0.2

A cautionary note, genetic selection for low MFA and high stiffness may induce an increased incidence of spiral grain as trees still need to sway in the wind. Species likes pine and spruce have large MFAs in the juvenile-corewood zone and selection for increased stiffness may induce feedback that generates an increased incidence of spiral grain. Anecdotally there is a greater incidence of spiral grain the further north one goes in New Zealand, paralleling the natural decrease in MFA and the increase in density.

Most softwood plantations produce wood of passably quality, reasonably quickly and at acceptable cost. Mediocrity should not be equated with something shoddy. Wal-Mart sells goods cheaply, but they need not be shoddy. However, there has to be much greater attention to quality control (of which tree breeding is a crucial filter). The boundary between acceptable and unacceptable is more finely balanced and the choice of threshold values for performance is critical. A mediocre material is vulnerable, not because it is mediocre, but because part of the population is less than mediocre. The desire to off load, i.e. to sell as acceptable, that portion of the wood supply that is less than mediocre is an irresistible temptation (Gaunt, 1998).

Finally it is worth reiterating that for hardwoods the key characteristics are different – brittleheart, growth stress, splitting and tension wood; and collapse, high shrinkage and slow-drying.

6. ACOUSTICS TO SELECT FOR STIFFNESS

Current log grading considers external, visible features: diameter, sweep, branch sizes and distribution. Once graded all logs within a particular sort are taken to be representative of that grade. However, two working rules – at least for softwoods – would be (i) that the corewood of the best 10% in the sort will be as good as the outerwood of the poorest 10% of the population; and (ii) that 90% of all processing problems will arise from the poorest 10% of the population. Any differentiation within a population or batch of logs will require knowledge of the intrinsic qualities of the individual logs. Therein lies an opportunity.

Acoustics, NIR and microwave technologies unveil the huge differences hidden in intrinsic properties that are found in a pile of visually similar logs.

Tapping a log excites sonic waves in it that can be many metres long. Such waves no more 'see' the microfibrils (or indeed knots) than does a rolling ocean swell 'see' and break around the piles of a pier: only very short-wavelength, high-frequency microwaves can observe knots. However the microfibrils acting together with the lignin-hemicellulosic matrix contribute to the macro-property of elasticity, and it is the average stiffness of the entire woody medium that determines the velocity of sound along the log.

The advent of robust, cheap, field-based acoustic tools has allowed the sorting of logs (harvesting), of standing trees (forest evaluations after thinning a stand) and of young seedlings (tree breeding). Harvesting is of immediate interest as it impacts on profits. Therefore the majority of interest has been on logs, invariably using acoustic resonance techniques (Aratake, 1992; Huang *et al.*, 2003; Ross *et al.*, 1997; Wang *et al.*, 2001, 2004).

For logs, the mean log stiffness (GPa) is given by the universal equation:

$$E_{log} = \rho . V^2 \tag{1}$$

where ρ is the green density (kg m^{-3}) and V is the acoustic velocity (m s^{-1}).

The beauty of this equation is that green density is relatively constant between logs and, generally, a constant value can be presumed rather than measuring individual log densities. Thus for radiata pine a green density of 1000 kg m^{-3} has proved acceptable – because the trees are felled young and have little heartwood whose green density is nearer 600 kg m^{-3}. Further the observed changes in measured acoustic velocities amongst logs are much greater than any possible change in green density, so measuring acoustic velocity is sufficient to clearly distinguish between logs of varying stiffnesses. The validity of this approach is demonstrated in a Weyerhaeuser patent (Figure 6.11). Thus logs could be broadly sorted into stiffness classes according to their acoustic velocity: all logs were freshly harvested so the green log densities would have been close to 1000 kg m^{-3}.

Figure 6.11 demonstrates the ability to broadly segregate logs as they come onto a skid site, a central processing yard or mill. A structural plywood mill or LVL plant might seek logs having a sonic velocity of at least, say, 3.6-3.7 km s^{-1} from which to peel stiff outerwood.

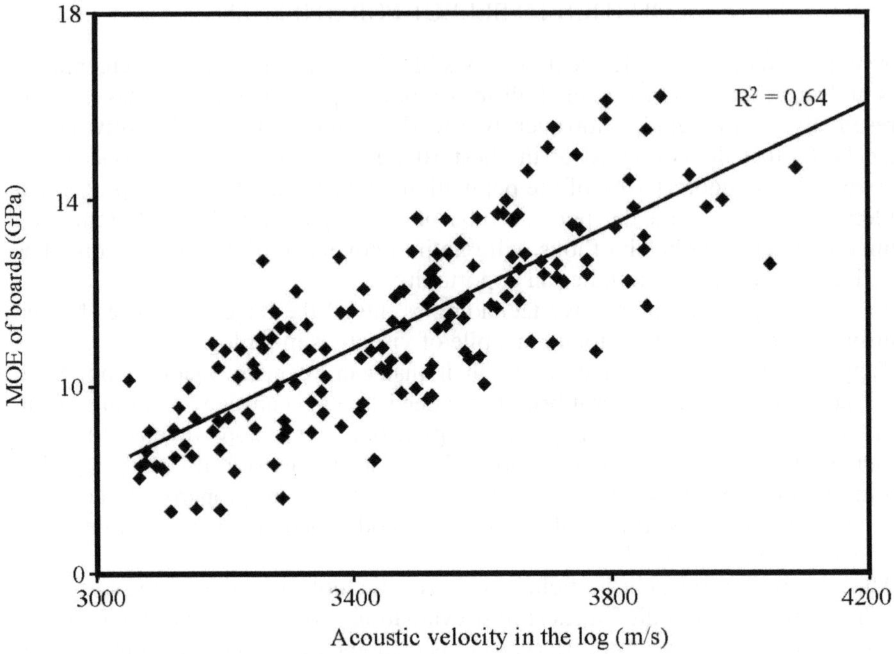

Figure 6.11. Comparison between the velocity of sound along a loblolly pine log and the mean stiffness of all lumber cut from that log (US Patent 6,006,689, 2000 filed by Weyerhaeuser).

Acoustic sorting can be used to identify the relative stiffness of timber within a forest region, or it can be used to monitor the change in stiffness with age. In Figure 6.12, the interesting features are that there is a small proportion of the 14 yr-old logs which exceeds the stiffness of a small proportion of the 35 yr-old logs, and within any age class the stiffest logs are twice as stiff at the poorest logs (there is huge genetic diversity).The influence of environment or of the particular seedlot could be judged by noting how the distributions shift to left or right when comparing stands of similar age.

Finally, acoustic properties can be measured on standing trees. An acoustic impulse is timed over a known distance between two piezoelectric sensors with pins/nail-like probes that penetrate through the cambium. This time-of-flight approach, pioneered by Fakopp Enterprise in Hungary, provides a non-destructive measurement of outerwood stiffness in standing trees (Grabianowski *et al.*, 2004; Lasserre *et al.*, 2004) that can be related to tree age, stocking and other variables. Again within-stand variations of stiffness at young ages provide an early assessment of prospective wood quality. Using adjacent stands of different ages a Russian Doll model (*matroschka,* dolls-inside-dolls) for stiffness groupings reveals what can be expected when the trees are felled (Figure 6.13). The outerwood stiffness, determined from some 300 trees in each of three stands, aged 11, 17 and 25, were

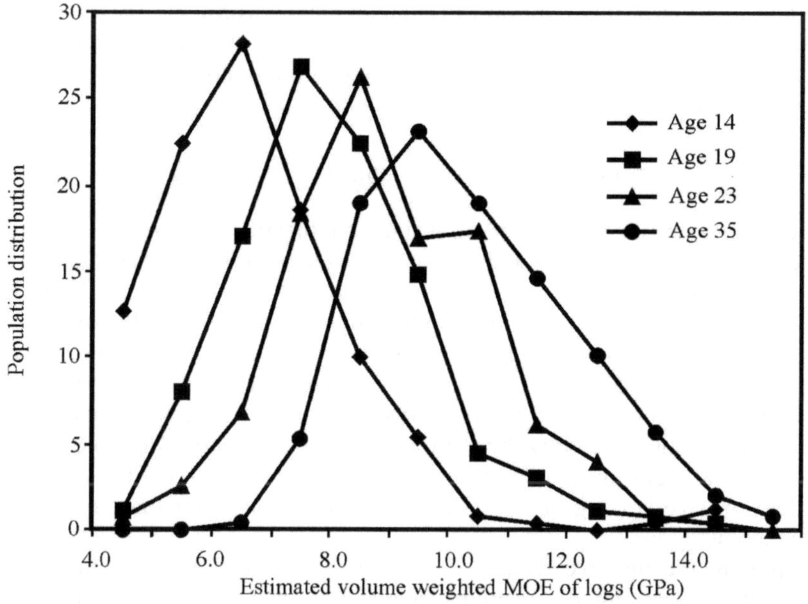

Figure 6.12. Stiffness in radiata pine stands in Tasmania (unpublished, Norske Skog Australia).

	Stiffness of trees in the various quintiles (GPa)				
Age	1st	2nd	3rd	4th	5th
25 yr	9.2	10.7	11.8	13.1	14.8
17 yr	6.8	8.1	9.0	9.9	11.3
11 yr	4.6	5.5	6.0	6.6	7.7

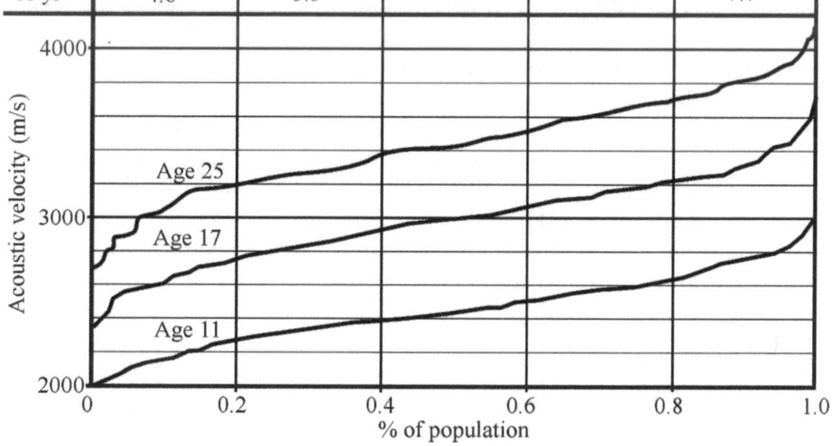

Figure 6.13. Within and between tree variations for sub-populations of logs, constructed from standing tree measurements in stands of differing ages (Bascuñán, 2005).

ranked and so grouped into quintiles. Presuming that on average trees will remain in the same quintiles as the stand ages, then the internal stiffness properties of the eventual butt logs can be anticipated by sorting them into quintiles and applying the equation, $E_{log} = \rho.V^2$.

7. NEAR INFRARED TO PREDICT WOOD QUALITY

While acoustics provides a direct determination of stiffness, near-infrared offers useful correlations with many wood characteristics and properties (So *et al.*, 2004).

In near-infrared (NIR) a sample is bathed in diffuse energy and the reflected spectra is examined across a wide bandwidth extending from 800 to 2500 nm. Energy of particular wavelengths is strongly absorbed. These strong absorption bands, corresponding to fundamental (in the infrared) and overtones (in the NIR) of various modes of vibration and rotation, arise from C-H, O-H, N-H bonds (that are moderated and distinguished by their ability to hydrogen bond). The spectra provide an enormous amount of structural information, but this is confounded by the multiplicity of overlapping absorption bands that are the signatures of individual vibrating atoms/ groups.

Interpreting the massive data is the key issue given the processing speed of current mills. Examining by discrete wavelengths may solve the communication problems but the robustness and accuracy of a parsimonious model may not be adequate. In this procedure narrow 'windows' or 'frequency bands' in the spectra may be (in large part) attributed to specific characteristic vibrations of bonds or molecules, e.g. galactans/compression wood, condensed lignin content *etc*. In these specific examples the intensity of the spectra over these narrow frequencies may be interpreted in terms of the quantity of galactan or lignin type, which in turn are related to the presence of compression wood (and so the risk of the board warping) or to difficulties in pulping. Less obviously and more empirically NIR predicts wood characteristics such as density, fibre length, microfibril angle and spiral grain, as well as properties such as stiffness, pulp yield *etc*. Many interesting correlations can be found using multivariate analysis and other statistical procedures.

Traditionally, these characteristics have been measured directly by expensive, labour-intensive wet chemistry and sophisticated technologies. With NIR, calibration models are first derived from processing NIR spectra derived from the same samples as those for wet chemistry. Then NIR can be used routinely, using those pre-determined statistical relationships between the readily measured spectra and underlying chemical and physical characteristics. NIR spectroscopy offers the prospect of a cheap, field robust technology. It will provide chemical and property analyses (Figure 6.4) that can be gathered as routinely as density is from increment cores. Intriguingly, So *et al.* (2004) reveal a much broader range of wood composition in *Pinus taeda* – cellulose 40-55%; lignin 22-32%; hemicelluloses 18-28%; and extractives 2-11% – than that indicated in Table 2.2. Probably some of this variation arises from their selection of a very wide range of wood tissue, but there are obvious opportunities for breeding specific populations for particular industries, e.g. low extractives and lignin for chemical pulping.

8. STRENGTH AND ADSORPTION OF ENERGY

As one might anticipate, the two key characteristics influencing the strength of clearwood are density and MFA. However, a further effect is that of branches/knots on the local strength in their immediate vicinity. Discussion on the role of knots is deferred to Chapter 10.

The stress-strain curves in Figure 6.1 show the stiffness/elasticity of wood (initial slope), its extensibility (strain to the point of failure), work to failure (the area under the curves), and the failure stress or strength (the maximum stress). The crucial feature is the enormous increase in irrecoverable longitudinal extensibility beyond the elastic limit once the MFA exceeds 20°.

From these observations, Reiterer *et al.* (1999) conclude that trees and branches require:

- A high elastic modulus to prevent the crushing of fibres under the dead-load of the tree itself (more important in outerwood as the tree becomes more massive), and to withstand strong periodic wind gusts.
- A large fracture strain so it can bend without breaking.
- A high work to failure or toughness, so absorbing a large amount of energy.
- A high fracture stress to resist high lateral loads.

It is not possible to maximize all these criteria at once (Table 6.2). Thus the tree is optimized (compromised) to achieve different goals at certain times during its growth, while concurrently laying down wood of different densities and MFAs in different parts of the tree.

Table 6.2. Influence of microfibril angle on mechanical properties essential for standing trees (Reiterer *et al.,* 1999; 2001). The data has been normalized, i.e. values have been adjusted by multiplying by cell wall density (1500 kg m^{-3}) and dividing by the actual density of the sample. Maximum property values (highlighted in grey) occur at different MFAs.

	Microfibril angle, θ			
	5°	20°	27°	50°
Tensile strength/fracture stress [MPa]	219	193	135	37
Tensile strain/extensibility [%]	0.5	1.0	3.5	11
Work to failure [MJ m^{-3}]	0.7	1.4	4.1	2.8
Stiffness/modulus of elasticity [GPa]	54	44	17	1.5

Strength is a key requirement of all materials. Where lumber is used in construction each piece carries the legacy of its role in the live tree. Here outerwood is sought with its high density and a small MFA (Table 6.2). However, the intrinsic strength of clearwood is generally compromised by the presence of defects – knots, spiral grain, splits *etc*. This is discussed further in Chapter 10.

Grading for structural markets is generally non-destructive and relies on correlations with wood stiffness. First the stiffness of the member is measured, and from that the strength is estimated.

Under tension a small microfibril angle results in high stiffness and strength, whereas a large MFA results in high extensibility and toughness (Lichtenegger *et al.*, 1999; Färber *et al.*, 2001). At low MFAs the failure appears brittle (a smooth, clean fracture surface), with the microfibrils kinking under high load before failing catastrophically. At large MFAs extensive shearing, either at the cellulose-hemicellulosic interphase or within the lignin-hemicellulosic phase, results in deformed and torn cell wall fragments spiralling out of the fracture zone (Figure 6.1): also, lateral separation at the middle lamella is likely.

It is known that wood is two to three times as strong in tension than it is in compression, so simple bending will result in compressive failure well before tensile failure. This low compressive strength relative to tensile strength is due to bucking of the hollow cells under load. Thus optimization should favour a larger MFA under compressive than tensile stress and favour an increased density as occurs in compression wood in contrast to opposite wood in a leaning softwood stem.

Thus, young, thin branches have largish microfibril angles even on the upper side (35-27° for opposite wood) as well as on the lower side (45° in compression wood). The young branch needs to be flexible. As the branch enlarges and extends – becoming more massive and exerting a progressively larger bending moment on the branchwood close to the stem – the MFA on the upper side falls to 0° only a few rings from the pith as the new function of the opposite wood is to stiffen efficiently the growing branch and prevent the branch sagging. The design seeks to avoid the bending stresses exceeding the (lower) critical compressive stress on the underside side (Färber *et al.*, 2001).

Trees need to absorb large amounts of energy. It is not enough to sway in the wind. If the periodicity of sway co-insides with that of the wind gusts, sway becomes increasingly exaggerated and risks catastrophic failure. A damping mechanism is needed. The actual mechanism probably involves shearing within the hemicellulose-lignin matrix. Damping is optimized where the microfibril angle is *c.* 30° (Table 6.2). This explains the MFA/stiffness profile in young trees (Figures 6.2, 6.4 and 6.6).

However in timber construction work to failure is of less concern except with pit props, poles and fencing. In construction the metal connectors – bolts, nails and metal plates – are designed to absorb any large deformations that arise in cyclones or earthquakes. The timber members are 'protected' by the ductility of the connectors so wood stiffness and strength are more critical than work to failure.

9. FIBRE LENGTH

Tracheid length in softwoods varies both within growth rings and throughout the stem. For *Pinus radiata* tracheid length and cell wall thickness increase gradually with ring number from the pith, with tracheid length in the first few rings being typically half that found a further 10 rings out. The transition between corewood and

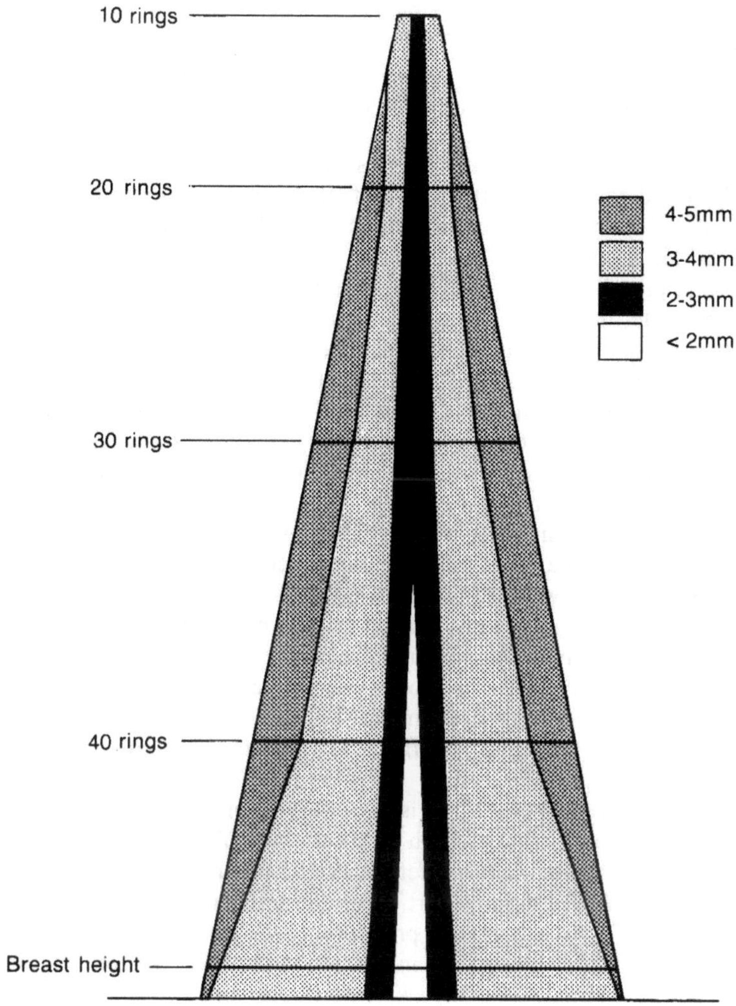

Typical tracheid dimensions

		Earlywood	Latewood
Corewood:	Length (mm)	2.0	2.5
	Diameter (μm)	30.0	20.0
	Wall thickness (μm)	3.0	4.0
Outerwood:	Length (mm)	3.5	4.0
	Diameter (μm)	40.0	30.0
	Wall thickness (μm)	4.0	5.0

Figure 6.14. Variation in tracheid length within radiata pine from the central North Island of New Zealand: typical values for corewood and outerwood are included (Cown, 1992).

outerwood is taken to occur at about ring 12-15. The longest tracheids are found at mid-tree height (Figure 6.14).

Tracheid length declines by 0.75 mm in going from the north to the south of New Zealand, being influenced by simple variables of latitude, altitude, temperature and rainfall.

These trends are valid for many species both hardwoods and softwoods. Bao *et al.* (2001) found for seven softwoods that tracheid length in corewood was 21-52% shorter than in outerwood, and for the three hardwoods studied the fibres were about 24% shorter in corewood.

Anticlinal or pseudotransverse division of a cell results in two new cambial cells that are much shorter than the original cell. Anticlinal division allows for rapid radial growth of the tree by providing additional radial files of cells, which replicate by subsequent periclinal division. The subsequent periclinal divisions of a succession of daughter cells and their initial rapid elongation mean that after about three years these 'progenitor' daughter cells at the cambium regain their normal length. Trees with rapid radial and diameter growth are subject to more frequent anticlinal division and this tends to reduce average fibre length.

Significantly the radial increase in tracheid length is quite gradual in the juvenile zone at the base of the tree. This pith-to-bark trajectory increases over the first 4-6 metres up the stem (Megraw, 1985 p. 42). Megraw associated this corewood-juvenile zone with the enlarged ring widths due to butt swell. The difference in tracheid length up the stem for a given ring position is of the order of 10-30% for a number of softwoods. As already noted, the microfibril angle parallels these trends and since cell elongation has ceased by the time microfibrils are deposited in the S_2 layer it is logical that the microfibril angle is influenced by tracheid length rather than *vice versa*.

A minimum length of about 2 mm is necessary to produce acceptable softwood kraft pulp, and improved tear strength is most noticeable on increasing tracheid length to about 3 mm (Zobel and Buijtenen, 1989). Where young thinnings and top logs are destined for kraft pulping short corewood tracheids will predominate and a case may be made for tree breeding programmes to consider increasing tracheid length. Similarly there may be benefits in seeking to improve the fibre length of long fibred hardwoods (> 1 mm), but there is little to be gained in improving the fibre length of very short fibred hardwoods. Where good printing grades are sought fibre length is less relevant and slender, short fibred hardwood pulps have proved admirable. It is the combination of short fibre length and narrow fibre diameter that leads to the production of sheets that are both smooth and even textured.

9.1. Integration with other desired characteristics

For over 40 years there have been intense efforts to develop improved eucalypt hybrids for 7 yr-old pulpwood crops. Elite clones of *E. urophylla x grandis* have both desired 'long' fibre characteristics, i.e. long for a hardwood, and less variation despite consisting solely of 'variable corewood':

Physical properties

- Basic density, increasing from about 300 at age 2 to 500 kg m^{-3} at age 5

Fibre characteristics

- Fibre length, ranging 0.95-1.20 mm
- Fibre diameter, ranging 14.5-20.5 μm
- Wall thickness, ranging 3.5-5.0 μm

Chemical analysis

- Lignin, 26-28%
- Holocellulose, 71-69%
- Extractives, around 3%

Production

- An unbleached yield of *c.* 54% at a kappa 15.

Short rotation hardwoods have a high lignin content in part because the wood in a 5-7 yr pulp rotation consists entirely of corewood.

There is interest in the trade off between volume production (> 40 m^3/ha/yr) and an increased basic density as a higher density means more 'fibre' in the digester and so more pulp production. This could be achieved by crossing another species with the traditional *E. urograndis* hybrid.

10. SPIRAL GRAIN (HARRIS, 1989)

Harris (1989) concludes 'that the propensity for spiral grain is controlled by heritable factors, yet its expression is dependent, at least in part, on environment', i.e. environmental extremes do not appear to be directly causative but may provide conditions that favour the expression of genetically determined factors. He illustrated his argument with examples of seed from trees of good form producing trees with spiral habit when grown in a different environment, a simple result of genotype *x* environment interaction. Altitude, growth rate, soil, temperature and wind have been implicated in one study or another. Unfortunately it is hard to move from specific instances to broad generalized conclusions, even for a single species. There are, or may be, within-tree trends over which one can superimpose broad, between-tree variations – but predicting behaviour across sites is at best tenuous.

> Spiral grain values [in *Pinus radiata*] showed significant regional variation and a suggestion of a decreasing north-south trend [in New Zealand]. All families showed the expected pith-to-bark trend of decreasing spiral grain. [However] family ranking of spiral grain across sites was relatively variable, suggesting that genotype *x* site interaction for this trait is large (Kumar *et al.*, 2001).

Typically in softwoods the wood immediately adjacent to the pith is straight grained with the tracheids aligned parallel to the stem axis, but over the first few growth layers (< 10) the tracheids become increasingly inclined to the stem axis, forming a left-hand spiral: by convention spiral grain is left-handed if the grain winds up the stem moving across the face from the lower right to upper left. Thereafter the grain angle decreases until the wood becomes straight grained, before inclining to a right-hand spiral whose angle gradually increases with age. This pattern of spiral grain is not invariable. For example, with the Indian chir pine, *Pinus roxburghii*, the left-hand spiral may continue to develop until the angle becomes almost horizontal in some trees. Whatever the pattern, the direction and spiral angle in the cambium are liable to change markedly during the life of the tree. The large spiral angle sometimes encountered in the outerwood of senescent trees is not a problem in plantations as there is insufficient time for it to develop. Instead, particular attention has to be paid to the severity and pattern of spiral grain near the pith.

Despite – or as a consequence of – the unpredictability between trees and across sites spiral grain is an important cause of degrade in sawn lumber of many species. For convenience, the pattern in radiata pine is examined. Here the broad trends within-tree (Figure 6.15) are:

- A rapid increase in spiral grain angle in the corewood on ascending the stem (from the juvenile zone in the butt log to the mature zone in the upper logs) that is completed over the first 6 m. Spiral grain increases from about 3 to 6°; and
- A maximum spiral grain angle occurs around ring 3, thereafter there is a steady decline on moving from corewood to outerwood.

Many product specifications require wood to be straight grained because spiral grain angles exceeding 5° result in noticeable twisting of unrestrained lumber on drying. With a mean angle in the upper logs of around 5°, half the lumber assessed in Figure 6.15 exceeds the critical threshold for twist instability and requires drying under heavy restraint and subsequently care in sorting and allocating to various end uses.

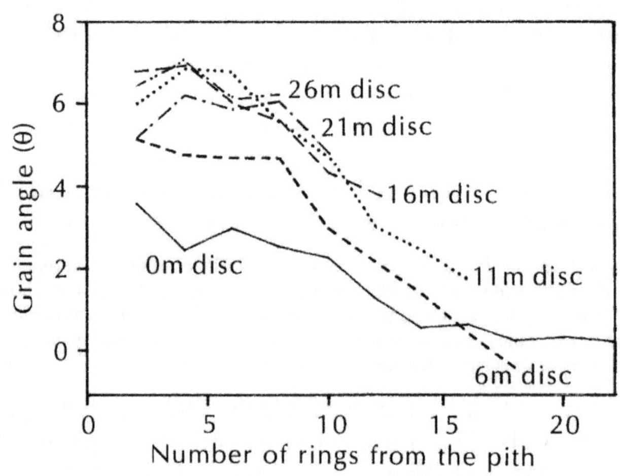

Figure 6.15. Mean grain angles within trees, for fifty 26 yr-old *Pinus radiata* from the central North Island of New Zealand (Cown *et al.*, 1991).

Fortunately there is very considerable between-tree variation (Nicholls and Dadswell, 1965) with corewood values ranging from 1 to 20° and the trait is strongly heritable, so there is the prospect of genetic selection to screen the worst material. Spiral grain is not such a problem in the South Island of New Zealand, supporting the belief that there are pronounced genotype *x* environment interactions.

It is hard to generalise about spiral grain in hardwoods. Many commercial timbers have a reputation for being straight grained (< 2°) although a small proportion of the stems may show spiral grain. Many tropical hardwoods, e.g. *Entandrophragma* sp. and *Eucalyptus* sp., and a few temperate genera display interlocked grain. Interlocked grain appears to involve a switching of grain direction at more or less regular intervals across the stem radius, although a change in direction *per se* is not always observed. Interlocked grain reflects a regular switching in grain angle, and where this is well developed the change in grain angle can be as great as 60-80°.

The effects of spiral grain on wood properties arise principally from the anisotropy of wood. Since the ratio of crushing strength perpendicular and parallel to the grain is less than the ratio for tensile strength, spiral grain has less influence on crushing strength (Chapter 10). Similarly, the greater longitudinal shrinkage associated with spiral grain merely arises from the component of shrinkage that is transverse to the fibre axis being directed parallel to the axis of the stem.

Where spiral grain is moderate to severe it effects utilization. In general the loss in strength is less significant than the loss of function. Warp of lumber is a particular problem where equilibrium moisture content values vary widely within a country. For example, in South Africa the equilibrium moisture content in the narrow coastal fringe is around 15% whereas in the majority of the country it is between 6 and 10%. Material that is adequately dried for use in the coastal strip will be inadequately dried and liable to twist if used inland. *Pinus patula* often has excessive spiral grain (de Villiers, 1973), although the mean value adjacent to the pith is only 3.5°, emphasizing that the range in values is more important than the mean.

This particular example from South Africa leads to the generalisation that for most plantation softwoods there are two strategies. The initial left-hand spiral grain, which is of the greatest concern, reaches its maximum angle in the first few annual rings and that segregation of the corewood allows it to be sawn, dried and marketed while bearing in mind its distortion-prone characteristics: either the corewood is sawn into thin boards and dried under restraint or it should be sawn into large baulks, relying on internal restraint to reduce distortion. Fast growth means that the development of large angles in the outerwood will not occur before the trees are large enough to fell. In contrasts the greater spiral grain in second-growth Douglas fir growing on poor sites is largely due to the greater age of the trees by the time that they reached a merchantable size compared to stands on better sites (Elliott, 1958).

Spiral grain in combination with heart shake lowers both volume conversion and grade recovery of lumber where either by itself would not be significant. For example a check in a 5 m log will rotate through 96° as a result of spiral grain of only 2°, degrading the output from a whole quadrant.

Finally both spiral and interlocked grain present difficulties in finishing. The grain runs out of the surface in opposite directions on either face of the board (spiral grain) or on the same face (interlocked grain) and the grain tends to tear out of the wood when machining against the fibre: the same problem arises with wild grain in the vicinity of large knots.

11. HEARTWOOD (HARRIS, 1991)

The initiation of heartwood occurs early at about age 10 with *Pinus elliottii* and *P. caribaea*, a little later with *P. radiata*, and later still with *P. taeda* and *P. oocarpa*, at about age 30. With *P. radiata* the percentage heartwood in the stem increases from zero at age 10 to 10% at age 20 and 20% at age 30, whereas with *Pseudotsuga menziesii* heartwood forms earlier averaging about 35% at age 30 and 60% at age 50 (Cown, 1992). The southern pines are fortunate in forming relatively little heartwood at ages less than 50 years, *c.* 15% or less. For those softwoods whose heartwood offers only modest decay-resistance, heartwood formation results in a lower green density (desirable), but also (undesirable) reduced penetration by preservatives and pulping liquors and troublesome resinous extractives.

Predictive equations for the percent heartwood in radiata pine according to stand age have limited value because tree-to-tree and site-to-site variations about the means are considerable. Dominant trees form significantly less heartwood than suppressed trees in the same stand. In Australia heartwood formation begins around age 14 and advances at a rate of about 0.4 rings per year in Victoria and 0.6 rings per year in South Australia compared to New Zealand where heartwood formation begins around ring 10 and expands at 0.33 to 0.5 rings per year on going from south to north. In pine, the heartwood zone (with its resinous aspirated pits) is surrounded by a narrow impermeable drywood zone ('intermediate wood' between sapwood and heartwood), which has progressively fewer live ray parenchyma cells and where the bordered pits of earlywood tracheids progressively aspirate in anticipation of becoming heartwood. This band of intermediate wood is often wider than the region that turns to heartwood in any year.

Until recently, for many plantation pines and spruces any interest in breeding has been to delay the onset of heartwood formation and to rely on preservative treatment to provide protection: or in the short term to take advantage of the reduced amount of heartwood in fast grown, short rotation operations. Yet even today, heartwood of older southern pine is used for flooring in the US south.

Heartwood is a desirable feature for a number of high value species having some natural durability, and attractive colour and/or odour – the more so today as preservative treatments are subject to ever more critical and rigorous scrutiny. A 20 year-old progeny trial with half-sib families for teak, *Tectona grandis,* revealed significant variations in heartwood content and basic density: basic density recorded a narrow sense heritability of 0.31 and 0.47 at the individual and family levels while for heartwood the figures were 0.64 and 0.81 respectively (Mandal and Chawan, 2003). Future breeding programmes can achieve improved durability and stability and aesthetics values, i.e. the dark colour associated with its extractives.

Cultural preferences and tradition are powerful influences. Hinoki, *Chamaecyparis obtusa*, is a highly desired plantation species in Japan. Its pinkish-yellow to pale reddish-pink colour makes for attractive wall sidings and is used for bath-tubs where the odour adds to the physical pleasures. Japanese red pine (*Pinus densiflora*) while having limited heartwood – but more so when compared to many plantation pines – forms a deeper coloured heartwood – and discolouration from blue stain does not spread into the heartwood. In the Pacific Northwest of the US, western redcedar (*Thuja plicata*) is used mostly for roof singles due to its natural durability, but colour and odour are also desired for furniture and interior finish.

Among hardwoods, Udai-kanba (*Betula maximowicziana*) can be separated into two groups according to the proportion of heartwood, with those trees having abundant heartwood being twice as valuable as those with little heartwood. Ebony and teak are ultimate, universal exemplars of heartwood.

12. GROWTH STRESS AND REACTION WOOD (YAMAMOTO ET AL., 2002)

Wind is a principal agent in determining tree form. Taper, i.e. the increased girth down the stem, counters the increased bending moment imposed by wind loading on the green crown, such that a plot of (stem diameter)^3 against height is roughly linear. In this way the maximum strain and stress, near the cambium, along the length of the stem remain approximately constant (Wilson and Archer, 1979) and there is no obvious weak point where breakage is more likely – the stem swells in the vicinity of branch whorls so reinforcing a potential point of weakness. Some feed-back mechanism is ensuring that there is more growth where the strain at the cambium is greatest, i.e. the swaying stem is stimulating the rate of cell division. Thus guying a tree to prevent it swaying reduces diameter growth below the point at which the stem is guyed.

The efficiency of the tapered stem is improved by natural pre-tensioning of the fibres near the cambium while the fibres near the pith are in compression, with the neutral plane occurring about one-half to two-thirds of the distance from the pith to cambium (Figure 6.16a). Even straight trees, growing without much exposure to wind and producing normal wood, are prestressed in this way. These growth stresses originate in the tendency of cells to change shape during development: normal wood and tension wood cells contract longitudinally while compression wood cells expand longitudinally (Boyd, 1972). In normal wood tensile strains develop near the cambium as the newly maturing cells attempt to contract longitudinally but are restrained from doing so by the rest of the wood within the cross-section. It is presumed (without much evidence) that the strain and corresponding tensile growth stress at the periphery of the stem remain constant as the tree's diameter increases. Matching compressive stresses develop in the vicinity of the pith. Each new layer of cells at the cambium contributes to the cumulative compressive stress acting on the centre of the section.

In practice growth stresses are not measured directly (Archer, 1987). It is easier to release the stress and measure the strain relief (ε), together with the elastic modulus or stiffness, E. The growth stress (σ) is calculated assuming simple elastic theory, $E = \sigma/\varepsilon$. Where growth stresses are severe the longitudinal tensile strain at

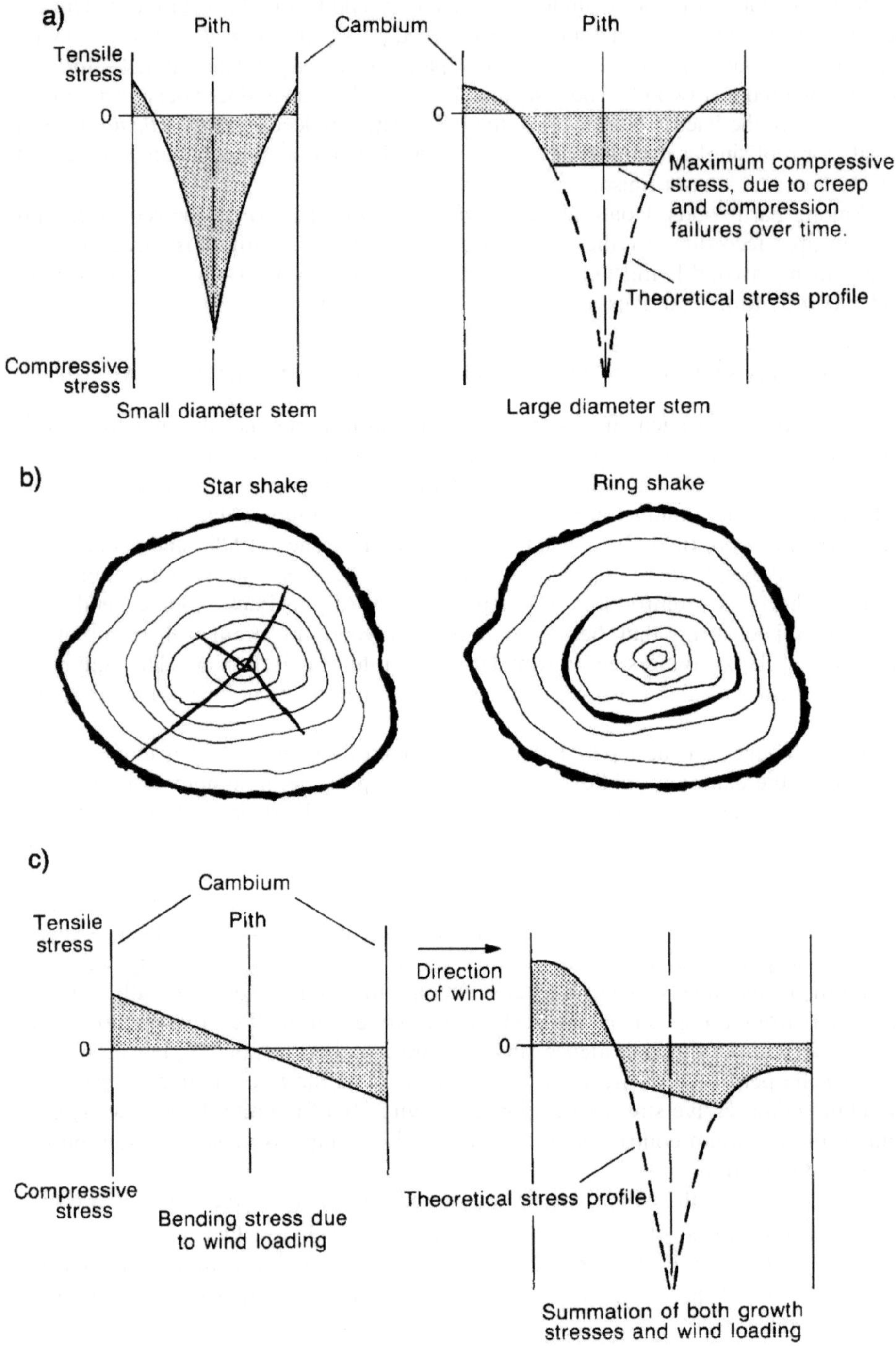
a)
Pith
Cambium
Pith
Tensile stress
0
0
Maximum compressive stress, due to creep and compression failures over time.
Theoretical stress profile
Compressive stress
Small diameter stem
Large diameter stem
b)
Star shake
Ring shake
c)
Cambium
Tensile stress
Pith
Direction of wind
0
0
Compressive stress
Bending stress due to wind loading
Theoretical stress profile
Summation of both growth stresses and wind loading

the periphery is of the order of 0.1%, but near the pith the theoretical accumulated longitudinal compressive strains are much greater. In reality growth stresses at the centre of the log are not as large as predicted: where the compressive growth stress exceeds the elastic limit the cells deform and behave like a viscoelastic body (Figure 10.17a) so that the imposed stress gradually abates and is redistributed (Figure 6.16a). Consequently the smaller observed stresses are only the residual stresses as creep over time has led to some permanent deformation (strain).

The pretensioning of outerwood allows the flexing stem to withstand much greater windloads. This is necessary because timber is much weaker in compression – where the wood buckles and fails at a lower stress – than it is in tension (Figure 10.3). The actual stress at any point within the stem is the combination of the local growth stress and the bending stress induced by wind. As a stem bends with the wind, the outerwood on one side of the stem will experience a tensile bending stress (superimposed on a tensile growth stress – no problem as wood is strong in tension) while the other side will experience a compressive bending stress (but here the pre-existing tensile growth stress mitigates and off-sets the wind-induced compressive bending stress).

Also, wind is the principal cause of leaning stems, although poor root development, soil creep, and factors such as phototropism can play a part. Once leaning, enormous forces are necessary to correct the lean. Not only must the stresses sustain the bending moment of the leaning tree, but also they must eventually overcome the rigidity of the stem itself, forcing it to bend so that it regains an upright position. These stresses are generated by the formation at the cambium of reaction wood in response to the lean. Even a large tree can be straightened given enough time for the stem to accumulate a large amount of reaction wood.

In softwoods reaction wood is found on the underside of the stem or branch. The tissue is called compression wood (Figure 1.17). In hardwoods reaction wood forms on the upper side and the tissue is called tension wood (Figure 1.34). Both types of reaction wood act to correct the lean of the stem. During formation of compression wood the softwood tracheids on the underside expand longitudinally so pushing the stem or branch up, whereas with tension wood the hardwood fibres on the upper side seek to contract longitudinally and pull the stem or branch up. In general the stem is enlarged on the underside in compression wood and on the topside in tension wood. The presence of reaction wood can be deduced if the stem is somewhat elliptical as a consequence of increased growth in the reaction wood zone. In leaning stems the wood on the opposite side of the stem to either compression or tension wood is called opposite wood while wood on the two vertical faces/sides are called side wood (and approximates to normal wood in a straight stem).

←

Figure 6.16. Growth stresses (Boyd, 1950). (a) In large logs compressive stress near the pith is less than predicted as the fibres deform through creep and compression failure. (b) Star and ring shake failures arise from transverse growth stresses and from the relief of longitudinal growth stresses as the tree is cross-cut. (c) Superimposing bending stresses from wind loading on growth stresses results in higher tensile stresses at the cambium on the windward side and lower compressive stresses on the downwind side. Likely brittleheart and compression failure is limited to the central 20% of the cross-section, within 70-100 mm of the pith.

Reaction wood is not confined to obviously leaning trees. If a stem eventually straightens the presence of reaction wood near the pith may not be suspected. A high incidence of reaction wood near the pith is understandable as a thin stem is misaligned easily and seeks to straighten itself. Instability – or a wandering leader as individual stems manoeuvre to compete for light – is evident in fast grown exotics and these generate reaction wood as the stem moves to counteract any lean. The lean can be over corrected, with reaction wood forming first in one side and then on the other side and so on. Indeed very vigorous growth in conifers can produce mild compression wood all round the stem, as has been noted after heavily thinning or fertilising of stands on good sites. Fast growth and wide spacing encourage the formation of reaction wood, but this can be countered by breeding for stem straightness and by removing the worst formed stems during a thinning operation.

If the stem of a young sapling is bent into a loop in the vertical plane, then at the top of the loop the uppermost part of the looped stem is stressed in tension and its underside is in compression whereas at the bottom of the loop the stresses are reversed. In such a looped stem softwoods form compression wood tissue on the underside at the top of the loop (where the bending stress is compressive) and on the underside at the bottom of the loop (where the bending stress is tensile), whereas hardwoods form tension wood on the upper part of the stem in both positions, i.e. compression wood can form in that part of the stem that is in tension as well as that part of the stem that is in compression, and similarly for tension wood.

Compression wood forms on the underside of all softwood branches and tension wood on the upper side of hardwood branches. Thus reaction wood is responding to and resisting the additional incremental mass of branch wood that is weighing it down and which would otherwise result in the branch drooping from its preferred inclination. Most curious, if a branch is forced upwards compression wood forms on the upper side with softwoods, while tension wood forms on the lower side with hardwoods. Reaction wood is seeking to maintain some pre-determined, natural branch angle.

In summary and confusingly, under certain circumstances compression wood (in softwoods) and tension wood (in hardwoods) can each be found on the underside or upper surface of a branch/stem, and they can also be found where the local bending stress is tensile or compressive.

Reaction wood also forms in the stem immediate below branches. It is a continuation of the reaction wood tissue in the branch downwards into the stem. In softwoods the volume of this associated compression wood can be from one to several times the knot volume (Von Wedel *et al.*, 1968).

Reaction wood varies from mild to severe, and the adverse effects of reaction wood really relate to the more severe manifestations. These are summarised in Table 6.3. Foresters can minimise its presence and severity by selecting straight-stemmed trees with small, wide-angled limbs and by silviculture.

Growth stresses and reaction wood, while normal in the sense that they occur with some frequency, are generally considered atypical and their presence can create problems in utilization. The very large growth stresses in severe compression wood or tension wood, contrasting with reduced growth stresses in opposite wood, are extreme examples of the asymmetric distribution of growth stress around the periphery. Such characteristics should not really be regarded as abnormal but rather they are extreme variants of normal wood.

Table 6.3. Characteristics of severe reaction wood.

	Softwoods: compression wood	Hardwoods: tension wood
Curious features	Found in all genera	Completely absent in some genera
	Significantly higher galactan content than in normal wood	
	15-40% greater than normal wood	10-30% higher
Physical features:	Darker coloured with extra, atypical latewood	Hard to distinguish from normal wood, sometimes appearing 'wet' or silvery. Woolly when sawn
Fibre length:	10-25% shorter	Variable relative to normal fibres
Longitudinal shrinkage:	3-5% and even greater, as against 0.1-0.3% in normal wood	0.5-1.5%
Drying:	Liable to warp badly	Can warp and is liable to collapse
Strength:	Strength is comparable to that of normal wood of that species despite the 20% increase in density. Brittle in tension	Generally superior strength and superior toughness
Stiffness:	Variable: an increase in basic density offsets an increase in MFA	Similar to or somewhat better than normal wood
Chemical pulp:	Cellulose content is low (40 *vs* 50%). Loss of yield is more significant than the reduction in pulp quality. The thicker walled tracheids are harder to bleach, having high residual lignin	Cellulose content is high (65 *vs* 55%). Higher yield if not over cooked. After beating pulp quality is acceptable
Mechanical pulp:	Yields fibre fragments and paper has a low burst and tear strength. High lignin content (40 *vs* 30%) means more bleaching is needed	Produces good mechanical pulp. Low lignin content (15 *vs* 20%) means that it yields a brighter pulp

The occurrence of growth stresses, and the parallel deduction that particular cells in reaction wood tissue must generate stresses, is not in doubt. Analytical models have been developed that offer a plausible description of these phenomena. However, the fundamental biophysics and biochemistry remain controversial and unresolved.

> Two theories have been put forward to explain the origin of these growth stresses. The first theory explains their origin in terms of longitudinal contraction of the cells as they swell laterally due to the deposition of lignin in the interstices of the wall (Boyd, 1985b), whereas the second theory states that the longitudinal tensile stress in the cells is generated naturally from the contraction of the cellulosic microfibrils due probably to crystallization of the cellulose and that it is subsequently reduced or modified by the deposition of lignin (Bamber, 1979,1981), (Dinwoodie, 2000, pg 30).

This quote considers processes (lignification and crystallization) rather than causality. The lignification hypothesis is most relevant when discussing compression wood (where the MFA is large), while 'crystallization' – a term that needs careful defining – is most relevant when discussing tension wood (where the MFA is small). Particular differences in both process and causality between tension wood (upper side of stem) and growth stresses (around the entire circumference of hardwood and softwoods) have yet to be made clear. Underlying all three processes lie the speculative roles of hormones (auxin), growth regulators (gibberellins), stress (and microtubule alignment) *etc.* (Barnett and Jeronimidis, 2003).

12.1. Compression wood in softwood (Wilson, 1981).

Burdon (1975) in a study of 12 yr-old *Pinus radiata* growing on four sites estimated that 30-45% of the stems contained mild to severe compression wood; although Timell (1986) suggested that 15% would be a representative figure for virgin spruce forests and for plantations of southern pines and *Pinus radiata*. Lindström *et al.* (2004) recorded compression wood occupying from 6-30% of the stem cross-section in 3 yr-old clones of radiata. Whichever number is taken, there is a lot of compression wood in young softwoods.

In the elliptical cross-section the growth rings are narrow on the upper face and much broader underneath so that the pith is located nearer the upper face of the leaning stem. In the compression wood zone the wood is denser, with more thick-walled cells (all the better to carry compressive loads). Clear differences between normal tracheids and cells with severe compression wood are visible in the scanning electron microscope (Figures 1.18 and 1.19). The cells are more rounded with intercellular spaces between tracheids. Usually the S_3 layer is absent. The S_2 layer has an outer highly lignified zone and an inner less lignified zone in which there are deep helical splits whose inclination correspond to that of the microfibrils (30-50°). The large microfibril angle in the S_2 layer means that the tracheids shrink more in the longitudinal direction on drying compared to normal wood. Further the tracheids are somewhat shorter and their tips tend to be distorted and bifurcated.

The key to understanding the role of compression wood lies in the microfibril angle of the S_2 layer that, because of its dominant size, over-rides the behaviour of the other cell layers. The effect of microfibril angle on shrinkage is examined in Chapter 4. The essence of that argument is that longitudinal shrinkage in wood on drying is not of great consequence when the microfibril angle is less than 30° but that longitudinal shrinkage increases dramatically once the microfibril angle exceeds about 40°, with shrinkage of 3-5% being possible.

Compression wood has its origin in the cambial layer. The dimensional changes in the cell wall that occur during cell lignification are the converse of those that develop during drying wood (Boyd, 1972, 1985). In the latter case removal of water molecules from the S_2 layer results in volumetric shrinkage, but significant longitudinal shrinkage occurs only when the microfibril angle exceeds about 40°. In contrast, lignin deposition between the cellulose microfibrils in the secondary wall results in swelling transverse to the microfibrils: the greater the quantity of lignin deposited in the cell wall the greater the swelling. Where the microfibril angle exceeds about 40° – and this is only likely in corewood wood and in compression wood – lignification will result in an axial expansion and an increase the cell length, inducing compressive growth stresses at the periphery. In this way the formation of compression wood cells on the underside of a leaning stem gradually bends the stem back upright. This process is the opposite of what happens when wood dries (Figure 4.5). Other factors are involved: the microfibril angle in compression wood can be quite modest in mature-outerwood, and the large MFA of compression wood in juvenile-corewood may not be significantly greater than that of normal juvenile-corewood (the MFA cannot get much worse).

The lignin and hemicellulosic contents in compression wood are higher than average. In chemical pulps yield is reduced (less cellulose), tear strength is reduced (shorter tracheids) and bleaching is more difficult (more lignin which is more highly condensed). Not only is the lignin content of compression wood high, but it differs chemically from that in normal wood having a noticeable proportion of *p*-hydroxyphenylpropane units (which have no methoxyl groups on the C_3 and C_5 positions), and having more carbon-carbon linkages and fewer β-alkyl-aryl ether linkages (Figure 2.9b): both of these make the lignin more resistant to chemical attack. Further, while the increased level of galactans in compression wood tissue has long being noted (Panshin and de Zeeuw, 1980 Table 8.1), only recently has any practical significance been attached to these observations. Floyd (2005) has observed (Figure 6.17) that the worst effects of compression wood occur only where the tissue has both a large microfibril angle (MFA) and a high content of galactans (neither by itself is sufficient). As discussed earlier (Chapter 2), unlike the other hemicelluloses the C_1-O-C_4 interunit link in galactans is axial rather than equatorial so encouraging the molecule to coil, and thus exert a transverse shrinkage or swelling force between microfibrils that is disproportionate to its physical volume. In severe compression wood the presence of abundant galactans results in enhanced volumetric swelling while the large MFA channels this swelling in the axial direction.

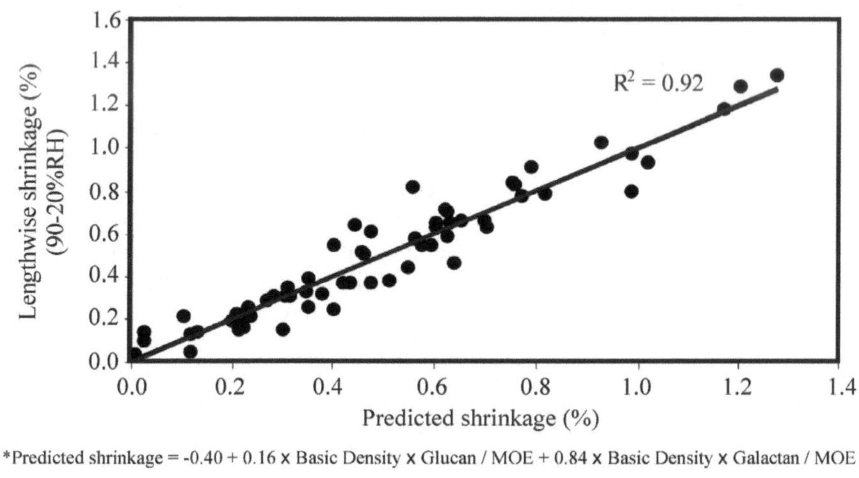

Figure 6.17. Large axial shrinkage depends on a high microfibril angle (equivalent to a low MOE that is more easily measured) and an atypically high galactan content (Floyd, 2005).

The principal problem with severe compression wood in sawn timber is not so much its excessive longitudinal shrinkage (Table 6.3): if all parts of a board shrank longitudinally by the same amount there would be no distortion. The problems are local variability and, where present, the gradient of severity. Where a board has compression wood on all or part of one edge or face and not on the opposite edge or face it will shrink unevenly and distort badly on drying and the grade and value of the board will be reduced significantly.

12.2. Tension wood in hardwoods

A review by Barnett and Jeronimidis (2003) notes that tension wood fibres are absent from branches of many hardwoods, while occasional pockets of tension wood fibres have been found amongst normal tissue in vertical stems (in which growth stresses might reasonably be expected). Then there are two important statements (Robards, 1969) about tension wood that deserve attention: 'tension wood is completely absent from some genera', and both compression wood and tension wood have 'significantly higher galactan content than normal [wood]'. Also Sigglekow (1974) concluded that true gelatinous fibres are comparatively rare in New Zealand hardwoods. Where then does the stem-righting mechanism arise? The most primitive hardwoods (in the evolutionary sense) are vessel-less and produce compression wood. Others produce a kind of tension wood that lacks both a S_3 layer and a G-layer, but the innermost part of the secondary wall has a very low microfibril angle (Yoshizawa *et al.*, 2000). Then there are those that produce 'classic' tension wood.

First, some general observations. The longitudinal shrinkage of severe tension wood (0.5-1.5%), while not as great as that of compression wood, is nevertheless much higher than that of normal wood. As with compression wood excessive and uneven shrinkage will result in warp. Severe tension wood is less easy to identify than compression wood. In dry dressed temperate hardwoods it can have a silvery sheen; in tropical woods it appears as darker streaks; while in green sawn lumber the fibres get pulled out, resulting in a woolly surface. Unlike compression wood, individual tension wood fibres tend to be less heavily lignified than normal, principally because these cells are characterised by the presence of a gelatinous (G) layer that usually replaces the S_3. The G-layer is unlignified and is mostly hydrated (loosely bound) crystalline cellulose whose microfibrils lie at a low angle (5°) to the axis of the fibre. At times the G-layer largely fills the lumen. While tension wood fibres can be found over a sizeable area of tissue, more usually they tend to be scattered amongst other fibres. Tension wood fibres are frequently absent in latewood. Boyd (1985) has emphasised that even in localised areas of severe tension wood only a proportion of the fibres (about 30%) have the non-lignified G-layer that distinguishes them as tension wood fibres whilst a larger proportion of the fibres have thick fully lignified cell walls.

The cellulose tension hypothesis presumes that the microfibrils are deposited in a partially crystalline or paracrystalline state and that during cell maturation they experience further crystallization and in so doing they contract axially. A conceptual difficulty with this hypothesis is that there is a strong presumption that cellulose microfibrils are crystalline when 'spun out' from the plasma membrane. The analogy is with prestressed concrete where steel rods are embedded in concrete. Only after the concrete has set can the tightening of threaded nuts – against washers and the hardened concrete – tension the steel rods. This stretches the steel rods and in reaction compresses the concrete matrix. In the case of tension wood the microfibrils are equivalent to the steel rods. To be effective, the microfibrils must be sufficiently anchored to other material in the cell wall and the cell itself sufficiently bonded to the adjacent xylem cells: only then can any axial contraction of the microfibrils result in their going into tension, in reaction to the restraint offered by the matrix material and by other cells.

Discussion of the cellulose tension hypothesis of tension wood began with the suggestion that the crystalline microfibril can incorporate point defects that are annealed (removed or redistributed) over a significant period of time (hours, days, weeks?). Stöckmann (1972) recognised three types of point defects – wrongly matching hydrogen bonds, inadvertent inclusions (a water molecule) and locally distorted chains (tiny misalignments). Stöckmann envisaged that these defects might be swept out of 10-30 nm lengths of the microfibril and dumped or aggregated in short fully disordered zones (< 0.5 nm long):

Unannealed cellulose → crystalline segments + very short disordered segments

Stöckmann suggested that this slow sweeping out of point defects could be a powerful force even if the disordered segments are infrequent and short.

This argument can be restated: long lengths of annealed fully crystalline material alternate with very short fully disordered segments. The latter account for perhaps 1.5% of the microfibril length in ramie (Nishiyama *et al.*, 2003) in which region there is considerable local misalignment of residues, i.e. 200-600 perfectly ordered residues (a 10-30 nm long crystallite) and then a few totally disordered residues *etc.* Newman (*pers. comm.*) suggests that this local misalignment might represent a local contraction of 33%, in which case annealing would result in a contraction in ramie of 0.5% (33% within 1.5%). The eventual outcome is not too different from the very early 'micelle' model of cellulose (Preston, 1974 Chapter 7). To generate tension wood or growth stresses this contraction must be resisted both by the matrix material of the cell wall and by adjacent xylem cells.

However, Marchessault and Sundararajan (1983) found no evidence of periodic disordered regions using dark field electron microscopy with *Valonia ventricosa*; so they can be no larger than the limit of resolution (< 0.5 nm). Further, this hypothesis has contradictions. There is no logical reason why defects should be swept some distance along the microfibril rather than escaping laterally to the surface.

We need to revisit current thinking about the cellulose microfibril (Chapter 2). The cellulose in the interior of the microfibril (where wood cellulose is a discrete mixture of I_{α} and 1_{β} allomorphs) ought to have a unit cell repeat distance along the microfibril of 1.043 nm. These are surrounded by the surface chains – also of pure cellulose – that do not have the same degree of hydrogen bonding found in the interior. The surface chains differ principally in that the $-CH_2OH$ group of the C_6 carbon is exposed and adjacent units are twisted about 30° (Figure 2.4b). As a consequence the unit cell repeat distance for the surface cellulose chains is slightly less, 1.029 nm. Spectroscopic tools clearly distinguish the slight differences between the cellulose chains on the surface and the cellulose I_α and 1_β allomorphs of the interior (Newman, 1997). Overall, for microfibrils in wood with roughly equal proportions of interior and surface chains the compromise unit cell repeat distance is between 1.034 and 1.038 nm (Figure 6.18). Thus the surface chains are in tension and the interior chains in compression (Davidson *et al.*, 2004).

Consider the possibility that the synthesizing complex spins out a thin metastable fully crystalline microfibril with a unit cell repeat distance of 1.043 nm for all cellulose chains (both surface and the interior). Subsequently, over a period of time (hours, days, weeks?) the surface chains reorganise to produce the realigned hydrogen bonding network recognised by NMR which seeks a reduced unit cell repeat length of 1.029 nm. A compromise between interior and surface chains would result in a small contraction and a repeat length of 1.034-1.038 nm for both:

Metastable crystalline microfibril → crystalline interior + annealed/weathered surface
… with a spacing of 1.043 nm → 1.043 nm + 1.029 nm

The ‘annealed’ surface layer may be stabilized by lateral association with those hemicelluloses that bond strongly to the microfibrils. Also in the G-layer, which is mostly cellulose, the presence of any (uncertain) hemicelluloses and the layer’s strongly hydrated nature (gelatinous implies a gel) must prevent strong lateral bonding of the individual microfibrils to form larger macrofibrils.

In summary, the axial contraction and within-microfibril tensioning are not consequences of paracrystalline microfibrils becoming fully crystalline as originally

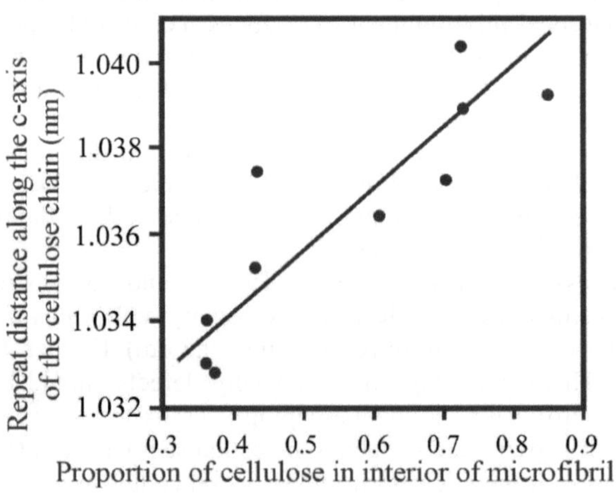

Figure 6.18. Davidson *et al.* (2004) collected samples of cellulose I from a wide diversity of sources. They observe that the larger the cross-section of the microfibril the greater the repeat distance along the cellulose chain (c-direction).

proposed. The suggestion is that axial contraction occurs due to the slow annealing that creates the distinct and reoriented surface layer. The slow annealing means that cell wall lignification and maturation can be completed ahead of delayed surface realignments and eventual contraction of the microfibrils, and so the matrix and the adjacent xylem are cohesively connected to resist this contraction – hence the development of growth stresses. Further, this in part explains why microfibrils in wood cells are so narrow, only 3 nm. A high ratio of surface to interior chains results in a strong axial contraction and high stress in tension wood, and also accounts for the more general phenomenon of growth stress. There is no other logical reason for wood cellulose forming a 'surface-modified nanofibril'. Other cellulose producing organisms form larger microfibrils.

While it is logical to presume that the presence of the G-layer in some hardwood fibres is involved in the straightening of leaning stems, not all hardwoods form tension wood. Further, a similar phenomenon must be involved in generating peripheral growth stresses around the entire circumference so it is equally logical to implicate the microfibrils of the S_2 layer when and where the MFA is also small. Growth stress provides the unifying link between normal and reaction wood.

Yamamoto (1998) compared measurements of growth stress in *Cryptomeria japonica* at low MFAs with theoretical values generated in his model and in this case found excellent agreement by presuming an axial strain of 0.15% (Figure 6.19). This is equivalent to tensioning – to prevent the microfibrils from contracting axially. A strain of 0.15% may seem insignificant but the cellulose microfibril is exceptionally stiff (134 GPa) so large forces – stresses – would be involved.

12.3. Growth stresses (Kubler, 1987)

Growth stresses occur in softwoods, although they are more severe in hardwoods. They are particularly marked in a few genera, for instance *Eucalyptus*, *Fagus* and *Shorea* sp. Growth stresses are subject to genetic control and improvement through breeding can be made with species of economic importance. Within a stand there are great variations; the highest mean value for the longitudinal tensile stress in one tree can be three times that in another tree. Growth stresses vary around the periphery and the range is far greater in leaning stems – with the greatest difference between the reaction wood and opposite wood.

The discussion of tension wood introduced the unified hypothesis of Yamamoto (1998) that incorporates lignin swelling (effective at large MFAs) and cellulose tensioning (effective at small MFAs). This model was developed using data from *Cryptomeria japonica* that – obviously – does not form tension wood but does generate tensile growth stresses around the periphery of the stem. Here the growth stresses generated at low MFAs must be attributed to the prestressing of the microfibrils in the S_2 layer as softwoods do not have a gelatinous layer. With judicious selection of realistic constants Yamamoto achieves an excellent match between observations and theory (Figure 6.19a).

Many trees have low microfibril angles yet large growth stresses are not universal. This implies that it is subject to regulation (as is reaction wood), perhaps by controlling the rate of 'annealing'. With most trees the rate at which the microfibril surface is restructured must be rapid so permitting axial contraction ahead of lignification, whereas for trees with high growth stresses the process must be delayed and completed only after lignification.

In very young softwoods (< 4-yrs) there is a tendency for the stems to develop longitudinal compressive growth stresses in the outer layers and corresponding tensile stresses adjacent to the pith (Boyd, 1950), but as the trees get older these stresses reverse. This is to be expected since the MFA is very large in corewood (Figure 6.19a). For young fast growing hardwoods the initial MFA is generally smaller (< 30°) and the growth stresses in the cambial tissue will be zero or tensile and the corewood will experience compressive stress from the beginning.

One might speculate that the large microfibril angle in the corewood of softwoods is an intrinsic feature of corewood – quite independent of issues concerning compression wood. The high MFA is there to generate peripheral compressive stresses to stabilize the tree and to be there as a precursor for compression wood formation. Compression wood might then be associated with the production of galactan that rapidly reorients the stem by amplifying – or building on – the pre-existing high MFA. One might also speculate that the smaller initial MFA in young hardwoods is to generate peripheral tensile growth stresses to stabilize the tree and to be there in anticipation of the need to form tension wood, which when needed results in formation of the gelatinous layer in some genera. In both cases the MFA has the latent potential to reorient the stem.

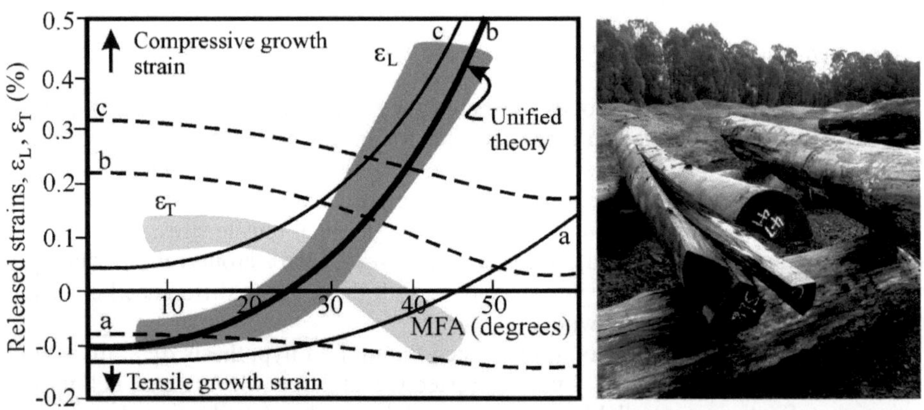

Figure 6.19. (a) Relationship between MFA in the S_2 layer and the released growth strains, comparing experimental data for *Cryptomeria japonica* with outcomes from the unified model (Yamamoto, 1998). ε_L and ε_T are the axial and tangential growth strains with the shaded zones representing the scatter of experimental data; the solid and dashed lines 'a' are the predicted strains assuming the microfibrils contract by 0.15%, while lines 'c' are the predicted strains assuming heavy liginification – as in compression wood. Lines 'b' are the predicted outcome where microfibril contraction and heavy lignification occur together. (b) Atypically severe end-splitting of *Eucalyptus nitens*.

Stem correcting tension wood in some hardwoods, with its very low MFA, reinforced by a general low MFA in very young trees contrasts with compression wood in softwoods and some primitive hardwoods, with its high MFA, reinforced by a generally high MFA. This is a double negative for corewood quality of softwoods and somewhat positive for hardwoods. This implies that some hardwoods are intrinsically better candidates for very short rotation plantation species (Figure 5.11). For live trees both strategies are equally valid, it is just that once the trees have been felled and the lumber dried the practical outcomes are very different for the end user.

One could argue that in young trees the microfibril angle of first choice is centred around 30°. This provides maximum resilience and work to failure (Table 6.2) whereby the tree absorbs and rapidly dissipates the energy from the wind. The departure to a new equilibrium, to a smaller MFA for some hardwoods or to a larger MFA in softwoods and some hardwoods, depends on the environmental pressures – whether the tree is at the forest edge or growing under the canopy, whether dominant or sub-dominant, its relative stem diameter *etc.* – and on genetic factors.

A further level of complexity is provided by the basic density of the wood that also affects the capacity of woody cells to absorb energy. The stem has to have woody mass to absorb this energy, disadvantaging balsa. On reflection, the need for a tree to absorb energy provides a tentative explanation as to why Jozsa and Middleton (1994) observed for Canadian softwoods that the density in the first few growth rings declines before recovering to form high density outerwood (Figure 6.20): see also Figures 5.4 and 5.12. For low density woods damping may arise from some hydraulic pumping that forces sapflow under stress between adjacent vessels.

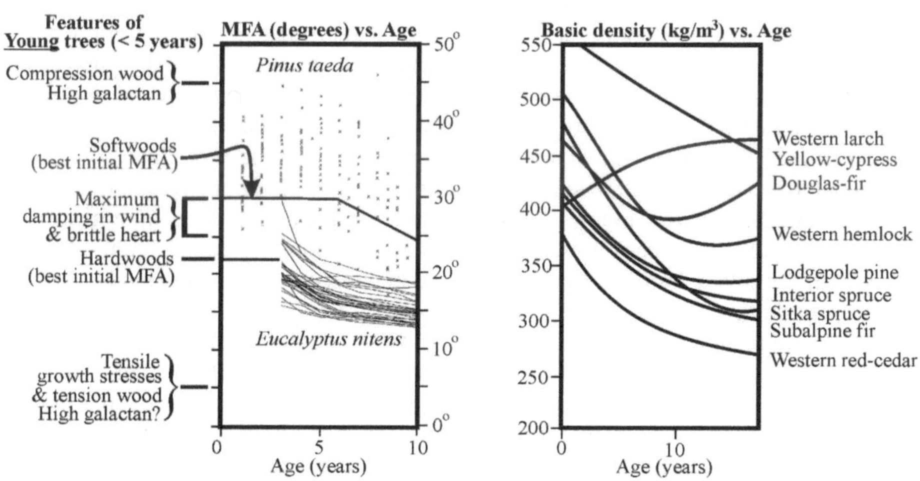

Figure 6.20. Hardwoods are favoured for sawlog production in very short rotation plantations (< 10-15 yrs). The high density in the innermost rings may increase the ability of a tree to absorb energy: one might speculate that the first 10-15 yrs for second-growth Canadian softwoods might be compressed into only 1-3 yrs for the fastest growing plantation species. See Yang *et al.* (1994); Evans *et al.* (2000); Jozsa and Middleton (1994).

These three ideas – the universality of growth stresses; a modest MFA in some very young hardwoods and a high MFA in very young softwoods; and the need to absorb the energy of the wind – are represented in Figure 6.20.

In older trees, the assumption must be that these influences become less critical, resulting in a steady decline in MFA with increasing ring number. The contribution of density to a tree's survival strategy remains uncertain – with air-dry densities ranging from 200 to 1200 kg m^{-3} (Figure 5.11).

For species having large growth stresses, the fibres near the pith will experience compression failure if the compressive stress becomes too great. When the tree is felled compression failure in the corewood may show up as brittleheart (Figure 6.21), which in sawn lumber breaks at unexpectedly low loads. While brittleheart is more prevalent in tropical and semi-tropical hardwoods it is found occasionally in softwoods. For example Plumtre (1984) notes that it may occur in *Pinus caribaea*, where its formation can be accentuated by the presence of very low density corewood both in absolute terms and relative to the density of the outerwood. Brittleheart occurs as a result of the superposition of symmetric growth stresses and asymmetric bending stresses due to wind. Fibres near the pith may be weakened already by the imposition of intense compressive growth stresses that gradually abate as the corewood experiences permanent strain. When subject to wind loading fibres in this region would be susceptible to failure under the combined stress system (Figure 6.16c). Towards the cambium the intensive compressive stress due to wind loading is mitigated by the tensile growth stresses and consequently the fibres in this region are more able to resist failure under the same loading conditions.

The discussion has been simplified as growth stresses occur in the longitudinal, radial and tangential directions: a material that is deformed in one direction will experience counter-balancing strains in the transverse plane. Although transverse growth stresses are about a tenth of the longitudinal value, because wood is much weaker and less stiff in these directions relatively small stresses can cause internal splitting (Figure 6.16b). A typical tangential strain at the cambium of about 0.15% is approximately twice the mean longitudinal strain at the cambium (Kubler, 1987). Tangential stresses are the opposite of longitudinal stresses: near the cambium wood is under tangential compression, which induces a corresponding circumferential ring tension near the pith. From pith to cambium, radial stresses are weakly tensile.

On cross-cutting the stem, longitudinal stresses relax in the vicinity of the cut face. There the rebalancing of strain energy imposes considerable transverse stresses – further increasing the transverse tensile stresses near the pith – that may be sufficient to cause end-splitting of the log during or after felling. In extreme cases growth stresses are so severe that whole stems can split along their length (Figure 6.19b). Hence when felling, stems should be laid down and not dropped.

Small stems are more likely to warp on sawing: assuming the peripheral longitudinal growth stress varies little with the size of the tree, then with small trees there is a much steeper stress gradient from cambium to pith that results in greater distortion where young trees are milled. However, small stems are less likely to develop end-splits or heart checks as the total stored strain energy within the log is proportional to the log volume. Further, small logs are less likely to have brittleheart: perhaps because in young trees there has been less time for compressive creep, and so the wood is able to store more recoverable strain energy. Large growth stress gradients are more often present in young hardwood trees, especially some eucalypts, and for uses other than for fibre production some see this as a strong reason for prolonging the rotation age until the diameter at breast height is 0.6 metres or more. Rather than accept a policy of failure – long rotations vitiate production economics – it is more logical to select intensively for clones with little growth stress. Chauhan (2004) measured growth strains in *Eucalyptus nitens* ranging from 430 to 1645 μm (CV 37%) so there is enormous natural diversity within an unimproved population from which to make selections. Favoured fast grown plantation hardwoods are generally medium density species that offer the broadest utility. Of these, eucalypts have been the most successful.

Growth stresses are less severe if the trees are subject to less competition. Heavy thinning reduces growth stresses, provided the stand is not destabilized and reducing competition does not encourage the crown of the tree to reorientate (Kubler, 1987). The most intractable logs are liable to come from unthinned stands, while the least stressed are open grown trees on farmland. Some abatement of stress in sawlogs is possible with long-term wet storage (3-6 months). Alternatively logs can be heated in vats ($T \leq 100^{\circ}C$) where viscoelastic strain energy is released: although this increases the amount of immediately available elastic energy so end checks can enlarge, they will not propagate into the log so rapidly over time as the slow release of viscoelastic energy has been avoided. Milling full length logs is preferable to short lengths, and if stems are to be cut to merchantable lengths this should be delayed until immediately before milling. Drying presents further difficulties, as collapse, distortion and splitting are likely. Some of the practical aspects in milling and drying such material are discussed in Chapters 7 and 8.

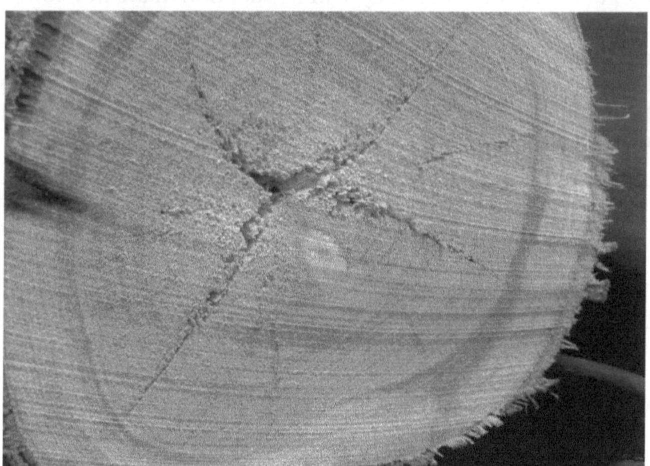

Figure 6.21. Brittleheart in which the wood near the centre of the stem fractures on cross-cutting and the end-grain has a ragged appearance despite cutting with a sharp saw. Rebalancing of growth stresses has induced star checks.

13. ENDGAME

Empirical studies of wood quality carry the risk of unintended consequences, so the aim should not so much be to 'eliminate' any problem as to avoid the worst outcomes: uncertain side effects cannot be quantified with any confidence; at least not until there is greater understanding of the links with cell biology.

> Biochemistry, biophysics and cell biology are difficult roads to walk, but there can be no doubt that they, in conjunction with tree genetics and wood science capabilities, hold answers to understanding how trees make wood, and thus to manage forests for wood and all of their other benefits (Savage, 2003).

Wood technologists do not have the ability and few cell biologists have shown an inclination to draw the connections between these disciplines. Wood technologists have largely played with easily measurable properties and characteristics, while cell biologists have sought to understand how and why the cell wall comes together in the way it does. By default the task of integration has fallen to bioengineers who are more interested in building models that describe structural performance. These models build towards but do not attend to fundamental causes.

As a generalization little has been achieved in wood quality for sawlog production over the last 50 years, other than improved stem growth and form producing over-obese trees rather than toughened grid-iron champions. Figure 6.20 is a challenge to those with long established and well-managed softwood plantations. It is hard to see how such unimproved plantations grown on 25-45 year rotations can compete in commodity markets with the new wave of plantings of intrinsically stiffer, more stable, and more decorative short rotation hardwoods (Table 6.3). There are political stones to hide under, but markets provide a persuasive force compelling industries to reorganise in ways that they will neither anticipate nor comprehend.

Table 6.3. A broad indication of grade recovery for *Pinus radiata* as compared with those from traditional forest practices in Canada and Sweden (van Wyk, 1990): unimproved pine has generated a disproportionate amount of low grade wood, with upgrading that material being the immediate challenge. Anticipated prospects for hardwoods have been included.

	Canada	Sweden	Past imperfect pine[1]	Future perfect hardwoods
Finishing (clears, dressings, cuttings)	35%	30%	10%	40%
Construction grades	50%	40%	40%	40%
Industrial/utility	15%	30%	50%	20%

[1]With intensive silviculture the proportion of finishing grades could rise to 15-20%

Studies on wood quality, with a heavy empirical emphasis on a few topics of interest – checking, growth stresses, resinification (heartwood and resin pockets), shrinkage (warp) and stiffness – reflect this tragedy. In this chapter, the over-long commentary on growth stresses provides a tentative example of the potential for better understanding that ought to be achievable where there is greater integration of disciplines. The case is incomplete and there remain logical inconsistencies.

Many challenges. Many opportunities.

CHAPTER 7

SAWMILLING

PHILIP BLACKWELL[1] AND JOHN WALKER

[1]*School of Forest and Ecosystem Science, University of Melbourne, Australia.*

1. INTRODUCTION

Four factors most influence the design of a sawmill:

- The wood resource in terms of species, its quality, log supply and predicted log size. The species, whether it is old-growth, second-growth or plantation (whether managed or not, and if managed the silvicultural treatments that have been applied), and its age and condition. These are all factors that determine the grade of material likely to be cut, because they affect the log quality (density, growth rate, knottiness, stiffness) and log form (straightness, taper). The certainty of supply over time – and any expected changes in log volume and log size with time – determine the economic size of the mill and influence the most profitable level of investment. The predicted mean diameter of the logs and the lineal throughput of the mill determine the amount of timber that can be cut from a given volume of logs and the likely timber sizes.
- The markets. The complexity of a mill and its design depend on the market that is targeted. A mill can be sawing for pallets, for the framing (or stud) market, for high value board grades and furniture components, for export so requiring non-standard sizes, or for a general local market. Product diversification reduces market risk, but tends to require greater capitalisation: the mill needs more flexibility in processing, especially in the transfer and buffering of flows between machines. Small specialist mills can be as profitable as larger more diversified entities since economies of scale are less pressing than in most other industries.
- Mill location. The mill is best located near the geographic centre of the resource to minimize log haulage costs. Alternatively it can be sited by a river or railhead for efficient receipt of logs and dispatch of timber. In timber importing regions mills have located near ports, e.g. Bristol, Liverpool and London in England, but in the 1950s-1970s controlled dock and stevedore labour markets and high port charges distorted such rational decisions, with mills at the dockside in Liverpool importing through Felixstowe on the other side of the country. Sawmilling is a noisy, traditionally labour intensive industry that has been unable to pay high wages: both reasons for avoiding prosperous metropolitan areas.
- The available capital. The last decade has been characterized by neutral to positive real interest rates, i.e. above the rate of inflation, which means that the

pay back time on investment has to be shortened. Mills are designed to cut particular products for specific markets, and while some flexibility is desirable, flexibility costs money. As a general statement mills in North America, Northern Europe and Australasia produce an assortment of sizes, lengths and grades that are too varied for production to be quite rational.

Sawmills are characterized by the timber resource they cut, by operational size, by the type of machinery used to breakdown the logs, and by the degree of automation. Every sawmill is unique. There can be no standard design. A mill should be judged on its operational efficiency and profitability, which are as much a result of good management as of good mill design.

Good design is seen in the smooth flow of wood through the mill with no bottle-necks and with no machine waiting for material to cut. This necessitates a piece by piece analysis of material through the mill as logs are progressively broken down and the timber is cut to the required sizes. Mill design involves repeated simulation of various design options, varying the resource characteristics, the saws and mill layout, and the market demand for different sawn products.

Throughput, volume conversion and grade recovery determine the profitability of a mill. The conversion is defined as the ratio of the volume of green sawn timber that can be cut from a given volume of debarked logs. Usually the conversion is based on the nominal sizes being cut (i.e. 100 x 50 mm) rather than the actual green dimensions (*c.* 102 x 52 mm). The nominal size allows for sawing variation, shrinkage and planing loss. Alternatively conversion is based on the dry dressed sizes (*c.* 94 x 45 mm). Green conversion ranges from as low as 40% to 65% in large-log mills cutting for grade. In modern mills conversion is typically about 55-65% and the sawdust generated should not exceed 10%, with the balance of the material being chips. In old inefficient mills sawdust can amount to 25%. There are further volume losses as part of the production is dried and some is dressed. Indeed in the United States the volume of lumber shipped from all mills is only 42% of the log input. An Australian sawmill cutting a range of high value furniture and flooring grade eucalyptus hardwood timber is profitable using small low quality logs while only achieving an average recovery of 26%.

Grade recovery is important where milling valuable logs. Grade recovery is concerned with maximizing profits by cutting the more valuable grades or sizes rather than trying to achieve profitability by maximizing throughput or volume conversion. It involves turning and careful examination of each log as it is being sawn to determine the best sawing strategy. A log carriage is needed.

2. BASIC SAW TYPES AND BLADES

A variety of saws is used to break logs into boards or larger dimension timber: circular saws, bandsaws, framesaws and chipper canters. The first three saws generate a saw kerf. The saw kerf is the width of cut that a sawblade produces when sawing lumber (Figure 7.1). The wood in the kerf is reduced to coarse sawdust. Chipper canters function differently. They chip the edges of logs, cants or flitches to generate two parallel faces while reducing the waste material to chips that can be sold to the pulp and paper industry. The choice of machinery is influenced by the log resource (quality, size and volume).

A variety of saws are used to progressively cut the logs into timber of the desired dimensions. The first saw to cut a log as it enters the mill is the headrig. Other saws are resaws, which further process material coming from the headrig, and edgers, which edge material. Once the wood is faced on four sides it only needs to be cross-cut with circular docking saws, either trimming to length or, where necessary, cutting out defects such as large knots.

The knowledge and experience of the sawdoctor is crucial to an efficient sawmill, if only because a change in any one of a number of factors affects the ability of the sawblade to produce a straight cut. Only when the balance of factors is optimized does the saw run at its most efficient, with minimum power consumption, highest production, and high quality cuts for the maximum time between blade sharpenings. The sawdoctor has to consider a number of interactions concerning:

blade material,	sharpness or tooth angle,	feed speed,
blade speed,	face angle,	face hardening,
blade tension,	gullet depth,	spring setting, or
blade thickness,	gullet shape,	swage profiling,
blade width,	side clearance,	tungsten tipping,
clearance angle,	tooth design,	stellite tipping,
hook or rake angle,	tooth pitch or spacing,	*etc.*

2.1. Tooth geometry and blade steel

Koch (1964), Quelch (1972, 1977) and Williston (1989) provide detailed accounts of tooth geometry, saw doctoring and saw characteristics.

When cutting along the grain the process is termed 'ripping'. When cutting across the grain the process is termed 'cross-cutting'. In ripping the removal of the chips is akin to chiselling parallel to the grain, whereas when cross-cutting the individual fibres are being severed by side-pointing teeth moving across the grain (Figure 7a,b).

Consider a back of the envelope calculation. Each tooth on a 10 metre long bandsaw blade with a rim speed of 50 m s^{-1} will engage with the logs roughly 50 000 times in an 8 hour shift (assuming that the saw is cutting wood for a third of that time). If the bite per tooth is about 2.5 mm, if the kerf is 5 mm and if the average depth of cut is 200 mm, then each tooth chisels out about $1/8^{th}$ of a cubic metre of solid wood per shift. Teeth are subject to relentless impact loading and fatigue.

As a rule of thumb, the bite per tooth (Figure 7.1) should be roughly equal to half the width of the kerf, corresponding to a bite of 2-4 mm. If the bite is too small, the tooth generates fine sawdust that is small enough to escape from the gullet and spill into the clearance between the face of the blade and the wood, where rubbing friction overheats the blade. If the bite is too great the volume of coarse sawdust created as the tooth moves through the log will exceed the space in the gullet. Again frictional heating will be considerable and in extreme circumstances the compacted sawdust can choke the saw. Ideally sawdust should be retained in the gullet and discharged as the tooth exits the log, but a small amount of spillage is inevitable.

The maximum and optimal feed speed, f_{max}, is given by:

$$f_{max} = \text{GFI}.\ cA/pd \qquad (1)$$

where GFI (gullet feed index) is the fraction of the gullet that should be filled (70% or 0.7 for bandsaws), c is the speed of the blade, A is the gullet area, p is the pitch and d the depth of cut. For the tooth geometry in Figure 7.1 and a blade speed of 50 m s^{-1} the maximum feed speed is about 1.2 m s^{-1}.

Tooth geometry is subject to a number of interactions. Increasing the pitch increases the bite per tooth and the necessary power of the saw. Deepening the gullet would increase gullet capacity and so allow a higher feed speed, but lengthening the tooth increases the strain on the tooth and the likelihood of cracking in the gullet. Increasing the rake angle and reducing the sharpness of the tooth will reduce power consumption but the tooth will wear faster and is more likely to break. A positive clearance angle ($>>0^o$) and instantaneous clearance angle ($>0^o$) prevent the back of the saw rubbing against the freshly cut wood.

Sawblades are manufactured from the highest quality steel, where required characteristics include good durability, resistance to wear and fatigue, retention of tension (so the blade does not stretch during sawing), and an ability to remain flat and straight. Blade materials range from carbon steels with the carbon content around 0.75%, to steels alloyed with manganese or nickel to increase toughness. Tempering and plastic working of blades is common. These induce phase changes and develop dislocation networks in the steel, and amongst other things harden the blade (Neely and Bertone, 2003). Bandsaw steel blades are robust and easy to repair and sharpen.

Steel for bandsaws comes in rolls. The desired length is cut and teeth are punched before welding into a continuous band. This invisible joint has to sustain endless flexing on the wheels of the bandsaw.

2.2. Setting of teeth

The set provides clearance between the gauge of the blade and the wood.

Figure 7.1. Geometry of a sawblade. (a) Spring set teeth. For cross-cutting the face is angled (as shown) so the tips of the teeth can sever the fibres; for ripping the face is perpendicular or angled only slightly to the direction of cut. (b) Most modern saws are swage set. The cutting edges of the teeth extend beyond the gauge of the metal to provide clearance between saw and timber, so preventing excessive rubbing. (c) A typical sawtooth profile for cutting pine. The gullet must be large enough to retain the bulk of the unconsolidated sawdust until the tooth emerges from the log: the volume of unconsolidated sawdust is about three times that of the solid wood. (d) The bite per tooth equals the product of the pitch, p, and the log feed speed, f, divided by the speed of the saw, c, i.e. the bite = pf/c. Typically it is 2-4 mm and roughly half the width of the kerf.

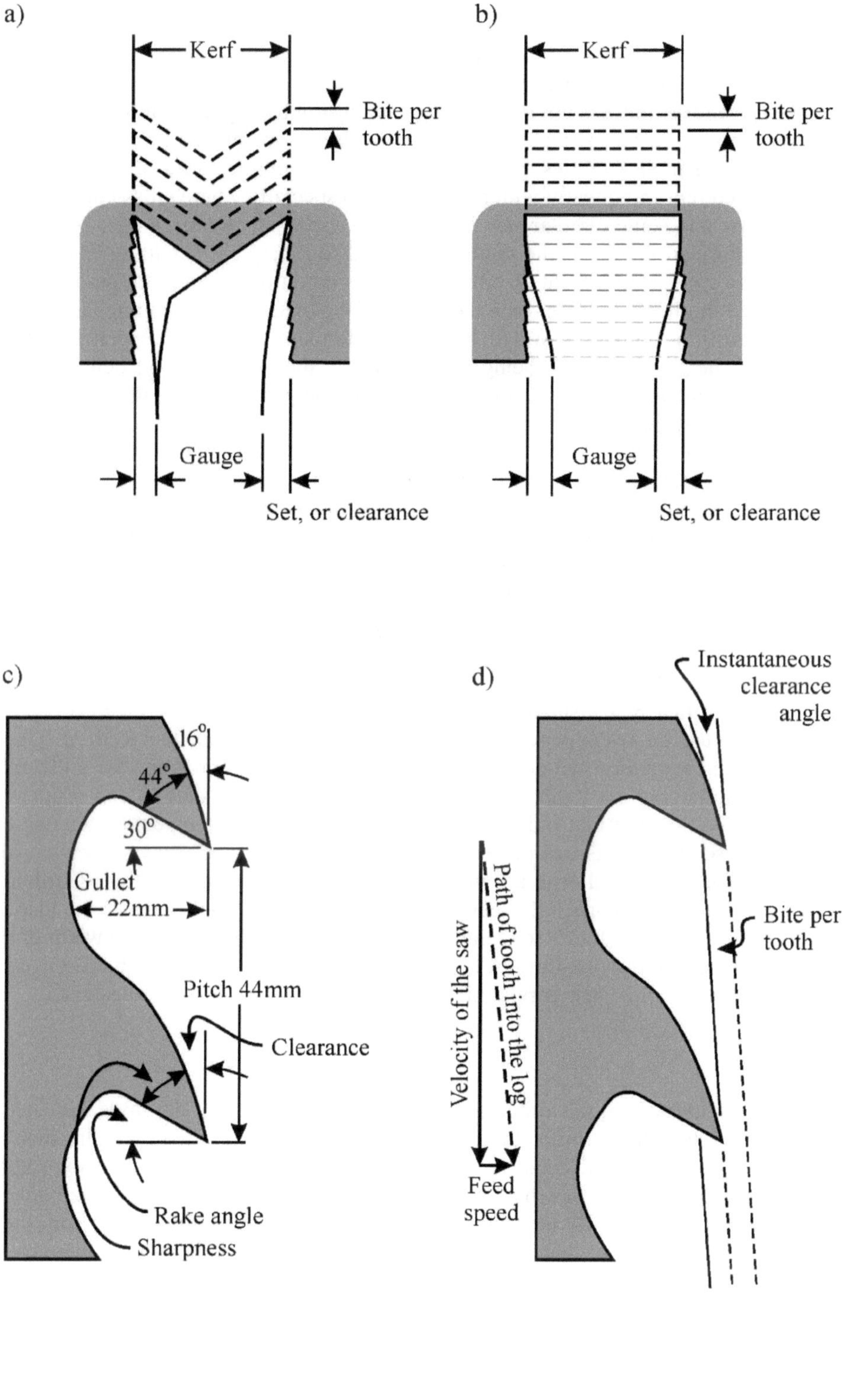
a)
Kerf
Bite per
tooth
Gauge
Set, or clearance
b)
Kerf
Bite per
tooth
Gauge
Set, or clearance
c)
16°
44°
30°
Gullet
22mm
Pitch 44mm
Clearance
Rake angle
Sharpness
d)
Instantaneous
clearance
angle
Path of tooth into the log
Velocity of the saw
Feed
speed
Bite per
tooth

The traditional method is 'spring set' where the tooth or tip of the tooth is bent to one side. For circular saws adjacent teeth are bent in opposite directions, alternately set left-right-left *etc.* (Figure 7.1a). For bandsaws a left-right-raker sequence is common: the raker tooth is not bent and may stand slightly higher than the bent teeth to keep the saw cutting straight. Spring set bandsaws are more common in small sawmills with simple facilities and in secondary processing (*ex*-sawmill). Spring set teeth are used more frequently with circular saws, especially where cross-cutting.

Bandsaw and some circular saw blades are 'swage set' in large mills. Here the metal that formed the tip of the tooth is punched/squeezed back; with nowhere for this metal to go it is spread out on either side. This is sharpened and side ground into a broadened tooth that is wider than the gauge of the blade, thus providing clearance (Figure 7.1b). The amount of swage, before and after side dressing, is specified, with the size of the finished tooth being determined by the type of wood being sawn. Swaged saws cut a smaller kerf, some 0.2 mm less compared to a spring set blade.

Superior sawing can be achieved by tipping the teeth. Wear resistance is substantially increased by having tungsten carbide (very hard but brittle) or cobalt/chrome stellite tips (not so hard, but resistant to chemical corrosion, e.g. acid from western red cedar or eucalypts). The advantages are particularly noticeable when sawing hard and abrasive woods of species with a high silica content.

Stellite is annealed, brazed, resistance welded or plasma melted onto the saw teeth by means of automated machines (but can be welded manually). Stellite tipped sawteeth should be reground every 8-16 hours even though they can cut for much longer. Failure to do so will result in fatigue gullet cracks and impairment in the sawing quality. Stellite tipping offers a narrower kerf (↓8%), overcut reduction (↓10%) due primarily to a superior sawn finish, and an increased saw-life (↑50%).

Blades must be resharpened regularly. A dull edge means the teeth will be stressed and saw quality will deteriorate. Grinding involves reprofiling the gullet as well as resharpening the tooth. Careless grinding in the gullet bottom can result in superficial overheating that induces a phase change, forming tempered martensitic steel. Microscopic checks can nucleate in this surface layer and grow subsequently during milling as the gullet is a region of moderate stress-concentration (Gordon, 1978). Regrinding to sharpen the teeth and reprofiling of the gullet reduce the blade width (bandsaws) or diameter (circular saws). This may continue until the width of a bandsaw blade is about one-third of the original but with a circular saw blade the rip speed will reduce with each grind until the blade no longer performs as designed.

2.3. Tensioning

Blades are tensioned. In this process the metal a little back from the tip is stretched thus allowing the blade to run straight when centrifugal forces (with circular saws) and the heating of the teeth cause the metal on the outer edge of the blade to expand.

Tensioning of circular saw blades involves carefully hammering or squeezing the blade between narrow rolls in a zone about half to two-thirds of the way from the centre to the periphery so that the metal there is very slightly thinner with the metal being spread out sideways. The blade is pre-strained. However the metal in the hammered region is largely restrained from spreading by the surrounding metal, so that in the plane of the saw the metal in the tensioned region is actually in compression while the metal in the rim of the saw is slightly stretched and is in tension. The blade becomes very slightly dished (like a saucer) when placed on the arbor: the greater the degree of tensioning the greater the dishing effect. When the saw is running at the desired peripheral speed the metal at the rim stretches under the centrifugal force that permits the metal nearer the collar to expand correspondingly so counterbalancing the strains induced by hammering. The saw blade straightens and can cut accurately when running at the correct peripheral speed (30-45 m s^{-1}). Tensioning must take account also of the thermal expansion of the metal at the rim due to frictional heating of the teeth, where the temperature at the cutting tip can reach 700°C. Clearly the amount of tensioning depends on the desired peripheral speed: the faster the desired rim speed the greater the tensioning required. If the saw revolves too slowly the blade will remain slightly dished and will have been over-tensioned. If it revolves too fast it is liable to vibrate as it will have been under-tensioned. At some critical speed the saw starts to resonate and 'snake' with the teeth of the saw moving laterally out of the plane of the saw (Schajer, 1989; Valadez, 1990), somewhat akin to a vibrating guitar string. The optimal peripheral speed is slightly less than that associated with the onset of snaking. Some flutter is inevitable when a tooth hits a hard knot and is deflected slightly. This vibration is dampened with guides. The clearance between saw-guides (Figure 7.2a) and the blade is only a few tenths of a millimetre. A thin film of oil or water keeps the surfaces apart. The film needs to be a little greater than the surface roughness of the blade so that asperities are prevented from rubbing against one another. The film shears readily but resists lateral movement: the analogy is with the ease of sliding two wet flat glass surfaces past one another (no hindrance to the rotation of the saw) and the difficulty in trying to pull the glass sheets apart (dampening vibrations).

Bandsaws also require tensioning as the teeth get warm as they cut. In one approach the metal at the centre and back of the saw is slightly stretched by passing between tension rolls. This means that the back of the saw is initially slack when the blade is first strained between the two wheels. Only as the front of the saw heats up and the metal there expands slightly does the whole width of the saw feel the straining forces applied between the two wheels, but the front of the saw still remains in tension after it has heated up and can cut accurately without wandering. Further the teeth stretch as they are pulled through the log.

There is a trend to high strain bandsaws that use thinner gauge steel and to strain the blade by applying significantly higher loads between the two wheels. These saws cut a narrower kerf but are harder to tension. If poorly tensioned the blade will wander, cutting a wavy kerf and the timber will have to be dressed afterwards to get a good finish – more will be planed off than would have been lost as kerf had the saw being less highly strained, and the blade been of thicker gauge.

2.4. Circular saws

The circular blade rotates on an arbor (spindle) and is self-supporting.

Generally it is undesirable to use circular saws to process large diameter logs or thick timbers because large saws have to be of thick gauge metal necessitating a very wide saw kerf and an excessive amount of sawdust: the width of the cut (the kerf) and the stiffness of the blade, and hence the accuracy with which it cuts, all depend on the thickness (the gauge) of the blade. Where circular saws are used to breakdown large logs (>500 mm) the kerf is typically 7-8 mm and sometimes even 10 mm. Despite the large kerf many small mills throughout the world have circular saw headrigs principally because they are cheap and robust.

On the other hand manufacturers now produce excellent headrig saw blades with an operational but consistently narrower kerf of about 4 mm.

A double arbor circular saw allows a greater depth of cut without such excessive kerf. The two saws should cut in exactly the same plane with the upper saw off-set (ahead of the lower saw) so the teeth of the two saws do not mesh. The slightest lateral offset between the saws will produce a small visible step on the cut faces that has to be planed off later and this partially negates any benefit of a narrower kerf. Accurate alignment should not be difficult to maintain, but even with prefect matching the fact that the two blades rotate in different directions leaves a visible 'line' along the timber which discriminating consumers find unacceptable, especially in Japan.

It is desirable to keep the diameter of the saw to the absolute minimum because as a general rule saw kerfs are approximately twice the thickness of the saw blade steel. Therefore for deep cuts double arbor systems have advantages, e.g. as double arbor edgers and circular gangs. Minimizing the diameter of a circular saw blade provides real benefits. The further the saw tip is away from the arbor the greater the flexibility and movement. To compensate for this a thicker saw gauge needs to be used to maintain stability.

Small circular saws are used in most mills to edge the boards and timbers, i.e. to remove all or some of the curved waney edges, which correspond to the cambial surface. Circular saws are favoured because the depth of cut is not large (typically 25-100 mm) and so the kerf can be kept small (2-4 mm). An edger has a number of adjustable sawblades on a single arbor each of which can be moved along the arbor to cut material of any desired width. The saws are indexed relative to one another, so that they piggy-back on the position of the previously set blade. Wide slabs, flitches or cants can be ripped simultaneously into a number of pieces, while also removing some wane from the edges. The gang-edger usually has two sets of blades positioned on either side of the wide throat (*c.* 1.5 m). On one side the spacing between blades is close (*c.* 25 mm) to cut boards, while on the other side the spacing is wider (*c.* 50 mm) to cut softwood dimension With softwoods the terms board, dimension and timber generally refer to the thickness of the material cut. Boards are less than 50 mm thick, dimension is between 50 and 100 mm thick and timbers are greater than 100 to 125 mm thick. The terminology varies between counties.

A high rate of production is achieved as the saws do not need to be reset. Where deep cuts (>150 mm) are required as in a gang-edger a double arbor may be used, in which two identical sets of saws are positioned above one another and the timber is passed between them. The previous comment on surface finish applies.

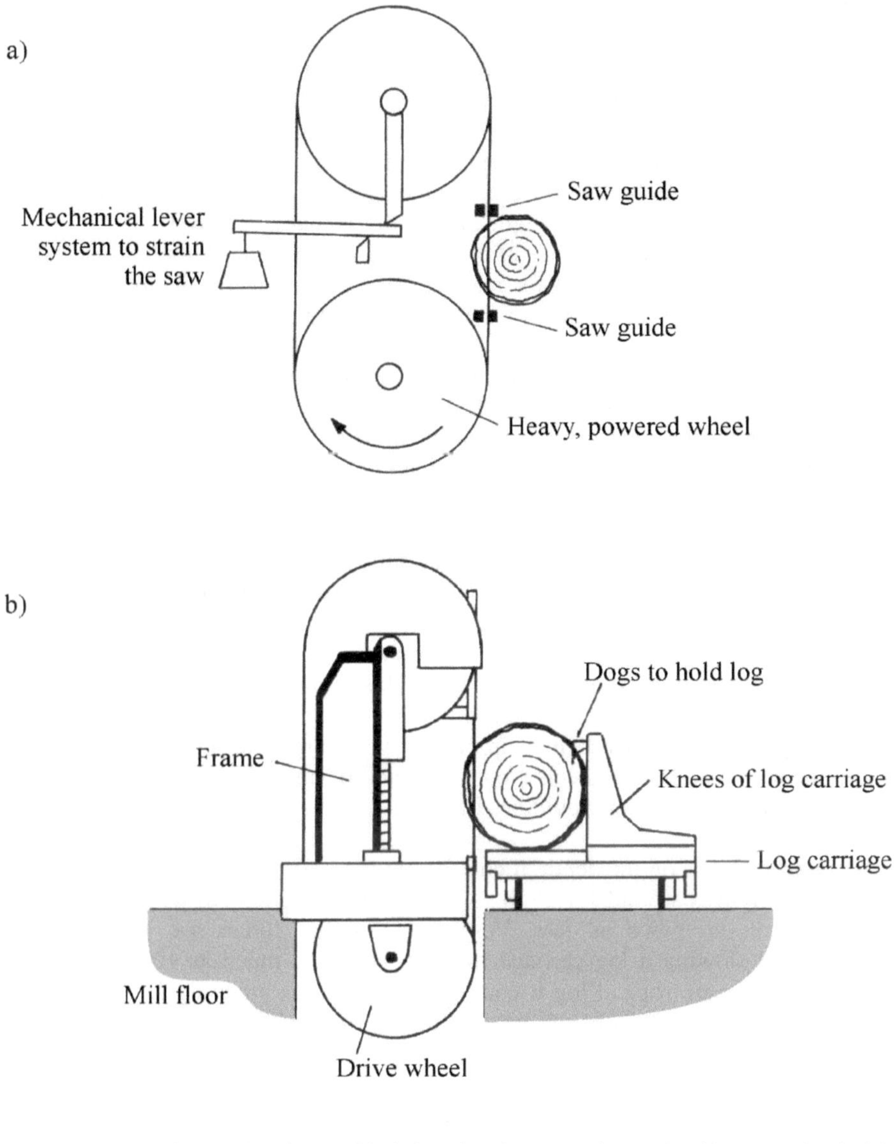

Figure 7.2. Bandsaws. (a) The sawblade in a bandsaw may be strained by a mechanical lever system. Bandsaws cut fast and accurately provided the blade is correctly tensioned and strained between the two wheels. Saw guides dampen slight vibrations, where the teeth strike a hard knot. (b) End view of bandsaw with log carriage. The log is rotated on the log carriage to present the desired face to the saw.

2.5. Bandsaws

A bandsaw has an endless steel band that is mounted between two large wheels (Figure 7.2a). These wheels can be 1.5 to 3 metres in diameter, with larger wheels being used on the headrig and smaller wheels being used on resaws. The lower, heavier wheel is powered and pulls the blade down through the log as it is fed into the saw. The blade is not self-supporting. Instead the saw is strained between the two wheels permitting the blade to be of thinner gauge metal than that used in a circular saw. The saw kerf, 1-5 mm (typically 4 mm), is much narrower.

Modern strain systems and the use of cartridge pressure guides both help the saw cut accurately. Bandsaws are ideal for making deep cuts with little kerf. They cut accurately and because of the length of the saw, 10-20 m, can run for long periods between sharpening. They are found at the headrig and as resaws.

A bandsaw with a log carriage is used in the breakdown of medium or large size logs (Figure 7.2b). This combination is ideal for logs of variable quality as well as those of high quality since it offers versatility in sawing patterns and a deep cut while keeping the kerf to a minimum. The logs are firmly and accurately held on the log carriage before being fed into the saw that makes a single cut on each pass. The cut material is dropped off onto outfeed rollers and the remainder of the log is taken back past the saw before being repositioned for another cut. With a log carriage the log can be turned between cuts to maximize the quality and value of timber cut. For example a log can be turned to explore the grain or to separate off the sapwood. The vertical knees of the log carriage, against which the log is firmly secured (dogged), can move independently to allow for log taper. This allows a full length slab to be cut parallel to the cambium on any or all four sides, or the log can be cut parallel to the pith (Figure 7.3). A laser light helps the sawyer align the log correctly with respect to the saw. The laser projects a narrow pencil of light along the length of the log showing where the sawcut would be if the log were to be fed into the saw when in that position.

Feed speed depends on the depth of cut, the hardness of the wood and the size of the saw: generally it is around 1-1.25 m s^{-1} in a high production softwood mill. The return speed can be twice as fast. Very large inertial forces are involved in accelerating and slowing a log carriage that can weigh as much as 10 tonnes and when carrying a green tropical log it could easily weigh as much again. Timber is only being cut about 25-35% of the time, allowing for loading and positioning the log on the carriage, and in bringing the log carriage back past the saw after each cut. Obviously the volume of timber cut is limited by the unproductive time handling logs and a bandsaw with log carriage is uneconomic for milling small logs (<350 mm): with large logs the volume throughput can be sustained as each sawcut releases a large piece of timber for the secondary breakdown saws.

Slant bandsaws where both the saw and log carriage are tilted about 17-27 degrees have found increasing popularity. Gravity aids fast, accurate location of the log against the knees of the log carriage, and the slabs and flitches fall from the saw in a well-disciplined manner, sawn face down.

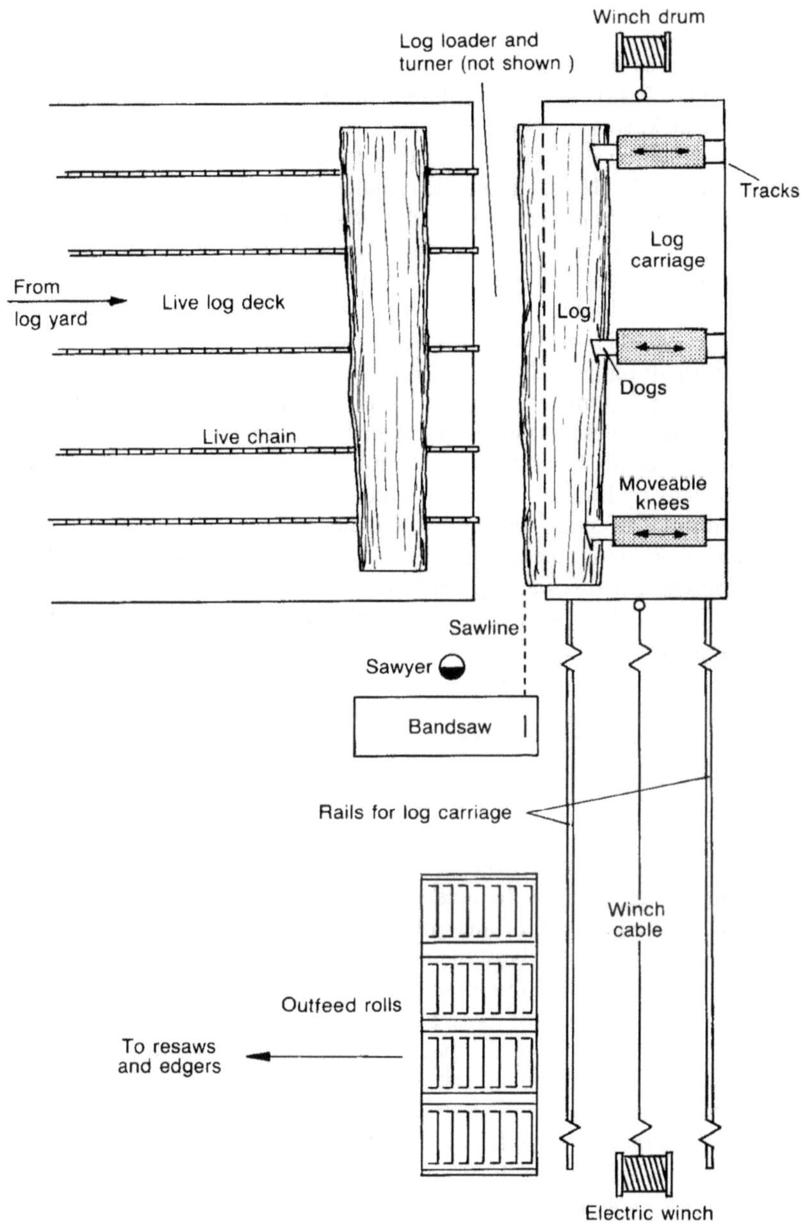

Figure 7.3. Headrig bandsaw with log carriage. The log is feed into the saw by a powered winch. The position of the log with respect to the saw is adjusted by moving the knees of the log carriage. The more the knees are moved forward the deeper the cut. With some log carriages it is possible to move the knees independently and so to cut parallel to the cambium or parallel to the pith.

2.6. Chipper canter

The chipper canter, as its name implies, chips two faces from a small log leaving a central cant (Figure 7.4a). The cutting tools consist of two rotating truncated cones fitted with chipper knives arranged in concentric circles or in a spiral. These chip opposite sides of the log or cant as it passes through the workstation. The length of the chip along the grain (16-25 mm) is determined by the feed speed (0.4-2.0 m s^{-1}), the number of revolutions per second (6-18) of the chipper and the number of knives (6-12) on the face of the cone. The chipper canter was developed to chip material during the normal course of log conversion instead of having to undertake a secondary slabwood chipping operation. No sawdust is produced and the material that would have been in the kerf is turned into profitable pulpwood chips. The system is ideal for straight small logs, 100-300 mm small-end diameter, from which conventional sawing systems would generate too much sawdust. The twin discs can move sideways quite quickly to alter the thickness of the central cant, which can be cut to ± 0.1 mm. Even so, batching logs of similar diameter is desirable to minimize the time needed to hydraulically readjust the position of the discs. Ideally there should be a continuous flow of logs butted together.

The primary workstation need not be a single saw. A twin or quad band reducer saw mounts the chipper discs and either two or four bandsaws symmetrically on a single frame (Figure 7.4b). Such units suit high production mills. By placing all the saws on a single frame and holding the log firmly while passing it through these saws the pieces are cut more accurately and to better tolerances than when mounting the saws individually. Circular saws can be mounted behind the chipper discs if a cheaper, more robust system is desired.

2.7. Frame saws

Frame or gang saws are currently out of favour, largely due to their inflexibility and slow lineal processing speeds. In essence multiple blades at fixed spacings are assembled and strained within a frame allowing multiple cuts of the log in a single pass. The thickness of the two slabs, the side flitches and central cant are determined by the spacing of blades on the frame, so that a batch of similarly sized logs must be fed into the saw for two hours or more before stopping production and resetting the saw pattern: with a change of blade spacings in the frame another batch of logs of different diameter class can be milled. These mills can have a framesaw as the headrig and downstream for processing the central cants.

Mills batch logs into some 40 log sorts, in 20 mm diameter classes. Logs must be of uniform quality as it is impossible to turn the log and explore ways of maximizing recovery of the better grades of timber. The system favours medium-sized logs of about 360 mm diameter from naturally regenerated, second-growth forest.

→

Figure 7.4. Chipper systems. (a) The chipper canter reduces the slabwood to chips and generates no sawdust. (b) Chipper canter with a quad-band as the headrig. In this mill the log input is insufficient to justify a second unit and instead the cant is taken around and refed into the headrig.

a)

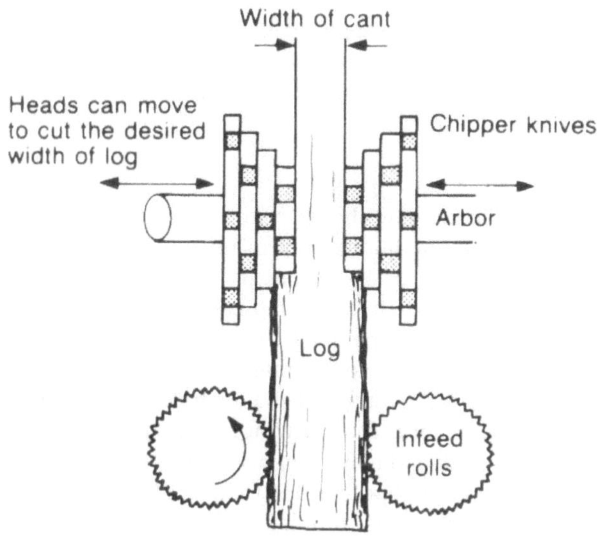

b)

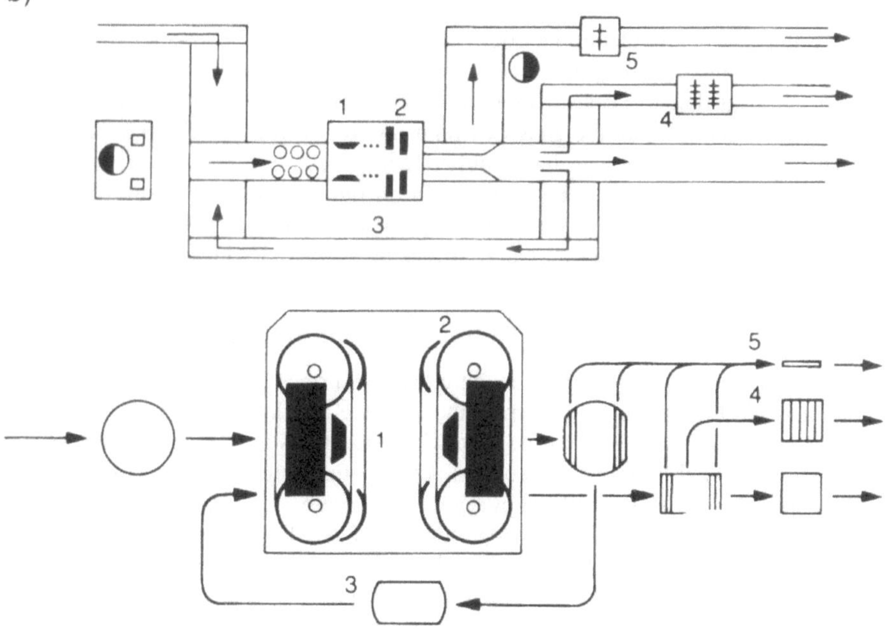

1 Chipper canter
2 Quad-bandsaw
3 Merry-go-round
4 Double arbor circular gang saw
5 Board edger
◒ Operator

Framesaw mills place emphasis on low production costs and high volume throughput, with an almost continuous supply of wood, butted end-to-end, moving through the saw at about 0.2-0.3 m s^{-1} and cutting some 30 m^3 per hour. More details on frame saws can be found in the first edition (Walker, 1993).

2.8. Secondary breakdown systems

Logs are not processed by a single saw. If the primary breakdown saw has a log carriage the logs can be passed through a number of times while slabs, flitches and cants are cut. These are then processed by secondary breakdown saws to produce lumber of the desired sizes. The alternative to a log carriage could be a frame saw, twin/quad band or circular saw, or a chipper canter or some combination. With these headrigs the logs make a single pass before being passed on to be broken down further by other saws. It is possible to drop off the central cant and bring the cant back to the in-feed side of the headrig for resawing, using a merry-go-round: this allows the headrig to do more work and may be appropriate where the timber supply could not sustain the high lineal throughput needed for a single pass headrig.

Crooked or swept cants present special problems. Rather than cross-cutting logs into ever shorter lengths, special cant-guiding systems are available where some lateral/rotational movement is permitted that allows the resaws to cut with the curve of the cant. The bowed timbers can be straightened by drying under heavy weights.

Secondary breakdown saws operate on the same principles as already outlined. Since the log has been cut into smaller pieces the saws can be smaller than that at the headrig. Secondary breakdown saws include amongst others the band-resaw for handling material that has one or more cut faces, the edger for trimming the wane from slabs and flitches, and the gang or frame saw for cutting the cant into a number of boards. These machines can have higher feed speeds and cut thinner kerfs than the headrig. Further there is greater emphasis on the use of circular saws since the depth of cut is less and small circular saws cut a smaller kerf. Circular saws are cheaper and more robust than other alternatives in edging and docking operations.

2.9. Relocatable sawing systems

Relocatable sawmills are appropriate where the resource is limited or the demand for the output is not high enough to warrant the expense of a larger fixed sawmill. They can cut wood close to or even in the area where the trees are being felled. Portable sawmills are a valuable option in isolated small communities, avoiding the need to transport logs to a distant mill. They cannot compete economically with a large volume sawmill producing commodity products but satisfy local needs.

A niche specialist market for such mills exists where special purpose timbers yield a very high valued product despite there being only a limited volume of logs.

Mobile mills range in size and complexity from a single operator chainsaw mill to a large hydraulically controlled trailer mounted bandsaw. The mobility and setup time varies with the size and complexity of the system. An outline of the production capacities, weaknesses and strengths is described in Table 7.1.

Table 7.1. Advantages and disadvantages of various mobile sawmills (Waugh, *pers. comm.*).

Sawing System	Mobility, Set-up	Capability (sawn output) Resource	Productivity	Product	Strength	Weakness
Travelling head units on truck or 4WD						
Chainsaw	Set-up < 1 hr	Wide range of log sizes	Very low, 2m^3/day	Slabs need resawing	Versatile Low cost	Small logs Large kerf Poor sizing
Horizontal bandsaw	Set-up 1-2 hr	Best on medium sized logs	Low-medium, 4-6 m^3/day	Slabs resawn on unit	Narrow kerf Low cost Accuracy	Small eucalypt & dense wood
Singular circular saw units	Set-up 1-2 hr	Best on larger logs	Low-medium, 4-6 m^3/day	Flitches & product sizes off saw	Product cut without extra handling	Small eucalypt Wide kerf
Twin circular saw units	Set-up 2-3 hr	Capable of handling very large logs	Medium, 6-12 m^3/day	Flitches & product sizes off saw	Large logs. Handles log & product. Rugged	Small eucalypt Wide kerf Saw cut stability
Bench-type units						
One-man unit	Truck or 4WD Set-up 2-3 hr	Flitches & small logs	Medium, 10-15 m^3/day 2-man crew	Accurately sawn product	Good productivity Handles log & product	Expensive Less mobile External power
One-off design	Truck Up to 2 days to set-up	Designed for resource	High 20 m^3/day 2-3 person crew	Accurately sawn product	Good throughput Standard sawmill	Expensive Large crew Lack of technical support

3. MILL DESIGN

Williston (1981, 1988) gives comprehensive reviews of sawmilling that repay reading. Further a study by Kockums CanCar (Hall, 1983) provides an excellent introduction to mill design: the analysis of a softwood mill processing small logs and a hardwood mill processing large logs illustrates the need for close attention to the smooth flow of material through the mill. This approach is summarised below.

3.1. A small-log softwood mill (Hall, 1983)

An inventory is essential to reveal the quantity of timber available, the projected log sizes and log quality. The latter strongly determines the appropriate market strategy. Once the expected log size distribution is known the log supply can be segmented into a number of log diameter classes and in the first instance a set of typical sawing patterns assumed for each diameter group. In the first example a small-log softwood mill is to be supplied with 110 000 m^3 of roundwood per year (Table 7.2). This supply equates to 580 000 logs with an average small-end diameter of 245 mm and an average length of 4.0 m. The mill is designed as a high production mill with emphasis on accurately cutting a large volume while achieving a high conversion factor.

Initially sawing patterns are developed for the various log diameter classes bearing in mind the market for specified end products and favoured product dimensions. Figure 7.5 illustrates a breakdown pattern that is deemed appropriate for sawing a log of 255 mm small-end diameter. On the first pass through the reducer bandsaw (RBS 1) the log is broken into three pieces while at the same time chipping the outer slabs to expose two faces somewhat greater than 60 mm wide. The 150 mm wide central cant is turned through 90 degrees and passed a second time through the reducer bandsaw (RBS 2) that faces the two rounded edges of the cant as well as producing two flitches and chips for pulp. The rectangular cant can now go to the gang-edger (GE) to give 7 boards, 24 x 150 mm. Of the four flitches, two go direct to the chipper board-edger (BE) while the two thick flitches (from RBS 1) pass through a horizontal band-resaw (HRS) to give four pieces before also going to the board-edger (BE). With this particular breakdown pattern the piece count through the various saws is: reducer bandsaw two pieces, gang-edger one piece, horizontal resaw two pieces and board-edger six pieces. All thirteen pieces have to be trimmed to length. This analytical procedure is repeated for each diameter class and the total flow of material through the mill can be estimated (Table 7.3).

The mill layout is developed by summing all the piece counts through the various saws and the appropriate flow paths for material through the mill. In this case study the reducer twin bandsaw has insufficient lineal throughput to justify a second such workstation so a ‘merry-go-round’ takes the central cants round to refeed them into the headrig (Figure 7.6). Although the throughput is about 2.6 logs

Table 7.2. Projected log supply for a small-log softwood mill, by small-end diameter (Hall, 1983). Operating 250 days/year, 2 shifts/day and effective production time 900 min/day.

Log diameter class, mm	Mean S.E.D. mm	Mean log vol. m^3	% of total volume	Volume per year m^3	Logs per year	% of total logs	Logs per day	Logs per minute
116-155	135	.077	6	6 600	85 714	14.78	343	.381
156-175	165	.109	9	9 900	90 826	15.66	363	.404
176-205	185	.133	10	11 000	82 707	14.26	331	.368
206-225	215	.175	13	14 300	81 714	14.09	327	.363
226-245	235	.206	13	14 300	69 417	11.97	278	.309
246-265	255	.239	12	13 200	55 230	9.52	221	.245
266-285	275	.275	10	11 000	40 000	6.89	160	.178
286-305	295	.313	8	8 800	28 115	4.85	112	.125
306-325	315	.354	6	6 600	18 644	3.21	75	.083
326-345	335	.397	4	4 400	11 083	1.91	44	.049
346-365	355	.443	3	3 300	7 449	1.28	30	.033
366-385	375	.491	2	2 200	4 481	.77	18	.020
385-660	525	.934	4	4 400	4 711	.81	19	.021
Summary	245	.190	100	110 000	580 091	100.00	2 321	2.579

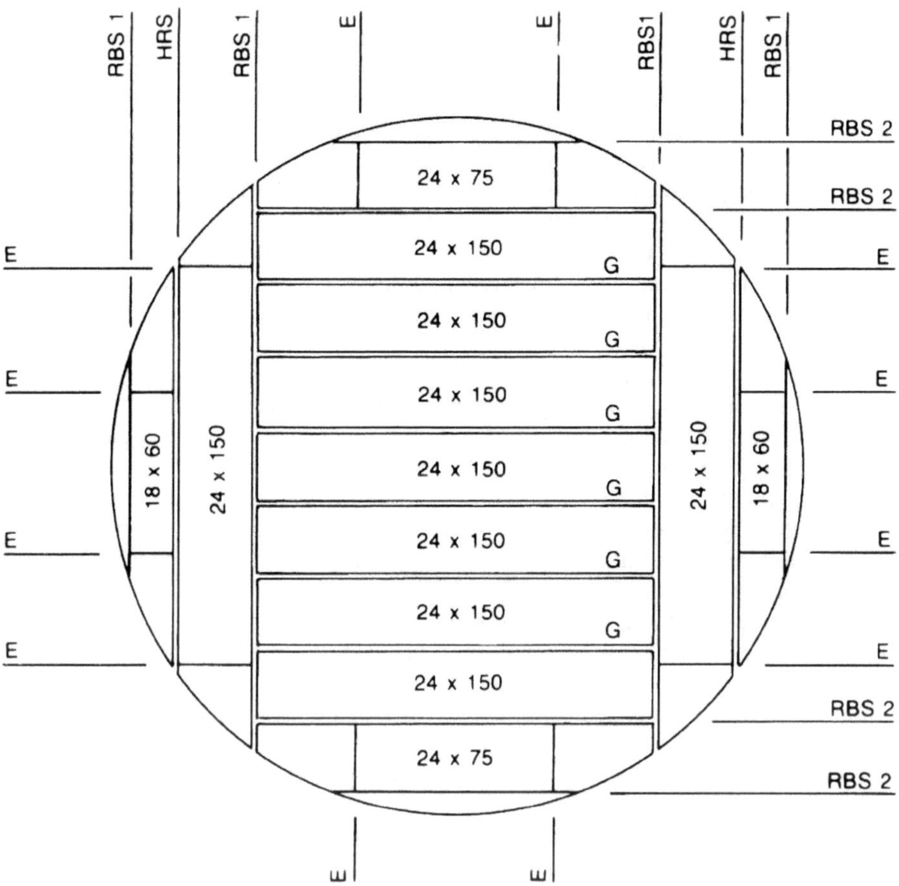

Sawlines:
RBS 1 - Reducer twin bandsaw first pass
RBS 2 - Reducer twin bandsaw second pass
HRS - Horizontal resaw
G - Gang edger
E - Optimising chipping edger

Figure 7.5. Sawing pattern for 255 mm small-end diameter class softwood logs (Hall, 1983).

per minute (Table 7.2) the headrig actually processes material at twice that rate, which emphasizes the high production capacity of reducer bandsaws. The circular saw gang-edger is equipped with two saw clusters on either end of the arbor allowing two different thicknesses of timber to be produced without resetting the machine. The horizontal band-resaw has a good depth of cut and so can handle wide flitches. Only one sawn face is necessary which is held down on the table/bed feed rolls by powered toothed rollers pressing down from above (this feature of horizontal resaws is used in other mills to process slabwood). Finally the waney-edged boards pass to the board-edger. Here a chipper-edger with two movable ripping saws maximizes grade recovery and volume conversion. The projected conversion for the mill is 63% (Table 7.4). As one would expect the conversion factor increases with log diameter. Good conversion with small logs can only be achieved if the logs have little sweep or taper, and if boards as well as dimension lumber are cut. Where sweep or taper is a problem it is preferable to cut the logs to as short a length as possible: for example by cross-cutting a swept log in half the effect of sweep – the curvature of the stem – is reduced by a factor of four.

Various mill layouts are simulated before deciding on the mill design. There is some of choice in machinery and breakdown patterns, e.g. a frame saw is an option for handling the central cant. Critical factors include matching flows through the various workstations to ensure that the saws have adequate capacity and none are underutilized. They should be capable of responding to peak loads and have adequate storage areas both ahead and behind. Changes in log size have a dramatic effect on volume production. For example, a 4% decrease in average log diameter from say 245 to 235 mm would have to be compensated for by a 9% increase in lineal throughput to maintain volume output. Such a change in log supply could arise from intensifying competition for wood, from over-cutting or as a consequence of fire or windblow. The mill design must be flexible enough to cope with changes in wood supply and markets.

In parts of the world, e.g. Scandinavia, delivered sawlogs account for 70-80% of total production costs and it is essential to ensure that the smaller logs, especially if of low grade or with sweep, at least recover the variable costs of production (wood, labour and power). Small-log mills in North America and Scandinavia are economic because the small logs are generally straight, with little or moderate taper, and slow grown, having small branches and narrow growth rings (not too much corewood and

Table 7.3. Small-log softwood mill: flow of material through the various machines based on breakdown patterns for small-end diameter log classes (Hall, 1983).

	No. of logs			Pieces per minute to machines				
Small-end diameter	Year	Day	Minute	Reducer bandsaw	Gang edger	Horizontal resaw	Board edger	Docking saws
135	85 714	343	.381	.762	.381	-	1.524	3.048
165	90 826	363	.404	.808	.404	-	1.616	3.232
185	82 707	331	.368	.736	.368	-	1.472	3.312
215	81 714	327	.363	.726	.363	.726	2.178	4.356
235	69 417	278	.309	.618	.309	-	1.236	3.708
255	55 230	221	.245	.490	.245	.490	1.470	3.185
275	40 000	160	.178	.356	.178	-	.712	1.958
295	28 115	112	.125	.250	.125	.250	.750	2.000
315	18 644	75	.083	.166	.083	.166	.498	1.411
335	11 083	44	.049	.098	.049	.098	.294	.784
355	7 449	30	.033	.066	.033	.132	.264	.627
375	4 481	18	.020	.040	.020	.080	.160	.420
525	4 711	19	.021	.042	.021	.084	.126	.357
Totals	580 091	2 321	2.579	5.158	2.579	2.026	12.300	28.398

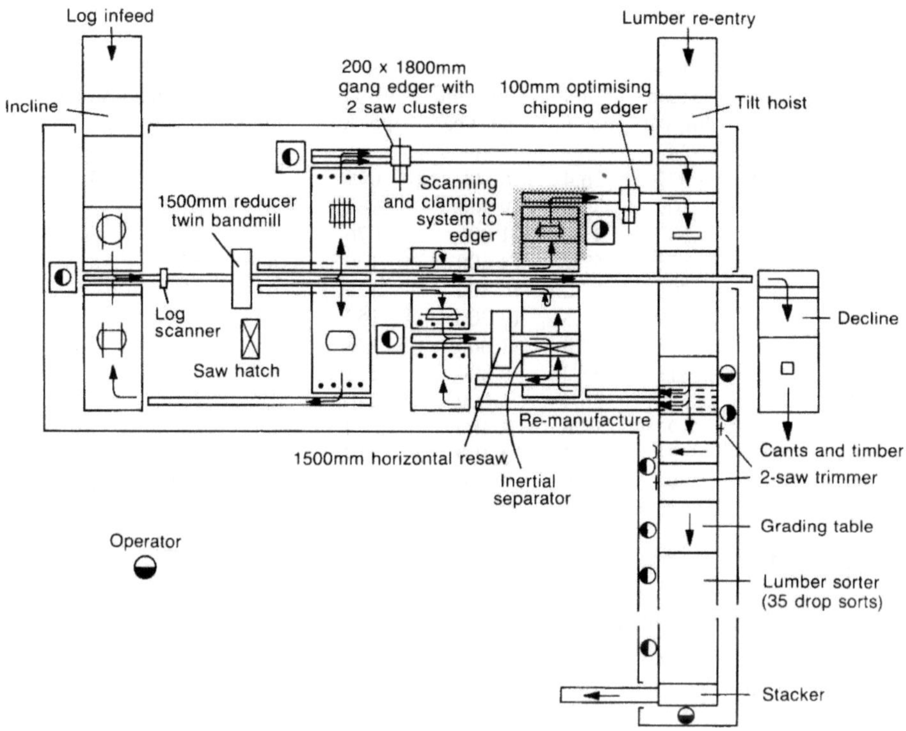

Figure 7.6. Small-log softwood sawmill layout, with an anticipated output of 70,000 m^3 yr^{-1} (Hall, 1983).

so still acceptable for studs). In such cases defects within the log do not pose great problems and overall the logs give an acceptable grade outturn. Often these mills cut a limited product line. They were developed in response to declining production from old-growth forests and the exploitation of smaller diameter second-growth stands, the exploitation of stands on harsher sites and the increased demand for pulp chips.

Such a generalized statement about wood quality is liable to qualification. For example second-growth Douglas fir contains a noticeable amount of corewood wood, ranging from 5-40% depending on age. Fast grown *Pinus radiata*, at least in New Zealand, presents a far more complex optimization problem and associated algorithms. The markets are varied (both local and export) as are the required sizes for boards and dimension. Furthermore, log quality is much more variable. Fast-grown, thinned and pruned stands give clearwood from the pruned butt logs and timber with very large knots from further up the stem. With fast-grown, wide-spaced and untended stands the best returns may be obtained from the clear cuttings between very large nodes. Heavily stocked and untended stands with dead, loose-knots in the butt log should give some structural timber, while some board grades

Table 7.4. Small-log softwood mill: theoretical conversion of green timber (Hall, 1983).

	Log data				Theoretical green recovery			
Log diameter class, mm	Mean S.E.D. mm	Mean log vol. M^3	Volume per day m^3	Volume per year m^3	Volume per log m^3	% of logs	Volume per day m^3	Volume per year m^3
116-155	135	.077	26.4	6 600	.039	51	13.4	3 343
156-175	165	.109	39.6	9 900	.061	56	22.2	5 540
176-205	185	.133	44.0	11 000	.076	57	25.1	6 286
206-225	215	.175	57.2	14 300	.109	62	35.6	8 907
226-245	235	.206	57.2	14 300	.129	63	35.8	8 955
246-265	255	.239	52.8	13 200	.153	64	33.8	8 450
266-285	275	.275	44.0	11 000	.178	65	28.5	7 120
286-305	295	.313	35.2	8 800	.211	67	23.7	5 932
306-325	315	.354	26.4	6 600	.240	68	17.9	4 475
326-345	335	.397	17.6	4 400	.274	69	12.1	3 037
346-365	355	.443	13.2	3 300	.317	72	9.4	2 361
366-385	375	.491	8.8	2 200	.341	69	6.1	1 528
386-660	525	.934	17.6	4 400	.688	74	13.0	3.241
Summary	245	.190	440.0	110 000	.119	63	276.6	69 175

should be cut from further up the stem, despite the presence of live knots. This wide diversity in wood quality is further compounded by the fact that the age at felling is likely to be variable and many logs have considerable sweep, so lowering conversion. In Australasia there has been a tendency to consider 300 mm small-end diameter to be the minimum size for a sawlog (O'Dea, 1983), which with its 150 mm diameter fast-grown knotty core generates little enthusiasm. Small-log mills are probably the best choice for logs up to 450 mm in diameter, and an increasing proportion of Europe and North America's log supply can be described as small logs.

Small-log mills can incorporate a variety of headrigs. With the smallest logs high production mills often use chipper canters, with two pairs of heads set at 90 degrees to each other to produce a rectangular cant or a profiled cant that is subsequently ripped into material of different widths. Such profile chipping is only viable where edging decisions at the headrig can be made as efficiently as when edging subsequently, i.e. the logs must be very straight. With slightly larger logs twin or quad reducer saws can be used: the saws can be either bandsaws or circular saws. Most small-log operations need a market for their chipwood. These mills need not be stand alone operations. The mill can be integrated with a large-log softwood mill using a band headrig with log carriage, with log allocation being determined operationally in the log yard. The advantage of integration does not lie wholly or fundamentally in the use of common facilities or in sharing overheads and marketing expertise. It lies in the allocation to each plant of that proportion of the raw material, which if converted by another plant, would be used less profitably. However these advantages should not be overstated. A profitable small scale operation might use one or two pairs of circular saws mounted on a single arbor (known as a scragg saw). Two cuts are made, generally symmetrically about the centre of the log, yielding a cant and two slabs. A second scragg saw can take the central cant and edge this, taking off two further slabs. Small-log mills by their very nature produce a limited product line so there is less sorting and grading, and the marketing of a limited range of products can be advantageous. There is a resurgence in proprietary inline, lineal sawmill systems using narrow kerf circular saws to rapidly process very small logs into standard sections.

3.2. A tropical hardwood mill (Hall, 1983)

The wood supply to a tropical hardwood mill is totally different to that just examined. The log size is extremely variable: in this case it ranges from 0.4-1.8 m in diameter, and has a mean small-end diameter of 690 mm (Table 7.5). The effective production time, 400 minutes per shift, is less than that for the small-log mill. The greater amount of downtime is associated with handling large, heavy logs, sometimes with pipe rot in the centre, and the increased time required for maintenance. The 300 operating days per year reflects different social conditions and traditions. Conversion is only 48%, which relates to a number of factors – rot at the centre of the logs, brittleheart, buttresses, flanges and fluting of the logs, a lack of markets for smaller sizes and lower operational efficiency. A typical breakdown pattern is shown in Figure 7.7. The aim is to cut quarter-sawn boards (25 mm) of maximum possible width and of the highest grade. The logs are orientated on the log carriage to maximize grade recovery. The headrig quarters the logs, first sawing the log in half and then reloading behind the saw and taking each half log through the saw a second time. All material goes to a pony-rig which does some of the work that could be done on a headrig, breaking down the flitches to smaller piece sizes. The pony-rig cuts material faster because the depth of cut is reduced and the carriage is lighter and accelerates more quickly. It is a smaller saw that produces both unedged boards and flitches. A resaw breaks the thick flitches into boards, and all boards go to the edger. The edger removes one or two waney edges and where desired makes a single saw cut in the board to yield two narrower boards. The piece count through the saws is shown in Table 7.6 and the flow through the mill is shown in Figure 7.8.

Table 7.5. Log supply for a tropical hardwood mill (Hall, 1983). Log input of 150,00 m^3 per year (4.9 m mean log length), for a mill operating two shifts a day (400 minutes effective time per shift) and 300 days a year.

Log diameter class, mm	Mean log S.E.D. mm	Mean log volume m^3	% of total volume	Volume per year m^3	No of logs per year	% of total logs	No of logs per minute
400-500	450	.887	1.5	2 250	2 537	3.34	.011
500-600	550	1.295	19.8	29 700	22 943	30.22	.096
600-700	650	1.780	31.1	46 650	26 208	34.54	.109
800-900	850	2.980	9.9	14 850	4 983	6.57	.021
900-1000	950	3.696	8.6	12 900	3 490	4.60	.015
1000-1100	1050	4.489	3.7	5 550	1 236	1.63	.005
1100-1800	1450	8.091	3.9	5850	723	.95	.003
Summary	690	1.977	100.0	150 000	75 896	100.0	.316

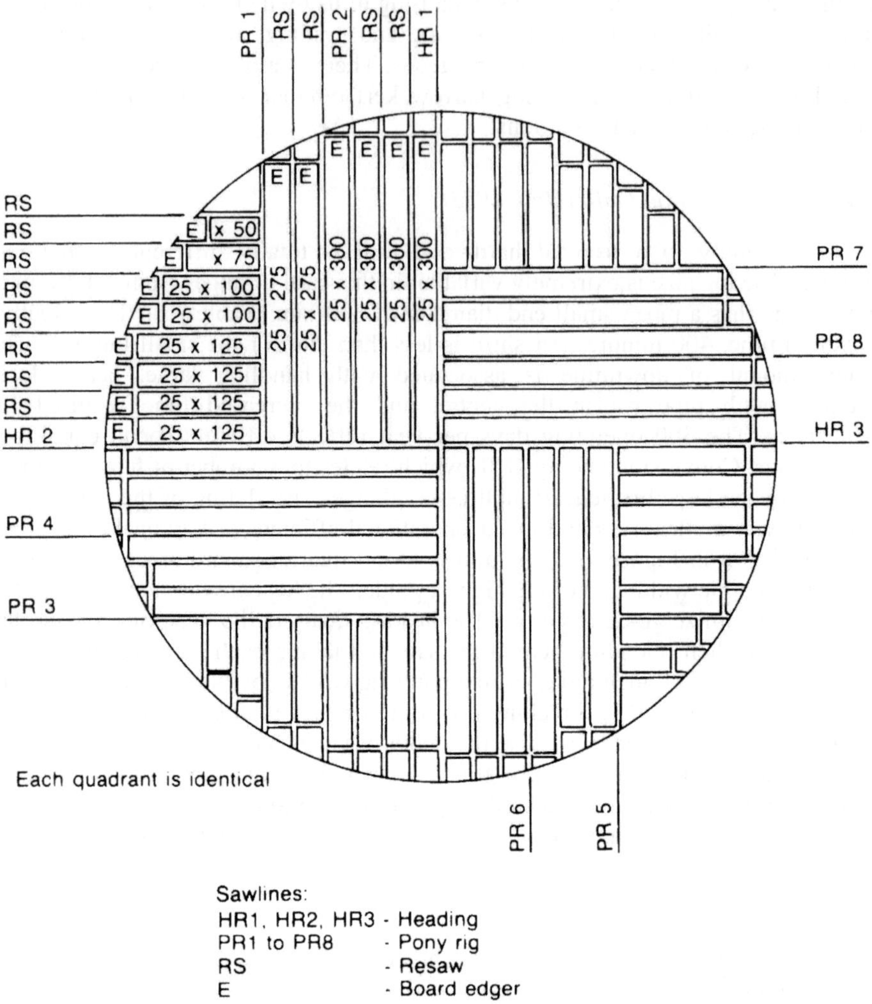

Figure 7.7. Quarter-sawing of a tropical hardwood log with a band headrig (Hall, 1983). With a 650 mm small-end diameter log the headrig makes three cuts, the band pony-rig makes eight cuts, the two resaws 48 cuts between them, and the two board-edgers 56 cuts between them.

These two studies illustrate some of the analytical procedures in mill design. Again it is important to emphasize that there are as many mill designs as there are sawmills. Other manufacturers might offer alternative solutions (Mason, 1975). Differences in log characteristics and quality can result in a radically different approach. For instance another approach to hardwood milling involves cutting a thin slab parallel to the outer face of the log, turning the log through 90 degrees and cutting another thin slab. By turning and gradually opening up the log the natural

Table 7.6. Tropical hardwood mill: piece count summary through the various saws on a 25 mm board basis (Hall, 1983).

Mean S.E.D. of logs mm	Logs per minute	Lines cut on headrig	Lines cut on pony-rig	Lines cut on resaws	Pieces to board-edger	Pieces to trimmer
450	.011	.033	.088	.352	.440	.445
550	.096	.2888	.768	3.840	4.608	4.608
650	.109	.327	.872	5.232	6.104	6.104
750	.057	.171	.456	3.420	3.876	3.876
850	.021	.063	.168	1.428	1.596	2.100
950	.015	.045	.120	1.140	1.260	1.740
1050	.005	.015	.040	.440	.480	.640
1450	.003	.018	.048	.462	.318	.516
Summary	.316	.960	2.560	16.314	18.682	20.024

defects such as knots and shake are disclosed while at the same time the sapwood is being cut out and segregated. In tropical hardwoods the best timber is in the outer heartwood as the centre of the log can suffer from brittleheart or heart rot. By turning the log the best grades can be cut leaving a tapered piece in the middle containing the least desirable and lowest valued material: this material is removed at the edger in the previous mill layout (Figure 7.8). Growth stresses in some species mean that the wood is liable to move in the saw. The saw kerf can close behind the teeth and the wood then pinches the back of the saw and may even cause the saw to seize up. For these reasons gang or multisaw edgers are inappropriate for most hardwoods. Where growth stresses are severe it is often advisable to flat/back saw as this will result in bowed boards rather than ones with crook (also called spring) which would arise with quarter-sawing.

Temperate hardwood logs tend to be smaller than tropical hardwood logs. Although the tradition has been to turn the log to recover the best grades, this may not be the best approach after all. Richards *et al.* (1980) found that live-sawing followed by ripping the boards to segregate the quality outerwood from the defective wood in the centre performed best with logs having a small defect core. These are precisely the logs where tradition has maintained that live-sawing is inappropriate and where turning the log constantly will yield the best return. Such simulation studies are supported by mill studies on yellow-poplar, *Liriodendron tulipifera,* in the south-eastern United States (Peter, 1967) and on hard maple, *Acer saccharum* in Eastern Canada (Pnevmaticos and Bousquet, 1972). The advantages of live-sawing, especially where ripping to segregate the defect core is practiced, lie in skilful edging. It is this skill that is most lacking in hardwood mills which explains why the benefits have not been appreciated in practice (Richards *et al.*, 1980). Live-sawing or cant-sawing would allow the use of frame saws. Frame saws are used in Scandinavia for milling oak and birch.

However, four-sided grade sawing still performs better with small logs having a large defect core.

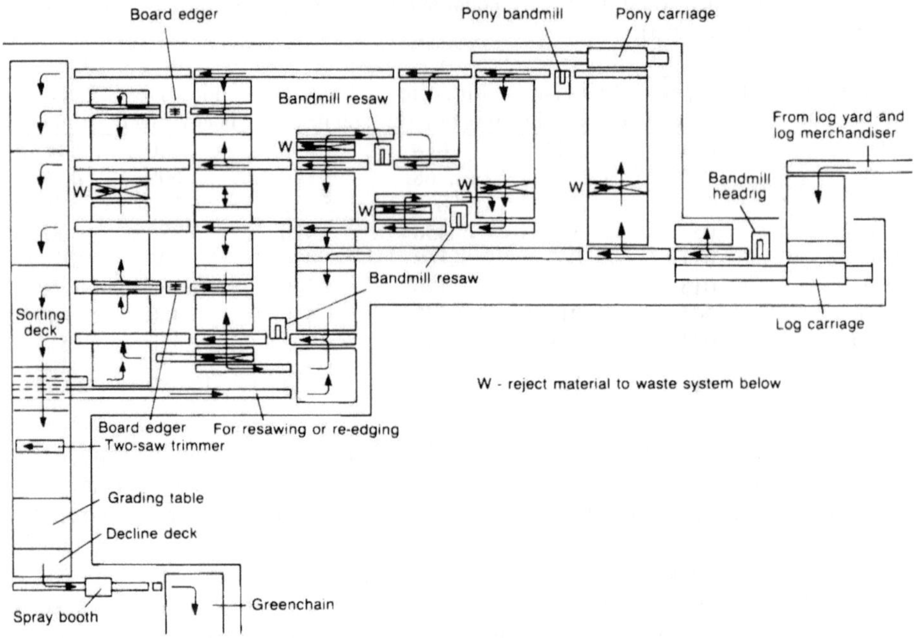

Figure 7.8. Tropical hardwood mill with an anticipated output of 72 000 $m^3 yr^{-1}$ (Hall, 1983).

3.3. A large-log softwood mill

The headrig in large-log mills is invariably a bandsaw with a log carriage that produces large slabs, flitches and cants for the secondary saws. For example, mills in the Pacific Northwest sawing large coastal Douglas fir logs sought to cut clear, high grade timber from the outside of the log. The knees of the log carriage must be able to move independently so that the side of the log presented to the saw is parallel to the line of the saw cut and all the taper of the log is taken up by movement of the knees. Once the saw exposes the poorer grades within the log, the knees are moved back so they are in alignment with the carriage and in line with the saw. The back of the log will lie parallel to the cutting plane of the saw. A tapered wedge of low-grade wood can then be taken off. On turning the log through 180 degrees another untapered full length slab will come off the saw. If the log is turned through 90 degrees it must be skewed again so that the first slab or flitch has minimal taper. This cutting strategy applies to quality hardwoods and pruned plantation softwoods. The emphasis is on grade recovery where the price differential between the highest and lowest grades can be as much as 800%.

The bandsaw with log carriage can handle logs of variable size and quality and log sorting is unnecessary. When a large log is presented to the saw a number of cuts can be taken while still giving plenty of work to the secondary saws, whereas when a smaller diameter log is presented the headrig will make one or at most two cuts before turning to another log. Where logs are of poorer quality adjustment of the knees is unnecessary. Saw cuts can be parallel to one another to give flitches and a large cant. In the latter case a framesaw is as suitable provided the logs are pre-sorted into various diameter categories: such saws are most efficient when the log sizes are intermediate, i.e. 300-400 mm small-end diameter.

3.4. Milling small hardwood logs

3.4.1. Eucalypts

It is reported that peripheral growth stresses are broadly the same in small logs as in large logs (Archer, 1986), so the internal stress gradient is more severe in small hardwood logs as the forces are distributed over a smaller radius, and boards cut from small logs will distort more. Halving the log diameter increases the amount of distortion in that log when sawn by about a factor of 8 (Chapter 10). Hence the perception that it is desirable to grow eucalypts until they achieve a diameter at breast height of at least 750 mm. The butt log will then be large enough to quarter-saw efficiently. However small logs will always be available for milling, especially from second-growth forest. The growth stresses and the occasional presence of brittleheart in small logs of many hardwoods generally means that it is not economic to grade saw such logs where the small-end diameter is under 400 mm when quarter-sawing, or 300 mm when flat-sawing.

Haslett (1988) has reviewed strategies for sawing plantation-grown eucalypts, where the problems of brittleheart, growth stresses and shake can be acute, especially with small logs. Here only an abbreviated and partial discussion is possible. Within the log longitudinal growth stresses are symmetric: tensile at the periphery and compressive around the pith. If the log were to be cut through the centre the two halves would bend apart as the stresses within the two halves seek to achieve a new equilibrium. The longitudinal bending of both pieces is toward the bark (the same effect is seen in cutting celery 'along the stick'). When using a log carriage and cutting only a thin slab most of the movement shows up in the bowing of the slab away from the saw (Figure 7.9). However it is difficult to hold the log firmly enough on the carriage and the sawn face also bows out slightly at mid-length. If the bowing of the sawn face exceeds about 3 mm in a 4.8 m log a thin non-productive straightening cut is required before further sized timber can be taken from that face. The straightening cut shaves off the slight curvature in the face of the log. This approach suits large logs that do not distort much in the saw. With small logs distortion can be severe and frequent straightening cuts may be necessary which lowers both conversion and productivity. Haslett (1988) discusses appropriate cutting strategies that are essential to minimize distortion and to maximize grade recovery of defect-free wood. This requires a proper appreciation of the stress distribution within round and partially sawn logs. If the log is flat-sawn the stresses in the log result in bow in the sawn timber, but for the ash group of eucalypts flat-sawing results in an unacceptably high level of surface checking (high tangential shrinkage) and collapse (high moisture content and low basic density) during drying.

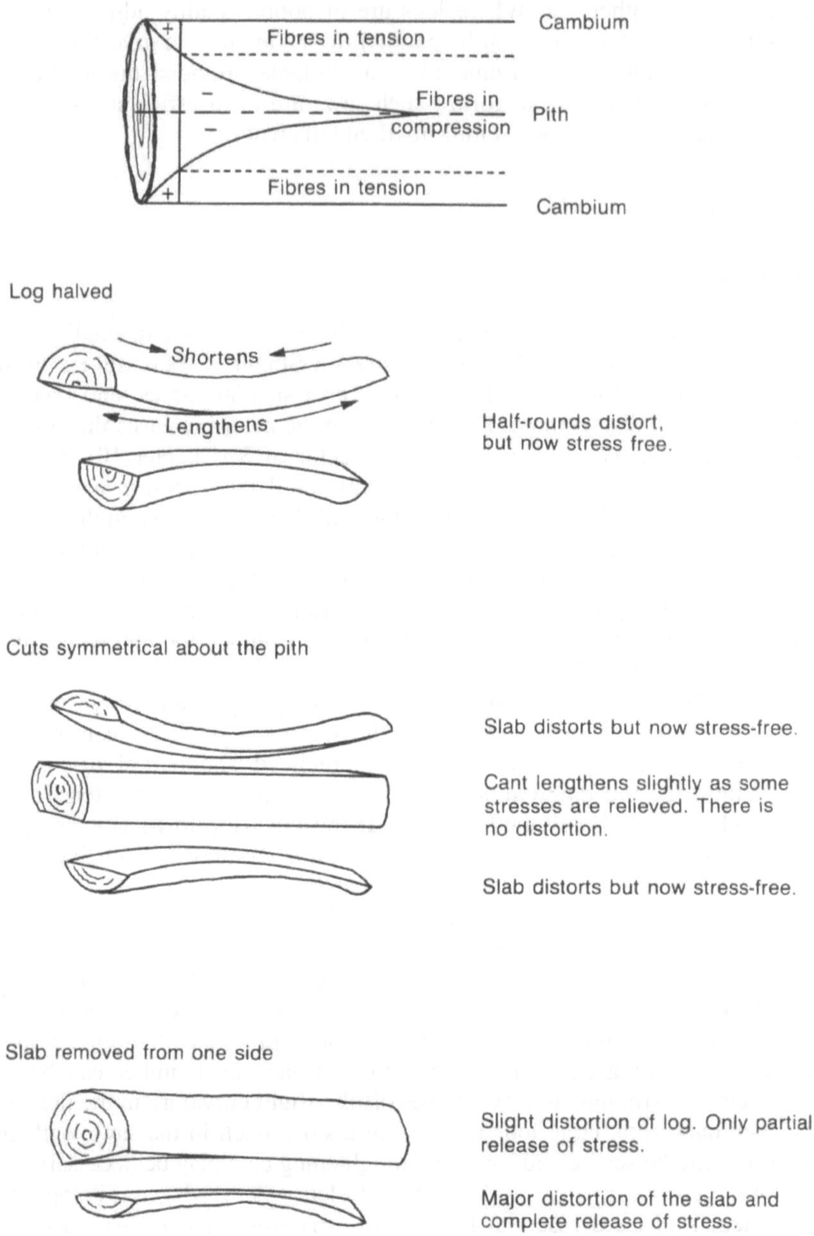

Figure 7.9. Distortion is a consequence of the release of growth stresses (Figure 6.16) on sawing. In a scragg mill two circular saws cut symmetrically about the centre so it is possible to produce an undistorted central cant. Unfortunately the centre of the log has the lowest grade material: maximizing recovery from the centre of the log may be less desirable than cutting round the log and accepting the distortions that arise (Haslett, 1988).

Figure 7.10. Milling of a partly sawn eucalypt log on a band headrig fitted with log carriage and linebar (Haslett, 1988). The partial relief of growth stresses has resulted in distortion of the face that is being presented to the saw.

Quarter-sawing has its problems too, the lumber shows crook (Figure 8.14), a problem that is harder to remedy. However the attractive ribbon figure of quarter-sawn boards of a number of hardwoods means that a premium is paid for quarter-sawing.

In Australia an alternative approach is to use a carriage together with a linebar to mill small eucalypt logs. Originally the linebar was developed for resaws to provide a long fence against which the face of the lumber could be pressed to provide good alignment into the saw and for accurate referencing and sizing. Where adapted to the headrig the linebar is placed on the infeed side of the saw and is hydraulically positioned to set the thickness of the cut. For the opening cut the linebar is used only to align the log to ensure that the first cut is parallel to the outer face of the log. On the second cut the sawn face of the log is pressed firmly against the linebar by the knees of the log carriage which can be individually operated (Figures 7.10 and 7.11). Because of the release of growth stresses the sawn face of the log is slightly bowed outwards and can only make contact with the linebar at one point (tangential contact). The sawyer adjusts the individual head pressures on the knees of the carriage in order to maintain contact with the linebar at a point just ahead of the saw as the log moves through the saw. The technique calls for judgement and practice. A linebar can cope with approximately 30 mm distortion in a 4.8 m log so fewer straightening cuts are needed and conversion is enhanced, although there is the penalty of lower throughput compared to a traditional carriage. Both traditional and linebar carriage sawing equipment and strategies are fully discussed in Haslett (1988).

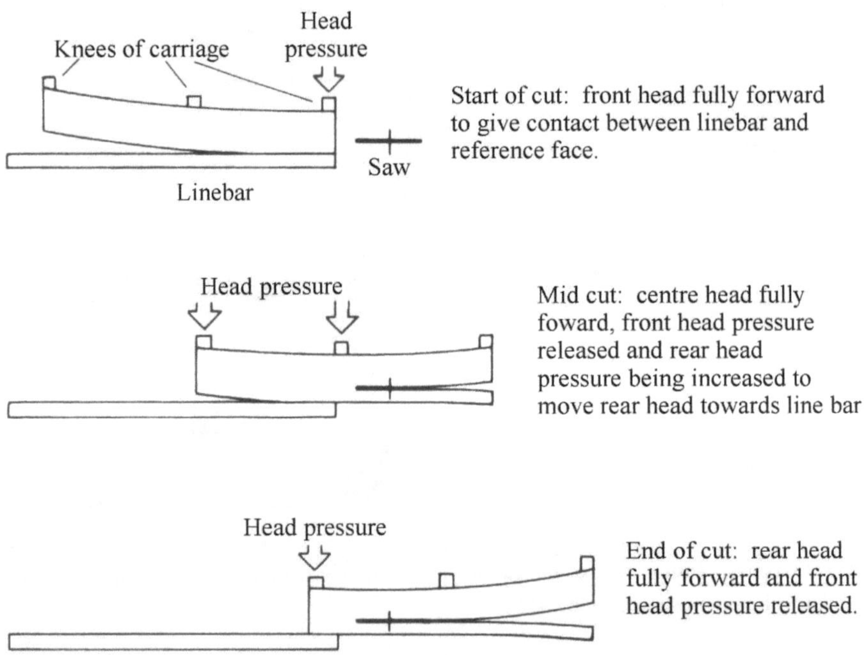

Figure 7.11. Sketch of a linebar operating in conjunction with log carriage (Haslett, 1988). The log is held against the linebar as it travels through the bandsaw with the thickness of the material cut being determined by the gap between the linebar and the saw. The curvature especially of the log due to the release of growth stresses is exaggerated.

Another obvious solution for hardwoods is a scragg mill, where twin circular saws make two cuts symmetrically about a central cant that will not distort (Figure 7.9). This cant has proportionately more wood from the centre of the log (compared to the original log) and there will be some readjustment of the initial stresses. Normally there is sufficient wood in the cant to sustain the severe stresses without splitting. The cant can be recut sequentially parallel to the first cuts or set at 90 degrees. When the cant is ripped in a second scragg saw the stress gradient across the ripped boards is less severe as the central cant has experienced already some stress relief. Unfortunately cutting for grade is not feasible in a scragg mill since the best wood from the outside of the log is removed as tapered slabs while the central brittleheart is recovered in full. Scragg sawing suits low quality logs, where quarter-sawing and grade recovery are not primary objectives, as well as thinnings from plantations. More sophisticated, larger mills may use twin-saw reducer headrigs.

Storage of full length logs under water sprinklers for long periods (>3 months) greatly reduces stresses, and end-splitting which is a manifestation of growth stresses, and offers a means to control some of the worst effects. However storage cost are high (Clifton, 1978). High growth stresses are not necessarily associated with fast growth rates. The severity of growth stresses appears to be subject to both genetic and environmental control (Hillis, 1978) and both genetic improvement and specific silvicultural treatments are being pursued in several countries.

3.4.2. Other hardwoods

Although most small-log hardwood mills do not encounter growth stresses as severe as those in some eucalypts, the economics of milling small hardwood logs has never been very promising, especially when they are of low-grade. Nearly 87% of all hardwood sawlogs in the United States are of secondary quality, No. 2 and below (Rast *et al.*, 1979). However techniques to make the milling of such logs more viable have been promoted in the North America. The philosophy is very simple (Maeglin and Boone, 1983; 1988). It involves live-sawing flitches, which are lightly edged and kiln-dried, preferable using a high temperature schedule. Only after drying are the flitches ripped. Leaving the flitches in full-width until they are dried overcomes the problems associated with the release of residual stresses which, if the flitches had been immediately resawn, would cause crook. Some end-splitting may be unavoidable in full-width flitches because steep stress gradients will still be locked into flitches cut from logs which had severe growth stresses. By kiln-drying green timber at elevated temperatures thermal softening and flow of the lignin can occur (Goring, 1965). This permits some physical rearrangement and displacement of material so relieving the stresses within the cell wall. Then the dry, virtually stress-free flitches can be ripped with minimal distortion occurring. This approach (saw-dry-rip also known as SDR) has proved viable for both the construction and furniture industries, which have very different wood procurement objectives.

Hardwoods are used for plywood, for structural applications including pallets, in furniture and in the remanufacturing sector. It is important not to discount the lower grade pallet and container market as this sector absorbs about half of all US hardwood lumber, with dimension, furniture and cabinet markets absorbing a further 30%. The pallet and container markets favour medium-to-high density timbers because of their high strength and impact resistance.

Hardwood mills cut both structural and board grades. Board grades are intended to be used either in the size sawn, e.g. flooring, or for factory lumber that is recut subsequently into smaller pieces. Hardwoods are usually sold rough sawn in random widths and lengths, being edged to give the maximum possible width rather than being cut to a standard width. The remanufacturer, e.g. the furniture factory, subsequently cuts up the boards to maximize the number of small pieces of the desired size (dimension stock) and quality (defect-free if visible in the finished product or tolerating certain defects if the face is not exposed). The term dimension stock refers to the pre-cut components of the desired size, tolerances and grade ready for assembly.

Hardwood mills require a reasonable quality and size of log to be economic. However hardwood forests contain large volumes of low-grade material that cannot be milled economically by traditional techniques. The Eastern United States illustrates the problem (Table 7.7). According to Araman (1987a) the principal markets for FAS & Sel (First-and-Seconds and Select) are mouldings, millwork and export (Pacific and Europe). Second line material (No. 1 & 2 Common) is used for domestic dimension, furniture, cabinet, flooring and other manufacturing. The poorest material is used in rail ties, mine work, for the production of pallet parts and flooring. The top grades (FAS & Sel) are most profitable although output is limited, while producers are satisfied if they can recover their costs with the lowest grades (below No. 2 Commons). Thus a sawmill's profits hinge on having adequate outlets for No. 1 and No. 2 Commons that account for about half of all production.

Table 7.7. Standing timber sawlog volume and grade recovery of select hardwoods in the Eastern United States (Araman, 1987a). Mill grade outturn should be somewhat better as lumber grades were derived from log grade inventory and many of the small low-grade logs would never be milled.

Species	Volume	Lumber grade (percentage)			
	M m^3	FAS&Sel	No. 1C	No. 2C	Below 2C
Select oaks	323	12	24	27	37
Hard maple	102	11	21	26	42
Ash, walnut, cherry	104	19	25	29	27
Yellow birch	21	12	21	24	43
All select hardwoods	550	12%	23%	27%	38%

Much of the material graded for appearance is not needed in long lengths. Factory and shop grades assume that each board will be further processed to yield clear cuttings or sound pieces (blanks) for the furniture and cabinet dimension stock, for various forms of millwork, and for finger-jointing. The US NHLA (1991) sets out the number and minimum size of cuttings (6 in x 6 ft to as little as 1-1/2 in x 2 ft for No. 3B Commons). Grading is based on the percentage recovery of material of the desired quality (Figure 10.1). Recovery ranges from 83-1/3% to as low as 25%.

A novel strategy to utilize the very large volumes of No. 2 Common or lower grades of hardwood lies in producing standard length blanks, from 300 mm and increasing in 100 mm increments (Araman and Hansen, 1983; Hansen and Araman, 1985). This approach deserves emphasis because traditional mills cannot cope with small low-grade logs that predominate in unmanaged hardwood stands, and there is an acute shortage of quality logs. The strategy (Reynolds and Gatchell, 1982) differs from conventional hardwood milling in that:

- A new, non-lumber product is produced: standard-sized blanks.
- Log diameters are restricted and conversion is simplified: no cutting for grade.
- Every board containing a minimum sized cutting is processed and no other product is produced.
- Operator decision making is minimized and recutting options are strictly limited.

Interest has centred on the manufacture of furniture and cabinet blanks (Reynolds and Gatchell, 1982; Reynolds *et al.*, 1983) from 1.9-2.5 m hardwood logs with small-end diameters of 190-320 mm. Milling short lengths alleviates the worst effects of sweep and allows the poorest sections of the stem to be assigned immediately to fuelwood or pulp chips. The recommended log diameters is a compromise to avoid the need for large circular saws with wide kerfs, which would seriously affect the recovery from small logs, and too much corewood if log sizes are too small. Over half the US hardwood growing stock falls in this size range. The problem in North America is that hardwood forest owners cannot afford to pay for thinning operations which yield small, poorly-formed logs that have little value as sawlogs. Yet this thinning operation is needed to improve stand quality and yield larger logs when the stands mature and are felled. The inability to thin these forests economically means that the forests are being undercut and future timber quality will show little improvement. The concept of furniture and kitchen cabinet blanks and associated technology was developed to make use of abundant low-grade hardwoods in ways that would see the wood end up in much more valuable end-uses than would have been possible with traditional technology and thinking, and for the first time make thinning of the US hardwood resource economically viable.

The key is to minimize complex decision making and rely instead on standard cutting procedures using simple technology (Figure 7.12). Short-length harvesting is possible because long length material is rarely needed (80% of pieces needed are less than 1.2 m and over 50% are less than 0.9 m). There is no cutting for grade. Logs are live-sawn to give just two cants, either 82.5 or 101.5 mm thick. The cants are gang-ripped to produce boards that are 25.4 or 31.7 mm thick. All boards with at least one minimum clear sized cutting (38 x 380 mm) are partially air-dried before kiln-drying to 6% moisture content. Gang-ripping of the cants and drying between smooth stickers permits the boards to crook and no effort is made to prevent this. However, the stacks are top-weighted to minimize cup and twist. Badly crooked boards are rejected. The remaining boards are stress-free although still containing other defects. The worst defects in these standard width boards (82.5 or 101.5 mm)

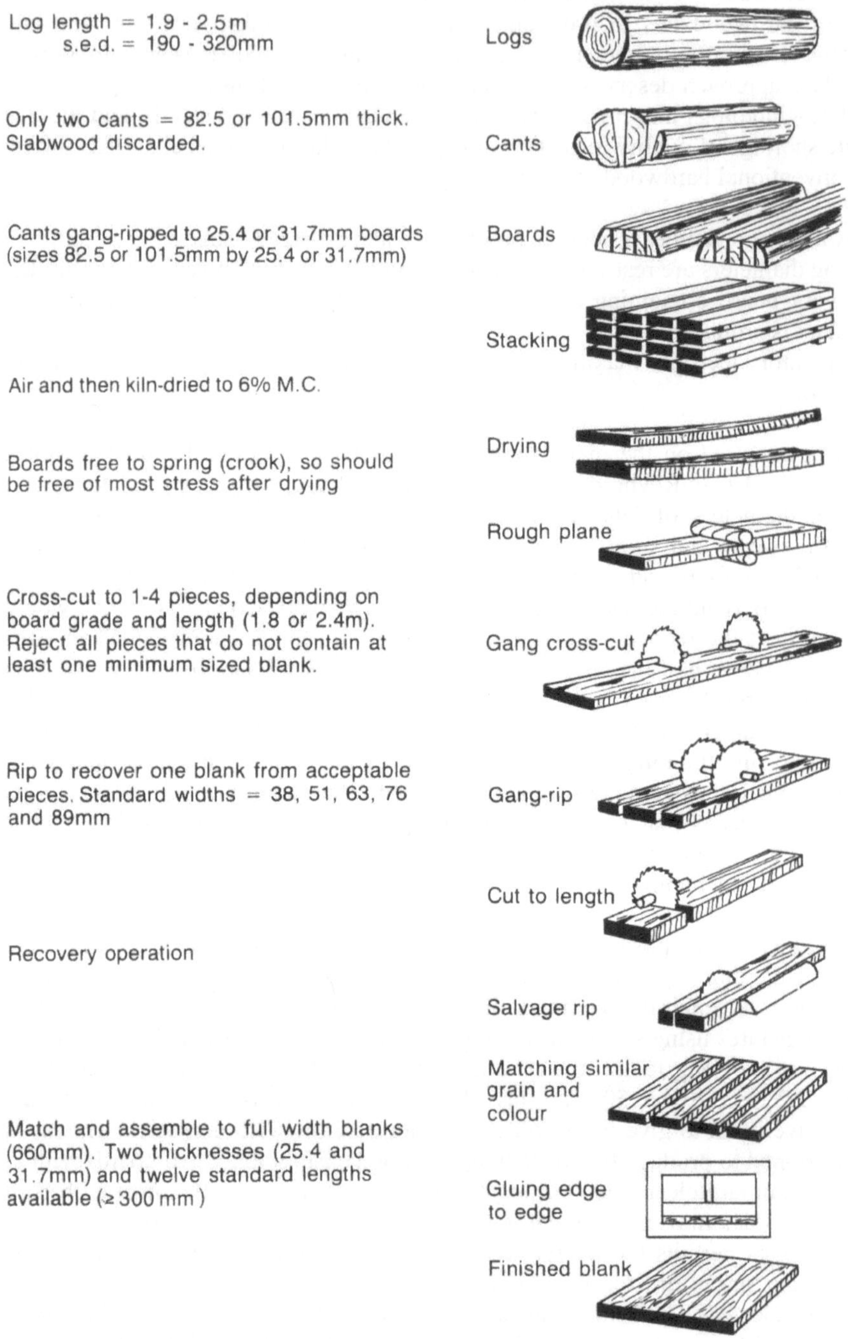

Figure 7.12. A new way to process low-grade hardwoods (Reynolds and Gatchell, 1982).

are removed by cross-cutting to give one to four pieces of standard length. To simplify decision making it is suggested that only 4 out of 12 standard lengths are cut at any one time. Finally each piece is ripped to yield a single cutting of a standard width (38, 51, 63, 76 and 89 mm). Alternatively the material can be gang-ripped first and then cross-cut to give blanks which are slightly longer but narrower. The cuttings can be clear (defect-free) or of frame quality (admitting certain small defects).

While standard widths meet the majority of needs of furniture and kitchen cabinet manufacturers, there are advantages in edge-gluing the standard width blanks to full-width (660 mm) blanks. Then the purchaser can rip full-width blanks to the precise, narrower sections desired in his secondary manufacturing operation. There is one decisive marketing advantage in this approach, the buyer does not need to know or understand the vagaries of processing and grading, which differ markedly between countries. The buyer merely purchases edge-glued clear blanks of such standard dimensions as meets his stock requirements. Although furniture manufacturers use literally thousands of different component sizes and grades, by ordering an appropriate mix of standard blanks trimming losses to convert to any final component should be less than 10%.

The development of such a strategy in the United States is significant for a number of reasons. First, the United States is the biggest producer of sawn hardwoods in the world, and approximately half of its hardwood resource is amenable to such technology. Secondly, the United States hardwood resource is being undercut so there is potential to lift exports significantly. Finally, the processing strategy is similar to that operating in Japan. This suggests a long term coincidence of interests.

3.5. Eucalypt sawmilling revisited

A major problem with small re-growth eucalypt is the distortion arising from the release of growth stresses. If the wood moves on the saw the cut material will not be accurately sized. Also the cut surface of the part-sawn log retained on the carriage may not stay flat. Here, the flexibility of a log carriage with a linebar lies in its ability to accurately size the majority of pieces even from distortion-prone small logs, and its ability to re-reference and accurately recut large dimensioned flitches to final product size. This reduces the work-load on the resaw. Depending on log size and the product mix, a line-bar carriage followed by a two-man resaw cutting for grade should produce 30-40 m^3 of green lumber per day, working a single shift.

In the above example of a single resaw in a small-log eucalypt mill, the amount of resizing required at the resaw increases as log diameter decreases and growth stresses become more severe. A piece that needs resizing effectively has no flat/plane face to work from and the first job at the resaw is merely to create a flat face from which to reference subsequent cuts This is slow and labour demanding.

Higher sawing rates can be achieved on a conventional carriage without a linebar but to work well it needs two resaws in which case it should produce 40-60 m^3 of green lumber per day in a single shift. The head saw can cut more wood but, without a linebar and with young small-log eucalypts that can be highly stressed, more pieces are poorly sized. With two resaws, the two-man resaw concentrates on poorly sized, difficult-to-handle pieces, that require face-cutting (including slabs and round-backs), while a high production one-man resaw handles the accurately sized flitches that only need ripping and edging to final product sizes.

Other issues relate to the physical location of the sawyer and not being able to gauge on either side of the saw. With small eucalypts, there is the need for a lot of face-cutting. With a one-man resaw the sawyer is located next to the saw and relies on remote controlled equipment to feed the saw. This operation is imprecise and slow. The sawyer cannot orientate a flitch with the same degree of precision as with a two-man resaw; consequently such saws are best fed with accurately sized material. With a two-man resaw the sawyer located in line and on the in-feed side of the saw is well placed to carry out minor corrections to the flitch orientation prior to sawing and where necessary to turn/prepare material for resizing.

Single-bladed saws are right or left handed, e.g. a resaw can stand on one side or the other of the infeed rolls. When observed from the infeed side, if dimensioned products are removed from the left-hand side of the saw, it is referred to as a left-handed sawing unit, and likewise, a right-handed unit would have products sized on the right-hand side of the saw. Ideally, resawing should work from a dressed, flat face. Where the headrig cuts two faces are right angles to each other, one face will lie flat on the feed/transfer rolls and the other is pulled by the sawyer to lie against the linebar of the resaw to cut a sized board (emerging from between the linebar and the saw) leaving the residual piece of variable thickness that can be returned by merry-go-round transfer belts and rolls to the same saw. If the resaw is on the other side of the infeed rolls to that of the headrig, the face that lies on the feed rolls will have to be lifted against the linebar while the face that was originally vertical will now lie on the feed rolls. But what if only one face is straight?

With two sawn faces, whether parallel to one another (a live-sawn flitch) or at right angles, growth stresses may result in only one face that is straight enough for the resaw to cut accurately-sized, dimension products from. If the warped face is presented to the linebar then either the flitch has to be turned or a fresh, straightening face cut must be made. Such practices are either wasteful or cause loss of production or both.

These problems are peculiar to Australia. The native hardwood industry has been trapped in a legacy of cheap wood to be processed, ideally, somewhat less inefficiently. Equally civic authorities have reluctantly supported a continuation of a rural industry that offered work of a kind without much investment and without energetically driving change. Thus the log carriage with linebar has been an intelligent, low-cost response to high growth stresses evident in much unmanaged forest.

Elsewhere in the world eucalypts are planted. It is worth repeating the observation, that although Australia is the home of almost all eucalyptus species, the 3.1 million m^3 of sawlog production comes almost entirely from native forests. This roughly matches the estimated 2.9 million m^3 coming from plantations worldwide (with 37% from Brazil alone). Indeed in Australia the plantation area is only about 15% of that in Brazil, which has *c.* 3 million hectares (Donnelly *et al.*, 2003).

4. MILL EFFICIENCY

The fragmented nature of the sawmilling industry in most countries has resulted in undercapitalisation and weak management. In other industries equipment 5-10 years old is obsolete, yet in sawmills the machinery can be substantially older. Some of these mills are uneconomic and inefficient, others are highly profitable. Because of a shortage of or lack of access to finance as much emphasis is placed on further investment in existing mills as in building new mills. Incremental improvements – replacing a single machine, improved sawing patterns and flow through the mill, and better maintenance – can all result in improved grade and volume conversion, and mill profit. For example, during the 1970s sawmill improvement programmes in the United States resulted in an average increase of about 4% in the amount of timber recovered (Lunstrum, 1982). Only simple improvements were involved, mainly aimed at reducing kerf, sawing variation, shrinkage and planing allowances, and simply cutting oversize to allow for uncertainty. Adequate surge areas to hold material between saws and unscrambling devices greatly improve efficiency.

Bryan (1977) highlighted the obvious fact that in a competitive environment the percentage of total revenue that represents profit is quite small. This lead to the conclusion that there is considerable leverage for profit improvement through improved performance.

> In every business environment involving more than a few variables there exists a profit gap – the difference between actual profits recorded for an accounting period, and that which would have been achieved had all the resources and opportunities been utilised in an optimal manner. Because of the complexities and large number of variables influencing financial performance in sawmilling, the profit gap in this industry is usually very large. Even with the best run operations, there are usually many ways in which things can be done differently to improve earnings.

Further,

> In a commodity market, where no single producer can have a significant effect on the market, product prices over a full market cycle tend to seek levels that keep the average producer in business – barely. As a result, any company that can lift their performance level above that of the average producer will prosper as it enjoys prices in the market place which indirectly reflect the capabilities of less efficient organisations (Bryan, 1977).

The attraction of a small mill lies in its low capitalisation which, coupled with ingenuity and skill, good management and operational efficiency, hopefully arising from good planning and cooperation, should be capable of good profits. Indeed Richardson (1978) has long argued that sawmilling is an industry which is not particularly amenable to economies of scale. The principal difficulty with that view lies in matching the unit sizes of ancillary facilities (boiler, kiln, treatment plant, machine stress grader and the sawshop) which are available in discrete sizes or capacities that may not match the scale of the operation, so that all may not be fully utilized or alternatively demand may exceed capacity. Small-scale industries are successful because they are generally more efficient and more flexible than similar large-scale operations. The successful deployment of capital has little to do with its availability. Small industry requires a high level of managerial and technical skill.

5. ASPECTS OF OPTIMIZING SAWLOG BREAKDOWN

5.1. Log debarking (Wingate-Hill and MacArthur, 1991)

There are many benefits in debarking logs:

- Bark picks up sand and dirt during extraction. Its removal reduces tooth damage and wear resulting in reduced maintenance of saws and less down time.
- Better exposure of log shape and defects. The sawyer is in a better position to make the correct decision which lead to better conversion and grade recovery.
- Easier handling of material with reduced fouling of saws and transfer systems.
- Clean, bark-free chips and slabwood command a better price. The segregated bark can be used for fuel or disposed of.

When debarking there are three points to consider:

- The object is to shear the bark from the wood but the bark-wood bond strength varies greatly, depending on species, age, the time of the year when felling and the time between felling and debarking.
- Some bark can remain tenaciously attached and an exposed face is needed to work from.
- Some barks break into small sized fragments whilst others with strong, long bark fibres pull away in very long strands that can block and otherwise foul up the debarker.

The commonest debarkers in softwood sawmills are ring debarkers in which the log is held between spiked rollers and moved by them longitudinally through the debarking ring (Figure 7.13a). The rotating ring carries a number of blunt knives that pivot and press against the log, shearing the bark off at the cambium. Logs from 650 to 100 mm diameter can be accommodated at feed speeds ranging from 0.25-1 m s^{-1}. Stringy-barked hardwoods present a problem. The bark tends to pull away easily in long strands that wrap round the debarking arms in a tangled mass, so blocking the machine. In other species the bark clings tenaciously. One approach has been to separate the cutting and shearing functions of the debarker by using two cutterheads. The first of which cuts helical grooves in the tight bark while the traditional debarking head removes the bark between the grooves.

A Rosserhead debarker has a much lower throughput and is more suited to smaller mills (Figure 7.13b). It has certain advantages: it can accommodate poorly shaped logs and logs too large for a ring debarker, and can handle most bark types except those which tear away in long strands. The log is rotated while at the same time a rotating cutter head is lowered onto the log. Either the log is static and the cutterhead traverses the log or *vice versa*. Either way the bark is removed in a helical pattern. Some Rosserhead type debarkers remove substantial amounts of wood fibre and peel the nodal swelling around branches, which may be undesirable for high

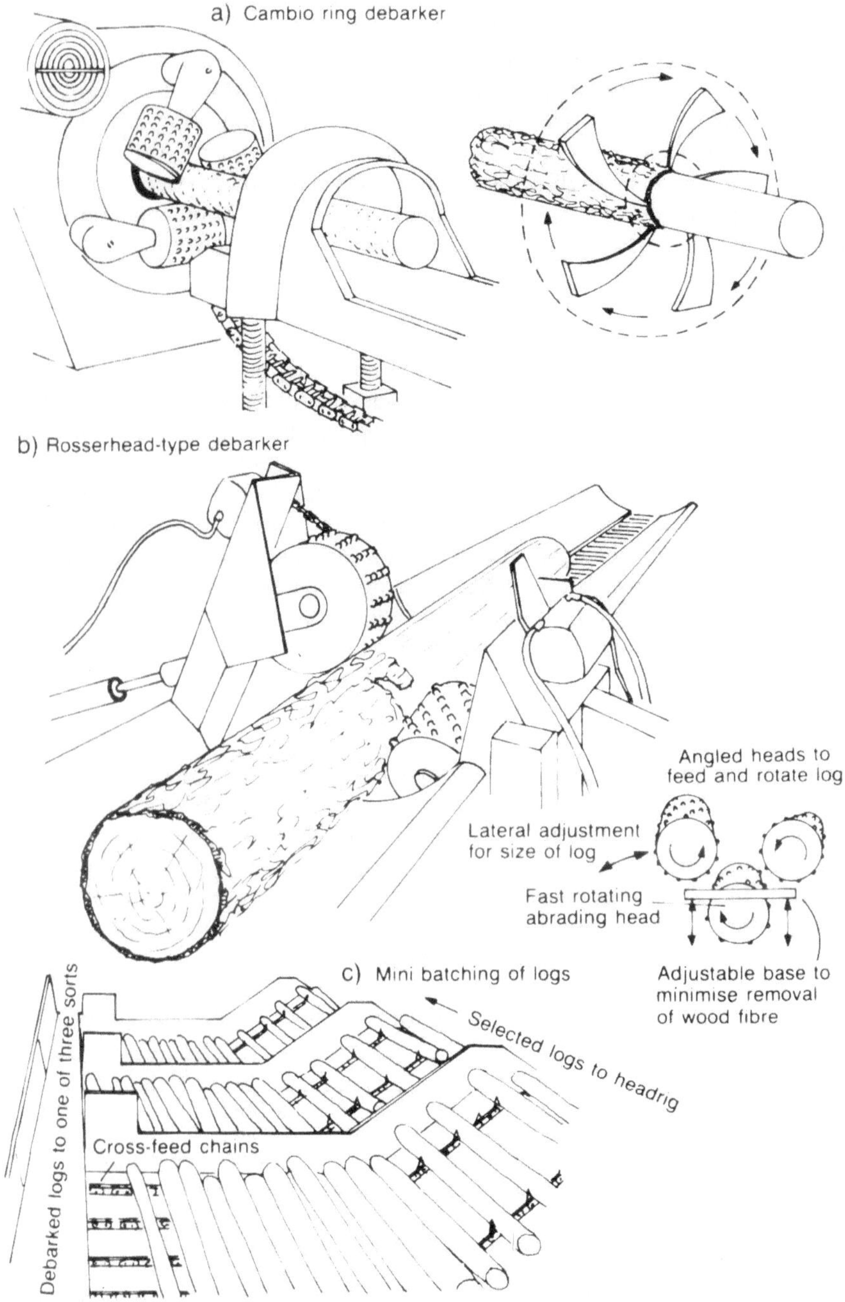

Figure 7.13. Log preparation before entering the mill. (a) Cambio ring debarker. (b) Rosserhead type debarker. (c). Mini-batching immediately prior to milling.

strength pole material. Other versions have moveable abrading heads that float around the nodes, giving good debarking with little fibre removal.

Logs of poor form or heavily fluted are effectively debarked with high pressure water jets (10 MPa). Hydraulic debarking uses large quantities of water and can erode the wood as well as blasting off the bark if traversing the log too slowly. Effective filtration and recycling of water and the disposal of wet bark are two causes for concern. Very high pressure water jets have lower water consumption. Such jets have been developed experimentally for cutting various materials but the penalty for going to higher pressures is the greater electrical power demand.

5.2. Log sorting

This is a costly operation that appears to add no value, but is essential to efficient processing. Sorting by species, size and grade facilitates handling later on, and permits provisional allocation of logs to appropriate headrigs if more than one is available. Framesaws require a very large number of sorts (as many as 60) in order to maximize conversion. Where saws can be reset rapidly and the log diameters are not too variable, the number of sorts can be much reduced. Sorting is still desirable, but feeding a single size category into a mill for long periods can result in underuse of some saw(s) because with a particular breakdown patterns only certain piece sizes will be cut if volume conversion is to be maximized. Mini-batching may help as it ensures an appropriate mix of logs to keep all the saws in wood (Figure 7-13c).

5.3. Sawing patterns

These interact unpredictably with log form and size. There is no single best sawing method for all logs. The four basic sawing patterns are live-sawing, sawing around, cant and quarter-sawing (Figure 7.14). Sawing around and quarter-sawing are only appropriate for large logs (Table 7.8) with quarter-sawing largely confined to hardwoods. In general cant-sawing gives higher volume yields than live-sawing (Hallock *et al.*, 1976) because in cant-sawing some of the taper in the cant can be recovered as short boards whereas in live-sawing this taper is lost as edgings.

Table 7.8. Generalizing on the consequences of flat and quarter sawing in eucalypts.

	Quarter sawn	Back/flat/live sawn
Productivity	Reduced, up to 30% less	Greater
Recovery	Reduced, typically by 10%	Greater
Dimensions	Limited by log diameter	Wider products possible
Knotty core	Affects a high proportion of products	Lesser impact
Distortion	Crook/spring: problematic	Bow
Drying degrade	Easier to control	Much harder to control
Growth rings	More uniform	A feature on face of boards
Kino veins	Less obvious	Prominent
Stability/movement	Good stability	More movement

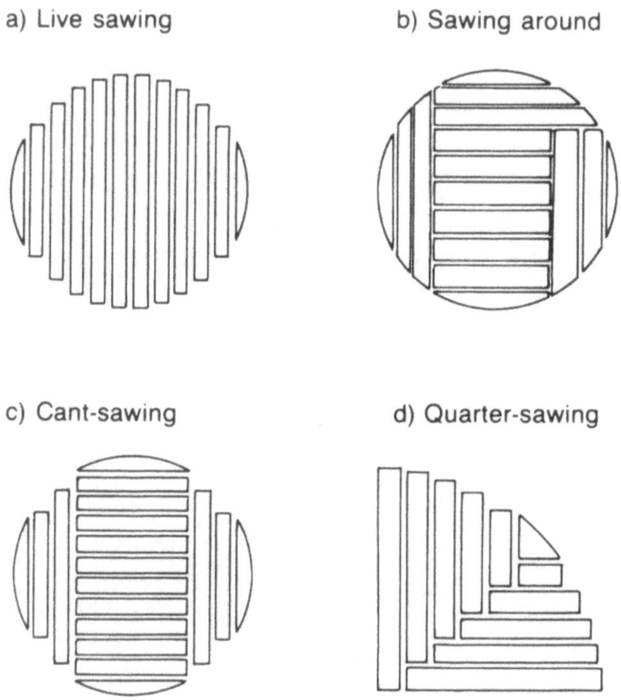

Figure 7.14. Basic cutting patterns. (a) Live-sawing. (b) Sawing around or sawing for grade. (c) Cant-sawing (see also Figure 7.5). (d) Quarter-sawing (see also Figure 7.7).

Further there is an increased incidence of large spike knots when live-sawing, which results in a lower recovery of better grades in softwoods.

Sawing (Figure 7.15) can involve split-taper (sawing parallel to the pith) or full-taper (sawing parallel to the cambium). In general full-taper gives a higher conversion with short logs having little taper, where there is the opportunity to recover an extra piece of short lumber from the side. On the other hand with long logs having significant taper split-taper sawing gives a reasonable conversion of short side boards and a better conversion from the central cant compared to full-taper sawing. With split-taper sawing the width of the central cant is approximately the same on both its faces. With full-taper sawing the cant has one face of constant width along its length whereas the opposite face is heavily tapered, resulting in a lower conversion. In this situation with full-taper sawing the higher conversion from the tapered side slab is insufficient to offset the lower conversion from the cant, and split-taper sawing gives a better yield. Full-taper sawing on all four faces as the log is turned maximizes the conversion and recovery from the outside of the log and yields a tapered trapesoid of boxed-pith, which in softwoods would never be of much value. Grade recovery will be better with pruned logs.

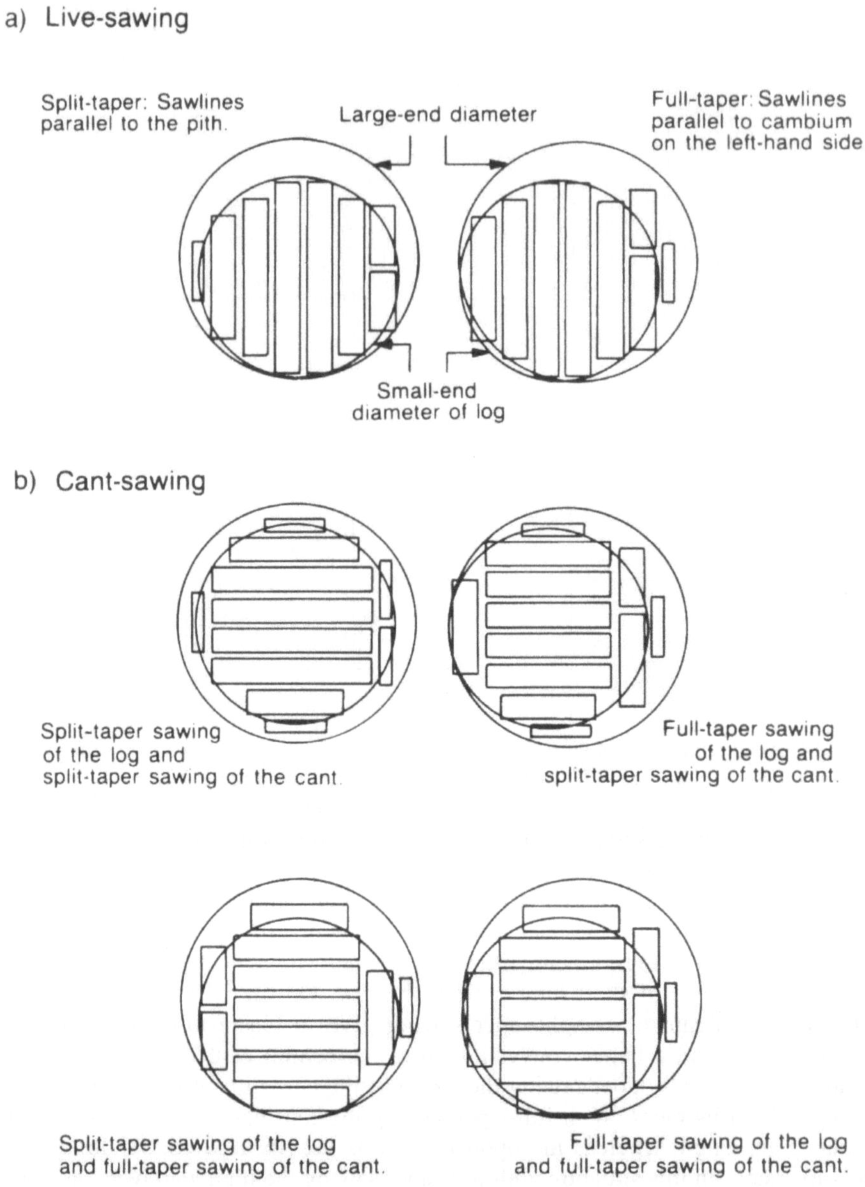

Figure 7.15. Optimal sawing patterns using various sawing strategies for a 310 mm diameter, 6 m log with a taper of 56 mm (Hallock *et al.*, 1976).

Sawing patterns can be centred or variable which offers a further set of possibilities to consider. A variable opening face gives a consistently better yield but not all sawmills are able to cut their logs in this manner.

In practice the cutting pattern is determined by the available saws, the log quality and size, the market demand, and the sawyer. The interactions are not easily understood and the effects are not apparent from casual considerations. For this reason computer control is universal in new mills. Optimal cutting solutions can be achieved once the log geometry is known, and saw and mill characteristics, and timber sizes have been entered into the simulation. Equally important the logs must be held very accurately. The log handling system must mechanically hold each log so that its position is maintained from scanning until it is automatically orientated, realigned by computer-controlled setworks and passed through the saws. The setworks are the hydraulic or mechanical devices for adjusting the positions of the knees on the log carriage (Figure 7.3) relative to the position of the saw, so controlling the thickness of the timber to be cut. In small-log continuous feed mills the setworks adjust the positions of the reducer heads and saws. The Best Opening Face (BOF) programs developed by scientists at the United States Forest Products Laboratory (Hallock and Lewis, 1971; Hallock, 1973) recognised that in converting logs the positioning of the opening cut is crucial, as this fixes the position of subsequent sawlines (Figure 7.16).

The benefits are most noticeable in small, straight logs. Again, if a cant is produced its orientation (skewness) relative to the sawlines and the position of its opening cut also needs to be simulated, the cant correctly repositioned and the sawcuts set. The first BOF simulation system was based on predetermined cutting patterns for logs of various sizes. These cutting patterns were stored in a computer. They were derived on the simplistic assumption that a log is a truncated cone. After scanning a log all possible positions for the sawlines were considered, knowing the saw kerfs and the green target dimensions for the various timber sizes. The BOF simulation proceeded to look at a myriad of predetermined cutting patterns and selected the best solution. The advantages of BOF systems are greatest when processing small logs, of about 150 mm diameter, and declines substantially with diameters over 275 mm (Figure 7.16). Occasionally the sawyer over-rides the settings if defects are seen which the scanning system leaves out of account, e.g. knots, rot *etc.*

For optimization to be effective three inter-related elements are necessary:

- A non-contact measuring system to provide a three-dimensional image of the log.
- A simulation program to determine the best sawing pattern.
- An infeed system capable of quickly and accurately turning the log while it is being scanned, adjusting the positions of the ends of the logs both horizontally and vertically for optimal sawing, and then holding the log firmly in that optimal position as it is fed through the saw.

5.5. Scanning

The most important requirement for scanner performance is consistent, reliable operation. In their simplest form scanners merely operate as log scalers (measuring

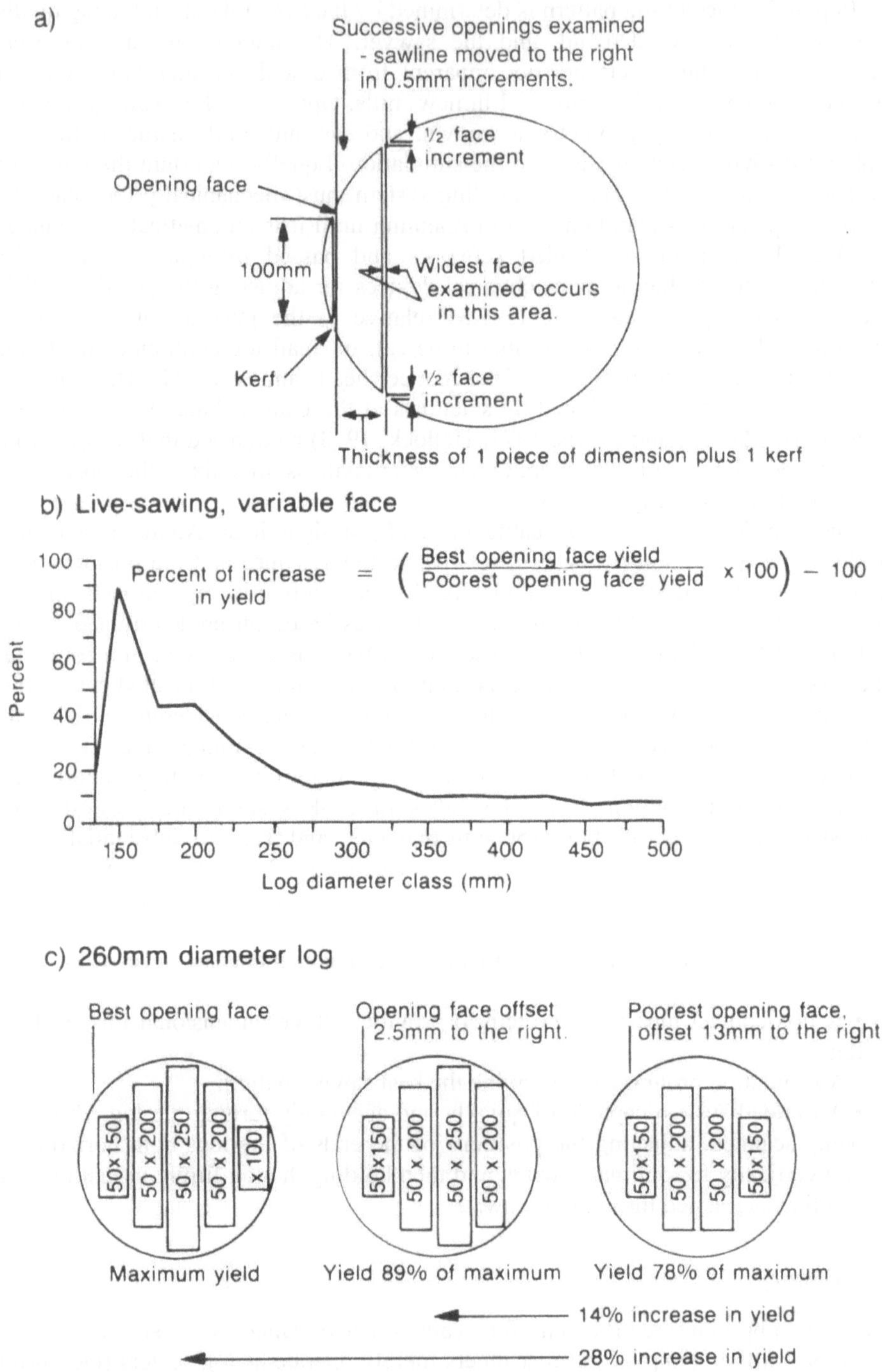
a)
Successive openings examined
- sawline moved to the right
in 0.5mm increments.
½ face
increment
Opening face
100mm
Widest face
examined occurs
in this area.
Kerf
½ face
increment
Thickness of 1 piece of dimension plus 1 kerf
b) Live-sawing, variable face
Percent of increase in yield = (Best opening face yield / Poorest opening face yield x 100) − 100
Percent
Log diameter class (mm)
c) 260mm diameter log
Best opening face
Opening face offset 2.5mm to the right.
Poorest opening face, offset 13mm to the right.
50x150
50 x 200
50 x 250
50 x 200
x 100
Maximum yield
Yield 89% of maximum
Yield 78% of maximum
14% increase in yield
28% increase in yield

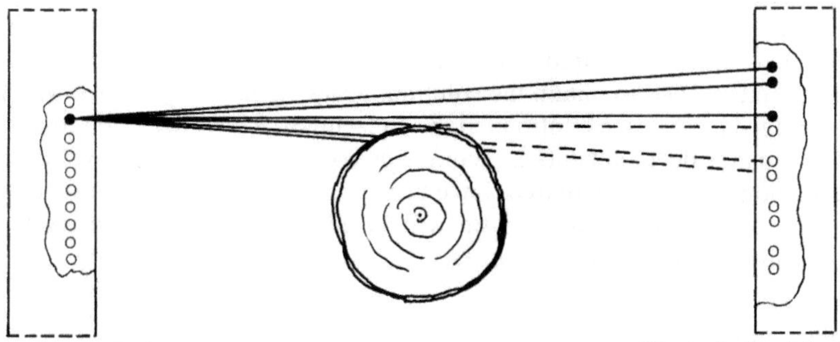

Figure 7.17. By lighting up each light emitting diode (LED) successively and noting which photodiodes grouped opposite detect the light the position of the limiting surfaces of the log can be determined very accurately, ±1 mm. Typically there are three pairs of scanners around the log. Scans are repeated at 100 mm intervals along the log to build up a three-dimensional image. This is a robust system with no moving parts (*from* Kockums Rema).

log volume). The scanning data can be used to develop optimal cutting strategies when the operation is integrated with the log feed system and the positioning of the saws. Scanning frames employ light sensitive devices that measure the shadow cast by the log as it passes through the scanner (Figure 7.17). These can be quite simple with batteries of lights and detectors positioned on either side of the log. Multiple scans from different angles provide data which can be built up into a comprehensive three-dimensional picture by repeatedly measuring the profiles at close intervals along the length of the log, so yielding its diameter, ovality, taper and sweep.

5.6. Real-time simulation

This is the ideal method for determining the log breakdown pattern. This requires scanning each log comprehensively to determine its three-dimensional shape for which the optimal cutting is sought. The best sawing strategy must be obtained before the log reaches the infeed to the headrig. The information needed includes the three-dimension image of the log along with mill parameters (kerf, sawing variation and saw characteristics) and the market situation (product demand and product price list). The time available for real-time simulation can be as little as one second. Here

←

Figure 7.16. Best Opening Face (Hallock and Lewis, 1971). (a) The yield from a small log depends on where the opening cut is placed: as this determines the position of subsequent cuts in the same plane. (b) Estimates of the increase in yield when live-sawing softwood dimension (50 mm stock), assuming no wane. Similar benefits are possible when cant-sawing (not shown), which requires a BOF to open the log and another to open the cant. The benefits of optimizing cutting patterns are greatest with small logs. (c) Live-sawing of a 260 mm diameter log. A small error in positioning the opening cut results in a significant drop in yield.

the principal challenge lies in writing computer algorithms that adequately represent the log to the sensitivity and accuracy required and which can be solved in the time available. It is essential to trade off some accuracy in locating defects in order to reduce the time needed to obtain the solution. Scanning technology in sawmilling is fast changing. The widescale introduction of machines able to detect internal defects in logs is a major development, e.g. computer tomography using x-rays. Interest in such technology is keen where trees have been pruned or where defects such as heart rot may be present. Installation costs will be high and workers are likely to be concerned about perceived health issues.

5.7. Infeed systems

Generally the log is rotated so that it can be correctly presented to the scanner, usually with the sweep of the log lying in the vertical plane. The log is picked up by a charger and passed through the scanner. The computer then instructs the setworks to correctly position the log before passing it on to the infeed mechanism of the

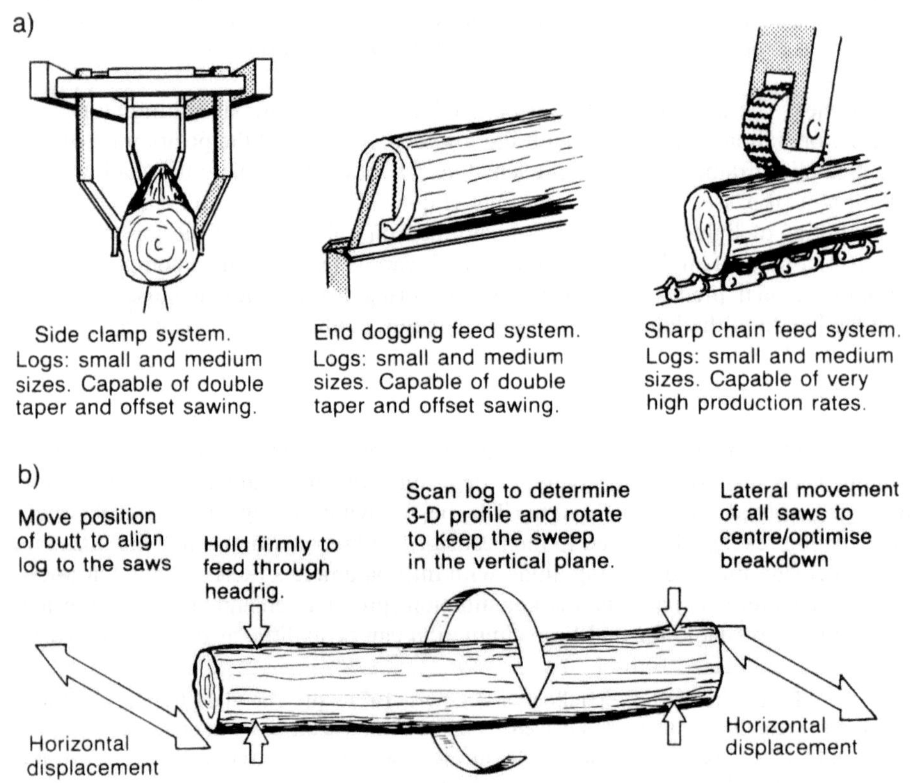

Figure 7.18. Log feed systems for small and medium sized logs. (a) Side-clamping, sharp chain and end-dogging feed systems. (b) Alignment of the log before dogging.

headrig. Some proprietary systems differ somewhat from that just described. For example logs can be rotated while being scanned to determine the most favourable rotational position of the log, and only then are the ends of the log held and adjusted. Systems vary according to the lineal through put desired, log size and form (Figure 7.18). The log must not move once it has been correctly positioned. Movement of the log either before or while being sawn would result in inaccuracies in sawing. BOF simulations show that a slight misplacement of the sawline can dramatically influence conversion (Figure 7.16). Reversing that argument, a slight error in the scanning and transport system will also dramatically lower the predicted conversion.

With a sharp chain feed or V-track the log is centred (split-taper) which gives a slightly lower theoretical conversion. It is best suited to straight, well formed logs whereas poorly-formed logs cannot be held firmly enough. Logs with sweep need to be orientated with the ends pointing up. Side-clamping or end-dogging systems cope better with poorly formed logs and where BOF full-taper sawing or offset sawing is practiced. With end-dogging the clamping arms hold the log as it passes through the saw.

5.8. Material flow

The smooth flow of wood is an essential prerequisite of an efficient mill. For this reason mills are elevated with space beneath so that waste material from each saw can be collected and conveyed away at a lower level than the timber that moves from saw to saw. The headrig is elevated relative to the other saws and the timber moves through the mill on powered (live) roll conveyors, or continuous belts or chain conveyors with cleats.

The capacity to accumulate timber (surge areas) ahead of every saw and to transfer between saws avoids the problem of down-time when particular sizes or products are being cut which can result in specific flows through the mill, overloading one saw and starving another. Transfer systems between saws and to an external outfeed may allow part of the mill to operate in isolation while one machine is not operating.

The transfer and holding areas occupy a disproportionate amount of space in the sawmill. Unscrambling devices are very necessary ahead of workstations with high piece counts, e.g. the edger, the trimmer and stacker. Unsorted scrambled material passes through a broad U/V-shaped trough from which pieces escape one at a time by being lifted out while balanced on short cleats projecting from the live chain.

5.9. Grade and volume optimization at the multisaw edger and trimmer

Grade recovery and volume conversion at the edger are not optimized where the operator manually aligns and then determines the best cutting pattern for each board. The throughput of these machines is too high. The best operator makes frequent mistakes, and with fatigue performance declines during the shift. With manual edging conversion can drop to 65-75% of the theoretical value. With optimizing edgers 97-99% of theoretical conversion is claimed, and at least 92% is achievable

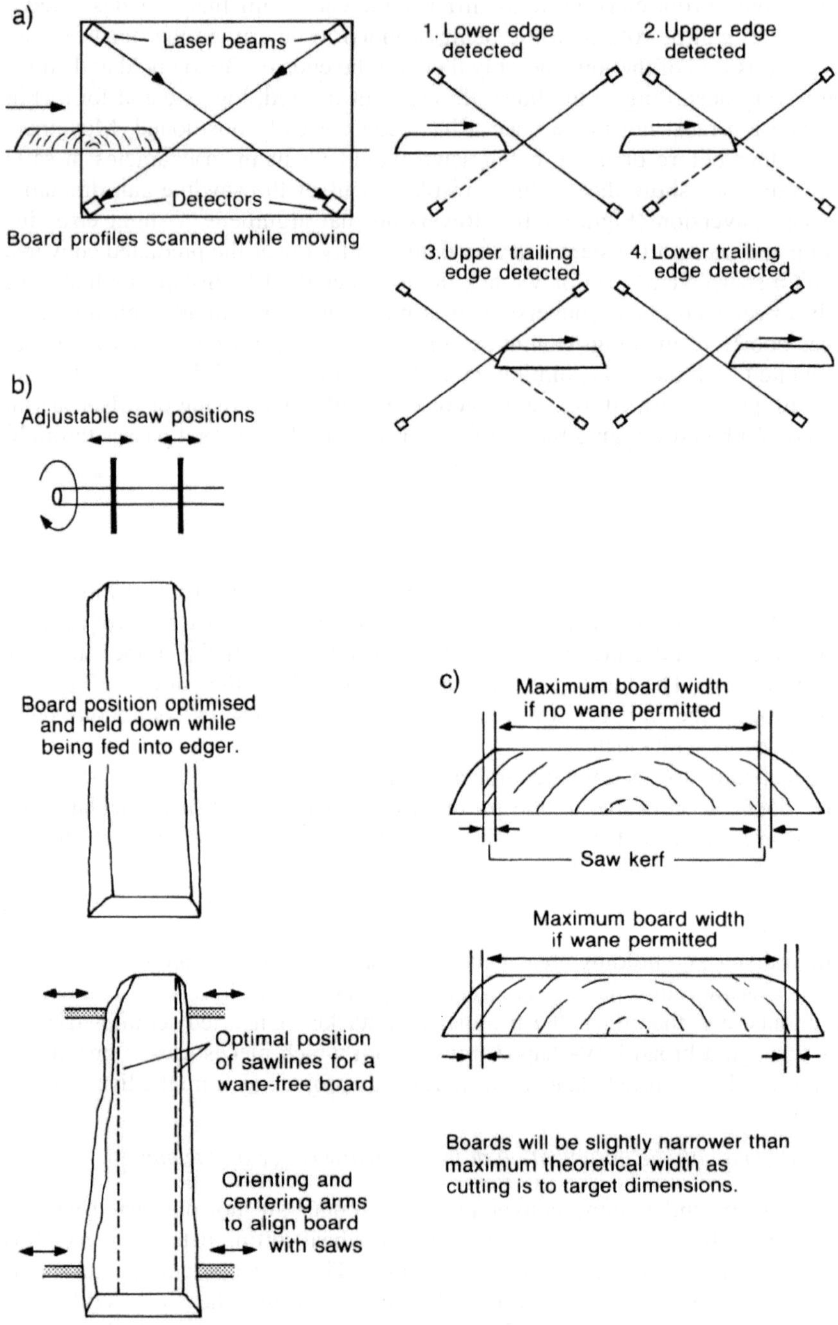

Figure 7.19. Optimization at the edger, boards are fed sideways into the scanning frame.

in practice. If one recovers 20% more at the edger and 30% of total production passes through the edger this gives an additional 6% conversion. Optimizing edgers use optical scanners to determine the upper and lower profiles of unedged boards: a resolution of 1 mm in width and 0.5 mm in thickness is claimed. A computer then aligns the board with index pins (functioning in a manner that is similar to the knees of a log carriage), adjusts the positions of the saws and the board is held down firmly as it moves through the edger (Figure 7.19).

The maximum amount of wane can be taken into account where grade and markets permit: for example in some structural uses wane is accepted as long as there is an adequate residual edge that can be nailed. Scanning rates of 30-45 boards a minute are possible compared to 10-12 pieces a minute in a manual operation. Most optimizing edgers maximize volume and take little account of the board grade so the operator can over-ride and alter cutting patterns to improve grade recovery. Optimizing for grade at the edger has been practiced on a limited scale in Scandinavia (Karonen, 1985). Most attention has been directed to the detection of knots as they are the reason for lowering the grade 80-90% of the time. Optical scanners can detect knots on the basis of colour disparity. Such systems can be supported by microwave, x-ray scanning and infra-red sensors. Edging decisions and the subsequent cross-cutting (docking) of the boards to remove particular defects must be integrated and optimized together.

5.10. Board sorting and stacking

Traditionally the sorting and stacking of timber occurs on the green-chain. In smaller sized mills a circular table may be preferred, but in larger mills the linear green chain permits, in principle, an unlimited number of sorts. An enormous number of sorts may be needed, especially in large-log mills with their flexible cutting patterns and ability to cut large timbers. The material needs be sorted to length, by thickness, and also for species and grade. A long green chain is labour intensive and for this reason automatic sorters and stackers are in use in many countries: these are capital intensive so most mills cope with 20 or less sorts although they would prefer more. Extra sorts can be provided by a short green chain handling the low volume piece sizes. In Australia and New Zealand 40-60 bin sorts are common, because milling is less specialised and mills supply diverse markets.

Recently some mills have introduced automatic density scanning of boards from within bin sorts. This is used to segregate material into different density groups to aid with kiln drying. This process reduces the variation within a 'kiln charge' and therefore can reduce the time needed to dry the lumber.

In summary, Hall (1983) states that the use of state-of-the-art technology can increase the overall yield, by:

- 3-5% through accurate positioning and feeding of logs through the headrig.
- 2% through cant optimizing;
- 6% with the use of an optimizing edger;
- 5% with optimized trimming.

These figures are based on the small-log softwood mill design that was examined earlier. The application of some of this technology is only appropriate in new mills or where volume production is high.

6. FLEXIBILITY

The scale of any business depends on the markets that one is in. Volume throughput dominates consideration where products are commodities – and where an annual throughput of 300,000 m^3 would be ideal. Here vertical integration is desirable to capture for oneself some of the profits that are to be found in the distribution chain as commodity production by itself is notoriously fickle.

There are examples of price-competitive, modern mills being designed and built in countries that do not have a long tradition in sawmilling. These are being built at a fraction of the price for a turnkey project. The success of these ventures does not lie exclusively in low cost labour although that plays a part. The saws may be imported, but for everything else the secret lies in identifying those components that are better made off-shore and incorporating these in local products - bearings, rolls, chains, sprockets, bushes etc.

At the other end of the scale there are the high quality, value adding enterprises, where the log is turned and opened up with thought. Here value rules; not volume.

CHAPTER 8

DRYING OF TIMBER

TIM LANGRISH[1] AND JOHN WALKER

[1]*Department of Chemical Engineering, University of Sydney, Sydney, New South Wales, Australia*

1. INTRODUCTION

The moisture content of freshly felled timber varies enormously from over 200% to as little as 40%. Once felled timber starts to dry and, provided it is not in contact with a moist body and is protected from rain, it dries eventually to a moisture content that is in equilibrium with the surrounding air: this might be as high as 20% in a humid environment and as low as 6% in a hot dry climate. Everywhere timber dries to below the fibre saturation point and shrinkage is to be expected. The yard or kiln manager seeks to regulate drying in order to:

- Dry the stack of lumber as fast as possible but without excessive degrade.
- Dry to the desired moisture content and within set limits, e.g. ±2% MC.
- Produce lumber that is relatively free of internal stresses, e.g. relieve inevitable drying stresses by conditioning at the end of the schedule.
- Produce lumber that is relatively free of visual blemishes, e.g. checks, collapse and stains.

This must be done as cheaply and efficiently as possible. Economics queries the high cost of holding air-drying stock for months – and even years – and favours fast kiln-drying schedules. In both situations the ideal is to dry wood as fast as one dares without it suffering too much degrade. It is a fine balance that can be achieved only with sophisticated process control and gnesio-models (modelling the 'real thing' and minimizing empiricism).

The elements that control the drying rate are the relative humidity of the air, the air temperature, and the airflow across the timber surfaces. In a kiln the temperature and (paradoxically) relative humidity are maintained at higher levels than in the open air, while powerful fans control the air velocity. In air-drying the same elements cannot be controlled or manipulated nearly as effectively.

Lumber drying is a two-stage process involving:

- The migration of moisture from the interior to the surface of the board.
- The evaporation of moisture from the surfaces into the moving air stream.

As a broad generalization, kiln conditions are designed to ensure that the rate of evaporation is adjusted to match the limiting supply of moisture from the interior to the wood surface, i.e. water should be removed from the surface not much faster than it can be replenished. The drying elements are manipulated to maintain that balance. If evaporation is too rapid, excessively steep moisture gradients will result within the wood, and this will be accompanied by drying stresses that may exceed the transverse tensile strength causing checking (splitting) and related damage.

2. THE DRYING ELEMENTS

Relative humidity of air is defined as the vapour pressure of water vapour in a given volume of air divided by the saturated vapour pressure. The absolute humidity at saturation (in g m^{-3}) and the saturated vapour pressure of water are shown in Table 8.1. Thus a cubic metre of saturated air can hold 17 g of water vapour at 20°C, 130 g at 60°C and 600 g at 100°C. In the wider drying literature, absolute humidity is normally given as a mass ratio of water vapour to the mass of dry air (kg kg^{-1} dry basis), since the flow rates of dry air and dry solids through any process are normally constant.

An absolute humidity of 50 g m^{-3} corresponds to a relative humidity of 98% at 40°C but only 8% relative humidity at 100°C. The drying capacity of such air would be minimal at 40°C being able to absorb only a further 1.1 g m^{-3}, but is enormous at 100°C being able to absorb a further 548 g m^{-3} before becoming saturated. The drying capacity of air is a function of its temperature and relative humidity.

Table 8.1. Absolute humidity and saturated vapour pressure of air varies with temperature.

Temperature (°C)	Absolute humidity at saturation (g m^{-3})	Saturated vapour pressure (kPa)
0	4.8	0.61
20	17.3	2.34
40	51.0	7.38
60	129	19.9
80	290	47.4
100	588	101
120	1090	199
140	1890	361
160	3090	618

Relative humidity is difficult to measure reliably and instead it is determined indirectly from the wet- and dry-bulb thermometers of a hygrometer. The wet-bulb thermometer is kept moist with a fabric sleeve whose other end is in a reservoir of clean water. As air passes over the wet sleeve water is evaporated and cools the wet-bulb thermometer: the drier the air the greater the cooling effect. The dry-bulb measures the air temperature: there is no cooling effect on the dry-bulb thermometer. The difference between the dry-bulb and wet-bulb temperatures, the wet-bulb depression (ΔT), and the dry-bulb temperature are the parameters used to control the relative humidity in the kiln. The equilibrium moisture content of lumber also varies with temperature and humidity of the surrounding air (Figure 3.2). It is read from tables in terms of the dry-bulb temperature and the wet-bulb depression (Table 8.2).

Table 8.2. The equilibrium moisture content of wood depends on the kiln condition, i.e. the wet-bulb depression and the dry-bulb temperature. The equilibrium moisture contents of individual species can deviate ±3% from these values (Brunner-Hilderbrand, 1987).

Wet-bulb depression (°C)	Dry-bulb temperature (°C) 0	10	20	30	40	50	60	70	80	90	100	110	120
2	12.2	15.5	17.0	17.9	18.1	18.1	17.6	16.8	15.9	15.2	14.6		
3	9.0	12.0	14.2	15.4	16.0	15.8	15.3	14.7	14.1	13.4	13.0		
4	6.6	10.4	12.2	13.4	14.0	14.1	13.8	13.3	12.8	12.3	11.8		
5	3.8	8.5	10.6	11.8	12.4	12.7	12.5	12.1	11.	11.1	10.8		
6		7.0	9.2	10.6	11.2	11.5	11.4	11.1	10.7	10.2	9.9		
7		5.3	8.2	9.6	10.3	10.7	10.6	10.3	9.9	9.5	9.1		
8		3.6	7.2	8.8	9.5	9.8	9.8	9.6	9.3	9.0	8.6		
9		1.7	6.1	8.0	8.8	9.2	9.2	9.0	8.7	8.4	8.1		
10			5.0	7.2	8.2	8.6	8.7	8.5	8.2	7.9	7.5		
11			4.0	6.5	7.6	8.0	8.1	8.0	7.7	7.4	7.1	6.9	
12			2.9	5.8	7.0	7.5	7.7	7.5	7.2	7.0	6.7	6.5	
13			1.7	5.0	6.4	7.0	7.2	7.0	6.8	6.6	6.4	6.1	
14				4.3	5.9	6.6	6.7	6.7	6.5	6.3	6.0	5.8	
15				3.6	5.3	6.2	6.4	6.4	6.2	6.0	5.8	5.6	
16				2.9	4.9	5.7	6.0	6.0	5.9	5.7	5.5	5.3	
18				1.1	3.9	4.9	5.4	5.4	5.4	5.2	5.0	4.9	
20					3.0	4.2	4.8	4.9	4.9	4.8	4.7	4.6	
22					1.8	3.5	4.2	4.4	4.4	4.4	4.3	4.2	4.1
24						2.8	3.7	4.0	4.0	4.0	4.0	3.9	3.8
26						2.1	3.1	3.5	3.7	3.7	3.6	3.6	3.5
28						1.4	2.6	3.1	3.3	3.3	3.3	3.3	3.2
30							2.1	2.7	2.9	3.0	3.0	3.0	3.0

Temperature influences the rate of drying in a number of ways. The principal reason for kiln drying at high temperatures is to increase the rate of moisture transfer to the wood surface. Raising the temperature dramatically enhances the rate of diffusion of water molecules across cell walls. The rate of diffusion increases with temperature at approximately the same rate, as does the saturated vapour pressure (Table 8.1; Figure 8.9b).

The absorptive capacity of the airstream increases with temperature (Table 8.1). Hence when kiln drying at high temperatures a much smaller volume of humid air needs to be vented to be replaced by cold air from outside through another vent.

Raising the temperature of wet wood marginally reduces the energy required to evaporate the moisture from wood. The energy needed to evaporate a kilogram of water at 20°C, i.e. the heat required to raise the temperature of the water from 20°C to 100°C plus the latent heat of evaporation = 4.2 x (100-20) + 2255 = 2590 kJ kg^{-1}, is only about 15% more than the 2255 kJ kg^{-1} required to evaporate the same amount of water at 100°C (assume that the specific heat of water = 4.2 kJ kg^{-1} $°C^{-1}$ and the latent heat of evaporation = 2255 kJ kg^{-1}).

Airflow performs two roles, as a carrier of heat and as a medium to absorb evaporating moisture. Drying time and wood quality depend on the air velocity and its uniform circulation. If the fan speed is unnecessarily high air will pass through the stack taking up only a fraction of the moisture it is capable of absorbing: this results in excessive power consumption. If the fan speed is too low the air as it passes across the stack cools and becomes very humid, and the boards at the 'back' of the stack dry very slowly. Early in the drying of permeable lumber, Ashworth (1977) observed that the humidity potential or driving force for drying at the 'back' might be only two-thirds of that at the 'front' for the air velocities used at the time, which meant that the drying rate at the inlet side was initially 50% faster than that at the back. Airflow reversals are used to counter the progressive humidification of air across the stack. A single switchover about one-quarter of the way through the schedule may be sufficient although more usually operators reverse the air-flow every 4 hours – to mitigate developing drying stresses.

With a slow-drying impermeable species a fan speed as low as 1.5-2.0 m s^{-1} may be sufficient as the moisture content at the surface drops quickly below the fibre saturation point – no mass flow – and there is no point in installing overly powerful fans just to strip off surface moisture for the first few minutes of a long kiln schedule (>14 days): thereafter the slow rate of transfer of moisture from the centres of the boards to their surfaces becomes more important than the rate of evaporation.

An air velocity of 5-15 m s^{-1} is desirable for fast drying of permeable species at high temperatures (>100°C). The benefits of high velocities are retained throughout the entire schedule because the diffusion coefficient rises more rapidly with temperature than does the external mass-transfer coefficient for evaporation.

The choice of optimum air velocity is an economic decision.

3. SURFACE TEMPERATURE

Consider the surface temperature of lumber during the course of a kiln schedule in which the dry-bulb and wet-bulb temperatures, and so the wet-bulb depression, are held constant. The initial rate of evaporation is independent of the air temperature (dry-bulb temperature) but is proportional to the wet-bulb depression. This is because evaporation is sustained by the rate of heat transfer, and the rate of heat transfer from the warm air to the moist wood surface is proportional to the temperature difference between the air and the wood surface which is at the wet-bulb temperature. Thus, provided the wet-bulb depression is the same, say $\Delta T = 5°C$, the rate of evaporation from a wet timber surface is essentially the same whether the kiln air temperature is 40, 70 or 100°C.

Only when the supply of moisture from the interior is no longer able to sustain the maximum cooling effect due to evaporation does the surface begin to dry out, its moisture content drops below the fibre saturation point and the temperature at the surface begins to rise above the wet-bulb temperature. The steep, superficial moisture-content gradient observed for much of the schedule when drying permeable timbers is associated with a static evaporative front. As a result the temperature at any point inside the board only rises significantly when the evaporative front recedes past that point late in the schedule (Figure 8.1). Then the

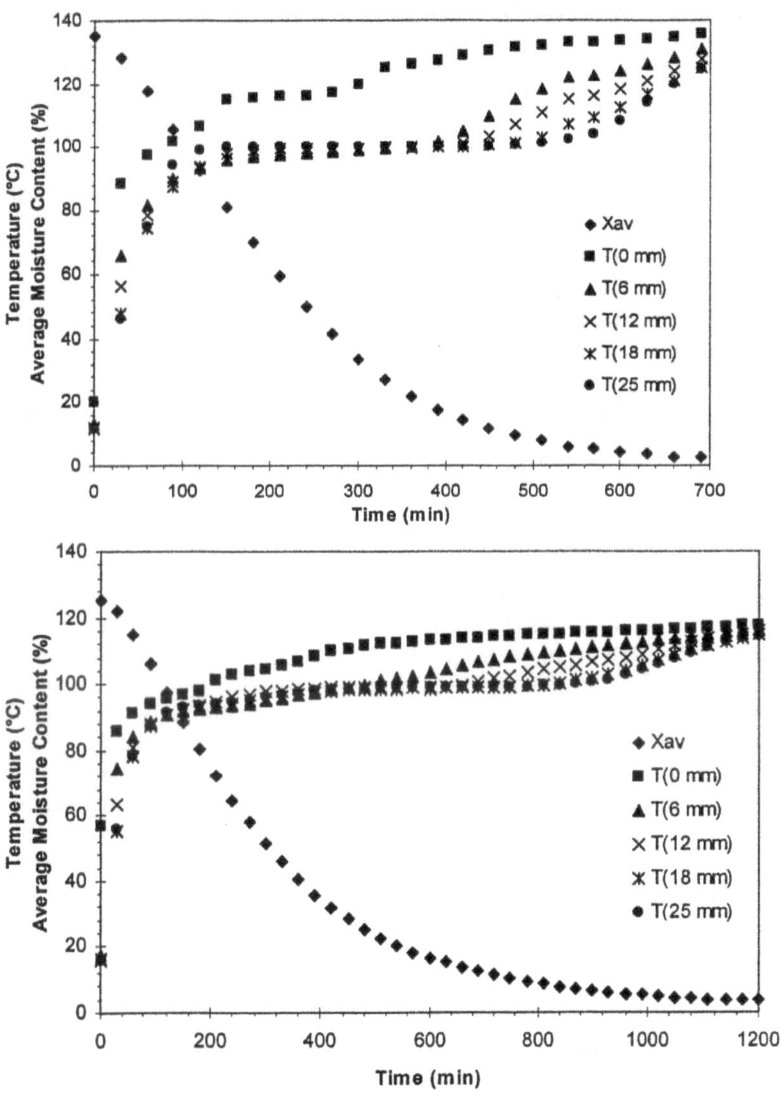

Figure 8.1. Temperature and average moisture content profiles in 50-mm thick sapwood boards (Pang, 1994). (a) 140/65°C, airflow 5 m s^{-1}, dry density 438 kg m^{-3}. (b) 120/85°C, airflow 5 m^{-1}, dry density 469 kg m^{-3}.

diminished air-to-surface temperature differential results in a lower rate of heat transfer to the surface and drying progresses more slowly. Eventually, as the surface moisture content approaches the equilibrium moisture content the temperature at the surface approaches the dry-bulb temperature and heat transfer and drying will approach zero. The total heat transferred must equal that needed for evaporation plus the heat required to raise the temperature of the wood to the dry-bulb temperature.

In summary: the temperature of the wood surface is determined by the cooling effect of evaporation, rising from the initial wet-bulb temperature when the surface is wet to the dry-bulb temperature as the wood approaches the equilibrium moisture content. With permeable woods the surface temperature remains at the wet-bulb for a considerable proportion of the schedule. With impermeable woods the surface temperature soon begins to rise toward the dry-bulb temperature, as there is no mass flow of water from the interior to keep the surface moist and above fibre saturation.

4. THE MOVEMENT OF FLUIDS THROUGH WOOD

A digression is required to appreciate the complexities of the 'inside, outside' aspects of drying:

- To understand influence of wood structure on the movement of moisture within the drying board, and
- To learn more about the external drying environment.

The process is made more complex because each board is intrinsically different carrying with it its own individuality and case history. At the same time there are difficulties in minimizing the different drying conditions associated with a board's position within a stack. In reality the drying process treats boards as a population, acknowledging that the drying experiences – both the immediate external conditions and the board's individual response – will give somewhat different outcomes. The desired outcome and challenge is to minimizing between and within board variations in final moisture content, without excessive degrade and in a reasonable time.

The movement of fluids (gases or liquids) through wood is complicated by the fact that the coarse capillary system is interconnected via smaller openings (Figure 8.2). The favoured path for the movement of fluids varies according to the fluid, the nature of the driving force (pressure gradient or chemical potential/concentration gradient) as well as being sensitive to variations in wood structure.

4.1. Permeability of softwoods

Two terms, permeable and porous, can be distinguished. Porosity relates to the proportion of free space in a material. A low-density wood or cellular foam is porous in that it contains a large void volume. A permeable material, on the other hand, is defined in terms of the ease of fluid flow. If the cells are interconnected then air/water can escape when compressed and the material is porous and permeable (a sponge). A material is porous and impermeable where the cells are closed and the

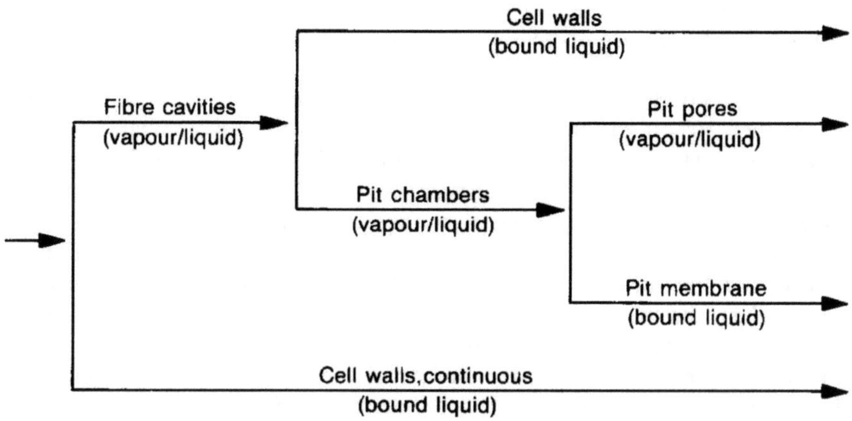

Figure 8.2. Various flow paths through softwoods (Stamm, 1967a).

air/water can escape only by rupturing the cell walls (bubble wrap). Timbers having the same porosity can have widely differing permeabilities: some are relatively permeable, some highly impermeable. The size of the openings connecting various wood cells determines the degree of permeability according to Poiseuille's law.

Poiseuille's law for laminar flow of fluids states that the flow rate is proportional to the radius (R) of the flow passage to the fourth power, indicating that longitudinal flow through pits and lumens is limited by the small openings in the pit membrane:

$$\text{Rate of flow} = \frac{\pi(p_1 - p_2)R^4}{8\eta L} \tag{1}$$

where $(p_1 - p_2)$ is the pressure drop over a distance L, and η is the fluid viscosity (Massey, 1986). This equation indicates that halving the radius will result in a 16-fold decrease in the flow rate even though the cross-section of the capillary only decreases 4-fold. The relative importance of the possible flow paths in Figure 8.2 is sensitive to the geometry of the obstructions to flow. In terms of permeability, the longitudinal flow of fluids in sapwood occurs almost exclusively along the tracheids, with communication between one tracheid and another occurring through the restrictive openings in the pit membrane.

4.2. Pit aspiration in softwoods

In sapwood, the pit membrane of the bordered pit is suspended in the centre of the pit chamber between two over-arching pit borders (Figures 1.7 to 1.10). The membrane is composed of a cobweb-like network of microfibrils, known as the margo, with a central thickened area, known as the torus. In a living tree these pits protect the continuity of the ascending sap stream. Where the sap column is ruptured, the pressure differential across the sap-vapour meniscus forces the pit membranes towards their pit borders and the tori seal the pit apertures, so isolating these vapour-filled cells from the sap-conducting column – a process known as aspiration.

In green sapwood, the great majority of pits are unaspirated (>85-95%), but once dried, fewer than 10% in the earlywood pits remain unaspirated, compared to about 50% in latewood. There are two reasons for this difference. First, in latewood the strands in the margo are generally thicker and the flexural stiffness of the pit membrane is correspondingly greater. Secondly, a latewood bordered pit is more spherical while an earlywood pit is more saucer-shaped as the cell wall is thinner. Thus, in latewood the microfibrillar strands of the margo are shorter (the diameter of the pit is smaller), and the displacement distance to the pit border is greater (the cell wall is thicker) than in an earlywood pit. Consequently, in order for the torus to deflect until it touches the border, the strands of the margo have to stretch much more than is required in earlywood pits; and the force needed to aspirate the pit must be much greater, as must be the adhesive force holding the aspirated torus against the pit border. Once pits aspirate, they remain aspirated, so the permeability of air-dried sapwood is much reduced. Then earlywood permeability can be less than a hundredth of that for the green earlywood sapwood, and latewood permeability exceeds that of its earlywood – but the permeability of both is considerably reduced compared to the green condition.

The moisture content of the heartwood is only a little higher than the fibre saturation point (Table 3.1), so most pits in heartwood are aspirated. These pits are also clogged to some extent with extractives. Heartwood is impermeable.

4.3. Liquid-vapour interfaces

The presence of water vapour interfaces, for example a bubble, within a capillary disrupts the process of permeation. Such interfaces significantly influence the drying and preservative treatment of wood.

The pressure drop (p_1 - p_2) across the interface is reflected in the curvature of that interface (r), as shown in a simplified version of the Laplace equation:

$$(p_1 - p_2) = \frac{2\gamma}{r} \tag{2}$$

where γ is the surface tension. Note that the pressure difference across the meniscus decreases as the bubble expands. This agrees with experience, since it is harder starting to blow up a child's balloon (when r is small) than to expand the balloon, i.e. as r gets larger, $(p_1 - p_2)$ gets smaller.

Consider the case of evaporation from a container having a single opening of radius R. As water evaporates, the curvature of the concave meniscus across the opening must decrease, and it follows from the Laplace equation that there will be a pressure differential across the meniscus equal to $2\gamma/r$. The pressure difference means that the water will be in tension and by adhering to the container walls will pull them in unless these are sufficiently rigid. The radius of curvature of the meniscus, r, can only decrease to a value corresponding to the radius of the opening, R. When it reaches this value, the meniscus is hemispherical and cannot decrease further (the meniscus could not bridge the opening if its radius were smaller than that of the opening). At this point, the curvature of the meniscus can increase again, forming an expanding bubble within the vessel, and at the same time the capillary tension within the water will decrease again as r increases. The capillary tension is a maximum when the curvature of the meniscus, r, equals the radius of the opening, R. While the size of the openings in the margo (Figures 1.8 and 1.9) range from 0.1 to 1.0 μm it is the largest pores that are most important when considering capillary tension.

Pit aspiration is a complex phenomenon. The old view held that capillary tension was responsible for pit aspiration, gradually drawing the torus into contact with the pit border as evaporation proceeded. However, Booker (1989) noted that cells are either totally saturated or they are well drained. The distribution is strongly bimodal, i.e. there can be no small vapour bubbles in the sap stream. Booker's mechanism for pit aspiration involves cavitation. During drying, evaporation generates a capillary tension that is felt throughout the entire network of saturated cells. Removal of water by evaporation results in the water within the network being stretched very, very slightly, and the volume of the capillary system must shrink very, very slightly: the cell walls are being pulled into the lumens. The forces pulling the cell walls inwards are generated in response to the capillary tension of the water. In this manner, the entire wood-water system can store a large amount of strain energy. As evaporation proceeds, capillary tension increases until it reaches a point at which cavitation occurs in a single cell lumen, most likely through loss of adhesion between the water and the cell wall. Booker (1989) argues that the vapour bubble, created suddenly, expands so rapidly that the escaping sap or the bubble itself hammers the torus against the pit border, giving the aspirated pit membrane its characteristic dished depression (Figure 8.3). Cavitation generates a shock wave that can be detected with acoustic emission equipment (Tyree and Dixon, 1983). The energy storage capacity of many millions of saturated tracheids is hardly affected by the drainage of a single tracheid, so the capillary tension drops little and quickly builds up again, and the cycle is repeated endlessly.

Thomas and Kringstad (1971) provide convincing support for this theory. They examined the affect of a number of liquids on pit aspiration. They found no relationship between pit aspiration and the surface tension of the evaporating liquid: liquids having a moderate surface tension may not result in pit aspiration, whereas liquids with a low surface tension may do so. Obviously properties other than surface tension are important. They suggest that hydrogen bonding is necessary both to transmit the capillary tension force required to aspirate the pit and to allow the formation of hydrogen bonds between the pit membrane and border that maintain the pit in the aspirated state (Figure 8.4). Pit aspiration was not observed with pyridine, despite its moderately high surface tension. They suggest that when the

Figure 8.3. Characteristic dished appearance of the torus after being hammered against the over-arching border, in radiata pine (unpubl. Dr J.A. Kininmonth).

capillary tension reached only a modest value the liquid cavitated since pyridine, unlike water and diethylamine, is unable to cross-link with other pyridine molecules and so form a hydrogen bonding network (a number of one-armed individuals cannot form and dance the conga). This means that the energy storage capacity in pyridine saturated cells would be insufficient to develop the large dynamic forces necessary to aspirate pits. On the other hand, the surface tension of diethylamine is low yet pit aspiration is observed because a hydrogen bond network is possible.

Figure 8.4. Water and diethylamine have both hydrogen bond donor and acceptor properties through the –OH or = NH groups. They cross-link through hydrogen bonds and can withstand considerable capillary tension. Pyridine has only hydrogen bond acceptor properties and cannot cross-link with itself. Pyridine cannot withstand large capillary forces: the energy storage capacity of the pyridine-saturated system is small (Thomas and Kringstad, 1971).

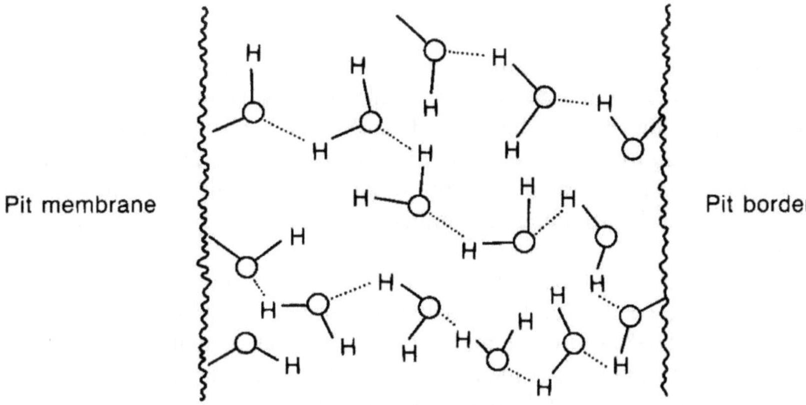

Water, H_2O

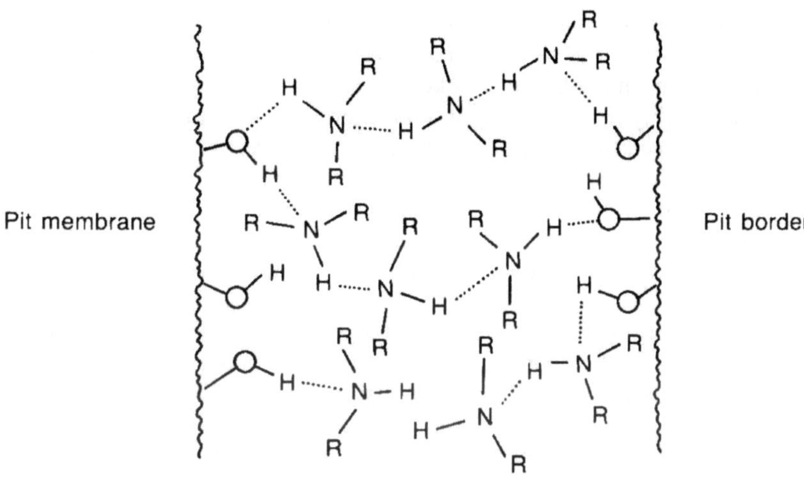

Diethylamine, $(C_2H_5)_2$ NH
R = C_2 H_5

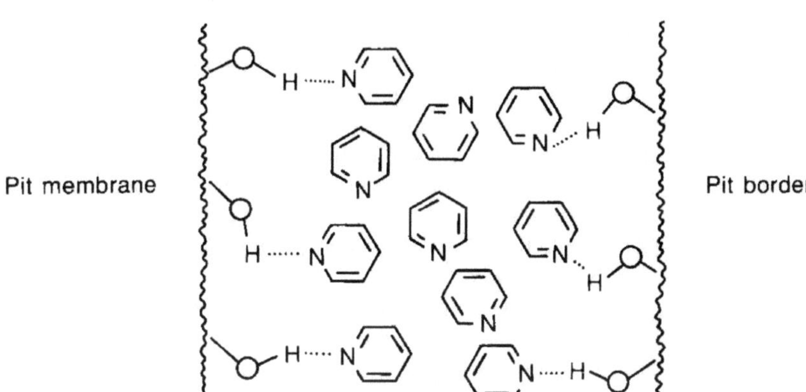

Pyridine, C_5H_5N

4.4. Permeability of softwoods

Using a modified version of Poiseuille's law, Petty and Puritch (1970) estimate that the lumen accounts for about 40% of the resistance to sapflow in *Abies grandis*, with the unaspirated pits contributing the other 60%. For *Pinus radiata,* Booker (unpub.) found the contribution to resistance to flow by the lumen to be 26% and that for the pits to be 74%. Thus unaspirated pits only moderately impede the natural flow of sap in a living tree while providing vital protection to the sap-capillary system where the tree is injured.

The living tree has only modest needs for lateral redistribution of sap throughout its conducting tissue so it is understandable that transverse permeability in green sapwood is only *c.* 0.01% of that along the grain. In the tangential direction, flow is through bordered pits and lumens in series. Bordered pits in tracheid cross-walls are encountered much more frequently than in the longitudinal direction, and resistance to flow is dominated by these pits. When drying wood, the longitudinal permeability is dominant even though it would appear at first sight that in long, thin boards the transverse permeability should totally dominate. Indeed, with such high anisotropy it is technically very difficult to measure transverse permeabilities accurately, and few reliable values have been published (Booker, 1990).

Radial permeability is not as well understood as longitudinal or tangential permeability and indeed the tree has no path for – or interest in – radial flow except within ray tissue. However, after drying (and pit aspiration) the minimal radial permeability in some species can exceed tangential permeability. The greater radial air-dry permeability for *Pinus sylvestris* compared to either *Picea sitchensis* or *Pseudotsuga menziesii* is attributed, in part, to the pine having many more ray tracheids.

Radial permeability of *Pinus radiata* sapwood actually increases on drying due to the collapse of thin-walled parenchymous tissue surrounding both horizontal and vertical resin canals (Booker, 1990). In contrast, resin canals and ray tracheids are absent in *Dacrydium cupressinum*, and its radial permeability is reduced on drying. Booker (1990) found a decrease in permeability on drying *Pinus radiata* sapwood of about 200-fold in the axial direction, 100-fold in the tangential direction and an increase of about 20-fold in the radial direction. The decrease in tangential permeability is similar to that found with *Dacrydium cupressinum* and is due to pit aspiration. The longitudinal permeability of *Pinus radiata* is reduced as well: but now it is dominated by flow along infrequent axial resin canals, whereas in green sapwood it is almost entirely due to flow along earlywood tracheids.

In the sapwood of pines, thin-walled epithelium cells surround the resin canals, whereas in heartwood these may be thicker-walled, lignified and occluded with tylosoids and so resistant to collapse on drying. However, the opportunity for collapse of the thin-walled parenchyma cells on drying explains why the sapwood of pines is generally more treatable than that of spruces, larches and Douglas fir, which are thick-walled and probably lignified.

Approximate longitudinal permeabilities for a variety of species are shown in Figure 8.5. There can be large differences in the values obtained, depending on the test procedures (whether tested green or dry and the method of drying), on the fluid used, and on the length of the specimen. The utility of Figure 8.5 lies in the relative rankings of their permeabilities rather than in the absolute values. Transverse permeabilities are approximately 10^4 times less.

4.5. Permeability in hardwoods

For hardwoods the resistance to flow is due to intervascular pits between contiguous vessels (Figure 1.25). In diffuse-porous hardwoods vessels are of limited length and flow through these intervascular pits must be efficient. Such pit membranes appear impermeable when viewed in the scanning electron microscope (Figure 1.26) but in fact tortuous flow paths across the membrane occur between the interwoven networks of microfibrils in the primary wall layers of the two cells. Petty (1981) examined fluid flow through vessels and intervascular pits in sycamore, *Acer pseudoplatanus,* and calculated that the cross-wall between two vessels would have approximately 60 pits, each with some 180 openings with a pit pore radius of 90 nm; while according to Siau (1984) the openings are at least an order of magnitude smaller than for softwoods, with a logarithmic mean of 30 nm.

Tyloses greatly increase the resistance to flow along vessels (Figures 1.31 and 1.32). They account for the low permeability of white oaks such as *Quercus alba,* which is used in tight cooperage, e.g. whisky barrels. Tyloses occur in both heartwood and on drying in the sapwood of some but by no means all hardwoods. Species that do not normally form tyloses may form them in response to injury. Alternatively, such species can seal vessel cavities by secreting gums and resins.

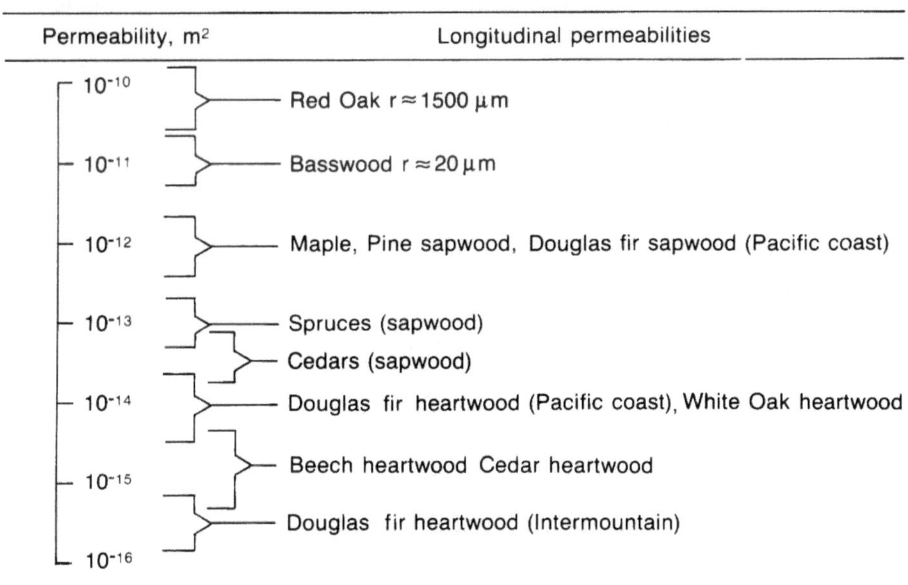

Figure 8.5. Estimate of air-dry, longitudinal permeabilities of various timbers (Siau, 1984).

Little is known about the mechanism of lateral permeability other than that transverse flow is very small by comparison to longitudinal flow. It is generally less than in softwoods. Surprisingly there is little difference in radial and tangential permeability (Comstock, unpub., see Siau, 1984). Ray tissue, despite the generally higher volume fraction in hardwoods, is not particularly efficient in radial flow. Neither are the pits on the radial surfaces of fibres efficient in tangential flow.

4.6. Impermeable timbers and diffusion

For permeable species there is an *irreducible moisture content*, occurring around 50-60% MC, below which diffusion takes over from permeability as dominating moisture migration processes. At this moisture content there are still numerous saturated tracheids but these are presumed to lie in isolated groups with no hydraulic continuity, such that noticeable sap flow toward the surface is no longer possible. At this point the evaporative front, or wet line, starts to move into the wood from its original position a millimetre or so below the surface. Further, as individual cells cavitate the stored strain energy within such limited volumes of saturated cells is insufficient to result in dynamic pit aspiration – so explaining why some 5-10% of earlywood cells and 50% of latewood cells in dry sapwood have unaspirated pits.

With impermeable woods – and heartwood – the supply of moisture from the interior cannot keep pace with evaporation of water vapour from the surface, because mass flow of water is not possible and diffusion is a much slower process. Thus the surface moisture content quickly falls below fibre saturation and the evaporative front starts receding into the wood. Figure 8.6 shows the parabolic moisture content profile for a slowly air-dried impermeable hardwood. Similarly for permeable softwoods that have been dried below the irreducible moisture content, Stamm (1964, 1967b) reported parabolic moisture profiles that are consistent with diffusion of both water vapour and bound water.

Poiseuille's law describes the flow of water through the capillary network in response to an applied pressure gradient. However, water and other small molecules can migrate across swollen cell walls even where the wood is impermeable and there is no pressure gradient. This other process is called diffusion. The rate of bound water diffusion is the product of a diffusion coefficient and a driving force (Stamm, 1959). In classical thinking, the driving force is a concentration gradient or more strictly the chemical potential. Diffusion is a molecular process, directing the otherwise haphazard wanderings of individual molecules.

The same pathways are available for diffusion of water molecules as are available for flow of water down a pressure gradient, but their relative contributions are very different. The reason is that the rate of diffusion along a capillary of radius R is proportional to its cross-sectional area, πR^2 (Eq. 3), whereas the rate of flow is proportional to πR^4 (Eq. 1), so the small capillaries in a distribution of capillaries will make a proportionally greater contribution to diffusion processes than to permeability.

Diffusion along 100 tubes of radius R – where $A = \sum\pi R^2 = 100\,\pi R^2$ – will be identical to the diffusion along a single tube of radius $10\,R$ – where $A = \pi(10\,R)^2 = 100\pi R^2$. On the other hand, the permeability of the array of a 100 small capillaries will be proportional to 100 R^4, which is 1/100th of the permeability of the large capillary as the permeability of the latter will be proportional to $(10\ R)^4$, i.e. 10 000 R^4.

Hence, the diffusion of water molecules through the swollen cell wall is of far greater significance than diffusion of water molecules across pits, simply because the total area occupied by the openings in the pit membranes is small compared to the area available for diffusion through the cell wall itself (Chapter 3).

In Fickian diffusion (Treybal, 1980) water molecules diffuse individually from a region of high moisture content to a region of low moisture content, so reducing the moisture gradient and equalizing the moisture content. Under steady state conditions the diffusion rate of water molecules through a piece of wood becomes:

$$J = -D \cdot A \cdot \frac{\delta M}{\delta x} = -D \cdot A \cdot \frac{M_1 - M_2}{x} \qquad (3)$$

where M_1 and M_2 are the moisture contents ($M_1 > M_2$) on either side of the wood of thickness x across which diffusion is occurring; A is the cross-sectional area across which diffusion is occurring. The moisture content cannot exceed the fibre saturation point simply because at this point the cell wall is saturated.

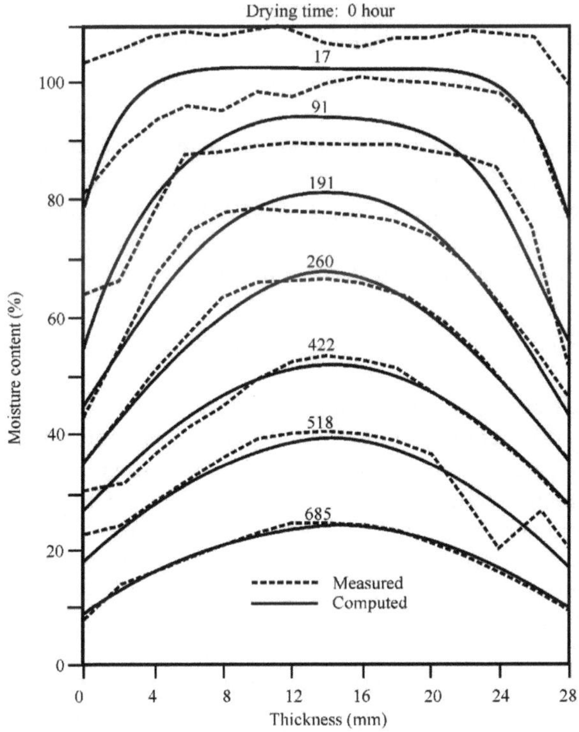

Figure 8.6. Moisture content profiles for slowly air-dried Tasmanian oak (Wu, 1989).

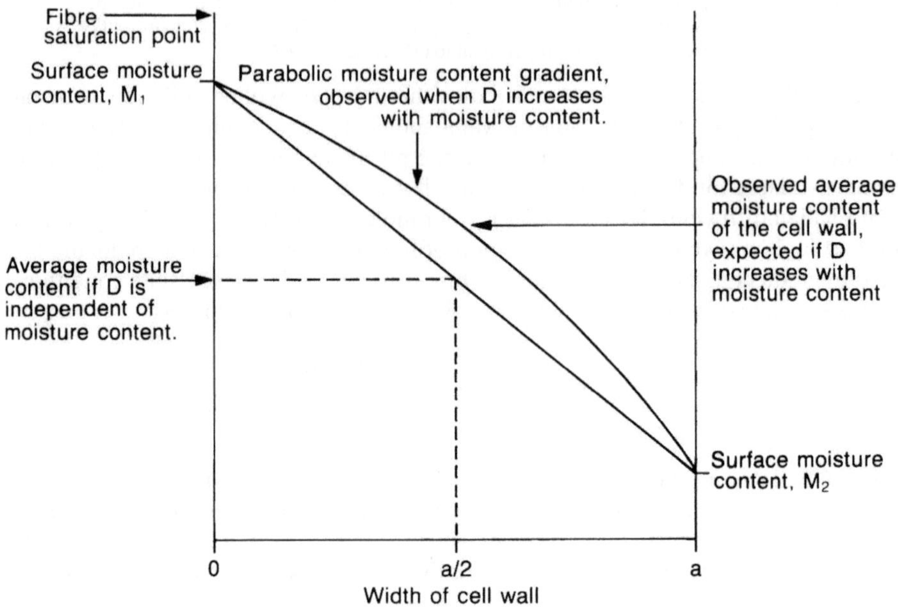

Figure 8.7. Diffusion of moisture across a timber slab. The parabolic gradient reflects the fact that the diffusion coefficient increases with moisture content.

The diffusion coefficient is sensitive to moisture content. The amount of water entering the slab ($J_{M1} = D_{M1}. A . (\delta M/\delta x)_{M1}$) on one side of Figure 8.7 must equal the amount of water leaving the slab ($J_{M2} = D_{M2} A . (\delta M/\delta x)_{M2}$) on the other side under steady state conditions, otherwise the moisture profile within the wood would have to change and the system would not be in steady state. Thus:

$$D_{M1} \cdot A \cdot (\delta M/\delta x)_{M1} = D_{M2} \cdot A \cdot (\delta M/\delta x)_{M2} \tag{4}$$

Since the diffusion coefficient is greater at high moisture contents, $D_{M1} > D_{M2}$, then to compensate the moisture gradient must be steeper at lower moisture contents, $(\delta M/\delta x)_{M2} > (\delta M/\delta x)_{M2}$. A parabolic moisture gradient results. A feature of a parabolic moisture content profile is that the average moisture content in a slab is $M_{\mathrm{Av}} = 2/3M_1 + 1/3M_2$ rather than $1/2M_1 + 1/2M_2$, where the moisture contents on each side are given by M_1 and M_2 (with $M_1 > M_2$). In other words, the average value for the moisture content of the slab is greater when the diffusion coefficient varies with moisture content than when the diffusion coefficient is a constant.

It would be more correct to describe the moisture gradient in terms of the partial pressure of the surrounding air that maintains the surfaces at moisture contents M_1 and M_2 but redefining the equations in this way, while technically appropriate, does not alter the thrust of the argument.

At very low moisture contents (<5%) most water molecules are hydrogen bonded directly to the hydroxyls of the cell wall and are very strongly held. These molecules require considerable thermal energy to break their hydrogen bonds and so lack mobility. This is reflected in a very low diffusion coefficient (Figure 8.8a). As the moisture content in the cell wall increases multilayers build up and the water molecules in these multilayers will be less strongly adsorbed. At fibre saturation, those water molecules sufficiently removed from the underlying substrate have a mobility approaching that of free water, due to the attractive forces holding these water molecules to the cell wall material being small, as demonstrated by NMR and by the minimal heat of wetting. The higher diffusion coefficient at higher moisture contents is due to both the higher molecular mobility and the greater proportion of the swollen wall volume that is occupied by water and so participating in the diffusion process.

Temperature has a crucial influence on the economics of drying. The principal reason for drying at elevated temperatures is to increase the rate of diffusion (Figure 8.8b). Diffusion plays a vital role in the drying of impermeable timbers at all moisture contents, and for permeable timbers wherever the moisture content is below the irreducible moisture content. Raising the temperature from 25° to 100°C increases the rate of diffusion about 37-fold (Stamm, 1964): a month's air-drying is roughly equivalent to a day in a kiln at 100°C. The diffusion rate approximately parallels the increase in the vapour pressure of water with temperature. The saturation vapour pressure increases from 2.3 kN m^{-2} at 20°C to 20 kN m^{-2} at 60°C and to 101 kN m^{-2} at 100°C (Table 8.1).

Even with a permeable wood diffusion assumes increasing importance as the average moisture content approaches the irreducible moisture content: indeed, in every part of the board where the moisture content approaches this value drying is diffusion controlled. Permeable and impermeable timbers of similar densities should dry from fibre saturation at about the same rate. The behaviour of mixed heart/sapwood boards is complicated since, at first, there is both an evaporative interface near the sapwood surface and one in the interior at the zonal boundary between heart and sapwood. For a board with only a slither of heartwood along one face, mass flow can only move to the sapwood face so in effect the board appears to be twice the width than it actually is. Pang *et al.* (1994) predicted that such a 50 mm thick board would dry from green to 6% moisture content using a 140°C/90°C schedule in 14 hours, compared to 10 hours for sapwood and 11 hours for heartwood.

4.7. Longitudinal vs. transverse diffusion

When wood is green the diffusion coefficient of water vapour in the lumens is about 10 times greater than that of adsorbed water in the cell wall, but when the wood is dry, *c.* 5% moisture content, the diffusion coefficient of water vapour in the lumen is about 1000 times greater than that of water in the cell wall: the difference is least when the moisture content of the cell wall is close to the fibre saturation point and greatest when the moisture content approaches oven-dry.

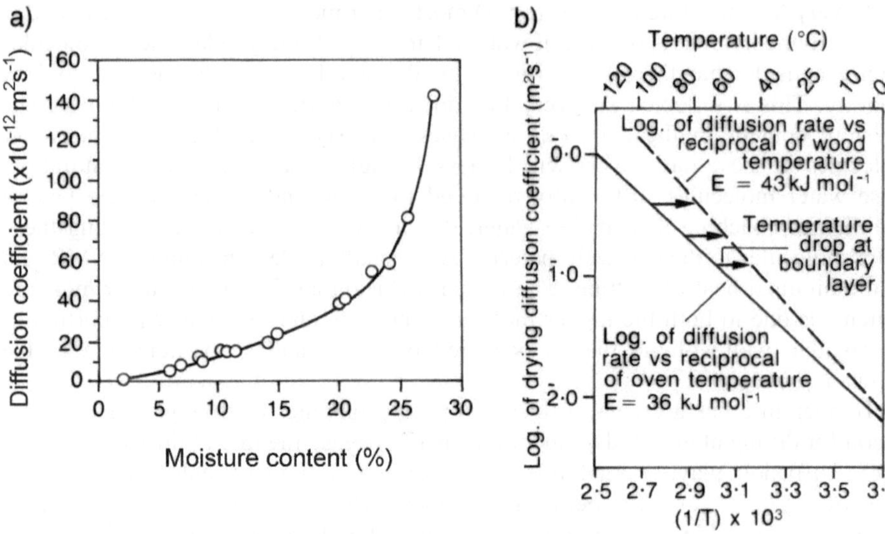

Figure 8.8. (a) Bound water diffusion in Sitka spruce at 26.7°C as a function of moisture content (Stamm, 1959). (b) Diffusion rate as a function of temperature normalised to a basic density of 400 kg m^{-3} and at an average moisture content of 20%. Measuring air temperature under-estimates the actual diffusion rate (Stamm, 1964; Bramhall, 1979).

Since the cell walls offer more resistance to diffusion than do the lumens, it is evident that transverse diffusion is essentially determined by the diffusion coefficient of the moisture through the cell walls and by the thickness of cell wall traversed per unit distance, i.e. the basic density of the wood. The presence of pits and the condition of their pit membranes do not influence diffusion very much: if no diffusion occurred through the pits the transverse diffusion coefficient would be reduced by only 10% and if the pit membranes were removed entirely the diffusion coefficient would increase only three-fold (Stamm, 1967b).

The ratio of the longitudinal to the transverse diffusion coefficient varies from 100:1 at 5% moisture content to 2-4:1 at 25%.

5. THE EXTERNAL DRYING ENVIRONMENT

A kiln (Figure 8.9) is in essence a large box-shaped chamber with a pitched roof, incorporating overhead fans above a false ceiling (to circulate the air), heating coils (to maintain the dry-bulb temperature) and a water bath/sparge (to provide low-pressure steam to raise humidity). The humidity is controlled by opening vents, thus governing the amount of humid air to be recirculated (>95%): once the kiln is at temperature moisture evaporation from the lumber provides the humidity, the excess of which has to be bled off by venting.

Lumber is prestacked and moved into the kiln on flatbed trolleys. The individual stacks are usually 2.4 m wide. A kiln may have a single track or double track (with two stacks side-by-side). In the stack boards are packed tight – without gaps – within each layer since the air is forced across the stack from one side to the other and there is no natural vertical air drainage (as in air-drying). Stickers between layers must be of uniform thickness and sufficiently wide and frequent to avoid indenting the boards, as wet wood softens at elevated temperatures. Stacks should have only 50-100 mm clearance with the false ceiling. Stacks should be square-ended and gaps at either end between stack and kiln walls minimized. Baffles and curtains seek to direct the airflow through the stack, and reduce the amount of air flowing around the stacks. Nijdam and Keey (1996) note that with 100% bypass – with equal amounts of air moving around and through the stacks – drying is only 75% of what would be achieved if all the air moved through the stack. Even with well-aligned stack ends bypassing of 10-25% is likely. Ragged-ends and missing board lengths will result in uneven, slower drying.

When drying permeable species kiln productivity increases linearly with the velocity of the airstream, for as long as the external environment controls the drying rate. Optimal airflow is just below the value at which the permeability of the wood itself begins limiting the rate of drying – and this in turn is temperature dependent. Higher kiln temperatures allow the use of higher airflows and so increased productivity and lower drying costs. Unintended consequences such as the development of kiln brown stain impose limitations on excessive temperature depending on the grade of material being dried and the intended market.

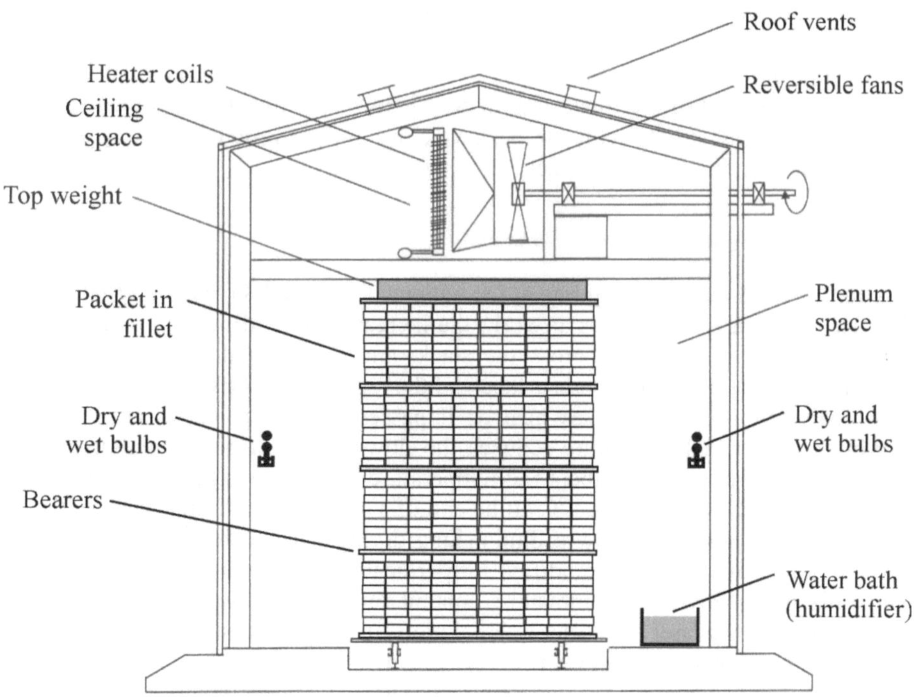

Figure 8.9. Schematic plan of a traditional kiln.

When drying impermeable species, the imposed external drying conditions determine the moisture-content profile near the surface of the wood. These profiles influence the development of stresses in the wood, the severity of which restricts the severity of drying schedules that may be used. High humidity and a small wet-bulb depression early in the schedule are desirable as this results in a relatively high equilibrium moisture content, less surface shrinkage, smaller differential strains and so reduced degrade. Thus while the external drying conditions control and limit the rate of drying, that limit is also dictated by slow diffusion within the wood.

5.1. Maldistribution of airflow with height in the stack

Recirculating air has to change direction – like an athlete on a running track – and so the inner radii within the kiln ought to be rounded to ease airflow around the bends. For example, where the airstream enters the plenum space a sharp edge tends to shed vortices, as the airstream breaks free from the surface. Such a vortex is similar to the recirculating eddies that can form downstream of rocks in a fast-flowing river. These effects were investigated using an hydraulic kiln, scaled to achieve dynamic similarity (Figure 8.10). A fast moving airstream, an unradiused edge to the false ceiling and a narrow plenum space encourage the development of a vortex at the top of the stack that impedes airflow into the upper sticker spaces – indeed under severe conditions the air in these upper spaces can flow in the opposite direction to the overall airflow (Nijdam and Keey, 2002). Radiused guides and a wide plenum space (>1 metre) ensure a more uniform air velocity through all the sticker spaces. One criterion for a well-designed kiln is to minimize the difference in air velocities throughout the height of the stack. In practice airflow velocities in the sticker spaces can vary by at least ±10% with the peak velocity appearing two to four layers of boards down from the top of the stack.

5.2. Uneven drying potential across the stack

Early in the schedule the drying rate at the outlet side can be less than half that at the inlet face. There are two factors at work: the increasing humidification and cooling of the airstream as it passes across the stack; and the gradual thickening of the boundary layer between the wood surface and the rapidly moving airstream.

Air gains moisture and loses heat as it passes through a stack. This rise in humidity and fall in temperature reduces the downstream drying potential of the airstream. With permeable timbers the temperature drop across the stack can be as high as 15°C initially, falling to 4°C at the end of a conventional kiln schedule. Both hardwoods and softwoods show a diminishing drying rates with distance from the inlet – although this profile is modified for permeable softwoods once the upstream boards reach the irreducible moisture content and must now dry by diffusion whereas for boards nearer the outlet internal moisture transfer is still by mass flow.

A boundary layer of retarded flow is zero at the leading edge and builds up gradually with distance as flow is hindered by skin friction. The mass-transfer

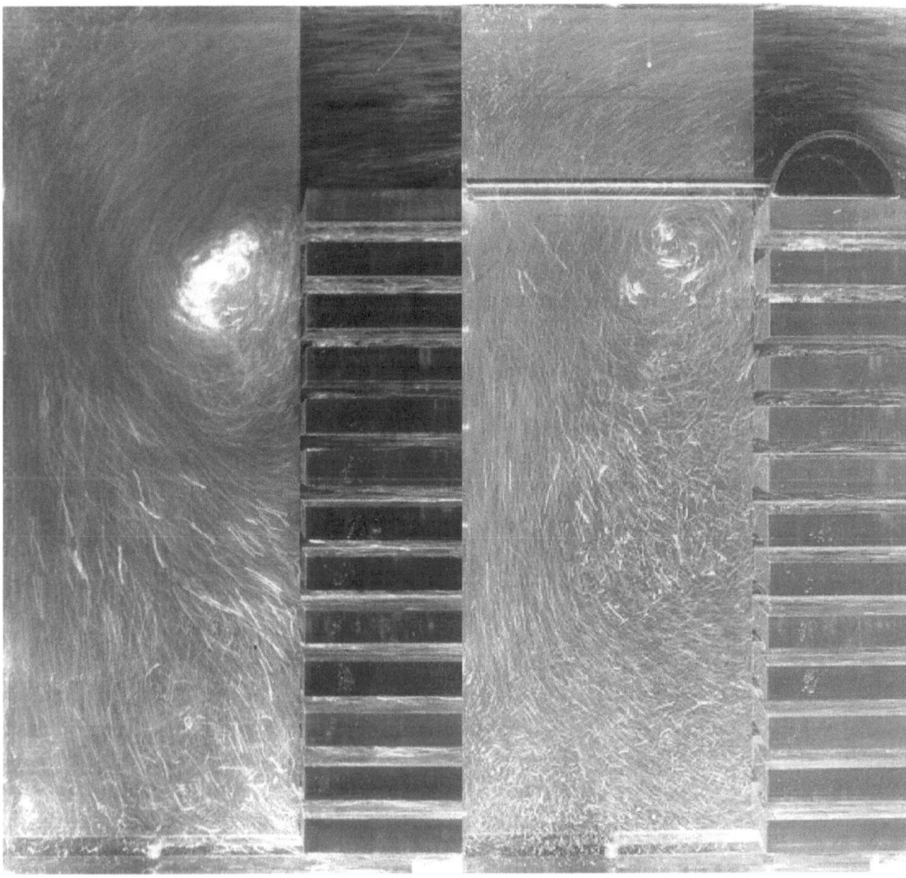

Figure 8.10. 'Airflow' in a hydraulic kiln. (a) On the left, the recirculating air from the fans has to take a right-angle bend to enter the plenum space and the airstream breaks free from the surface to form a vortex. This hinders air moving thought the upper layers of the stack. (b) On the right, more uniform airflow with a radiused edge and wide plenum space (Nijdam, 1998).

coefficient is inversely proportional to the thickness of the boundary layer through which the moisture vapour must diffuse to escape into the rapidly moving airstream. Thus the mass-transfer coefficient dwindles in the direction of airflow (Salin, 1996).

Without any airflow reversals there is a very wide variation in moisture content across the stack mid-way into the schedule (Figure 8.11) and although the absolute variation then diminishes the relative variance increases. Flow reversals reduce the moisture content variability across the stack by trimming excess over- and under-drying in the outer portions of the stack, such that toward the end of the schedule the slowest drying boards are in the centre of the stack. Nijdam and Keey (1996) suggest that when drying a permeable species at moderate temperatures (< 80°C) a single reversal about one-quarter of the way into the schedule is enough.

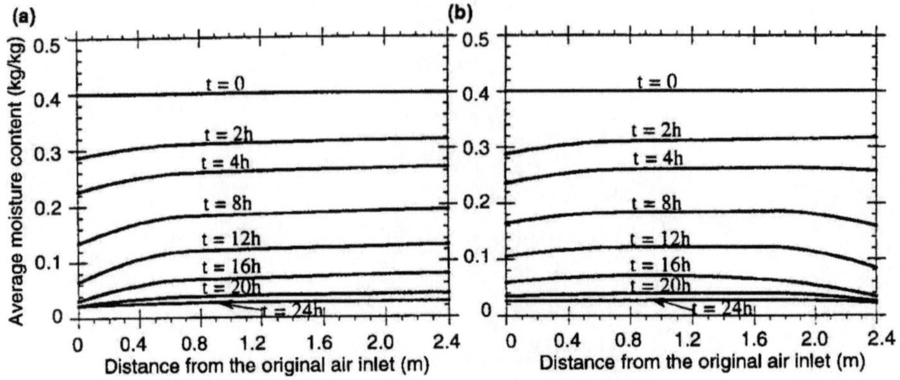

Figure 8.11. Drying heartwood of radiata pine at 120/70°C. With no reversal of airflow there is uneven drying across a stack with the wettest wood on the outlet side. This contrasts with the more even drying with periodic airflow reversals where the centre of the stack dries more slowly (Pang, 1994).

5.3. Uneven heat transfer across individual boards

Even when boards are tightly butted together on stacking, inevitably small gaps develop due to shrinkage. These gaps, even where as small as 1 mm, become recirculation zones in which vortices develop low frequency oscillations, 'spilling out' every 1-7 seconds (Lee, 1990). These emerging vortices create turbulence in the airstream, disrupt the boundary layer at the leading edge of each board, and this results in enhanced transfer coefficients. Kho *et al.* (1989) using naphthalene-coated test boards and Langrish *et al.* (1993) using numerical simulations demonstrate the greatly enhanced mass-transfer (Figure 8.12) that is occurring close to the leading edge of each board. In practice this means that individual boards do not dry evenly and that the upstream portion of the board will dry faster than the downstream part.

The overall consequence is that despite over-drying a load of lumber some boards will still not achieve the desired end-moisture content. Merely extending the drying schedule does not solve the problem of airflow maldistribution and uneven drying; this is especially true when drying permeable species using low air velocities of ≤ 3.5 m s^{-1}.

6. DRYING METHODS

6.1. Air-drying (Rietz and Page, 1971)

In small mills the entire production may be air-dried while in larger operations only part of production may be air-dried, generally the lower grades of lumber. Air-drying can suit larger sized members and items for exterior use that do not require a low final moisture content. Then again, air-drying can be an essential precursor to kiln-drying many valuable, impermeable or collapse-prone species.

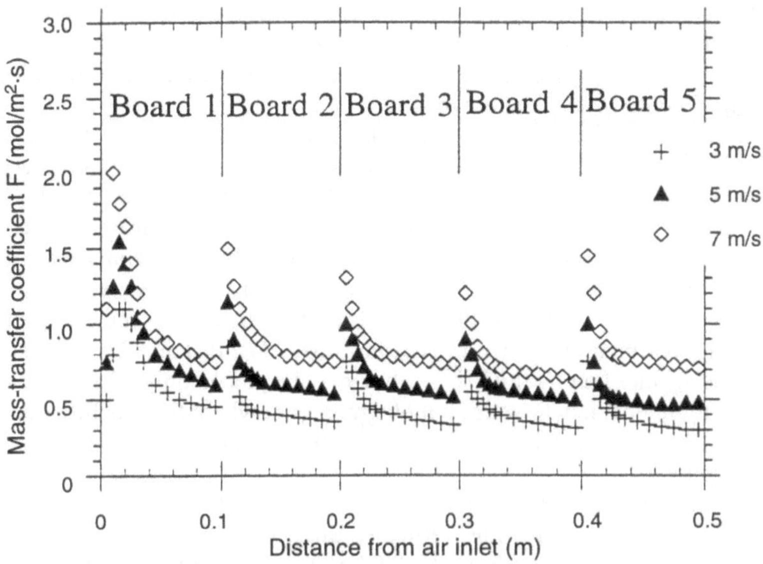

Figure 8.12. Experimental local mass-transfer coefficients, from the leading edge of stack of 100 x 25 mm boards with side-gaps of 0.5 to 1.0 mm, for an airflow of 3-7 m s^{-1} (Kho, 1993).

When air-drying, there is no control over the drying elements (wind and wind direction, temperature and sunshine, humidity and rain), and the drying time is very variable depending on location, season and species. As a broad generalization, 25 mm boards of American hardwoods can be dried in 50-200 days, while softwoods can be dried in 30-150 days, depending on the time of year. Thicker timbers dry even more slowly: doubling the board thickness more than doubles the drying time. Lumber dries quicker in spring and early summer than wood going into stacks in late summer. For example, in the upper midwest of the United States, Rietz (1972) indicates that there are only 15 effective air-drying days between 1st December and 28th February, and only 55 between mid-October and mid-April, but 165 in the remaining six months. Denig and Wengert (1982) have developed a simple predictive air-drying calendar equation based on historical weather data (relative humidity and temperature) and the initial moisture content of the wood.

The drying yard should have good air movement (a gently sloping site) and the surface should be sealed or free draining. A yard needs a lot of space: to hold 6-12 months stock; for access and to manoeuvre a forklift; for good air movement; and for fire control and to meet insurance requirements. The traditional layout has the stacks aligned parallel to the wind for good all-round drying characteristics, namely high average evaporative loss coupled with a small variation in moisture content between stacks (Finighan and Liversidge, 1972a,b). The alternative of aligning stacks perpendicular to the wind results in the windward stacks drying more rapidly than the downwind stacks: drying is less uniform. There is no general, ideal yard layout. Wind direction is not constant, and features such as aspect (to take advantage of the sun) and the constraints of the site itself must be considered. For example, the width of the alleys can be reduced if side loading forklifts are used, or stacks can be four deep in the rows. Uniform drying conditions may not be an overriding objective of management: some species or thick timbers require milder drying conditions, in which case they could be stacked downwind.

When stacking, stickers (also called fillets) are placed between the layers of timber. Stickers should be well seasoned, of uniform dimension, and preferably preservative treated or heartwood timber. Stickers are laid across the layers of timber to allow air movement and drainage through the stack. Their thickness (15-25 mm) determines the gap between the layers. They should be sufficiently wide (20-50 mm) to avoid crushing/marking the boards. They should be placed vertically above one another to carry the weight of the stack directly to bearers. Accurate alignment prevents the lumber bending under an eccentric load. The distance between stickers depends on the species and its thickness. They may be placed only 300 mm apart for boards with a high tendency to warp, but 600 mm apart where there is little tendency to distort on drying. Uniform, well-aligned stickers keep the timber flat and minimize warping.

When air-drying it is important to encourage vertical air drainage within the stack. This is achieved by having vertically aligned gaps (25-50 mm) between boards. As air picks up moisture from the lumber it cools. Cool, moister air is denser than the surrounding air and so drains downwards through the stack. Consequently the packets of lumber near the base of the stack dry more slowly than those at the top. The stacks should rest on level, firm foundations some 0.5 m above ground level. Stacks are generally not more than two metres wide, with the height of the stack no more than three times stack width (Figure 8.13).

Lumber should be sorted by grade, size of cross-section and length before stacking. If short lengths are included in a stack they may need further support from an extra line of stickers immediately below the unsupported ends. Within a layer short boards should be placed flush with alternate ends of the stack which should be maintained square at both ends.

If the timber is prone to degrade due to too rapid drying the thickness of the stickers can be reduced, to say 15 mm, to reduce the airflow through the stack. Stacks can be moved closer together and stored at the leeward side of the yard. To further inhibit airflow, stacks can be covered with horticultural shade-cloth (50% open mesh), but the boards will need to have had an anti-sapstain treatment.

Where end splitting is a problem – due to excessive drying of the end-grain – it can be minimized by placing wide stickers flush with the ends of the stack and, with valuable slow-drying species, by coating the ends with a sealant. Wax emulsions reduce the rate of evaporation from the end grain and hence end-splitting.

Degrade can be severe in the upper layers due to exposure to sun and re-wetting from rain. Permeable timber absorbs rainwater much more rapidly by capillary action than it can subsequently escape by evaporation. Consequently a brief period of wetting retards drying and creates favourable conditions for decay that will remain for some time. Also, lumber near the top is liable to distort, as there is no great weight of wood above to hold it flat as happens lower in the stack. Ideally packs of the least valuable grades should be placed at the top of the stack. Stack covers improve the quality and speed of drying.

Figure 8.13. Air drying with some features of poor practice: perfection is rarely attained.

6.2. Predriers

Forced air-driers merely require a roof and fans to circulate air (1-2 m s^{-1}). A cut-out switch turns the fans off when the relative humidity is exceeds 85-90% (to save power when drying will be ineffectual) or when the relative humidity falls below about 40% (when surface checking is likely).

A more robust method involves preheating the air 10-20°C above ambient. This increases the low adsorptive capacity of the air and ensures that drying continues throughout winter. Less basic versions recycle the heated air within an enclosed space and in effect function as a uninsulated, low-temperature kiln. Such an operation has attractions where damp winters make air-drying very slow. They are cheap, and capital is often limited for small companies. Low-temperature, low-cost driers provide some control over the drying elements that is lacking in air-drying.

Since the drying conditions are mild compared with those of a kiln there is less difficulty in drying mixed loads of species and thicknesses. Significant cost savings can be achieved with slow-to-dry quality hardwoods: the savings are both in time and in reduced degrade (Wengert, 1979).

Some green, impermeable hardwoods can be very slow drying. If dried from the green condition in a kiln the low temperature schedule would be far too long and the cost prohibitive. Therefore the timber is first air-dried to around the fibre saturation point before being kiln-dried. Paradoxically, predriers are essential often to slow down the rate of drying – to avoid collapse (see later). Rather than accelerating air-drying, the aim is to control the environment and achieve measured, uniform drying.

6.3. *Kiln-drying*

While collapse-prone and impermeable lumber may require long, slow air-drying, elsewhere the trend is overwhelmingly toward investment in fast, controlled drying with kilns. Kiln-drying requires technical skill and considerable capital investment. However it offers distinctive benefits over air-drying.

- Air-drying is more expensive than it seems. Land is required (capital + rates) and large quantities of stock have to be held (capital + insurance). With hardwoods the volume of stock may equal the annual turnover.
- Kiln-drying is much faster and turnover is quicker. Many hardwoods can be dried in three weeks or less, although impermeable species must be partially pre-dried beforehand. Most softwoods can be kiln-dried in less than a week, and some can be dried in a matter of hours where high-temperature schedules are appropriate.
- Orders can be filled at short notice. The operation is more flexible because drying is independent of the weather. In winter air-drying is very slow.
- Kiln-drying is essential if drying to low moisture contents (<15-18%). Many markets require wood dried to 12% and as low as 6%.
- Accurate control of the drying elements reduces the amount of degrade (distortion, checking) in the stack.

6.4. *Solar kilns*

The use of solar radiation to deliberately dry materials is as old as civilisation. Imré (1984) noted some general advantages of solar energy as its 'free' nature (although there are costs in collecting and using it), its availability in remote locations, the absence of monopoly in its use, and low environmental impact. Disadvantages include its intermittent nature, dependence on time, season and weather, and the relatively low energy flux compared with conventional energy sources. The world average solar insolation on a horizontal surface is 3.82 kWh/m^2/day (McDaniels, 1984), while the maximum flux is around 1 kW m^{-2}. Plumptre (1979) reviewed 35 solar kiln designs that were spread almost uniformly over the range of latitudes from 0-50°. This distribution lends weight to the suggestion that the rate of solar drying relative to that in the open air is likely to be similar over a wide range of latitudes.

Plumptre (1979), classified solar kilns as follows:

- *Greenhouse kilns* where the solar collectors are within the same chamber as the wood. These may be further grouped according to:
  (i) whether air circulation in the kiln is by natural convection or driven by fans;
  (ii) whether the vents are opened and closed manually, or venting is powered and governed by humidity, temperature and/or external conditions such as sunlight;
  (iii) whether the roof, walls and floor are insulated or glazed to reduce heat loss.
- *External collector kilns* with an insulated collector in a separate structure that is connected to the kiln by insulated ducts.

Greenhouse kilns appear to be more popular, possibly because of their simplicity – an important factor as these kilns are usually operated by unskilled labour.

Most heat absorbers in solar kilns use metal surfaces, painted matt black, to absorb solar energy. Configurations including flat, corrugated and vee surfaces are used, together with perforated metal screens or a honeycomb patterned surface, to transfer this heat to the air flowing over the absorber surface.

Other features have been added to some designs, including heat storage systems (rock piles, masonry partition walls or floors, and heat storage tanks, including water and gels), microprocessor control of vents and circulation, and dehumidifiers.

Materials for glazing include glass and a variety of plastic films or rigid sheets. Glass is heavy, costly and durable, but easily broken. Tempered or laminated glass is stronger and, in the case of tempered glass, not much more expensive if it is available. Optically, glass is much better than most plastics since it is transparent to most in-coming shortwave radiation and opaque to out-going longwave radiation. Plastics are less opaque to longwave radiation. Horticultural polythenes are very poor relative to glass, while PVC, PVF and polyesters are intermediate between polythene and glass, with some polyesters approaching the performance of glass. Plastics deteriorate and embrittle in ultra-violet light. Horticultural polythene has a life of about a year in the tropics (two to three years in temperate regions), while the others vary up to a maximum of five to six years in the tropics.

6.5. Dielectric and microwave drying

Schiffmann (1995) defines dielectric/radiofrequency drying as working in the range 1-100 MHz, while microwave frequencies range from 300 MHz to 300 GHz.

The considerable advantage of these techniques is that the penetration and dispersion of drying energy avoids the formation of steep moisture gradients associated with conventional drying. In practice the temperature is higher within the wood than on the surface, and highest where there is most moisture.

Water molecules are dipolar by nature (that is, they have an asymmetric charge centre), and are normally randomly orientated. These dipoles attempt to remain aligned with the rapidly changing polarity of an alternating field. As the field changes polarity, the dipoles return to a random orientation before being pulled the other way. This build-up and decay of the field, and the resulting stress on the molecules, causes a conversion of electric field energy into stored potential energy, and then to random kinetic or thermal energy. The power developed per unit volume (P_v) by this mechanism is:

$$P_v = k\,E^2 f\,\varepsilon'\,tan\,\delta = k\,E^2 f\,\varepsilon'' \tag{5}$$

where k is a dielectric constant, depending on the units of measurement, E is the electric field strength (V m^{-3}), f is the frequency (s^{-1}), ε' is the relative dielectric constant or relative permeability, tan δ is the loss tangent or dissipation factor, and ε'' is the loss factor. Field strength and frequency are dependent on the equipment, while the dielectric constant, dissipation factor and loss factor are material dependent. The electric field strength is also dependent on the location of the material within the microwave/radiofrequency cavity (Turner and Ferguson, 1995), which is one reason why domestic microwave ovens have rotating turntables (so that the food is exposed to a range of microwave intensities).

The dielectric constant of water is at least an order of magnitude higher than most underlying materials (Schiffmann, 1995), and overall the dielectric constant of most materials is nearly proportional to moisture content up to a critical value, often around 20-30%. Hence microwave and radiofrequency methods preferentially heat and dry wetter areas in lumber, a process that results in more uniform final moisture content. For water and other small molecules, the effect of increasing temperature is to decrease the heating rate slightly, hence leading to a self-limiting effect, so that spontaneous thermal runaway is less likely to occur. However, there is still a need to limit the temperature rise in the wet core of boards, since microwave power is absorbed preferentially by the wet centre of boards.

A secondary heating effect is due to ionic conduction, since the ions (sodium, chloride and hydroxyl) in the water are accelerated and decelerated by the changing electric field. This type of heating is not significantly dependent on either temperature or frequency. It is most noticeable with preservative treated wood.

A frequent feature in the literature is the use of radiofrequency or microwave power to augment vacuum drying. This practice stems from the difficulty in getting convective heat transfer when wood is under vacuum. RF/microwave energy when superimposed on vacuum drying not only warms the entire board but also generates a pressure gradient – along with a mild temperature gradient – from the centre to the surfaces. The low absolute pressure in the drying chamber reduces the boiling point (Table 8.1) enabling water to evaporate at temperatures far below 100°C, while the RF/microwave energy generates a modest overpressure within the wood that sustains a flow of water vapour down these pressure and temperature gradients. This is an attractive system for valuable, slow-to-dry, green hardwoods that can be susceptible to collapse if dried above 35°C, and certainly above 50°C, early in the schedule. Here, microwave power needs are modest as it is necessary to limit the temperature rise in the interior of the boards.

Rozsa (1994), when drying 30 mm *Eucalyptus regnans* (>100% MC) under vacuum, found that radiofrequency drying at an absolute pressure of 1 kPa increased the initial drying rate twofold compared to working at 10 kPa. At 10 kPa there was a rapid rise in the core temperatures even at low power inputs, a situation that might lead to damaging pressures inside the wood. Rosza concluded that control of the core temperature below 40°C to be necessary for good product quality. The estimated energy costs of A\$130/m^3 (50W) and A\$175/m^3 (100W) were compared with A\$230/m^3 for conventional hardwood drying on a small scale (Schaffner, 1991), suggesting that heating costs would not be excessive at these low microwave energy inputs. Conventional predrying (35°C, 85% RH) resulted in a 1-2% moisture loss per day (8 weeks to the fibre saturation point) compared to 5% per hour or 7.5%/hr for the two dielectic-vacuum drying treatments (50W or 100W).

Quoted ratios of drying times (Table 8.3) to achieve the same final wood quality using microwaves (and radiofrequencies) to those from conventional schedules range from 0.25 for white spruce, *Picea glauca*, (Miller, 1971) to 0.03 for Douglas fir (Barnes *et al.*, 1976). These accelerated drying rates are an important factor when evaluating the economic viability of such processes. Schiffmann (1995) suggests that, as a rule of thumb, radiofrequency or microwave systems become economically attractive if the drying rate can be increased fourfold over that from conventional drying for the same product quality. Since these enhancements have been noted by most workers, the economics appear to be promising, particularly for new construction.

Barnes *et al.* (1976) reported trials in a kiln with a 25 kW magnetron. They concluded that the cost of drying per unit volume of timber for microwave drying is influenced mainly by the initial moisture content and density, while for conventional drying, the cost per unit volume is mainly dependent on the length of the drying cycle required to achieve acceptable degrade levels. They felt that microwave drying is most likely to be appropriate where the wood species has a low initial moisture content, has problems with degrade in conventional drying, and/or is relatively valuable (so that stock-holding charges are significant). For example, they found that microwave drying was economic for Douglas fir, which had a low initial moisture content (40%), but not for western hemlock, which had a higher initial moisture content (86%). On this basis, the authors considered that microwave drying would find most applications for high-value hardwood species.

Table 8.3. Typical reductions in drying times with microwave and radiofrequency drying.

Author	Species, size of timber	Moisture content (%)		Relative drying times microwave/ conventional
		Initial	Final	
Avramidis *et al.* (1996)	Mixture of western red cedar, western hemlock and amabilis fir, 71-152 mm thick	80	16	0.12
Barnes *et al.* (1994)	Western hemlock, 50 mm thick	86	9-11	0.04
	Douglas fir, 50 mm thick	36	11-13	0.03
Harris and Taras (1984)	Red oak, 25 mm thick	57-67	7-8	0.06
Miller (1971)	White spruce, 50 mm thick	55-65	15	0.25
Smith and Smith (1994)	Red oak, 57 mm square	85	8	0.25

A consistent feature has been the concern to keep core temperatures and pressures quite low to avoid damaging pressures building up inside the wood. However a radically different approach, namely the use of high microwave power and high internal pressures, aims to disrupt the fine structure of the wood. Further, under such conditions the use of a vacuum is irrelevant to severe microwave drying. Vinden and Togornikov (2002) at the University of Melbourne in Australia have progressed vigorously the construction and commissioning of such microwave drying plants including continuous conveyor-belt systems, taking the project through to market readiness. High intensity microwaves with green or partially green wood generates high internal steam pressures that selectively rupture and modify wood cells, creating pathways for liquid movement through unlignified ray tissue in pine species, and micro-checks at ray/fibre interfaces in hardwoods. Both result in greatly improved radial and longitudinal permeabilities. The permeability of *Eucalyptus obliqua* increased 170-1200 fold. The destructuring of wood after a few seconds exposure to intense microwaves and subsequent drying is exemplified by physical expansion, up to 15% for radiata pine, 9% for Douglas fir, 13% for *E. obliqua* and 11% for *E. regnans*, i.e. a 10-15% expansion rather than the expected 10-15% shrinkage.

The end product of the microwave pre-treatment is a material with frequent checks/splits. The impermeable wood structure is 'blown' open. Subsequent resin impregnation by cold soaking and press curing offers the prospect of a product with high structural performance and great aesthetic value – resealed and glued checks do not generate the hairline stain marks usually associated with drying checks.

6.6. Dehumidifiers and heat pumps (Bannister et al., 1994, 2002)

Negative misperceptions of dehumidifiers are that they dry too slowly, operate at too low a temperature (<50°C) to relieve drying stresses, and use electricity rather than cheap process fuels such as wood residues. In reality inefficiencies have arisen from poor design, over-sizing of equipment and poor operational management. For example, compared to a conventional kiln an efficient dehumidifier requires far superior seals for door/vents and better insulation since it is energy frugal in its recovery of waste heat, and in part because of the relatively high cost of electricity. However, dehumidifiers are less efficient at low moisture contents (<20%).

Such kilns are inappropriate for industrial and structural softwood lumber that may be dried at over 100°C in less than a day, but they stand comparison with a two-day 90/60°C schedule for 25 mm appearance grade softwood lumber. Dehumidifiers can dry such wood in 6-7 days at 40-60/35-50°C (dry/wet-bulb temperatures) – but down to only 20%, from 150% MC. Further they are particularly well-suited for slow-to-dry, collapse-prone, check-prone hardwoods that must be dried at low temperatures. These can be dried in about 15 days to 20% MC.

Benefits include energy conservation using 40% less energy overall (even accepting the 33% 'inefficiency' in generating electricity), reduced greenhouse gas emissions (a 80-150 kg of CO_2 'saving' for every m^3 of green pine that is dried compared to using wood or coal as fuel), superior effluent trapping (condensate) and avoidance of volatile emissions, a paler surface colour and superior quality of lumber, e.g. no kiln brown stain or within-ring checking. Further, dehumidifier technology is immature with the potential to increase the specific moisture extraction rate of between 2.8 and 3.5 kg to over 5.0 kg of water per kWh.

The choice of environmentally benign refrigerant, e.g. CO_2, NH_3, hydrocarbons, its physical properties (liquifaction temperature, heat capacity) and the compressor unit determine the efficiency of these driers and the temperature range over which they operate: most operate at 40-50°C but some can work efficiently up to 80°C.

A heat pump dehumidifiers is operationally equivalent to a refrigerator except that both evaporator and condenser coils are inside the kiln. Energy efficiency is achieved by condensing moisture from the circulating air and recovering its latent heat of vaporization, rather than venting moist air to the atmosphere as in a conventional kiln. With a heat pump the refrigerant is forced round the system by a compressor. This low boiling point liquid passes through a pressure-reducing device (an expansion valve) to the evaporator coils where the liquid absorbs heat from the surrounding warm, moist air and boils off. The vapour is then recompressed to a high pressure and pumped into the condenser coils where it is cooled by passing cold, dry air over the coils. When cooled sufficiently the vapour condenses back to a liquid. Thus the working fluid absorbs heat at the evaporator coils and gives out heat at the condenser coils.

Moist warm air from the stacks is drawn through the evaporator coils and is cooled until the absolute humidity of the air corresponds to the absolute humidity at saturation (Table 8.1), the dew point. On further cooling moisture condenses on the cold evaporator coils. This water is collected and drained from the kiln. The cold air then passes through the warm condenser coils where it is reheated. The dry warm air then passes through the timber stacks again.

There are limitations with dehumidifiers. First, dehumidifiers may need an auxiliary heat source to warm up the kiln charge to the optimum operating temperature. Secondly, they operate most efficiently at high humidities, which results in extended drying times, and become progressively less efficient as the relative humidity declines towards the end of the schedule. Finally, more heat is released than is needed to dry the timber. This can be abstracted by venting as required (for temperature control) or removed by coupling with an external heat exchanger. The latter allows the system to operate in a fully-closed, non-vented mode which means that the lumber can be dried in a controlled or modified atmosphere, e.g. oxygen-free to control colour.

7. A CONVENTIONAL KILN SCHEDULE

7.1. Warming up

The initial warm up stage aims to heat the wood thoroughly throughout its cross-section before drying commences. To do this the kiln operator maintains a small wet-bulb depression. Too small a wet-bulb depression results in excessive condensation on the lumber but, if too great, drying will occur while the wood is still cold and severe surface checking may result – a consequence of a high rate of evaporation (a large ΔT) and a low rate of diffusion (low T) resulting in a very steep moisture gradient close to the surface. Impermeable woods are likely to check if the wet-bulb depression exceeds about 2°C. Once the air temperature reaches the kiln operating temperature the wood is allowed to warm through: an hour per 10 mm timber thickness would be typical. This warm up treatment helps relieve any stresses present in the wood, especially likely if it has been partially air-dried.

7.2. A traditional schedule

It is not good practice to mix species or timber sizes, as the schedule must be that for the slowest-to-dry material, i.e. the slowest drying or most degrade-prone boards limit the drying rate. Thus for plantation-grown material it is usual to presort material by dimension, by grade, and into sapwood, heartwood and mixed. Each is dried by its own specific schedule. Drying schedules for various species, thicknesses and moisture contents are available in most countries (Boone *et al.*, 1988). These schedules (Table 8.4) tend to be conservative and experienced operators can dry a little faster without causing degrade.

Accelerated schedules for fast drying softwoods – a single step 90/60°C – compared to that offered in Table 8.4a produce evenly dried, stress and strain free wood in two days rather than 4-5 days. The lumber will also be free of kiln brown stain (see later).

Table 8.4. Conservative kiln schedules of the 1970s that were developed empirically.

(a). Conventional schedule for drying and equalizing of 50 mm *Pinus radiata* framing timber.

Moisture content of wettest lumber in stack	Dry-bulb (0°C)	Wet-bulb (0°C)	Relative humidity (%)	Equilibrium moisture content (%)
Green	71	60	58	9.0
50	75	60	49	6.3
20	80	60	39	4.9
Equalizing	80	73	73	9.9
Conditioning	85	84	96	18.5

(b). Drying schedule for teak[1] up to 38 mm thick[2] (Boone *et al.*, 1988).

Moisture content of wettest lumber in stack	Dry-bulb (0°C)	Wet-bulb (0°C)	Relative humidity (%)	Equilibrium moisture content (%)
Green	60	56	82	14.2
50	60	54.5	75	12.0
40	60	51.5	64	9.6
35	60	49	55	8.0
30	65.5	51.5	49	6.8
25	71	54.5	43	5.8
20	76.5	57	30	5.1
15	82	54.5	26	3.5
Equalizing	82	69	57	7.0
Conditioning	82	79	87	13.7

[1] Curiously this schedule also applies to balsa, *Ochroma lagopus*, as well as to some tropical pines, *Pinus caribaea* and *P. oocarpa*.

[2] Thicker material requires a milder schedule. The humidity is increased by 5-6% and the dry-bulb temperature reduced by 5-6°C.

In one sense, kilns maintain a relatively high humidity in order to slow down the rate of evaporation, particularly early in the schedule when drying impermeable species. This means a high equilibrium moisture content, less surface shrinkage and differential strains early in the schedule – less risk of checking.

The kiln operator selects the appropriate schedule and adjusts the temperature and humidity as necessary by manipulation of the heating coils (dry heat) and spray valves (steam), with occasional adjustments to the air inlet and outlet vents, which exchange hot, moist air from the kiln for cold, drier air from outside. The steam line raises and maintains the humidity while the kiln warms up. It should be used only infrequently thereafter as economy in operation depends on making maximum use of the moisture extracted from the lumber to maintain the required humidity in the kiln. As moisture evaporates the humidity rises above the schedule value and the vents open a little to bleed off some of the hot damp air.

During the schedule automatic controls (or manual operator) intermittently adjust the dry and wet-bulb temperatures so that the kiln conditions correspond to the next stage of the schedule and so on until the schedule is complete. The wet-bulb depression must be controlled accurately as it has a disproportionate effect on the severity of the schedule, compared to any drift in both dry and wet-bulb temperatures (Table 8.2). If the wet-bulb depression is too great, the rate of drying will be too fast, and the lumber may suffer degrade.

Pines such as *P. radiata* are permeable and can be dried quickly (Table 8.4a). However their pits aspirate and if subsequently pressure impregnated with an aqueous preservative such as copper-chrome-arsenate using the full cell process they cannot be redried nearly as rapidly. If such resaturated, preservative treated timber were to be dried using the schedule in Table 8.4a steep moisture gradients would develop leading to severe checking. A milder schedule is required. Treated pine takes approximately twice as long to dry.

7.3. Equalizing and conditioning

The final step of the schedule would, if left long enough, dry the load more than desired. Indeed when drying refractory hardwoods it is advisable to wait until the majority of the timber has dried well below the specified moisture content, to ensure that no excessively wet material exists within the load, before applying an equalizing treatment. The final step in the hardwood schedule in Table 8.4b corresponds to an equilibrium moisture content of 3.5%, which is much lower than the desired final moisture content of around 10%. When the driest pieces are 3% below the desired moisture content the kiln schedule should be terminated. Assuming a moisture content of 10% is sought the driest pieces would have a moisture content of 7% while the wettest sample might be at 13% moisture content. To reduce the variation in moisture content between boards the humidity in the kiln is now increased. The dry-bulb temperature is kept at 82°C while the wet-bulb is raised to 69°C giving a relative humidity of 57% and an equilibrium moisture content of 7.0%. Under these equalizing conditions the drier boards are unable to dry further but the wetter boards continue to do so, albeit more slowly.

When the moisture content of the wettest sample board drops to 10% moisture content the humidity is raised again for a final conditioning period in which the conditioning equilibrium moisture content is set about 3-4% above the desired moisture content. In this particular case the dry-bulb temperature is still 82°C and the wet-bulb temperature is raised to 79°C, giving a relative humidity of 87% and an equilibrium moisture content of 13.7%. The equalizing and conditioning periods reduce the variability in moisture content between boards as well as reducing moisture gradients within the boards (target moisture content ±2%). They also relieve drying stresses generated during kilning as discussed later, e.g. case hardening.

Separate conditioning chambers with fully saturated, low-pressure steam are favoured if kiln throughput is large enough to justify the investment. There are no fans and no environmental control. High-temperature dried ($T > 100°C$) charges should be left to cool in the conditioning chamber for 1-2 hrs, until the temperature of the wood falls to 70-90°C. The wood is now cool enough for steam to condense on the boards. With the wood able to absorb moisture reconditioning can commence.

7.4. Cooling down and storage

If lumber were taken out of the kiln immediately there would be a risk that the hot wood will heat the cool air around the stack, making the air warmer and much drier. The warm dry air would then lead to further drying and checking at the surface of the boards – cool saturated air at 20°C has a relative humidity of only 12% if heated to 60°C and the moisture content of the wood in equilibrium with that air would be only 2%. For the better grades of timber the heat is turned off and the load cooled under a constant wet-bulb depression of about 5°C until the temperature is within 15-20°C of that outside. Only then can the stacks be removed safely from the kiln.

Once dried the lumber should be block stacked, without stickers, and wrapped in a plastic covering. It should not be left unprotected in the mill or on the building site.

7.5. Kiln control

Drying is a compromise between the ideal and the practical. Problems arise from the steep moisture gradients generated if drying is too fast, especially with impermeable timbers. These can occur early in the schedule, which is why emphasis is placed on a low initial temperature (the timber is stronger) and high humidity (slower drying, a higher equilibrium moisture content and so less shrinkage stress). This minimizes the danger of checking and case hardening. Once past this critical stage more severe drying conditions can be imposed.

Traditional schedules result in a drying rate that decreases with time, only partly countered by increases in the dry-bulb temperature and the wet-bulb depression as the schedule proceeds. Modern automatic process control means that the kiln schedule can be adjusted continuously so ensuring a more constant rate of heat transfer and evaporation. Also this avoids any 'shock' that an abrupt change in the schedule imposes on the timber.

Strain-limited schedules have been developed by a number of workers (Doe *et al.*, 1994; Langrish *et al.*, 1997). With this approach, the drying rate is maximized subject to a limit on the maximum instantaneous strain experienced anywhere in the wood. In one example (Langrish *et al.*, 1997) the drying time of a hardwood, *Eucalyptus paniculata*, was reduced from 155 hours for a conventional schedule to 115 hours for a strain-limited one, with a reduction of 75% in the number of cracks from the strain-limited schedule compared with the conventional one. Doe *et al.* (1994) included sensor-based monitoring of acoustic emissions to override the mathematically-developed strain-limited schedule. The development of strain-limited schedules that allow for biological variability has been described by Pordage and Langrish (2000). A key outcome was the prediction that the productivity (amount of good quality timber divided by drying time) appears to be maximized when the schedule is such that 90% of the wood produced is good quality.

8. HIGH-TEMPERATURE DRYING ABOVE 100°C

Drying with a dry-bulb temperature above 100°C is termed high-temperature drying. It differs from normal kiln-drying in that there is a slight steam pressure within the wood wherever the temperature exceeds 100°C. In permeable wood this results in mass flow of water vapour from the wet-line to the wood surface, i.e. the movement of the water vapour is no longer entirely diffusion controlled despite the local moisture content being below fibre saturation. It is questionable whether high-temperature drying is best viewed as being fundamentally different to drying at lower temperatures or whether it should be seen as an extreme example of the latter. The processes that contribute to drying of a permeable species at moderate temperatures also contribute very effectively to high-temperature drying. The vapour flow through the pits acts in parallel with moisture diffusion through the cell walls (which becomes increasingly efficient since diffusion is so temperature sensitive). The enormous wet-bulb depression sustains the high evaporative load.

Most high-temperature kiln schedules operate in an air/steam environment, for example with a dry-bulb at 120-140°C and a wet-bulb at 70-85°C, and use high fan speeds (10-15 m s^{-1}). Stickers are wider, up to 45 mm: the timber is softer at high temperatures and stickers will indent into the wood unless the bearing area is increased. They are also thicker, 25-32 mm: the wider gaps between layers encourage better airflow and more rapid heat transfer. With permeable softwoods the drying of 50 mm thick material can be completed in less than 24 hours whereas with a conventional kiln the schedule might last 5-7 days. It is used successfully with the southern pines and with *Pinus radiata* and hoop pine, *Araucaria cunninghamii*. For example, in Australia 45 mm radiata pine is dried in purpose designed kilns using a 150°/75°C schedule in about 14 hours. These high-temperature schedules include steam conditioning at 100°C for 2-4 hours depending on thickness and careful attention to this procedure is essential to minimize degrade from internal checking or honeycombing (Williams and Kininmonth, 1984).

Potential problems with high-temperature drying, apart from limitations regarding species, are that it can be difficult to achieve a uniform moisture content, the wood is commonly discoloured or darkened, and its mechanical strength is slightly reduced. Also warp is very prevalent unless the stacks are heavily weighted to prevent movement. Distortion-prone corewood of radiata pine can be successfully dried by restraining with thick reinforced concrete slabs (<1000 kg m^{-2}).

Some success has been reported with hardwoods, but only for those species that are not particularly difficult to dry using conventional schedules. Overall drying time can be halved but at the expense of additional degrade such as honeycombing and collapse. Some recalcitrant timbers may be dried successfully but only if they have been pre-dried to the fibre saturation point. Other hardwoods cannot be high-temperature dried at all.

9. DRYING DEGRADE

Apart from stain, virtually all seasoning degrade is due to shrinkage or to differential shrinkage. Moisture gradients cause the most difficulties. These are minimized by drying slowly, but that would be uneconomic. In practice the kiln or yard supervisor seeks to dry the timber as fast as possible without causing excessive degrade.

9.1. Staining

Heavily sapstained pine has a dull blue colour. Indeed the sapwood of maritime pine, *Pinus pinaster*, when imported to Britain from the Continent early in the last century was described as such: no just in time delivery 100 years ago.

Sapstain and fungal attack is controlled by minimizing the time between felling and milling, and by getting the lumber into stack as quickly as possible so that the surface of the boards can dry quickly to below 20% moisture content. Alternatively, a prophylactic anti-sapstain dip immediately after milling provides effective protection for a few months. Rather than using broad-spectrum, long-life fungicides, sapstain is controlled by combinations of less-effective chemicals having a number of active ingredients with different modes of action – each targeting a specific range of organisms.

9.2. Warping

This arises simply because of the anisotropic shrinkage of wood (Figure 8.14). Spiral grain, cross-grain and reaction wood contribute, and warping is especially likely in corewood. Drying under heavy restraint (1000 kg m^{-2}) mitigates the problem.

Diamonding. Square cross-sections that are sawn with the growth rings running diagonally become diamond shaped simply because tangential shrinkage is greater than radial shrinkage. To remedy, the material is dressed on all four sides.

Cupping. Flat-sawn boards cup (Figure 8.14). The outcome of cupping is such that the growth rings 'straighten out' a little. Only boards at the top of timber stacks can cup in this way as the others are held down flat by the weight of the timber above – and may crack instead.

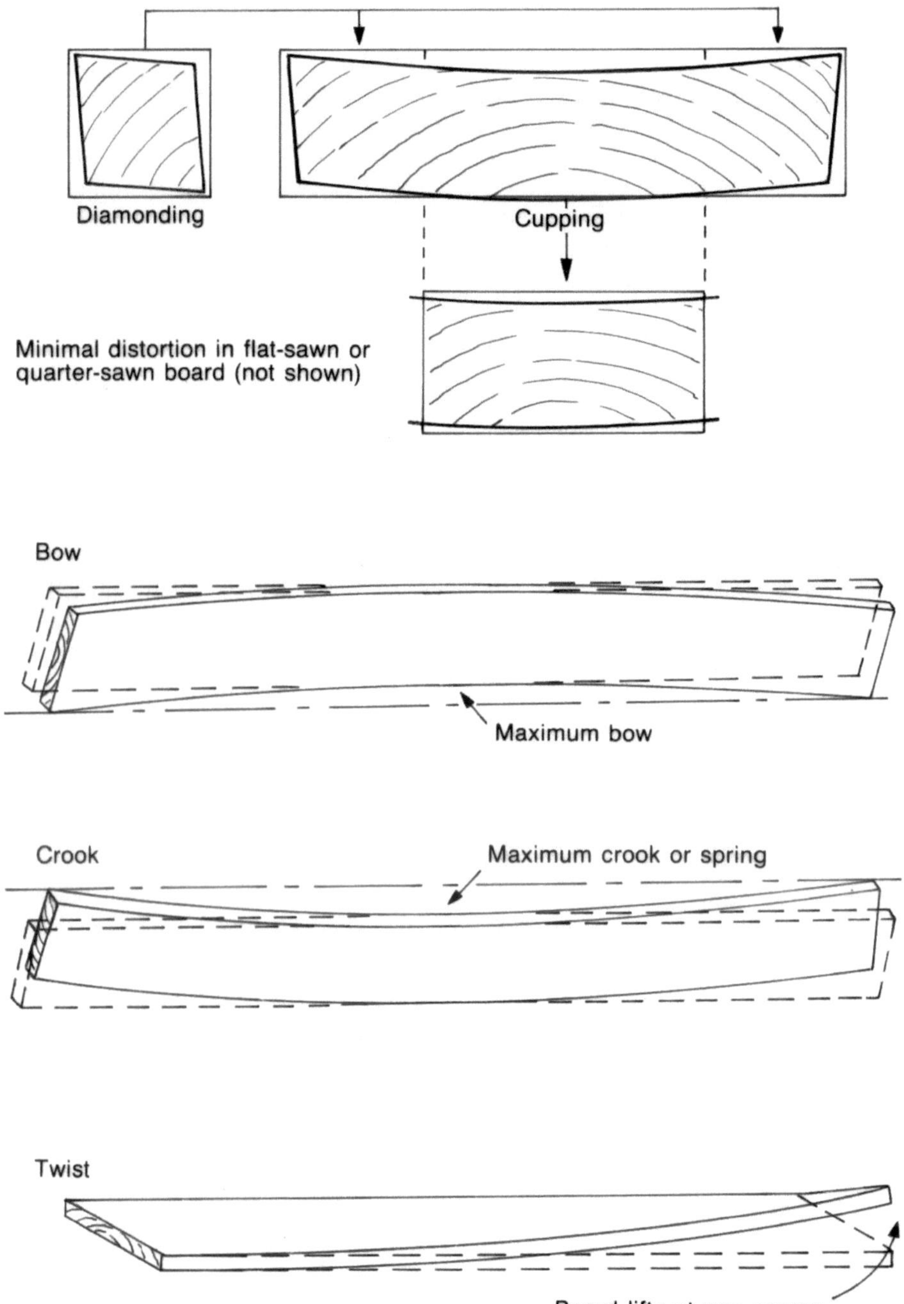

Figure 8.14. Distortion in lumber has various causes.

Bowing is the distortion of the board, with the face no longer planar, but curved about an axis lying parallel to the width of the board.

Crook or spring is similar to cupping and bowing except that it occurs on the edge. Far larger forces are involved (it is far easier to bend a ruler about its face than about its edge). The weight of the stack, or concrete weights placed on top, provide only limited restraint. Crook is encountered with reaction wood or in pithy timber where the fibres on the edge adjacent to the pith may have a large microfibril angle. Crook arises from differences in longitudinal shrinkage between opposite edges of the board.

Twist is a spiral distortion along the length of a piece of wood. It is found in species with spiral grain. It arises because the angle of the grain varies with the position of the fibres within the tree, and this is reflected when sawn into lumber. It is also associated with cross-grain.

9.3. Kiln brown stain and collapse

Sawing creates a rough-sawn surface layer 1-2 mm thick. When high-temperature drying (>100°C) a permeable sapwood, sap flows to the edge of this layer for a considerable part of the schedule and on evaporating leaves behind an accumulation of water-soluble plant foods and pectic materials. Chemical or microbial action results in the formation of complex brown-coloured, tannin-like molecules (Figure 8.15). Unfortunately, kiln brown stain is revealed only subsequently, once the wood is dressed. It is a particular problem as the market prefers light-coloured woods and mills are sawing closer to the desired dimension – and are reluctant to dress 1-2 mm off each face. Kiln brown stain is much less acute with milder schedules using lower dry-bulb temperatures, less than 90°C for radiata pine. With heartwood there is no liquid flow, no accumulation immediately below the surface and no kiln brown stain.

Collapse precedes conventional shrinkage, only occurring within saturated cells, whereas normal shrinkage occurs below the fibre saturation point. Where present, severe collapse is revealed by a series of corrugated depressions matching earlywood

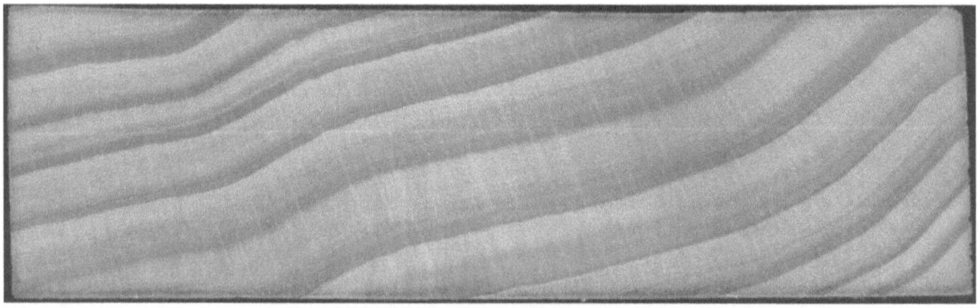

Figure 8.15. Kiln brown stain, which appears as only a feint line in the cross-section above, can be a costly and unsightly discolouration. When dressed the face, which should be a bright and white, has a blotchy brown appearance.

bands on the radial surfaces of boards. Collapse occurs predominantly in the radial direction, and radial shrinkage can be as great as 50%.

The Laplace equation (Eq. 2) determines the capillary tension that is experienced by all waterlogged cells within the capillary network, and not just those at the wet-line. On drying the negative tension within the sap stream increases until the onset of cavitation and the flow of sap from the emptying, cavitated cells to the wet-line. Alternatively, where the capillary tension exceeds the compressive strength of the cell wall – and this can occur with the thinnest and weakest cell walls – the walls can collapse inwards and the sap flows out (there is no cavitation). Collapse is associated particularly with saturated, thin-walled earlywood. In permeable wood this results in the flow of sap to the wet line, whereas in impermeable wood collapse results in the gradual local redistribution of the sap as it escapes from the cell through the wall by diffusion.

The compressive strength of the saturated cell walls decreases with increasing temperature, so raising the temperature increases the likelihood of collapse. Thus collapse is most likely when a waterlogged impermeable timber of low density (and so of low strength) is kiln-dried above 50°C. Collapse is best avoided by drying at low temperatures.

When drying radiata pine at 105/72.5°C, Deyev and Keey (2001) observed that collapse in earlywood occurred suddenly, in less than a minute, once the moisture content had fallen to around 80% of the original green value. These collapsing earlywood bands were surrounded by air-filled cells, both in the adjacent latewood bands and in the 1 mm deep saw-damaged zone at the surface. The tightly aspirated pits of these surrounding cells isolated the saturated earlywood allowing the capillary tension to build until the tension reached *c.* 6-7 MPa. This value was estimated by creating similar depressions by pressing a cylindrical indenter into the hot surface (Deyev and Keey, 2001).

Lumber that has suffered from collapse can be conditioned by steaming at 100°C and 100% relative humidity for 4-8 hours, depending on the degree of collapse. Steaming can commence once the moisture content is below 15%, or below 10% in the case of high-temperature drying. During steaming the wood reabsorbs a certain amount of moisture (1-4%). The cell walls re-swell and become sufficiently plastic (they are hot and moist) to permit the latent or residual stresses within the walls, which opposed collapse, to reassert themselves and force the distorted and compressed cells to return to their normal shape. No collapse occurs on re-drying after steaming since the lumens are not saturated. The reconditioning schedule is very severe on the kiln and it is usual to build special concrete reconditioning chambers.

Unless collapse is a common occurrence and of a serious nature, it may not be worthwhile curing it, in which case it is best prevented by partial air-drying or kiln-drying at a low temperature (<50°C).

Eucalyptus regnans is the classic example of a collapse-prone timber. Once considered suitable only as fuelwood, today it is in demand for flooring, furniture and in other prestige items. Collapse-prone timbers include certain other eucalypts, oak (*Quercus* sp.), black walnut (*Juglans nigra*), western red cedar (*Thuja plicata*) and redwood (*Sequoia sempervirens*).

9.4 Case hardening

A moisture gradient is essential for drying. Consequently the wood shrinks unevenly with the surface drying first. Initially surface shrinkage is restrained because the inner fibres are above the fibre saturation point and do not want to shrink yet. Because of this restraint, the outer fibres dry in tension perpendicular to the grain (shrinkage is principally in the transverse directions) and, as a reaction, the inner fibres are in compression. If the moisture gradient is excessively steep during the first stage of drying, the tensile stresses experienced by the surface fibres may exceed the failure stress perpendicular to the grain. Surface checks will form and the stresses will relax again. If the moisture gradient is a little less steep, the surface fibres will be subjected to a tensile stress that is in excess of the elastic limit but is less than the failure stress. The surface fibres will be stretched. As drying continues the surface fibres dry further and the zone in which the wood is attempting to shrink gets larger. The magnitude of the tensile stresses decreases while the compressive stresses in the centre of the board increase: the proportion of cross-section drying and wanting to shrink becomes greater and the counterbalancing compressive forces are concentrated in a smaller wet core. Because the surface fibres were restrained from shrinking they do not shrink as much as expected.

Later on the interior fibres begin to dry below fibre saturation point and in turn want to shrink but now they are prevented from doing so by the outer fibres, which are set in a stretched condition – stretched in the sense that they have not shrunk as much as expected had they been free to dry and shrink without restraint. At this stage of drying the stresses within the wood have reversed. The interior is in tension while the surface zone experiences a compressive stress. This condition is called case hardening. Typical strains and moisture gradients within the timber during drying are shown in Figure 8.16. Note that compressive strains in the fibres result in elongation when cut free as thin sections, and tensile strains in the fibres are revealed as a contraction when cut free.

McMillen (1955) measured strains in oak by cutting thin sections from samples at various times during air-drying (Figure 8.16). The maximum tensile strain in the surface fibres occurs within 5 days when the average moisture content is high. The corresponding tensile stresses exceed 15 MPa. The 2,9 slices reach their maximum tensile strain after 18 days and at that time the 5,6 slices sustain their maximum compressive strain. Stress reversal occurs after 28 days. By day 36 the compressive strain in the outer slices is 80% of the tensile strain in the same slices on day five, but the dry surface fibres easily sustain this stress without crushing. At this time slices 3 to 8 attain their maximum tensile strain and the corresponding tensile stresses in this zone may exceed the ultimate tensile strength of the wet wood (the average moisture content is 17% but the core is still above fibre saturation point). Internal checking (honeycombing) is most likely in the period between days 28 and 36. At the end of air-drying, day 50, the lumber is case hardened.

Case hardening is one condition that the kiln operator prefers to live with and cure rather than prevent. It is not economic to dry timber too slowly. In practice the aim is to dry lumber as fast as one dare without getting excessive checking and to

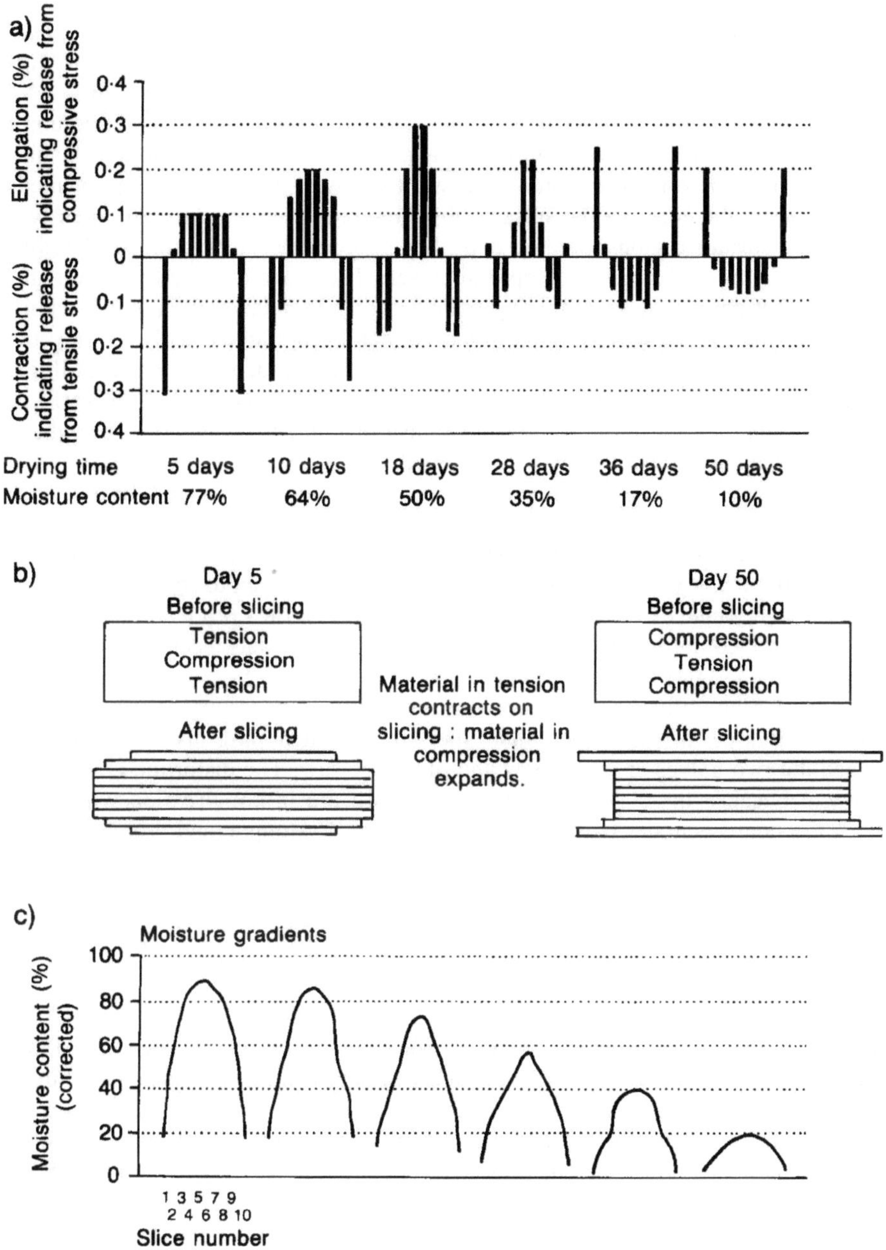

Figure 8.16. (a) Strains in 50 mm red oak during air-drying (McMillen, 1955). (b) Transverse movement in partially dried timber when stresses are released by cutting in sections parallel to the surface. (c) Moisture profile through the timber during drying (McMillen, 1955).

condition out the stresses found in case hardened wood. Species requiring the most care during drying are those that are impermeable and those that possess poor dimensional stability. As noted, mild case hardening also occurs during air-drying and this is not relieved unless the lumber is subsequently conditioned in a kiln.

Case hardening is relieved during the final stages of kiln-drying by conditioning. The relative humidity and temperature are raised to make the wood more plastic. The surface fibres re-adsorb moisture and want to swell again, but are prevented from doing so by the core. As the surface fibres re-adsorb moisture the compressive stresses in the surface regions increase until they exceed the elastic limit for the hot, soft surface fibres and induce compression set. The trick is to induce enough compression set during the conditioning period to counterbalance the initial tension set during the early days of drying. If that is achieved lumber will dry free of stresses. At the end of the conditioning period the surface fibres are still in compression and the core is still in tension, but the moisture content in the surface now exceeds that of the core. When the moisture content throughout the wood finally equalizes internal stresses will be relieved.

The conditioning treatment can be overdone. If the wood is steamed for too long and the surface fibres re-adsorb too much water this will induce an excessive compression set in the surface fibres. At the end of drying, when the surface fibres return to the natural equilibrium moisture content they will want to shrink more than normal wood (because of the excessive compression set), thus producing tension at the surface and compression in the interior. The lumber has been reverse case hardened. This condition cannot be treated.

Case hardening only becomes obvious when wood is recut or when more material is dressed off one face than the other. When case hardened boards are ripped the two halves crook or bow and pinch the saw. The distortion is towards the saw cut because the surface zones of the board are in compression while the centre of the board is under tension. An edge groove, for tongue and grooving, will pinch too. Similarly a board that is machined more heavily on one face will bow towards that face. Kiln-dried lumber ought to be dimensionally stable, and such distortion on processing is undesirable.

9.5. Honeycombing

During the later stages of drying, when the surface fibres are in compression and the core in tension, the moist interior of the wood can check internally (honeycombing) with the checks running in the radial direction, following the rays. Ray tissue is weaker and fails in tension across the grain more readily than other tissue. Honeycombing is insidious since its presence is not obvious until the wood is further processed.

Honeycombing can usually be avoided by ensuring that large stresses do not develop early in drying and by avoiding too high a temperature during the final stages of kilning. The tensile strength of green wood is very temperature dependent, decreasing with increasing temperature; the compressive strength of dry wood, is less temperature dependent. Honeycombing is also associated with collapse.

9.6. Checking and knots

A check is a split parallel to the grain, normally a few centimetres long. Surface checking occurs during the early stages of drying when using too severe a schedule.

Species such as oak and beech check quite readily, and to avoid this problem the humidity is kept high early in the kiln schedule. Also the temperature is kept low in order to maintain the timber's strength. Only as the lumber dries and becomes stronger can the humidity be lowered and the temperature raised to provide more rapid drying conditions. Surface checks forming early in the kiln schedule may close up later when the surface fibres go into compression and the core into tension, although the failure plane remains in the tissue. Such checks can be revealed subsequently as hairline streaks if the wood is stained.

Checks develop at points of weakness, e.g. along rays. Hence flat-sawn boards face-check and quarter-sawn boards edge-check. Probably the rays dry faster than the surrounding tissue but are restrained from shrinking, and are stretched tangentially, in the sense that they are unable to shrink as much as they would like. A steep moisture gradient favours checking.

Very rapid end drying will cause the ends of boards to shrink ahead of the rest of the board and so end-check. End-coatings and wide stickers minimize this.

A knot with its exposed end grain dries rapidly, but the surrounding wood that dries more slowly restrains its shrinkage. In the board the fibres are aligned longitudinally whereas in the knot the fibres are transversely orientated relative to the board, so the board will shrink little longitudinally whereas the transverse shrinkage of the knotty tissue will be much greater in this direction. Live (intergrown) knots are firmly bound to the surrounding wood. These knots are liable to crack because of differential shrinkage between the knot and the surrounding wood. Such degrade is comparatively serious in fast growing plantation species. Dead or bark-encased knots in lumber can loosen and fall out during drying.

10. PRACTICAL IMPLICATONS OF DRYING MODELS

The development of early kiln schedules and estimates of drying times provided the rudimentary skills for successful drying of lumber. To be viable they had to be conservative as they were empirical, and they could not be extended outside the range of data used in their development. One early empirical approach assumed the time to dry to be proportional to $(\text{density})^n$ and $(\text{thickness})^n$, where n ranged between 1 and 2. At the two extremes, the sapwood of *Pinus radiata* is highly permeable and drying times are proportional to thickness whereas the heartwood of *Nothofagus fusca* is highly impermeable and drying time is proportional to the square of the thickness. Empirical models provided no information about moisture-content gradients, information that is required for stress analysis, which in turn is needed to understand and predict the development of collapse, internal checking *etc*.

This introduction has followed the easy inclination to conceive of drying in terms of processes within the wood, based on the questionable intuitive belief that it is the wood itself that controls drying.

In contrast realistic, functional drying models provide the basis for optimizing kiln schedules that then can be automated. They need at least three close-coupled modular elements: the airflow around the kiln; the boundary layer above the lumber surface; and the processes of moisture movement within the wood itself. In turn each of these has subsidiary complexities. Here, the intention is to reiterate the ideas and features that must be incorporated in these models – and to emphasize the importance for managers of getting quality technical advice before investing in drying facilities.

In this section the complexity of the fluid dynamics is ignored and the emphasis is on explaining the outcomes and on understanding the utility of these models. The parameters that are woven into such mathematical models will have been derived from a variety of fundamental studies – and validated in extensive trials.

For example, regarding transport processes occurring within the wood, there are a number of parameters that are needed to develop a complex model. These are gathered though a series of iterative studies examining the drying of single boards in a wind tunnel or small stacks in experimental kilns. Boards are monitored continuously, and intermittently sample boards are removed and destructively tested – by sectioning and measuring individual 'slices' taken from the board – to experimentally determine the moisture gradients and stresses as they evolve during the course of the drying schedule. Diffusion coefficients, which vary widely between species (10^{-8} to 10^{-10} m^2 s^{-1} for hardwood) and with temperature (activation energies in the range 23-43 kJ mol^{-1}), and developing elastic strains within the board are obviously needed if one is to dry a wood rapidly without excessive degrade: to which should be input the initial, irreducible and final moisture contents as well as green and basic densities of the lumber *etc.*

The boundary layer between the wood and the air stream has proved particularly complex, with vortices being spat out of tiny gaps between boards, with the leading edge of a board drying faster than the trailing edge, with turbulence being generated by boards having slightly different thicknesses (±0.1 mm) and the increased humidification and cooling across the stack.

Uniform airflow across the stack is influenced by the size and shape of the plenum spaces, the sticker gaps, air velocity, and effective baffling to avoid bypassing.

Finally the kiln type and design has to be included in the model: size, insulation, operating temperatures and throughput in order to correctly dimension operational units (fans, heat exchanges) to optimize mill drying – and this involves a trade-off between flexibility, costs and befits.

Some time ago Kamke and Vanek (1994) compared the performance of a number of within-the-timber drying models, representing mainly diffusion-like and multiple-transport mechanism approaches, for predicting average moisture contents and moisture-content profiles. Four data sets were used, with three sets representing idealised problems. The fourth data set was the experimental results of drying 40 mm boards of Norway spruce, *Picea abies*, from initial moisture contents of 29-66% at a dry-bulb temperature of 60°C, wet-bulb depressions of 8-25°C, and an air velocity of 6 m s^{-1}. The required inputs for the models, including physical properties of the timber and initial and boundary conditions, were supplied to the authors of the models. Requests for computer simulations were submitted to 31 authors in 16 countries, and simulation results were obtained for 12 models.

There was significant variability between the predictions of most models. Many models did not predict the data well, and there was no single 'best' model. Uncertain coefficients for the models, different degrees of simplifications and different methods for solving the heat and mass-transfer problems were given as possible reasons for the variations between the predictions of the models and for the discrepancies compared with the measurements. Kamke and Vanek concluded:

> Prediction results from drying models are only as good as the input data supplied. The more 'sophisticated' models do not perform any better than the simple models if the physical property data is inadequate...A simple model will work quite well if good physical property data is available for the species and the drying conditions are within the range for which the model was developed. For research purposes where detailed heat and mass transfer information is required (such as predicting stress and strain behaviour) the models that separate the transport mechanisms may prove more useful.

Understanding of drying processes has advanced considerable since then as has the speed and capacity of data analysis. Despite the comment above lumber drying is hugely better than 15-20 years ago: then radiata pine framing would have been dried in 5 days as against 12 hours today. However the point to reiterate is that mill managers are simply not equipped to make investment decisions on drying without wide technical consultations. Further the industrial support base is fragmented with consulting parties only familiar with one technology – dehumidifier heat pumps, heat-and-vent kilns, or microwave drying. Yet there will be numerous instances where the optimized solution will lie at the boundaries of these alternative approaches.

CHAPTER 9

WOOD PRESERVATION

KEVIN ARCHER[1] AND STAN LEBOW[2]

[1]*Chemical Specialties, Inc., Charlotte, North Carolina, USA;* [2]*USDA Forest Products Laboratory, Madison, Wisconsin, USA*

1. INTRODUCTION

Wood preservation can be interpreted to mean protection from fire, chemical degradation, mechanical wear, weathering, as well as biological attack. In this chapter, the term preservation is applied more restrictively to protection from biological hazards and the reader is directed to one of several references (Feist and Hon 1984; Hon and Shiraishi, 2000; USDA, 1999) for a more extensive discussion of non-biological aspects of wood protection.

Most people accept that because wood is of biological origin it must be a perishable material. In contrast, man made materials such as concrete and steel are generally considered to be more durable and permanent. The non-durability of wood is often cited as being one of its greatest disadvantages when compared to other building materials. The premature degradation of solid timber and wood-based composite products costs the consumer substantial amounts of money. Indeed in the United States alone the annual financial losses attributed to fungal decay of timber have been estimated to be well in excess of five billion dollars (Lee *et al.*, 2004). Estimates of the damage just caused by termites in the United States range from 750-3,400 million dollars, and these estimates can be doubled if the damage caused by other wood-destroying insects and fungi are included (Williams, 1990). Much of this loss is avoidable. The first line of defense is the use of construction techniques that minimize the exposure of wood to conditions that favour biodeterioration. Usually this means keeping it dry. Where such construction is not practical, wood preservation techniques can greatly extend the service life of wood.

The use of preservative chemicals and treated wood has been and still is sometimes criticized on the basis of health or environmental concerns. Ignorance on the part of the treating industry, poor work practices and lax environmental regulation all share part of the blame for that negative perception. Innovation in the first half of the 20th century led to the development of more effective wood protecting chemicals and processing techniques that turned a specialty industry into a commodity business (Preston, 2000). As can happen in all commodity businesses, research and development was not sustained when profit margins began to fall and the door was opened for competitive products such as plastics, concrete and steel.

Some countries, such as New Zealand, have a well established and regulated timber preservation industry and the benefits of construction with treated timber are well appreciated by the public at large. This is not so true of the United States or Europe where treated wood for residential decking and other consumer applications is losing market share to man made materials such as plastics (Clemons, 2002).

The old adage 'familiarity breeds contempt' might certainly be applied to wood preservation in recent years. The construction industry, building code enforcement and the public at large have come to expect extended lifetimes for wood-based building components while forgetting how that longevity is achieved. In New Zealand in 1998 an ill-advised decision to allow untreated house frames coincided with a trend toward monolithic cladding systems which aided by inadequate design/ detailing and coupled with poor construction practices resulted in a 'leaky building' crisis. The failure was in the weather tightness of the external envelope arising from the rigidity of the panels and the movement of the underlying timber, from poor detailing or even the absence of flashings around openings, and in poor performance of jointing materials. This allowed egress of water with no means of drying out any wet elements within the enclosed wall cavity. The problem was systemic, with the way the components were put together rather than poor performance of an identified product. While blame was diffuse the reputation of timber framing suffered. The inference is that timber treatment is not a solitary activity and needs to be seen in the context of building design and construction practices. Preservative treatment should not be used to compensate for loss of eaves, omission of flashings, abuse of sealants, moisture entry into concealed spaces with nowhere to drain *etc*. Sadly the problem has been evident elsewhere, in Canada, the U.S. and Europe.

Moving into the 21st century the wood preservation industry is of necessity facing a major overhaul. Health and safety concerns are being alleviated through a transition to less toxic chemicals. Environmental concerns with preservative treatments are counter-balanced by their ability to extend the durability of wood products, allowing conservation of forest resources. New preservative chemistries have been developed to target specific wood biodeteriogens. While other construction materials could substitute for wood in many applications, such materials are generally more expensive and require more energy to produce (Cassens *et al*., 1995). In that regard, life cycle analysis concepts are being used to promote the virtues of wood preservation (Hillier and Murphy, 2000). Best management practice concepts are also being adopted by the wood preservation industry (Anon., 1996). Wood is no longer being over treated, efforts are being taken to minimize dripping after treatment and surface residues are no longer an issue. Innovative processes and preservative chemistries are being developed to protect wood-based composites such as oriented strand board and medium density fibre board further expanding the universe of wood protection. In short the future for preservative treated wood is a positive one.

2. ORGANISMS THAT DEGRADE WOOD

Depending on where and how they are used, wood products may be attacked by a range of biodeteriorogens that include fungi, insects, marine borers and bacteria.

Fortunately, wood may be protected from biological degradation in a number of ways. The optimal choice depends on the local environment and organisms present. Accordingly, it is important to have some understanding of the biology of these organisms. To do justice to such an interesting topic deserves a dedicated discussion in its own right, but only a brief review of the key points can be provided here. More extensive reviews of the biology of wood-degrading organisms are available (Daniel, 2003; Eaton and Hale, 1993; Highley, 1999; Nicholas, 1973a; Rayner and Boddy, 1988; Zabel and Morrell, 1992).

2.1. Wood inhabiting fungi

Fungi require air, moisture and nutrients in order to invade and colonize wood.

Fungi are micro-organisms that depend on organic matter for nutrients. Above ground and out of soil contact, fungi typically infect or colonize wood either via reproductive spores carried on air currents or in liquid water. Where timber is in contact with the ground or immersed in water it may be infected by fungal spores but more commonly the fungal invasion is in the form of microscopic, threadlike structures each of which is known as a hypha, or collectively as mycelium. Fungi spread within wood only where there is a source of water and where environmental conditions favour growth.

Fungi need adequate moisture, not only to prevent desiccation but also to provide a medium for the outward diffusion of the extracellular enzymes and other degradative systems produced by the fungus and for the movement of mineral nutrients and degradation products in the opposite direction. The optimal moisture condition for decay by the most active rot fungi is above the wood's fiber saturation point where free water is available for the transport of enzymes and nutrients, but also there is plenty of oxygen in the lumens for fungal metabolism. Below 20-22% moisture content infected wood will generally not decay because the fungus cannot grow. However, some fungi may persist for years under dry conditions and if the moisture content later rises above that critical level the fungus may reactivate and attack the wood again.

Fungi are facultative aerobic organisms; they need oxygen to survive. Decay is retarded and may even be completely inhibited by an excess of moisture because it can limit the supply of oxygen needed for fungal respiration. Decay of wood is most severe at or just below the ground line in power poles, fence posts *etc.* for the simple reason that the amount of oxygen and moisture is optimal. As the depth of soil increases the oxygen supply becomes reduced while the moisture content generally increases. Where buried in the ground timber can survive for hundreds of years provided either moisture or oxygen is lacking.

A temperature of 25-30°C is optimal for the growth of most fungi. Below 12°C decay is usually very slow and few fungi are active above 40°C. In general fungi are not killed by low temperature but they are somewhat more sensitive to elevated temperatures. That sensitivity to heat can be utilized to advantage for sterilizing infected wood in a conventional kiln, provided high temperatures are applied for long enough to ensure heating of all infected parts of the wood. Such treatment is therefore appropriate for timber known to be susceptible to decay or where decay is only at the incipient stage, i.e. the wood is infected but not yet decayed. It is pointless to kiln sterilize even slightly decayed wood as the material will have lost much of its strength, particularly its toughness.

2.1.1. Mould and stain fungi

Mould fungi can be broadly classified as being saprophytic organisms that utilize simple sugars and other carbohydrates derived from cell lumens. Since they do not attack the wood cell wall structure they do not cause significant decreases in wood mechanical properties. Moulds are noticeable as fuzzy or powdery growths with colours ranging from white to black. They primarily affect the aesthetic appearance of the wood.

Unfortunately, to the layman all or any fungal growth associated with sawn or round wood is of considerable concern. Not only is there a misconception that the structure is in danger of premature collapse but in extreme cases hysteria ensues out of concern about exposure to mould spores (Uzonovic *et al.*, 2003). Moulds can cause allergic or asthmatic reactions in some sensitive people and a few moulds produce potentially toxic substances; however anything more serious than allergic or irritant symptoms is rare.

Sapstain fungi are similar to mould fungi, with the primary distinction being on the depth of the discolouration in the wood. Sapstain results where fungi with pigmented hyphae grow within the sapwood which can become badly discoloured as a result. As with the moulds these fungi derive their nourishment principally from cell contents, and therefore attack parenchyma-rich ray tissue. As a result the discoloured wood in softwoods is often wedge-shaped when seen in cross section, although in hardwoods a more diffuse staining distribution may result. This discolouration can be unsightly and is undesirable under natural finishes. Sapstain fungi are also significant because their hyphae can break down pit membranes and make fine holes as they pass through cell walls. This increases wood permeability and can create a number of problems when the wood is used. It makes the timber more susceptible to rewetting which in turn favours decay and if the wood is treated it can lead to over treatment and subsequent bleeding of the excess preservative in service. Sapstain fungi grow best in warm, moist conditions and so are particularly common in the wet tropics, especially as suitable insect vectors are very numerous.

If harvesting and milling is undertaken efficiently a prophylactic dip or spray immediately after sawing may provide the necessary short term protection against mould and sapstain during seasoning, storage or export.

Mould fungi can sometimes be a problem in preservative treated wood during prolonged storage especially if the wood is prevented from drying quickly after treatment. While this might seem counterintuitive because the wood is preservative treated in reality many mould fungi are not susceptible to the same preservative chemicals that are effective against decay fungi, To address this problem, preservative formulations may include mouldicide additives to provide short term protection against mould growth.

2.1.2. Decay fungi

Decay is the most destructive form of fungal attack on wood and occurs in three forms that are generally described as brown, white and soft rots. The terminology relates to the physical appearance of the wood after it has been extensively attacked. Brown and white rots result from the growth of highly specialized higher fungi (of the *Basidiomycotina*). The hyphae of Basisdiomycetes are able to ramify through the three-dimensional structure of wood creating large bore holes in the cell walls. These fungi utilize extracellular enzymes to degrade the wood cell walls to derive their nourishment. Under optimal conditions the process quickly weakens infected areas. Soft rot is caused by another group of higher fungi (*Ascomycotina* and many *Deuteromycotina*) which produce fine bore holes without the extensive enlargement seen with the Basidiomycetes.

Brown rots are more commonly associated with softwoods. The fungi attack primarily the cell wall carbohydrates (cellulose and hemicelluloses) and change the structure of lignin only slightly. As a consequence, the decayed wood develops a brown colour that will eventually exhibit extensive cubical cracking as it dries. Dry rot (a particular form of brown rot caused principally by *Serpula lacrymans*) is so-called because it is capable of colonizing, transporting water to and subsequently destroying sound, initially dry wood. The fungi can wet wood by transporting water over considerable distances along macroscopic root-like structures formed by aggregations of hyphae. In many respects the use of the word 'dry' is a misnomer because the wood was in fact moistened at some point and subsequently dried after decaying, creating the illusion that dry rot occurred (Bech-Andersen and Elborne, 1999).

White rot affects both softwoods and hardwoods. Cellulose, hemicelluloses and lignin are degraded. Progressive erosion by hyphae in the cell lumen as well as bore holes weaken the cell walls. Wood affected by white rot may darken in the early stages of decay but as the decay advances bleaching may occur. It does not split into cubical fragments but, because the breakdown of the lignin weakens interfibre bonding, the wood becomes spongy or stringy in texture.

Soft rot is a form of decay caused by a quite different group of fungi that is more closely related to moulds. They usually attack wood in wetter conditions than those favoured by brown and white rot fungi. Soft rot fungi characteristically attack the surface of the wood, gradually eroding inward at the rate of a few millimetres per year. The principal distinguishing microscopic feature of soft rot is the production of chains of geometrically shaped cavities oriented with their long axis following the microfibrils of the cell wall layer in which they are located, typically in the S_2 layer. Generally these cavities are cylinders with biconical ends or they are diamond-shaped. In many hardwoods an additional form of attack occurs with erosion of the cell wall lumen surface caused by hyphae. In softwoods erosion may be less severe because the S_3 layer is more developed and more highly lignified.

Soft rot is of economic significance mainly under conditions that retard or inhibit the activities of brown and white rot fungi, e.g. in preservative treated wood, in thermophilic situations and aquatic environments. This slow and initially superficial rot is sometimes more significant than might appear at first sight for several reasons:

- The outerwood contributes disproportionately to the bending strength of timber e.g. in a stressed pole or corner post.
- In some species heartwood is attacked as rapidly as sapwood.
- Many of the fungi involved are tolerant of high levels of commonly used wood preservatives.

2.2. Wood destroying insects

Wood destroying insects are of major significance in most regions of the world, although the number of species involved is relatively small (Creffield, 1996; Lenz, 2002). They damage wood by chewing it with their mandibles, although in many cases they derive no direct nourishment from it. For some, such as longhorn borers, only the insect larvae tunnel within the wood; in other cases, such as pinhole borers, all stages occur there. From a wood durability perspective, insect attack is less predictable than decay because some insects can bore into sound dry wood, and because insect populations are not uniformly distributed. However, most insects are similar to fungi in attacking only moist wood.

In the natural environment most wood decomposes as a result of both insect and microbial activity. Most insect pests of wood are either termites or beetles. Other insects such as wood wasps, moths, carpenter ants *etc*. are sometimes significant locally but by and large the termites (order *Isoptera*) and beetles (order *Coleoptera*) are the wood destroying insects of greatest importance.

2.2.1. Termites

All termites feed on cellulosic materials (Creffield, 1996). The most important are the subterranean termites that are found throughout the world within 40-45° of the equator. The number of species and total termite biomass increases nearer the equator, and they are generally regarded as a more serious threat in tropical and sub-tropical regions. Like all *Isoptera*, subterranean termites are social insects that live in colonies that are established in the soil. In their quest for food, subterranean termites may enter buildings and other above ground structures through enclosed galleries which they construct to protect themselves from desiccation and which connect to the soil and ultimately to the colony. Once inside a piece of wood, termites tunnel along the grain often leaving only a thin shell of sound wood to conceal their activities. Traditionally wooden structures have been protected by treating the soil under and around the building with an insecticide: subsequent soil treatments are necessary to maintain protection. Physical barriers such as metal caps between building and foundation supports have some limited value in that they force the colony to construct an enclosed gallery across both faces of the cap and thereby warn the home owner of their presence. Soil barriers such as graded gravel and steel mesh show some promise, as do toxic bait systems (Lenz and Runko, 1993). The bait systems use slow acting insecticides, allowing foraging termites to return to the nest to feed the colony (Su and Scheffrahn, 1991). Building with preservative treated timber provides another layer of protection if other protection mechanisms fail.

Drywood termites are the other group which sometimes attack wooden structures. They do not require access to soil as the queen actually invades the wood and her progeny become established there. Fortunately, such colonies are rarely as large as those of subterranean termites so that the damage is seldom as extensive. Where they occur they are nevertheless a serious pest and control measures are required. The best control is achieved by using preservative treated wood.

2.2.2. Wood boring beetles

The beetles infesting wood fall into three groups:

- Bark beetles and the related pin hole borers.
- Other beetles found in green wood.
- Borers found in dry wood (<25% moisture content).

A few species of bark beetle and pin hole borer are able to attack live trees, but most species prefer to invade green logs or stumps after felling. Wood damaged by bark beetles is largely discarded in slab wood. In lumber, the loss of strength associated with the 'holes' is minimal and the impact largely cosmetic. However these insects sometimes carry sapstain fungi that can result in very visual aesthetic degrade. Many other beetles such as flat-headed borers can infest green logs and timber but usually do not cause extensive damage in wooden structures. Under normal circumstances the wood is removed from the forest, processed and dried too quickly for these insects to have much effect.

The most destructive beetle pests are those which attack seasoned wood in service, e.g. *Anobium punctatum*, *Hylotrupes bajulus*, *Lyctus brunneus*. Only a few species are capable of doing this, but those that do can cause serious problems. They include long-horn beetles, the common house borer or furniture beetle and powder post beetles. Given susceptible lumber and suitable conditions for development, all of the above insects are difficult and expensive, or in some cases impossible, to control. The use of preservative treated wood obviates the necessity for control.

2.2.3. Marine borers

Marine borers damage wood structures in salt or brackish water throughout most of the world, although the severity of attack generally increases in warmer waters. The most damaging marine boring organisms are shipworms, pholads and gribbles.

Shipworms, i.e. *Bankia* and *Teredo* spp., are molluscs. Their minute free-swimming larvae move around until they lodge on the timber surface prior to gaining entry. Once within the timber they proceed to elongate and grow as they tunnel through the wood creating an extensive honeycombed structure: superficially the timber appears sound. Treatment with creosote or with waterborne preservatives containing copper and arsenic can protect wood from shipworm attack.

Pholads are clam-like molluscs i.e. *Martesia* or *Xylophaga* that create pear-shaped cavities near the surface of the wood. Pholads are limited to warmer waters, and can cause severe problems in tropical ports. Pholads also have some resistance to copper and arsenic based wood preservatives.

By contrast gribbles, i.e. *Limnoria* spp., are small crustaceans that attack the surface of the wood, and tunnelling seldom extends far from the surface. The combined action of water movement, gribble and microbial attack effectively wears away the wood. Damage is concentrated on exposed timber between low and high tide. Related crustaceans (*Sphaeroma* spp.) are somewhat larger than *Limnoria* and have a similar attack pattern: however, *Sphaeroma* spp. are less numerous and less damaging than *Limnoria* spp. In warm waters a species of Limnoria (*Limnoria tripunctata*) is able to attack creosote treated wood. A more detailed discussion of marine borer biology can be found in Cragg (2003) or Distel (2003).

3. NATURAL DURABILITY

Although sapwood is rarely durable, the heartwood of many tree species exhibits some degree of resistance to attack by decay fungi and insects (Table 9.1). This natural durability can be attributed to a combination of toxic extractives present in the wood and low inherent permeability. As a result of this natural durability such woods can be used outdoors and in some cases in ground contact or submersed in water. Wood from naturally durable species is sometimes viewed as being environmentally preferable to chemically treated wood, and many of these species have an attractive appearance. In addition, some species such as black locust, greenheart and ipe also have excellent strength properties (Green *et al.*, 1999). As might be expected such a combination of desirable attributes has led to increasing interest in use of durable species from the tropical countries for construction in North America and Europe. However, several factors limit the use of naturally durable species. In developed countries the volume of growing stock of naturally durable species is relatively low compared to the demand for durable wood

Table 9.1. Natural heartwood durability of certain timbers in ground contact, based on 50 × 50 mm stakes: indicative figures only. Hardwoods (HMSO, 1969); softwoods (Hughes, 1982).

Perishable (<5 yrs)	Non-durable (5-10 yrs)	Moderately durable (10-15 yrs)	Durable (15-25 yrs)	Very durable (>25 yrs)
Hardwoods				
Alder	Elm	Keruing	Kempas	Afrormosia
Beech	*Eucalyptus regnans*	Sapele Seraya, red	Meranti Oak	Iroko Teak
Birch	Obeche			
Poplar, black	Seraya, white	Sepetir		
Ramin				
Softwoods				
Corsican pine	Douglas fir	*Cupressus macrocarpa*		*Podocarpus totara*
Ponderosa pine	European larch			
	Radiata pine	Redwood		
	Western red cedar	Sitka spruce		

Figure 9.1. Field tests, also known as graveyard trials, as used to establish the durability of untreated heartwood of various timbers and also to determine the effectiveness of a variety of preservative systems (unpubl. courtesy New Zealand Forest Research Institute).

products. The felling and export of tropical species from developing countries to industrialized nations raises concerns about exploitation, deforestation and destruction of habitat. On the other hand, woodlots of fast growing species such as black locust *Robinia pseudoacacia* and some eucalypts whose heartwood is rated moderately to very durable may be a viable proposition for on-farm commodities such as posts and rails and even for simple farm buildings. Elsewhere durable heartwood is a scarce commodity.

While the durability of many species has been evaluated with post or stake tests (Figure 9.1), evidence for durability of other species is largely anecdotal. A comprehensive review by Scheffer and Morrell (1998) has helped to collate the literature related to durability for a wide range of wood species. Further, usage is also limited by variability in durability. For some species there are wide differences in heartwood durability between adjacent trees and even between boards cut from the same tree. Also boards can contain both sapwood and heartwood as it is often not economic or practical to cut timber so as to exclude all sapwood. Thus only broad estimates of durability can be developed (Table 9.1). As a result of these sources of variability the use of naturally durable species is often restricted to above-ground applications where the biodeterioration hazard is lower and the consequences of an early failure are less severe.

Very few wood species have sufficient natural durability to allow their use in marine environments without additional protection. Two species that have provided excellent performance as marine piles are turpentine (*Syncarpia glomulifera*) from Australia and greenheart (*Ocotea rodiaei*) from Guyana. The uncertain supply and the high cost of naturally borer-resistant timbers has led to the successful development of a marine construction industry throughout the world that relies on preservative treated wood.

One should state the obvious. Naturally durable timbers contain various extractives that are able to inhibit decay, so one should expect some of these timbers to have the potential, at the very least, of inducing allergic reactions in people that handle and process them (Woods and Calnan, 1976).

Finally, there are numerous instances of wood remaining in sound condition for hundreds and even thousands of years, but this is usually a result of construction practices and favorable environmental conditions. Norwegian Stave Churches have survived from the early Middle Ages because for much of the year the air is dry and very cold (being below freezing for up to eight months) while in summer it is hot, the relative humidity is low and the level of ultraviolet radiation is high. These structures also benefited from designs that minimized trapping of moisture and that kept timber out of ground contact.

4. PHILOSOPHY OF PROTECTION

During the nineteenth century the demand for durable construction particularly for rail road tracks and bridges so necessary for the industrial revolution, the scarcity of naturally durable timbers and an inability to control and regulate the immediate environment led to the development of a timber preservation industry (Freeman *et al.*, 2003). Spurred on by initial successes it was surmised that provided the timber, the preservative and the treatment process were all appropriate, it should be possible to ensure that treated timber retains its integrity for as long as is desired. In practice wood is exposed to a wide spectrum of hazards that vary with end-use, geographic location, and construction practice. It was soon recognized that no single preservative treatment was optimal for all situations. It is inappropriate to use a high concentration of a relatively toxic preservative for applications such as millwork where a lower concentration of less toxic preservative would provide an adequate service life. Similarly, a water-soluble preservative such as a borate that may provide excellent protection for wood used indoors will not provide long-term protection for wood used outdoors. Again, some preservatives are effective in preventing attack by fungi but not insects. Others may offer little protection against mould fungi. Failure to put the potential hazards into perspective tended to create uncertainty with the result that preservative treatments used were stronger than necessary. Today, increasing emphasis is placed on using preservatives that are targeted more specifically to particular applications (Goodell *et al.*, 2003). Such preservatives are safer to use and potentially less damaging to the environment.

A key but vexing question in any consideration of the philosophy of wood preservation must be how long a piece of treated wood should last. It is apparent that no one specific time frame exists.

The efficacy of a preservative treatment in wood is a function of:

- Type of organisms present and environmental conditions.
- The preservative's intrinsic toxicity to or efficacy against the target organism(s).
- The preservative's ability to resist leaching, UV degradation or other forms of environmental degradation.
- The degree of penetration and uniformity of distribution of preservative within the treated wood.
- The retention, or concentration, of the preservative within the treated wood.

In recognition that the deterioration hazard varies with end-use, many countries have developed 'hazard class' or 'use category' systems that specify those preservative formulations that are suitable in particular situations, the amount of preservative to be used (its 'retention'), and the depth to which the preservative must penetrate the wood (Morrell and Preston, 1995) (Table 9.2). As might be expected there is considerable overlap between these end-use categories.

Table 9.2. General guidelines for the specification of treated timber.

End use, relative hazard	Principal hazard	Choice of timber	Condition of timber	Choice of preservative	Quantity of preservative uptake	Treatment process
Marine	Marine borers	Hardwood or softwood	Incised or otherwise modified	Oil or water based	High or low chemical uptake	Pressure treatment
Ground contact	Fungi	Permeable or impermeable	Treat dry or green	Environmental hazard level: broad toxicity	Deep treatment or envelope	Sap Displacement
Exposed exterior	Fungi/ insects	Wide or narrow sapwood band		Fixed or leachable		Vapour phase
Interior of buildings	Wood-boring insects			Clean or staining		Diffusion

Wood preservatives are generally classified or grouped by the type of application or exposure environment in which they are expected to provide long term protection. Some preservatives have sufficient leach resistance and broad-spectrum efficacy to protect wood that is exposed directly to soil and water. These preservatives will also protect wood exposed above ground, and may be used in those applications with lower retentions (lower concentrations in the wood). Yet other preservatives have intermediate toxicity or leach resistance. This allows them to protect wood that is fully exposed to the weather, but not in contact with the ground. Other preservatives lack the permanence or toxicity to withstand continued exposure to precipitation, but are effective with occasional wetting. Finally, there are formulations that are so readily leachable that they can only withstand very occasional, superficial wetting.

It is not possible to evaluate a preservative's long term efficacy in all exposure environments. Preservatives have been tested most extensively in ground contact only, and there is no perfect formula for adjusting or predicting how well a wood preservative might perform in another situation. This is especially true for above-ground applications. To compensate for this uncertainty, there is a tendency to be conservative when selecting a preservative for a particular application.

5. PRESERVATIVE FORMULATIONS

Historically, wood preservatives have been thought of in terms of their solubility in either water or oil-type solvents (Ibach, 1999). Thus we have so called oil-borne and water-borne preservative systems. More recently that classification has become less relevant, because, with advances in formulation chemistry active ingredients can be formulated with either type of solvent, while others may be emulsions or suspensions. Water-based preservatives often include some type of co-solvent such as an amine or ammonia to keep one or more of the active ingredients in solution. Each solvent has advantages and disadvantages depending on the application.

Oil-type systems in medium to heavy oils are among the oldest and most effective preservatives. These systems usually leave the wood surface dark brown in colour although some lighter solvents can minimize colour changes. Oil-type systems are widely believed to reduce checking and splitting, although this can be difficult to document (Ibach, 1999). Oil-type preservatives are commonly used for applications such as utility poles, bridge timbers, railroad ties and piling. They are less likely to be used for applications that involve frequent human contact or for inside dwellings because they may be oily or have a strong odour.

Water-based preservatives are often used where cleanliness and paintability of the treated wood are required. Typically, wood treated with a water-based preservative has little or no odour when compared to oil-based preservatives. However, unless supplemented with a water repellent, the water-based systems do not confer any dimensional stability to the treated wood. In addition, water-based preservatives that utilize copper as a fungicide may not adequately protect hardwood species under conditions that favour soft rot attack. Some water-based preservatives can increase the rate of corrosion of mild steel fasteners.

The original water-based preservatives were simple salts, e.g. $ZnCl_2$ and NaF, but it was found that they had a tendency to leach out when exposed to liquid water and so were unsuitable for many exterior situations. Some recent preservatives are initially soluble in acidic or alkaline solutions but after pressure impregnation they are designed to chemically bind or 'fix' with the wood or form insoluble compounds. These are versatile preservatives. By varying the solution strength or the treatment process the amount of chemical deposited in the wood, i.e. the retention, can be adjusted according to the degree of hazard likely to be encountered in service. The lowest retentions are used to combat insect attack and the highest are used against marine borers.

Copper has been a primary ingredient in wood preservative formulations for over a century because of its excellent broad-spectrum fungicidal properties, low mammalian toxicity and relatively low price (Evans, 2003). A few fungi are tolerant of high levels of copper (Barnes, 1995; Choi *et al.*, 2002), and under some unusual circumstances copper treated wood exposed to copper tolerant fungi can decay faster than untreated wood placed in the same environment. The existence of 'tolerant' fungal species is not confined to copper. Fungal species tolerant to arsenic and creosote are well known. Preservative formulators will often include a co-biocide to provide further protection against such tolerant species.

Historically, chromated copper arsenate (CCA) has been the most widely used water-borne treatment. CCA is a mixture of chromic acid, cupric oxide, and arsenic pentoxide. CCA is strongly fixed to the wood and for the last 70 years has provided excellent protection in a variety of environments. The primary drawback to CCA is the perceived human health concerns associated with arsenic and hexavalent chromium: both are recognized as potential human carcinogens. As a result of these concerns CCA is no longer available for use in a number of countries, and its usage is severely restricted in others (Freeman *et al.*, 2003). Non-chrome, non-arsenic alternatives to CCA have been developed and several of these alternatives have gained wide commercial acceptance. For the most part the alternatives still rely on copper as the primary biocide, but the chromium and arsenic has been replaced with other components. In some places, particularly in Europe, even copper is coming under environmental scrutiny. In Europe there has been considerable interest in developing wood preservatives that do not contain copper or other heavy metals (Goodell *et al.*, 2003). Such preservatives must of necessity depend on combinations of relatively low toxicity organic fungicides and insecticides originally screened for agricultural uses. Developing new wood preservative systems presents technical difficulties because bacteria or other non-wood attacking organisms may degrade these organic compounds. This challenge is particularly acute where wood is in contact with the ground.

Each preservative has unique characteristics that might affect its suitability for a particular application. These include factors such as appearance, odour, toxicity, wood species compatibility and availability. The discussion that follows provides a basic background to a wide range of preservative systems. Some of these systems are still in use today, while others have been phased out and others are currently under development. Further discussion of preservative systems can be found elsewhere (Ibach, 1999; Nicholas, 1973b; Richardson, 1993; Schultz and Nicholas, 2003). It will be readily apparent from this section that the transition away from traditional heavy metal broad spectrum biocidal compounds to organic chemistries has added significant complication to the wood preserving industry as a whole.

5.1. Preservatives used in marine environments

Marine borers present a severe challenge. Some preservatives that are very effective against decay fungi and insects do not provide protection in seawater.Thus, despite severe reservations about the continued use of creosote and CCA these remain the only viable treatments currently available. Creosote is most commonly used, preventing attack by all marine borers except *Limnoria tripunctata*.

Waterborne preservatives containing copper and/or arsenic such as CCA have also proved to be efficacious either alone or in combination with creosote. Waterborne preservatives such as CCA or ACZA protect against attack by shipworms and *Limnoria* spp, but they do not protect against pholads.

Much higher preservative retentions are required to protect against marine borers than are needed to protect wood in terrestrial or fresh water applications. Further, with no single preservative effective against all marine borers, more expensive dual treatments involving an initial treatment with a waterborne preservative followed by a conventional creosote treatment may be required in some locations.

Physical barriers such as plastic sleeves or wraps have been used to protect piling, but they are vulnerable to breaches arising from mechanical damage. These are most effective when applied to pressure preservative treated piles.

5.2. Heavy duty preservatives designed for use in high deterioration hazard areas

Soil contact and fresh water immersion applications present a high deterioration hazard to wood and wood-based products. Preservatives used in these environments must exhibit sufficient toxicity and leach resistance to protect the wood for the intended lifetime of the building or structure, as building components in such environments typically have a structural or support function and can be difficult to replace *in situ*. The preservative's active ingredients should penetrate deep into the wood for maximum performance. Thus, almost without exception, only pressure treated materials find their way into high deterioration hazard end uses.

Broad-spectrum biocides with relatively high retention levels are the preservatives of choice. The sections that follow are not intended to be all inclusive rather they provide a brief summary of the major historically important systems and the products currently in use around the world.

5.2.1. Coal-tar creosote

Creosote is the oldest 'modern' wood preservative. It is formulated by fractionating coal tar distillate that in turn is a by-product of high temperature carbonization of coal. Creosote is a complex mixture of polycyclic aromatic hydrocarbons (PAHs), tar acids and tar bases that makes it such an effective broad-spectrum preservative. Difficult-to-treat woods can be pressure impregnated with hot creosote for lengthy periods. The wood has improved dimensional stability. However treated wood sometimes bleeds and has an oily surface, so it is not the first choice for applications where there is a high probability of human contact. Workers may dislike creosote treated wood as it soils their clothes and on contact photosensitises the skin.

Creosote treated wood has a lengthy record of satisfactory use in a wide range of industrial activities – as telegraph poles, on wharfs and with the railroads. Treating facilities using creosote are widely distributed in many parts of the world, making it one of the more readily available preservative treatments. The ease with which workers can climb creosote treated wooden utility poles is a significant advantage.

Concern over toxicity of creosote has limited or curtailed its use in many places.

5.2.2. Pentachlorophenol (PCP) in heavy oil

PCP was first introduced in the 1940s as a substitute for creosote. The active ingredient, a chlorinated phenol, is a crystalline solid that dissolves in a variety of organic solvents. The performance of PCP and the properties of the treated wood are influenced by the choice of solvent. A heavy oil solvent is preferred where treated wood is to be used in ground contact – wood treated with lighter solvents is not as durable. PCP treated wood has many characteristics and properties that mimic those of creosote, except that it is ineffective against marine borers.

Long-standing concern about broad and persistent toxicity (from contaminants) has curtailed the use of PCP in many countries and severely restricted use elsewhere.

5.2.3. Chromated copper arsenate (CCA)

CCA, developed in the 1930s, was once by far the most commonly used of all wood preservatives and until very recently represented over 90% of the sales of waterborne wood preservatives in the United States – as the preservative of choice for most ground and marine applications. There were numerous formulations with varying ratios of copper, chromium and arsenic. One of the most common formulations is 47.5% chromium trioxide, 18.5% copper oxide, and 34.0% arsenic pentoxide dissolved in water (CCA Type C). Typical retentions of active elements are several kilograms per cubic metre of wood, with yearly production of around 20 million cubic metres in the mid-1990s (Clausen and Smith, 1998).

CCA has decades of proven performance in field trials and in-service. With the correct species and treatment CCA provided an assured in-ground service life in excess of 50 years. Recently Bull (2001) has proposed that the fixation products of CCA are dominated by chromium (III) arsenate, chromium (III) hydroxide and wood-carboxylate-copper (II) complexes. CCA is potent precisely because it is bioavailable – and persistent. Significantly the separation of copper from chromium and arsenic is consistent with the observation that acetic acid and chelating organic acids – and silage or compost – under certain circumstances can promote leaching and early failure (Cooper and Ung, 1995; Kazi and Cooper, 1998).

With difficult-to-treat species it may be hard to obtain adequate penetration. There is an upper limitation to the temperature during impregnation and the rapid reaction of chromium within the wood structure can hinder penetration during longer pressure periods.

Today CCA is longer used in most jurisdictions; elsewhere its use is severely restricted. However, for accelerated testing, CCA is still the reference preservative used to evaluate the performance of other waterborne wood preservatives.

5.2.4. Copper naphthenate in heavy oil

The efficacy of copper naphthenate has been known since the early 1900s, and various formulations have been used commercially since the 1940s. It is an organometallic compound formed as a reaction product of copper salts and petroleum derived naphthenic acids. Like pentachlorophenol, copper naphthenate can be dissolved in a variety of solvents, but is more durable when dissolved in heavy oil. Although not as widely standardized as creosote and PCP treatments, copper naphthenate is used increasingly in the treatment of utility poles.

More generally, it is recommended for field treatment of cut ends and drilled holes (that expose untreated wood) made during construction using pressure treated wood. With the right solvent and treatment procedure, it is possible to paint copper naphthenate treated wood after it has been allowed to weather for a few weeks.

Copper naphthenate has been formulated as a water-based system, and sold in this form for consumer use. The waterborne formulation minimizes concerns about odour and surface oils. Water-based formulations are not used in pressure treatment.

5.2.5. Acid copper chromate (ACC)

ACC is an acidic water-based preservative currently in limited use in Europe but at the time of writing is under a no sell regulatory moratorium in the United States. It was originally developed in the 1920s but could not compete effectively with CCA on either price or performance so was largely relegated to small niche markets such as cooling tower components. ACC contains 31.8% copper oxide and 68.2% chromium trioxide. The treated wood has a light greenish-brown colour, and little noticeable odour. Tests on stakes and posts exposed to decay and termite attack indicate that wood well-impregnated with ACC gives acceptable service. However it is susceptible to attack by copper-tolerant fungi, and because of this its use has largely been limited to above-ground applications. It can be difficult to obtain adequate penetration of ACC in some of the more refractory wood species such as white oak or Douglas fir. Since it does not contain arsenic ACC is perceived to offer certain environmental and handling advantages over CCA. However, from a practical perspective the arsenic is replaced by a higher proportion of hexavalent chromium. In principle the hexavalent chromium should be converted to the more benign trivalent state during treatment and subsequent storage of the wood but recent unpublished studies seem to indicate that the time frame for full conversion is exceedingly long. Given the potential for the product to expose consumers to hexavalent chromium it seems unlikely that acid copper will receive widespread acceptance in the United States except perhaps for industrial applications where human contact is minimal.

5.2.6. Ammoniacal copper zinc arsenate (ACZA)

ACZA or Chemonite® is a water-based preservative containing copper oxide (50%), zinc oxide (25%) and arsenic pentoxide (25%). It is a refinement of an earlier formulation, ACA, that is no longer available. The ammonia in the treating solution, in combination with processing techniques such as steaming and extended pressure periods, allow ACZA to obtain better penetration of difficult-to-treat species than many other water-based preservatives. Treating facilities using ACZA are currently located in western United States, where many of the native timbers are difficult to treat with other waterborne preservatives. The primary biocidal activity can be ascribed to both the presence of copper and arsenic although zinc exhibits some fungicidal properties.

5.2.7. Copper-chromium-boron (CCB) and copper-chromium-phosphate CCP

CCB and CCP are similar in many respects to CCA except for the fact that the arsenic is replaced by boron in CCB and by phosphate in CCP. Most commonly used in Europe, both formulations were developed in part to address concerns about the toxicity of the arsenic in CCA. CCB and CCP are less efficacious than CCA and in the absence of arsenic the fixation processes are compromised. The systems still contain significant levels of chromium, which faces significant regulatory pressure from the Biocidal Products Directive in Europe. In the longer term the future for preservative formulations containing chromium is questionable.

5.2.8. Alkaline copper quat (ACQ)

ACQ is one of a number of recent water-based preservatives developed to address environmental concerns about the use of arsenic and chromium in treated wood. Several formulations of ACQ have been developed and marketed but all share a similar composition. The active fungicide and insecticide components in all ACQ formulations are copper and the quaternary ammonium compounds ('quats'). Copper provides the primary fungicide and insecticide activity in ACQ formulations, while the quaternary ammonium compounds ('quats') provide additional protection against copper tolerant fungi and insects. The type of quat may vary as can the copper-to-quat ratio in the formulation. The copper solublilizing agent may be ammonia in ACQ type B or ethanolamine in ACQ types C or D. Alkaline formulating agents, particularly ammonia, have the ability to swell wood cell walls and so improve the penetration of chemicals into wood. This characteristic has proved useful for improving the treatment of the refractory timbers such as Douglas fir prevalent on the West Coast of the United States.

At the time of writing ACQ based technology has secured the lion's share of the wood preservative market in Canada and the United States.

5.2.9. Copper azole

Copper azole is another recently developed water-based preservative formulation that relies primarily on copper solubilized in ethanolamine and an organic trizaole co-biocide. The first copper azole formulation developed contained 49% copper, 49% boric acid, and 2% tebuconazole. More recently, a formulation containing 96% copper and 4% tebuconazole has been used. As with ACQ formulations the copper in copper azole systems provides the primary fungicide and insecticide activity. The azole component provides protection against copper tolerant fungi.

Copper azole has gained widespread use in Europe, North America, Australia and New Zealand.

5.2.10. Copper HDO

Copper HDO is an amine copper water-based preservative that has been used in Europe and is currently is being registered for use as an above ground wood preservative in the United States. The active ingredients are copper oxide, boric acid, and copper-HDO (Bis-N-cyclohexyldiazeniumdioxy copper). The appearance and handling characteristics of wood treated with Copper HDO are similar to the other amine copper-based treatments. It is also referred to as copper xyligen.

5.3. Preservatives used above-ground and fully exposed to the weather

In volume terms the majority of preservative treated wood is used above ground – not in contact with soil or immersed in water. Typical examples might be residential decking or fencing. Logically the heavy duty preservatives mentioned in the last section can also be expected to perform well above ground and many are in current commercial use for that purpose, albeit with a reduced retention of active ingredient.

In many respects a ground contact or fresh water immersion environment represents a very consistent and high exposure hazard: the same cannot necessarily be said of all above ground applications. In certain situations, for example where moisture or organic debris can collect, the above ground environment may present a deterioration hazard similar to a ground contact exposure. This can be particularly problematic to the wood preservative formulator and treater. Here, the heavy duty preservatives discussed in the previous section may be more appropriate for such applications, especially in critical structural members.

Most of the preservatives listed here have not demonstrated the ability to provide long-term protection against a broad range of decay organisms in high decay hazard applications. However, they provide adequate protection for wood that is above ground and occasionally exposed to wetting. Examples of such use include members that may be subjected to wetting from wind-blown rain or from splashing during heavy rainfall, such as millwork. Many applications in this category involve dwellings or inhabited structures for which there has been a steady move in the past few decades to use preservatives with low mammalian toxicity.

There is an increasing move way from treating millwork *etc.* using light solvent carriers because of economic and environmental concerns. The attraction of such solvents was in the dimension stability of the product – no need to redry and remachine – so dressed final product could be so treated. More recently one of the larger millwork producers in the United States successfully developed and marketed millwork components that are pressure treated with a waterborne formulation containing a water repellent emulsion. Rough sawn timbers are treated, dried and then machined into profiles suitable for millwork components. The machined waste can be recycled to make preservative treated wood composite door cores.

In this category the distinction between oil and water-based preservatives has been blurred, as many of these components can be delivered either with solvents or micro-emulsions. The triazole fungicides, such as tebuconazole and propiconazole are being used more widely. Other azoles, including cyproconazole and azaconazole are used in more limited quantities.

5.3.1. Oxine copper (Copper-8-quinolinolate)

Oxine copper is an organometallic preservative comprising 10% copper-8-quinolinolate and 10% nickel-2-ethylhexoate that offers protection against sapstain and moulds. It has low mammalian toxicity. The treated wood has a greenish brown colour and little or no odour. Of particular interest, when used alone it is permitted by the U.S. Food and Drug Administration for treatment of wood used in direct contact with food.

It dissolves in a range of hydrocarbon solvents, but provides much longer protection when delivered in heavy oil. Oxine copper is sometimes used for treatment of the above-ground portions of wooden bridges and deck railings, protecting against both fungi and insects. Adequate penetration of difficult-to-treat species can be achieved, despite the treatment solution being somewhat heat sensitive, which limits the use of heat to increase preservative penetration. Oilborne oxine copper does not accelerate corrosion of metal fasteners relative to untreated wood.

However oxine copper is not widely used by pressure treatment facilities.

5.3.2. Tributyltin compounds

A number of related chemistries belonging to the tributyl tin family of compounds e.g. Bis (tri-n-butyltin) oxide (TBTO) and Tributyl tin naphthenate (TBTN) have been used as wood preservatives. They are colourless to slightly yellow liquids that are soluble in organic solvents, but insoluble in water. They have proved to be most efficacious as anti-fouling agents in marine paints (to be phased out completely by 2008), as preservatives in paint finishes, and in dip treatments for wood used in millwork. Used alone, tributyl tin is not effective in protecting wood used in ground contact, but it can protect wood products that are above-ground and partially exposed to the weather. While cost effective TBTO use has declined steadily due to concerns about the environmental and health effects of tin.

5.3.3. Triazoles

The development costs of biocide ingredients are exceedingly high. Most of the currently available organic fungicide and insecticide compounds being used as wood preservatives and those being considered for future wood preservative applications can trace their origins back to agricultural use. The triazole family of compounds is a good example of this process in action. Some of the more widely used triazoles include propiconazole and tebuconazole. They tend to be sparingly soluble in water but reasonably soluble in light organic solvents. As a consequence most formulations containing these compounds are emulsion systems. As might be expected from their agricultural usage their mammalian and environmental toxicity profiles are quite benign. From an efficacy perspective they do not have as broad a spectrum of fungicidal activity as might be desired and little if any insecticidal activity. For this reason most of the wood preservative formulations in use today contain mixtures of triazoles or other fungicides with or without the addition of insecticides. For example tebuconazole is used as co-biocide component in the ground-contact copper azole wood preservative discussed previously. Triazoles are also relatively poor performing compounds against mould and stain fungi.

Their efficacy against soft rot fungi is weak and as a result they are not usually used as the primary biocide in applications where softrot is a concern.

5.3.4. Quats: DDAC and ADBAC

Didecyldimethylammonium chloride (DDAC) and alkyldimethylbenzyl ammonium chloride (ADBAC) are quaternary ammonium compounds that are widely used as bactericides, antiseptics and fungicides. More recently the mainstream quaternary ammonium compounds used in wood preservative formulations have transitioned to chloride free products such as didecyl dimethyl ammonium carbonate ('carboquat'). The removal of the chloride ion reduces the corrosion characteristics of the quat. ADBAC, DDAC and DDAcarbonate can all be used as the 'quat' component of ACQ wood preservative formulations. DDAC is used as a component of anti-sapstain formulations. They are colourless, nearly odourless, and can be formulated for use with either water or oil-based carriers, although solvency is diminished in lighter aliphatic hydrocarbons such as mineral spirits.

Although quats can be used as stand alone wood preservatives in other parts of the world – especially in Japan – it is more common to see them used in combination with other fungicide or insecticide components for example with triazole fungicides or nicotinyl insecticides.

5.3.5. IPBC

3-Iodo-2-propynyl butyl carbamate (IPBC) is used in anti-sapstain formulations, or as a fungicide in water-repellent finishes for decks or siding. It is also used to treat millwork, and may be combined with azoles to enhance efficacy against mould fungi. IPBC may be used as either a solvent or water-based formulation. IPBC is colourless, and depending on the solvent and formulation, the treated wood may be painted. Protecting IPBC treated wood from direct sun light helps prolong its longevity as it appears that IPBC is somewhat susceptible to UV breakdown. Some formulations may have noticeable odour, but formulations with little or no odour are possible. IPBC is not an effective insecticide, and is not used as a stand-alone treatment for critical structural members.

Some pressure treating facilities use a mixture of IPBC and an insecticide such as permethrin or chlorpyrifos to treat structural members for above-ground end-uses that are largely protected from the weather. The advantage of this treatment is that it is colourless and allows the wood to maintain its natural appearance.

5.3.6. Zinc naphthenate

Zinc naphthenate is used as a component in over-the-counter wood preservative products. It can be formulated as either a solvent borne or waterborne preservative.

Unlike copper naphthenate, zinc napthenate imparts little colour, and thus is more compatible with finishes. When formulated in light solvent, the treated wood may be painted. However, wood treated with zinc naphthenate may have a noticeable odour and as a result it is not recommended for interior uses. Zinc is not as effective a fungicide as copper, and zinc naphthenate is not typically used as a stand-alone preservative for exposed structural members. However, zinc naphthenate does have some preservative efficacy, and may be sufficient to protect wood used above ground and partially protected from the weather. Zinc naphthenate pressure treatments have been shown to extend the life of treated stakes exposed in Mississippi, and brush treatments of a waterborne zinc naphthenate significantly improved the performance of pine fully exposed to the weather (above-ground) in Mississippi (Barnes *et al.*, 2004). However, zinc naphthenate was less effective in protecting hardwoods in that above-ground study. The addition of a water repellent component to the treating solution appears to increase the efficacy of zinc naphthenate treatments.

5.4. Preservatives used in applications above ground and protected from liquid water

In general the primary threat in this end use category is insect attack, but protection against mould fungi or even decay fungi from occasional wetting may be desirable. Since liquid water is not an issue preservatives that do not fix in the wood can be used. They can be expected to provide adequate protection as long as the wood is not subjected to liquid water that could leach the active ingredients.

5.4.1. Borates

Boron has some exceptional performance characteristics including low mammalian toxicity, activity against fungi and insects, and low cost. A further advantage of boron is its ability to diffuse into green timbers that normally would resist traditional pressure treatment. Also, wood treated with borates has no colour and no odour

Borates are the most commonly used and versatile of the unfixed waterborne preservatives. They include formulations prepared from sodium tetraborate, sodium pentaborate and boric acid but the most common form is disodium octaborate tetrahydrate (DOT). DOT has higher water solubility than other forms of borate. Glycol is also used to increase solubility in some formulations.

Frequently wood is treated in the green condition. With greater mobility through heating and with higher solution concentrations, adequate core loadings of borate can be achieved by diffusion within a reasonable time frame (weeks). To avoid the long green diffusion times, there is a trend to partial drying before pressure impregnation. In conjunction with partial drying, DOT is able to penetrate relatively refractory species such as spruce using heated solutions, extended pressure impregnation periods and a subsequent diffusion soak. Pressure treated framing is used in areas of high termite hazard. With high retentions borates provide fire-retardant treatments for wood.

While boron has many potential applications, it is not suitable for applications where it is exposed to the weather, because borates are readily leachable. Therefore care must be taken to ensure that where borate treated wood is stored on site it is protected from precipitation. Research continues to develop borate formulations that have increased resistance to leaching while maintaining biocidal efficacy. Various combinations of silica and boron have been developed that appear to somewhat retard boron depletion, but the degree of permanence and applicability of the treated wood to outdoor exposures have not been well defined.

Also it is used as a surface treatment for a wide range of existing wood products such as log cabins, and the interiors of wood structures. Borates are also applied as internal or as remedial treatments using rods or pastes.

Another form of borate, zinc borate (ZB), is used as a preservative for wood-composite products. ZB as defined in American Wood Preservers' Association Standard P18 is 38.2% ZnO, 48.2% B_2O_3, and 13.6% H_2O. Zinc borate is a white, odourless powder with low water solubility that is added directly to the furnish or wax during panel manufacture. Zinc borate concentrations in the panel usually range from 0.75 to 1.5%. Because of its low solubility it does have some leach resistance once incorporated into the panel, and can be used in conditions with slight exposure to the weather if the panel is coated.

5.4.2. Insecticides

For interior uses protected from the weather decay or mould protection may not needed and wood may be treated with an insecticide only. Historically, insecticides with unnecessarily high mammalian toxicities such as lindane, dieldrin, aldrin and chlorpyrifos were used. More recently these have been largely replaced with pyrethroids such as permethrin and cypermethrin, as well as chloronicotinyl and neonicotinoids, pyroles, and insect growth regulators. These insecticides may also be incorporated with a fungicide, such as IPBC or the triazoles, to provide a greater degree of protection.

5.5. Non-biocidal approaches (see Chapter 4)

It is possible to impart a degree of durability to wood without the use of toxic components. Such treatments use strategies that limit water movement into the wood and/or render the wood structure unusable to degrading organisms. The simplest example is that of water repellents. Pressure treatments with high concentrations of wax greatly extend the service life of wood, even in ground contact (Crawford *et al.*, 2002), but the loadings required are uneconomic.

Instead, interest is largely focused in two areas: wood modification and heat treatment. Both approaches are more expensive than conventional pressure treatments and historically their use has been limited. However increasing concerns about the environmental effects of biocides, and the increasing costs of the biocides themselves has made these alternative approaches more attractive. Apart from Europe, currently they are limited to niche markets.

5.5.1. *Wood modification*

The idea of wood modification is to make the wood both more moisture resistant and less attractive as a food source by replacing the cellulose and hemicellulose hydroxyl groups with other moieties (Evans, 2003). Various reactants have been considered, but the most common is acetylation. Acetylation, when applied with weight gains of 15-30%, results in more dimensionally stable wod. The ability of the wood to resist insect attack is less clear, and there is little or no protection against the growth of mould fungi. Due to the high weight gain required wood modification has not proved to be economically viable for broad scale usage, although in some niche markets such as flooring it has found some utility.

5.5.2. *Heat treatment*

The goal of heat treatment is to both volatilize wood components that are used as food by fungi and to alter the wood structure. Typically the wood is heated to 160-260°C. In one process the vessel is flooded with nitrogen, while another uses vegetable oil for rapid heat transfer. Decay resistance and dimensional stability are increased and the wood darkens to a brownish colour, making it suitable for some above-ground applications where appearance is important. Depending on the process, the wood suffers some loss in mechanical properties and so is not appropriate for critical structural applications. Heat treatments have gained popularity in Europe, where some of the most common wood species such as spruce are difficult to treat with preservatives. Research continues to optimize the trade-off between an increase in durability and losses in strength properties (Militz, 2002).

6. TREATMENT PROCESSES

A key to effective wood protection is to ensure that the active ingredient is present in a sufficient quantity and is well distributed within the treated wood. With some permeable softwoods this is a relatively simple exercise but with certain refractory softwoods and hardwoods getting the active components sufficiently deep into the wood to afford long term protection is a significant challenge. The treatment process used depends on the end use, the wood species, preservative characteristics, and the technology available. It is generally desirable that the wood is permeable so that the preservatives can penetrate readily.

6.1. *Preparing wood for treatment*

6.1.1. *Green preconditioning*

The ideal forest operation sees the lapse time between felling and milling reduced to a week at most. For somewhat longer periods limited protection can be provided by brushing or spraying the exposed end-grain of logs with a biocide such as copper-8 quinolinolate.

In some regions it is difficult to ensure a stable log supply due seasonal weather *etc.* Here short-to-medium term storage under sprinklers is a viable merchandising operation in the normal management of a forest. In more extreme instances, e.g. after a major storm, fresh windblown logs can be kept for some years submerged in ponds or under sprinklers (Figure 9.2a) that minimize oxygen and prevent growth of sapstain or decay (Liese, 1984). Anaerobic bacteria rapidly colonize these log piles and can selectively attack pit membranes, so improving permeability (Figure 9.2b). Impermeable Douglas fir sapwood can be treated with waterborne preservatives after sprinkling with a bacterial inoculum for a couple of weeks (Archer, 1985). Optimal conditions required incising when green to give the bacteria radial access to the full depth of the sapwood band at which point the bacteria migrated tangentially degrading pit membranes (Figure 9.2c). In many species however, the increased permeability is undesirable because it causes excessive preservative uptake.

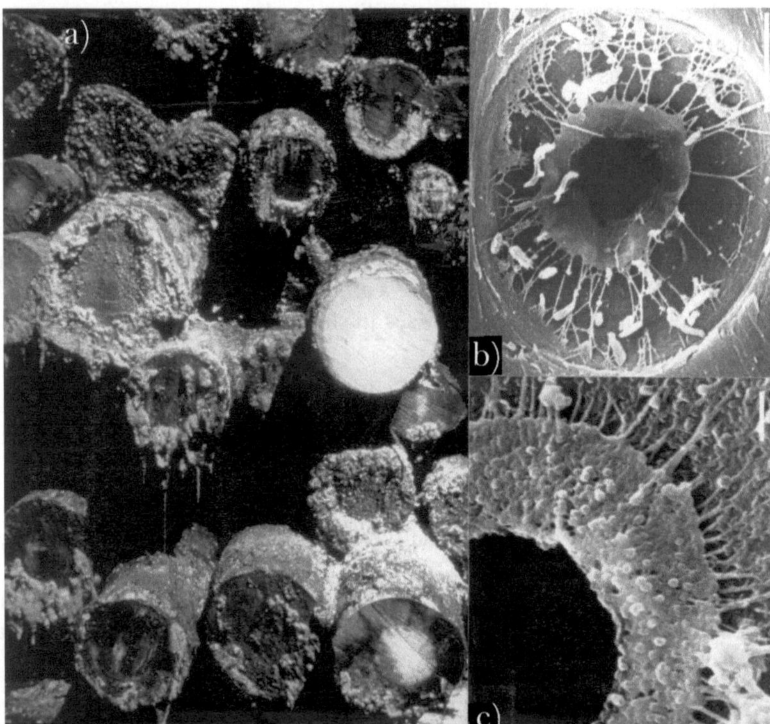

Figure 9.2. (a). Logpile in Balmoral Forest, New Zealand, five years after windblow (Liese 1984). A fresh exposed face, cut 100 mm from a log end, shows no stain or decay despite extensive surface colonization by microflora. (b) Douglas fir wood after several weeks under sprinklers show the complete disappearance of a central torus region: note the rod-shaped bacteria adhering to the relatively intact margo microfibrils (Archer, 1985). (c) Same material as in (b) emphasizing the doughnut appearance of the remaining torus, the intact margo and the granular material encrusting the pit chamber and torus (Archer, 1985).

6.1.2. Drying

As a rule, wood should be dried to its fiber saturation point or below before preservative treatment. Kiln-drying is common for dimension lumber, but the method of drying varies with climate and capital resources. For large timbers and railroad ties air-drying is used, despite the increased time required. However, in some climates it is difficult to air-dry material before it begins to suffer attack by stain fungi or even decay fungi, and alternative approaches must be considered.

Drying increases preservative penetration and also ensures, for larger timbers and roundstock, that much of the checking occurs before treatment. If timber is not adequately dried there is the risk that these checks might subsequently extend into untreated wood when the timber is in service. An alternative is to control subsequent checking through pre-treatments. One method for sawn or roundwood is to cut a saw kerf to the centre of the timber prior to drying and treating. As the wood shrinks, the kerf opens like a hinge to relieve the drying stresses.

Not all material needs to be dried, for example where the treatment relies on the diffusion of active ingredients through the green wood, or uses a pre-treatment schedule that removes water, e.g. steaming.

6.1.3. Incising

Some species, such as Douglas fir, larch and spruce, are very resistant to the penetration of preservatives and can only be pressure treated effectively if incised. In this case the wood is passed between toothed rollers (lumber) or through a cylindrical collar (poles) that contain adjustable steel knives (or needles) that incise the wood parallel to the grain (Ibach, 1999). The incisions are 6-20 mm long, about 3 mm wide and 12-24 mm deep (Figure 9.3), with the trend towards use of smaller, thinner teeth at closer intervals (Ruddick, 1987). Under pressure the preservative enters through the exposed end-grain in each incision and forms an envelope of treated wood that is slightly deeper than the incisions.

When treating poles, incisions can be concentrated on the region close to the groundline, so putting the preservative where it is most needed. Incising also promotes a more uniform checking pattern, with many small shallow checks spreading from the incisions rather than a few deep checks. The process causes a slight reduction in strength, especially if applied to dry wood or used on small dimension material (Winandy and Morrell, 1998).

6.1.4. Steaming or Boultonizing processes

With large members such as poles or piles thorough drying may be uneconomic and/or the members may get infected and begin to decay while drying. Steaming or Boultonizing is sometimes used to condition the green wood as part of the treatment process (Ibach, 1999).

In the steaming process, green wood is steamed in a pressurised treating cylinder for several hours, usually at a maximum temperature of 118°C (245°F) so that the

Figure 9.3. In the incising ring shown the needles can penetrate 20-60 mm and on subsequent treatment a preservative envelope of that depth forms in the impermeable timber. Deep incising is needed for demanding end uses, e.g. utility poles in the vicinity of the groundline.

outer annulus of wood is heated above 100°C. A sufficiently long steaming period also sterilizes the wood. Once steaming is completed a prolonged vacuum is applied. This generates a pressure gradient within the wood as moisture escapes as steam – largely through the ray tissue. The duration of the steaming and vacuum periods depend upon the size of members, the species, and moisture content. The boiling off of the superheated sap not only reduces the moisture content in the heated outer sapwood zone but also blows out unlignified ray tissue in some pines so that rays provide uninterrupted pathways for easy radial movement of the preservative solution: virtually every tracheid is connected to ray tissue. Steaming is much less effective where the ray tissue is lignified as in *Pinus elliottii*. The timber is left for a while to cool: this allows for moisture to redistribute: furthermore traditional CCA salts will precipitate out prematurely if the wood is too hot when pressure treated.

In the Boulton or boiling-under-vacuum method of partial seasoning, the wood is heated in the oil preservative under vacuum, usually at about 82°C to 104°C (180°F to 220°F). This temperature range, lower than that for the steaming process, is a considerable advantage in treating woods that are especially susceptible to collapse at high temperatures. The Boulton method removes much less moisture from heartwood than from sapwood. Both processes can result in strength losses to the treated wood if strict temperature and time limitations are not followed. Most countries have such limitations included in their treatment standards.

6.2. Vacuum/pressure impregnation treatments

Combined vacuum and pressure treatments are the most common methods of applying preservatives to wood. These techniques result in deep penetration of permeable timbers while at the same time controlling the amount of preservative retained. The process requires large heavy-gauge cylindrical pressure vessels up to 2 x 30 metres in size (Figure 9.4). There are a number of variations in the treatment schedules depending on the timber, preservative and intended end-use of the treated product (Hunt and Garrett, 1967; Nicholas, 1973b; Richardson, 1993).

6.2.1. Bethell (full cell) treatment

The distinctive feature of this treatment is the application of an initial vacuum (not less than -85 kPa) to draw much of the air out of the timber (Figure 9.5). The vacuum is held for at least 15 minutes. Then the preservative solution is drawn into the cylinder while maintaining the vacuum and when filled a hydraulic pressure is applied. Pressures up to 1575 kPa (225 psi) are common, with pressure periods varying from as little as 15 minutes to many hours. The pressure is maintained until the charge of timber is fully impregnated and/or the rate of absorption of preservative by the timber becomes negligible. At this point the preservative is drained from the cylinder and pumped back into the storage tanks. Since most of the air was removed during the initial vacuum high net preservative retentions are attainable with the full cell process. With a permeable timber the uptake of preservative can be in excess of 550 litres m^{-3} of timber, although a lower uptake is

Figure 9.4. CCA pressure treatment plant. The chemical storage tanks are out of sight.

common in refractory woods or in charges containing significant volumes of heartwood. Because the initial vacuum is unable to draw all of the air from the permeable wood a small amount will be trapped and compressed during treatment. When the timber is removed from the cylinder the compressed air can expand again gradually displacing some of the preservative from the timber charge. To avoid excessive kickback or bleeding a final vacuum (-85 kPa) is drawn for a few minutes before removing the timber from the cylinder. This process is most commonly used with water-based preservatives because the carrier (water) is inexpensive and because the solution concentration can be adjusted to achieve the desired retention of active ingredient within the wood.

6.2.2. Lowry (empty cell) treatment

With this method the aim is to achieve maximum penetration with a low net retention of preservative. No preliminary vacuum is applied before flooding the cylinder and an hydraulic pressure of up to 1575 kPa (225 psi) is maintained until the timber is fully treated (Figure 9.5). The pressure is released and a vacuum pulled to prevent excessive bleeding of preservative once the timber is removed from the cylinder. The compressed air re-expands displacing some of the preservative. With a permeable timber the net retention may only be 60% of the gross uptake, about 300 litres m^{-3} of timber. This process is useful for treating permeable timbers such as pine for exterior joinery and framing timber in low hazard situations. Subsequent drying is much shorter compared to the full cell treatment as considerably less moisture must be removed. The lower weight after treatment also reduces transport charges from the treating plant to the retailer or jobsite.

With some preservatives the temporary residence of the solution within the wood can result in partial fixation and in some cases selective absorption of one or more of the chemical components in the formulation such that the expelled solution ('kick-back') is no longer correctly balanced. Imbalance in the preservative solution needs to be monitored. Another undesirable characteristic of empty cell cycles is the fact the kick-back solution can contain dissolved wood sugars. These sugars can react with preservative components leading to the accumulation and deposition of insoluble precipitates, commonly referred to as sludge, in the bulk storage tanks.

6.2.3. Modified full cell or 'low weight' method

A method that combines aspects of both the full and empty cell treatment methods is now commonly used for treatment of permeable species with water-based preservatives. In a modified full cell treatment, the initial vacuum is of lower intensity and shorter duration than with a true full cell treatment. The pressure period is also shortened, while the final vacuum is of greater intensity and longer duration than the initial vacuum. This method yields adequate treatment with lower solution uptake than a full cell treatment. The wood gains less weight, reducing shipping costs. It is also less likely to drip preservative and has a much drier surface.

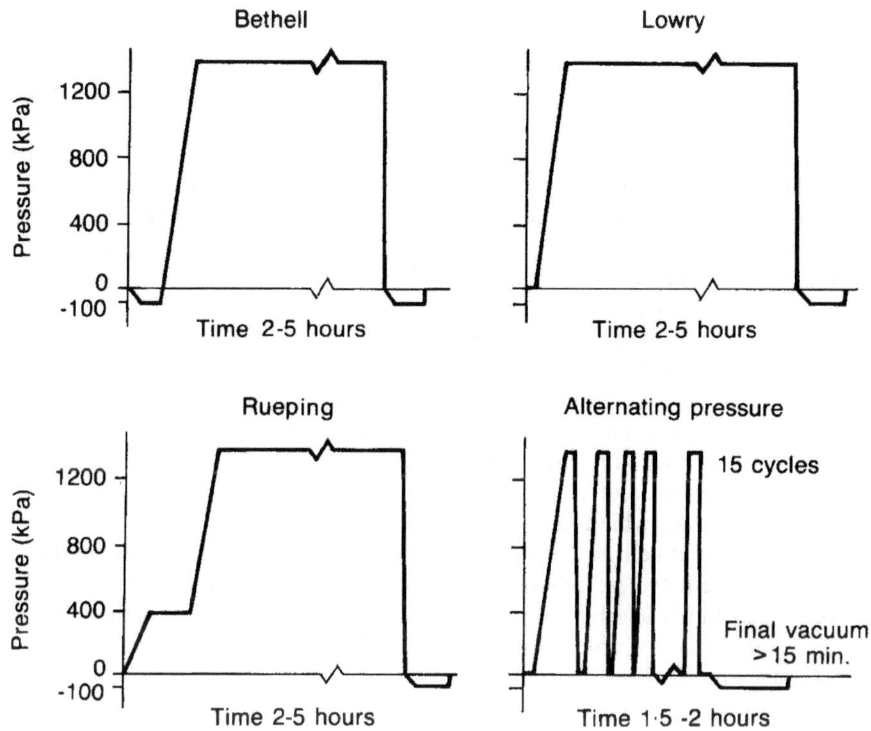

Figure 9.5. Time-pressure impregnation treatments.

6.2.4. Rueping process

This treatment is used principally with hot (>82°C) oil-type preservatives such as creosote and PCP where a low net retention is desired for some hazard categories. The treatment cycle begins with pressurizing the cylinder with air, no more than 700 kPa (100 psi) for creosote and PCP in oil (Figure 9.5). The preservative is pumped into the cylinder whilst maintaining pressure and when flooded the hydraulic pressure is increased to 1400 kPa (200 psi): species such as Douglas fir and larch are prone to collapse when the hot moist cells are subject to high pressures and the working pressure may have to be reduced somewhat (but still greater than 860 kPa). After the desired treatment time the pressure is released, the preservative is pumped back into the storage tank and a final vacuum pulled, again to minimize weeping. With a permeable timber the net retention is as low as 40-50% of the theoretical uptake, or about 220 litres m^{-3} of timber. Because creosote and PCP solutions are not usually diluted, adjustment of the initial air pressure and other treatment parameters is the primary method of obtaining a desired retention. This is an inexact method and it is difficult to produce material treated to a specified retention level.

6.2.5. Oscillating pressure method

Pressure treatments using waterborne preservatives require drying the wood before treatment and, in some case, again after treatment. Many pits aspirate when dried prior to treatment and the timber becomes less permeable. Redrying treated timber requires milder conditions as there is greater risk of steep moisture gradients and of checking. The oscillating pressure method utilizes repeated applications of high pressure and vacuum to force preservative into green wood so circumventing the problems arising from pit aspiration (Hudson and Henriksson, 1956). There is no pit aspiration prior to treatment and the timber need only be dried once – after treatment.

When a vacuum is applied the air in the tracheids expands and displaces some sap out through the rays to mix with the treatment solution in the cylinder. Some air is also expelled. When the hydraulic pressure is applied the air in the lumens is compressed and preservative solution is forced through the rays into the tracheids to mix with the sap. The cycle time is gradually extended to allow for the slower response deeper in the wood to the fluctuating pressure. This process was originally developed in Europe to treat unseasoned Norway spruce, *Picea abies*, and Scots pine, *Pinus sylvestris*, which are difficult to pressure treat with water-based preservatives. The treatment of large pole material took about 20 hours and involved numerous of treatment cycles; although where applied to more permeable species far fewer cycles and much shorter treatments times were needed. The method is not well suited to most current water-based formulations because the preservative reacts with sap displaced from the wood, causing sludging and surface deposits.

6.2.6. Vacuum treatments

With permeable wood species and members with small dimensions, a short vacuum or a double vacuum treatment may be sufficient to achieve the desired penetration (Table 9.3). In this process atmospheric pressure may be though of as the pressure period. Vacuum treatments have been commonly used for treatment of dry profiled or machined components (millwork) using preservatives carried in light organic solvents. The use of a volatile organic solvent avoids the dimensional swelling associated with aqueous treatments, and allows finishing within a short time after treatment. Although complete sapwood penetration is possible, this method emphasizes treatment of the end-grain where decay is mostly likely to occur in the exposed joinery. Organic preservatives containing azoles or IPBC are commonly used with this method. With permeable sapwood the uptake would be around 50 litres per m^3 of timber and with an impermeable hardwood using a more intensive schedule the solution uptake would be no more than 20 litres per m^3 of timber.

6.2.7. Other pressure treatment approaches

Certain timbers, such as some eucalypts which are highly impermeable to pressure impregnation, have been treated with varying degrees of success by resorting to very long treatment schedules or to the application of very high pressures, up to 7,000 kPa (1000 psi). Very high pressure treatments could only be considered for dense timbers, otherwise the wood cells will collapse before the preservative penetrates the lumens (Tamblyn, 1978). The capital cost of such a treatment plant would be high.

There has also been research to evaluate the use of wood preservative treatment chemicals with supercritical CO_2 combined with appropriate co-solvents (Acda *et al.*, 1997). Although promising, this method would also require substantial capital investments in treatment plant equipment (Evans, 2003).

Another alternative is to re-examine the type of solvents used as carriers. Pressure treatments with a liquefied hydrocarbon gas can achieve much better penetrations especially in refractory timbers, because the viscosity of the liquefied gas is so low, about one fifth that of water in the case of butane. After impregnation the liquefied gas can be drained from the cylinder and that part which is retained in the wood can be evaporated off under reduced pressure. This process has the advantage of almost complete solvent recovery so that it is economic to select an expensive solvent which has optimum technical properties. The treatment gives a clean finish, except with certain timbers where there can be excessive exudation of resin which is solubilized in the butane. This treatment was originally conceived for treating with PCP but it is no longer used after significant in service failures were reported. While it was not anticipated at the time we now know that the oil carrier in traditional PCP treatments enhances the overall performance. The explosive flammability of the liquified gas was also a hazard, requiring the treatment cylinder to be flushed with nitrogen to remove any air. In some respects the underlying approach remains attractive but significant technical hurdles remain unresolved.

Table 9.3. Vacuum treatments using light organic solvents as carriers of the preservative (BWPA, 1986). The two schedules shown represent the extremes of treatment. The choice of a particular schedule is a function of the species, dimension of the material and the end use.

Increasing resistance of timber to impregnation requires a severer, more prolonged treatment ↓	Initial vacuum		Pressure phase		Final vacuum	
	(kPa)	(min)	(kPa)	(min)	(kPa)	(min)
	-33	3	0	3	-67	20
	-83	10	100	60	-83	20

6.3. Non-pressure treatments

6.3.1. Brushing, dipping and soaking

The simplest treatment is an application by brush or spray. Although penetration across the grain is minimal, some penetration along the grain is possible. The additional life obtained by such treatments over that of untreated wood will be affected greatly by the conditions of service, e.g. just brushing untreated wood with a simple wax water-repellent is surprisingly effective for rustic joinery (Feist and Mraz, 1978; Feist, 1984).

Dip applications provide very limited protection to wood used in contact with the ground or under very moist conditions, and they provide very limited protection against attack by termites. However, they do have value for exterior woodwork and millwork that is painted, not in contact with the ground, and exposed to moisture only for brief periods.

Dipping wood for even a few seconds will increase end-grain penetration somewhat beyond that achieved with brushing. In some cases, preservative in light solvent may penetrate the end-grain of pine sapwood by as much as 25 to 75 mm. Good end-grain penetration is especially advantageous for joints that are the most vulnerable point for decay in millwork products. However, if the wood is subsequently cut untreated end-grain will be exposed that needs retreating.

Soaking differs from dipping only in the amount of time that the wood is immersed. Members may be soaked for several hours and ever for many days, yielding substantial end-grain penetration. This process is still used in many parts of the world for the treatment of dried fence posts and small poles. Pine posts treated by soaking for 24 to 48 h or longer in a solution containing 5% of PCP in No. 2 fuel oil have shown an average life of 16 to 20 years or longer. The sapwood in these posts was well penetrated, and preservative solution retention levels ranged from 32 to 96 kg/m^3. Preservative penetration and retention levels obtained by soaking lumber for several hours are considerably better than those obtained by brief dipping of similar species, but still well below that obtained by pressure treatment.

6.3.2. Diffusion

Traditionally rough-sawn lumber is treated green off the saw where the moisture content is well in excess of fiber saturation (>50%). The moisture content is critical: even if only the surface has dried out briefly it becomes hydrophobic and does not pick up the solution (Dickinson and Murphy, 1989).

The boards are box piled, loosely strapped and immersed in a highly concentrated solution of boron salts for a few of minutes (Figure 9.6a). Alternatively timber on the green chain can be passed through a boron mist-spray tunnel or chain dip and then block stacked. The salt retention is a function of the surface area to volume ratio of the timber. Consequently thicker members may require a second dip 2-4 days later to fortify the salt concentration in the surface film. Once treated the timber is tightly wrapped and left for a number of weeks (Figure 9.6b). During this period the boron salts diffuse into the wood. The holding time varies from 4 to 6 weeks for 25 mm boards and up to 12 weeks for 50 mm stock, the time depending on the green moisture content and basic density of the timber (Barnes *et al.*, 1989; Dickinson and Murphy, 1989).

After the holding period there is still a moderate concentration gradient across the material and a high overall loading of salt is needed in order to achieve a minimum core loading of 0.1% boric acid equivalent for softwoods and 0.2% boric acid equivalent for hardwoods in the centre of the timber. The eventual uptake of salts is controlled by such factors as:

- The concentration of the treating solution.
- The surface area to volume ratio of the timber.

• The temperature of the treating solution (the solubility of the boron salts increases with temperature, allowing more concentrated solutions to be used).
• The thickness of the solution film: for rough-sawn softwoods this is assumed to be about 0.2 mm, but with hardwoods and dressed softwoods the film is thinner.

Timber species can be grouped to take account of the fact that those having a high basic density and low green moisture content need to be immersed in stronger solutions in order to obtain the correct amount of preservative (wt/wt basis). Solution strengths vary from 15% to 45% of boric acid equivalent, but the more concentrated solutions can be achieved only by heating the solution above 50°C.

Figure 9.6. (a) Timber about to be immersed in a boron dip tank. Concrete drip storage area to the right. (b) Block stacked and covered timber is held for 4-8 weeks to allow salts to diffuse into the core.

The use of high molecular weight branched polymers as thickening agents results in a marked increase in the viscosity of the treatment solution (Vinden and Drysdale, 1990). In consequence a thicker film of boron salts clings to the timber and the vertical drainage of the salts through the block stacked timber is reduced. With thickened solutions there is much less within charge variability, less concentrated solutions are necessary and treatment times are reduced. Further it becomes possible to treat gauged timber so that there is no chemical loss or waste disposal problem as where rough-sawn timber is subsequently dressed.

The emphasis in Australia and New Zealand is on treating permeable pine, but diffusion treatments are effective with impermeable green hardwoods and softwoods such as hemlock and spruce. In tropical countries boron diffusion offers many advantages: no health hazard to operators, simple technology and the ability to treat local timbers locally. The major disadvantage is the stock holding period for diffusion and subsequent air-drying.

Today, just in time stock control favours a totally different approach, that of kiln-drying followed by pressure impregnation to obviate the long diffusion holding period.

6.3.3. Double diffusion

This process was suggested as an appropriate technology for developing countries. The double diffusion process consists of soaking green wood first in one chemical solution and then in a second solution (Johnson and Gonzalez, 1976). Because the chemicals are each water-soluble, they diffuse into the green wood, where they react to form leach-resistant compounds. In one scheme the wood is first soaked in a solution of copper sulphate ($CuSO_4$) for 1-3 days, and then soaked in a mixture of sodium dichromate ($Na_2Cr_2O_7$) or sodium chromate (Na_2CrO_4) and sodium arsenate (Na_2HAsO_4) for the same period (Markstrom *et al.*, 1999). In another scheme the wood is soaked first in sodium fluoride and then in copper sulphate. In theory, the first salt starts diffusing into the timber and as the other salts follow later they react with the first salt to precipitate out the non-leachable preservatives. However recent research with $CuSO_4$/NaF combination indictes that much of the fluoride remains leachanble after treatment (Morrell *et al.*, 2005). Another recent development involves partial air-drying and an initial hot soak (80-90°C) with the first salt, so that as the timber cools the partial vacuum encourages deeper initial penetration as the solution is drawn in by capillary tension. Consequently the salts used in the second dip have to diffuse further into the timber before the two chemicals react and precipitate out. With a hot soak or with thickening agents there will be less contamination of the second solution by the residues of the first solution still clinging to the wood surfaces. A negative to this method is the handling and dripping of preservative.

Despite the simplicity and elegance of the process it is hard to justify when used with such chemicals that have been restricted or withdrawn from general use in many developed countries.

6.3.4. Sap displacement (Boucherie process)

In the live tree there is a continuous conduction system within the outer sapwood. Thus water-soluble preservative solutions can be drawn up the tree after felling by immersing the butt in a solution of preservative – and relying on transpiration from the needles. Or, a freshly felled log can have its butt end elevated so that preservative can be introduced via a charge cap – a minimal hydrostatic head is all that is needed provided no air-water menisci intrude. More efficient systems use either vacuum caps to draw the preservative through the timber or pressure caps to force the preservative into the timber. No end-grain drying is permitted as air-water menisci require much greater forces to displace them through the capillary network in wood – dry ends of logs should be precut to re-expose green wood. These processes result in a preservative gradient within the roundwood, with the one end having a higher chemical loading unless the direction of flow is reversed. These processes are not commercial as there are problems of quality control, but they have uses in remote locations and where an on-farm treatment is desired. The displaced sap will contain some salts that are partially precipitated by reaction with the wood sugars. The expressed solution can be recycled or mixed with sawdust (to fix any residual chemical) and buried.

6.4. Treatment of wood composites

Some wood composite products such as plywood, glue-laminated beams, laminated veneer lumber, and parallel strand lumber can be treated using conventional pressure-treatment techniques. However, products made from smaller particles such as oriented strand board (OSB) or particle board may suffer significant losses in mechanical properties when pressure-treated. Even though they are used typically in dry environments, there is increased interest in protecting these panels from termite attack as well as from mould and decay fungi that may occur after unexpected moisture problems, for example in cases of building envelope failure (Gardner *et al.*, 2003). Treated versions of these products incorporate preservatives such as zinc borate or copper ammonium acetate into the furnish or wax (Laks, 2004). In other cases an organic mouldicide such as IPBC/azole mixture is simply sprayed on the surface to provide temporary protection against mould during construction.

Another approach proposed for protection of composites is a vapour phase treatment (Murphy *et al.*, 2002; Vinden *et al.*, 1990). Certain esters of boron have high vapour pressures making them readily volatile and suitable for vapour phase treatment. For example trimethyl borate boils at 65°C so the treatment requires both timber and pressure vessel to be heated to at least this temperature. Trimethyl borate will react with the adsorbed moisture in the wood to yield methyl alcohol (which is recovered) and boric acid that remains in the wood:

$$B(OCH_3)_3 + 3H_2O \rightarrow H_3BO_3 + 3\ CH_3OH$$

Hydrolysis is virtually instantaneous, so in order to get deep penetration the wood must be very dry (<5-6% moisture content) otherwise most of the trimethyl borate will react with the adsorbed moisture near the surface and the core will be deficient in boric acid. Such a low uniform moisture content is very hard to achieve, even in a kiln.

6.6. Wood properties affecting treatment

A basic knowledge of wood anatomy is helpful in understanding how wood structure affects the movement of preservatives through wood. The primary cell types in wood are tracheid/fibers (softwoods and hardwoods) or vessels (hardwoods) that can be thought of as collections of tubes oriented along the grain (Siau, 1984). Movement through these tubes (along the grain) is relatively rapid. Paths for movement across the grain are more limited, in which preservative must move through the relatively small pit openings between axial cells, or along the transversely oriented ray cells. Because ray cell are less numerous and shorter than the longitudinal cells, they do not provide for rapid movement across the grain of the wood. As a result, penetration of preservatives is usually many times greater along the grain than across the grain. But, because most wood products are very much longer than they are wide, adequate penetration is largely dependent on movement across the grain. Thus, it is the differences in paths of flow across the grain that causes differences in treatability. Much of this difference is attributable to the size, number and condition of the pit openings. Generally pines are easy to treat because the ray cells have very large openings between cells, whereas spruces and Douglas fir have very small openings. Between ray cells and longitudinal fibers, pines can have very large window pits (pinoid) whereas spruces have very small pitting between ray cells and longitudinal fibers (Panshin and deZeeuw, 1980).

Some species have notable differences in penetration between earlywood and latewood bands of the annual growth ring. Latewood cells with thicker walls mean the pit membranes are less likely to aspirate and permeability can remain high. This differential treatability sometimes results in a 'zebra' treatment with alternating bands of treated latewood and untreated earlywood.

The ratio of sapwood to heartwood volume in a tree species is also a key to its treatability. In most species sapwood is more permeable than heartwood; and in some species, such as many pines, the difference is very great. In the heartwood there is a higher proportion of extractives, which block the ray cells and encrust pit membranes. The pit membranes are also lignified and often aspirated. Thus the perceived treatability of a species may be largely a function of the proportion of sapwood in lumber cut from that tree. Many pine species, such as the southern pines, have a large sapwood band that results in a larger proportion of treatable sapwood in most lumber dimensions. Conversely, Douglas fir has only a thin sapwood band and most material cut from this species contains substantial heartwood. In other species, such as spruce, the sapwood and heartwood are both difficult to treat with the heartwood being only slightly more impermeable than the sapwood. Although heartwood is often more naturally durable than sapwood, a wide permeable sapwood band is preferred for many uses since the durability of treated sapwood can be considerably greater than that of untreated heartwood. The difficulty in treating heartwood has led to the practice of calculating the preservative retention on the basis of the volume of sapwood in the treatment charge. The sapwood content can vary widely and is often much less than the volume of untreatable heartwood. In some cases it has been recommended that the specified retention should consider not just the volume of treatable wood but also the amount of treatable wood (volume x basic density), with denser material requiring a higher preservative loadings.

Hardwoods have a more complex structure than softwoods, and penetration and distribution of preservative is often adversely affected. The main flow paths are provided by vessels. Connections between vessel elements are efficient but the vessels themselves have limited length. Some species have a very intensive branching and interconnecting system (*Fagus* spp.), in others vessels are very straight with few interconnections (*Eucalyptus* spp.). Further there is limited flow to adjacent fibres. The proportion of vessel tissue in hardwoods is also variable, ranging from 15-50%. Although tyloses can occur in sapwood they are much more abundant in heartwood and dramatically reduce its permeability. Tyloses are found in about half of all hardwoods. Other species secrete resin and gum exudates to seal the vessels.

Penetration will be poor if the vessels are blocked by tyloses, if there are too few vessels, or if the vessels are too small. Ring-porous hardwoods have much larger vessels in the earlywood than in the latewood. For example, *Eucalyptus delegatensis* has no vessels in the latewood in which to adsorb preservative. There is little evidence of lateral movement of creosote within eucalypt wood and the vessels are sharply defined by their preservative content. Also with CCA salts the distribution is non-uniform with copper salts tending to remain in or near the vessels. Such material can fail in ground contact despite having high preservative loadings as the poor preservative distribution means that fungi can attack the untreated fibres away from the immediate vicinity of the vessels. However the susceptibility of hardwoods to soft rot fungi is not simply a matter of poor distribution of preservative, rather hardwoods are better utilized by these fungi. Soft rots tolerate greater amounts of preservative where the substrate is highly nutritive and can support good growth.

It should be emphasized that world-wide the treatment industry is based on comparatively few moderately permeable timbers. Problems can arise when there is commercial interest in using a timber that is somewhat less than ideal, perhaps because it is the main plantation species of that country (for example in the use of eucalypts and spruce). Although treatment of refractory species is not ideal, by drying to a low moisture content and with a high preservative loading in the surface layer, adequate service life may be achievable for certain end uses.

6.7. Remedial treatments

There is substantial interest in using preservatives to extend the life of treated wood that is already in service (Barnes *et al.*, 1995; Morrell *et al.*, 1996). These remedial treatments are most economic for high-value products that are expensive to replace, such as utility poles, piles, and bridge timbers. However, they are also used to protect log cabins, fence posts and millwork.

In utility poles the treatments fall into two categories: those intended to protect the untreated heartwood; and those intended to fortify and replenish the preservative in the sapwood around the groundline area. Treatment of the internal areas in a pole are usually accomplished by drilling holes at a 45° angle downward into the pole. A liquid or solid preservative is then placed in the hole and the hole is plugged. Preservatives for internal treatment of poles commonly contain a fumigant ingredient such as methylisothiocyanate (MITC), although boron and fluoride rods are also used. Piles and bridge timbers may be treated internally in a similar manner. External treatments are applied to poles by digging the soil away from the base of the pole and applying a paste or bandage to the groundline area. Copper and boron are the most common ingredients in these groundline treatments.

Remedial treatments for log cabins and millwork are applied generally by drilling holes into the member and adding a diffusible borate preservative. Borates have been formulated as rods, pastes, thickened solutions and powders for this type of application.

7. HEALTH AND ENVIRONMENTAL ISSUES

Wood preservatives must strike a balance between beneficial toxicity towards wood-attacking organisms and potential harm to non-target organisms. Because a wide range of organisms can attack wood, the most versatile wood preservatives must have a broad-spectrum toxicity. It is almost inevitable that preservatives that protect against a wider range of wood attacking organisms also have the greatest potential for harming non-target organisms. This is the situation with the traditional broad-spectrum preservatives such as creosote, PCP and CCA.

The shift to preservatives based on copper, azoles, and quaternary ammonium compounds has lessened the risk associated with wood preservatives. However, all wood preservatives contain ingredients that pose some degree of risk to non-target organisms, and the public and regulatory perception of a proper balance between risk and benefit is steadily changing (Brooks, 2002; Lebow *et al*., 2002). Preservative ingredients that are considered acceptable today may be considered as less desirable in the future.

Perhaps the greatest health and environmental risks with wood preservatives occur at the treatment plant. Here, improvements in handling and containment technologies have greatly lessened these risks at modern treatment facilities. Now, more concern has shifted to end-use, where risks may be encountered by construction personnel, consumers and the environment. Where still allowed, the use of creosote, PCP and inorganic arsenical preservatives is usually limited to high degradation hazard applications where direct human contact is minimized. In high contact areas such as residential decks or buildings, these preservatives have been replaced with formulations containing ingredients with lower mammalian toxicity such as copper, azoles and borates. Concerns about environmental impacts, especially in aquatic environments, are also associated with treated wood applications.

Restrictions on use of creosote and arsenical preservatives have been proposed in some areas despite relatively little evidence of environmental impacts (Brown *et al.*, 2003). Because there is little evidence of traditional preservatives causing harm to the environment, it is difficult to establish that the alternative treatments are less harmful. However, it is apparent that some release of preservative occurs from all types of treated wood and that treatment and processing practices can be adapted to minimize these releases (Cooper, 2003; Lebow and Tippie, 2001).

7.1. Over-treatment and re-treatment

In most parts of the world preservative retentions in common use are specified by wood preservative standards which are in turn backed by scientific studies. Wood treated to a standard, combined with third party quality audit inspection schemes, can be expected to provide consistent performance appropriate for the intended application. It is common practice for standards to prescribe minimum retention and penetration levels as opposed to maxima. If the goal is to maximise the longevity of preservative treated wood this approach might at first seem counterintuitive. But where the broader picture is taken into consideration, increasing the retention based on the premise that 'more must be better' needlessly increases the amount of leachable chemical in the wood without necessarily providing a durability benefit. It is rarely good practice to ask for a retention higher than that specified in wood treatment standards. A similar concern arises with the practice of retreatment of charges that originally failed to meet penetration or retention requirements. Although retreatment of failed charges is acceptable in some situations, it can lead to increased bleeding or leaching of excess preservative. The modern approach to these issues relies on best management practice concepts that define pretreatment, treatment and post treatment handling of treated wood products.

7.2. Bleeding of oil preservatives

Oil-type preservatives sometimes bleed or ooze to the surface of the treated wood. This may be apparent immediately after treatment. More problematic bleeding may occur in service in a location where it is exposed to direct sunlight: dark wood can get very hot. Now the problem is harder to remedy. This issue is best addressed through strict control of treatment processes. Processes used to reduce bleeding include:

- Maintaining clean facilities and working solutions.
- Avoiding over-treatment.
- Using post-treatment conditioning techniques such as final vacuum, steaming, and expansion baths.

Typically the volume of preservative that oozes out of the wood into the environment is quite small, but it can appear much larger if it spreads on the surface of standing water. Wood with a visibly oily surface should not be used for projects in sensitive environments or in applications likely to involve human contact, i.e. decking and handrails.

7.3. Fixation of water-based preservatives

The active ingredients of various waterborne wood preservatives, i.e. copper, chromium, arsenic and/or zinc, are initially water-soluble in the treating solution but become resistant to leaching when absorbed in the wood. This leach-resistance is a result of the chemical 'fixation' reactions that render the toxic ingredients insoluble in water. The mechanism and requirements for these fixation reactions differ depending on the type of wood preservative (Bull, 1998). For each type of preservative, some reactions occur very rapidly during pressure treatment, while others may take days or even weeks to reach completion, depending on post-treatment storage and process conditions. If the treated wood is placed in service before these reactions are completed, the initial release of preservative into the environment may be many times greater than for wood that has been adequately conditioned. Concerns about inadequate fixation have led Canada and European countries to develop standards or guidelines for 'fixing' treated wood, and similar efforts are underway in the United States (Cooper, 2002; Pasek, 2003).

The essence of CCA-C fixation is the reduction of chromium from the hexavalent to the trivalent state, and the subsequent precipitation or adsorption of chromium, copper and arsenic complexes in the wood substrate. Some of these reactions, such as the adsorption of copper and chromium onto the wood components, occur within minutes or hour while others are completed during the ensuing days or weeks. The length of time needed for fixation is greatly dependent on temperature, and the reactions may proceed slowly when the treated wood is stored out of doors in cool weather (Cooper, 2000). Because fixation at ambient temperatures may be unacceptably lengthy, several techniques are used or have been proposed to elevate the wood temperature and accelerate fixation, including various forms of kiln-drying, hot water baths and steaming. These accelerated fixation methods are quite effective, although care must be taken not to dry the wood too quickly or to elevate the temperature to a level that may harm the mechanical properties of the wood.

In ammoniacal systems the metals are solubilized by ammonia, and become insoluble as the ammonia evaporates. Some of the metals appear to simply precipitate within the wood, while others react with the wood structure (Lebow and Morrell, 1995). Volatilization of ammonia appears to be a key factor in fixation with ammoniacal preservatives, and this can be accomplished by air-drying, kiln-drying, or a combination of both. Placing stickers between layers of wood greatly increases the rate of drying of the treated wood. Until recently the fixation processes of the amine wood preservatives were poorly understood but ongoing research in North American university laboratories is beginning to expand the knowledge base considerably. At low retentions the bulk of fixation appears to occur very rapidly, within a few hours after treatment. At higher retentions, however, fixation is slower and temperature dependent (Ung and Cooper, 2005).

7.4. Recycling and disposal

A significant challenge facing treated wood products is the lack of an effective strategy for handling treated wood that has been removed from service (Connell, 1999). Currently, much treated waste wood is either placed in landfills or stockpiled waiting disposal. Land filling certain types of treated wood is restricted in some countries and under close scrutiny in others because of concerns about groundwater contamination. The potential environmental impact from treated wood in landfills is debatable; but the lack of strategies for reuse or recycling treated wood is clearly a legitimate concern. Several obstacles have been difficult to overcome in managing treated wood waste. For treated wood used in residential construction, one of the greatest difficulties is the lack of an efficient process for collecting and sorting treated wood (Smith *et al.*, 2002; Solo-Gabriele and Townsend, 1999). This is less of a problem for products such as railroad ties and utility poles.

Once collected, a number of options have been proposed for reuse or recycling of treated wood. Reuse is a desirable option as long as the secondary use is appropriate for that product. Used railroad ties are often reused as fence posts or landscape timbers, and utility pole are reused for fence posts or bridge supports. The proportion of wood treated with heavy metals that is reused is smaller, again in part because of problems with collecting and sorting. Appearance is also an issue, because many of these products are used in residential applications.

Researchers have demonstrated that wood treated with heavy metals can be chipped or flaked and reused to form durable panel products or wood-cement composites. However, this type of reuse has not gained commercial acceptance because of concerns with processing the treated wood, with the introduction of pesticides into the panel fabrication process, and with the leaching or environmental impacts from the final product (Kartal and Clausen, 2001).

Another viable option for products treated with creosote and PCP (and presumable other organic treatments in the future) is burning to generate power (cogeneration). When added as a small percentage of the overall fuel load these types of treated wood can be burned without unduly increasing air emissions. As fuel costs and energy demands increase, disposal of treated wood in this manner becomes more attractive.

The direct extraction and reuse of the metals from treated wood has been proposed. These include acid extraction, fungal degradation, bacterial degradation, digestion, steam explosion, or some combination of these techniques. All of these approaches show some potential, but none are currently economic (Helsen and Van den Belk, 2005).

Cogeneration poses additional challenges for wood with heavy metals – particularly for wood treated with arsenic. As well as concerns with emissions, the concentration of metals in the ash requires further processing (Solo-Gabriele *et al.*, 2002). Various processes have been proposed to extract and reuse the metals from the ash, but when combined with challenges in collection and sorting, the economics of these processes become daunting (Bull, 1998).

Nurmi and Lindros (1994) had the ingenious scheme of feeding treated wood chips into the smelting furnace at a copper smelter. This causes no difficulties since copper ores contain arsenic and other heavy metals, and both copper and arsenic are recovered.

In most situations disposal in designated landfills is deemed sufficient – as well as being the least expensive option – but others may require immobilization in concrete.

CHAPTER 10

GRADING TIMBER AND GLUED STRUCTURAL MEMBERS

DAVID E. KRETSCHMANN AND ROLAND HERNANDEZ

USDA Forest Service, Forest Products Laboratory, One Gifford Pinchot Drive, Madison, Wisconsin

1. INTRODUCTION

Because of its biological nature, which is influenced by many factors as discussed in Chapter 6, the quality of timber is enormously variable. Therefore, some sort of arrangement or classification must be undertaken prior to its use, in order to get the most out of this valuable resource. It is obvious that not all lumber can be used for the same purposes because not all of it has the same properties. It is the intension of this chapter to discuss the various classification or 'grading' techniques that are applied to timber and glued structural members.

A grading rule is a set of definitions of timber characteristics together with methods for measuring them. Grading rules determine, in an orderly manner, the way in which a given piece of timber from a certain species or group of species will be designated and located into any of a number of groups or categories. Grading of timber has evolved into two major categories: material graded for structural use and material graded for appearance. Strength and stiffness of timber are primary considerations for structural use, whereas appearance grades are dictated by the size of the largest clear cutting. Different grading rules apply to these two situations. Both categories will be discussed, but the majority of this chapter will be concerned with the grading rules appropriate for timber as a material for structural uses. Also, in this chapter a distinction is made between the properties of 'wood', that is, clear, defect-free material, and 'timber' (called lumber in the U.S.) with all its natural defects of knots, splits, cross-grain, and distortion.

The grading of timber should be viewed as part of a marketing strategy, designed to ensure that timber buyers obtain the quality of timber appropriate for their needs and timber sellers receive an optimal price for their product. Unfortunately, grading suffers from conflicting objectives and can be best described as an attempt to bring some order out of what would otherwise be a chaotic situation.

Timber, being a natural material, is very variable in strength and appearance. This is compounded by the enormous number of commercial species and by the multiplicity of grading rules that evolved in isolation to take account of the vagaries of each species. There is strong historical justification for such practice. Heart rot and brittleheart may be particular problems with certain over-mature tropical hardwoods, while corewood is encountered in softwood plantations. In theory, rationalization of grading rules ought to be simple. However, rationalization will never be easy as timber grading can be a powerful tool in non-tariff protection of local interests. Naturally local grading rules are written with local timber in mind. It would be unrealistic to disadvantage home-grown material. In recent years, more objective and rational procedures for the grading and property assignment of timber have been developed, driven by increased international trade. Unfortunately, there has been only limited success to date in adopting these procedures.

2. THEORETICAL STRENGTH OF WOOD

The theoretical limits of wood strength are impressive. The strength of a cellulose molecule dwarfs the values associated with high strength steel (Table 10.1). Individual fibres are incredibly strong in tension, but once assembled into solid wood, much of this potential is lost because of weakness across the grain.

The real advantage of timber structures over steel and concrete equivalents usually lies in their strength to weight ratio. Wood has a very high strength to weight ratio and therefore requires a less massive foundation for an equivalent load. Timber is also aesthetically pleasing and non-corrosive.

Table 10.1. Indicative values for mechanical properties of an air-dried softwood and other materials (Gordon, 1978; Mark, 1967; Marra, 1975). Densities for timber, concrete and steel are taken as 500 kg m^{-3}, 2300 kg m^{-3}, and 7800 kg m^{-3} respectively.

Strength	MPa
A cellulose molecule in tension	7000
Individual delignified fibres in tension	700
Clearwood in tension along the grain	140
Clearwood in compression along the grain	50
Clearwood in tension across the grain	3
Construction lumber in tension along the grain	30
Timber, allowable working stress in tension along the grain	10
High tensile engineering steel	1600
Concrete in compression	40
Concrete in tension	4
Stiffness	**GPa**
C-C covalent bond	1200
Clearwood	14
Timber	10
Steel	210
Concrete	25

The variability in the quality of timber is large, even within a selected grade of a particular species assessed by visual techniques. Thus one distinctive feature of timber grading is the inability to assess reliably the strength of a piece of timber: to date, the best that can be done is to estimate how weak the piece might be. Historically, the quality of wood required for construction depended on tradition and local experience for the most part. Trade guilds and craftsmen applied judgements that resulted in magnificent structures, such as cathedrals in the Middle Ages, shipbuilding from the fifteenth to nineteenth centuries, and railroads and bridges during the nineteenth century. More recently, machine-grading techniques have successfully reduced the variation in the properties of lumber and related products.

3. TIMBER GRADING FOR NON-STRUCTURAL PURPOSES

Timber sawn from logs varies widely in quality. Some pieces are clear, others have a few knots, and still others are very knotty. Some contain checks or splits, and others have bark on the edges or can have areas of decay. Because of this variety, it is necessary to separate timber into classes or grades based on the number, condition, and size of defects. In 1764 in Stockholm, Sweden, Swan Alversdon published the first set of grading rules of which there is an authentic record (Ivory *et al.*, 1923). These rules recognized four grades of lumber: (i) bests or clear lumber, (ii) good or select lumber, (iii) common or lumber containing numerous sound knots, and (iv) culls or usable lumber containing coarse defects.

These early Swedish grades were based on the appearance of each piece, and lines of demarcation between grades were drawn on the basis of the character and position of the admitted defects. The rules were applied to lumber regardless of potential use. Use of these rules spread during the nineteenth century. For example, these grading rules accompanied Swedish loggers to the United States and followed them as they progressed west.

While all lumber was originally graded on an appearance basis only, two major categories of grading developed with time: appearance grading and structural grading. Material technology developments and different requirements of wood consumers have further subdivided grading into three distinct systems of use: the two appearance use categories of Factory or Shop timber, and of Yard timber; and a Structural use category of structural timber. Factory and Shop refers to timber that will undergo a number of further manufacturing steps and reach the consumer in a significantly different form. Yard and Structural timber relate principally to timber expected to function as graded and sized after primary processing (sawing and planing). Appearance qualities are important for Yard lumber, whereas Structural lumber is material that is graded for structural use with assigned design properties.

The grading rules described and discussed relate to those applicable in various parts of the United States. They are illustrative of the principles that apply. Other countries adopt different rules and the terminology may be different, but the underlying principles are generally the same.

3.1. Hardwood grades

A principal use for hardwood lumber is for floors, siding/weatherboards, scaffold planks, stair treads, truck decks, rail-ties/retaining walls *etc.* For these end-uses the lumber sizes and grades relate directly to the final product. The major constraint is the restricted availability of high grade large dimension hardwood lumber, and this necessitates finding and retaining markets for the predominant amounts of lower grade material.

The alternative strategy recognizes that small blemishes severely downgrade whole boards. Hardwood factory and shop grades provide very profitable markets for such material by cutting out such defects and using the smaller pieces in the manufacture of furniture, cabinetwork, and pallets, or directly for flooring, panelling, moulding and millwork. The rules adopted in the U.S. by the National Hardwood Lumber Association are discussed here as an example of a standard in grading hardwood timber intended for cutting into smaller pieces to make furniture or other fabricated products (NHLA, 1991). In these rules, the grade of the timber is determined by the size of the piece and the proportion of the piece that can be cut into a certain number of smaller pieces, commonly called cuttings, which are generally clear on one side, have a sound reverse face, and are not smaller than a specified size. Figure 10.1 shows typical examples of such cuttings and their grades.

The best grade in the Factory timber category is termed First and Seconds (FAS). The second grade is FAS one face (F1F). The third grade is Selects, which is followed by No. 1 Common, No. 2A Common, No. 2B Common (sound wormy), No. 3A Common, and No. 3B Common. These admit progressively larger defects/blemishes. This summary is only given as an example of the thought processes used to separate out Factory and Shop grade timber.

3.2. Softwood factory and shop grades

As with hardwoods, here softwood lumber is sorted at the sawmill into factory or shop grades on the basis that these boards can be recut to yield a certain proportion of smaller pieces of specific quality and size. This lumber provides the basic raw material for many secondary manufacturing operations.

The variety of species available has led to numerous grade names and grade definitions that seek to reflect as accurately as possible the grade and yield that can be obtained in the subsequent cutting operations. During secondary manufacturing processes, the quality, size, and often the entire appearance of the pieces are changed (as defects are chopped out) and these transformed cuttings become integral parts of the final end-product. The names of these factory grades can reflect end-use, e.g. industrial clears, box lumber, moulding stock, and ladder stock. Here, some of the more common classifications are outlined briefly. Such grading procedures are largely the responsibility of manufacturers' associations and, because of the wide variety of wood species, industrial practices, and customer needs, different lumber grading practices coexist. For details, reference must be made to industry sources for the certified grading rules for that country, region, and species.

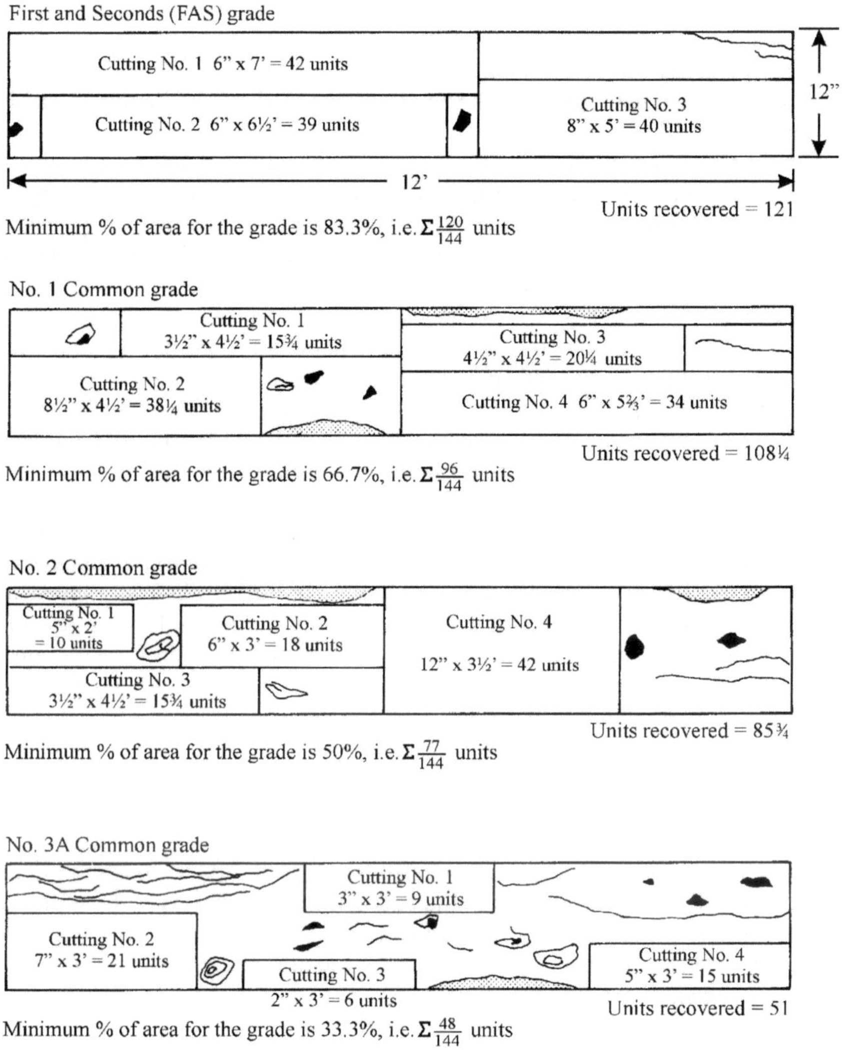

Figure 10.1. Grading of hardwoods: cutting strategies for boards that vary greatly in quality (NHLA, 1991). In this example the 12 foot x 12 inch board has a surface area of 144 units.

Traditionally, softwood timber used for cuttings has been called Factory or Shop. Some grading rules refer to these softwood grades as Factory, while others refer to them as Shop. All impose a somewhat similar nomenclature in the grade structure. Shop timber is graded on the basis of characteristics that affect its use for general cut-up purposes or on the basis of the size of cuttings required for sash and doors. Factory Select and Select Shop are typical high grades, followed by No. 1 Shop, No. 2 Shop, and No. 3 Shop.

Grades and grade rules take account of the width, length, and thickness of the original piece and of the amount of high quality material that can be obtained by cutting. Factory Select and Select Shop lumber is required to contain 70% of cuttings of specified size, clear on both sides. No. 1 Shop is required to have 50% cuttings, and No. 2 Shop only 33-1/3%.

Industrial Clears are used for trim, cabinet stock, garage door stock, and other product components where excellent appearance, mechanical and physical properties, and finishing characteristics are important. The principal grades are B&BTR (B and better), C, and D Industrial. Grading is primarily on the best face, although the influence of edge characteristics is important and varies depending upon piece width and thickness. In redwood, the Industrial Clear All Heart grade includes an 'all heart' requirement for decay resistance: it is used in the manufacture of cooling towers, tanks, pipe, and similar products.

Moulding, ladder, pole, tank, and pencil stock requirements relate to specific consumer products. Custom and the characteristics of the wood supply have led to different grade descriptions and terminology. For example, with U.S. West Coast species, the ladder industry can choose between 'ladder and pole stock' grade plus two ladder rail grades and one ladder rail stock grade. With southern pine, ladder stock is available as Select and Industrial. Moulding, tank, pole, stave and stadium seat stock, as well as box timber, and pencil stock are other classes oriented to a specific final product. Some product classes have only one grade level; a few offer two or three levels. Special features of these grades may include a restriction on sapwood related to desired decay resistance; specific requirements for slope of grain and growth ring orientation for high-stress use such as ladders; and particular cutting requirements as in pencil stock.

3.3. Softwood yard timber

The grading requirements of yard timber relate specifically to the construction uses intended, and little or no further grading occurs once the piece leaves the mill. Yard timber falls into two categories, Select and Common, and encompass those products in which appearance is of primary importance; structural integrity, while sometimes important, is a secondary feature.

3.3.1. Select timber

Select timber is intended for natural, stain and paint finishes. This category of timber includes material that has been machined to a pattern, and further processing of these items is limited to on-site fitting, i.e. cutting to length and mitering. The Select category includes trim, siding, flooring, ceiling, panelling, casing, stepping, and finish boards.

In the United States, most Select grades are described by names (Superior, Prime) or by letters and combinations of letters (B&BTR, C&BTR, D) depending on the species and grading rules under which the timber is graded. The specifications

FG (flat grain), VG (vertical grain), and MG (mixed grain) are offered as a purchase option for some Select timber products. Select timber grades emphasize the quality of one face and in consequence the reverse side may be of lower quality.

Colour can play a role in grade descriptions. For cedar and redwood, there is a pronounced difference in colour between heartwood and sapwood. Heartwood also has high natural resistance to decay, so some grades are denoted as 'heart'.

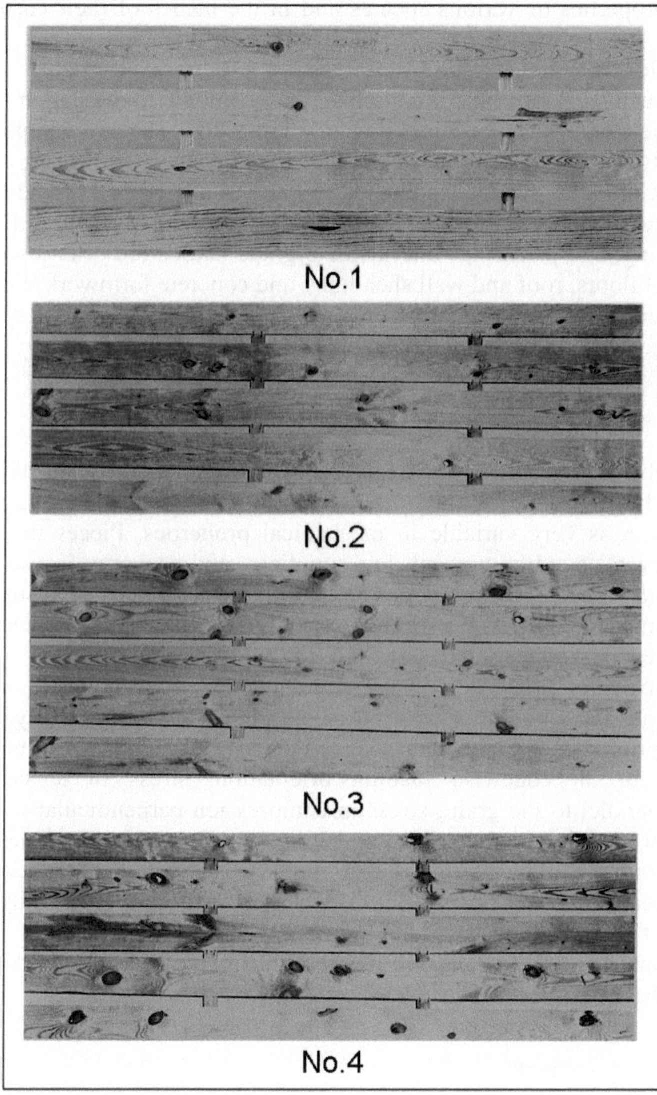

Figure 10.2. Typical examples of softwood boards in lower Common grades.

3.3.2. Common timber

Common timber is generally not stress-graded. The grades reflect suitability for general construction and utility purposes. Common timber is segregated into three to five grades depending on the species and grading rules involved. These may be described by number (No. 1, No. 2; or No. 1 Common, No. 2 Common) or by descriptive terms ('Select Merchantable', Construction, Standard). Differences in the inherent properties of various species and in the names of their corresponding grades means that the grades and grade names for different species are not always interchangeable.

Features such as knots and knotholes are permitted to be larger and more frequent as the grade level becomes lower. Figure 10.2 shows examples of four grades of softwood boards produced in the United States. The top grade boards (No. 1, No. 1 Common, 'Select Merchantable') are usually graded for serviceability, but appearance is also considered. These grades are used for things like siding, cornice, shelving, and panelling. Intermediate grade boards are often used for such purposes as subfloors, roof and wall sheathing, and concrete formwork. Lower grade boards are selected for adequate strength, not appearance. They are used for roof and wall sheathing, subfloor, and rough concrete formwork.

3.4. Softwood structural lumber

Structural timber is material graded for use as a structural member. What makes structural grading so challenging is that lumber sawn from a log, regardless of species and size, is very variable in mechanical properties. Pieces may differ in strength by several hundred percent. For simplicity and economy in use, pieces of lumber of similar mechanical properties are placed in categories called stress grades, which are characterized by (i) one or more visual or mechanical characteristic, (ii) a set of properties for engineering design, and (iii) a unique grade name.

As is the case for properties of any structural material, allowable engineering design properties must be either inferred or measured non-destructively. From one to six design properties are usually associated with a stress grade: bending modulus of elasticity for an edgewise loading orientation; stress in tension and in compression parallel to the grain; stress in compression perpendicular to the grain; stress in shear parallel to the grain; and extreme fibre stress in bending. With lumber, the properties may be inferred through visual grading criteria, or by non-destructive measurement such as flat-wise bending stiffness or density, or by a combination of these methods.

At this point, it is best to consider the mechanical properties of wood and timber and the potential of timber as a structural material.

3.4.1. Introductory concepts

Everyone has been exposed to wood and its use in everyday life. The terms 'strong', 'stiff' and 'tough' are familiar words. However, such familiarity should not be confused with knowledge of what is technically meant by these terms. Strength is defined in terms of the ability of a material to sustain a load. The load that can be carried by an object depends on the size, shape, and properties of the object. Stress is the magnitude of a load distributed over a unit of area. For all materials there is a critical stress at which they will fail. At less than the critical stress, the material will simply be compressed, stretched, or bent, often by an almost imperceptible amount. The load can be applied in tension, compression, shear, or some combination of these. With wood, the situation is even more complicated. Wood is orthotropic (properties differ in the radial, tangential, and longitudinal directions), so it is necessary to define the direction of the stress with respect to the grain of the wood. Wood tested in tension or compression and loaded parallel to the grain is considerably stronger than wood loaded perpendicular to the grain, but the reverse applies in shear. To add to the complexity, strength is a function of moisture content.

The first systematic testing procedures to determine the properties of wood used small specimens, either 20 by 20 mm or 2 by 2 in. in cross-section, free of all defects (BSI, 1986; ASTM, 2005a). The mechanical properties of numerous woods available to the British timber trade are listed in Lavers (1969), and those of North America are found in the Wood Handbook (USDA, 1999). An enormous amount of work is required to fully characterize the mechanical properties of small clearwood of a single species. It is important to appreciate that these values relate to the timber sampled and the standard rate of loading applied. The data are representative of only the population from which the samples were taken.

Stress applied to clearwood specimens results in some distortion or deformation of the body. This deformation is known as strain. In a tensile test, the sample is very slightly stretched and the strain, ε, is defined as the change in length divided by the length, $\delta L/L$. Most materials fail in tension after they have experienced a strain of about 1%. Figure 10.3a illustrates the green and dry tensile behaviour of wood. Initially, the strain, ε, increases proportionally with the stress, σ, up to the limit of proportionality (*PL*), i.e., $\sigma = E\varepsilon$, where E is known as the modulus of elasticity (E or MOE). The MOE is the ratio of the stress to the strain for the initial slope of the stress–strain diagram. The modulus of elasticity for wood is a quantitative measure of its resistance to deformation. A high modulus of elasticity indicates a stiff material, necessitating a greater applied stress to achieve a given amount of strain. Beyond the limit of proportionality, the tensile stress–strain curve becomes slightly non-linear and the specimen soon fails. For standard testing speeds, the limit of proportionality occurs at about 80% of the failure stress (known as the ultimate tensile stress). The behaviour of wood in compression parallel to the grain is also shown (Figure 10.3a). When dry, the maximum crushing strength of clearwood is on average no more than half the ultimate tensile strength, and the ratio is even smaller when wood is tested green. Compression failure is initiated by buckling and separation of individual fibres adjacent to ray tissue, probably in the earlywood, which throws a disproportionately large load on the adjacent fibres. These in turn become unstable over a portion of their length, and a buckled failure band eventually spreads across the specimen. The failure band is only about 0.2 mm

(0.008 in.) wide. and, typically, lies perpendicular to the grain on the radial face (Figure 10.4a).

Tensile testing of small clearwood specimens parallel to the grain has never been part of these systematic studies and, until recently, bending strength has been taken as an adequate surrogate measure of tensile strength. Indeed, it is extremely difficult to get clearwood specimens to fail in tension because the tensile strength of wood parallel to the grain is so much greater than is its shear strength parallel to the grain or its crushing strength perpendicular to the grain. Thus, it is difficult to pull a specimen in tension without getting premature shear failure or crushing in the grips used to hold the specimen. The tensile specimen must be gradually necked down to a narrow waist some distance from the grips (Figure 10.3b), resulting in a much smaller necked area that has to sustain a much higher stress than occurs in the wood in the vicinity of the grips. Also, test specimen alignment is critical to avoid the introduction of unknown bending stresses.

If tension is about pulling and compression is about pushing, then shear is about sliding (Gordon, 1978). In contrast with compressive and tensile strength in which values along the grain are much higher than values across the grain, the shear strength of wood is much higher across than along the grain. Shearing wood across the grain involves severing the fibres, whereas shear parallel to the grain merely displaces the fibres relative to one another (Figure 10.4b). The low shear stress parallel to the grain presents design problems: a simple example would be the

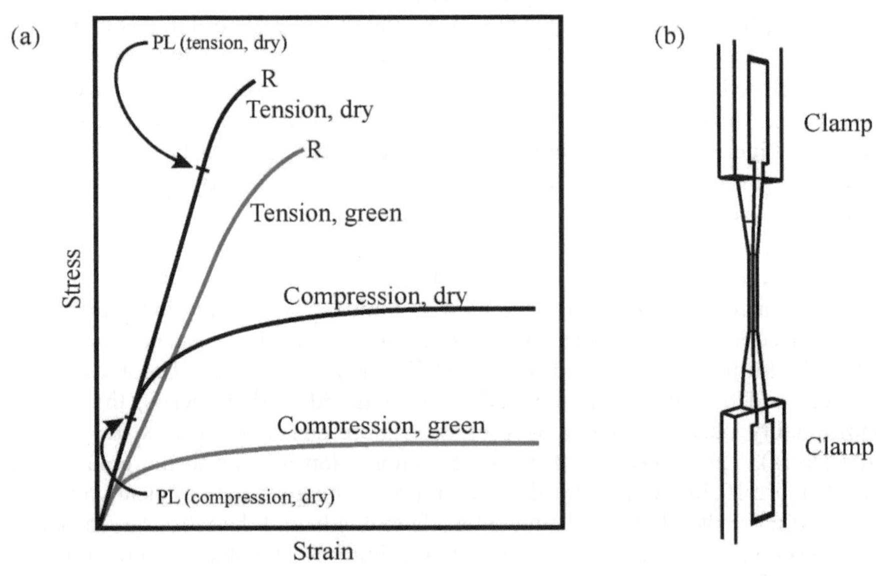

Figure 10.3. (a) Tensile and compressive behaviour of small clearwood samples in green and dry conditions: PL = proportional limit; R = rupture or failure stress. Wood in compression does not rupture suddenly under load but buckles. (b) Geometry of a necked-down tensile specimen.

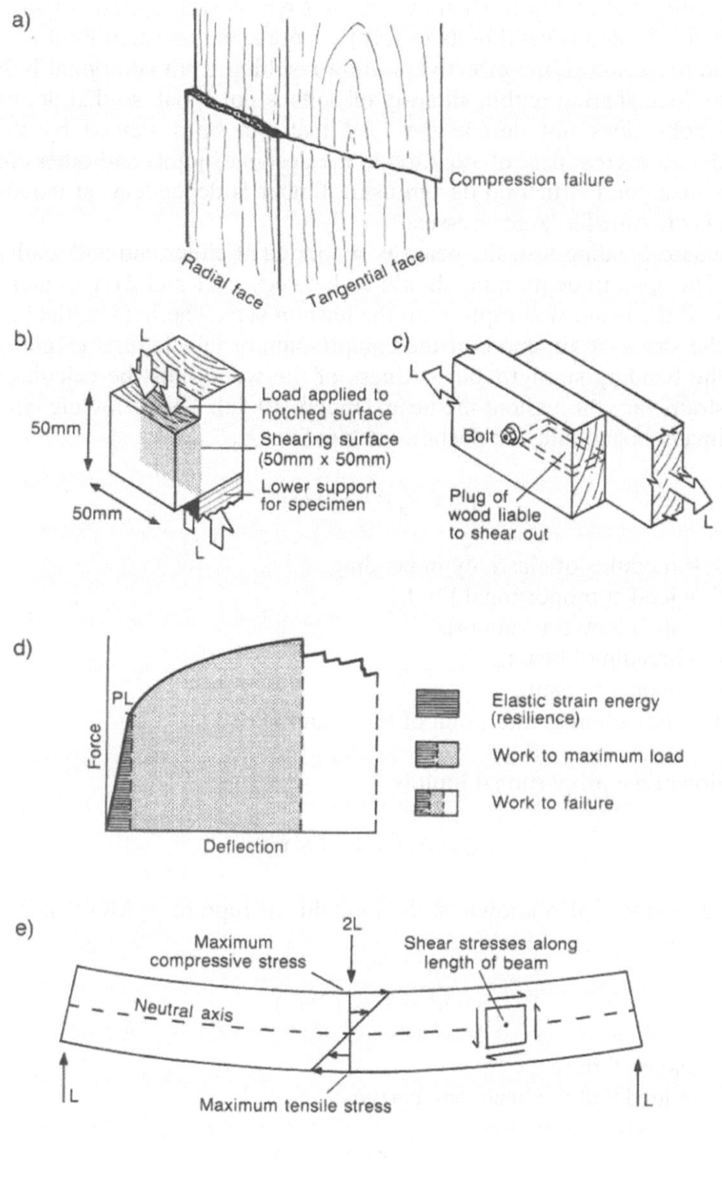

Figure 10.4. Examples of failure modes and bending tests. (a) Buckling failure of short specimens compressed parallel to grain (ASTM, 2005a). (b) Shear test of notched specimen: wood has low shear strength parallel to grain. (c) Low shear strength: timber members in tension parallel to grain cannot be readily connected with bolts, which tend to pull out under comparatively small loads. (d) Load–deflection curve during a bending test. (e) A three-point bending test, with maximum deflection at mid-span.

premature pulling out of a bolt when the tensile stress is still modest (Figure 10.4c). Any benefit from using several bolts to form a more shear resistant joint is partially offset by the reduction in the effective section resulting from additional bolt holes. Furthermore, load sharing within an array of bolts is not equal, so that doubling the number of bolts does not double the load that can be sustained by the joint. Engineers devote a great deal of attention to the design of joints and other connector systems because good structural design using timber is dependent on the ability of these elements to transfer large stresses.

In a standard bending test, the beam is supported at either end and loaded at the mid-point. The span to depth ratio should be between 14:1 and 21:1 to increase the likelihood that the beam will rupture on the tension side. The load–deflection curve resembles the stress–strain curve of the compression or tensile tests (Figure 10.4d). However, the bending strength and stiffness of the wood must be calculated since stress and strain vary throughout the beam (Figure 10.4e). The following equation is used to estimate the modulus of elasticity in bending, E_b:

$$E_b \approx F_p l^3 /(4\Delta bh^3) \tag{1}$$

where E_b = modulus of elasticity in bending,
F_p = load at proportional limit,
L = span between supports,
b = breadth of beam,
h = depth of beam,
and Δ = deflection at mid-point of beam under load F_p.

The deflection at the proportional limit is:

$$\Delta \approx F_p l^3 /(4E_b bh^3). \tag{2}$$

The bending strength, also known as the modulus of rupture or MOR, is calculated using:

$$\text{MOR} \approx 3F_R l/(2bh^2), \tag{3}$$

where F_R = load at failure.
The maximum load that the beam can carry is:

$$F_R \approx (2\ \text{MOR}\ bh^2)/3l \tag{4}$$

The horizontal shear stress is a maximum at the mid-depth of the beam (known as the neutral plane because here the bending stress is zero) and the shear stress falls to zero at the upper and lower surfaces. The shear stress in the neutral plane is:

$$S \approx 3\ F_R/2bh \tag{5}$$

The modulus of rupture is an estimate of the stress at the upper and lower surfaces of the beam. It is calculated on the assumption that wood behaves elastically up to the point of failure, and it assumes that the maximum crushing strength and ultimate tensile strength have the same value. As Figure 10.6b illustrates, this is not true, but the analysis is an acceptable approximation for many purposes. The modulus of rupture (MOR) overestimates the crushing strength parallel to the grain at the upper surface and underestimates the tensile strength parallel to the grain at the lower surface of the beam. The equations given here assume a rectangular beam: the numerical constants and the form of the equation change somewhat if a member of a different cross-section carries the load, e.g. an I-beam or pole.

3.4.2. Simple beam theory

A beam is really two cantilever beams joined back to back (Figure 10.5a) and then turned upside down (Figure 10.5b). Thus, the largest compressive stress occurs on the concave surface of the beam at the mid-span, while the largest tensile stress occurs on the convex surface at the mid-point. Although it is obvious that tensile and compressive forces are generated within a beam when it is bent, it may not be self-evident that shear forces also exist. In a loaded beam, shear stresses act in both the horizontal and vertical directions.

The diagonal compressive and tensile stresses in a girder are equivalent to the two shear forces in a solid beam (Gordon, 1973). The shear forces act at right angles to each other and are oriented at 45° to the equivalent compressive and tensile forces. Both stress systems will deform a square piece of material into a diamond (Figure 10.5c). In a beam, the shear forces lie horizontally and vertically. The imposed weight, 2L, seeks to shear the beam in the vertical direction. The shear force acting at a point within the beam is defined as the algebraic sum of all the perpendicular forces acting on that portion of the beam that are either to the right or to the left of the point considered. Thus, the vertical shear stress acting at any point within the beam between A and B is always L (Figure 10.5b). In addition, an equally strong shear stress must act in the horizontal plane; otherwise the beam would rotate.

In a solid beam, the compressive and tensile stresses are not confined to the surfaces. The compressive stress in a section is highest at the upper surface and gradually diminishes to zero at the neutral plane. Similarly, the tensile stress is highest on the lower surface and diminishes to zero at the neutral plane (Figure 10.6a). While the beam deforms elastically, the compressive and tensile stresses increase proportionately with distance from the neutral plane. The compressive stress at a distance, d, above the neutral plane will be the same as the tensile stress at a distance, d, below the neutral plane. Further, as the modulus of elasticity is the same in compression and tension, the strain at both positions will be similar. Simple beam theory assumes that the beam behaves elastically until failure. However, the limit of proportionality in compression is quite low and once exceeded the fibres near the upper surface will start to buckle, crush, and strain at a greater rate: while

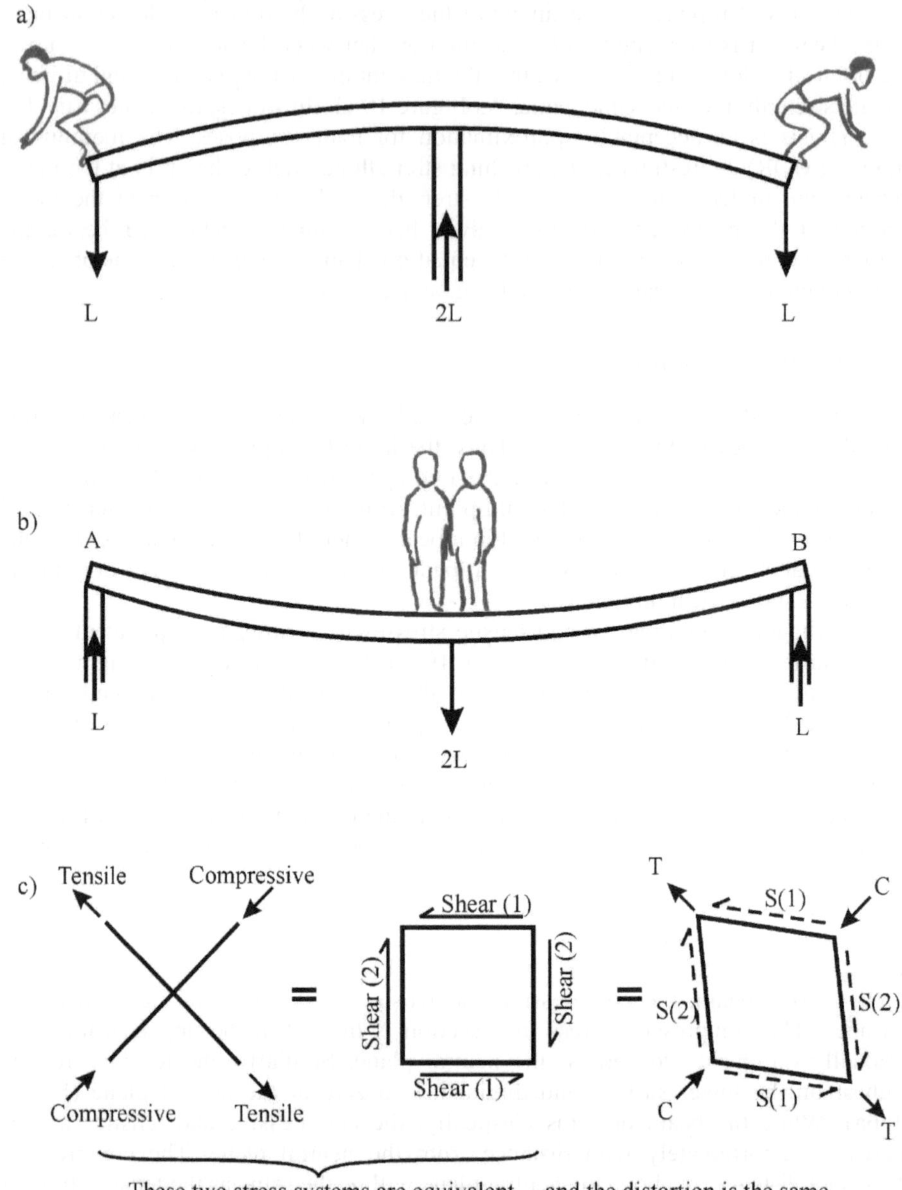

Figure 10.5. A beam under load is equivalent to two cantilevers placed back-to-back and rotated through 180°. Diagonal tensile and compressive stresses in a cantilevered girder are identical to two shear forces acting horizontally and vertically within a solid cantilever or beam. The effect of both stress systems is the same, deforming a square of material into a diamond (Gordon, 1973).

fibres in tension near the lower surface will still be deforming elastically (Figure 10.3a). An increased compressive load can be carried by moving the region of compression towards the lower face. In other words, the location of the neutral plane shifts. There is a little non-elastic deformation when the stress on the lower surface reaches the ultimate tensile strength value (Figure 10.6b). The ultimate tensile strength, calculated from simple elastic beam theory, underestimates the true tensile strength of 'wood', but not, as we shall see later, that of 'timber'.

3.4.3. Some implications of beam theory

With large spans, excessive deflection under load can be a problem. Referring to the simple beam equations (equations 1-5), it can be seen that doubling the length of the beam will result in an eight-fold increase in the deflection at the mid-point of the beam (Eq. 2), while reducing the load capacity of the beam by half (Eq. 4). Doubling the width of the beam only halves the deflection (Eq. 2) and doubles the load that can be sustained without the risk of rupture or shear failure (Eqs. 4 and 5). However, doubling the depth of the beam reduces the deflection by a factor of eight (Eq. 2), while the load that the beam can sustain before failure is quadrupled (Eq. 4). This means that it is more economic to increase the stiffness and strength of a beam by increasing the depth of the beam than to increase its width, because less material is needed in the cross-section. However, the shear stress within a beam decreases as the cross-section increases but it is not affected by the geometry of the cross-section or the length of the beam (Eq. 5). Shear failure can be a problem in short or deep beams that otherwise would sustain very heavy loads without excessive bending or failure. Shear failure is usually initiated by end splitting. Increasing the depth of the beam only results in a linear increase in the horizontal shear load that the beam can sustain, whereas the bending strength increases proportionate to the $(\text{depth})^3$.

Frequently, shear stresses do not govern in the design of beams; either the stiffness or the bending strength limit the carrying capacity of the beam. This means that material in the centre of the beam depth is not highly stressed. Engineers can design more efficiently by using an I-beam, which has wide flanges to carry the high stresses on the upper and lower surfaces and a thin diaphragm in the middle to bear the shear stresses and to keep the flanges apart.

High strength and stiffness are not the only criteria to judge structural efficiency. A third criterion is toughness, the ability of the material to absorb energy. Lignin, both in the middle lamella between fibres and as part of the matrix material between microfibrils, is able to absorb large amounts of energy without irreversible damage. Toughness is important in pit props, which have to sustain enormous loads without risk of sudden failure, and in highway fencing or power poles, which can be subject to impact in a traffic accident. For toughness the optimal MFA is about 25-30°.

Toughness can be defined and measured in a number of ways. A simple impact test can be used to measure the energy absorbed in the high strain rate fracture of a standard notched specimen. Another approach is to measure the strain energy

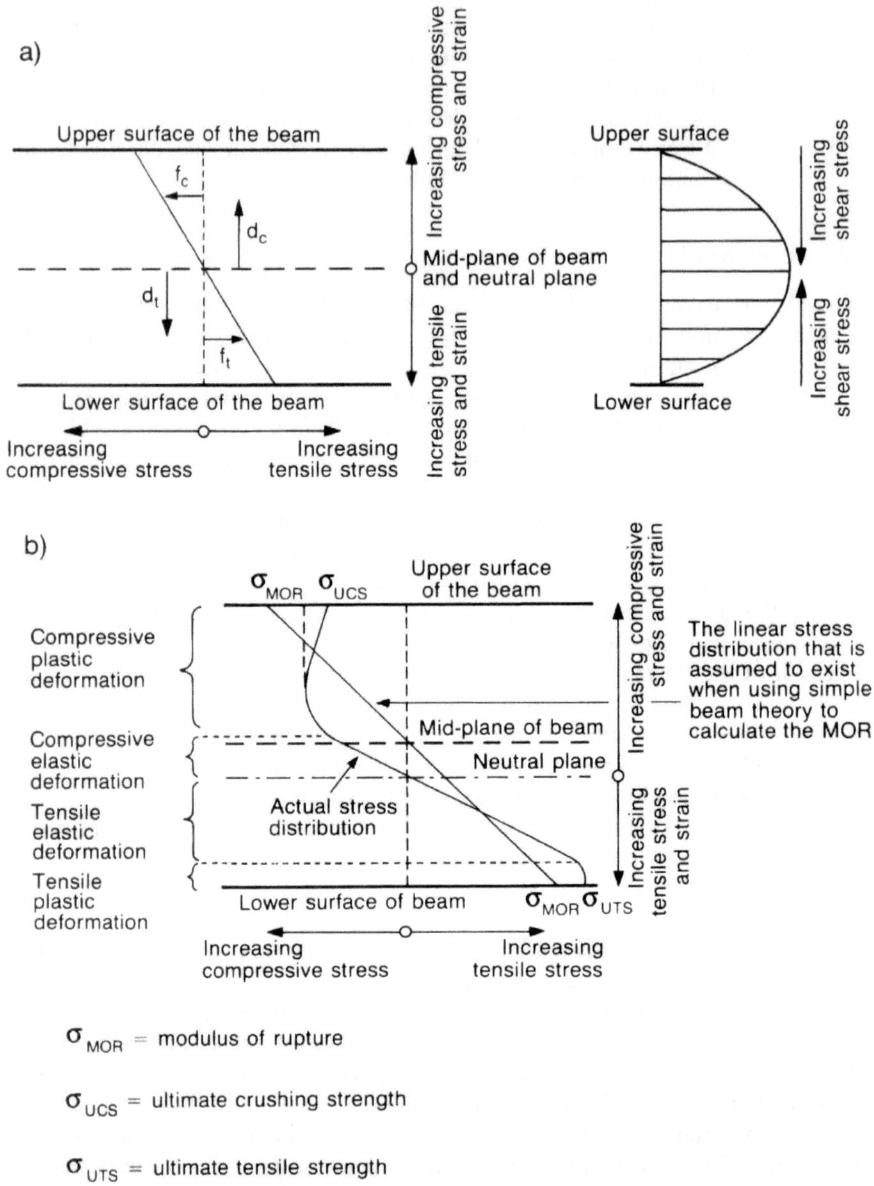

Figure 10.6. Bending of timber beams. (a) Schematic stress distribution at the mid-span of the beam, assuming that the beam deforms elastically. (b) In practice, clearwood starts buckling due to compression failures at the upper surface, whereas knotty timber is more likely to fail in a brittle manner in the vicinity of a knot lying in the tension zone. The behaviour of knotty timber is more akin to a beam deforming elastically up to brittle failure (rupture).

absorbed as timber under flexural loading is bent (Figure 10.4d). The elastic strain energy, also termed resilience, corresponds to the energy absorbed when timber is loaded to the elastic limit: it is equal to the area beneath the load–deflection curve up to the proportional limit (PL). The other two strain energy terms are work to maximum load and work to failure, which correspond to the area beneath the load–deflection curve from zero to the maximum load and to the failure point, respectively. Energy absorption is slightly better in green timber relative to the usual dry service moisture conditions, but it may increase with extreme drying. Round wood is tougher than sawn timber, as the sweep of grain around knots may be severed in sawn timber whereas it is preserved around branches in round wood.

Simple beam theory is a good starting point to understand the major implications of changes in dimensions in wood members, but lumber has further complexities that simple beam theory cannot interpret. Lumber is anisotropic and has many natural defects. Grading is a formal attempt to allow for the impact of these natural defects. The weakening effect of knots is very critical to determining the performance of particular timber products. However, before discussing grading further, it is critical to develop an understanding of the variability present in wood.

3.4.4. Effect of specimen variability on wood properties

Wood as a material inherently has a great deal of variability since it is naturally formed. Even with small clearwood specimens there is considerable variability in the strength (Table 10.2). This is to be expected. It reflects differences in density and microfibril angle within a tree, between trees in a particular stand, and between trees from contrasting locations and growing under difference management systems. For example, the bending strength of European redwood, *Pinus sylvestris*, based on a representative sample collected from material imported into Britain, is approximately Gaussian (Figure 10.7a). The distribution of strength values about the mean is very important. From this information, an estimate can be made about the probability of a piece withstanding a particular load. In this example of European redwood, there is a 1 in 100 chance of the piece failing under a bending stress of 26.1 MPa (378.6 lb/in^2) during the test; this value is significantly less than the mean value of 44.4 MPa (644 lb/in^2). The coefficient of variation (CV), the ratio of the standard deviation to the sample mean, provides a measure of the range of this distribution. A small coefficient of variation is highly desirable as it means less variation in the measured values of the property.

Unlike the strength of clearwood specimens, the strength distribution of material with defects is typically not symmetric. The distribution of material with many defects can be positively skewed, resulting in a lower mean value and a high strength tail (Figure 10.7b).

Typical coefficients of variation for clearwood, for graded kiln-dried Douglas fir, Hem-fir and Southern Pine, and for alternative materials is given in Table 10.2. The variability in properties increases on going from small clearwood specimens to commercially graded material. The poorer grades of timber display more variable

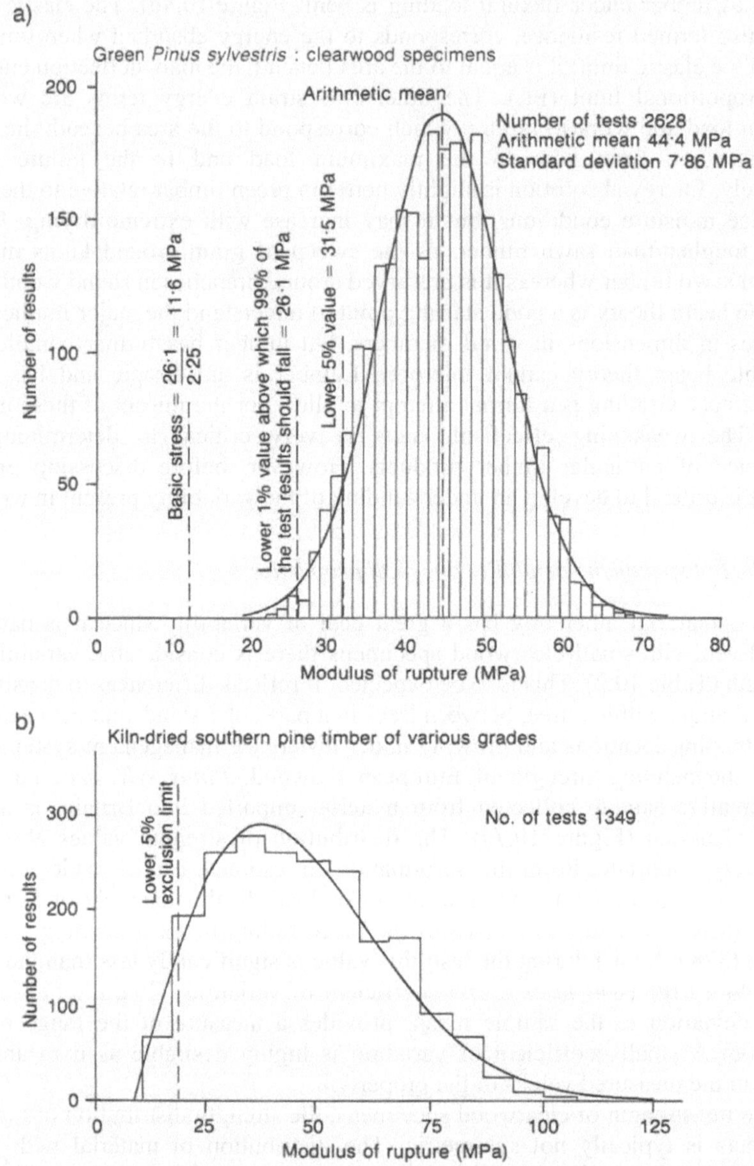

Figure 10.7. Bending strength of clearwood and timber. (a) Small clear green specimens of *Pinus sylvestris* (Sunley, 1968). Data approximates to a normal distribution with a coefficient of variation of 17.7%. Thus 1% of population fails before the stress reaches 26.1 MPa (44.4 – 2.33 x 7.86) and 5% fails below 31.5 MPa (44.4 – 1.645 x 7.86). (b) Variability in bending strength of kiln-dried southern pine of various grades (Doyle and Markwardt, 1966).

Table 10.2. Coefficients of variation for mechanical properties of clearwood and kiln-dried visually graded lumber compared to those of competitive structural materials. Clearwood values are based on tests of approximately 50 species (USDA, 1999) and lumber properties are from Green and Evans (1989). MOR is modulus of rupture, MOE is modulus of elasticity.

	Coefficient of variation (%)					
Material	Properties of other materials	MOR	MOE	Tension parallel to grain	Compression parallel to grain	Shear parallel to grain
Clearwood		16	22	—	18	14
Select Structural						
Douglas Fir		22	24	34	17	—
Hem–Fir		22	21	31	17	—
Southern Pine		23	18	32	18	—
No. 1 & Better						
Douglas Fir		30	25	41	20	—
Hem–Fir		38	23	36	20	—
Southern Pine		25	19	38	20	—
No. 1						
Douglas Fir		28	22	34	14	—
Hem–Fir		28	21	30	16	—
Southern Pine		28	23	36	20	—
No. 2						
Douglas Fir		34	27	47	22	—
Hem–Fir		35	26	40	20	—
Southern Pine		37	24	48	23	—
Concrete	10–20					
Structural steel	5–15					
Metal connectors	10–15					

mechanical properties. The gradual exhaustion of the prime virgin forests of the world and the utilization of second-growth and plantation forests have forced manufacturers to use lower grade and more variable material. This in turn has necessitated the development of different grading methods that take account of this greater variability in timber properties.

There are three main methods to stress-grade timber for structural use: visual stress grading, mechanical stress grading, and proof grading.

4. VISUAL GRADING OF STRUCTURAL LUMBER

Visual grading of structural lumber imposes limits on the maximum size of knots and slope of grain and the minimum density permitted for a given grade. It involves turning the timber to examine all four faces. Yet the relatively low value of the product does not allow for a slow, deliberate examination. A piece of timber is typically graded in two or three seconds. Visual grading has to be able to tolerate errors; a packet of graded timber is deemed to meet the grade if, on re-examination, at least 95% of the pieces are of that grade. Destructive testing of visually graded timber shows that very few pieces fail below the grade stress value, but many pieces are far stronger than the grade stress would indicate. The characteristics of visually grading are that:

- It does not require great technical skill,
- It is safe but inefficient,
- It is labour-intensive but fast,
- It is ideal for small mills and local markets,
- It permits a quick primary sort (to remove visually unacceptable material that might have inadequate strength) prior to other structural grading techniques, and
- It retains market acceptability.

4.1. Visual sorting criteria

Visual grading is the original method for stress grading. Originally it was based on the premise that mechanical properties of lumber differ from mechanical properties of clearwood only because of the many growth characteristics that affect properties, and these characteristics can be seen and judged by eye. Growth characteristics are used to sort lumber into stress grades. The typical visual sorting criteria for softwoods are knots, slope of grain, shake, checks and splits, density, decay, pitch pockets, wane, growth rate and pith. Other features may be characteristic of a great number of tropical species and a few temperate ones, such as interlocked grain, brittleheart, and severe growth stresses.

4.1.1. Knots

Knots cause localized cross-grain with steep slopes. A damaging aspect of knots is that the continuity of the grain around the knot is interrupted by the sawing process. The weakening effect of knots depends on: knot size; mode of testing; and position of the knot within the piece of timber (Figure 10.8).

There is a modest correlation between the strength of timber and knot size. The adverse effect of a knot is primarily attributed to the presence of cross-grain in the immediate vicinity of the knot, rather than to the size of the knot itself. Knots have a greater effect on strength in tension than in compression; in bending the effect depends on whether a knot is on the tension or compression side of a beam (knots along the centre line only affect shear). Intergrown (or live) knots resist compression and transmit some tension, but neither bark-encased knots (unless very tight) nor knotholes can carry a tensile stress. On the other hand, distortion of grain is greater around an intergrown knot than around an encased (or dead) knot of equivalent size. As a result, overall strength effects are roughly equalized, and often no distinction is made in stress grading between intergrown knots, dead knots, and knotholes.

The distorted cross-grain around a knot is not 'parallel to piece' so its stiffness is less than that of straight-grained wood; thus, knots are associated with local areas of low stiffness. However, such zones generally constitute only a minor part of the total volume of a piece of lumber and, because overall stiffness of a board reflects the sum of all its parts, 'whole piece' stiffness is not greatly influenced by knots.

The effect on strength depends roughly on the proportion of the cross-section occupied by the knot, the knot's location, and the distribution of stress in the piece. Limits on knot sizes are therefore made in relation to the width of the face and the location of the knot within the face. Compression members are stressed about equally throughout, and no limitation related to the location of knots is imposed. In tension, knots along the edge cause an eccentricity that induces bending stresses, and they should therefore be more restricted than knots away from the edge. In simply supported structural members subject to bending, stresses are greater in the middle of the length and at the top and bottom edges than at mid-height. These facts are recognized in some grades by differing limits to the sizes of knots in different locations. Thus some grading rules distinguish between knots lying at the edge/margin and in the central part of the wide face (Figure 10.8). Other rules apply the 'knot area ratio' concept to determine the impact of knots on overall properties

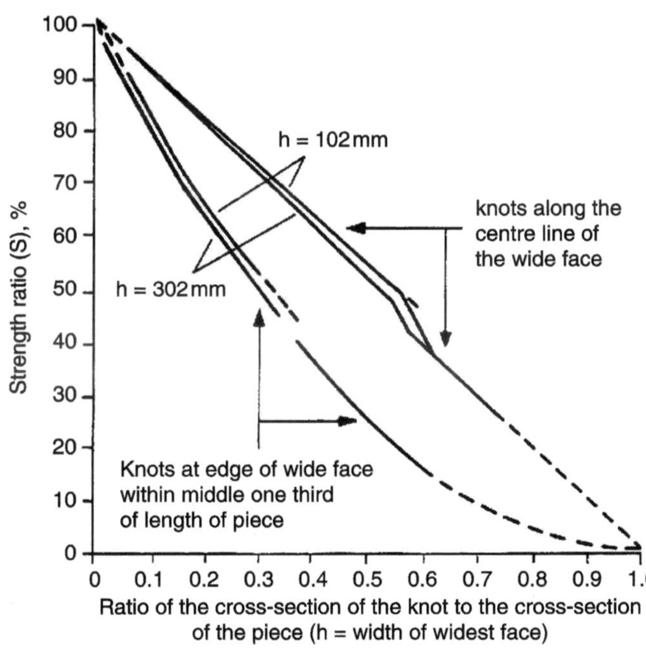

Figure 10.8. The loss in strength due to knots is a function of their size and location. The equations here are from an appendix in ASTM D245 (ASTM, 2005b). In essence, the strength ratio for a timber with a knot on the edge/margin of the wide face is deemed to be equivalent to the square of the strength ratio of a timber with an identical knot that lies at the centre of that face.

(Figure 10.9). Bending strength is more sensitive to margin knots than to centre-face knots, while the effect of knots confined to the narrow edge of the member is less than but similar to that due to a knot in the edge/margin of the wider face. The results are expressed in terms of a strength ratio that relates the strength of the timber with knots occupying a proportion of the cross-section to the strength of an equivalent knot-free member.

4.1.2. Slope of grain

An empirical, elliptical relationship, originally developed by Hankinson (USDA, 1999), is often used to describe the effect of grain orientation on strength properties:

$$\sigma_\theta = \frac{\sigma_{//}\,\sigma_\perp}{\sigma_{//}\sin^n\theta + \sigma_\perp\cos^n\theta} \tag{6}$$

where σ_θ is the strength property at an angle θ to the fibre direction, and $\sigma_{//}$ and $\sigma_\perp$ are the strength parallel and perpendicular to the grain respectively, while *n* is an experimentally determined constant. Slope of grain reduces the mechanical properties of lumber because the fibres are not parallel to the edges. Representative values for *n* and $\sigma_\perp/\sigma_{//}$ are given in Table 10.3. Note that the lowest $\sigma_\perp/\sigma_{//}$ value of 0.03 for compressive strength applies to very low density wood: a ratio of 0.1 or more is more typical.

Table 10.3. Constants for Hankinson's formula for the effect of grain orientation on strength properties (USDA, 1999).

Property	n	$\sigma_\perp/\sigma_{11}$
Tensile strength	1.5–2.0	0.04–0.07
Compressive strength	2.0–2.5	0.03–0.40
Bending strength	1.5–2.0	0.04–0.10
Modulus of elasticity	2.0	0.04–0.12
Toughness	1.5–2.0	0.06–0.10

Lumber with highly localized cross-grained (arising from knots) are especially undesirable as they tend to warp with changes in moisture content. Stresses caused by shrinkage during drying are greater in structural lumber than in small, clear, straight-grained specimens and are exaggerated in zones of distorted grain. To provide a margin of safety, the reduction in design properties resulting from cross-grain in visually graded structural lumber is considerably greater than that observed in small, clear specimens that contain similar cross-grain.

→

Figure 10.9. Grading rules must take knots into account. (a) Typical knot area ratios (KAR) and resulting grades (BSI, 1986); (b) Knots are considered to act together and occupy the same cross-section if the grain disturbance around the knot has not recovered before the grain starts to deviate around the next knot (TRADA, 1974).

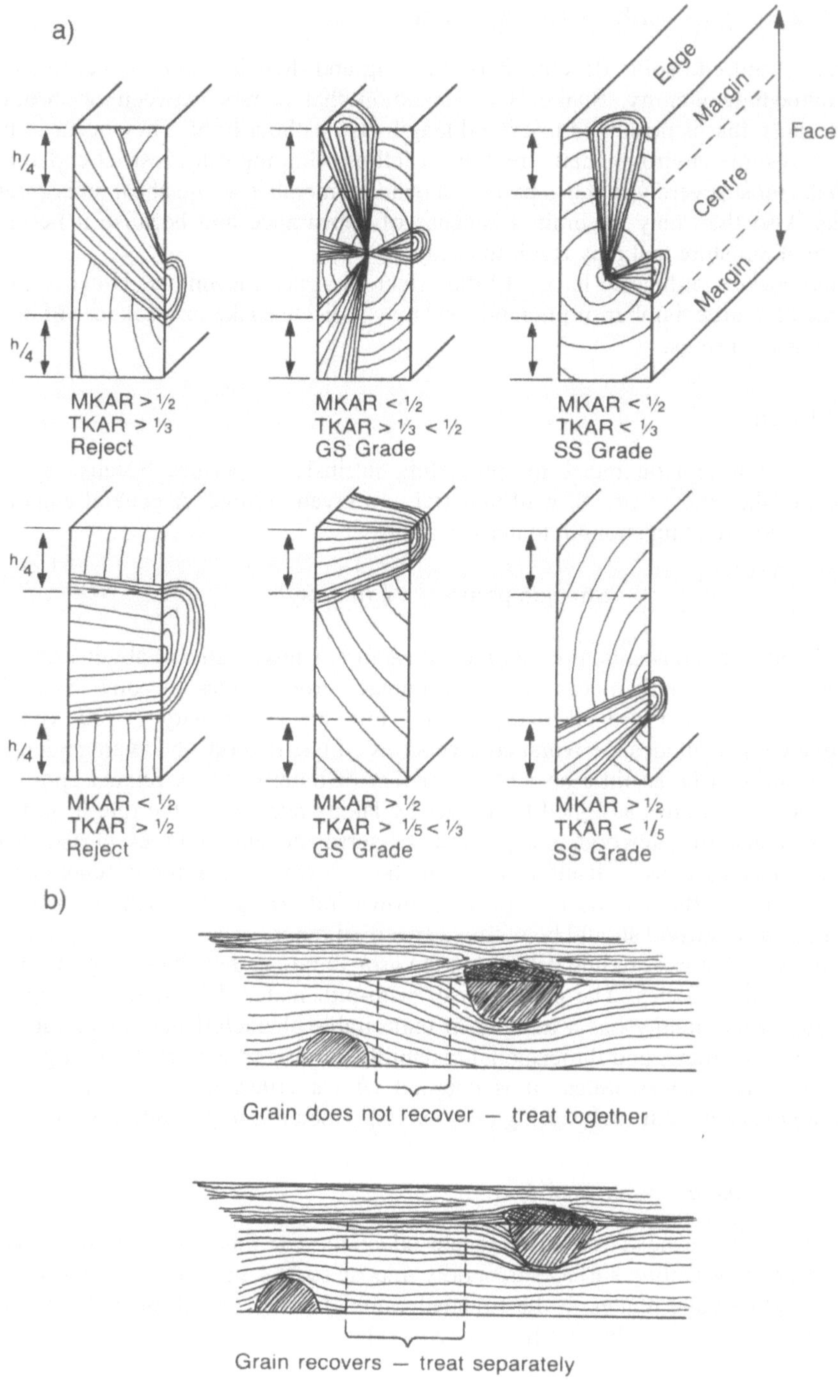
a)
Edge
Margin
Centre
Margin
Face
h/4
h/4
MKAR > 1/2
TKAR > 1/3
Reject
MKAR < 1/2
TKAR > 1/3 < 1/2
GS Grade
MKAR < 1/2
TKAR < 1/3
SS Grade
MKAR < 1/2
TKAR > 1/2
Reject
MKAR > 1/2
TKAR > 1/5 < 1/3
GS Grade
MKAR > 1/2
TKAR < 1/5
SS Grade
b)
Grain does not recover – treat together
Grain recovers – treat separately

4.1.3. Checks, splits, shake and pitch pockets

Checks and end-splits develop during drying and their influence is confined to their immediate vicinity. Shake is a separation that occurs between or through growth rings, that is presumed to extend lengthwise without limit. They occur in the tree due to some environmental stress or on felling. Grading rules restrict end-splits and shake most severely in those parts of a bending member where shear stresses are highest. Also they may be limited because of appearance and because it permits entrance of moisture that may result in decay.

Pitch pockets ordinarily have so little effect on structural lumber. However, the presence of a large number of pitch pockets may indicate shake or weakness of bond between annual rings.

4.1.4. Density

Density is a pragmatic index for predicting intrinsic properties, because it is a measure of the amount of cell wall material in a given volume. A general equation (USDA, 1999) relating strength to density is:

$$\text{Strength property} = k\,(\text{density})^{n} \tag{7}$$

where k and n are constants. For modulus of rupture, n has a value of about 1.05.

The lowest strength pieces can be eliminated from grades in some codes by excluding those that are of exceptionally low air-dry density (<400 kg/m^3). Effectively this excludes the worst corewood, as with softwoods the average density of corewood can be as little as 60-65% of that of outerwood. A related approach adjusts property values assigned to lumber by taking account of the rate of growth and percentage of latewood, using these as surrogate and indirect measures of density – just as density itself is taken to be a surrogate for more fundamental characteristics of the cell wall. Typically, growth rate (rings per inch/cm) and the percentage of latewood should be within a specified range.

Some visual stress grading rules limit the growth rate and exclude all pith. There are two separate issues. The exclusion of with-pith material is understandable as corewood in its vicinity has a number of undesirable characteristics, including low density, low stiffness, and dimensional instability. However, a restriction on growth rate is less satisfactory unless it is coupled with a criterion that addresses ring curvature to ensure that such sorting only excludes material in proximity to the pith.

4.1.5. Miscellaneous: wane and decay

Aesthetics and the need for an adequate edge/face for fabrication or to provide ample bearing or nailing surfaces generally impose stricter limitations on wane than does strength. Wane relates to the rounded cambial surface (with or without bark) that may be found in lumber cut from outerwood.

Decay in most forms is prohibited or severely restricted in stress grades because the extent of decay is difficult to determine and its effect on strength may be greater than visual observation would indicate. Further without preservative treatment arrested decay in dry wood will resume if the material were to be rewetted. There are circumscribed exceptions. Limited decay of the pocket type, e.g. *Fomes pini*, can be permitted in stress grades, as it occurs in knots but not in the surrounding wood.

4.2. Responsibilities and standards for visual stress grading

An orderly, voluntary, but circuitous system of responsibilities has evolved in most countries to handle visual stress grading. Most visually stress graded lumber is dimension lumber produced in sizes from 35 to 90 mm (1-1/2 to 3-1/2 in.) thick and 35 to 200 mm (1-1/2 to 8 in.) wide. Each country has its own series of grades that are applied to their products. Table 10.4 provides a partial list.

In North America, stress grading is handled under the auspices of the American Lumber Standard Committee (USDC, 1999). In Europe, individual countries have their own grade classifications but a visual grading standard EN 518 (ECS, 1995) is used as the overarching guideline for property development. Elsewhere the following apply: South Africa SANS 1783 (SANS, 1997); Australia AS 2858 (SAA, 1986a); New Zealand NZS3631 (SANZ, 1988); Japan JAS 143 (JAS, 1997).

Attempts are currently under way to develop a more standardized international approach through the International Organization for Standardization (ISO).

Table 10.4. Partial list of various grades for visually graded timber throughout the world.

European visual grades										
Belgium	France	Germany	Greece	Holland	Ireland	Italy	Nordic	Portugal	Spain	UK
S10	B	S13	S10	A	SS	S10	T3	S10	Eng	SS
S8	S	S10	S8	B	GS	S8	T2	S8	P2	GS
S6		S7	S6	C		S6	T1	S6	P1	
			S4	D			T0			

Elsewhere							
Australia		China	Japan	New Zealand	South Africa	U.S. & Canada	
F14	F7	Grade1	No. 1	Eng	S14	SS	Const
F11	F5	Grade2	No. 2	No. 1	210	No. 1	Std
F8	F4	Grade3	No. 3	No. 2	S7	No. 2	Util
					S5	No. 3	

4.3. Classification into strength groups

In grouping species into strength classes, MOE and MOR are assigned conservative values. Species are usually grouped together where they have roughly the same mechanical properties and so can be treated as equivalent, or where two or more species are very similar in appearance, or for marketing convenience.

The advantages of strength groupings were first recognized in Australia where an enormous number of indigenous species have been evaluated for structural purposes (Leicester, 1988). In tropical regions, there are literally thousands of individual timbers, which are often difficult or impossible to identify once sawn. Similar looking species can have very different strength characteristics. In Malaysia only 400 of 2500 species are deemed to be of commercial value. To achieve orderly marketing, these 400 species are grouped so that, in effect, 70% of the commercial timbers of Malaysia are actually species mixtures.

The Australian standard AS 2878 (SAA, 1986b) offers two approaches to categorize a timber into strength groups. Where adequate but limited data of mean strength characteristics are available, a positive strength grouping is possible. Where these data are not available, the mean air-dry density of the timber is used instead to give a more conservative, provisional strength grouping (Table 10.5). Application of visual grading rules (SAA, 1986a) leads to four or more stress grades for each of the 600 species listed (SAA, 1986b), giving a total of 2500 individual stress grades.

Table 10.5. Lower limits for strength groups for seasoned timber (SAA, 1986b).

Strength group	Testing of small clearwood specimens			Density at
	Bending (MPa)	Compression (MPa)	Stiffness (GPa)	12% MC (kg m^{-3})
SD1	150	80	21.5	1200
SD2	130	70	18.5	1080
SD3	110	61	16.0	960
SD4	94	54	14.0	840
SD5	78	47	12.1	730
SD6	65	41	10.5	620
SD7	55	35	9.1	520
SD8	45	30	7.9	420

These stress grades have been reconciled and rationalized into only 12 interlocking stress grades in a geometric preferred number series (Table 10.6). For structural purpose, it is sufficient for the engineer to specify the desired stress grade, i.e. F14. Selection may be tempered by considerations such as cost, natural durability, or amenability to preservation, seasoning, or gluing characteristics.

Strength groups avoid any confusion that can arise where it is presumed all material of a given grade has equivalent properties. No. 1 dark red meranti does not have the equivalent strength of No. 1 balau: dark red meranti falls into SD6 group and its No. 1 lumber is F14, whereas balau is in SD3 and its No. 1 lumber is F27. Balau and dark red meranti both contain species groupings of the genus *Shorea*.

A number of grading systems are employed in Europe. To remove barriers to trade, the European Union (EU) instituted a strength class approach to species groupings as part of the development of a EU lumber building code, Eurocode 5 (EC5, 1995), to provide member countries with a unified framework of harmonized

Table 10.6. Basic working stresses and stiffness for structural timber (SAA, 1988b). Each grade stress is 25% greater than that below. MOE includes an allowance for shear.

	Basic working stresses (MPa)				MOE
Stress grade	Bending	Tension	Shear	Compression	(GPa)
F34	34.5	20.7	2.45	26.0	21.5
F27	27.5	16.5	2.05	20.5	18.5
F22	22.0	13.2	1.70	16.5	16.0
F17	17.0	10.2	1.45	13.0	14.0
F14	14.0	8.4	1.25	10.2	12.0
F11	11.0	6.6	1.05	8.4	10.5
F8	8.6	5.2	0.85	6.6	9.1
F7	6.9	4.1	0.70	5.2	7.9
F5	5.5	3.3	0.60	4.1	6.9
F4	4.3	2.6	0.50	3.3	6.1
F3	3.4	2.0	0.45	2.6	5.2
F2	2.7	1.6	0.35	2.1	4.5

standards and regulations (Cooke, 1988; Sunley, 1979). Table 10.7 shows the strength classes that have been established in the Eurocode.

The ISO Technical Committee 165 is developing a global strength class system that will be compatible with the existing European and other systems.

A strength class system does not exist for visual stress grades in the United States and Canada. However, ASTM D2555 (ASTM, 2005e) has procedures for calculating clearwood properties for groups of species to be used with D245 (ASTM, 2005c). Also, D1990 (ASTM, 2005d) contains procedures for calculating design properties for groups of species tested as full-size members. The properties assigned to a group by such procedures will often be different from those of any species in that group. The group will have a unique identity, with nomenclature approved by a given country's regulatory agency.

Table 10.7. Minimum strength requirements for European strength classes (ECS, 2003).

Mechanical	European (EN 338) strength classes (C-Grades)								
property	C14	C16	C18	C22	C24	C27	C30	C35	C40
F_b, MPa	14	16	18	22	24	27	30	35	40
F_t, MPa	8	10	11	13	14	16	18	21	24
F_c, MPa	4.3	4.6	4.8	5.1	5.3	5.6	5.7	6.0	6.3
E, MPa	7000	8000	9000	10 000	11 000	12 000	12 000	13 000	14 000

F_b is bending stress; F_t, tension parallel to grain; F_c, compression parallel to grain; E, modulus of elasticity.

4.4. Revision of the visual stress grading philosophy

In the mid-1960s, some allowable properties for lumber began to be questioned as more data on full-size lumber specimens became available (Bohannan, 1966; Doyle and Markwardt, 1966, 1967, 1968). This was especially true for tension parallel to grain. In 1968, a new provision was adopted in ASTM D245 that set tension values at 55% of bending values for visually graded timber. By the 1970s, additional work on full-size lumber (Johnson and Kunesh, 1975; Kunesh and Johnson, 1972, 1974; Littleford, 1978; Littleford and Abbott, 1978) provided enough evidence to show that the visual stress grading procedures used at that time were no longer appropriate or particularly efficient. Also, during this period there was a major liability suit centred on the failure of a large cooling tower constructed of wood, which heightened the concerns of the North American lumber industry. The final straw came from a series of studies on full-size lumber by Borg Madsen at the University of British Columbia (Madsen, 1975, 1976, 1978). Madsen described these studies as 'in-grade' tests because they were tests of material representative of typical commercial lumber that was within grade. These tests implied that published bending stress values could be overstated by as much as 25% to 35%. Also, Madsen called into question the current moisture adjustment procedures for commercial structural timber. Another study at Colorado State University seemed to confirm that bending stress values were overstated (Bodig, 1977).

The basis for the grading of small clear timber had been developed when it was relatively easy to obtain large timber members with few, if any, serious defects. These members could be cut from enormous trees in the virgin forests of most continents. Today, such timber is much more difficult to obtain. This is significant since the failure mode of clearwood in bending is quite different from the failure mode of pieces in graded structural timber (Figure 10.10). Tensile failure perpendicular to the grain, caused by localized grain disturbances around knots, weakens structural timber considerably. Madsen argued that because of different failure mechanisms, strength characteristics of structural timbers should be obtained by evaluating graded material directly rather than using clearwood strength to estimate the allowable grade stresses. A large testing program known as the North American In-Grade Testing Program evaluated material in the grades as produced in the United States and Canada (Green *et al.* 1989).

It should be emphasized that all visual grading procedures are quite crude systems for categorizing the strength properties of timber. Thus, there will be a significant proportion of low grade material, e.g. No. 3, which would sustain a stress equivalent to that for a higher grade, e.g., Select Structural. Equally, a much smaller proportion of SS only has the strength equivalent to that expected of No. 1, No. 2, or even No. 3 lumber.

Grading rules for home construction timber have generally abandoned the clearwood approach outlined in Figure 10.7. In the old system clearwood tests established the lower percentile values for the population (either the lower 1% or 5%) which were reduced further to obtain a basic stress for clearwood, by applying

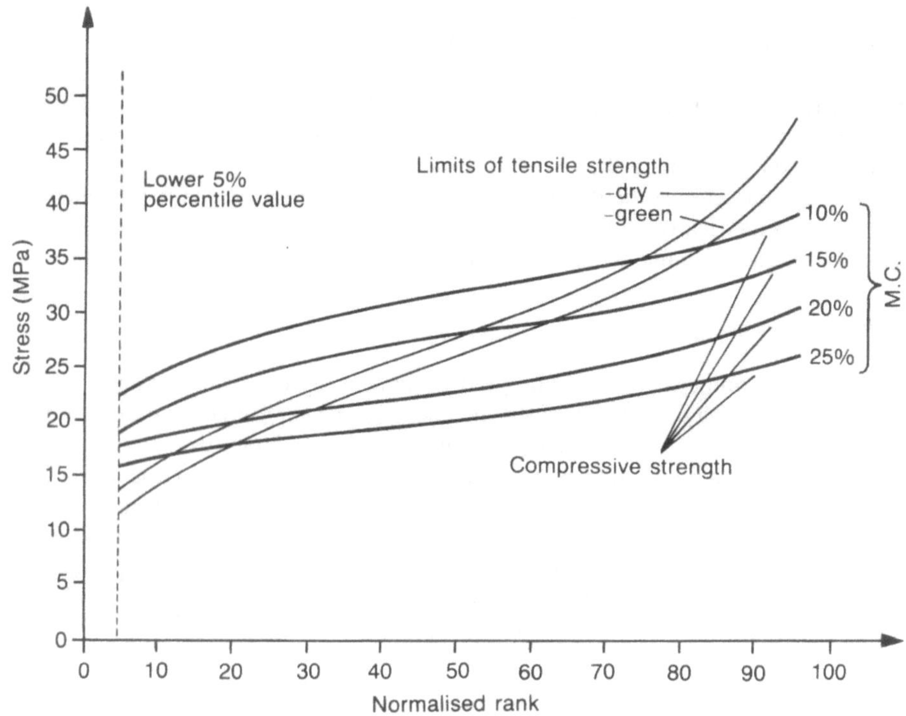

Figure 10.10. Ranked data of compressive and tensile tests with 38 x 140 mm (2" x 6") spruce-pine-fir (S-P-F) of No. 2 grade and better (Madsen, 1984b). The tensile strength is less sensitive to changes in moisture content than is the compressive strength. Consequently at the low strength end of the distribution pieces experience brittle failure in tension, in the immediate vicinity of cross-grain or knots. This contrasts with clearwood tests (equivalent to the strongest lumber, ranked to the right in this figure) where the tensile strength always exceeds the compressive strength.

corrective terms for long term loading and a factor of safety/ignorance – remember samples are broken in standard tests in a matter of minutes, yet a timber beam is expected to carry a load for many years. Finally grade stresses were obtained by considering the extent to which permitted defects in a particular grade reduced the properties of that grade relative to that expected of clearwood. The original British grades (Sunley, 1968), had strength ratios of 0.75, 0.65, 0.50 and 0.40 relative to the basic stress of clearwood

The updated BS 4978 (BSI, 1986) specification for softwood grades for structural uses replaced the four grades by just two visual stress grades, general structural (GS) and select structural (SS), which have strength ratios of the order of 0.3-0.35 and 0.5-0.6, respectively. However the strength ratio is no longer derived.

Instead the grade stress is determined directly by in-grade testing, and the link with clearwood values through the strength ratio has been broken.

Unlike clear wood specimens, the strength distribution for in-grade material is asymmetric: the population is positively skewed. This results in a lower mean value and a high strength tail (Figure 10.7b). Therefore, a Gaussian distribution function cannot be used to estimate the lower fifth percentile exclusion limit. Instead, it is preferable to characterize the distribution function and determine the fifth percentile value non-parametrically. A three-parameter Weibull distribution has been found to accurately characterize these skewed distributions.

4.5. Procedures for deriving design properties

The mechanical properties of visually graded lumber may be established by (i) tests of a representative sample of full-sized members, or (ii) appropriate modification of test results conducted on small clear specimens. Design properties for the major commercial species in a country may be a mixture of these two methods. For example, values for softwood dimension lumber species listed in current design specifications and codes in the United States have been derived from full-size member test results using D1990 (ASTM, 2005d). In contrast, design properties for most hardwood dimension and structural timbers are still derived using results of tests on small clear samples under standard D245 (ASTM, 2005c).

4.5.1. Application of strength ratios

The influence of knots still needs to be assessed; in the United States with the strength ratio method (Figures 10.11 and 10.12); in the UK by looking at the proportion of the cross-section that they occupy, their interaction with one another, the total knot area ratio (TKAR), and their location within a member (Figure 10.9). Smaller knots are permitted in the margins than elsewhere in the member. The permissible knots are presumed not to lower the strength of the material below the stress assigned to that grade. The weakening effect of adjacent knots, even if not lying in the same cross-section, is recognized: in such cases: both knots are deemed to be in the same cross-section (Figure 10.9b). This takes account of any serious

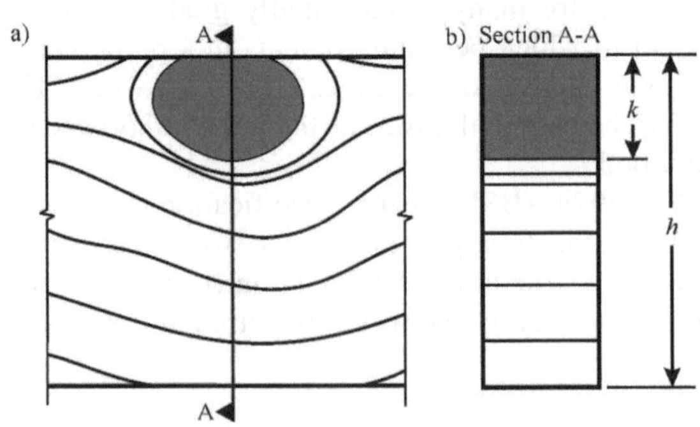

Figure 10.11. (a) Edge knot in timber. (b) Assumed loss of cross-section (shaded area).

cross-grain in the vicinity of one knot interacting with the grain around an adjacent knot. Other defects are prescribed, e.g. the slope of grain must not exceed 1:6 for general structural and 1:10 for select structural.

In the United States, sorting criteria that influence mechanical properties of hardwoods are handled with 'strength ratios' for the strength properties and with 'quality factors' for the modulus of elasticity. Conceptually each strength property of a member is derived from the product of the clearwood strength for a population representing the species and a measured limiting strength ratio for that member, i.e. the strength ratio is the hypothetical ratio of the strength of the piece with visible strength-reducing growth characteristics to its strength if those characteristics had been absent. The true strength ratio is never known and can only be estimated.

Estimated strength ratios for cross-grain and density have been obtained empirically; strength ratios for other growth characteristics have been derived theoretically. For example, to account for the weakening effect of knots, the assumption is made that the knot is effectively a hole through the piece, reducing the cross-section, as shown in Figure 10.11. For a hardwood beam containing an edge knot, the bending strength ratio can be idealized as the ratio of the bending moment that can be resisted by a beam with a reduced cross-section to that of a beam with a full cross-section:

$$\mathrm{SR} = 1 - (k/h)^2 \tag{8}$$

where SR is the strength ratio, k the knot size, and h the width of face containing the knot: effectively comparing two beams of depth h and $(h\text{-}k)$ using Eq. (2) assuming that depth of the beam is reduced by the size of the knot. Figure 10.12 shows how the strength ratio changes with knot size according to this formula. Strength ratio formulae are given in D245 (ASTM, 2005c). The worst defect in an individual piece having several defects is used to derive the strength ratio.

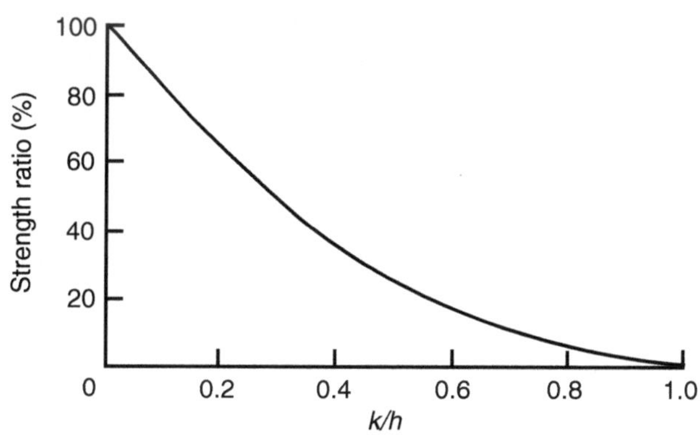

Figure 10.12. Relationship between bending strength ratio and the size of an edge knot expressed as fraction of the face width: k is knot size; h, the width of face containing the knot.

The range of strength ratios in a grade and the natural variation in clearwood strength values give rise to variation in strength between pieces in the grade. To account for this variation and to ensure safety in design, it is intended that the actual strength of at least 95% of the pieces in a grade exceed the design properties assigned to that grade (before a reduction for duration of load and safety). In visual grading, according to D245, this is handled by using a near-minimum clearwood strength as a base value and multiplying it by the minimum strength ratio permitted in the grade to obtain the grade strength property. The near-minimum value is the 5% exclusion limit (Figure 10.7a). D2555 (ASTM, 2005e) provides clearwood strength data and a method for estimating the 5% exclusion limit.

The assigned modulus of elasticity, E, is an estimate of the average modulus of clearwood when tested in static bending, adjusted for shear. The average modulus of elasticity for clearwood of North American species is recorded in D2555. The clearwood average is multiplied by empirically derived 'quality factors' to represent the reduction in modulus of elasticity that occurs by timber grade for pieces tested in an edgewise orientation. This procedure is outlined in D245 (ASTM, 2005e).

4.5.2. In-grade procedure

First, the natural population of lumber is visually segregated into sub-populations (grades). The properties of these grades have broad distributions that overlap one another, but the properties of each grade are more discrete and less imprecise than run of mill lumber.

The mechanical properties for specified grades are established from tests of full-size specimens of those grades, testing representative samples of the population following procedures outlined by a given country's standard. In the United States, those standards are D2915 and D1990 (ASTM, 2005f,d). The specimens are tested using appropriate test procedures for the United States, D198 or D4761 (ASTM, 2005b,h). The 5% exclusion limit for the strength of the grade being tested is established directly and further modified for design use by consideration of service moisture content, duration of load, and safety. Adjustment of the data for strength ratios is not needed. In Europe, the development of characteristic properties is covered in EN 384 (ECS, 2004).

4.5.3. Visual stress grade stamps

Visually graded timber should be marked with a grade stamp (Figure 10.13). With few exceptions, the grade stamp should include five features: (i) trademark, indicating identity of agency with quality supervision; (ii) mill identification, giving product manufacturer's name brand or mill number; (iii) grade designation number or abbreviation; (iv) species or species group identification; and (v) condition of seasoning, indicating moisture content at time of surfacing.

Figure 10.13. An example of a grade stamp for visually graded No. 2 Douglas fir, surfaced dry and produced by mill #12 of the Western Wood Products Association. Reproduced with permission.

5. MACHINE-GRADED STRUCTURAL TIMBER

Machine-graded timber is graded by machine using a non-destructive test followed by visual grading to screen out certain characteristics that the machine cannot or may not properly evaluate. In New Zealand and Australia, machine-graded structural timber is described as machine stress-graded (MSG). Terms common in the United States and Canada to describe machine-graded structural timber are machine-stress-rated (MSR), machine-evaluated lumber (MEL), and *E*-rated timber. No matter the terminology, machine-graded timber allows for better sorting of material, with reduced variability, for specific applications in engineered structures.

The attraction of machine grading lies in the fact that individual pieces are tested so that the deflection in bending (the usual method for allocating a piece to a grade) reflects both the natural defects (knots *etc.*) and the intrinsic features (density, MFA *etc.*) of that piece.

The basic components of machine-grading systems are sorting and prediction of strength through machine-measured non-destructive determination of properties coupled with visual over-ride; the assignment of design properties based on strength prediction; and quality control to ensure that assigned properties are being obtained.

The quality control procedures ensure:

- proper operation of the machine used to make non-destructive measurements,
- appropriateness of predictive parameter–bending strength relationship, and
- appropriateness of properties assigned for tension and compression.

5.1. Machine sorting criteria

Generally timber is machine-graded using the modulus of elasticity, *E*, as the sorting criterion for mechanical properties. The modulus of elasticity can be measured in a variety of ways. Usually, the apparent *E*, derived from three-point bending, is measured (Figure 10.14). Because timber is heterogeneous, the apparent *E* depends on span, orientation (edge or flatwise in bending), load speed of test (static or dynamic), and method of loading (tension, bending, concentrated, or uniform). As long as the grading machine is properly calibrated, any of the apparent *E* values can be used to assign the graded piece to a 'not to exceed' grade category. Most grading machines are designed to detect the lowest local stiffness in flatwise bending, E_{min}, that occurs in any 1.2-m (4-ft) length as well as the average flatwise bending, E_{av}, for the entire length of the piece. Here the timber is fed continuously through the stress grader. The machine flexes each piece as a plank between two supports applying a fixed deflection (Figure 10.14) and measuring a load, or measuring the deflection under a particular load. Thus the piece is tested as a plank (to obtain a significant measurable deflection) but is used as a joist. The visual over-ride considers edge knots that have a far greater influence on strength when in service as a joist than they have when the piece is tested in bending as a plank.

In the United States and Canada, MSR and MEL timber are subject to a visual override because the size of edge knots in combination with E is a better predictor of strength than is E alone. Maximum edge knots are limited to a specified proportion of the cross-section, depending on grade level. Other visual restrictions, which are primarily for appearance, are placed on checks, shake, skip (parts of the board 'skipped' by the planer), splits, wane, and warp.

Another method of sorting machine-graded timber is to use density to estimate knot size and frequency. X-ray sources in conjunction with a series of detectors gather density profiles in the timber, which are sensitive to knots. This information is then used to assign the graded piece to a 'not to exceed' grade category.

For over 40 years, it has been known that strength correlates with stiffness, and appropriate regression equations have been established for commercially important timbers. In Britain, for example, regression equations were derived by taking a representative sample of the timber, lightly loading each piece in bending as a plank to determine its stiffness, E_{plank}, and then turning the piece on edge and loading to

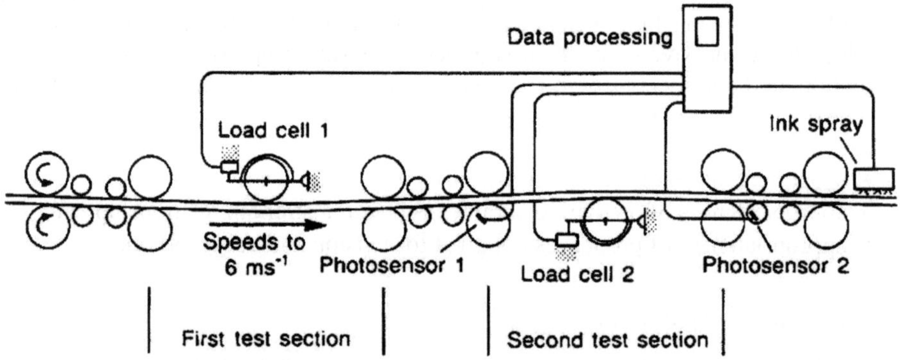

Figure 10.14. The continuous lumber tester (CLT) measures the load required to achieve a fixed deflection, continuously along the length of the piece. Measurements from both bending directions are combined to compensate for distortions in the timber (Metriguard Inc., Pullman, Washington). Deflections are set so that pieces experience a stress level roughly equivalent to the 10-year design bending strength for the grade. A few pieces break during their passage through the machine (normally not more than three pieces per shift), which helps to rid the machine-grades of 'rogues' that have relatively high stiffness but low strength.

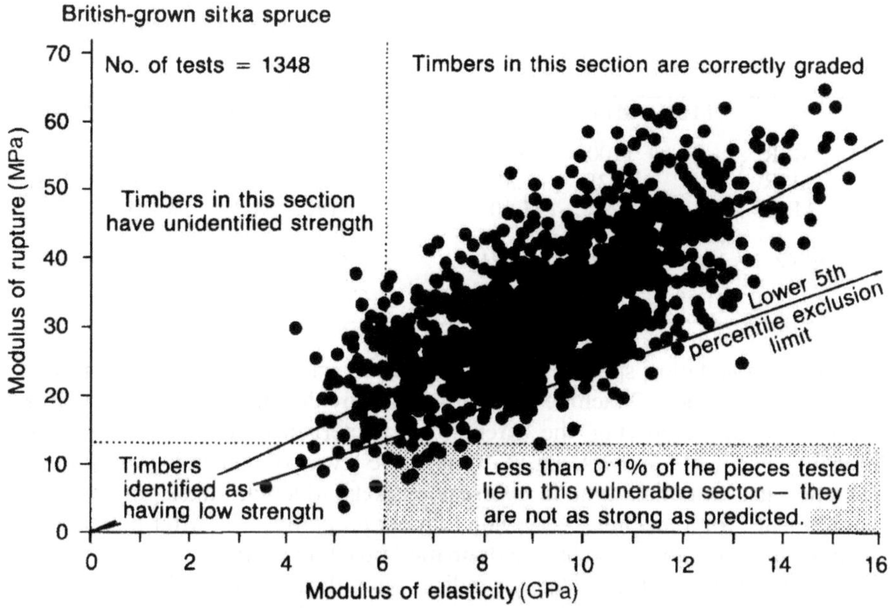

Figure 10.15. Relationship between MOR_{joist} and E_{plank} for British-grown Sitka spruce, *Picea sitchensis*. MOR = 0.002065 $(MOE)^{1.0573}$ with correlation coefficient of 0.702.

determine the bending strength, MOR_{joist}. The MOR_{joist} is then plotted against E_{plank} and the regression calculated and superimposed over the data points (Figure 10.15).

The scatter about the regression line is attributed both to the inherent variability of the timber due to its density and MFA and to the presence of defects. The regression coefficient (r), which describes how tightly the data cluster about the regression line, is generally found to fall between 0.5 and 0.85. The square of the regression coefficient, known as the coefficient of determination, indicates the percentage of variability of one variable, MOR_{joist}, that is accounted for by the other variable, E_{plank}. Thus, a correlation coefficient of 0.707 means that 50% of the variability in bending strength can be accounted for by E_{plank}.

Like the designations for visually graded material, a number of different grade designations are used in various countries. A partial list of grade names for machine-graded timber available from different countries is shown in Table 10.8.

Machine stress grading is more objective and efficient than visual stress grading. A typical regression of MOR_{joist} on E_{plank} has a regression coefficient of 0.6 to 0.7, which compares favourably with a regression coefficient of around 0.3 to 0.4 for MOR_{joist} based on visual grading, using knot area ratio. Even so, the correlation between stiffness and strength is still not that good as 60-70% of the material in a grade can sustain a stress twice that allocated to timber of that grade. In essence grading is focussed on the weakest pieces in the grade – how weak the poorest pieces might be – and says nothing about the strength of better pieces in the grade.

Table 10.8. Examples of machine-graded timber designations from different countries.

Country		Machine-graded timber designation							
UK		M10	M14	M18	M22	M26	M30	M35	M40
Canada & U.S.	MSR / MSR		1200f –1.2E	1450f –1.3E	1650f –1.5E	1950f –1.7E,	2400f –1.7E,	2700f –2.2E,	2850f –2.3E
	MEL			M-10		M-19	M-23	M-24	
Australia & NZ		MGP6	MGP8	MGP10		MGP12		MGP15	
Europe			C14	C18	C22	C26	C30	C35	C40

Fast-grown plantation species have been viewed with suspicion by sawmillers, merchants and specifiers. Machine stress grading provides the forest owner with an objective way of comparing the strength characteristics of their timber with traditionally acceptable timbers, e.g. it offers Chile a way of enhancing the status of its plantation-grown radiata pine in overseas markets. Machine stress grading is the current ultimate 'truth machine' – revealed timber properties can differ dramatically between populations, and these have a dramatic impact on profits.

The results shown in Figure 6.5 for radiata pine illustrate this point. Changes in silvicultural management, especially the age of clearfelling, can have a dramatic impact on timber quality. Figure 6.5 compares the strength–stiffness characteristics of timber coming from more mature plantations with those of material harvested at a younger age. The loss of the low MFA, high density outerwood is the major reason for the lack of high strength material in the younger population. Unimproved pine has an abundance of low quality wood, and the failure to meet certain grading thresholds devastates profits. Tree breeding offers a realistic opportunity to trans–form profitability if superior stock is selected.

Independent statutory agencies formalize these grading rules and oversee quality control procedures. Depending on the country, visually and mechanically graded timber may be branded or grade stamped. The stamp usually includes details such as grade agency, grade, species, mill identification, or company number.

An advantage of machine stress grading is that fewer grades are required. All species can be graded to a limited number of stress settings. There are more than 80 different design values for 2 by 4 lumber of visually graded softwood in the Unites States, compared with some 20 machine-grades. Benefits are reduced complexity for the designer and far fewer inventory items for the fabricator.

5.2. Procedures to derive design properties for machine-graded structural timber

Because modulus of elasticity, E, is a less than perfect predictor of strength, lumber sorted solely by average E falls into one of four 'accept-reject' categories (Figure 10.16a):

- Category 1—Material that has been accepted correctly, i.e. pieces have sufficient strength and stiffness as defined
- Category 2—Material that has been accepted incorrectly, i.e. pieces do not have sufficient strength
- Category 3—Material that has been rejected correctly because it does not have sufficient strength; nor sufficient stiffness
- Category 4—Material that has been rejected correctly because it does not have sufficient stiffness

Only Category 2 presents a real problem. These pieces are accepted as having sufficient strength but in reality they do not.

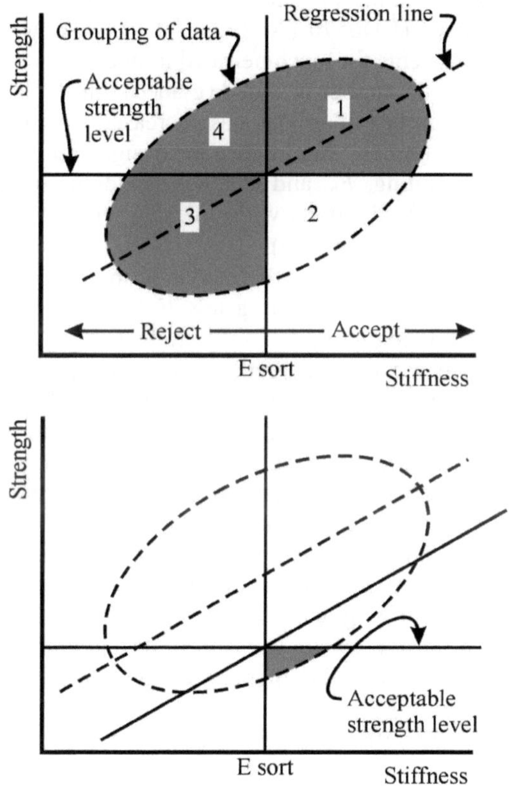

Figure 10.16. Schematic *E* sort. (a) A regression line used as the predictor for four categories: (1) accepted correctly; (2) accepted incorrectly; (3) rejected correctly; and (4) rejected correctly. (b) With the lower confidence line used as the predictor, then a relatively low proportion of material lies in the 'accepted incorrectly' category (lower right).

Engineers deal with this problem by minimizing the amount of material that falls into category 2. A lower confidence line, usually the lower 5th percentile exclusion limit, is used as the prediction model (Figure 10.16b). The number of pieces that fall into category 2 is now much lower compared with the mean regression line model. Furthermore, the probability of a piece (and thus the proportion of pieces) falling into category 2 is controlled by the confidence line selected.

Additional grading criteria (edge-knot limitations, for example) are also added to improve the efficiency of the sorting system.

5.3. Standards for machine-rated lumber

A number of standards outline the product requirements for machine-rated lumber. In Australia and New Zealand, the standard is AS/NZS 1748 (AS/NZS, 1997). In the United States, the standard is D6570 (ASTM, 2005k). In Europe, EN 519 (ECS, 1995) contains the requirements for machine-rated lumber.

In most MSR systems, the timber is sorted (graded) into E classes. In the United States and Canada, the number of grades has increased as specific market needs have developed for MSR timber. The grades are designated by the recommended extreme fibre stress in bending, F_b, and edgewise modulus of elasticity, E. Thus '2100F–1.8E' designates a MSR grade with a design stress F_b of 2,100 lb/in^2 (14 MPa) and E of 1.8×10^6 lb/in^2 (12.4 GPa).

In theory, any F–E combination can be marketed that can be supported by test data. In practice, a mill will produce only a few F–E classifications depending on the qualities of the timber being harvested, mill production capabilities, and product or market demand. Once the desired grades have been chosen, grade boundary machine settings are used to separate the timber into the F–E classifications. A qualification sample of timber is tested by a grading agency for strength and stiffness to verify that the proper machine settings are being used. After initial qualification, additional quality control tests are performed during production.

In Canada and the United States, the relationships between the 5th percentile 10-year bending stress and those in tension and compression are based on limited timber testing of the three properties but supported by years of successful experience in construction with visual stress grades of timber (ASTM, 2005j). For tension, it is assumed that the ratio of design tensile stress F_t to design bending stress F_b is 0.45 unless other test data are available; for compression the ratio is assumed to be

$$F_c = [0.338\,(2.1F_b) + 2060.7]/1.9 \qquad (9)$$

Strength in shear parallel to the grain and in compression perpendicular to the grain is poorly related to modulus of elasticity. Therefore, in machine stress grading these properties are assumed to be grade-independent and are assigned the same values as those for visual timber grades, except where otherwise predicted from specific gravity (basic density) on a mill-by-mill basis.

5.4. Quality control

Quality control procedures are necessary to ensure that stresses assigned by a machine-grading system reflect the actual properties of the timber graded. These procedures must check for correct machine operation. Two basic quality control systems are used; machine-controlled or output-controlled. In a machine-controlled system, which was mainly developed in Europe, each individual machine must be strictly assessed and controlled by a regulatory agency. The machine settings that control the operation of these machines are set by and are under the control of the third-party inspection agency. This system allows for the handling of a large number of sizes and species. In output control, the machine settings are controlled by feedback from daily testing of specimens. Statistical procedures are used to monitor and adjust the settings of the machines doing the grading. QC procedures are highly technical and local regulatory agencies are deeply involved.

5.5. Proof grading

Proof grading is practiced in Australia for situations where production is small. The multiplicity of timbers available encourages such a development. The procedure involves a quick visual segregation into two or more grades followed by passing each grade through a continuous proof testing machine that sorts out the exceptionally weak pieces at the tail of the grade distribution (Leicester, 1988). The proof load is set for each grade such that 1% to 3% of the pieces in that grade break. The unbroken pieces are deemed capable of sustaining over time a grade load that is equal to the short-term proof load divided by 2.1. Alternatively, sorting for grade could be based on stiffness, which would be preferable for high-strength tropical hardwoods where stiffness tends to control design. In theory any grading method can be used provided the presumed strength class is verified by proof testing.

Proof grading is also used in glued-laminated production plants to determine the quality of fingerjoints.

5.6. Non-destructive testing

It would be better if some wood quality assessment could be performed on sawlogs before they go into the sawmill. Significant savings can be made if poor quality structural material can be eliminated before the expense of sawing and drying occurs. Research in the area of non-destructive testing has been a most active area of wood research in the last two decades. Non-destructive testing techniques such as acoustic wave propagation, x-rays, lasers, microwaves, and near-infrared spectroscopy have been developed to the point where tools for early assessment of wood quality are available (Pellerin and Ross, 2002; Knowles *et al.*, 2004; So *et al.*, 2004). These techniques can determine or estimate the stiffness, density, or local slope of grain in logs and lumber prior to further processing. As mill operations become increasingly automated, non-destructive testing will become an integral part of the grading process.

6. ADJUSTING STRUCTURAL TIMBER PROPERTIES FOR DESIGN USE

The assigned values for grades of particular species may need further adjustments to take account of factors such as the effects of drying after grading, member size, duration of load, creep, chemical treatments, temperature and, in allowable stress design, an additional factor of safety. The principles underlying some of these adjustment factors are outlined briefly. Adjustment factors are primarily of interest to engineers. Specific adjustments are given in national standards and codes such as ASTM (2005c,d) designations D245 and D1990.

The strength of small clearwood specimens is noticeably improved on drying below fibre saturation point. For example, in D2555 (ASTM, 2005e) values for bending strength of Douglas fir at 12% MC are increased by 72%, stiffness by 22% and shear strength by 50%. However, when drying lumber its properties do not improve nearly as dramatically: presumable uneven, differential shrinkage around knots, for example, induces checking and distortion that weakens the drying wood. Thus for lumber, less than 102 mm (≤4 in.) only a small adjustment is permitted, whereas for thick lumber, greater than 102 mm (≥4 in.), green property values also apply to the dry condition – no improvement on drying is recognized. Green and Evans (1989, 2003b) and Kretschmann and Green (1996) have studied these effects.

An interesting feature of in-grade timber is that tensile strength is not changed significantly on drying, whereas compressive strength is (Figures 10.3 and 10.10).

In general, a size effect causes small members to have greater unit strength than that of large members. Brittle fracture theory (Bohannan, 1966; Madsen and Buchanan, 1986; Green and Evans, 2003a,b) interprets this in terms of the weakest link principle, which assumes that the member is as weak as its weakest part. It follows that the probability of finding a large strength-reducing defect within the member is greater the greater the volume of the member. As a result 100 x 50 mm members of a given grade are in general stronger than 200 x 50 mm members, which in turn are stronger than 300 x 50 mm members.

Strength properties are time dependent. The load that timber can sustain without failure decreases with time. If the short-term ultimate load in the 5-minute static bending test is taken as the reference point, then wood will fail, on average, at about 66% of that load after 1 year, at 62% after 10 years and at 56% after about 27 to 200 years depending on the curve used to fit the data: there is a dearth of long term experimental evidence (>10 yr).

Creep behaviour is particularly interesting. Under low stresses and for short loading periods deformation may be considered to be elastic. However, where a beam is held under constant load, over time, the deflection increases gradually although at a rate that diminishes (Figure 10.17a). When the load is removed there is

→

Figure 10.17. Time-dependent creep of timber. (a) The creep decelerates where the stress is less than 50% of the short-term failure stress but there is an inflection and accelerating creep where the stress is about 90% of short-term failure stress. (b) Estimated strength ratios for small clearwood specimens expressed as a ratio of the short-term strength, in the 5-minute bending test (Wood, 1951).

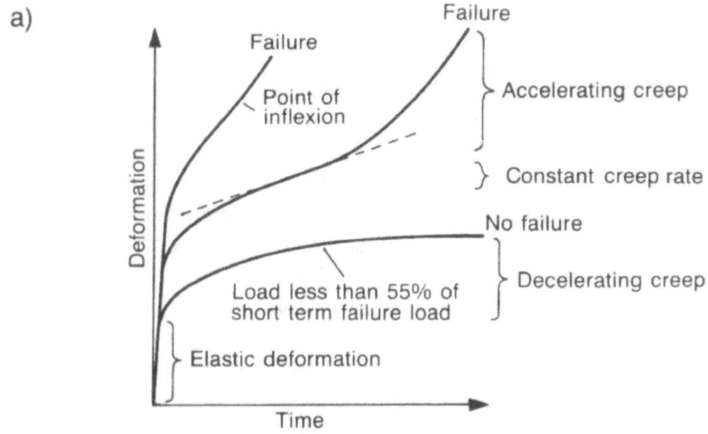
a)
Failure
Failure
Point of inflexion
Accelerating creep
Constant creep rate
No failure
Deformation
Decelerating creep
Load less than 55% of short term failure load
Elastic deformation
Time

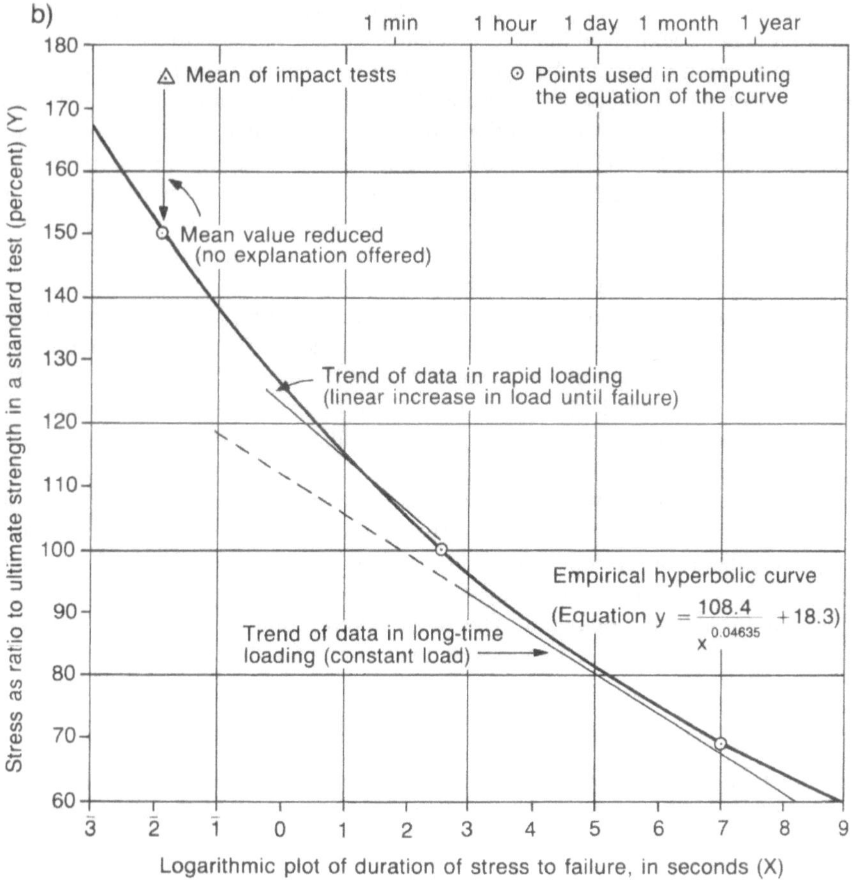
b)
1 min
1 hour
1 day
1 month
1 year
Mean of impact tests
Points used in computing the equation of the curve
Mean value reduced (no explanation offered)
Trend of data in rapid loading (linear increase in load until failure)
Empirical hyperbolic curve
(Equation y = 108.4 / x^0.04635 + 18.3)
Trend of data in long-time loading (constant load)
Stress as ratio to ultimate strength in a standard test (percent) (Y)
Logarithmic plot of duration of stress to failure, in seconds (X)

an immediate elastic recovery, a smaller delayed elastic recovery and an irrecoverable component. This time-dependent phenomenon is known as creep. Creep under a long-term constant load results in a long-term deflection that is roughly twice the initially observed elastic deflection. With higher stress levels, above about 55% of the short-term ultimate stress, the creep rate declines gradually before passing through a point of inflection and accelerating again (Figure 10.17a).

Evidence of creep is seen in the sag between supports along an old roofline. Musicians and archers only string their instruments tightly before use and release the tension immediately afterwards to prevent creep. Where green timber is used as the lintel, of a double garage door for example, it is advisable to provide some means of support, e.g. a strut, until the beam has partially dried: without initial support the deflection can easily increase to more than three times the initial short-term deflection. Even when dried, fluctuations in moisture content greatly accelerate creep: the long-term deflection can be as high as six times the short-term deflection. At the molecular level creep is the result of a rearrangement of molecules and bonds relative to one another. Such rearrangements occur more readily in the presence of water, so it is not unexpected that creep is much greater in green timber.

Certain treating processes have been shown to affect the final strength of wood (Winandy, 1995a,b,c). Incised and treated lumber has been shown to have a 5-10% reduction in E and a 15% reduction in strength. Reductions in energy-related properties are about 1.5 to 2 times those reported for static strength properties. There is no difference in long-term duration of load behaviour between treated and untreated material; however, current design standards in the U.S. prohibit increases in design stresses for short-term duration of load when considering impact-type loading for material treated with waterborne preservative.

Finally, as wood is cooled below normal temperatures, its properties increase. When heated, its properties decrease. The magnitude of the change depends upon moisture content. At temperatures up to 65°C (150°F) in the United States, the effect of temperature is assumed by design codes to be reversible. Prolonged exposure to elevated heat can lead to a significant permanent loss in strength (Green and Evans, 2003a, USDA, 1999).

In many design circumstances there are several loads on the structure, some acting simultaneously and each with a different duration. Intermittent loading causes cumulative effects on strength and is treated as a continuous load of equivalent duration. This 'cumulative' load may be more significant that a less severe constant load – and should be calculated. The most severe condition governs the design. Either the design stress or the total design load (but not both) can be adjusted.

7. GLUED STRUCTURAL MEMBERS

There are many types of glued structural members, with more products actively being developed. These use woody material that is reassembled with adhesives to make a wide range of structural products. The grading of glued structural members is either handled by a proprietary process internally by the various manufacturers or, with glulam, it is handled more like dimension lumber with third-party regulations.

Glued structural members are manufactured in a variety of configurations:

- Structural composite lumber (SCL) products consist of small pieces of wood glued together into sizes common for solid-sawn timber (Figure 10.18).
- Glued-laminated timber (glulam) is an engineered stress-rated product that consists of two or more layers of timber in which the grain of all layers is oriented parallel to the length of the timber.
- Glued structural members incorporate timber that is glued to panel products, to form box beams and I-beams, and structural sandwich construction.

7.1. Structural composite lumber (SCL)

Structural composite lumber was developed in response to increasing demand for high quality timber at a time when it was becoming difficult to obtain this from the forest resource. Structural composite lumber products are made from smaller pieces of wood glued together into sizes common for solid-sawn timber.

An example of SCL product is laminated veneer lumber (LVL), which is manufactured by laminating specially graded veneer with all plies parallel to the length. Another type of SCL product consists of strands of wood or strips of veneer glued together under high pressure and temperature. Depending upon the component material, this product is called laminated strand lumber (LSL), parallel strand lumber (PSL), or oriented strand lumber (OSL) (Figure 10.18). These types of structural composite lumber products can be manufactured from raw materials, such as aspen or any other under-utilized species. Different product widths can be ripped from SCL for various uses.

Figure 10.18. Three examples of structural composite lumber (top to bottom): laminated veneer lumber (LVL), parallel strand lumber (PSL), and oriented strand lumber (OSL).

Structural composite lumber is a growing segment of the engineered wood products industry. It is replacing timber in various applications and in the manufacture of other engineered wood products, such as prefabricated I-joists, that take advantage of engineering design values that are superior to those commonly assigned to sawn timber.

7.1.1. Laminated veneer lumber (LVL)

In early operations LVL was made in multidaylight presses using 3.2 to 2.5 mm (1/8 to 1/10 in.) veneer, which were hot pressed with phenol-formaldehyde adhesive into lengths from 2.4 to 18.3 m (8 to 60 ft) or more.

Commonly, LVL is produced in 0.6 to 1.2 m (2-4 ft) widths in a thickness of 38 mm (1.5 in.). New plants with continuous presses form a potentially endless sheet that is cut to the desired length. Various widths of product can be manufactured or recut in the retail facility.

Veneers are carefully selected to achieve the desired engineered properties. The visual plywood grading criteria of PS 1-95 (NIST, 1995) are inadequate of themselves and additional ultrasonic selection and sorting of veneer are needed to ensure that the finished product has the desired engineered properties. End joints between individual veneers may be staggered along the product to minimize their effect on strength. The end joints may be butt joints, or the veneer ends may overlap for some distance to provide load transfer. Some producers provide structural end-joints in the veneers using either scarf or fingerjoints. Laminated veneer lumber may also be made in 2.4-m (8-ft) lengths, having no end joints in the veneer; longer pieces are then formed by end jointing those pieces to create the desired length.

7.1.2. Parallel strand lumber (PSL)

Parallel strand lumber (PSL) is defined as a composite of wood strand elements with the wood elements primarily oriented along the length of the member: the average length of the strands must be a minimum of 150 times that of the least dimension.

Parallel strand lumber is manufactured from veneer about 3 mm (1/8 in.) thick, that is clipped into strands about 19 mm (3/4 in.) wide and 0.6 m (24 in.) long. The product was designed to use waste material from the roundup lathe as well as other less than full-width veneer arising from plywood manufacture (Chapter 11). Species commonly used for PSL include Douglas fir, southern pines, western hemlock and yellow poplar, but there is no restriction on the use of other species.

Strands are coated with a waterproof structural adhesive, e.g. phenol-resorcinol formaldehyde, and laid-up using special equipment to ensure proper orientation and distribution. The microwave continuous press both densifies the material and cures the adhesive. LVL is commonly produced in 0.28 by 0.48 m (11 by 19 in.) section – much thicker than that of LVL. Final product can be sawn into smaller dimension, if desired, while the length is limited only by freight and handling restrictions.

7.1.3. Laminated strand lumber and oriented strand lumber

Laminated strand lumber (LSL) and oriented strand lumber (OSL) products are an extension of the technology used to produce oriented strandboard (OSB) structural panels (Chapter 12). This product needs a greater degree of alignment of the strands than does OSB as well as higher pressure, which results in increased densification. Here the manufacturers control the grading process to produce products of certain load or span capacity. One type of LSL uses strands that are about 0.3 m (12 in.) long, which is somewhat longer than the strands commonly used for OSB. Waterproof adhesives are used in the manufacture of LSL. One type of product uses an isocyanate type of adhesive that is sprayed on the strands and cured by steam injection.

7.1.4. Advantages and uses

In contrast to sawn timber, strength-reducing characteristics of structural composite lumber are well dispersed within the veneer or strands and have much less effect on strength properties. Compared to solid the wood, these products have far smaller variations in property values, i.e. far smaller coefficients of variation (Figure 10.7 and Table 10.2). Thus high design values are assigned to the strength properties of LVL and PSL. Somewhat lower design values are assigned to LSL and OSL but they have the advantage of being produced from a cheaper raw material that need not be in a log size large enough to peel into veneer. All SCL products are made with structural adhesives and are dependent on a minimum level of bond strength. The final moisture content of SCL products is slightly lower than that of lumber for most service conditions and little change in moisture content will occur in many protected service conditions. When used indoors, the product is less likely to warp or shrink in service. However, the porous nature of both LVL and PSL means that these products need protection otherwise they can quickly absorb water.

All SCL products can substitute for sawn timber products in many applications. LVL is used extensively for scaffold planks and in the flanges of prefabricated I-joists, which takes advantage of its relatively high design properties. Both LVL and PSL beams are used as headers and major load-carrying elements in construction. LSL and OSL products are used for band/header joists in floor construction and as substitutes for studs and rafters in wall and roof construction. SCL products are used in non-structural applications, e.g. the manufacture of windows and doors.

7.1.5. Standards and specifications

While grading procedures by manufactures are proprietary, standards serve as guidelines for the minimum quality control of products produced. An example of such a standard is D5456 (ASTM, 2005j), which provides procedures to develop design properties for structural composite lumber products as well as requirements for quality assurance during production. Each manufacturer is responsible for gathering the required information on properties and ensuring that the minimum levels of quality are maintained during production. An independent inspection agency is required to monitor the quality assurance program. Unlike timber, no standard grades or design stresses have been established. The designer must consult the manufacturer regarding unique design properties and assembly procedures.

7.2. Prefabricated wood I-joists

Improved adhesives and manufacturing techniques have allowed prefabricated I-joists to replace larger timber sizes in floor and roof applications for both residential and commercial buildings (Figure 10.19).

Significant savings in materials are possible with prefabricated I-joists that use either plywood or OSB for the web material and small dimension timber or structural composite lumber for the flanges. High quality timber for the flanges is difficult to obtain from visual grading methods and, instead, several manufacturers use machine-graded timber or structural composite lumber. The details of fastening the flanges to the webs vary, but all flanges must be glued with a waterproof adhesive.

The ASTM standard D5055 (ASTM, 2005i) sets out procedures for establishing, monitoring, and re-evaluating structural capacities of prefabricated I-joists. However, each manufacturer is responsible for ensuring that the minimum levels of quality are maintained during production, while an independent inspection agency is required to monitor the quality assurance program.

These prefabricated I-joists are popular with builders because of their lightweight, dimensional stability, and ease of construction. Their accurate and consistent dimensions, as well as uniform depth, allow the rapid creation of a level floor. Utility lines pass easily through openings in the webs.

The prefabricated wood I-joists industry was one of the fastest growing segments of the wood products industry during the 1980s and 1990s. Today some 15 companies prefabricate I-joists in the United States and Canada. Often product is distributed through building material suppliers. Each manufacturer has its own catalogues with span tables, construction code and design information.

Recently, the APA–EWA (2004a) has issued a performance standard for prefabricated I-joists covering products used in residential floor construction.

7.3. Glued-laminated timber (glulam)

Structural glued-laminated timber (glulam) is a long-established stress-rated product. Typically, individual laminations are made from 38mm (nominal 2 in.) thick timber where used for straight or slightly cambered members, and thinner 19 mm (nominal 1 in.) thick timber where used for curved members. Individual laminations may be joined end-to-end to produce pieces that are much longer than the laminating stock itself. Pieces may also be placed or glued edge-to-edge to make a member that is wider than the input stock. Straight members up to 42 m (140 ft) long and more than 2.1 m (7 ft) deep have been manufactured: the only limitations are handling and transport. Curved members have been used in domed structures spanning more than 152 m (500 ft).

Timber of different grades can be arranged within a laminated cross-section depending on the anticipated loading, whether loaded parallel or perpendicular to the wide faces of the laminations. For loads applied perpendicular to the wide faces of the laminations, referred to as 'horizontally laminated', cross-sections are typically designed with higher-grade laminations in the outer top and bottom layers to carry the highest stresses and lower-grade laminations in the core layers where stresses are minimal. For loads applied parallel to the wide faces of the laminations, 'vertically laminated', all laminations are subjected to the same load, whether in compression, tension, or bending so placement of grades is less critical. Frequently glulam cross-sections have the same timber grade throughout the cross-section.

7.3.1. Manufacture of glulam

In the United States glulam is manufactured according to the national standard, ANSI/AITC A190.1 (ANSI/AITC, 2002). This standard contains requirements for production, testing, and certification. Plants meeting these requirements can place their product quality mark on the glulam, which contains key information regarding the type, species, and design values.

There are four steps in the manufacturing process: (i) drying and grading the timber, (ii) end jointing, (iii) face bonding, and (iv) finishing and fabrication.

Figure 10.19. Prefabricated I-joists with laminated veneer lumber flanges and structural panel webs. (a) With an experimental hardboard web. (b) With commercially produced oriented strandboard web. (c) With a plywood web.

Glulam is generally manufactured at moisture contents below 16%, with a maximum range amongst laminations of 5% MC. This minimizes dimensional changes following manufacture. End sealers, surface sealers, primer coats, and wrapping may be applied during manufacture to minimize changes in moisture content. Required protection will depend upon the final use and finish.

Grading standards of the regional timber grading associations describe those characteristics that are permitted in the various grades of timber, while manufacturing standards such as AITC 117 (AITC, 2004) describe the combination of timber grades necessary for specific design values. Both visual graded and *E*-rated lumber is permitted. For example for visually graded lumber there are three lamination grades, L_1, L_2 and L_3, which respectively limit knot size to 1/4, 1/3 and 1/2 of the section width. For *E*-rated timber various *E*-levels are permitted with a visual override, e.g. limiting the size of edge knots when grading as a plank. Where high design stresses are sought for some laminations, such as may be required for the outer 5% on the tension side of the beam, the grading criteria for these 'tension laminations' are given in AITC 117 (AITC, 2004).

7.3.2. End jointing

End jointing is necessary to manufacture long-length glulam. The most common end joint, a structural fingerjoint, is about 28 mm (1.11 in.) long (Figure 10.20). A key advantage of this process is that fingerjointing blanks can be cut from knotty stock provided the blanks have clearwood with straight grain in the immediate vicinity of the finger: in this way certain lower grades of lumber can be upgraded to lamination stock. Continuous production equipment with radio-frequency curing under end-pressure is standard for fingerjointing. Careful control of the process – determining timber quality, cutting the fingers, applying the adhesive, mating, applying end-pressure, and curing – is needed to produce consistent high strength joints.

Fingerjointed pieces can achieve 75% of the strength of clearwood. These joints are more than adequate for most applications as most timber grades used in the manufacture of glulam permit natural characteristics that result in reductions in strength of at least 25% compared to that of clearwood.

A continuing challenge in glulam production is to eliminate the occurrence of an occasional low-strength end-joint. Procedures regarding fingerjointing and glulam manufacture are covered in ANSI/AITC A190.1 (ANSI/AITC, 2002).

It should be noted as an aside, that there is a significant market for non-structural fingerjoints, only 3-8 mm long. Here the ideal joint has fingers that match perfectly with no visible gaps and so can be painted without visible blemishes. Their strength may be only 30-40% of that of clearwood, but they are not for structural use. Under certain circumstances the lowest, knottiest grades of timber can be cross-cut profitably to give some short-length blanks, as little as 150-200 mm long, and with a recovery of only 30-50%.

Figure 10.20. Typical fingerjoint used in glulam manufacture. The pin-holes at the tips of the fingers are typical of a structural fingerjoint.

7.3.3. Assembly, finishing and fabrication

Best practice is to dress the two wide faces of the laminations just prior to laying-up on the flat bed. This means that the beam will be rectangular and when glued and assembled the pressure will be applied evenly. Phenol resorcinol is most commonly used as adhesive. After the recommended open assembly time, the side clamping pressure is applied either hydraulically or mechanically and the adhesive is left to cure for some 6-24 hours, depending on the resin formulation and room temperature (Figure 10.21). Quality control includes shear tests on end trim samples, refer ANSI/ AITC A190.1 (ANSI/AITC, 2002). Test values should equal about 90% of the clearwood shear strength for the species. A very modest increase in glue bond shear strength (10%) occurs in the following few days.

The wide faces are planed to remove the adhesive that has squeezed out between adjacent laminations and to smooth out any small irregularities between the edges of adjacent laminations. Some additional finishing may be required following the industry standard, refer AITC 110 (AITC, 2001). Industrial appearance is applicable where appearance is not a primary concern, as in industrial plants and warehouses. Architectural appearance applies in most cases where appearance is important. Premium appearance is the highest classification.

Fabrication involves final cutting: holes are drilled, connectors added, and a finish or sealer is applied, if specified. Different degrees of prefabrication are undertaken at this point. Trusses may be partially or fully assembled. Moment splices can be fully fabricated, then disconnected for transportation and erection. Special precautions are necessary during handling, storage, and erection to prevent structural damage to glulam members, refer AITC 111 (AITC, 2005) or EWS R540 (APA–EWA, 2002).

Figure 10.21. After being assembled in the clamping bed, the laminations of this Tudor arch are forced together and into shape with a manual air-driven screw clamp.

Various countries have standard practices established to determine allowable properties for glued-laminated timber. In the United States, ASTM D3737 (ASTM, 2005g) guides the allowable stress development.

7.3.5. Advantages of glulam

Compared to sawn timber as well as other structural materials, glulam has several benefits to offer:

- Size. By laminating small-dimension timber in glulam, the production of very large structural elements is possible. Straight members up to 30 m (100 ft) long are common, with some spans up to 43 m (140 ft). Sections deeper than 2 m (7 ft) have been used. Thus, glulam produces large members from small trees.
- Architecture. By curving the timber during manufacture, a variety of architectural effects can be obtained from glulam that are impossible or difficult with other materials. The degree of curvature is determined by the thickness of the laminations. Beams with curvature are generally made with 19 mm (nominal 1 in.) timber, while 13 mm (1/2-in.) or thinner material may be required for very sharp curves.
- Large dry dimensions. Timber used in glulam must be seasoned or dried prior to use, so the effects of checking and other drying defects are minimized. Design

on the basis of seasoned wood permits greater design values than can be assigned to unseasoned large-dimension lumber.

• Varying cross-sections. Structural elements can be designed with varying cross-sections along their length as determined by strength and stiffness requirements. For example, the central section of beams can be made deeper to account for increased structural requirements in regions carrying higher moments. Similarly, arches often have varying cross-sections as determined by design requirements. Such beams offer great aesthetics.

• Mixing grades. A major advantage of glulam is that a large quantity of lower grade timber can be used within the less highly stressed laminations of the beams, placing the highest grades in the highly stressed laminations near the top and bottom and a lower grade around the neutral axis. Species can also be mixed to match the structural requirements of the laminations.

• Environmental consciousness. It is a modern material that emphasizes sustainability, its low energy demand during manufacture and its carbon storage capacity; and eventually it can be recycled. While aesthetic and economic considerations are major factors influencing material selection, these environmental advantages increasingly influence material selection.

8. FIRE

In a fire, the main hazard to life comes from the room contents. Wood smoke itself is relatively innocuous. Death arises most often from the choking toxic smoke and asphyxiation arising from burning plastics and synthetic fabrics. Regulatory authorities seek to eliminate such hazardous materials and to restrict the speed with which a flame spreads along a surface. Structural design aims to isolate areas and to ensure that fire is not able to spread from room to room. Thus, doors and other openings should seal properly, allowing no gaps through which fire might penetrate. Good design is an integral part of fire safety. An intumescent coating can seal small gaps and retard the spread of flame. Such coating acts by foaming when heated and insulates the substrate.

The fire rating of buildings relates to the nature of the building – home, school, hospital, or chemical warehouse. The greater the hazard to human life, the greater the required fire rating for doors, floors, walls, and ceilings. The fire rating refers to the time that the element must resist the fire without letting the flames spread further. Fire ratings range from 20 min to 3 h.

Fire damage is not directly related to the combustibility of the building materials but rather to the details of construction. Wood structures contribute very little to the fuel load in a building because large dimension material is consumed slowly and small dimension material is generally protected from fire by non-combustible cladding such as gypsum plasterboard. In an intense fire, structural failure is more likely where steel trusses or beams are exposed to the heat. With timber, the member can only char at a totally predicable rate of about 40 mm/h (1.5 in./h), so by modifying section dimensions of the timber a structure can be reliably designed to withstand a fire for as long as required. However, the metal connectors for timber require insulative coverings. This predictability of performance for large timber or glulam beams is a highly desirable characteristic in a major fire, allowing a fire crew to enter the building confident that sudden failure is unlikely. Further, because the centre of the section is well insulated from the heat it will not lose strength or shrink along the grain, whereas a steel beam will expand along its length on heating and can push the walls of a building outwards, resulting in collapse even though the heat is not intense enough for the beam to buckle.

9. TIMBER STRUCTURES

The discussion of this chapter has centred on the properties of individual pieces of timber. In many situations, the load in a structure is shared between a number of pieces, and the inability of a weak or less stiff member to carry its share of the load is offset by the carrying capacity of other members. Thus, some variation in mechanical properties can be tolerated. In only a few situations is load sharing not possible, e.g. in door lintels, trusses, and beams. It is in these situations that the benefits of machine stress grading and proof testing are greatest. Furthermore, in timber design the emphasis is on the efficiency and the effectiveness of structures rather than the individual parts. More profit and potential lie in the sale of prefabricated components and structures, and continued development in these areas is vital if timber is to remain competitive. Considerable attention is devoted to achieving efficient connections between individual pieces to build safe, stiff, and efficient structures. Timber connections determine the safety of structures in earthquakes, fires, hurricanes, and other natural disasters. The timber is only a part of the structure. Timber design and engineering are subjects in which few foresters have even a partial familiarity and little expertise. They are important subjects in maintaining a market place for timber and in keeping timber at the cutting edge of technological development.

CHAPTER 11

WOOD-BASED COMPOSITES: PLYWOOD AND VENEER-BASED PRODUCTS

SHELDON SHI[1] AND JOHN WALKER

[1]*Department of Forest Products, Mississippi State University, Mississippi, USA*

1. INTRODUCTION

A wood-based composite can be defined as a composite material mainly composed of wood elements. These wood elements are usually bonded together by a thermosetting adhesive (wood truss products could also be regarded as wood-based composites, but connected by metal connectors). The commonly used adhesives include urea-based adhesive (such as urea formaldehyde resin), phenolic-based adhesive (including phenol resorcinol adhesives), isocyanate-based adhesive, and adhesives from renewable resources (like soybean, lignin *etc*). The wood elements in wood composites can be in many different forms such as:

- Dimension lumber – for laminated glued timber (Glulam) and wood trusses;
- Veneers – for plywood, laminated veneer lumber (LVL), and parallel strand lumber (PSL);
- Fibres – for medium density fibreboard (MDF), high density fibreboard (hardboard), and other fibre-based products;
- Particles – for particleboard;
- Flakes or strands – for flakeboard, oriented strand board (OSB), oriented strand lumber (OSL), and laminated strand lumber (LSL); and
- Scrims – for scrim-based products, as in Scrimber.

Traditional composite panels are made from veneers and from mat-formed composites bonded by adhesive. More recently wood has also been combined (compression moulded or extruded) with synthetic polymers, e.g. thermoplastic polymers, to make wood-polymer composites (WPC). WPC products have been growing very rapidly in the recent years, especially in the decking market, where Wolcott (2004) observed that their market share has grown from 2% in 1997 to 14% in 2003. Further, much research work has explored the use of fibre-reinforced polymers (FRP) to enhance the structural performance of engineered wood composites, called FRP-wood hybrid composites (Dagher *et al*., 1998; Shi, 2002).

Some engineered wood composite products are made from combinations of other wood composites, such as wooden I-joists that can have flanges of sawn lumber, LVL, or other structural composite lumbers together with webs of OSB or plywood.

Table 11.1. Market demand (in Mm^3) of lumber and some major wood-based composites in the United States in 2004 (Adair and Camp, 2003; Adair, 2004).

	Residential [b]	Non-residential	Industrial [c]	Total
Lumber	99.58	5.57	29.99	135.14
Plywood	7.36	1.35	4.81	13.52
OSB	18.13	1.33	0.82	20.28
Glulam	0.51	0.21	0.04	0.76
LVL	1.93	NA	NA	1.93
I-joist [a]	269.93	22.88	NA	292.80

a) I-joist volume in million lineal feet. b) Includes remodelling. c) Furniture, pallets, transportation.

Wood-based composites fall into two categories from the end application standpoint: panel applications, such as plywood, OSB, particleboard and fibreboard; and beam or header applications, such as glulam, LVL, OSL, PSL and scrim-based lumber. Panel applications are mainly for sheathing and flooring in residential housing and other industrial applications. Beam and header applications are mainly for load-carrying members in the residential and commercial buildings, such as garage door headers, floor joists *etc.*

2. TRENDS

The long-term trend in wood use reflects social and economic development and the changes in resources. The diameters of the trees are getting smaller while the cost of labour has increased steadily, forcing industry to use it efficiently so that labour productivity keeps ahead of labour costs. Those production processes that break wood down into small pieces or fibres are most adaptable to continuous flow, to automation, to standardization of product and to large-scale operations. Such products are technologically progressive and can be manufactured while showing respect for the growing costliness of human effort. On the other hand, those products in which wood is kept more nearly in its original state and in which pieces are handled individually tend to be technologically backward and will decline in importance. Thus there is a natural progression from solid wood, through plywood to strand-based-based composites, particleboard, fibreboard and paper.

Wood-based composites are widely used in industrial applications (furniture, pallets, packaging materials and concrete formwork), and other outdoor applications, such as bridges. However, the biggest market for wood-based composites is with residential and commercial building applications. In the United States, about 95% of the residential housing is built with wood-based materials. As trees got smaller and new technologies developed for wood-based composites, so sawn lumber beams or joists in housing construction have been replaced gradually by engineered wood products (EWPs), such as glulam, LVL, I-joists, PSL *etc.* Solid wood floors, wall diaphragms, panelling and ceiling lining have instead being sheathed with structural composite panels such as plywood and OSB. Table 11.1 summarizes the demand for some major wood-based composites and lumber in the United States in 2004.

In North America, all plywood manufactures used to follow the prescriptive standard PS1 (APA, 1995). However, since the introduction of performance-based standards such as PS2 (NIST, 2004) and PRP-108 (APA, 2001) in 1990s other structural panel products, such as OSB, can and have been used interchangeably in structural panel applications.

Figure 11.1 shows structural panel production (plywood and OSB) from 1970 to 2004, and some major engineered wood products (glulam, I-joists, and LVL) from 1980 to 2004 in North America. Structural panel production was under 15 million m^3 in 1970, but has increased to 37 million m^3 by 2004. Although OSB only became significant in the structural panel market in the 1980s, it overtook plywood by the late 1990s because of its lower production costs. In turn plywood has sought more diversified industrial applications, such as the furniture and transportation markets. Veneer-based composites have penetrated other areas, such as in LVL applications where production has grown strongly in the past ten years.

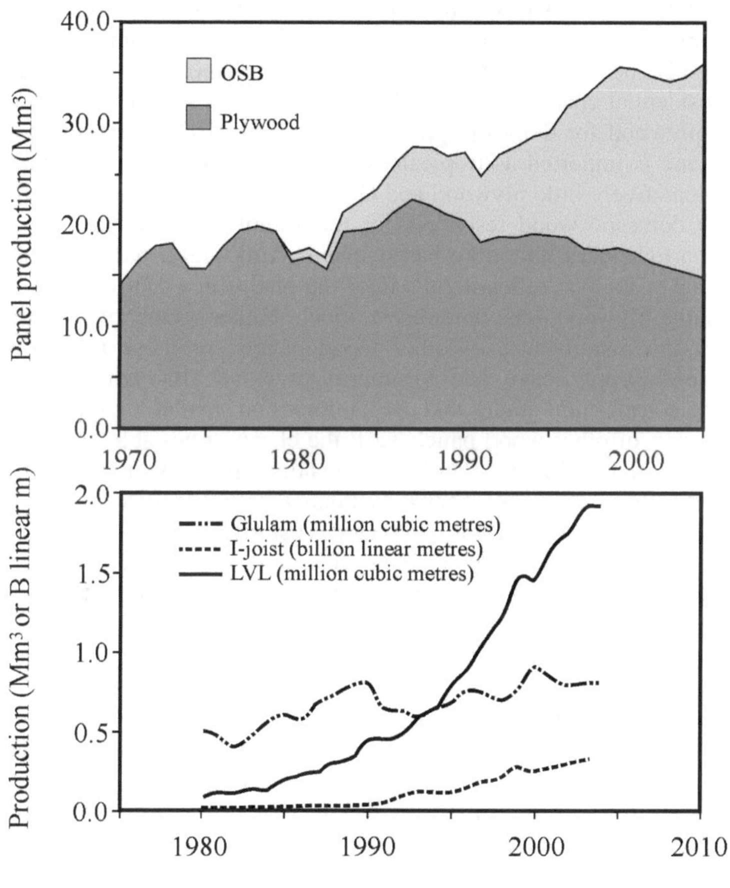

Figure 11.1. Production of structural panels and major engineered wood products in North America (Adair and Camp, 2003; Adair, 2004).

Table 11.2. Production (in Mm^3) of selected wood-based composites data in 2003 (FAO yearbook: forest products, 2005).

	Sawn wood[a]	Veneer sheets	Plywood[b]	Particleboard[c]	Fibre-based board[d]
Africa	7.7	0.88	0.69	0.47	0.23
N. & C. America	152.1	1.66	17.4	30.9	8.7
S. America	34.0	0.83	3.7	2.9	2.3
Asia	67.6	5.44	39.7	11.7	16.4
Europe	132.1	1.78	6.3	42.6	14.9
Oceania	8.6	0.72	0.58	1.2	1.7
World	402.0	11.31	68.4	89.7	44.1

a) All softwood and hardwood. b) Structural and decorative plywoods. c) Includes OSB but not those with inorganic binders. d) Insulation board, medium density fiberboard and hardboard.

Table 11.2 shows 2003 production of both sawn timber and wood panels for various regions of the world. Structural panel production is dominated by North America, mainly because structural panel products have enjoyed a dominant position in residential construction. The United States is a substantial manufacturer of softwood plywood for domestic production but only 10% is exported; whereas half that volume is imported as tropical hardwood plywood. Europe manufactures and uses comparatively little plywood and OSB. Instead the region relies on its own lower quality domestic wood resources for the manufacture of other wood-based panels, e.g. particleboard and fibre-based board. Amazingly, panel production in Asia equals that of lumber, reflecting the inter-regional China-centric supply chain.

Traditionally plywood has required a much higher grade of log than was necessary for the manufacture of other wood panels, so those nations with an unsuitable wood supply have had to import plywood although now there are manufacturing construction grades that use a poorer and smaller log type.

Manufacturers of other wood panels seek the cheapest possible wood. They are able to utilize lower quality logs and wood residues from other wood processing industries and still produce homogeneous boards with adequate mechanical and physical properties. Typically, the delivered cost of sawlogs, peeler logs and chipwood account for around 80-60%, 60-40% and 40-15% respectively of the production costs of lumber, plywood and boards made from comminuted wood.

3. PLYWOOD

Thousands of years ago Chinese and Egyptians shaved wood and glued it together to achieve special effects with veneered surfaces. In the 17th and 18th centuries, the English and French progressed the general principle of plywood, to where one or two veneers were overlaid on a plain, stable plank – or on narrow alternate wood strips jointed side-by-side to counteract wood's natural tendency to warp: the finest items would be counter-veneered and might be steam bent. However, Czarist Russia is credited for first making a form of plywood prior to the 20th century.

Typically early modern-era plywood was made from decorative hardwoods and was most commonly used in the manufacture of household items such as cabinets, chests, desktops and doors. Construction plywood made from softwood species did not appear in the market until the 20th century, although the first patent for plywood was issued in 1865, to John K. Mayo of New York City. The plywood industry really started in 1905 in the city of Portland, Oregon, USA.

Plywood is manufactured from sheets of cross-laminated veneers or plies, arranged in layers, and bonded with adhesives. Usually, the structure must be symmetric about the mid-point. Therefore, plywood has an odd number of layers, in which each layer may consist of one or more plies. Plywood construction is described by the number of plies and layers (i.e. 3-ply/3-layer). In the particular case of 4-ply/3-layer, this is manufactured using 4 plies of veneer sheets with the two inner veneers both orientated perpendicular to the face and back veneers (to maintain symmetry about the mid-point or neutral axis). Because of the way plywood is laid up, movement within the plane of the board is minimal because the wood grain lies at right angles to each other in alternate plies: the axial alignment of the grain in one sheet of veneer restrains the tangential movement in adjacent veneers. The resulting panel has similar shrinkage and strength properties in these two directions and thus the large dimensional changes and low strength values that occur across the grain in solid wood are eliminated. However, restraining the wood in the plane of the board results in greater than normal movement in the thickness of board. Further desirable features of plywood are its resistance to splitting, its availability in sheet form, and its ability to withstand large racking forces imposed on structures, for example by an earthquake.

Softwood plywood production in North America had grown to 345 000 m^3 by 1933 with major applications in industrial markets such as door panels, cabinets, trunks, and drawer bottoms. Later, in the 1940s and 1950s, plywood was promoted for residential construction and by 1960 softwood plywood production had reached 8 million m^3 of which nearly 50% was used as sheathing for residential construction. Plywood was also used for sub-flooring, siding, soffits, and stair treads and risers. The repair and remodelling and the non-residential building markets were also growing. By 1980, North America plywood production had reached 16 Mm^3. Then, OSB technology was introduced, and OSB has largely displaced plywood as structural sheathing in housing construction. Subsequently, plywood production has been static: in 1999, the production of OSB at 18 Mm^3 overtook plywood production at 17 Mm^3. Currently in 2005, the residential construction market accounts for only one-third of plywood market demand in the U.S. Instead plywood has gradually secured additional industrial markets, such as furniture, pallets, and others.

Over 70 wood species are used to manufacture plywood (APA, 1995). These species are divided into five groups on the basis of strength and stiffness. Strongest species are in Group 1, while weakest species are in Group 5. Veneer grades (A, B, C, $C_{plugged}$ or C_p, D) define veneer appearance in terms of natural features and the allowable number and size of repairs that may be made during manufacture (Table 11.3). A represents the highest grade and D the lowest. The grades of the face and back veneers in a sheet of plywood define the grade of the plywood (such as A-A, A-B, A-D, B-B, B-D, C-C_p, C-D, *etc.*). Grade A face veneer is necessary if a paintable surface is required, while B grade offers a solid face suitable for overlaying. The minimum grade of veneer for exterior applications is C-grade, while

D-grade veneer is used in plywood intended for interior applications. Lower grades may be permitted on the reverse face of a panel, e.g. C-D. The plywood will indicate whether it has been manufactured with an exterior or interior rated adhesive. Plywood can be used as rough sawn, unsanded, touch sanded, sanded, and overlaid. Plywood panels with rough sawn surfaces are used only for decorative purpose such as siding applications. Panels are unsanded if a smooth surface is not required, for subfloor, roof, and wall applications. Single floor and underlayment may require only touch sanded board for sizing to make the panel thickness more uniform. Plywood panels with B-grade or better veneer faces are always sanded smooth in manufacture since the intended end application is for cabinets, shelving, and furniture. There are two types of overlays: high density overlay (HDO) and medium density overlay (MDO). The overlays can be applied to the plywood at the same time as the panel is pressed, in one-step, or after the panel is pressed, in two-steps. The two-step process usually requires the panels to be sanded before overlaying.

Table 11.3. Veneer grades for plywood (APA, 1995).

A	Smooth, paintable. Not more than 18 neatly made repairs, boat, sled, or router type, and parallel to grain, permitted. Wood or synthetic repairs permitted. May be used for natural finish in less demanding applications.
B	Solid surface. Shims, sled or router repairs, and tight knots to 25.4 mm across grain permitted. Wood or synthetic repairs permitted. Some minor splits permitted.
C Plugged	Improved C veneer with splits limited to 3.2 mm width and knotholes or other open defects limited to 6.4 x 12.7 mm. Wood or synthetic repairs permitted. Admits some broken grain.
C	Tight knots to 38.1 mm. Knotholes to 25.4 mm across grain and some to 38.1 mm if total width of knots and knotholes is within specified limits. Wood or synthetic repairs permitted. Discoloration and sanding defects that do not impair strength permitted. Limited splits allowed. Stitching permitted.
D	Knots and knotholes to 63.5 mm width across grain and 12.7 mm larger within specified limits. Limited splits are permitted. Stitching permitted. Limited to Exposure 1 or Interior panels.

4. RAW MATERIAL REQUIREMENTS

The desired characteristics of the species for plywood include density, colour, ease of peeling or slicing, drying without wrinkling, bondability *etc*. However, only a few species have gained general acceptance as a sufficient volume of logs must be available on the international market on a continuing basis and these must be of sufficient size and adequate form.

Early mills in the Pacific Northwest of the United States made plywood from virtually flawless, old-growth, large diameter logs (>1.5 m) of Douglas fir. In the mid-1960s Douglas fir accounted for 90% of North American plywood production, falling to 55% ten years later. The declining availability of large, high quality

Douglas fir veneer logs has brought about a profound change in the United States plywood market. The main development has been in the construction and industrial (C and I) market, for sheathing, floor underlayment (with carpet, vinyl, or hardwood floor laid on top) and for containers. A key feature is the emphasis on physical and mechanical properties rather than on visual characteristics. Thus in the southern United States a new industry emerged in the 1960s producing relatively cheap 5- and 9-ply boards from the southern pines with C and C_p face veneers. These panels differ from those produced earlier in that knots as large as 75 mm in diameter and splits 25 mm wide are permitted and raw material requirements have shifted from traditional peeler grade logs to first and second grade sawlogs (Lutz, 1971). Such trends forced foresters to reassess their ideas on softwood plantation management. The largest, high quality logs used to be considered potential veneer logs and in a managed plantation these could only come from the final clearfell and even then only at the end of long rotations. Today, in the southern United States some of the first pine thinnings at age 12 are used for veneer. Log size is no longer of overwhelming importance. Here the better logs may go to the sawmill rather than the plywood plant. By the mid-1970s 30% of United States plywood was southern pine, rising to 50% by 1983. Elsewhere in North America the main plywood species are mostly Douglas fir and larch in Inland United States, mainly hemlock and Douglas fir on the West Coast, and in Canada mainly Douglas fir and spruce.

Commodity southern pine plywood requires three hours/m^3 and waferboard/OSB only one hr/m^3 (Spelter, 1988). By contrast in Finland the emphasis is on adding value to a basic commodity. Slow growing birch logs averaging only 200-250 mm in diameter are peeled down to a core diameter of 60-65 mm. Only efficient operations can hope to be profitable when peeling such small logs. The increased costs due to the lower yield (36%) and modest log quality are compounded by the fact that the lower yield also reduces output. The predominant use of a single species (pure birch or birch-faced plywood) and veneer thickness (1.5 mm) helps automation – although 35% of the raw material is 2.5-3.2 mm spruce veneer (Höglund, 1980). The veneer made from small diameter logs necessitates much repair work, e.g. patching and jointing represents about a third of the work input (Höglund, 1980). To add value two-thirds of plywood is processed: by scarf-jointing (with a 10-30% local loss of strength depending on board thickness 24-6.4 mm) into giant panels (up to 2.75 x 12.5 m); by preservative treatment; by overlying (for appearance or multi-pour concrete formwork – reused 10-80 times) or by adding a thick textured phenol-resin coat (2 mm; 200 g/m^2) providing a non-slip pattern for flooring, for use in van/trailer floors, warehouses, scaffolding and staging; or grooved ready-to-install wall panels.

Europe has over 40% of the 10 billion m^2 global market for panel surfacing materials, with market share for low pressure melamine (51%), veneer (18%), paper foils (13%), high pressure laminates and paint (7% each) *etc.* (O'Carroll, 2001). The prime substrate is particleboard and MDF, except for the uses discussed above, e.g. formwork and floating floors where plywood excels. The laminating paper is typically an absorbent kraft sheet that is saturated with resin.

Plywood is not a homogeneous commodity product (Todd, 1982). North America manufactures predominantly softwood plywood. Asian production is tropical hardwood plywood, while European production is a mix of softwood and temperate and tropical hardwood plywood. Currently, a 'combi' panel that combines both hardwood and softwood species is common. For example, many manufacturers in Europe use birch faces and spruce inners, while those in China are likely to use Russian larch faces and poplar inners. Some plywood mills in the United States are using radiata pine (from Australia) face and Douglas fir or hemlock back veneer: however, radiata pine veneer tends to show more linear expansion than the Douglas fir veneer because the wood's higher microfibril angle and grain irregularity and so within-plane warping is more problematic. Other mills use high density eucalypts for face veneer.

Hardwood plywood is sold in both decorative (thin boards, <6 mm) and construction (thick panels, >6mm) markets. Temperate hardwoods are used primarily for decorative purposes, although Finnish birch is an exception being used in specialized high value construction applications. Thin boards are manufactured from tropical hardwoods and are used for decorative or platform uses. Decorative uses include wall panelling and door faces. As a platform the thin board receives a decorative surface that is either printed or overlaid on the panel surface, at which point it is known as prefinished (ready to use), and these are the major items traded internationally. Thin tropical boards are manufactured with water resistant, interior grade adhesives, whereas the majority of other boards use phenolic-based resin that can be used in exterior situations.

5. PLYWOOD MANUFACTURE (BALDWIN, 1981; SELLERS, 1985)

Plywood production can typically be divided into three manufacturing stages (Figure 11.2): veneer manufacture; clipping, drying and up-grading; and panel layup, pressing and finishing.

Construction plywood panels (1.2 x 2.4 m) are made from rotary peeled softwood veneers of 2-6 mm thicknesses in grades generally admitting large defects. A typical mill would process 100 000 m^3 per year.

5.1. Veneer manufacture

5.1.1. Principles of rotary veneer cutting (Koch, 1964; Lutz, 1974)

The process of rotary veneer cutting is essentially to cut perpendicular to the grain with the knife lying parallel to the grain. The bolt is centred between two chucks on a lathe and then turned against the knife that extends the full length of the bolt. As the bolt turns, a thin sheet of veneer is peeled off through the gap between the nosebar and the face of the knife as a long continuous ribbon. The quality of the veneer is determined to a considerable extent by the precise set up of the lathe (Figure 11.3). It is important that the veneer does not break and that it should have a smooth finish. Uniform thickness is a sign of good control at the lathe.

As the knife cuts the wood, the veneer is bent or rotated like a cantilevered beam and is liable to break (Figure 11.3a). An obvious way to reduce the bending moment

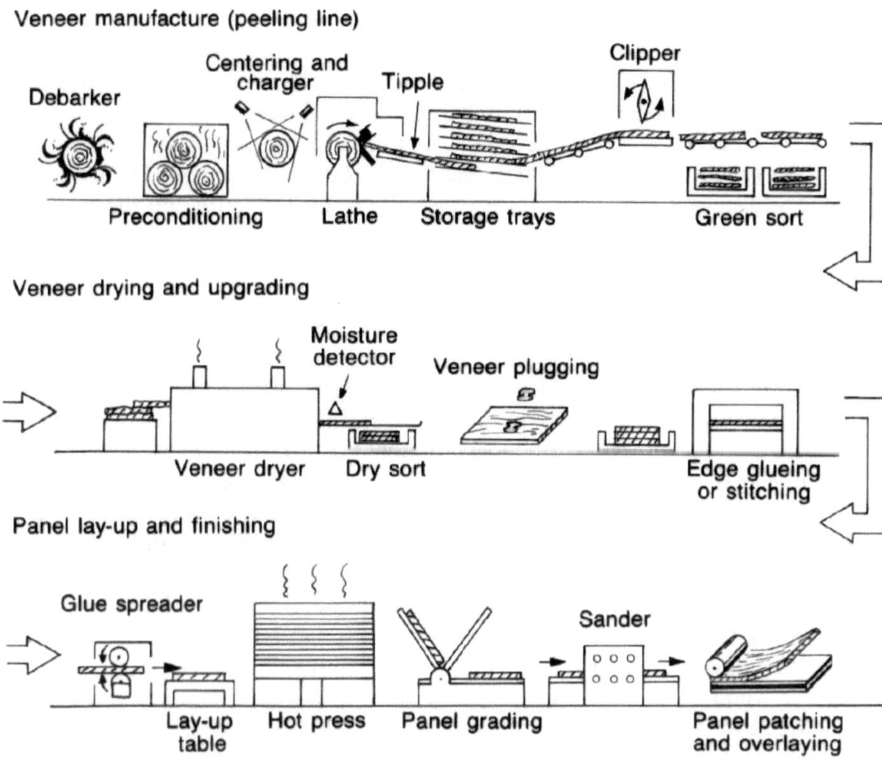

Figure 11.2. An outline of the principal features of plywood production.

is to increase the rake angle, α, and to keep to a minimum both the sharpness angle of the knife, β, and the clearance angle, γ (Figure 11.3b). Unfortunately it is not practical to use a knife with too fine an edge as this dulls rapidly. A microbevel (Figure 11.3d) on the knife gives a more durable cutting edge. Provided the joint or heel is not too large (<2.5 mm), a negative clearance angle is possible as only the small heel of the knife will press into the bolt. A nosebar is essential. Without a nosebar the veneer can split away from the bolt ahead of the knife and the surface of the veneer would be very rough. The nosebar compresses the wood perpendicular to the grain so that the veneer is cut at the knife and the knife edge itself defines the surface of the veneer. The nosebar pressure is achieved by reducing the gap between the nosebar and the knife so that it is less than the thickness of the veneer being cut. Adjusting the position of the nosebar does not affect the nominal thickness of the veneer but it does influence its quality. The nominal thickness of the veneer is determined by the rate of advance of the knife per revolution of the bolt, e.g. if 3 mm veneer is being cut the knife advances 3 mm per revolution. Typically the nosebar compression is between 10 and 20% so that the gap through which the veneer must escape is 90-80% of the nominal thickness of the veneer. If the nosebar opening is too large the veneer will be compressed insufficiently and will be loose and of uneven thickness. If the nosebar opening is too small the veneer will be compressed beyond the elastic limit for the wood and it will be very tight and over compressed: it will not recover to its nominal thickness. Further, the power required to peel a bolt increases steeply as the nosebar pressure is increased.

Veneer is characterized by the presence of small checks, called lathe checks, on the side of the veneer that was originally nearest the centre of the bolt. Lathe checks form as the veneer is bent sharply as it passes between the knife and the nosebar (Figure 11.3a inset). Checks on the knife side of the veneer are opened up as the sheet of veneer is unrolled from the bolt and flattened out for use. The knife side of the veneer is the loose side and the nosebar side is the tight side. Veneer that has many deep lathe checks is 'loose-cut' veneer while one having shallow checks is 'tight-cut'. Deep lathe checks will greatly affect the veneer quality.

A dull knife causes other problems. The fibres may bow and wrap themselves around the advancing knife rather than separating cleanly. This means that cutting at the knife edge can become intermittent as a plug builds up before being severed. At the same time that these fibres are being compressed the resistance to severance ahead of the knife will generate tensile stresses behind the cutting edge. These stresses may be sufficient to form checks in the veneer, resulting in groups of cells being torn from the veneer surfaces. Finally friction between the knife and the veneer may generate high shear stresses in the fibres adjacent to the knife resulting in a poor surface finish. Injecting superheated steam (up to 200°C) at the back of the knife and the introduction of a 'hot knife' has improved the cutting action by softening the wood fibres at the instant of contact with the knife and by reducing friction between the veneer and the back of the knife (Walser, 1978).

The nosebar performs a number of functions in maintaining veneer quality:

- It compresses the wood ahead of the cutting edge which reduces the chance of the wood splitting ahead of the knife. Cleavage ahead of the knife results in a very rough surface.
- Compressing the veneer permits it to bend more readily and with less risk of failure as the veneer escapes between the nosebar and the knife.
- By applying a steady pressure against the bolt the nosebar takes up any slack in the mechanical system due to wear, and guides the knife in relation to the outer surface of the bolt. This helps to ensure a veneer of constant thickness.
- A powered roller nosebar (Figure 11.3c) reduces frictional drag and clears slivers that stick in the gap. These spoil good veneer and interrupt peeling.

→

Figure 11.3. Rotary veneer lathe (Feihl and Godin, 1967). (a) Knife and roller nosebar mounted on a single carriage advance towards the chucks as the log rotates. The insert shows tension failures that are liable to form as the veneer is bent to pass between the nosebar and knife: the worst of the lathe checks are inhibited by slightly compressing the veneer. (b) Close-up of a fixed nosebar and knife. The gap between the two determines the degree of compression of the veneer, while the nominal thickness of the veneer is a function of the rate of advance of the knife carriage and the speed of rotation of the bolt. (c) Close-up of roller nosebar. (d) Microsharpening of the knife increases its resistance to damage from hard knots.

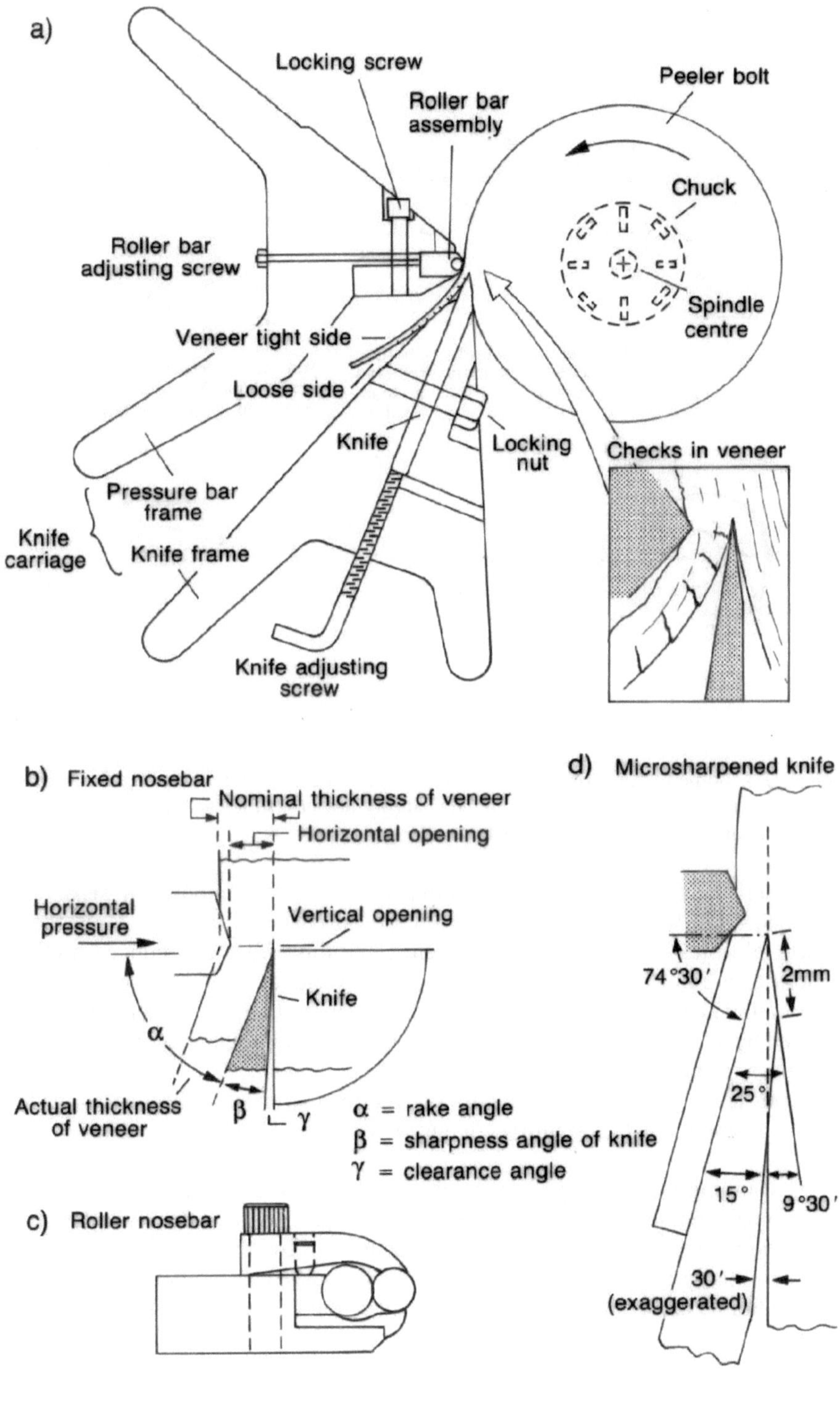
a)
Locking screw
Peeler bolt
Roller bar
assembly
Chuck
Roller bar
adjusting screw
Spindle
centre
Veneer tight side
Loose side
Knife
Locking
nut
Checks in veneer
Pressure bar
frame
Knife
carriage
Knife frame
Knife adjusting
screw
b) Fixed nosebar
Nominal thickness of veneer
Horizontal opening
Horizontal
pressure
Vertical opening
Knife
α
Actual thickness
of veneer
β
γ
α = rake angle
β = sharpness angle of knife
γ = clearance angle
c) Roller nosebar
d) Microsharpened knife
74°30′
2mm
25°
15°
9°30′
30′
(exaggerated)

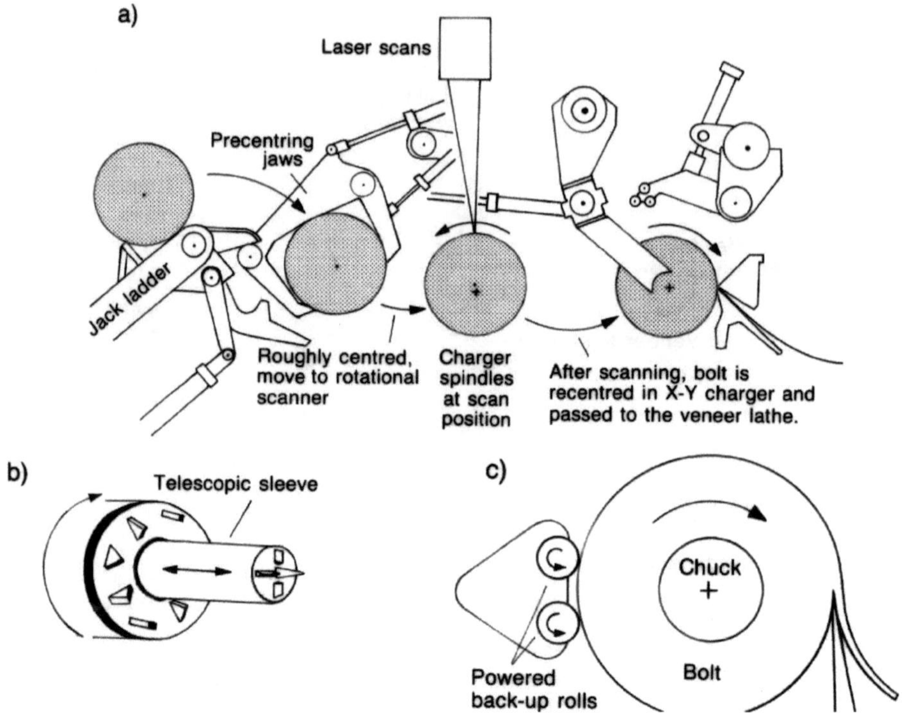

Figure 11.4. Some features of modern lathes (courtesy of Coe Manufacturing Co., Painsville, Ohio). (a) An automatic centring and lathe charging device. (b) telescopic, retractable chucks for peeling large logs. (c) powered backup rolls.

Walser (1978) examined the benefits of heating a contoured nosebar and of introducing hot water to the nip between the nosebar and the bolt. Only the most superficial fibres can be heated and softened, but this effect and reduced friction are sufficient to reduce power consumption and yield a smoother veneer of more uniform thickness. Since 1980, the small 16 mm diameter roller bar has been replaced by a larger roller bar ranging from 64–95 mm in diameter. The bigger roller bar allows for higher veneer productivity and recovery.

Other technical developments to conventional veneer lathes have made the peeling of small logs (<200 mm) economically viable. They include:

- Scanning bolts along their length to obtain a three-dimensional profile. The optimal spindle centre that would give the largest possible true cylinder of wood is computed. The bolt is then repositioned by displacing the ends both horizontally and vertically so that it can be passed to the lathe in that optimal position (Figure 11.4a). Traditional positioning devices only got within 25 mm of the optimal position. Precentring the bolts before loading the lathe has resulted in chucking rates in excess of five bolts a minute.

- The use of telescopic, retractable chucks allows the transfer of a large torque to the bolt when it is large, while the outer sleeve can be withdrawn to permit peeling down to a small diameter (Figure 11.4b). Spin out from the chucks sets a lower limit to the diameter that a bolt can be peeled to.
- Powered backup rolls augment the torque transmitted by the chucks and prevent the bolt bowing away from the knife (Figure 11.4c). The additional torque provided by coated, high friction backup rolls means that the torque transmitted by the chucks can be reduced. Backup rolls enable logs to be turned down to 130 mm cores without wood failure at the ends of the bolt that occurs where the chucks spin free. Smaller chucks are possible and the lathe can peel to a smaller core diameter.
- Fast peeling speeds, up to 6 m s^{-1}, allow large quantities of veneer to be peeled.
- Hydraulic control devices on the knife carriage have replaced gears and mechanical drives which can suffer from slack and wear. The use of hydraulics reduces the variability in veneer thickness and allows the thickness of the veneer to be adjusted very rapidly, for example peeling thick veneer during rounding up and then thin veneer when producing a continuous sheet.
- 1.2 m mini-lathes can be used to produce core veneer from large peeler cores coming from full-width veneer lathes or from smaller than average logs. Halving the length of the knife halves the torque needed to peel the shortened bolt while the torque that can be applied through the chucks remains unchanged. The short core lathe is a conventional response that has been superseded conceptually by the spindleless lathe.

The Durand-Raute spindleless lathe (Baldwin, 1987; Bland, 1990; Sorenson, 1985a) has transformed the economics and increased the productivity of small-log plywood mills. At the front end a round-up lathe peels the bolt until about 50% of its surface is dressed and the waste trim drops into the trash. The trimmed bolt is passed to the spindleless lathe (Figure 11.5a). The bolt is gravity fed onto the bottom rolls and then hydraulically lifted back up against the top fixed roll/nosebar and the adjacent knife that immediately peels a continuous ribbon of veneer. With a lineal output of 2.5 m s^{-1} a 165 mm diameter bolt can be peeled down to a 50 mm core in about one and a half seconds. Such machines are capable of peeling 15-20 bolts per minute, and 5,600 bolts per shift. The increase in efficiency is considerable as the time spent charging and rounding up a small diameter bolt in a conventional lathe can be half as long as the time to peel it.

The optimal clearance angle of the knife varies with bolt diameter and has to be adjusted continuously especially when peeling very small logs. If the clearance angle is too large, the knife tends to chatter while cutting, giving a short ripple (corrugation) in the veneer. If the clearance angle is too small the heel of the knife rubs on the bolt, and can force the knife out of the ideal spiral cutting line, giving a long, undulating wave in the veneer. They result in veneer of variable thickness.

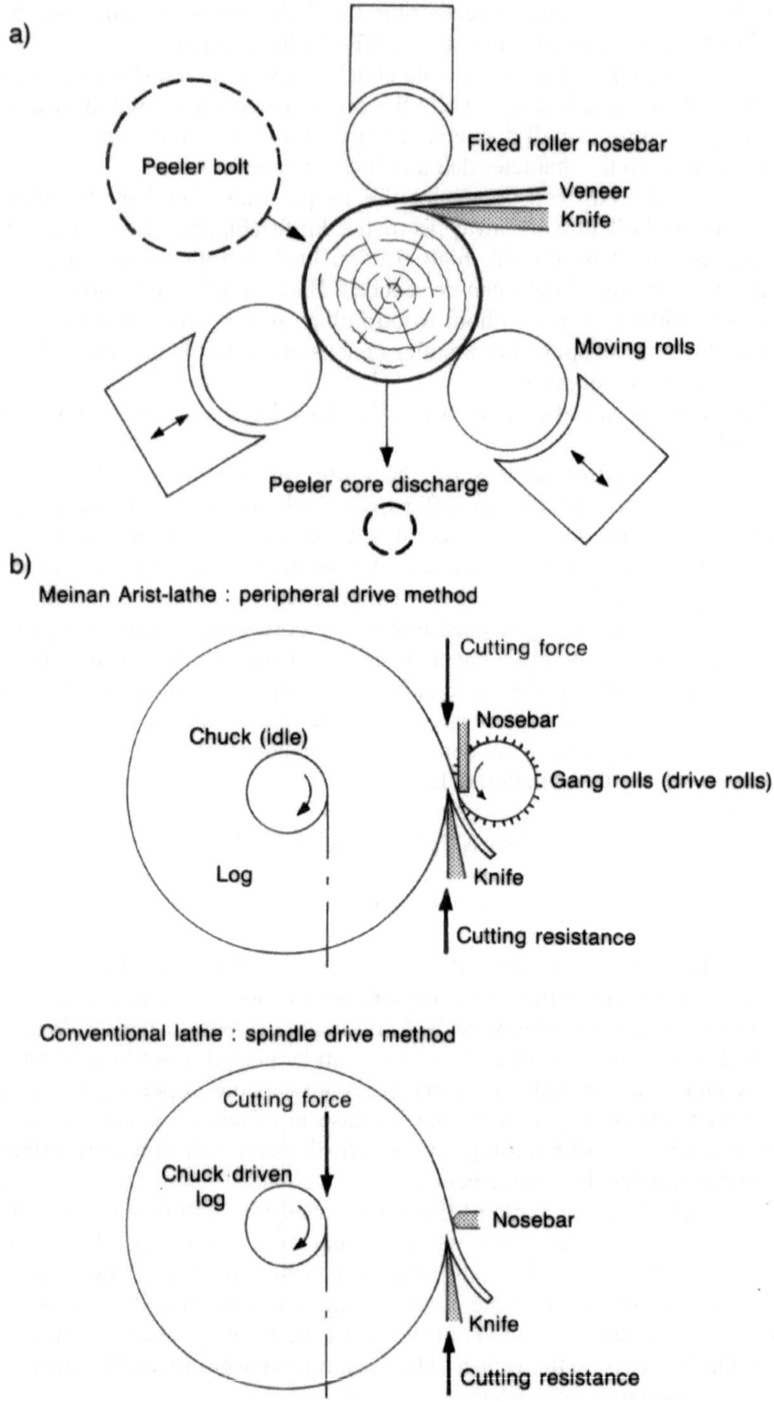
a)
Fixed roller nosebar
Peeler bolt
Veneer
Knife
Moving rolls
Peeler core discharge
b)
Meinan Arist-lathe : peripheral drive method
Cutting force
Nosebar
Chuck (idle)
Gang rolls (drive rolls)
Log
Knife
Cutting resistance
Conventional lathe : spindle drive method
Cutting force
Chuck driven log
Nosebar
Knife
Cutting resistance

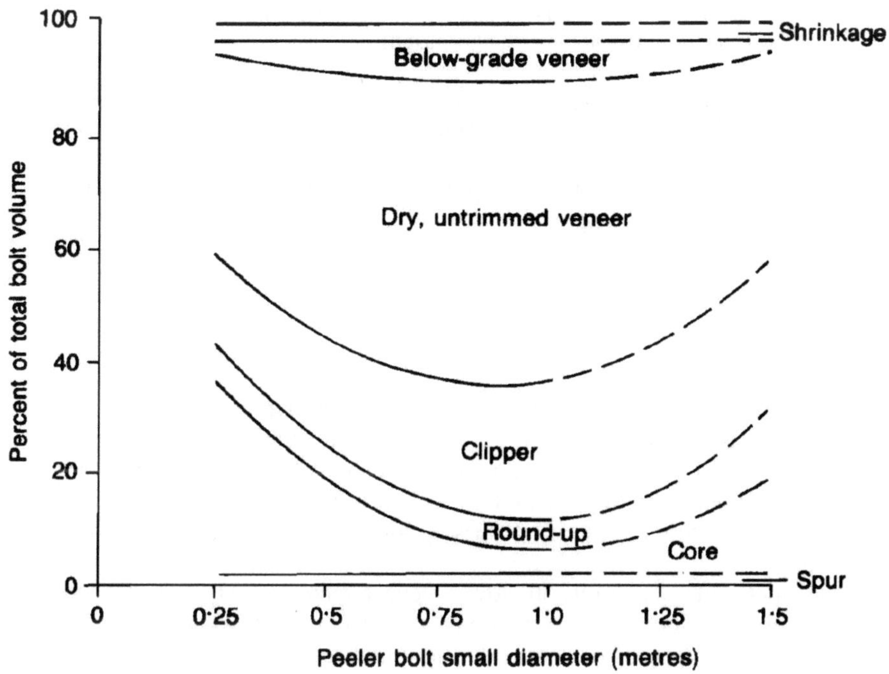

Figure 11.6. Veneer recovery from Douglas fir peeler logs (Woodfin, 1973).

Another unconventional approach is offered by the Meinan Arist-lathe (Figure 11.5b). The log is held in a modified lathe and a peripheral force is applied by a drive roll with a gang of spiked discs spaced at 50 mm intervals along its length to feed the bolt into the knife (Sakamoto, 1987). The spiked roller with its spacers functions as a peripheral drive unit and sectional nosebar. The cutting force is applied directly toward the knife eliminating the torque applied by a conventional chuck. There is far less tendency for badly split logs to break up and such material can now be peeled successfully. Further the spikes introduce small microchecks on the tight side of the veneer that becomes 'tenderized' so there is less likelihood of the veneer curling and breaking up. The veneer lies flat and dries more quickly.

In summary, Figure 11.6 demonstrates the high core losses with small logs and emphasizes the incentives to reduce these during veneer manufacture (Woodfin, 1973). Table 11.5 shows the efficiencies achieved by veneer manufacturers despite using much smaller softwood logs. Target core sizes and actual core sizes are far smaller than hitherto – and the recovery is actually improved.

Figure 11.5. Unconventional veneer lathes. (a) The spindleless lathe (courtesy Durand-Raute, Nastola, Finland). The lower rolls move up and in as the veneer is peeled, retracting again to release the peeler core and receive another bolt. Torque is provided by driving all three rolls. (b) Characteristics of the Arist-lathe compared to a conventional lathe (courtesy Meinan Machinery, Aichi-ken, Japan).

Table 11.5. The decline in both log quality and log supply on softwood plywood output has been mitigated by changes in technology (Spelter, 1988).

Technology of	Log size (mm)	Target size of core (mm)	Spin outs of bolts (%)	Core size after peeling (mm)	Recovery of green veneer (%)	Recovery of plywood (%)
Mid 1970s	350	135	8	240	65	51
Late 1980s	230	50	1	150	73	57

5.1.2. Log specifications

The poplar concept of the ideal, cylindrical log for veneer production relates more particularly to rotary peeled veneer (Lutz, 1971). Sliced veneer can be cut from logs having a degree of irregularity unacceptable in peelers. With sliced veneer irregular grain of various kinds is highly prized: burl walnut is a classic example. While large diameter logs are preferred, scarcity and rising log costs have forced mills to utilize smaller logs.

Sweep, taper and eccentricity all lower veneer recovery. The effect of sweep may be minimized by judicious cross-cutting of stems into individually straight logs. Taper produces short lengths of veneer during initial rounding up, and much of this material is unusable. Further the fibres in a tapered log do not lie parallel to the veneer knife so that the veneer is weak in bending, and can bleed from the glue line.

Eccentricity results in the production of narrow sheets during rounding up, which can be utilized but are not particularly valuable. More critically, eccentric or swept logs are indicative of the presence of reaction wood. Compression wood with its large longitudinal shrinkage (>1%) can result in imbalance and warping of softwood plywood, and board stability is one of the prized characteristics of plywood. Tension wood, on the other hand, can give a fuzzy surface after sanding because the fibres tend to pull out and bend rather than being severed cleanly. This is critical for decorative hardwood veneer.

Wood density is logically linked with hardness, machining characteristics and end use. Veneer species tend to have densities between 380 and 700 kg m^{-3} with a preference for those having a density near 500 kg m^{-3}. Lower density timbers are preferred for cores and crossbands because of ease of cutting, because they generally dry more easily with less tendency to warp and because they give lighter panels. However, low density species can be difficult to peel. They are liable to give a fuzzy surface though this can be counteracted to some extent by peeling when the wood is very wet and the cells are full of water: this gives some support to the cell walls during cutting. However with too high a moisture content there is no room for compression to occur until some water is forced out: if the bolt is peeled too fast the water is forced out at such a rate that the cells are ruptured. For this reason 'sinker' logs of species like redwood are not peeled. Species with uniform moisture content of about 50-60% cut best.

High density species are not necessary for structural grades of plywood, even though high density and strength are closely related. There are problems associated with peeling dense woods in that:

- They require more power to cut, and knives and machinery wear more.
- They tend to develop deep checks as they pass over the knife.
- They are not always easy to dry.
- They require an excellent glue bond as denser woods move more in service.

Knots are acceptable in structural plywood, and in core and in cross-plies of many other types of plywood. However, the knots should be sound and satisfy grade requirements. Steaming prior to peeling helps to soften knots and permits the bolts to be processed more easily.

Spiral grain can result in buckling or cracking during veneer drying. Even where such material is dried successfully thin plywood, in particular, can warp if the plies are not properly balanced.

Fast growth can result in peeling alternate strips of earlywood and latewood veneer. A 3-ply sheet manufactured from this material is liable to be unbalanced. Panels with seven or more plies are more stable.

5.1.3. Practical aspects concerning veneer manufacture

The primary objective is to maximize the recovery of veneer, in long ribbons with the minimum amount of veneer breakage, and with the veneer having a smooth surface and being of uniform thickness.

Veneer should be cut from logs as soon as possible after felling. If logs are stored they are best kept fully submerged or in a sprinkler system to prevent decay and seasoning checks.

Preconditioning. Veneer logs are first debarked and preconditioned (heated through) before peeling. Apart from the obvious need to heat frozen logs, conditioning is desirable as it improves veneer quality and recovery. Young, freshly felled, low density pine which is soft and pliable and Douglas fir are partial exceptions in that acceptable veneer can be obtained even if they are peeled cold. The object of preconditioning is to soften the bolts so that they peel more easily, with less sheet breakage and produce smooth veneer of uniform thickness. Power consumption at the lathe is reduced and the bolt can be peeled to a smaller core. Veneer quality is improved because hot wood is more plastic and will bend over the knife with minimum checking. Even large knots in lower grade logs are softened and can be cut more cleanly with less torn grain. Warm veneer can be handled with less chance of subsequent breakage: breakage results in more clipping to waste and fewer full width sheets of veneer are produced.

Efficient, uniform heat transfer is achieved when logs are heated in steam-water mixtures or immersed in hot water. Some form of segregation is desirable as the conditioning period ranges from a few hours to three or more days, depending on the log diameter and species characteristics. Logs can be conditioned in batches and in continuous-flow vats or tunnels: with tunnels the speed of the conveyor determines the conditioning period and individual sections in the tunnel allow batches of bolts to be treated separately. The most prominent characteristic of insufficient temperature is hard knots, which result in knife nicks, a reduction in knife life and rougher, looser veneer. In general dense hardwoods are heated to higher temperatures (80-100°C) than are softwoods. Excessive temperature causes earlywood/latewood delamination in softwoods that results in excessive roughness at low compression levels and fuzzy grain at high compression levels. Generally the centre of a softwood log needs to be heated to no more than 50°C: the lower the basic density of the timber the lower the necessary core temperature.

The round-up lathe. There are advantages in partially rounding up the log prior to passing it to the main lathe. In the round-up lathe the eccentricity, sweep and taper of the log are largely removed until recovery of short lengths or widths of veneer appears viable, at which point the bolt is prepared and ready for the main lathe. The round-up lathe reduces significantly the amount of unproductive time at the main lathe. Furthermore, any dirt or stones clinging to the log mark the round-up knife rather than the main veneer knife which must be kept in perfect condition if it is to cut quality veneer.

Lathe outfeed. Veneer is peeled at speeds between 2.5 and 6.0 m s^{-1}. Increased peeling speeds have put pressure on the clipper to enable the whole line to run faster. The traditional approach has been to separate the processes of peeling and clipping because they can rarely be synchronized: indeed the lathe may operate two shifts and the clipper(s) three shifts. Storage without damage and undue handling is needed. Veneer can be stored in reels (generally thin hardwood veneer) or most commonly directed automatically by a hinged tipple to multiple-deck storage trays (numbering 2 to 5 and up to 70 m long, enough to take the full length of veneer peeled from a single bolt). However, in a number of recent small-log mills the lathe has been directly coupled with a fast acting, computer-controlled rotary veneer clipper.

Clipper. One objective is to maximize veneer recovery, especially in full width sheets which minimize further handling. At the same time there is a call for the more valuable A and B grade veneers for faces or backs, which means cutting out defects such as decay, knot holes and splits. Management must determine the most profitable balance. Preference may be given to the production of A and B grade half widths or even random width (called strip), over full width C and D grade. The value added must exceed the cost of extra handling and jointing of narrow widths. Clipped veneer is sorted according to full sheet/random-width/fishtails, by sapwood and heartwood, and by grade before being fed into the dryer.

There are alternative mill configurations. Some modern mills direct the veneer from the lathe to the dryer and only then to the clipper. The advantage of clipping after drying is that veneer can be cut to size much more accurately than when green. The yield can be increased by some 3-5%.

Computerized clippers use scanners to detect breaks and defects in the veneer and then automatically clip these out. Typically 15-35% of veneer is in narrow strips and a clipper might spend 40% of its time trimming round-up and 60% of its time

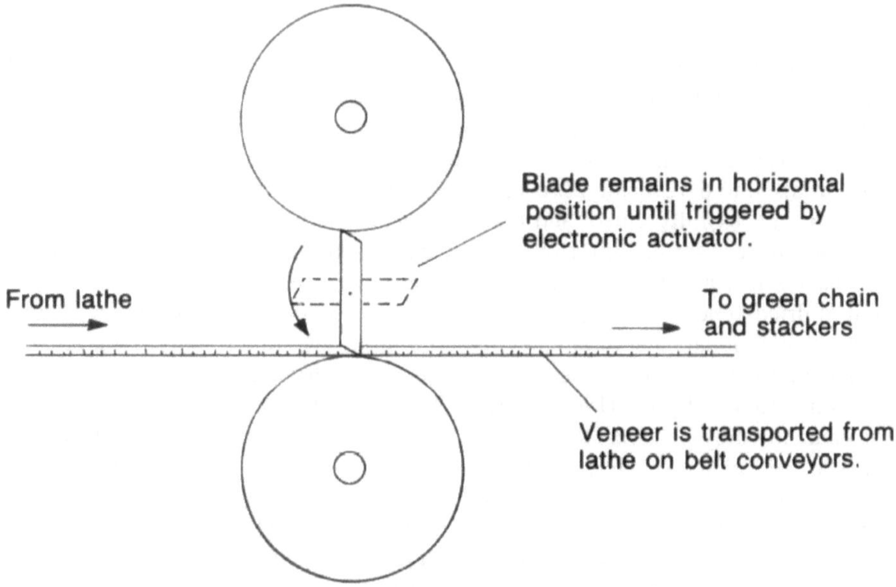

Figure 11.7. Principles of rotary veneer clipping (courtesy Durand-Raute, Nastola, Finland).

cutting full sheets from a continuous ribbon of veneer. Once the veneer is continuous, full sized sheets are clipped at a faster set rate. The up and down action of the old guillotine knife interrupted the steady flow of veneer, could cause the leading edge of the veneer to fold under, and cause pile ups and loss of veneer.

Today very fast clipping is achieved with a rotary clipper (Sorenson, 1985b). This device operates with the knife placed between two vertical rotating rollers (Figure 11.7). The bottom roll acts as an anvil and the top roll as a brace for the blade. The knife is electronically controlled and flips/spins between the rolls pressing and cutting the veneer against the bottom roller. The cut is very fast and does not cause any buckling of the veneer as both rolls and blade rotate with the flow of veneer. These clippers run at speeds of 1.8-3.0 m s^{-1} cutting with an accuracy of ±2.5 mm. They are quieter, more reliable, and require less maintenance than conventional clippers.

Veneer incising. Veneer can be incised when green before entering the veneer dryer (Dai *et al.*, 2003). The veneer sheets travel between incising rollers where a pattern of fine incisions are made on the tight side of the veneer in the grain direction. The incising rollers contain numerous chisel-shaped teeth. There are many benefits from veneer incising. It has been shown that the veneer drying time can be reduced by up to 10% and a less wrinkled, dry veneer obtained. Also the incised veneers can pick up 24% more preservative. Mill trials show that the incised veneer has a better bond quality and incising veneer can substantially reduce the panel blow (delamination), due to the greater permeability of the incised veneer.

5.2. Drying and upgrading veneer

Over half the mill's energy requirement is used in drying veneer. However, burning wood residues should be more than sufficient to meet the demand for process steam.

After peeling the veneer is too wet to glue and needs to be dried. Historically veneer was dried down to 2-5% moisture content, but today target moisture contents have been raised to 6-12% and even 15%. Veneer with a moisture content as high as 20% can be successfully glued, and this dramatically reduces dryer time and yields more plywood (less shrinkage). However, at the present time the objective would be for the panels to leave the press at 12%. Even with excellent control the moisture content of dried veneer varies quite widely, so it is essential to monitor the moisture content of individual sheets as they emerge from the dryer and to mark and segregate out the underdried material for subsequent redrying. Typically, 10-15% of production needs redrying (Figure 11.8). This is a desirable state of affairs as overdrying not only reduces throughput and increases dryer costs, but it causes unnecessary shrinkage of veneer, makes it more brittle and liable to degrade, and can cause gluing problems at the plywood press. For example McCarthy and Smart (1979) quote an increase in primary dryer throughput of 18% by raising the redry rate from 5 to 16%. Better utilization of the dryer is achieved by batching veneer to take account of large variations in initial moisture content. Moist sapwood, drier heartwood, and veneer to be redried require progressively milder drying schedules.

Continuous dryers. Very high throughputs can be achieved but it is essential to segregate the veneers according to species group, thickness, moisture content, e.g. heartwood or sapwood, and adjust the drying conditions accordingly. The veneer is restrained and kept flat by rollers or mesh wire and passed slowly through a long heated tunnel (15-100 m long) that has a number of independently controlled sections. The dryer can have multiple decks, typically four, which increases throughput. High temperatures (150-200°C) can be utilized in the first few sections of the dryer where the veneer is very wet. With jet impingement the hot gases are directed through small holes to impinge perpendicular to the veneer faces at very high velocities (20 m s^{-1}). The impact breaks up the thin stagnant boundary layer, that normally inhibits rapid heat transfer between the circulating air and the veneer surface, and drying is enhanced. Wet sapwood (>100% moisture content) can be dried in 6-8 minutes. The last section cools the veneer as the application of adhesive to hot veneer can create problems of excessive evaporation of water in the adhesive resulting in poor spreading and partial curing before pressing. The recorded temperature drop as the air passes over the veneer is indicative of the amount of evaporation taking place, which in turn provides a means of estimating the moisture content of the veneer. A large temperature drop indicates wet veneer with the air stream being cooled as a result of rapid evaporation. The dryer is adjusted to reflect the actual conditions within each section of the dryer.

Platen dryers. Drying systems have been considered that heat the surface of the veneer directly with heated platens (Loehnertz, 1988; Pease, 1980). A single veneer is placed on the lower plate in each daylight of the multiple daylight press and the platens closed (100-200 kPa pressure on the veneer) so keeping the veneer flat

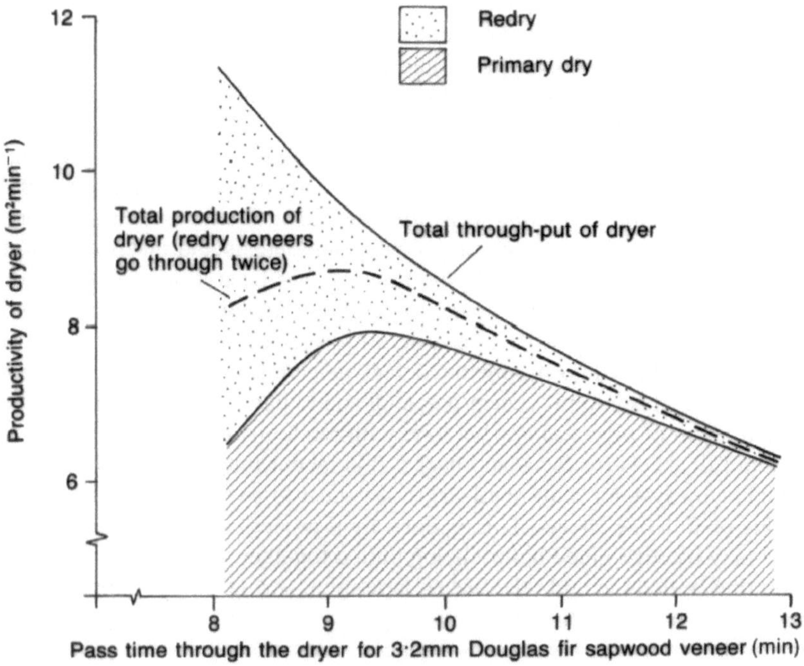

Increased productivity through redrying

Drying time	% redry	Production including redry
11·0min	5·2%	7·35 m^2min^{-1}
9·3	16·0	8·69
9·0	19·8	8·76

Figure 11.8. Effect of redry rate on throughput of dried veneer (McCarthy and Smart, 1979).

during drying. A platen dryer is unable to dry random widths (too much delay in loading) and the veneer has a pattern on the back as the platen has to be textured to allow the moisture to escape whilst pressing. The press dryer radically reduces steam and electrical consumption relative to a jet dryer of similar capacity, by 50%. Platen presses cost at least 25% more than a conventional jet dryer but a better return on investment is claimed as the running and maintenance costs are lower. Further it gives a flatter veneer having a more uniform moisture content. There is little shrinkage, which enables smaller sheet widths to be clipped. For example, full width sheets can be clipped to 1.285 m if press-dried rather than to 1.36 m. However much of the benefit of increased sheet width is offset by a corresponding decrease in veneer thickness. Wellons *et al.* (1983) noted that Douglas fir veneer loses only about 4% of its thickness during conventional drying compared to about 8% with platen drying. In other words restraining the individual veneers in the tangential direction reinforces shrinkage to the radial direction. Microchecks appear throughout the veneer, which is a consequence of this redistribution of the natural shrinkage pattern. Such veneer is fine for inner plies or for the outer plies of sheathing panels, but can cause appearance problems in sanded or siding panels.

Radio-frequency dryers are used to redry batches of veneer. These dryers are effective as the heat generated as internal friction by the vibrating water molecules is concentrated in areas where the moisture content is high. Some moisture is removed while the remainder is distributed more uniformly amongst and within veneers.

Veneer plugging and unitising. It is often desirable to recover a larger proportion of better grade veneer by cutting out/punching out defects and inserting a precisely fitting patch in the veneer sheets. In the past this has been a manual operation, but now automated systems feed the veneer to the router, cut out the defects, press in the patches and secure them with a couple of drops of hot-melt adhesive. These machines can be linked to automatic defect detectors that automate the process. However, the trend is to omit veneer patching and to patch the panels instead.

Manual layup of plywood can be done with random width cross-plies, two-piece centres and full sheets for faces and backs. It is labour intensive. On the other hand, automated layup systems require all strips of veneer to be jointed together into either continuous or full size sheets. This is done by a process known as unitizing where the strips may be edge glued before being butted together, usually by friction rollers. Edge jointing options include the use of tape, glue, string or a combination of glue and string. For example fibreglass strings pre-coated with hot-melt adhesive can be applied to the veneer in a number of places using a heated roller. Full sheets of jointed veneer can be readily handled as single pieces. Core unitizing simplifies mechanical layup and reduces the veneer losses and downtime at the layup.

Green veneer can be edge-stitched/sewn using either a zig-zag or looper stitch. A polyester thread is usually used as it shrinks by about the same amount as the veneer when dried. The jointed veneer is fed into a programmable clipper and onto an automatic stacker.

5.3 Panel layup, pressing and finishing

Structural plywood panels are manufactured with phenol formaldehyde resin, which is sufficiently durable to permit the panels to be used in exterior situations. With the traditional roll coater the amount of adhesive that is spread on the veneer is regulated by adjusting the gap between the steel doctor roll and the rubber applicator roll (Figure 11.9a). Glue coverage can be uneven if the veneer thickness is highly variable, with little glue coverage on the top face where the veneer is too thin to touch the adhesive film on the upper roll. While uneven coverage is undesirable it warns management that there is poor control of veneer thickness at the lathe. Unacceptably thin veneers are removed: their inclusion in a sheet of plywood would downgrade the board if that is not of the required thickness. Roll coaters remain popular in smaller operations making specialty and high quality plywoods.

With curtain coaters glue is forced from a reservoir through a narrow elongated gap or slit and falls as a continuous thin curtain across the entire width of the veneer

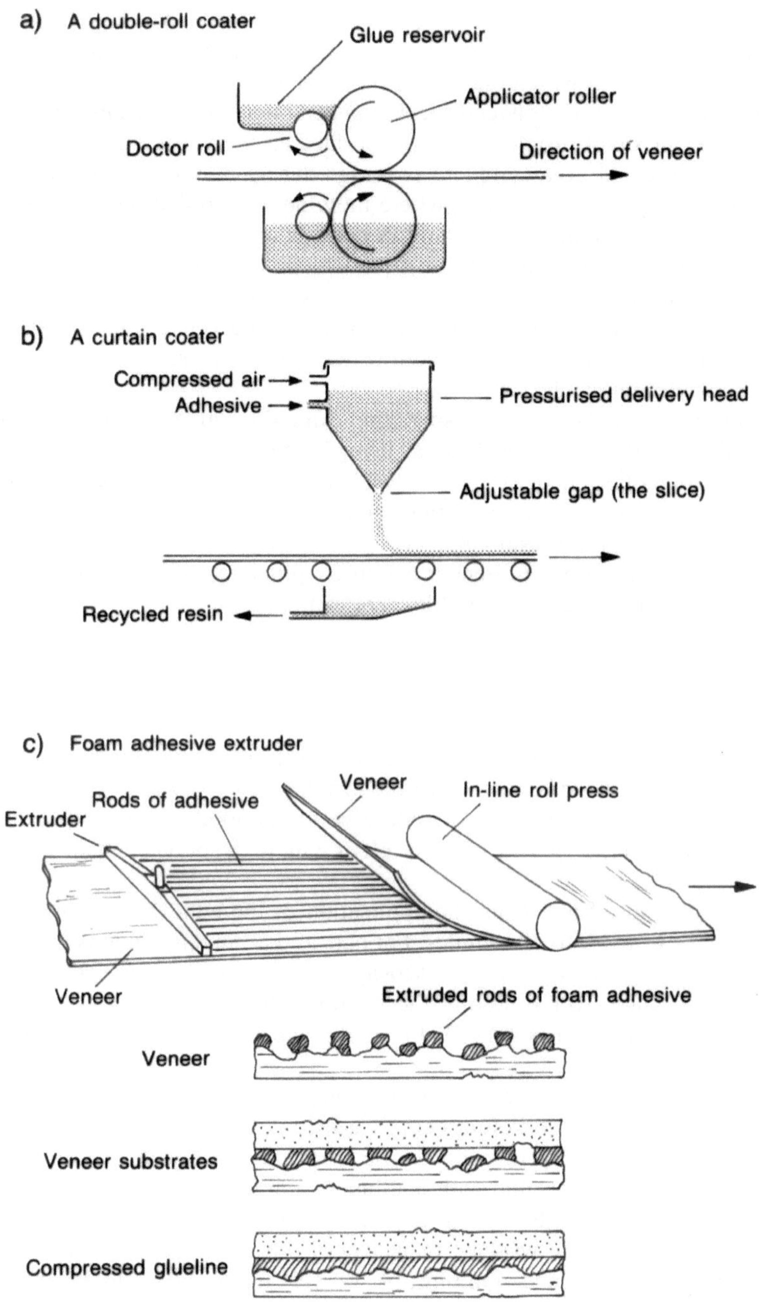

Figure 11.9. Various glue spreaders (Sellers, 1985). (a) Roller applicator. (b) Curtain coater. (c) Foam extrusion with in-line roll prepress.

which passes steadily underneath (Figure 11.9b). The amount of glue applied is controlled by the pressure head in the reservoir, the width of the gap, the viscosity of the glue and the speed at which the veneer passes under the coater. Glue that falls to either side of the veneers or between sheets can be collected and re-circulated.

Sprays can put small droplets onto the surface as the veneers pass underneath: the droplets spread and give complete coverage during pressing. Alternatively, the resin can applied by foam extrusion, where the glue is extruded through a series of holes spaced 10-15 mm apart and is laid in a series of continuous beads parallel to one another on the veneer (Figure 11.9c). The glue is foamed to five or six times its initial volume on being extruded in beads. These coaters use a continuous roller prepress to squeeze out the foam across the width of the veneer, to fill defects and holes and to ensure good coverage on both veneer faces. The re-circulated glue must be de-foamed before it can be recycled.

Such alternative methods of glue spreading can reduce glue consumption by 20% or more and are suitable for mechanical layup systems. Resin costs are contained further by adding fillers and extenders that both bulk and contribute to adhesion. They modify many resin characteristics such as viscosity and cure rate, and can contribute up to 50% of the resin volume.

Glue consumption (g m^{-2} of veneer per single glueline) is a function of the glue mix and a number of processing variables. For example:

- Rough veneer requires a higher than normal glue spread. Technical adjustments to the adhesive formulation are necessary to ensure that the gap-filling strength of the glueline is acceptable.
- Hot veneer (>35°C) requires more adhesive to counteract evaporation of moisture prior to closed assembly of the panel.
- Overdried veneer requires more adhesive to get good adhesive spread; while high moisture content gluing calls for glues with high solids content.

The total assembly time, from the application of the adhesive to entry into the hot press, ranges from 20 to 40 minutes. For part of that time the glue is exposed and can lose moisture rapidly. The viscosity of the glue and other adhesive characteristics change over time and the bonding strength after curing in the hot press is influenced by such factors. Automated layup minimizes open assembly and total assembly times.

Automated layup. Conveyors pass veneer under a series of gluing heads, cross-ply veneer is dropped on top before being passed to another gluing head where the next veneer is added, and so on until the desired number of plies are laid up. This process works for full sheets and with continuous jointed veneer that is cut to width at this point. There is a reduction in glue usage due to the ability to recycle and about a 4% decrease in the wastage of veneer at this stage due to better handling. However prior clipping and jointing of veneer is required. High production automated systems save skilled labour. They have better control and produce a more consistent product.

Cold pressing. Stacks of panels are pre-pressed cold for 3-5 minutes before being loaded into the hot press. The cold press ensures that the adhesive which is applied to one face of each veneer is transferred to the veneer on the other side of the glueline. Subsequent handling of panels is easier and more efficient.

Hot pressing. These presses are hydraulically operated and have 10-50 openings (daylights), each of which can hold one sheet of plywood. The trend is towards more daylights which means that hand loading is not fast enough. Instead the plywood is preloaded on racks and fed into the press in a single movement and simultaneously the hot-pressed panels are unloaded.

The press performs a number of functions. The initial pressure in the press, generally between 1200 and 1400 kPa, together with the plasticization (softening) of the veneers under the combined influence of heat and moisture, ensure intimate interfacial contact: the glueline film is less than 0.5 mm thick. The circulating medium heats the platens to around 140-165°C (for phenol formaldehyde resin) and as the heat migrates into the gluelines the resin polymerizes and hardens. Pressing is complete when the gluelines have been cured. Curing and moisture loss are rapid above 100°C and pressing is complete within two minutes of the innermost glueline reaching this temperature. Sellers (1985) provides indicative press times (Table 11.6). Wood veneer is a poor conductor of heat and this restricts the speed of glue cure at the centre of the board: heat transfer by evaporation of moisture from the surface veneers and its migration to the centre of the panel is not as easy as it is through a more open particleboard mat.

Table 11.6. Hot press schedules for phenol formaldehyde bonded southern pine plywood, 1980 (Sellers, 1985). Loading, press closure and unloading require a further minute or so.

Panel thickness[a] (mm)	Number of plies	Press time (in min.) from full pressure at:		
		140°C	150°C	160°C
9.5	3	3.5	3.0	2.5
12.5	3	4.0	3.5	3.0
12.5	4	4.5	4.0	3.5
12.5	5	5.0	4.5	-
15.5	5	6.0	5.5	5.0
19.0	5	7.5	7.0	6.5
19.0	7	8.5	8.0	7.5
22.0	7	10.0	9.5	9.0
25.0	7	12.0	11.5	11.0

a) Approximate metric equivalents, rounded down to next 0.5 mm.

Pressure is applied to consolidate the plywood and to ensure intimate contact at the gluelines. However, the hot moist plies can be densified, especially if the wood is a low density species, so as pressing progresses the pressure is reduced steadily to avoid unduly reducing the panel thickness. Wellons *et al.* (1983) observed a thickness loss of as much as 11% when Douglas fir veneer at 6% moisture content was pressed at 166°C and 1380 kPa pressure. To minimize thickness losses the closing pressure should be low and reduced further as quickly as possible. These losses are greater with rough veneer as greater pressures are needed to achieve intimate contact across all gluelines. There is some springback (2-5%) on unloading;

and by lightly spraying the panel surfaces with water a further 1% recovery is achievable. Low press pressures will be needed if the trend to high moisture content gluing, where boards leave the press at 12% moisture content, is to be achieved.

There are a number of variables that influence panel formation. They include press temperature, press time, moisture content, glue formulation, and modulation of the pressure. The influence of these variables is discussed further when considering the manufacture of fibreboard and particleboard (Chapter 12). The trend to pressing high moisture content veneer suggests that moisture migration through the veneers will contribute more to heat transfer than hitherto.

Inadequate cure or adhesion at the glueline due in part to high moisture can result in delamination when the boards are removed from the press. At the same time it is not economic to continue pressing until resin polymerization is complete, therefore curing of the phenolic resin continues after the panels have been removed from the press. Ultrasonic scanners at the outfeed of the press can detect air gaps (blisters) between veneers and suspect panels are marked and offloaded for further checking. This system reduces the number of defect boards being processed further and allows better feedback on the layup and pressing procedures.

From the press the phenol formaldehyde bonded boards are held in stack for further curing before edge and end trimming. The same does not hold for urea formaldehyde bonded boards. Urea formaldehyde panels are pressed at lower temperatures (<130°C) and should not be heated for prolonged periods as the resin is degraded by heat over 70°C. These panels should be cooled on leaving the press.

Panels are graded according to the veneer on both face and back. Splits, knots, knot holes and resin pockets are cut or routed out before being filled with putty or patched. Wood patches are being superseded by chemical patches, such as high density polyurethane foam which can be cured with heat lamps. It is not unusual to see sanded Douglas fir plywood with 20 plugs on the face veneer. Finally the panels are sanded. Modern high speed (0.75 m s^{-1} or more) widebelt sanders use a series of belts which give successively smoother surfaces.

Plywood can be processed further – with tongue-and-groove edges; overlaid, painted or given a textured surface finish; and edge-jointed to give oversized boards.

6. COMPETITION AND TECHNOLOGICAL CHANGE

The 1970s and 1980s, saw considerable innovation in the plywood industry, as it has responded to the decline in log size and quality, and to enormous competitive pressures from other panels such as waferboard and orientated strand board. The rate of application of new technology in North America is illustrated in Table 11.7. The effect of these changes on production costs and the competitive position of plywood *vis-à-vis* other structural wood panels are shown in another study by Spelter (1988). The interesting feature is the convergence of wood and resin costs for plywood and OSB/waferboard (Table 11.8). The loss in competitive position in the early 1970s was primarily the result of the policy of purchasing peeler logs, when the industry could have survived more profitably on a poorer mix of logs which would still have yielded the small proportion of A and B grades of veneer that was actually needed.

The misapprehension lay in the belief that all logs should be of peeler grade, whereas cheaper, lower grade logs would be adequate for core plies or even face veneers. Jointing of veneers and of plywood panels is significant as it allows the manufacture of enormous panels from low grade material, and offers greater flexibility in that the panels can be cut to the precise dimensions sought by assemblers of mobile homes, cabinet makers, *etc.* The challenge for the plywood industry today is to survive and profit from small logs at pulpwood prices. In Japan interest in the Arist-lathe and the spindleless lathe lay in the opportunity to peel short length veneer from low grade thinnings coming out of the indigenous forests. It is exemplified by the conceptual work of Okuma and Lee (1985) who examined the properties of laminated veneer board made from a 'patchwork of small, 450 x 900 mm, veneer elements' jointed into large sheets and then formed into plywood.

Wood composites today represent a matrix of opportunity. Each product seeks its own distinctive competitive advantage. A few of the commercially available products include:

- Blockboard, the original and still highly successful composite, with a side-butted core of timber overlaid by veneer.
- Plystran or Com-Ply with an orientated strand core aligned across the panel and overlaid with veneer aligned parallel to the panel length.
- Triboard with a thick orientated strand core overlaid with a layer of medium density fibreboard, so combining excellent strength with a smooth hard surface.

Table 11.7. Adoption of recent technological innovations by North American plywood mills.

Manufacturing technology	Number of units in service						
	1979	1980	1982	1984	1986	1988	1990
Charging:							
X-Y charger	1	3	70	120	135	140	>155
Peeling:							
powered backup roll			10	35	55	110	115
powered nosebar					70	120	140
peripheral drive lathe						1	2
hydraulic knife positioner			10	40	50	65	>100
spindleless lathe					1	1	4
Clipping:							
rotary clipper	1	2	6	45	90	105	>140
Drying:							
dryer control	-	-	-	1	5	7	>50
RF redryer	-	-	-	-	4	5	6
Gluing:							
foamed glue	--	-	-	-	6	7	8
Pressing:							
pressure controls	-	-	-	-	-	108	108
watering	-	-	-	34	43	36	36

Data from several plywood machinery suppliers (Spelter and Sleet, 1989). The 1990 estimate is supplied by Sleet (*pers. comm.*). With approximately 200 lathes in the United States and with major lathe changes too expensive for the smaller mills adaptation of 140 or so corresponds to effective market saturation.

Table 11.8. Impact of technological change on the estimated production costs of plywood and OSB/waferboard sheathing.

	Plywood		OSB/waferboard	
	Mid 1970s	Late 1980s	Mid 1970s	Late 1980s
Wood, net	66	37	32	28
Adhesives/wax	15	12	34	20
Energy	14	10	19	25
Labour	35	28	14	11
Overheads	26	26	26	26
Depreciation	7	7	12	12
Total	162	121	137	123

US dollar costs per m^3 for 9.5 mm (3/8 in) panels. The higher energy charges for OSB in the late 1980's is due to less wood waste and the need to buy more power. Assumptions include plywood bolts decreasing from 355 to 230 mm, with target core sizes of 135 and 50 mm respectively; and that OSB resin switching from 5.6% liquid to 2.1% powdered PF resin.

7. SLICED VENEER (LUTZ, 1974)

Sliced veneer operations are comparatively small scale with most mills processing less than 30 000 m^3 of logs per year. They cut a diversity of species. Slicing produces highly valued, figured veneers for face stock. The veneer is sliced very thin, 0.25-2.00 mm, typically *c.* 0.8 mm, to maximize the area of face veneer cut from expensive logs. Sliced veneer tends to be more brittle and to buckle and wrinkle on drying due to the wilder grain variation. Much of this thin sliced veneer is used for face veneer and may be laid on cheaper, non-decorative veneer. 5-, 9- and 15-layered plywood extends the surface coverage of valuable veneer since the consumer is not particularly interested in the material used as a substrate. For face veneer uniformity of colour is important and often requires the separation of sap and heartwood. Tradition or fashion dictates preferences for white or light-coloured woods or for darker woods. While many species would be suitable for slicing, selection is limited to those which are available in adequate volume, are of adequate diameter and free from excessive defects. Lutz (1971) indicated a minimum log diameter of 0.45 m for flat slicing and 0.6 m for quarter slicing, since the width of veneer strip that can be cut is limited by diameter. The visual characteristics that determine the value of a particular veneer relate to figure and colour of the wood and the manner in which the logs are sliced (Figure 11.10). Species with interlocked grain are best quarter sliced. Here the periodic reversals in the inclination of the fibres with respect to the axis of the log result in dark and light bands running the length of the veneer: if the veneer is reversed with respect to the lighting, the dark bands become light and *vice versa.* This is a characteristic of many tropical timbers, e.g. mahogany. Taste dictates that veneer with a narrow ribbon or stripe is more valuable. Wavy grain or irregular grain (Harris, 1989) is better flat cut, e.g. teak. Such veneer in violin backs is described as fiddleback, e.g. maple and walnut.

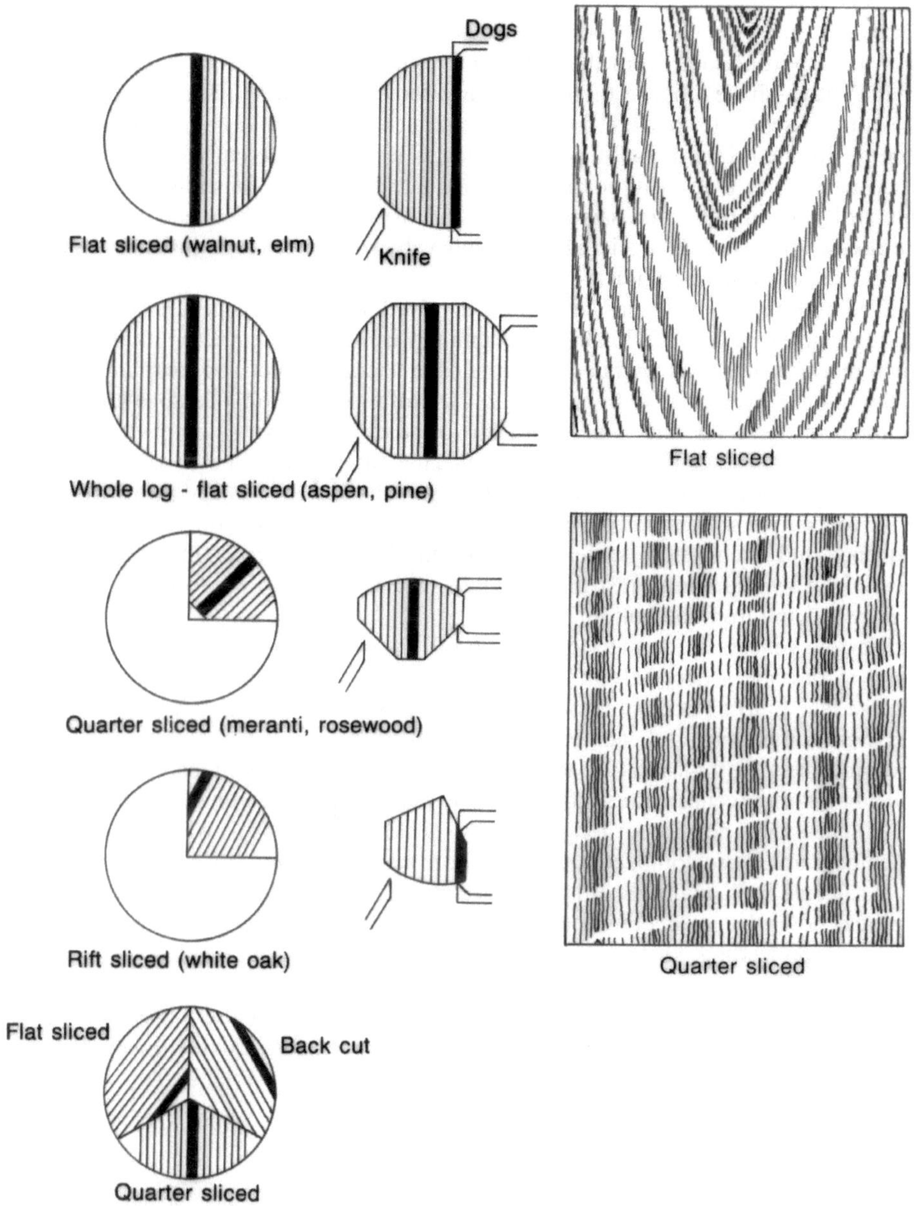

Figure 11.10. Veneer slicing (Lutz, 1974). Methods of breaking logs into flitches, and slicing strategies. Many species are flat or quarter sliced, the choice being determined by the log in hand and market demand. The wide dark bands represent the backboard left in the slicer after the veneer has been cut. Where a flitch is cut from pith-to-cambium the veneer is described as back-cut. It is used where the heartwood is narrow and highly prized, e.g. rosewood.

Generally logs are cut to length and sawn into flitches (Figure 11.10) before heating in vats. Some hardwoods are steamed as whole logs as this minimizes losses from end-splitting, and from the enlargement of existing ring (tangential) and star (radial) heart shake (Figure 6.16b). These wood failures arise in response to large growth stresses or to damage during felling: such trees should be laid down rather than be dropped. Internal splitting of the log is less of a problem when slicing than when peeling. It is usually possible to eliminate the effects of splits when cutting flitches by making the first saw cut along the worst split. Subsequently the flitches are dressed/planed before slicing. The high value of veneer logs justifies the high labour cost of these operations.

The cutting action is similar to that involved when rotary peeling except that cutting is intermittent (30-80 slices per minute) using an eccentric crank drive. The flitch must be firmly dogged and held against a vacuum table. The longitudinal axis of the flitch is skewed at a slight angle to the knife so that the blade does not impact simultaneously with the whole length of the log. There are two possible configurations. Either the knife is fixed and the flitch is fed into it or the flitch is fixed and the knife moves over it. Discharge belts lift the veneer away from the knife for stackers to handle in safety. The veneer from each flitch is stacked in sequence and each flitch is clearly identified. The character of the flitch can be ascertained by examining just three sample sheets: one from the top, one from the middle and one from the bottom.

Narrow veneers are jointed to make up full sheets that can be laminated onto any panel to provide a decorative finish: the panel requires a balancing veneer on its reverse face but this need not be of similar quality if it is not seen. Where alternate strips are turned over to become mirror images of one another the pattern is called book-matched. In this situation every second strip will have its loose face in view, and smooth tight veneer is essential if a high quality decorative finish is sought.

Flitches can be sliced longitudinally (Sakamoto, 1987), which makes the machine much more compact. With this slicer a small quantity of veneer can be cut from selected high quality boards/flitches coming from a sawmill. Individual flitches are held down by an overhead conveyor belt and driven over a flat machine bed having a protruding knife and nosebar. With a roundabout feed system 15-20 veneers can be cut per minute. The feed system allows flitches to be sliced down to a 5 mm backboard rather than to 30 mm as in a conventional slicer (Figure 11.10). The machine is not a high production unit but it recovers high quality veneer 0.2-0.5 mm thick from the best boards in some sawlogs.

8. TIMBER-LIKE PRODUCTS

There have been many interesting approaches to overcome the shortage of high quality, large dimension timber. Engineers have turned to trusses, I-beams, and space frames to achieve what had previously been accomplished by solid timber beams. Composites like glulam and the new products discussed below provide substitute elements that can be incorporated into the kinds of modern designs just alluded to. The thrust in the development of these new products has been improved reliability (strength and stiffness) and competitive marketing. There must be an overall advantage or financial saving. For example, the lightness of the timber/composite structure can reduce the cost of foundation work or it can reduce construction time, and so be an attractive solution even where the new product is more expensive.

8.1 Laminated veneer lumber (LVL)

The intrinsic efficiency of peeling logs to yield veneer has suggested that the same process should be applicable to the manufacture of laminated structural members. Laminated veneer lumber (LVL) was first produced in the early 1970s, and since then has been commercially available in many countries. It is now the most widely used structural composite lumber product in the US residential housing market.

In the manufacture of LVL, unlike plywood, all the veneers laid parallel to one another. It is produced by bonding layers of wood veneer using phenol formaldehyde resin (typically) in a large billet (1200 mm wide) under proper temperature and pressure in a stationary or staging hot press, or in a continuous hot press. To achieve the desired engineering design properties individual veneer sheets are transported through a veneer grade tester for measuring moisture content, density, and stiffness values. Acoustic or ultrasonic propagation time is used to grade veneer for stiffness in some LVL production lines. Scarf and lap joints are used to form an end jointed veneer sheet to a desired length. End joints between layers are staggered along the length to disperse their strength-reducing effects. The length of the scarf is typically 8-10 times the veneer thickness.

The thickness of the LVL can be from 19 to 75 mm and is available in lengths up to 25 m. Subsequently the material is cut to the required profile or dimension for beams/headers, I-joist flanges, scaffold planks, truss stock and for joinery work where its straightness and stability are positive characteristics. There is no limitation on the wood species for LVL. Any species used for plywood can be used for LVL. Low-grade or previously under utilized species can be used.

Laminating improves the strength of wood composites by reducing the defect area in any cross-section: by dispersing the defects, by averaging the wood densities of individual veneers, and by excluding the worst juvenile wood by confining it to the peeler core. Stiffness is a more tractable issue as grade stiffness values are taken as the mean of the grade population rather than the lower fifth percentile as for strength: hence the benefit of acoustic grading, sorting and screening of veneer. Indeed some companies are adding high stiffness hardwood veneer, e.g. plantation eucalypt, in their products. Research has considered incorporating a layer of fibre reinforced polymer or synthetic woven fabric into the LVL layup to improve the stiffness and strength (Laufenberg *et al.,* 1984; Dagher *et al.*, 1999). Typically, LVL is about 1.5-3 times stiffer and stronger than stress-rated timber.

Schaffer *et al.* (1972, 1977) examined the prospects for thick peeling southern pine, press drying, applying adhesive and then relying on the residual heat within the veneers to cure the laminated members. A phenol-resorcinol adhesive, which cures at moderate temperatures, was considered rather than a conventional plywood

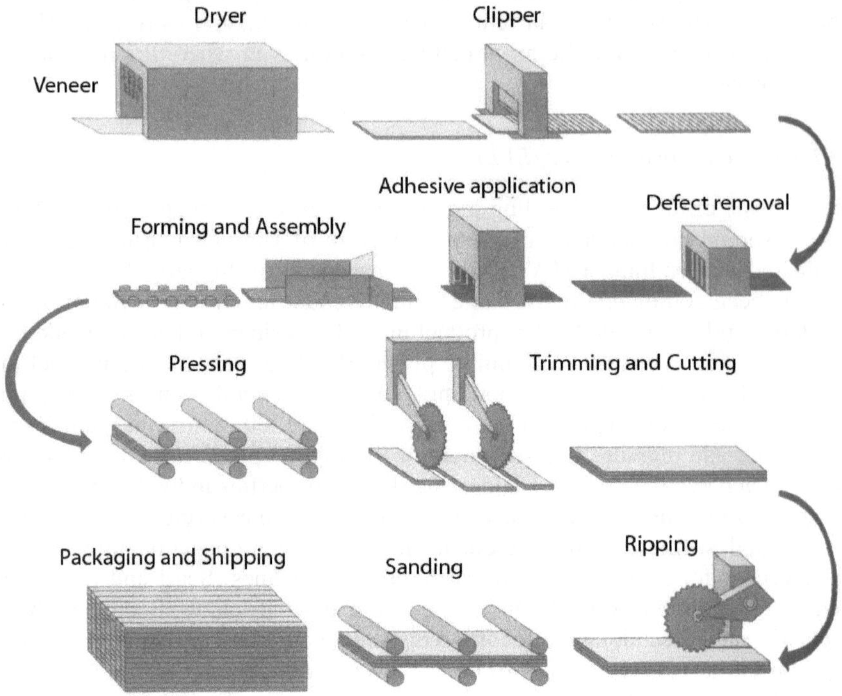

Figure 11.11. Manufacturing process for PSL (Williamson, 2002).

adhesive so that the laminated veneer would only need cold pressing. While the original scheme contemplated using thick veneer (up to 13 mm) to minimize the number of gluelines, these have very deep lathe checks and require much higher glue spreads (c. 50% more) than for plywood.

8.2. Parallel strand lumber (PSL)

Parallel strand lumber (PSL) was introduced to market by MacMillan Bloedel Ltd in the 1980s. Figure 11.11 shows the PSL manufacturing process. Residues from plywood and LVL plants can be used as the raw materials for the PSL – mainly Douglas fir, hemlock, southern pine, or yellow poplar.

It is manufactured from veneer (<6 mm) cut to a certain length (about 150 times the thickness of the strand), and a certain width (<18 mm), mixed with 4-6% phenol-resorcinol formaldehyde adhesive and cured by microwave. The long strands allow more complete transfer of load across the glue lines (Figure 11.12) so that the material approaches the ultimate strength of clearwood, partly because the requirement for long veneer lengths eliminates strands with knots and wild grain

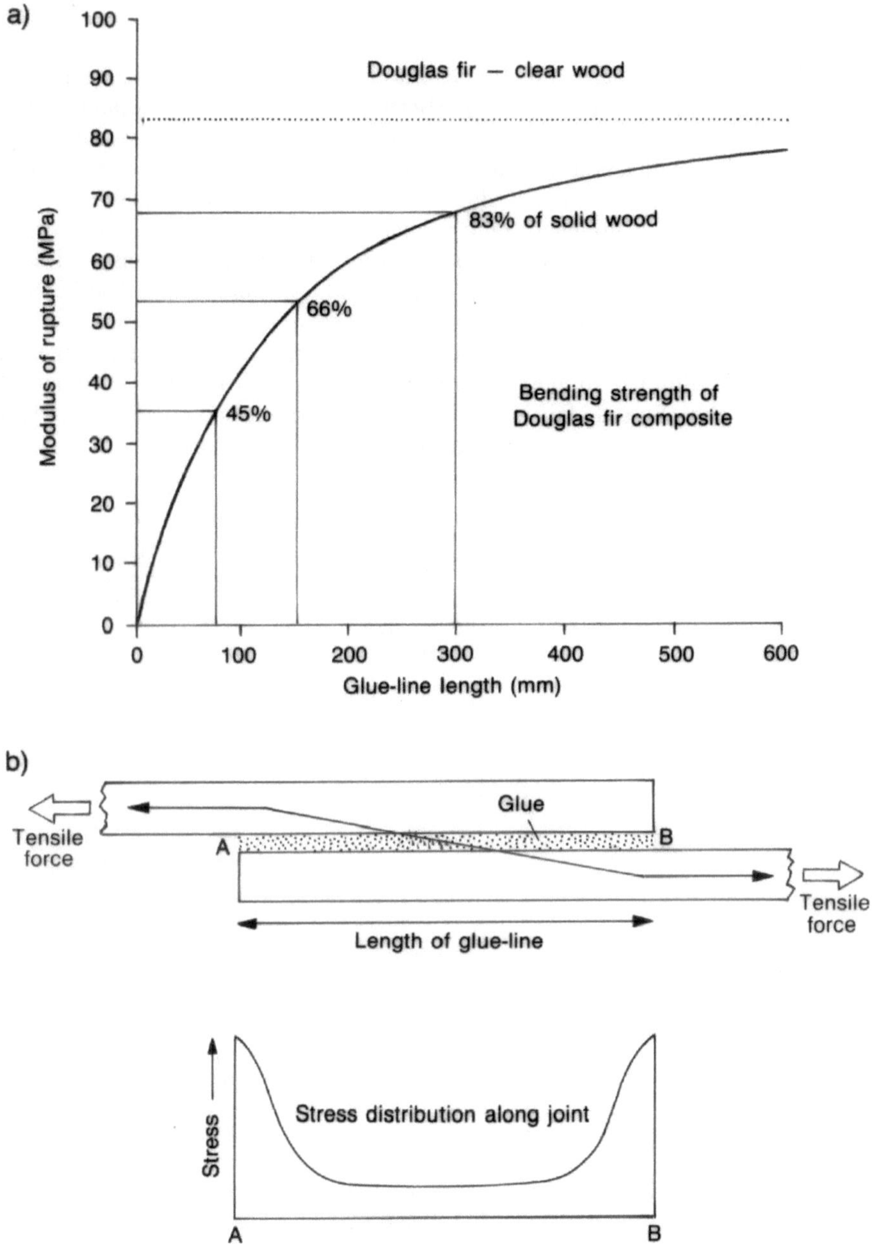

Figure 11.12. The strength of a glueline (Barnes, 1988). (a) Increasing the amount of strand overlap increases the joint strength. (b) Glueline strength is determined by the localized stresses at each 'glue joint' and the angle through which the load is transferred from strand to strand.

(Barnes, 1988). Maximum strength is achieved by accurately aligning straight-grained strands parallel to both the axis of production and the length of the eventual beams. Steel belts pull the mat of strands into a continuous press where the resin is cured with microwaves (Figure 11.13). Microwave energy penetrates and disperses uniformly across the large section which permits much faster curing of resin than would be possible with a conventional hot press: a hot press relies on heat transfer to cure the resin in the core and this is slow and uneconomic for thick members.

Technically, there is no length limit for parallel strand lumber since a continuous pressing operation is used. However, considering handling restrictions, PSL billets are usually cut to 20 m lengths, by up to 280 x 480 mm in section. The billets are recut to desired dimensions for use as beams, headers, columns, and studs.

PSL is very strong in its primary axis. The strength properties are higher than sawn lumber. Additional strength is gained from the 10% densification relative to the original timber. Strands fail in tension only because strand overlap is large and resistance to shear is greater than the tensile strength of the strand.

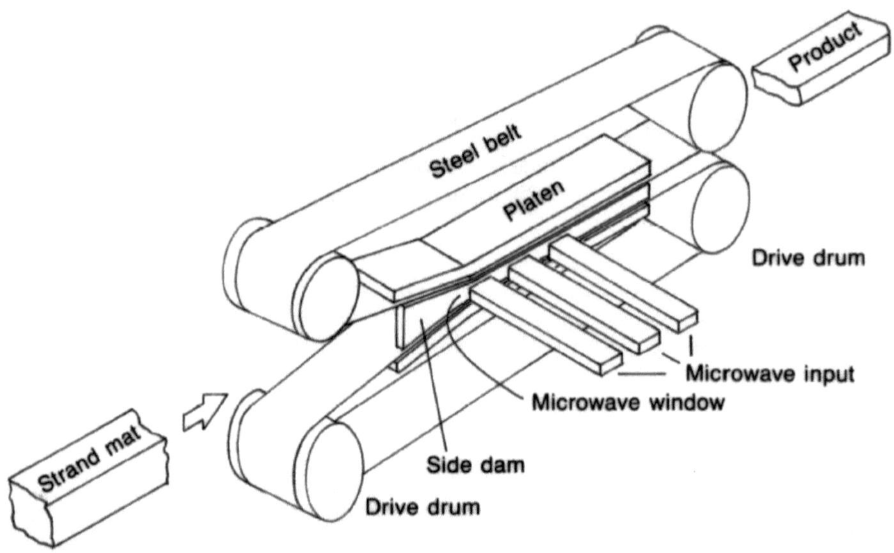

Figure 11.13. A 300 x 375 mm continuous press, for the manufacture of parallel strand lumber (Churchland, 1988). Four drive drums draw the strands into the press. The throat acts as a prepress reducing the thickness of the loose mat to 35-40% of its unconsolidated thickness. Both platens and side dams apply compressive forces to the mat that is cured by microwave energy admitted through ceramic windows in the side dams. These are transparent to microwave energy and yet sustain the full compressive forces on the edge of the product.

8.3. Scrim-based lumber

Scrim-based lumber (Hutchings and Leicester, 1988; Seale, 2004) is a product that utilizes low grade small diameter wood (76–203 mm). It was initially developed by

CSIRO in Australia and trademarked as Scrimber. Today it is being relaunched in the U.S. as TimTek. First, the roundwood is debarked, and each log flattened and crushed to form a mat of interconnected fibrous strands in a series of grooved rolling mills. The individual mats of flattened scrim (*tr.* a cloth) are passed through a continuous dryer before being collated and offset with respect to each other (as in LVL), coated with adhesive (such as phenol formaldehyde) and consolidated to a cross-section 180 x 1200 mm. The mat is passed through a steam injection press to be formed into beam product. Once cured, it is re-sawn into standard section sizes and cut to required lengths. This product is subject to a total quality management program throughout its manufacture. Its principal market has been identified as heavy section construction members and as such it competes directly with concrete, steel and timber beams. Figure 11.14 outlines the manufacturing process for TimTek scrim-based lumber.

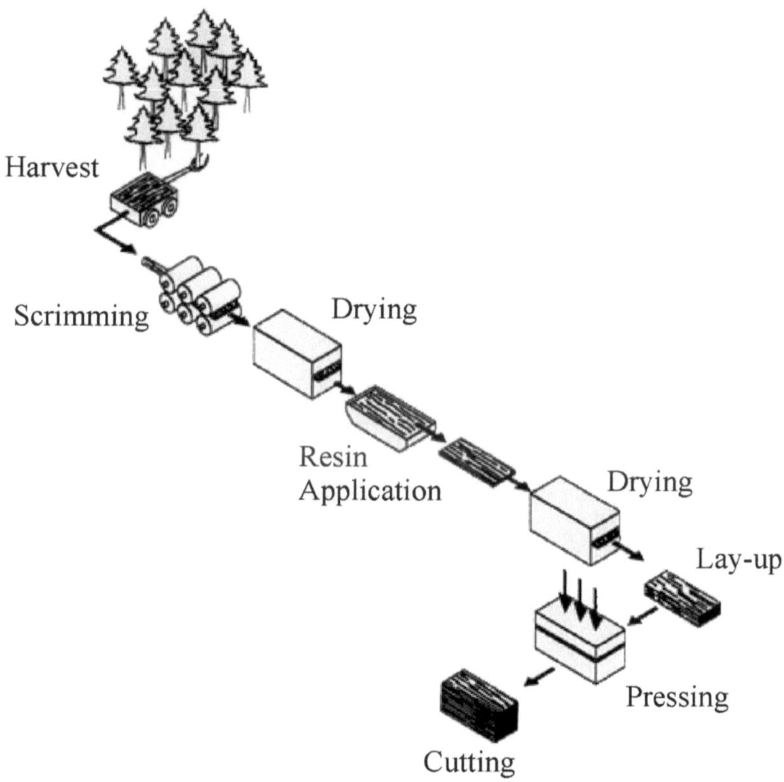

Figure 11.14. Scrim-based lumber uses young wood or thinnings from short-rotation plantations in a simple process to manufacture an engineered product of uniform quality (courtesy of TimTek Ltd).

The wood resource for scrim-based lumber is much inferior to that used in LVL or parallel strand lumber. The small diameter wood has much juvenile wood, but the split strands are very long and provide good transfer of stresses across the glue line. Scrim-based lumber has the potential to be made approximately twice as stiff and strong as the original (inferior) juvenile knotty timber. Scrim-based lumber is positioning itself in the general structural market where it aims to be competitive on price. LVL and parallel strand lumber have much superior properties and sell at a higher price for more specialized markets.

CHAPTER 12

WOOD-BASED PANELS: PARTICLEBOARD, FIBREBOARDS AND ORIENTED STRAND BOARD

KELVIN M. CHAPMAN

MDF Tech, Christchurch, New Zealand

1. INTRODUCTION

Wood-based panels have become an important component in the wood processing chain. From a materials viewpoint wood-based panels provide a way of realising value from materials that are not suitable for other uses, or that are residue streams from other processes. However these panels are now an integral part of the wood products market, meeting requirements that were once met by solid wood. The ability to produce panels of larger sizes reduces the number of components needed while the uniformity of panel properties give a more consistent performance. The panels are used extensively in construction and in the furniture, cabinet and joinery industries, while the ability to 'engineer' these products to meet specific performance requirements has been a significant factor in their growth. The technology for producing these panels has developed significantly, in particular with the introduction of continuous pressing providing new levels of product uniformity.

2. OVERVIEW

Particleboard, (PB), medium density fibreboard (MDF) and oriented strand board (OSB) developed in the latter half of the 20th century from technologies first applied to plywood manufacture, but required new approaches to particle preparation and drying, faster setting adhesives and press design. Particles, being smaller than veneer sheets used to manufacture plywood, allow a much wider range of raw material to be used. In fact these panels use raw material that is either a waste stream from other processes or a wood resource that cannot be used for other purposes: increasingly wood from urban waste streams is being used, particularly for particleboard. Another precursor to these panels was the wet-formed fibreboard panel known as hardboard, where a fibre mat was pressed to a high density (950-1200 kg/m^3). This panel, also known as 'Masonite', was self-bonded: this being achieved by reaching a temperature in the hot press (205-215°C) that caused thermal decomposition of the wood to produce an adhesive. Panel thickness was limited to about 6 mm by the necessity to achieve this temperature in the centre of the panel during pressing.

A wide range of adhesives are used. For the three panel types that are the focus of this discussion synthetic adhesives are used, although tannins extracted from bark have been used. However other binders such as gypsum and cement can be used with particles and with fibres. In this case the binder levels are much higher so that these products might be described as a wood-reinforced gypsum or cement matrix.

In general these products are pressed to consolidate the mat and to cure the thermoset adhesive. Whereas in plywood pressing the main object is to achieve good contact between the veneer layers with as little increase in density as possible, in the case of these products the average density is one of the key variables in determining panel properties and will normally be 30-100% greater than the bulk density of the wood from which the panels are made. A major advance in all these products is the ability to control the density profile through the thickness of the panel.

It is possible to characterise wood-based panels in terms of a plot of panel density against the major length dimension of the wood-elements used. In theory there is an infinite range of adhesive bonded products that can be defined within the range of particle size and density covered in Figure 12.1. In practice the range is limited by commercial requirements although the range indicated is by no means definitive, either for particle size or for density range.

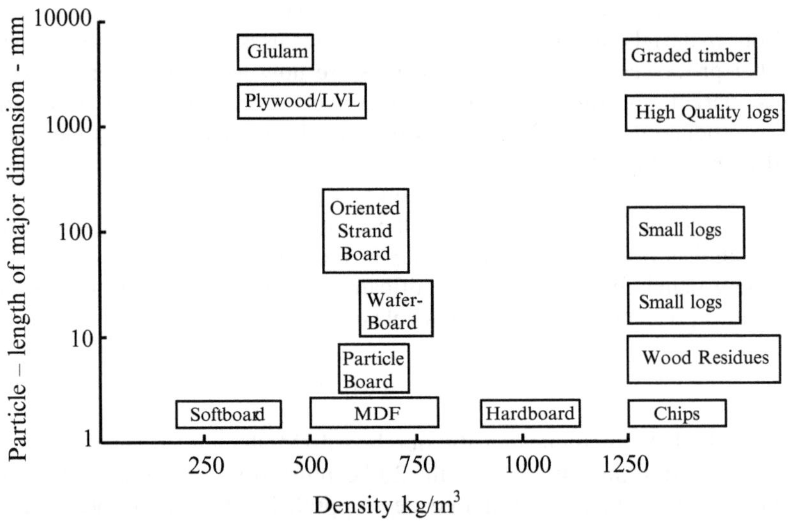

Figure 12.1. Classification of wood-based panels according to density of the panel and the size of the wood elements.

3. MARKET

Wood-based panels such as particleboard, MDF and OSB are a significant component in the wood industry. They satisfied performance requirements that are increasingly difficult to meet from solid wood sources and are able to meet these by using residues from other wood processing operations, or in the case of OSB, from small diameter logs from species that otherwise have limited potential, and at competitive prices. The technology to produce these panels has advanced to the point where these panels are made with a uniformity that allows processing on automated sawing, machining and finishing lines: increased automation of panel processing operations results in better quality components at a lower cost. Finishing systems range from paints and lacquers, grain printing, through to application of pre-printed or solid colour resin impregnated papers to provide wear resistance for working surfaces. To complete the cycle as it were, the composite panels can also be finished with a decorative veneer to provide a natural wood surface.

The market remains dynamic with newer panel types establishing markets by replacing other panel types. MDF grew initially in an area between particleboard and solid wood. The ability to create a machined or profiled edge enabled components to be manufactured from a single piece of MDF whereas particleboard required a wood clashing strip to provide such an edge. MDF also had much better ability to hold fastenings than did particleboard. This resulted in the introduction of new fixing systems for particleboard that overcame this limitation. Overall, particleboard has responded to the challenges posed by MDF by improving technology that raised its performance in a number of areas. Meanwhile the MDF sector has worked also to improve performance, not necessarily by increasing property levels, but rather by creating a more uniform product that performs more consistently. Lower panel densities (600 kg/m^3 as against 720-800 kg/m^3) have been used to increase plant capacity while maintaining performance and reducing costs.

OSB was developed explicitly as an alternative to plywood, at a time when that industry was faced with falling availability of modestly priced, large diameter logs for traditional construction plywood. OSB capacity has increased rapidly and now provides a competitive alternative to plywood sheathing over the timber frame of a North American house. While most OSB production is in North America, capacity is increasing elsewhere, particularly in Europe but also in South America. While initial growth was as a replacement for plywood, it has evolved to fill new market niches.

Panel volume data is available from several sources. While the FAO yearbooks provide some data, industry publications offer a more up to date perspective. The data are not always consistent or comprehensive. FAO data include OSB within the particleboard classification. Further FAO data cover world wood-panel production volumes whereas industry surveys (Wadsworth, 2005) are based on plant capacity of all operational units in each region. The difference between these two numbers reflects capacity utilization which varies as demand rises and falls. The volumes in this analysis are used to indicate trends rather than absolute levels (Table 12.1).

Table 12.1. World production of particleboard, OSB and MDF for 2004.

	million m^3	%	Average growth, 1995-2005 million m^3/yr
Particleboard	81.5	54.8	2.4
OSB	26.5	17.8	2.1
MDF	40.7	27.4	3.5
Total	148.6		

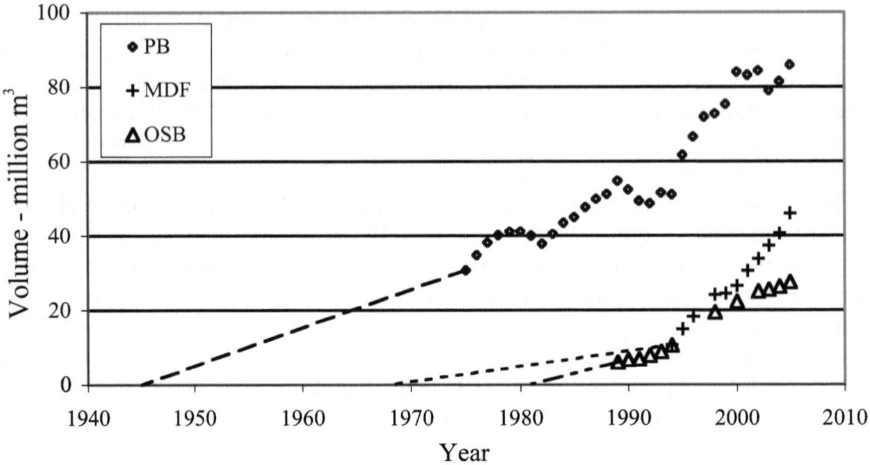

Figure 12.2. Annual world volume capacity for particleboard, MDF and OSB. The data are extrapolated back to the year when the products were first produced commercially.

These are comparatively new products with these volumes having been achieved in a relatively short time. The first particleboard was produced in 1941, with commercial production starting in the late 1940s. MDF was produced in 1965, with the first plants in USA in the early 1970s. Both particleboard and MDF moved quickly beyond the country of origin as the benefits of the new panels were realised. Commercial OSB production began in North America in 1981. World capacity is shown in Figure 12.2.

Product replacement is significant where two types of products compete within the same market area:

- Oriented strand board has grown as an alternative to plywood. For North America in 1987 OSB production was 29% of the plywood volume; by 1994 this had increased to 55%.
- Similarly in 1995 world MDF capacity was 24% of particleboard volume; by 2005 MDF had grown faster to reach 54% of particleboard volume.
- A greater proportion of MDF (43%) was exported from the country of manufacture over the period 1998-2002 than for particleboard (27%) although this varies significantly between different countries. Prices for particleboard are lower than for MDF by some 15-20% making this product more sensitive to freight costs, and therefore less likely to be sold in export markets at an acceptable price.

Growth has been sustained: annual increments over recent selected periods are shown in Table 12.1. The growth rate of MDF includes a major increment in China that may include some capacity from plants installed earlier.

A 1998 survey of manufacturing costs in the Canadian industry provides comparative figures for these panels. Table 12.2 shows the price advantage that has driven the use of OSB as a replacement for plywood, once satisfactory performance had been demonstrated. Waferboard was developed in 1962, but OSB had to await the optimization of flake geometry and the development of methods allowing the flakes to be oriented. Only then, in 1981, was OSB able to demonstrate a level of performance that allowed its use as an alternative to plywood.

Table 12.2. Canadian panel manufacturing costs (Poliquin, 1998). Note the relatively high costs for wood and labour in a plywood plant.

Cost $C/m^3	Softwood plywood (BC)	OSB (Quebec and Ontario)	Particleboard	MDF
Wood, net cost	151	58	28	40
Adhesive	17	23	35	36
Wax	0	4	3	4
Labour	103	29	22	20
Electricity	11	8	10	14
Misc.	22	18	17	22
Total Variable	304	140	115	136

The survey reinforces the benefits of large capacity automated plants producing OSB, particleboard and MDF have over smaller more labour intensive plywood plants. There are significant benefits of scale. These are achieved firstly by maximizing capacity of an individual press line, secondly by adding production lines in existing facilities. However increasing line capacity gives the greatest benefit, at least for new plants. Particleboard line capacities range up to 1800 m^3/day, while MDF press lines range up to 1500 m^3/day. The average line capacity for an OSB plant in North America is 1100 m^3/day, while new OSB lines proposed for 2005-6 have an average capacity of 1600 m^3/day.

4. CHARACTERISING WOOD-BASED PANELS

Wood-based panels such as particleboard, oriented strand board and medium density fibreboard consist of particles of widely varying shape and size bonded together with an adhesive system.

4.1. Wood particles

The variable shapes and sizes of the wood elements contribute significantly to the particular properties in any panel and in the way it can be used. These range from veneer sheets for plywood and LVL (Chapter 11), to individual fibres and fibre fragments used in fibreboards. The shape of the particles has become increasingly important as producers seek to maximize product properties while minimizing cost – for good structural properties long, thin particles and fibres are desired. The preparation of particles of the desired geometry and the separation of these from particles that do not conform is of increasing significance in quality control. Here, for convenience the term particle may refer to particles (chips, flakes and wafers) as well as fibre (fibres and fibre bundles).

4.2. Adhesive

The adhesive is selected to meet the specific strength and durability performance requirements of the panel, but also it has to meet other requirements. Generally the adhesives are thermosetting, in that they undergo permanent change with the application of heat. Cost is a major consideration, with the cost in the panel being determined by both the purchase price of the resin and its usage rate. An adhesive that requires higher temperatures to cure will also increase the cost of the panel by requiring longer times in the hot press.

Over the last 20 years, environmental concerns have become an important consideration in adhesive formulation and use. Firstly, in the plant the adhesive may require particular handling or the use of protective equipment. Secondly, volatile emissions arising from adhesive reactions both in the hot press and subsequently when the panels are in service are subject to tight regulatory control. In particular the reduction formaldehyde emissions from wood-based panels has been a major objective in adhesive development over the last few years, both because formaldehyde-based adhesives are the major type used and because these have been implicated in environmental and health concerns.

4.3. Pressing

The pressing of the particles – after being mixed with the adhesive – into a panel is the point where the composite panel is formed. Conceptually the press provides the desired clamping force that is used to increase the contact areas between the individual surfaces being glued, while heating the material in the press reduces the time necessary for the adhesive to develop sufficient strength. The development of panel presses has allowed the overall density of the composite panel to be increased, often significantly above the density of the wood from which the particles are prepared. This ability to change the density of the product has been further refined to the point where the density distribution through the thickness of the panel can be controlled.

4.4. Density

For plywood and LVL the objective in pressing the panel is to create only enough pressure necessary to form good bonds. However, the average density of the panel in the cases of OSB, particleboard and MDF is well above that of the wood from which the particles are derived. In the case of these panels, strength properties are as much a function of the resin system and the manufacturing processes as they are of the wood being incorporated into the panel.

However to generate good bonding through greater interfacial contact between particles and fibres it is necessary to densify the mat in the press. High density woods require greater force to form bonds of adequate strength and in turn produce undesirably heavy wood panels, a feature that can be countered by adding more adhesive and reducing the compression on the mat. The market prefers medium density woods (400-550 kg/m^3). It follows from beam theory (Chapter 10) that the reduction in thickness that results from increasing the density of the material in the hot press has a greater effect on the bending strength of the composite than does any improvement in bond strength. For OSB made from aspen/poplar mix the panel density ranges from 576-640 kg/m^3. When made from denser southern yellow pine, the common wood for U.S. plants, the panel density range is 608-672 kg/m^3.

The density of the panel is important in determining its properties, as is the ability to control the density distribution within the thickness of the panel. The bending strength of MDF depends on the tensile strength of the outer layers. Therefore increasing the density in the faces of the panel allows the bending strength to be improved. Concurrently a lower density core cuts fibre consumption compared to that that would be required if the core density were close to that of the face of the panel. Controlling the distribution of density through the thickness of the panel allows strength properties to be maintained at a lower overall panel weight.

4.5. Mat structure and lay-up

The method of assembly of the particles ahead of the press can also be used to manipulate the properties of the composite. For example, it is possible to use different particles (or adhesive systems) in different layers of the panel. Indeed, this approach allows composites with totally different structures in different layers of the panel to be produced. Such a product is 'Triboard' produced by Juken Nissho in New Zealand, where a product with MDF faces and a strand core is produced in a single pressing operation.

In a less dramatic form, separation of the fine and coarse particles in the mix prepared for particleboard during the mat forming process can produce panels with very different performance objectives. If the larger particles are concentrated in the outer surfaces of the panel, the bending strength of the panel will be improved (for flooring and structural uses). If the mat forming is configured to concentrate the fine particles in the outer surface of the panel, the surface will be much smoother, and respond better to paint and lacquer finishing systems, but the bending strength will be reduced. This approach is used where the panels are to be used for furniture or decorative face panels in joinery.

Different adhesives can be incorporated in different layers of the panel. The use of a faster setting resin in the centre of the mat will reduce the time needed to complete the resin cure in the hot press. The faster setting resin may be more expensive, or have other disadvantages, but these may be offset by using a slower curing, less expensive adhesive in the outer layers of the mat. This increases the complexity of the plant with separate lines needed to prepare material with different resin types or characteristics. This expense must be justified in terms of the increased production rates achieved in the press, in resin cost savings, or in improved product properties.

The lay-up or mat forming stage offers another opportunity to influence the directional properties of the panel. It is possible to orient the particles in one direction of the panel. A parallel alignment of members is inherent in glue laminated beams, while alternating the grain direction of veneers as they are laid up to prepare a plywood sheet for pressing is a way of achieving more uniform properties in the plane of the panel. While orientation is relatively straightforward for beams and veneer sheets, the technique has been applied also to smaller particles. As suggested by its name, orientation of the particles is important in manufacture of oriented strand board (OSB).

Orientation becomes more difficult as particles become smaller and consequently increase greatly in number. Some orientation does occur, even with the fibre used for MDF. In forming the mat the fibres are deposited on a moving belt that results in a preferential degree of orientation in the direction that the belt is moving. This effect shows in the bending strength of MDF which is greater in the direction of the mat travel than perpendicular to this direction. The difference depends of the speed of the forming belt and can be 20-25% higher in the belt direction than at right angles to this.

The shape of the space within the press influences the final form of the panel. For most applications a flat panel is preferred and the formed mat is pressed between flat platens. However there are plants using platen pairs formed as a three-dimensional die to produce a composite structure shaped for a particular application. The most successful of these 3-D shapes is the door skin, where the textured die is configured to produce the face of a wood-panelled door. The door is assembled by gluing the two faces to stiles and rails and trimmed to produce a finished product at a cost that has allowed this product to dominate the market for internal doors in buildings. Other successful three-dimensional wood-based composite products are shipping pallets, and circular table tops. The requirement is for a large volume. A door skin press may have 20 die sets, each able to produce the two matching faces of the door. If the total pressing and unload/loading time is 60 seconds, the press produces panels for 20 doors per minute, or in excess of 25 000 per day assuming the press line runs for 90% of a 24 hour day.

Early in the history of particleboard manufacture, extrusion was used to make profiled shaped material, but this process never became significant even where such a process might be expected to have advantages, e.g. for a product with an irregular cross-section and required in long lengths such as interior finishing mouldings. Instead, these are produced in large quantities, from MDF, not by an extrusion process, but are machined from strips sawn from flat panels. Extrusion has come back into fashion in the production of wood-plastic composites where the ratio of wood to polymer is close to 50:50 (Wolcott, 1996).

5. HISTORY OF WOOD-BASED COMPOSITES.

The range of wood-based composite panels available in the market is a result of the continuing development of processing techniques and adhesive technologies (Table 12.3). Wood glues have been used for centuries, but availability in quantity with consistency of performance was achieved only when synthetic formaldehyde-based adhesives became available.

Table 12.3. Evolutionary sequence for wood-based panels (Clark, 1991).

Year	Product	Country
1830	Mechanically sliced veneer	France
1896	Rotary peeled veneer	Estonia
1898	Fibreboard (wet formed)	UK
1906	Plywood	USA
1914	Insulating board	Germany
1925	Hardboard	USA
1941	Particleboard	Germany
1945	Dry process fibreboard	USA
1966	MDF	USA
1969	OSB	Germany

Plywood was originally produced using natural adhesives and later with phenol formaldehyde (PF), discovered by Dr Leo Baekeland in 1906. The first fibreboard panels were produced using papermaking technology. In this case the wet formed mat was made using a relatively crude fibre, and dried to form a low density panel called Softboard with a relatively low strength that limited its application. Subsequently this low density panel was pressed to a much higher density at platen temperatures that caused thermal decomposition of the wood to commence. This resulted in the panel known as hardboard. Hardboard is significant in that it was the first panel product made from fibres and fibre bundles to demonstrate property levels approaching those of plywood.

It was not until after World War II that particleboard appeared as the first product to use urea formaldehyde (UF) as an adhesive. Particleboard provided a low cost way of converting low grade raw materials into a useful product. Capacity increased rapidly after the World War II to meet the huge demand for building materials, especially in Europe and the technology quickly spread to other countries Experience with this product showed the benefits of controlling the particle size and geometry and lead to improved performance.

Early fibreboards were developed by taking fibre from a wet-process fibre line, drying this, mixing it with resin, and pressing it as had been done for particleboard. However, in the last three decades medium density fibreboard (MDF), a UF bonded and dry formed panel, has largely replaced the early fibreboard products.

Oriented strand board (OSB) was developed as an alternative to plywood and is now used in large volumes in construction, particularly in North America. It uses small thin flakes or strands which are oriented to provide directional properties.

These are prepared with specialist flaking equipment that can use small diameter logs. Initially this was from species having no other use, but as capacity has grown alternative wood sources have been used. The process evolved from the waferboard manufacture that used thin flakes or wafers to form a panel where the wafers performed as small veneers. The increased slenderness ratios of strands resulted in improvements in both strength and directional properties so that OSB became an alternative to plywood. The reduced width of the strands compared with wafers allowed them to be produced from smaller diameter wood.

The development of the wood-panel industry from 1966 has been chronicled in the proceedings of the annual Washington State University Symposia on Particleboard and Composites.

6. RAW MATERIALS

Wood-based composites require two major raw materials. These are:

- Particles, which are generally produced from wood although other cellulosic materials have been used, with straw being used in Europe early in the development of composites.
- An adhesive system that is compatible with the material to be bonded.

Other minor components may be included to impart specific properties, for example the addition of wax to improve short term response to wetting is almost universal. Insecticide and fungicides to impart resistance to fungal or insect attack may be added, while fire retardant materials can be incorporated to give panels with a specified fire performance.

6.1. Chips, fibres and sawdust

Each panel type requires particles of a particular quality and configuration. Veneer-based products such as plywood and LVL, and components for glue-lamination require the use of high quality logs or timber. The strands for OSB are engineered flakes that are prepared from logs. The flaking process is able to use logs that are too small for sawmilling, increasing the proportion of the tree that can be used, or alternatively using species such as alder and poplar that are unsuited to other uses.

Particleboard utilizes residue materials from other wood processing operations. Sawdust and shavings predominate, although most plants are able to use chips or flakes made from roundwood to meet specific strength requirements in the product. Cost is the major driver, with the proportion of high cost material such as chips or roundwood limited to that necessary to meet product requirements. Particle size distribution and shape have become more important as the effect of these variables on product quality has been recognised, leading to reductions in cost.

MDF is also produced from chips, but increasingly sawdust and shavings are used as fibre sources. As the name suggests the particles in this case are fibres or fibre bundles characterised by a high slenderness ratio, typically 80-100:1 for whole individual fibres. Most of the fibres in sawdust and shavings are damaged in the processes that lead to these residue materials, so that their average fibre length is reduced to the point where it is more difficult to meet strength requirements where these feed materials are used. Agricultural residues such as rice or wheat straw can also be used for both particleboard and MDF. However they often contain components such as waxes that require particular treatment, or the use of specific adhesives to achieve satisfactory bonding. Recently, reclaimed wood from urban waste streams has been included in the furnish for both particleboard and MDF.

Particleboard and MDF plants are thus tied closely to residue streams from other wood processing operations. The logs used for OSB may come from a specific small diameter resource or from top logs arising from harvesting of other forest resources. The processing of these residue materials into particles of the required shape and thickness is specific to each raw material.

6.2. Adhesive systems

A wide range of adhesive and other binder systems is used in the manufacture of composite panels, dominated by formaldehyde-based systems. Growth in the composite panels industry has depended on adhesives with the required properties being available in sufficient quantities at a suitable cost. The industry requires considerable economies of scale for synthetic adhesive production. The materials are derived from oil (phenol) or natural gas (urea and formaldehyde). Recently in the face of environmental constraints and the increasing cost of oil and natural gas there is renewed interest in adhesive systems derived from renewable resources.

The earliest wood adhesive for mass production of panels was phenol formaldehyde (PF). This is widely used for products designed to perform in severe weathering conditions and dominates the exterior plywood market. Resorcinol formaldehyde is a somewhat similar resin system that cures at ambient temperatures, but the cost is much higher, limiting its use in high-volume applications. These systems, although formaldehyde-based, do not have the continuing formaldehyde emission problem that is associated with other formaldehyde-based systems.

This other class of formaldehyde-based adhesives is the amine-based systems where the formaldehyde is reacted with urea (U) or with melamine (M).

The urea-based systems (UF) are the most commonly used and cure at relatively low temperatures (*c.*105°C), reducing time in the press. The glue bonds are less durable so these adhesives are used in products designed for applications where extreme weathering is not a consideration. The one problem with UF adhesives is that there is continuing emission of formaldehyde from the composite over the first year or so of its life. The UF polymer formed in the press can break down to formaldehyde and urea, at a rate that increases with temperature and moisture content. The formaldehyde lost during the pressing of UF bonded panel arises from this breakdown of the UF in the outer layers of the panel that reach temperatures close to that of the hot platen. Continuing formaldehyde emissions from the panel in service can affect the health of people working in built environments – causing various allergic reactions. A large reduction of formaldehyde emissions has been and continues to be a major development objective for UF resin manufacturers, with significant progress having been made. Formaldehyde emission classifications are specified in standards and are very often regulated by specifying the required emission classification. Roffael (1993) provides a useful review of UF resins and the factors influencing formaldehyde emissions, while Young (2005) provides a recent review of the tests used to measure formaldehyde emission levels and correlations between these.

Melamine also reacts with formaldehyde to form an adhesive system (MF). While the UF polymer is largely linear, the cyclic nature of the melamine molecule with three amine groups allows a cross-linked structure to develop. Melamine is 4-5 times as expensive as urea, and pure MF is not commonly used in composite panels. An attractive feature of MF resin is its miscibility with urea formaldehyde, allowing resins containing both melamine and urea formaldehyde components. In replacing a portion of the urea, one can achieve the level of cross-linking appropriate to the end use requirement. Low levels of melamine (2-3%) are incorporated in UF resins to improve strength properties for panels using UF resins with low formaldehyde emission characteristics. Higher levels (10-15% melamine) are used to improve the moisture resistance of the panels. The general term MUF is used for both these systems, but an indication of the melamine level is necessary to describe the performance attributes.

Three component formaldehyde systems have also been developed. In addition to melamine and urea these also incorporate phenol and are known as PMUF. The levels of each component can be varied to meet the specific resin performance required at minimum cost.

Tannins derived from tree bark also react with formaldehyde to form durable adhesive bonds. Tannins have been widely investigated, and the subject is well covered in the literature (Pizzi, 1989). The dark colour of the resin limits applications to those where a light colour is not important: indeed substrate colour can show through some overlay materials. Tannins are used commercially in a number of particleboard products notably a structural flooring particleboard panel used in Australia. Compared with the synthetic adhesives, the tannins, together with other adhesive systems derived from natural materials, are much more variable. This has limited their use, with the increased processing necessary to achieve uniformity increasing the cost to the point where there is no advantage. Other adhesive systems based on chemically modified plant derived raw materials such as oils have been proposed.

Isocyanate systems based on polymethylenediisocyanate (PMDI) are being used increasingly. This adhesive has a completely different chemistry to formaldehyde-based systems. While expensive, the absence of formaldehyde emissions is a determining factor in some applications. An attractive feature of isocyanates is the excellent wetting and coverage of the wood surface so requiring lower usage rates. Early on one problem with isocyanates was their ability to stick to the hot platens – a problem solved by use of release agents. Initially these were sprayed on the platens to prevent adhesion, but are now incorporated in the resin system.

PF and PMDI are classified as suitable for exterior rated panels while MUF (at the higher melamine level) is regarded as suitable for service in interior situations where some level of humidity or wetting may occur (bathrooms, kitchens, roof linings). UF resins are generally regarded as suitable for interior applications where contact with water would be unlikely. In addition to the resin system, many other factors such as particle size are significant in panel performance. Durability is discussed in more detail in a subsequent section.

Inorganic binder systems are also used. Cement bonded products both as particleboard and as fibreboard are used in exterior situations with the added advantage that these products resist fire. Generally binder levels in this case are quite high (*c.* 35% wt/wt basis), and the product can be regarded as a particle or fibre reinforced cement board. Gypsum is also used – both particleboard and fibreboard being manufactured with this binder. The inorganic nature of these binders gives panels with improved resistance to fire. Gypsum has the advantage of being a low cost residual byproduct created in the cleaning up of flue gases in coal-fired power plants. Indeed unreinforced gypsum board with paper overlays is the major contender with wood-panels for non-structural wall linings.

Chemical compatibility between the resin and the wood can also be an issue. UF and MUF systems require an acid pH environment (generally <5) to cure the resin at a reasonable rate. The acidity of wood species varies over a wide range and can affect resin cure speed. Acid generating catalysts are generally used in particleboard, these are less effective in MDF where the preheating of the wood prior to separating the fibres generates further acid groups on the fibres (Murton, 1998), and where the much greater surface area of the fibre allows the buffering capacity of the wood to determine the resin cure rate. This aspect is also an issue in cement bonded products where an alkaline environment is necessary to allow the cement to bond with the wood surface.

Within this wide range there are many possibilities. The adhesive purchase price and the required addition rate are major factors in determining the overall adhesive cost for a panel (Table 12.3). Performance requirements, particularly durability may limit the available choice. Overall UF systems provide the lowest cost solution for MDF and particleboard and are used in the greatest proportion of these panels, in excess of 95% for MDF.

7. GENERALISED PANEL PRODUCTION LINE

The production of composite panels (Figures 12.3 and 12.4) is based on the following stages that generally, but not exclusively, proceed in the following sequence:

- Preparation of the particles.
- Drying to the required moisture content.
- Spreading the adhesive.
- Forming the mat for pressing.
- Pressing the panel.
- Finishing the panel: sanding, trimming *etc.*

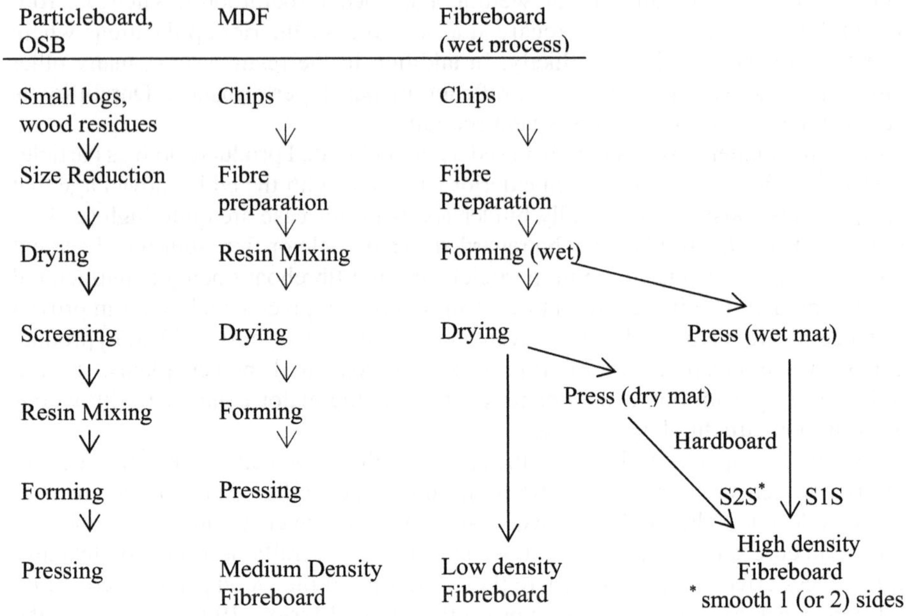

Figure 12.3. Production steps for particleboard/OSB, MDF and fibreboard (wet process).

7.1. Particle preparation

At first, particleboard was made with minimal preparation of the material, often using planer shavings and sawdust directly. As experience was gained, recognition of the importance of particle geometry and size in determining particular performance properties lead to much more attention to these aspects of wood preparation. Today, as particleboard producers have sought to maintain their market against competitors producing the same product and newer panel types, particularly MDF the size distribution and shape of the particles is closely controlled by a range of size reduction processes and size-based separation techniques. The process requirements are determined by the starting material and the particular product requirements. Broadly the size reduction process can be divided into two categories.

7.1.1. Impact mills

Impact mills fracture the material by striking it with a hammer, which is usually rotating at high speed. This approach induces a random failure in the material that probably starts near the impact point and continues either in the general direction of the stress induced by the impact or along the line of a weakness in the material effectively shearing parallel to the grain. The randomness of the process means that neither the alignment of the particle nor its position relative to the impact zone is controlled. The result of an impact operation is therefore an overall reduction in size

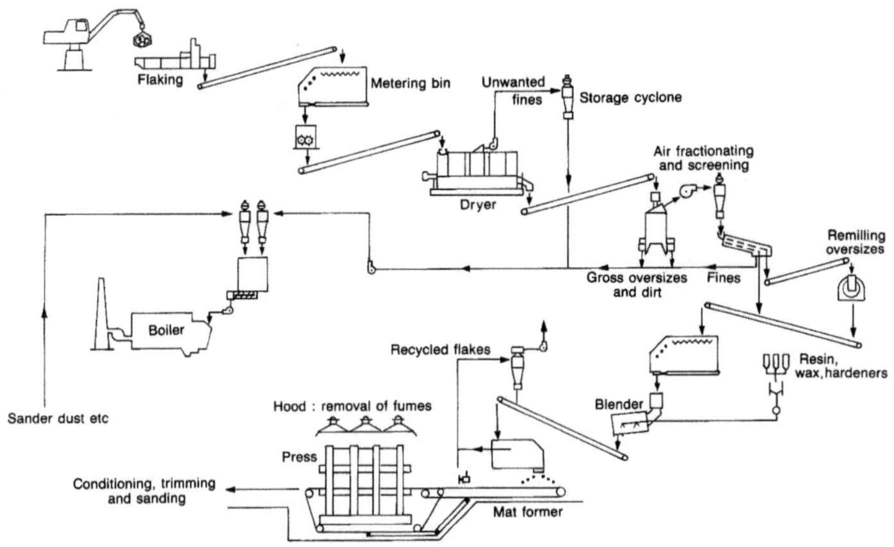

Figure 12.4. Material flows for the manufacture of OSB (Courtesy Siemplekamp, Germany).

of the incoming particles into a number of smaller particles with broad distribution of particle sizes that is determined by the impact energy and the original geometry and fracture characteristics of the material.

Impact devices (Figure 12.5) thus represent a relatively crude approach to size reduction. They are acceptable for processes where a range of particle sizes is required, perhaps to give a smooth surface as well as a coarser core layer. The main advantage is that they require little maintenance and are less susceptible to damage by foreign inclusions (metal, stones). However the random nature of the impact results in a wide range of particle sizes in the discharge some of which may be too small for the product. These machines are usually followed by a screening operation to separate oversize material (that may be processed a second time) and fines from the acceptable particle size range.

7.1.2. Engineered particles and flakes

The other approach to size reduction has been to use a knife to cut the material in a situation where the material is presented to the knife in a controlled manner so that the orientation of the cut and the distance between the cuts is controlled (Figure 12.6). Such devices include veneer slicers (which are not used in the preparation of particleboard furnish), chippers and flakers. These machines rely on close control of the relationship between the cutting knife, the backing plate that holds the wood in relation to the knife and the angle of presentation of the wood.

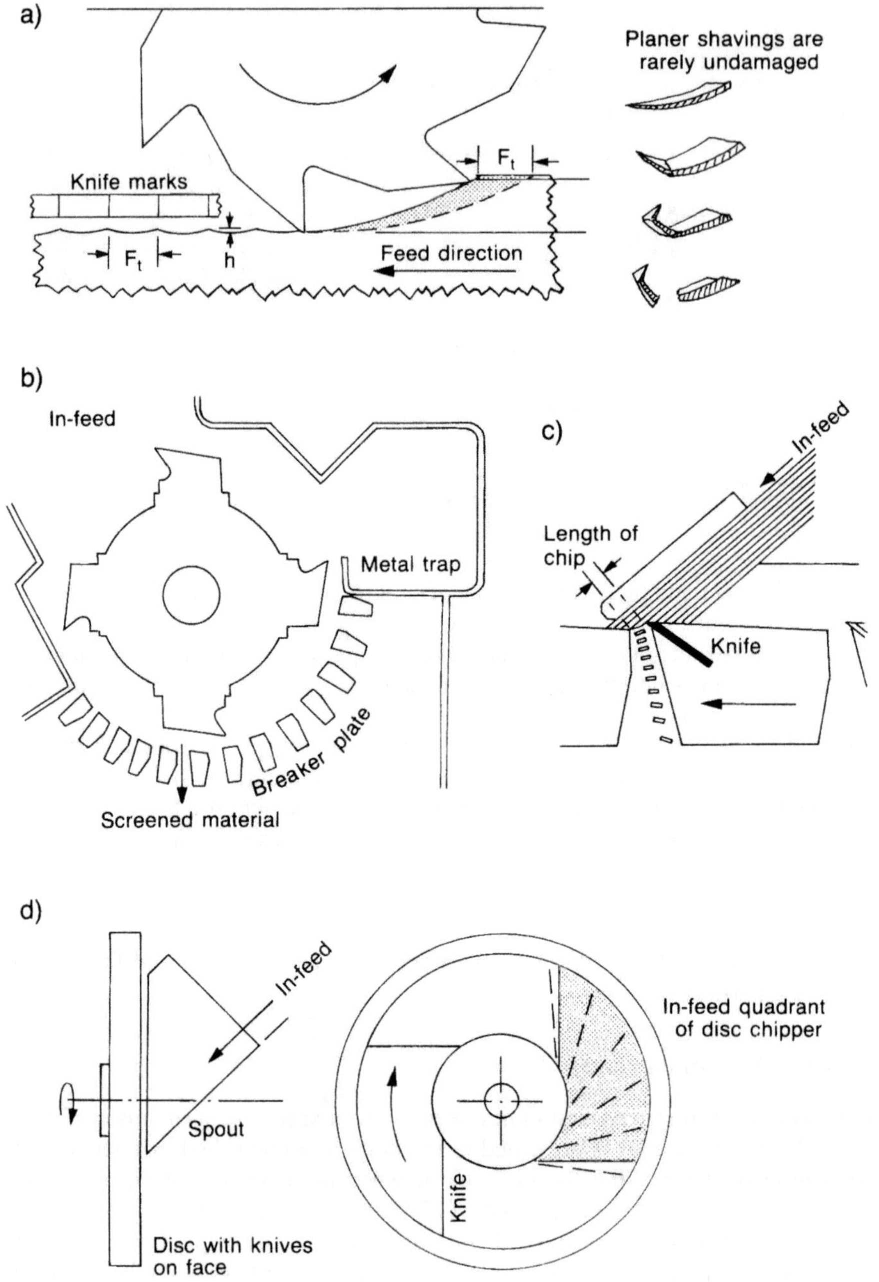

Figure 12.5. The generation of wood chips (Koch, 1964). (a) Planermills produce shavings of variable geometry. (b) A hammer hog reduces coarse material. (c) The cutting action of chipping knives. (d). A disc chipper.

The ring flaker is an example of this type of machine. It is designed to produce flakes for particleboard from chips. The knives are placed in an outer ring facing inwards. The projection of the knives beyond the backing plate of the knife ring determines the thickness of the flakes. A counter-rotating inner rotor with bars controls the presentation of the chips to the knives such that it both aligns the chip flat against the backing plate on the knife ring as well as providing the force necessary to move the chip past the knife. The distance between inner bars on the rotor and the knives must be less than the design flake thickness to maintain this for the last flake to be taken from the chip.

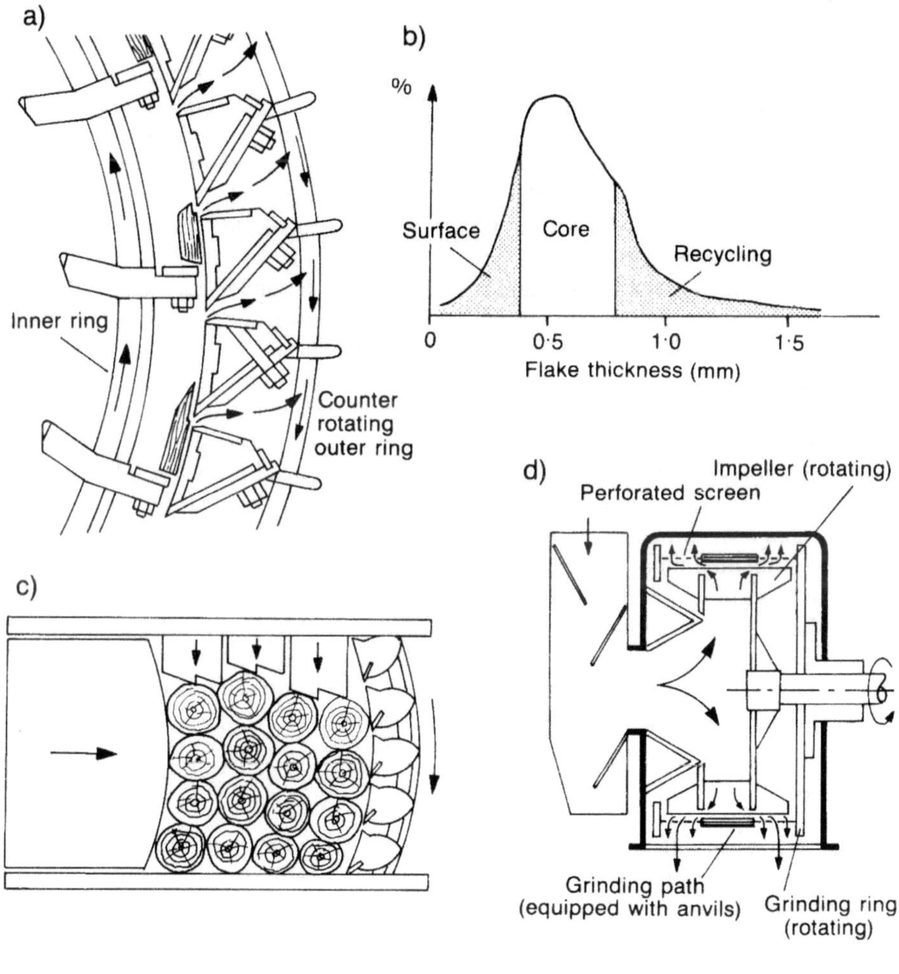

Figure 12.6. The generation of flakes. (a) Cutting action of a knife-ring flaker (Fisher, 1974). (b) Typical distribution of flake thickness (Jager, 1975) (c) Cutting action of a large drum flaker (courtesy Pallmann, Zweibrucken, Germany). (d) Ring refiner (Fisher, 1972).

Particleboard is unique among wood-based panels in that a range of particle sizes is desired. The requirement for particleboard varies, from applications where surface smoothness is the primary requirement to panels where strength requirements are more important. In the first case a range of particle sizes are needed to fill the gaps between the larger particles, to create a smooth surface. Where strength is the primary requirement then the material will be selected by rejecting the majority of the smaller particles at least from the surface of the panel that contributes most to the bending strength. The variable chip quality, arising from utilizing low grade wood resources in conjunction with a hammer mill, can be addressed by screening operations that separate oversize material for reprocessing and undersize particles for use in another part of the panel or as fuel. Screening leads to improved property values while at the same time reducing the amount of resin required. Engineered flakes prepared from chips are often used where improved strength is required, despite the much greater complexity and maintenance requirement of the ring flaker.

The preparation of fibre for MDF uses an impact device (the refiner) to separate the fibres but this follows a thermal pretreatment that softens the lignin. Consequently separation occurs in the lignin-rich middle lamella between the fibres so preserving as much of the cellulosic fibre structure as possible. The preparation of MDF fibre is considered later.

Stranding machines for OSB strand preparation use a similar geometry either in the form of knives placed radially on a disk, or in a cylindrical configuration similar to that ring flaker discussed above. With the input material in log form the force necessary to maintain the relationship between log and knife can be generated by mechanical clamps and pushers. This configuration will result in a small sheet of veneer with a length equal to that of the knife, or otherwise determined by circumferential scoring blades that cut across the grain ahead of the each knife. The path of the veneer sheet as it leaves the knife determines the width of the strand. A sharp change in direction a fixed distance after the knife will cause a fracture along the grain direction, determining the strand width. The stranding machine produces a flake or a strand with all three dimensions determined by the relationship between the knife, the support plate and the path available to the flake as it leaves the knife. Strands produced at the start and end of each log, or at the start or end of each cut may not reach the required quality level. These are separated by screening to be used in the core of the OSB panel, or used as fuel.

7.2. Drying

The approach to drying is constrained by particle size: the larger the particle the longer it takes to dry. Particles, flakes and strands typically pass through a rotating drum dryer, where the particles are lifted by projections (flights) on the drum surface to fall and tumble repeatedly through a hot gas stream. The drum may be arranged so that the particles have several passes, back and forth along the drum (Figure 12.7). The hot gas is usually obtained from a direct combustion source, and flows co-currently with the particles. Inlet gas temperatures can be quite high (up to 850°C) with the burner close to the dryer. The lightest particles are carried rapidly through the dryer in the combustion gas stream before being separated from this in cyclones. Larger particles such as the strands for OSB are too large and heavy to be pneumatically carried and tumble through the dryer to be discharged onto a conveyor: these particles remain in the dryer for up to 5 minutes. The exit temperature of the dryer is typically 105–120°C.

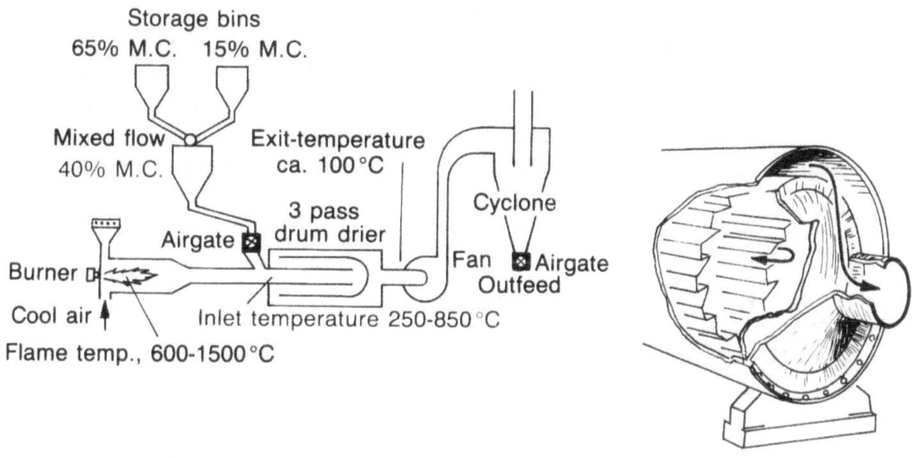

Figure 12.7. A three-pass rotary dryer that is often used in particleboard mills.

The high temperature used in particle dryers causes some breakdown of the wood. Although the evaporating moisture on the particle surface means the wood does not experience the high initial gas temperatures, this increases as drying proceeds. The exit temperature of the dryer is typically 105-120°C. The dryer gas flow is usually treated to remove particulate material as well as volatiles from the wood before discharge to the atmosphere. A range of treatment options are available with a wet electrostatic precipitator being commonly used.

The final moisture content of the particles leaving the dryer is in the range 2-5%. This is determined by the moisture content required in the press which is made up of that in the particles after drying and the moisture added with the resin.

Drying fibre for MDF takes a very different approach. The drying time for individual fibres is much shorter (<<5 seconds), allowing a flash tube dryer to be used. The fibre enters the stream of hot air that carries it to a cyclone that separates the dry fibre from the air. This short drying time has allowed the resin to be added before the dryer: the so-called blowline blending process overcomes the difficulties of mixing fibre and resin in a mechanical blender. The short drying time also changes the dynamics of the dryer, with the fibre being in equilibrium with the humid air stream that leaves the dryer. With the resin also travelling through the dryer, the moisture content of the material leaving the dryer is that required for the mat entering the press, typically 9-12%.

7.3. Resin blending: particleboard and OSB

Mixing the adhesive with the particles is critical in maximizing the resin performance and the approach again depends on the particle size. The aim is to have every particle contribute to the strength of the composite so that the resin particles need to be as small as possible and distributed uniformly over the surface. On contacting the dry particles moisture passes from the resin to the particles so the viscosity of the attached resin droplet increases.

The strands used for OSB are large enough to collect the resin particles – both solid and liquid resins are used for OSB – as they fall through a cloud of resin particles. The resin blenders used in OSB are large drums, typically 3 m or so in diameter rotating about a horizontal axis (Figure 12.8). Lifting bars carry the particles up to the point where they fall through the resin particles produced by the

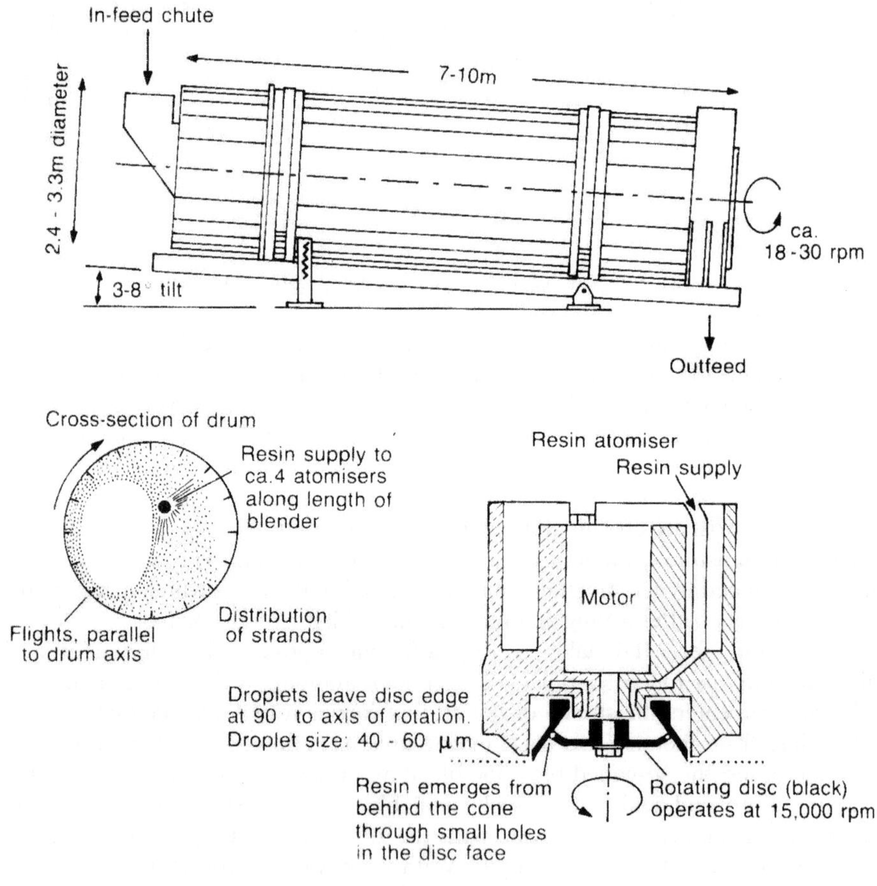

Figure 12.8. A long residence time blender (courtesy Coil Ind, Vancouver, BC).

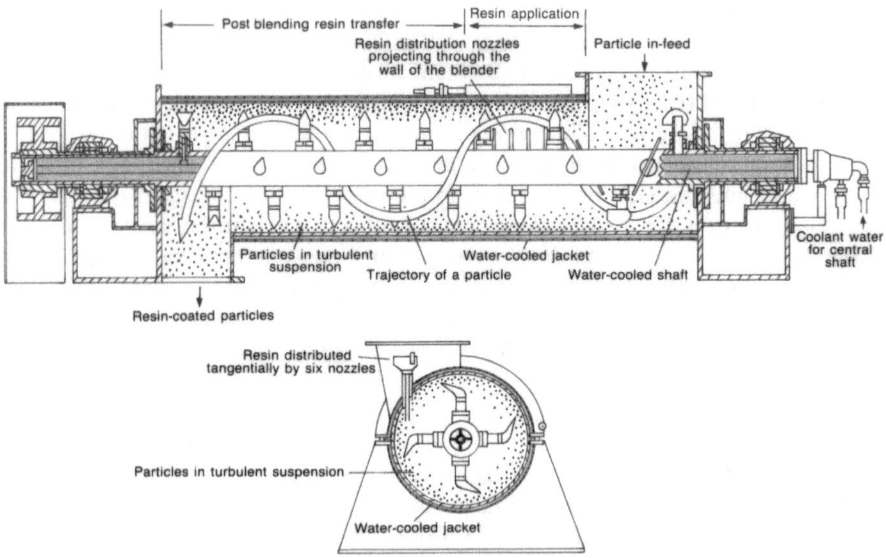

Figure 12.9. A short retention time blender (courtesy Draiswerke, Allendale, NJ).

nozzles. Solid resin is blown into the drum, while liquid resin is atomized in spinning disk atomizers installed along the length of the drum. Separate blenders for the surface and core layers allow different addition rates of resin to be used in the core and surface layers. Different resin formulations or types can also be used with separate face and core blending systems. The mixing for OSB strands is slow and gentle to minimize damage.

Generally particleboard blenders are small compact high speed mechanical mixers – also described as short residence time blenders – with chips remaining within the blender for less than 1 minute (Figure 12.9). The dry furnish enters at one end of a tube containing a central rotating arbor with agitator arms spaced along its length. These throw the particles out to the periphery of the tube while also moving the solid mass along the tube. The resin is injected through nozzles located at the entry end of the tube and projecting through the rotating particle mass around the periphery of the tube. Resin droplets are captured by the particles and further redistributed by the wiping action between particles induced by the movement of the particles within the blender. Fine particles tend to capture more of the resin than do the larger particles, due to their large surface to volume ratios. In some plants, particularly the larger ones, the fines are separated and mixed with a proportion of the resin to ensure that the resin is more evenly distributed between the larger and smaller particles. After blending the fines are mixed with the larger particles before reaching the mat former.

Blender design is essentially empirical, with particular aspects such as blade angle (that determines the direction and rate of material movement along the axis of the blender), rotational speed and nozzle location the basis of performance claims.

'Tack' is important in holding the particleboard mat together as it travels from the forming station to the press. It is difficult to quantify this aspect of resin performance with only subjective assessment of the property. Tack depends on the ability of the liquid resin to form surface tension bonds between adjacent particles in the low density, lightly consolidated mat that leaves the former. It is the result of a complex interaction between the water in the resin, the rate of absorption of this into the dry wood and the size of the resin droplets. The larger the resin droplets the easier it is to form bonds between adjacent particles that are further apart in a lower density mat. While larger resin droplets may contribute to improving the strength of the mat so that it can be transported from the former to the press, there are indications that smaller resin droplets improve bonding efficiency. Too much tack results in material building up on surfaces in the transport and forming areas.

7.4. MDF fibre preparation

With MDF the fibres are much smaller, but have a much greater slenderness ratio than those for particleboard. Fibre dimensions for most species are available, or can be readily determined, so that whole fibres can be characterized accurately. However production of the fibre requires a preliminary breakdown of the wood usually by chipping that will damage some fibres, so that fibre fragments will always be present. The level of these depends on the length of the chip in relation to the length of the fibre. A chip cut to an along-grain length of 25 mm from a material with an average fibre length of 3 mm will cut one fibre in 9 into two parts. Sawdust with a particle length of 3-5 mm will have almost every fibre cut into two parts.

It is these two characteristics – the slenderness ratio, and the overall small size of the particles – that have lead to the distinctive features of the MDF process.

The fibre for MDF is prepared using a low energy thermo-mechanical pulping process. This differs from the thermo-mechanical processes used to manufacture fibre for papermaking in that the thermal treatment is much more intense than for paper fibres, allowing the fibres to be separated while using much less mechanical energy. The thermal and mechanical treatments for chemical pulp, and for thermo-mechanical fibres for both paper and for fibreboard are compared in Figure 12.10.

The chips for MDF are treated thermally to a point where the lignin-rich layer between the fibres softens. This point is characterised by a glass transition point for lignin, generally taken as a temperature of 140°C. To achieve this, the material is heated in a steam atmosphere at a pressure of 8 bar (800 kPa), for a period sufficient to ensure that this condition is reached throughout the material. For chipped material this is achieved in 3-4 minutes. The thermally treated material is then transferred at the prevailing pressure to a refiner, an 'impact mill' that separates the fibres. This sequence in the fibre preparation system, involving the thermal pre-treatment followed by mechanical treatment in the refiner, is shown in Figure 12.11.

The raw material, prepared as chip-sized particles, is delivered to a bin above the preheater, where atmospheric steam is injected to start the heating process. From the bottom of the presteaming bin the chips enter a screw feed that forms a tapered plug

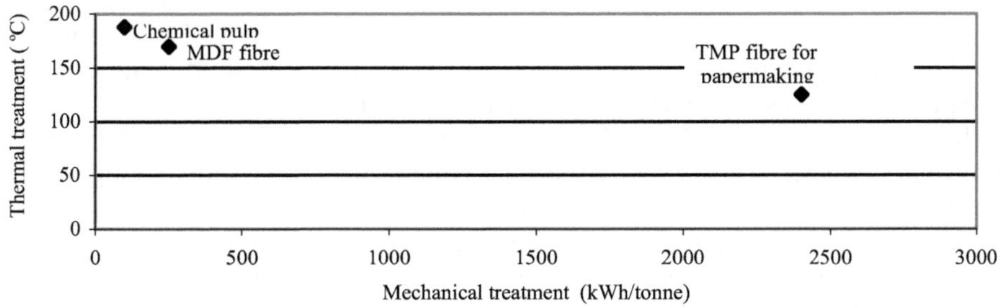

Figure 12.10. Fibre preparation conditions: contrasting primary processing procedures.

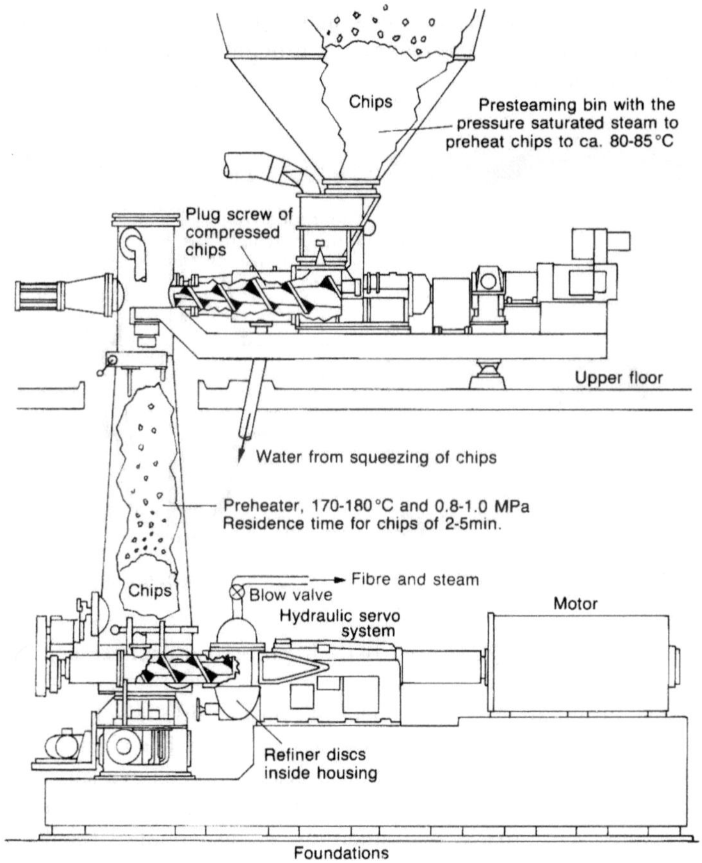

Figure 12.11. Pressurised disc refiner for MDF fibre (courtesy Sunds Defibrator, Sweden).

of material of increased density which creates the seal between the pressure in the preheater and the atmosphere. The speed of the plug screw is controlled to maintain a level of chip in the preheater that is selected to provide the desired treatment time in the preheater to ensure effective fibre separation in the refiner.

The chips move down the preheater pressure vessel and are removed from the bottom by a scraper arm that breaks up the column of material, delivering it to a metering screw conveyor. The speed of this conveyor determines the production rate while delivering the preheated material to the refiner.

Fibre separation is achieved between two discs, one stationary and the other rotating, each faced with a circumferential set of refining plates (Figure 12.12a). The rotating disc has a positioning system, usually hydraulic, that enables the distance between the fixed and rotating disc (0.2-0.4 mm) to be very precisely controlled. Fibre separation is achieved by contact with raised radial bars on the refining plates. The material moves in the grooves between the radial bars but circumferential dams in the refining zone ensure that the material is brought back again into the contact zone between the refiner plates.

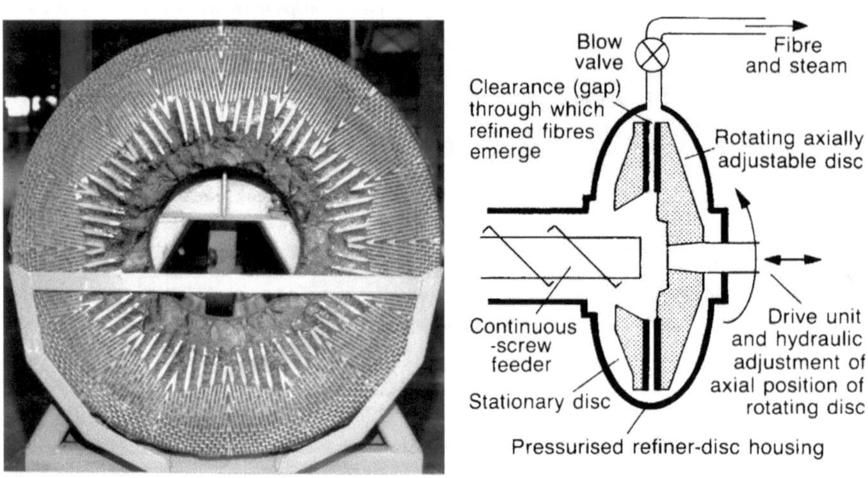

Figure 12.12. (a) Refiner plate. (b) Schematic of the pressurised disc refiner unit.

Refiner plate configuration is the subject of considerable development. Initially the focus was to achieve a satisfactory fibre. Recently this has changed slightly, to minimizing the refining energy consumption necessary to achieve satisfactory fibre quality. Typical refining energy levels are 125-200 kWh/tonne ODF (oven-dry fibre) for softwood chips, with hardwood energies being about 10% lower. New developments in refiner plate design claim to produce acceptable fibre using softwood chip at specific energies below 100 kWh/tonne. Specific energy levels to prepare fibre from sawmill wastes such as sawdust and shavings are higher, with typical refining energy levels of 300-400 kWh/tonne ODF for these materials: no reason for this increased energy requirement has been advanced. Generally the refiner plates used have been developed for chip as feed material and are not optimized for the smaller sawdust particles, or for the lower moisture content of the shavings. Refiner capacities range up to 60 tonne/h for single units with 1.5 m diameter discs driven by motors up to 12 MW in capacity. A typical refiner plate assembly is shown in Figure 12.12b.

7.5. MDF fibre quality

The fibre quality necessary for MDF is a subject of much debate. Reliable assessment of fibre quality is difficult and equipment to measure this is expensive. While panel properties are claimed to depend on fibre quality there are indications that this effect is more likely to be associated with changes in blowline blending conditions than directly attributable to changes in fibre quality. The fibre length for fibre derived from sawdust is probably less than 50% of that for fibre derived from chip. For equivalent strength properties the density may be increased by 5% (725 to 760 kg/m^3) while the resin addition rate may be 10-15% higher than for a panel made with fibre derived from fresh chip. Fibre quality can be mapped as in Figure 12.13, with each quadrant in the length/width axes defining a different area of departure from the ideal case of individual fibres separated without damage.

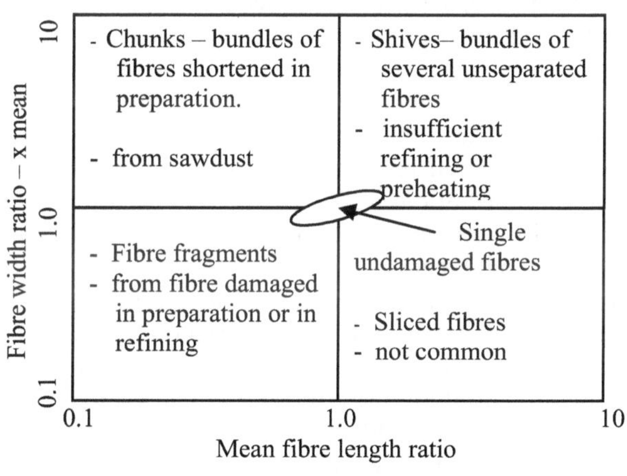

Figure 12.13. Fibre length – width map for defining fibre quality.

There will some fibre components from both the top right and lower left quadrants of Figure 12.13 in all MDF fibre. The shives are only a problem if they are noticeable in the surface of the panel, or if their different swelling characteristics affect overlay performance. Some fibres are always damaged in the preparation of the material before the refiner, so there will always be some fibre fragments. However the operation of the refiner must be such as to minimize these. If the feed material is sawdust or shavings there will be a much greater level of damaged fibres so that the average fibre length will be less.

7.6. Fibre-resin blending

The first MDF was made using particleboard equipment. The fibre was dried and mixed with resin in a particleboard blender. The product had excellent properties, but there were dark areas of varying sizes in the panel that on analysis were shown to have a resin content two or three times the average. These clumps of resin-rich fibre – known as resin spots – were sometimes sufficiently hard as to break saw teeth or to form lumps under overlays. A range of techniques was adopted to break up these lumps, including attrition mills, but with limited success.

The problem was reduced to manageable proportions by changing the point of resin addition to the blowline that carries the fibre from the refiner to the dryer. Resin spots were reduced to an acceptable level, but early experience with blowline resin addition established that higher resin levels were necessary to reach desired property levels. The additional resin requirement was estimated as between 15 and 30%, (Smith, 1998) but there is no published data that allows a direct comparison between the two approaches. One reason advanced for the increased resin consumption with blowline blending is that the thermoset resin has to pass through the blowline and the dryer, both operating at temperatures in the range needed to set the resin. It was claimed (Gran, 1982; Maxwell *et al.,*1984) that this exposure could be sufficient to start the resin cure process.

While dry fibre blending capacity was retained in some plants, the benefits of blowline blending meant that this approach has been adopted for an estimated 98% of MDF produced. Clearly the benefits of the blowline blending approach outweighed the cost of the extra resin needed where this approach was adopted.

Recently dry blending capacity has been installed in some plants with some (typically 20–25%) of the required resin being introduced at the blowline and the rest during the dry blending operation. Resin spotting limits the amount of the resin that can be added at the dry blending stage, but overall cost saving are claimed to justify the cost of the additional blender.

Chapman and Jordan (2003) presented an empirical model of the blowline blending process, based on a fibre-resin droplet interaction model that showed how blowline and resin atomization conditions could be controlled to improve resin efficiency as shown by the IB (internal bond) test. Reductions in resin requirements of 25% are claimed for an optimized blowline, suggesting that the reason for the apparent increase in resin consumption for blowline blending lies in the blowline conditions. The model shows that if the blowline is too large in diameter steam velocities are below those necessary to achieve a good blending outcome in the blowline. If too high a pressure in resin atomizing nozzle is used then the resin jet will hit the far wall of the blowline before atomization is complete. The boundaries for these two conditions limits the resin efficiency that can be achieved.

7.7. The fibre dryer and fibre moisture control

The blowline discharges the fibres, now mixed with resin together with the steam and moisture that accumulated in the refiner, directly into a straight section of the flash tube dryer. These have a straight tube, typically 100 m in length, followed by a cyclone to separate the fibre from the drying air that also conveyed the fibre through the dryer. Temperature measurements along the dryer tube show that most of the temperature drop between inlet and outlet temperatures occurs very shortly after the blowline discharges the fibre and resin into the dryer, typically within 20-30 m downstream from the end of the blowline, i.e. the fibre discharged from the blowline into the drying tube is rapidly dispersed within the dryer airflow. This indicates that the energy exchange between the hot gas and the fibre has been completed within this time. At a gas velocity of 30 m/s the drying time is probably less than one second. Consequently, provided the fibre dispersion condition is maintained, the dryer length can be reduced. In some cases the drying is achieved in a vertical tube rising from the refiner at ground level to the point of the entry to the cyclones that separate the fibre from the drying gas. Entry temperatures range from 100 to 300°C, with exit temperatures usually between 50 and 70°C. The hot gas for conveying and drying the fibre may be heated indirectly with steam or hot oil, but combustion gas, either from dedicated gas burners or from waste combustors is often used, after recovery of heat as steam and or hot oil for other plant requirements, with a consequent reduction in the thermal energy input to the plant.

The exit moisture content from the dryer depends on whether dry blending is used. If the resin is added at the blowline, then the moisture after the dryer is that required in the press, typically 9-13%. If resin is added after the dryer then the moisture leaving the dryer will be lower, so that the water added with the resin will raise the total moisture to that required in the press.

A fibre bin or bunker, sometimes incorporated in the mat former, provides a buffer between the refiner and the press. Bin capacity is quite limited as a tonne of fibre occupies 33 m^3, corresponding to a bulk density for the fibre of around 30 kg/m^3 for a softwood fibre derived from chip. With press lines requiring 10-50 tonnes/h of fibre, the provision of even a few minutes capacity between fibre production and panel manufacture involves a substantial bin volume. There is logic in making this bin as wide as the press line so that the process of distributing the fibre uniformly across the width of the press line can be partially accomplished in the fibre bin.

Press speeds can also be increased if the fibre mat is delivered into the press at an elevated temperature – so reducing the resin curing time. This approach is increasingly being adopted as a means of maximizing press capacity, particularly for thicker panels. The most efficient way of doing this is to maintain as much as possible of the temperature in the fibre as it leaves the dryer. To achieve this, the fibre processing and transport between the dryer and the press must take place under conditions that maintain both the temperature and moisture content of the fibre. The rapid response of the fibre as indicated in the time to complete the drying process requires that both temperature and humidity in processing and transport air must be controlled. The temperature-humidity relationship necessary to maintain a given fibre moisture content in the dryer is shown in Figure 12.14. The figure draws on timber drying data for *Pinus radiata* that is available from a number of sources.

The data applies to all wood particles, but only individual fibres are able to respond in the brief time the material is in a pneumatic transport system.

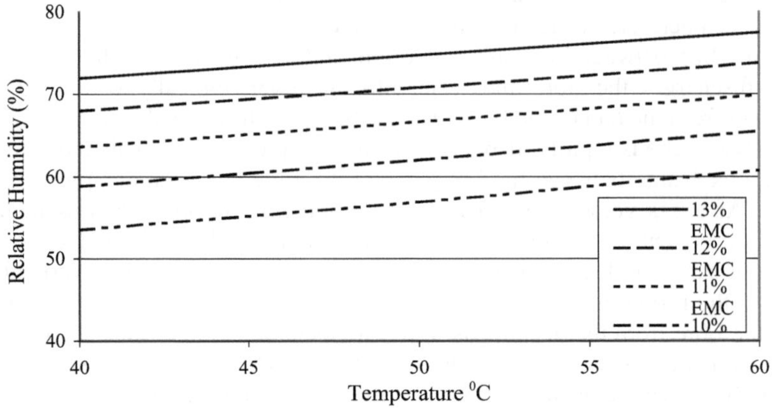

Figure 12.14. Equilibrium moisture content of fibre from the dryer, as function of relative humidity and temperature.

Microwave heaters have been installed on the forming line to preheat the mat (<70°C) prior to pressing. These require space that may not be available in a plant not initially designed for this. Other approaches include injecting steam into the mat, either sequentially from the top and bottom of the mat or in one case, by splitting the mat along its centreline to inject steam before reuniting the two sections of the mat. The mat splitting technique is limited to MDF, while the other approaches can be used with particleboard and OSB mats as well as MDF.

7.8. Mat forming

The major difference between the fibres used to make medium density fibreboard and the particles used to make particleboard is the greater slenderness ratio of the fibres. This is the reason why mechanical blenders designed for particleboard were not satisfactory for MDF. It is also the reason why the bulk density of fibre is low, particularly with fibres prepared from a chipped material. The fibres form networks that are quite strong, with fibres able to span considerable distances. The mats formed from such fibres hold together so that formed or even trimmed edges hold their shape. There is no requirement for tack from the resin to hold the fibre mat together.

There are two approaches to transporting material prepared for the press line. Belt conveyors are used for higher density materials, but the mat becomes very bulky when densities fall. For fibres, pneumatic transport systems, where the fibre is carried by a moving stream of air, are used: although belt conveyors are used for the moving floor of the fibre bin or bunker and for the conveyors that carry the MDF mat from the former to the press.

However pneumatic systems can cause size separation where there is a large range of particle sizes, as with a particleboard furnish. As particle sizes increase so the velocity necessary to carry the material pneumatically becomes greater, so that impact damage also increases. For this reason pneumatic conveying is not used for OSB strands after the screening to separate the strands onto face and core streams. While process considerations determine the type of conveyor selected, belt or chain conveyors have a higher capital cost but lower operating cost (electrical power).

Both approaches are used in the mat forming stage. The object is to form a mat the width of the press, with a weight per unit area such that when compressed to the required thickness in the press, it will have the density desired in the final product, at all points in the panel. Typically the density variation across the width of the mat should be within 2% of the mean value, with density measurements based on 100 mm square samples from the panel. To achieve this it is necessary to distribute the prepared particles uniformly across the width of the pressed mat (the cross direction) and also along the panel (the length direction). The mat former that performs this operation has at times used both mechanical and air transport systems to achieve this; sometimes in the same forming process.

7.8.1. Length direction

To achieve uniformity in the length direction the prepared furnish must be delivered into the mat at a uniform rate sufficient to meet the overall weight of material entering the press. A scale on the forming line weighs the mat as it passes over to provide a control variable that is used to adjust the volume metering rate to achieve the overall weight target for the mat. Two approaches are used. In MDF plants a scalping roll trims the top of the mat, with the material trimmed off being returned to the former. The scalping roll height is adjusted so that the material passing under the roll is that required for the press. The second approach is used for particleboard and for OSB and is being adopted in recent designs for MDF mat formers. In this case the material is discharged from a bin or silo in what is essentially a volume metering device at the rate required by the mat, together with an allowance for side trim.

7.8.2. Cross direction

Uniformity across the width of the mat requires that the prepared particles be distributed uniformly across the width of the mat. This can be measured off line as discussed above, but it is usual to provide online measurement of weight variability across the mat by a scanning density gauge that measures the density of the mat by absorption of radiation. These instruments indicate persistent areas where the weight per unit area is outside the desired limit. To achieve the weight profile across the mat requires some means of taking the uniform mass flow rate determined by the overall mat weight requirement and spreading this evenly across the width of the mat. With MDF mat formers this has been achieved by displacing the airstream carrying the fibre to the mat former from side to side. Both mechanically driven oscillating nozzles and air jets have been used to achieve this. To adjust the placement of MDF fibre across the width of the mat the nozzle mechanism can be adjusted, or a variable vacuum applied beneath the scalper roll to change the density of the mat at specific points across the width of the mat.

Recent designs for fibre mat formers have tended to move towards the mechanical formers used for particleboard and for OSB. As noted earlier, the low bulk density of prepared material requires large storage volumes. If the storage bin (or bunker) is made the same width as the forming line then it becomes part of the mat forming system. In this case the material is distributed across the width of the bin by an oscillating conveyor, moving across the width of the bin. By controlling the speed of the oscillating movement at each point as the conveyor moves across the width of the forming line the weight distribution across the mat can be controlled.

7.8.3. Orientation

OSB strands are oriented, along the mat on both surfaces, and across the mat for the core or central layer of the mat (Figure 12.15). To form the face layer a disk screen is used to align the strands lengthwise along the mat. Once aligned the strands fall through between the discs onto the mat. The disks align the strands until they are able to fall between the disks. By increasing the distance between the rotating shafts along the length of the former it is possible to achieve some sorting by length of the flakes.

For the core layer of the OSB mat the flakes must be able to fall free. They are caught in a series of rolls running across the width of the mat, with radial pockets that the strands fall into. The pockets then discharge the aligned flakes so that they fall across the width of the mat. There are three forming stations for OSB. First the bottom face is laid on the moving forming belt which then moves through the cross forming station that places the core flakes. Finally the top face of the mat is laid down to complete the process.

Particles falling onto a moving belt develop an inherent degree of alignment arising from the inertial forces involved on first contact with the mat. This alignment in reflected in a higher bending strength in the length direction of the mat than across the mat, a feature common to both particleboard and MDF. The edges of the mat are influenced to some extent by the board against which the edge is formed and it is usual to trim this after forming, returning this material to the forming station.

7.8.4. Particleboard mat forming

Particleboard formers use several techniques to achieve a smooth transition between the smallest and the largest particles in the mix. To achieve a symmetrical particle distribution from top to bottom of the panel the forming stations also need to be symmetrical, so that the top face of the panel is laid over the bottom face using a mirror image of that used for the bottom face. For the surface layers of the mat a wind sifting former is often used. This uses a controlled flow of air to separate the particles, the larger heavier ones dropping more quickly, the finer ones being carried further before dropping out of the air stream (Figure 12.16). The degree of separation is determined by the air velocity. The wind sifting former can be

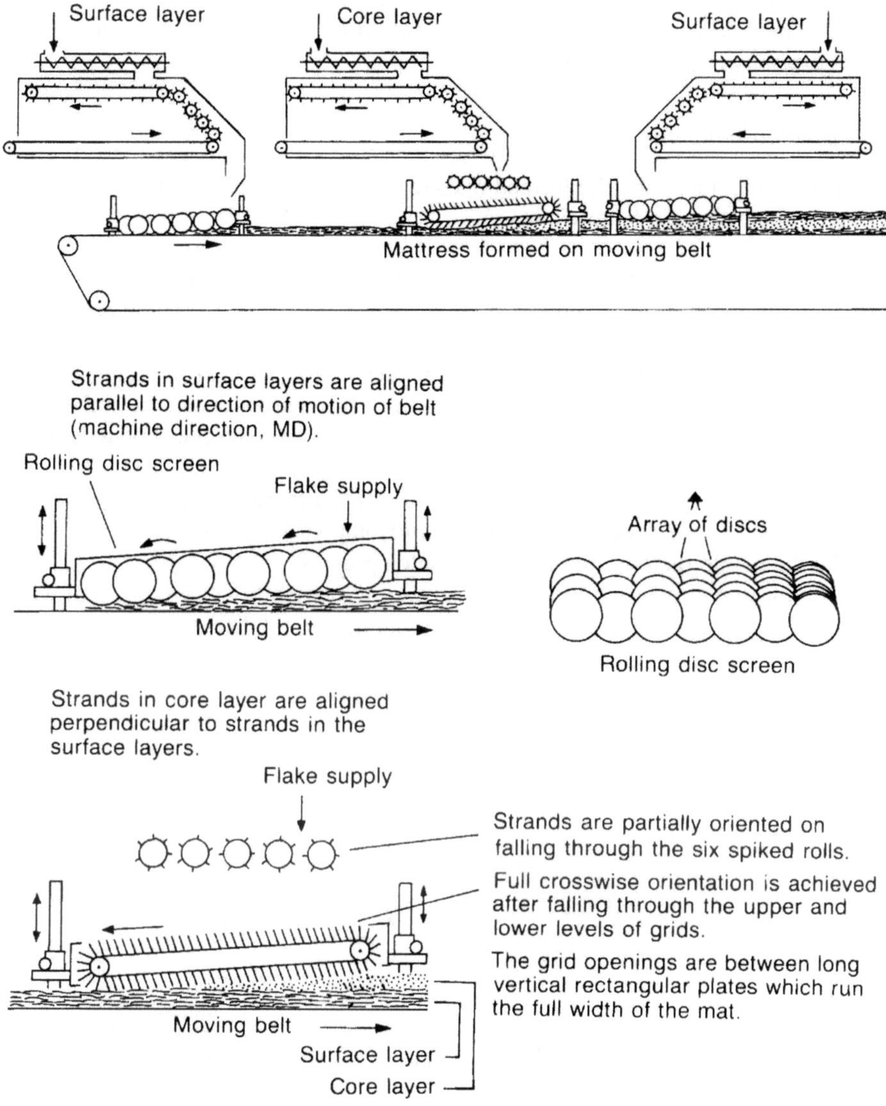

Figure 12.15. Orienting of strands along and across the mat for OSB. (courtesy Siemplekamp Krefeld, Germany).

configured to place the finest particles at the outer layers of the mat, or reversed so that the fines are placed in the centre of the mat thickness. Alternative forming arrangements use variations on the disc screen concept, with rotating elements agitating and aligning the particles while at the same time presenting an opening that particles of a given size are able to pass through. By varying the roll spacing and configuration a graduation of particle size is achieved. The mechanical formers are claimed to have lower power requirements, and do not have particulate-carrying air streams that may require cleaning.

7.9. The mat forming line.

The forming line transports the formed mats to the press. It is a conveyor system that may incorporate specific features to:

- Improve mat properties by cold pressing to increase its density. This increases its strength and makes it thinner so that the mat requires a smaller opening in order to enter the press.
- Apply moisture or other materials to the surfaces of the mat to increase initial heat transfer rates in the press or to modify mat characteristics in some other way: this includes release coatings to reduce sticking to the press platens or the steel bands of a continuous press.
- Trim the sides of the mat to remove the outer portions where the mat characteristics are affected by forming or precompressing operations.
- Cut the mat to the required length for batch presses.
- Monitor mat properties for overall mat weight, variability across the mat width and for the presence of metal or other material that could damage the press.
- Allow mats that contain foreign materials, or with weight variability outside density specifications to be removed before the press.

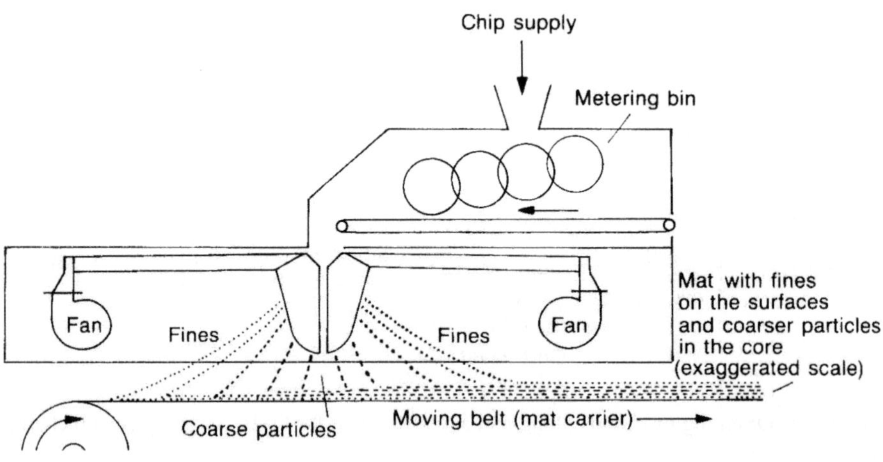

Figure 12.16. Schematic representation of an air sifting forming head producing a graded particle size distribution. The manifolds are alternately forward and backward facing vertical slits offset to allow the particles to be carried through by the airflow (courtesy Bison Werke, Springer, Germany).

7.9.1. The precompressor

The precompressor is a cold press that compacts the mat. Precompressors are used on almost all MDF lines, and also on some particleboard lines. The objective is to reduce the thickness of the mat from the former by increasing its density. As this occurs much of the air in the mat is expelled, reducing the effect of this on the compression of the mat and on heat transfer in the hot press. The mat becomes stronger as a result and is better able to traverse gaps between conveyors on the forming line, and pass through the side trim saws without damage. It also reduces the size of the press opening necessary to accommodate a mat of given weight. For an MDF mat, the formed density of the mat is very low (*c.* 30-50 kg/m^3), resulting in a very fragile mat. Batch precompressors have been used on batch press lines, following the example of the pre-presses used in plywood manufacture. However nowadays continuous inclined belt precompressors are standard.

As the mat travels down the line an inclined top belt makes contact, slowly compressing the mat as it moves through the press. The inclined belt is normally woven to allow air displaced from the mat to escape through the belt. As the force required to compress the mat increases the loading rolls become larger, and high strength belts are needed to move the mat through the zone of maximum compression. The precompressor is able to compress the mat to the density of *ex*-press panel. The thickness and density changes through the precompressor for an MDF mat are shown in Figure 12.17. The height of the MDF mat is reduced from 20-25 times *ex*-press panel thickness to 8-10 times this for a softwood fibre, measured after both the elastic and viscoelastic recovery of the fibre. Most of this reduction in height is due to a rearrangement of the fibres within the network laid down in the mat former. A study (Thorbjornsson, 1985) showed that the reduction in mat height was greater where a force was applied and released several times than where a greater force was applied in a single application. This suggests that the reduction in load allows the fibres to move relative to one another so that the lower mat height is achieved without the residual strain that arises where fibres are unable to move within the fibre network.

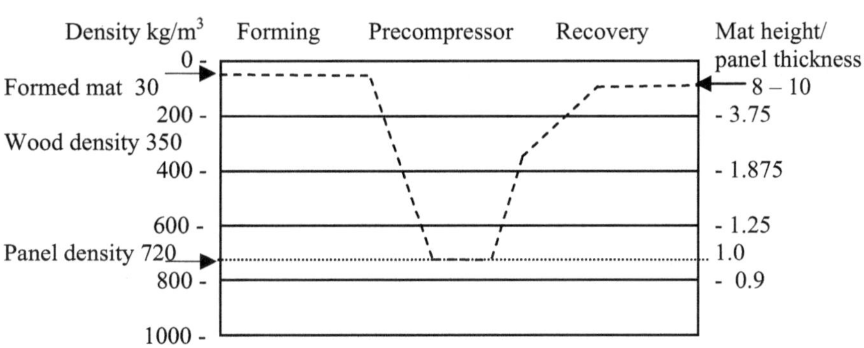

Figure 12.17. Density changes and mat height (as a multiple of the desired panel thickness) in a precompressor on an MDF forming line.

The stronger mat after the precompressor is better able to cope with the edge trimming saw and, in the case of a batch press, the cross saw that divides the mats into the length necessary for the press. It also enables the mat to bridge the space between the conveyors necessary to carry the mat into the hot press. One of these conveyors is able to retract to create an opening in the forming line that allows a mat containing foreign material to be removed ahead of the press.

The mats for particleboard are of higher initial density than those for MDF, typically 150-200 kg/m^3 on forming, so that a precompressor is not required to reduce mat height. However the particleboard mat tends to be fragile as a result of the reduced slenderness ratio of the particles and relies on 'tack' between the particles to maintain mat integrity. The precompressor is needed to improve the mat strength so that gaps between conveyors do not cause mat damage. As will be seen in the discussion of press types changes between conveyors are necessary in multidaylight presses and with continuous presses where the mat must be transferred between the forming belt and the steel belt that conveys the mat through the press. Precompressors are not used on OSB lines.

The forming line may also include the means for heating the mat ahead of the press. Microwave heaters have been used. Other systems apply steam or a steam/hot air mixture to heat the mat. Typically the mat can be heated to 50-60°C before the press. Temperatures in excess of this are likely to result in some resin cure ahead of the press.

7.10. The hot press

In the press, the prepared material is subjected to pressure to compress the material to the required thickness and to heat to cure the resin to form the bonds that give the panel the required strength and other properties. The press has a major impact on the plant design.

- The platen size determines the maximum size of the panels that can be produced and how these can be cut into the sizes required by specific markets. The press size can easily be specified for plants serving a local market with a preferred panel size. The decision is much more complex for plants seeking to meet the requirements of diverse international markets.
- The total area of the platen(s) determines the capacity of the plant. Once installed it is not easy to increase this, so that this limit needs careful consideration when equipment is sized.
- The rate of press closing, particularly under the high pressures that develop in the initial compression of the mat, has an important impact on the development of the density profile through the thickness of the panel. For a batch press the hydraulic pump capacity determines the closing rate. Increasing this rate reduces the quantity of material in which the resin has cured before the required face density has been achieved, and which must be removed in sanding.

7.10.1. The batch press

Initially particleboard was produced in batch presses developed from those used in the plywood industry. New approaches to transferring the fragile particleboard mats into the press were necessary. These required a greater distance between the platens to allow the thicker mat on its carrying tray to be loaded in the press, so reducing the number of openings that could be accommodated within a given press structure. On the other hand the particleboard mat was not subject to the size limitation of veneers, so that the platen could be both wider and longer than those used for plywood production. The benefits of larger ex-press panels were clear both in terms of minimising edge trim loss and in flexibility of cutting into panel sizes required by the market. Platen widths of 2.4 m are common in batch presses for particleboard and MDF, with continuous presses up to 3 m wide for MDF. Wider presses are used for OSB, with the widest press able to produce 3.6 m wide panels (Natus, 1996).

Batch presses are configured either as multidaylight, or single opening systems. The multidaylight press has a number of platens stacked vertically in a structural frame. The intermediate platens are held in position by a simultaneous closing system that ensures that the overall movement of the press is distributed evenly between the platens. This system of levers ensures that the weight of the platens above a particular opening in the press does not affect the loading. It is not strong enough to overcome the compressive load of the press, and incorporates springs or other mechanisms to allow the platen to move in response to the resistance of the mats on either side of the particular platen as loads increase.

The multidaylight press system requires a loading structure that accumulates the mats for the next press load while those in the press are being compressed. The loader has positions for each mat spaced according to the distance between the platens when the press is fully open. The loader moves vertically so that mats from the press line can be loaded into each position. When all mats for the next load are in position and the press operation is complete, the press opens. The loader trays carrying the mats move into the press, pushing the pressed panels out into an unloader that accepts the panels from all openings. As the loading trays retreat conveyors in the trays deposit the new mats directly onto the platen surfaces. When the loader trays are clear of the press it can close. Loading the press takes a fixed time, usually somewhere between 30 and 50 seconds.

The other type of batch press is the single opening type. As the name suggests this press has only one opening, but uses very large platens – up to 2.4 m wide and up to 27 m long. This system uses a steel conveyor belt that runs from the forming station through the press. The mat is formed on this conveyor while the panel is in the press. As the conveyor belt is stationary while the press is closed the mat former moves along the belt to lay the mat down. When the press operation is complete the press opens and the conveyor belt moves to discharge the pressed panel while moving the prepared mat into the press. With the mat travelling on a single conveyor mat integrity is less of an issue. The press is a downward acting one with the hydraulic cylinders acting to push the top platen down on the mat. This allows the press to open just sufficiently to allow the mat to enter the space between the platens, so that the press closing time is minimized.

A further variant of the batch press is the steam injected press. The platens have holes that allow steam to pass from headers through the platen into the material in the press, where it condenses to heat the material directly. When the resin has cured the headers can be connected to a vacuum system that reduces the pressure within the mat, so that steam flashes off and is drawn back out of the panel. The additional complication of a steam supply means that this approach has been implemented only in a single opening press. The approach is useful for very thick panels. Here press times would otherwise be very long because the heat must transfer from the platens to the centreline of the mat, but there are some limitations. While the steam travels rapidly through the mat, often with the flow from one side being delayed slightly to avoid the air in the mat accumulating in the centre, it has to pass through the faces of the panel. This limits the face density that can be achieved in the steam press.

7.10.2. Continuous presses

A continuous press was long an objective in the industry. There are a number of benefits of continuous pressing. From the panel perspective, the ability to determine the length of the panel after the press provides greater flexibility in panel size. Trim loss is also reduced.

Continuous presses also have no dead time associated with loading the mat and unloading the panel after the press, and are thus able to manufacture thin panel more economically than batch presses. There are at least four manufacturers of continuous presses, and today this type of press would normally be selected for a particleboard or MDF plant.

The first fully commercial continuous press was the Bison Mende calendar press (Figure 12.18.). This adopted the large diameter heated calendar rolls used in the paper industry to dry tissue papers. The heated calendar roll provided one 'platen'. The mat was formed on a continuous steel belt that carried the mat to the calendar roll and then around the roll. Additional heated rolls applied pressure and heat to the mat at points where maximum pressure was needed. The resin cured as the mat was heated and the finished panel left the calendar roll as a continuous strip that was taken to a sawing station. The steel belt returned to the forming station. In the Mende press the panel is formed and the resin cured in contact with a curved surface. The panel is straightened by inducing a reverse curve in the panel as it leaves the press, but this limits the thickness of the panel that can be produced. Calendar diameters of 3, 4 and 5 m are used. It is claimed that 12 mm thick panel can be produced on the 5 m diameter calendar roll. In practice this type of press is rarely used to manufacture panel above 6 mm in thickness.

The continuous production of thicker panels needed a straight line path through the press that required new solutions to moving the prepared mat through the heating and pressure zones needed to create the panel (Figure 12.19).

The first straight through press was the Küster press (now owned by Metso), this was followed by presses using a similar design by Siemplekamp and Dieffenbacher, both major suppliers of batch presses. The straight through press design uses two

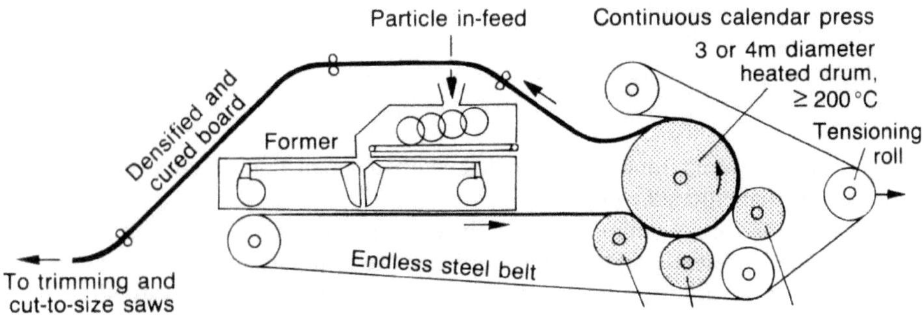

Figure 12.18. The Bison-Mende thin panel press cures the panel as it moves around a large diameter heated calendar roll (courtesy Bison Werke Springe, Germany).

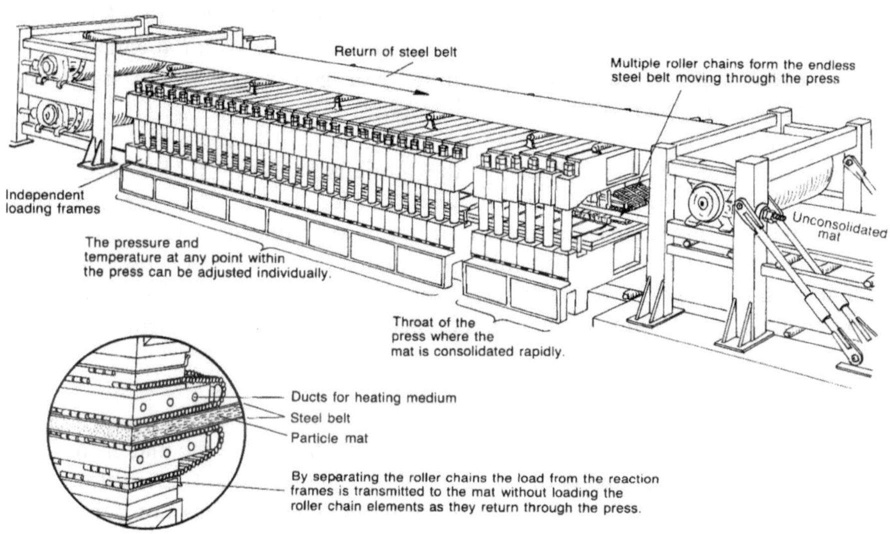

Figure 12.19. The Küsters continuous press. Similar presses are able to press a panel up to 3.1 m wide, with a press length up to 60 m. Maximum speed of the panel through the press can above 60 m/min. (courtesy Küsters, Krefeld, Germany).

steel bands to carry the material through the pressure and heating zones of the press. Some means is needed of controlling the friction between the moving steel bands and the stationary platen. In the case of the three suppliers just mentioned, a system of rollers between the steel band and the heated platens operates as a roller bearing to minimize the friction, while also transferring heat from the platen to the steel belt and then to the mat. Another straight through design by Bison uses a circulating oil film between the platen and the steel bands to control friction and transfer heat. This has not been particularly successful in the market.

There are other advantages in the continuous press that are specific to pressing panels.

- The press loading time is eliminated, with the mat compression starting as soon as the mat enters the press.
- Press design can be optimized, with the different stages of the pressing process occurring at different points along the press; the structure and configuration of the press can be adjusted to suit the pressure and temperatures required at that point in the press cycle. High pressures are only needed for the initial compression of the mat, perhaps 25% of the total press length. This level of pressure is not required in the middle part of the press so that the design pressure can be reduced. It is usual to provide a high pressure zone at the end of the press to allow the final thickness to be achieved. Platen temperatures can be varied along the length of the press. In one design, the last 25% of the press length can be configured to cool the mat, reducing the steam pressure within the mat.
- The platen distance can be set at each frame of the press so that a reproducible time – platen opening profile can be achieved.

Continuous presses have become the preferred approach for MDF and particleboard plants. The capital cost is higher than for an equivalent capacity batch press line but in most cases the benefits in terms of reduced material losses and improved product quality outweigh this. Continuous presses are also used for OSB production and for LVL manufacture, a product where the ability to produce material in a range of lengths is important in meeting market requirements.

7.10.3. Pressing time

The panel needs to be in the press long enough for the heat from the platens to reach the centreline of the panel and cure the resin, forming bonds of sufficient strength to allow the panel to be handled as it leaves the press. This depends not only on the platen temperature (or that of the steel transport bands in the case of a continuous press), but also on the rate of heat transfer through the mat. This is influenced by a number of factors such as the density of the panel and the resin type, and most significantly on the moisture content of the material in the press.

An empirical method is used to determine the pressing time and so the press size necessary to meet a particular capacity. Typical press factors for continuous presses are given in Figure 12.20 for particleboard, MDF and OSB.

Press factors are quoted as seconds/mm panel thickness. Generally press factors increase with thickness as the distance the heat has to move through the panel increases. But the curves are not smooth. The discontinuities result from changes in panel density or pressing strategy. Press factors are lower for continuous presses, a result that might not appear logical considering the need to transfer heat from the hot

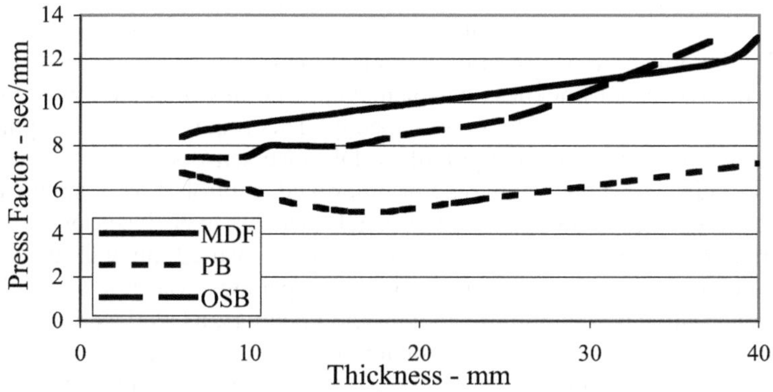

Figure 12.20. Typical continuous press factors for particleboard, MDF and OSB of various panel thicknesses using a continuous press (courtesy, Dieffenbacher, Eppingen, Germany).

platen, through the rollers and into the steel bands before it can move into the mat. However the rapid initial close of the press, together with the level of control of temperature and platen distance along the press result in reduced press times.

Press factors are used to determine the relationship between press size and plant capacity in the design phase. For an operational press, the press factors become a reference point with operational strategies generally resulting in press times below those determined by the press factors.

The press capacity (m^3/h) for panel of a given thickness is derived as follows:

$$\text{Press capacity} = 3.6\ n\ L\ W\ t/(t\ p_t + K)$$

where n is the number of daylights (n = 1 for a continuous press), L and W are the press length and width (m), t is the panel thickness (mm), p_t is the press factor (s/mm) for this particular press type/panel type, and K is the loading time (s) for a batch presses.

7.10.4. Press control

The objective in closing the press is to compress the material in the press to the required density while heating the material so that the resin cure will occur more rapidly. The press has the capability of compressing the material to a very high density, well beyond that required for the product, and must be controlled so that the desired final density of the panel, determined by the final thickness of the panel, is achieved.

With a plywood press, there is very little advantage in compressing the material beyond that point necessary to achieve a satisfactory level of bonding. In fact with the bending strength of the panel proportional to the square of the thickness: any reduction in thickness will adversely affect the bending strength of the panel.

However the property levels of both particleboard and MDF are density dependent, and this is used as one of the controllable variables for these properties.

Initially the objective was simply to control the closing of the press to the point where the desired panel thickness was achieved. The next objective was to reduce the closing time to maximize the press capacity. This required a more sophisticated approach to control of the press hydraulics. Distance sensors to measure the platen distance provided new opportunities for improved control. For a single opening press, position sensors allowed control of panel thickness at each sensor.

With multidaylight presses the distance sensors measured the aggregate distance over all openings. The thickness in individual openings was not controlled, and depended on relative mat conditions in adjacent openings, as well as the straightness of the platens at that point in the press: platen bending is not uncommon, particularly if a mat is incorrectly loaded in the press. To overcome the weight of the platens a system of simultaneous closing arms, designed to ensure that the weight of the platens does not contribute additional load on the mats below, is used. At first, these did not have the capability to impose additional loads on individual mats, but as multidaylight presses have responded to the improved control possible in a continuous press, thickness control on individual openings through a simultaneous closing system designed for greater loads has been offered by some manufacturers.

The straight through continuous presses enable very precise control of the process to be achieved. As the mat moves through the press, the time axis of the press cycle corresponds with the distance along the press, so that the different stages of the pressing process occur at different points along the length of the press. This allows the design of the press to be optimized for the particular process that is occurring at that point in the press. The platen temperature can be varied at different points along the press with one design, the Küster press, now offered with a cooling zone in the latter part of the press.

A number of studies have examined the relationship between press closing and panel properties, specifically bending strength. Overall, the average density of the panel is determined by the mass of material in the press as it closes and the thickness of the finished panel. However the distribution of material within the thickness of the panel can be controlled to give the desired density profile through the thickness of the panel. This profile, known as the vertical density profile (VDP), has a significant effect on the properties of the panel. Typical vertical density profiles for particleboard, MDF and for OSB are shown in Figure 12.21.

The vertical density profile provides a means of optimizing the press operation, as well as panel performance. While initially undertaken in the laboratory, the VDP can now be measured on the moving panel as it leaves the press, using an x-ray backscatter technique (Dueholm, 1996). This is an expensive tool, but the purchase is increasingly common as the value of real time VPD information is appreciated.

The density profiles reveal density areas at both surfaces, with a lower density in the centre of the panel. The high surface density gives good bending strength and surface finishing characteristics. A low density in the core of the panel allows these surface properties to be achieved with an average panel density that is only 60-70% of that at the surface. For particleboard and OSB the profile between the peaks is a

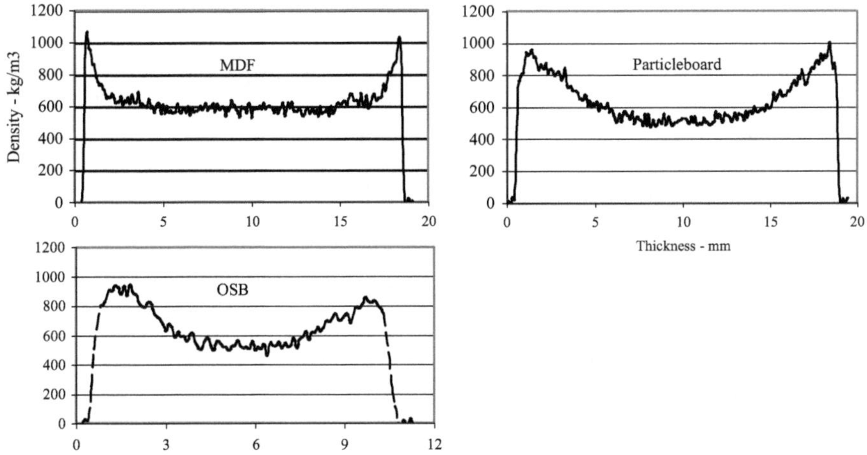

Figure 12.21. Vertical density profiles for particleboard, MDF and OSB.

relatively smooth curve. For MDF the density peak is more sharply defined, with a definite transition between this and the flat density in the core part of the profile.

The density profile results from differences in the compressive response of layers within the mat. Closing the press creates a strain on the mat; the mat responds to this strain depending on the local stress–strain relationship at each point. Apart from density, the only variables that change (both in time and position) during the closing of the press are the moisture and temperature within the mat (Figure 12.22). The degree to which gradients in these variables are established during the press closing determines the density differences between face and core that can be generated. The difference between the MDF profile and those for particleboard and OSB suggests that such gradients are more easily established within the fibres of the MDF mat. Preheating the mat ahead of the press also reduces the ability to create such gradients within the mat, making it more difficult to generate density profiles with sharply defined features. This situation has been modelled (Figure 12.21) with the objective of predicting the VDP of the panel from the press closing conditions, the initial conditions of the mat as it enters the press and the temperature and moisture changes within the mat as the press closes.

The VDP provides a means of identifying the contribution of the press in determining panel properties, separating this effect from other possible causes such as resin mixing and performance, or particle size/shape effects. It also recognises the significant effect of density on the properties of panel products.

7.10.5. Processing panels after the press

After the panel leaves the press it is cut to the required length (for continuous presses) and the edges trimmed. Panel thickness is measured at several points across

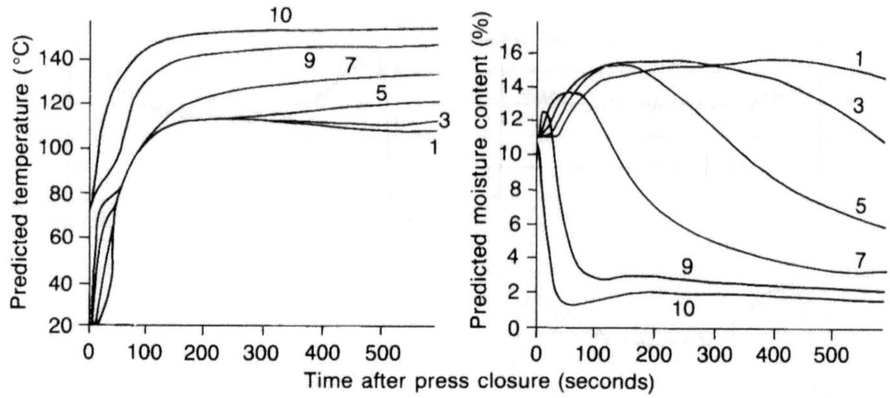

Figure 12.22. The effects of time on temperature and moisture content within a conventionally pressed panel are shown for different layers within the panel. Layer 1 spans the midpoint of the thickness of the mat while layer 10 is in contact with the hot platen (Humphrey and Bolton, 1989).

the width of the panel and acoustic transmission through the panel used to identify areas where delamination within the thickness of the panel has occurred. Panels bonded with UF resins must be cooled before stacking. If the temperature in the stack is above 65-70°C then hydrolysis of the resin can occur resulting in a loss of bond strength. Cooling is achieved by holding the panels in individual slots in a cooling wheel that allows each panel to lose heat to the surrounding atmosphere. During the cooling period the temperature of the panel falls until the temperature in the subsequent stack is below 60°C. During this time the moisture profile through the panel also equalises, with moisture moving from the centre of the panel back to the faces. If the cooling process is too rapid the *ex*-press moisture profile will be frozen in the panel leading to subsequent instability. Cooling times are usually based on 30-45 min in the cooler for the thickest panel produced on the press line. MDF and particleboard panels are then sanded to the finished thickness, with a finishing grit ranging from 120-180 to provide the necessary surface smoothness.

OSB is generally not sanded to a smooth surface, but one face is deliberately roughened, either by a patterned platen in the press, or by a coarse sanding, to provide a non-slip surface.

8. PRODUCT STANDARDS AND PANEL PERFORMANCE

Product standards provide established performance levels that have been found satisfactory for typical end-use applications for a given product. Standards evolve in that they build on methods and approaches that have been found to be suitable for similar products. Initially standards were local product specifications. These have been extended both in range and in scope firstly as national standards and increasingly as international standards to cover products in international trade. Often product standards are specified as a means of compliance with building codes.

For panel products these standards provide both test methods, together with the property levels required to meet specific performance levels. Product standards can be either be prescriptive, specifying the permitted means necessary to achieve a particular end, or performance based, where a particular property level must be reached but the means for achieving this are left to the manufacturer. Most standards are a mixture of both. Density is not generally specified, apart from a general range: MDF is specified in some standards as lying in the range from 400 to 900 kg/m^3.

Standards may cover:

- Tolerances on panel dimensions – length and width, squareness and thickness.
- Strength properties – bending strength, bond strength.
- Environmental issues – formaldehyde emission characteristics.
- Response to water.
- Durability assessment.

The strength properties required are generally the bending properties, measured as Modulus of Rupture (MOR) and less commonly as Modulus of Elasticity (MOE). Both these are significantly influenced by the surface density of the panel. The tensile strength across the thickness of the panel (also called internal bond or IB) is extensively used to measure the bond strength. In this test a sample from the panel (typically 50 mm square) has loading blocks glued to each face of the panel. When the glue has set a load is applied across the thickness of the panel until failure occurs. The calculated stress at failure is the IB value. The test is relatively simple, but has a high coefficient of variation. Failure generally occurs at the lowest density point in the thickness of the panel so that the IB correlates best with the minimum density point on the profile. Even so, correlation coefficients between IB and minimum core density for MDF are only in the range 0.6-0.7. Despite its limitations, the test is used extensively, and is the main test used to monitor the resin level in particleboard and MDF; this is adjusted to maintain the IB at the required level.

8.1. Formaldehyde emissions

Formaldehyde emissions are a significant factor in panel performance made with formaldehyde based resins. The formaldehyde is released from the panel at a rate that reduces over time, but this can be sufficient to raise formaldehyde to levels regarded as hazardous, particularly in living spaces. Various methods are used to assess the formaldehyde emission characteristics of a panel, ranging from the toluene extraction method used in Europe (the perforator test) to a number of chamber tests that measure the formaldehyde emitted from panel samples in an enclosed vessel. There are a range of rating classes, the most commonly used are the E ratings, where E3 is the highest emission class rating (rarely used today) through E2 to E1 levels, the most commonly specified level. E0 is a very low emission panel, while a new class of super E0 has a formaldehyde emission level close to that of wood itself. Formaldehyde levels in occupied spaces are generally controlled by regulation, with the use of a panel with a specified formaldehyde emission characteristic accepted as a means of complying. The development of formaldehyde – based resins able to meet the decreasing emission levels has been a major focus of UF resin development over the past few years. Initially UF resins with low emission characteristics required higher resin addition rates to maintain strength properties and proved to be more sensitive to plant and furnish conditions. Further development has recovered some of these disadvantages. Formaldehyde emission is limited to UF and MUF resins. PF, although a formaldehyde resin, is more stable and is classified as a low emission system. Isocyanate resin systems have no formaldehyde emission from the resin and can be used in situations where formaldehyde cannot be present at any level.

8.2. Response to water

A range of methods evaluating the response to water, either as liquid or as humid air are used in standards. The methods involve a short term response to immersion to water at a controlled temperature for a specified time with measurement of the resulting change in panel dimensions, particularly perpendicular to the plane of the panel. The swelling in this direction is a reversal of the compression locked by the formation of the adhesive bonds in the press. This response is density related and can lead to anomalous results if the test regime does not provide sufficient time for the water to completely penetrate the sample. Increasing density slows down the rate at which water travels through the sample, but because the compression is greater, a greater degree of swelling will result. Increasing density can give a lower thickness swell result in a short term test, but the long term swelling may in fact be greater. For products designed to operate in interior situations where exposure to moisture is not anticipated, resistance to water is not a major issue and the water response tests indicate whether wax added as a sizing agent is present.

For products intended to perform where a degree of water exposure is likely, such as kitchens or bathrooms a higher level of resistance to moisture is expected. In this case the permitted swelling level is lower, and additional tests may be required.

8.3. Durability

The response to water is critical in applications where water exposure occurs. Both particleboard and MDF ultimately fail unless moisture is excluded, limiting applications to those areas where the chance of this occurring is small. Most applications in furniture, fittings and cabinetry meet this requirement. There are also applications where structural performance is not usually critical, that is where a failure of the item does not endanger the structure as a whole.

Where a component contributes to structural performance in a building then a durability requirement, usually expressed in terms of exposure conditions and time, will apply unless the possibility of moisture affecting that part of the structure can be totally excluded. For established materials the durability performance is known.

For newer materials, particularly the adhesive bonded composite panels being considered, prediction of long term performance is much more difficult issue. There is an established history that suggests that some durability requirements can be met. OSB is used as a structural element in buildings, either as a bracing panel or as a component of a beam. Much of the time between the development of panels based on wafers or strands and the commercialisation of OSB was used to establish the durability of the material. Particleboard is used as a structural flooring panel in both Australia and New Zealand. The product formulations used are very different in the two countries, possibly reflecting the differing exposure conditions. In New Zealand a UF bonded particleboard has been used for some 50 years. It has proved to be satisfactory, even when exposed over a 2-3 month construction period. The Australian requirement is for a panel with a given MOR retention after a boil test, requiring the use of a PF adhesive, or one with similar response in a boil test.

There is a range of accelerated ageing techniques available that simulate the effect of moisture change in particular as well as ultraviolet radiation. For panel products it is the moisture change effect that is most significant. In order to simulate the effect of long term exposure, most accelerated tests use shortened time scales along with stress levels that are greater than those encountered in actual service conditions. There is a continuum between the stresses encountered in service conditions and the increasing levels of stress used in the accelerated tests and those used as quality control tests as shown in Figure 12.23.

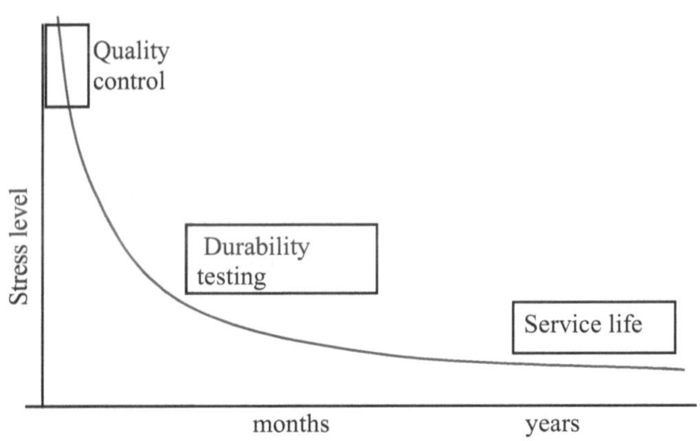

Figure 12.23. Stress-time relationship for durability assessment.

The extreme case is that induced by placing the material in boiling water. This test was originally used to distinguish between PF and other resin types. It has since been used in conjunction with IB or MOR tests on the sample after boiling to measure strength loss. The boil test generates a severe stress on the panel, one that has the advantage of being predictable, but the relevance of this test to actual exposure conditions has been questioned (Carroll, 1980). It has been used as a ranking test (Lehman, 1977, 1986) to screen a large number of samples quickly. Other tests use several cycles of stress, usually involving immersion or exposure to water, followed by drying and in some cases freezing. This test appears as the cyclic test (V313) in many standards (NZS, 4266.11) or in a similar test (ASTM, D1037-72a). In the cyclic tests the stress level is reduced, but is applied in several cycles (3 in European cyclic test, 6 in the US). There are a number of proprietary accelerated weathering tests that simulate selected climatic conditions in a shortened time period. The effect of the accelerated weathering on wood–based panels is evaluated using standard strength tests.

Fibreboards have been used as exterior sidings, a non-structural application, but with limited success. These examples indicate that while some applications where durability is a requirement are successful, these require control of the exposure conditions, particularly those that can lead to moisture ingress.

The durability assessment may also include measurement of the response to insect and fungal attack. There would appear to be little inherent resistance to these forms of attack beyond the characteristics of the wood. A range of insecticides and fungicides can be added. This is easiest if these can be mixed with the adhesive system provided that resin performance is not compromised. The size of the particles means that the selected material can be distributed more easily within the panel. However insect/fungal attack usually does not become a factor unless the moisture content increases, and it is this response that is most important.

8.4. Factors influencing panel durability

Selection of resin type is significant in the performance of wood–based panels. UF resins are generally not regarded as durable, while MUF systems with 10-15% of the urea replaced with melamine are considered suitable for moisture resistant grades of both particleboard and MDF. For OSB the resins used are generally PF or PMDI to meet the durability requirement. Most UF bonded products will disintegrate in a boil test, whereas other systems classified as moisture resistant will maintain some integrity. The best performance in the boil test is given by PF systems.

The mode of failure is significant. That adhesive bonded panels swell when exposed to water is clear. That is the expected response from the wood component of the composite. That this swelling is much greater in the direction of the compression applied by the press, in the thickness of the panel, is also clear. However in wood the loss of moisture is accompanied by a return largely to the original dimensions, so that the swelling is regarded as reversible. In the case of adhesive bonded panels the swelling in the thickness direction in particular is not completely reversed when the moisture returns to its original condition. This suggests that some structural change has occurred during the swelling, most likely the breaking of some of the adhesive bonds formed in the press. This is the starting point for the next moisture cycle where the process is repeated, with further bond disruption, ultimately leading to failure.

With the performance of both particleboard and MDF when exposed to cyclic moisture levels problematical, and OSB being less so, it might appear that particle size is significant in the response. Plywood responds to moisture variations in much the same way as does the wood from which it is made. Similarly OSB might be regarded as plywood made using a multitude of very small veneer slices. However the degree of compression necessary to create the level of bonding is higher. The resulting higher density, in terms of the model developed earlier, would be expected to increase the degree of swell. Particleboard and MDF are generally made to higher densities than for OSB, suggesting that the density effect is significant.

The conclusion of this discussion is that there is an inherent inevitability of the wood particles of the panel to expand when moisture content increases, particularly in the panel thickness direction: that is the most significant response of the panel to moisture. The response is influenced to some degree by the process conditions selected, with density, resin type and addition rate being the most significant. The effective limits on these variables for commercial products make the application of particleboard and MDF in exterior exposure situations unlikely to be successful in the increasingly regulated world of building products: but the goal remains. There are major applications for panel products that can be used in exterior applications, but the problem of the disruption and ultimate breakdown of these products in response to moisture changes must be solved in some other way.

There are three approaches that have demonstrated an ability to overcome this limitation. Two are based on modifying the wood, changing the way that it is affected by moisture. The third approach is to encapsulate the wood particles in a matrix that aims to exclude moisture.

Thermal modification is the earliest process. It was developed by Mason (1926) and was used in the first exterior rated product called Masonite. This was a high density fibreboard where the bonding agent was created by heating the wood to a high temperature, above 200°C. This lead to decomposition of the hemicellulosic material, reducing its ability to absorb water. In the Masonite process the high temperature was created in the press; this limited the thickness of the panel that could be made in this way to a maximum of 6 mm. There are other applications of the thermal process; Hsu *et al.* (1989) discussed the improved panel stability achieved by heating the panel after the conventional press in a second pressing stage. Others (Shen, 1988) have described processes for thermal treatment of cellulosic materials before pressing to produce autogenous bonding through thermal modification, while at the same time modifying the swelling response of the material.

The other approach is to modify the cellulose chemically. Rowell *et al.* (1994) discussed the application of both thermal and chemical modification techniques to modify fibre used to manufacture low density fibreboards. The acetylation technique appears to offer significant improvement in the stability of panels, but the technique has not achieved significant commercial success largely because of the cost of the chemicals required.

The third approach is to encapsulate the wood in a material that prevents water reaching the wood. This requires a high level of binder or coating agent. The cement bonded particleboard and fibreboards are examples of this approach, but it also applies to the extruded wood-plastic composite used as an exterior decking in USA. In this product the binder is a thermoplastic comprising 50-60% of the total mass. The wood in this case is also used in a finely divided form, so that the contribution to the structural strength of the product is less than it would be for larger particles.

However particle size is important in the durability performance of these materials. Some wood is always exposed as the material is cut to length or penetrated by fixings. This wood will respond to an increase in moisture by swelling. This is not an issue for small particles isolated within the plastic matrix, but would be more significant with larger particles used in the cement and gypsum matrix products.

9. CONCLUSION

Wood–based panels such as particleboard, MDF and OSB each represent a solution to a particular market demand, often for a product that can be handled like wood but without the limitations of that material. The ability to make panels in a range of sizes and thicknesses, with uniform, predictable properties tailored to particular applications has been a major factor in the significant growth of these materials. Each of these panel types uses a raw material that is either a waste stream from other wood processing operations, or is wood that is unsuitable for other uses. There is thus a ready source of raw material. In addition the processes have demonstrated an ability to use a range of raw materials, both in form, and in different wood species, extending this use to include recycled wood from urban waste streams and to agricultural residues.

This has been made possible by the development of suitable adhesive systems that are able to bond the particles together. The synthetic adhesives offer a consistency of performance that is difficult to achieve with natural products such as tannins, and at a cost that has enabled rapid growth to be achieved. At the same time the adhesive systems have shown a tolerance to a range of wood properties that has enabled most wood residues sources to be used. The environmental effects of formaldehyde emissions from panels made using UF adhesives have been successfully addressed. Initially these required higher resin addition rates to offset a loss in physical property levels, but further development has reclaimed much of this additional cost.

The technology to manufacture these panels has developed rapidly, both incrementally and by breakthroughs. The development of continuous pressing improved costs while offering significant benefits in panel uniformity. This allows large production lines with capacities in excess of 1500 m^3/day for both OSB and MDF, with even higher capacities in particleboard. The plants in Canada, on which the data in Table 12.2 is based, are generally newer plants at the top end of the capacity scale. Unit labour cost at the same wage rates would be higher for smaller plants on a capacity basis, with the benefits of increased automation giving further advantages in newer plants. The result of these developments is a dynamic sector of the wood processing industry, an integral part that contributes to the effective use of the forest resource. It has achieved this by developing new products and the processes that have enabled these to be made efficiently and at reasonable cost.

These processes provide a way of making products over the whole range of densities. While MDF started its life as a panel competing with particleboard as a wood substitute, it has become a new material in its own right. The ability to control density profile so that the surface density of the sanded panel is uniform has provided a consistent response to surface finishing operations. This, together with the thickness tolerance on the panels (typically below ± 0.15 mm) has given a panel that can be processed on automated component manufacturing lines with panels fed to the line, being cut to size, machined, and finished without operator intervention.

The MDF process is now used to produce panels well outside the density range envisaged as medium density. Low density panels in the 500-600 kg/m^3 range are now common in the market, while higher density products are produced, specifically as a substrate for prefinished flooring systems. The MDF process is also used to produce the fibre for moulded doorskin products. Thin MDF (<6 mm) became economic with continuous presses, firstly with the Mende design and subsequently in the straight through presses. Thin MDF grew rapidly as an alternative to hardwood plywood at a time when the availability of this material was declining. It is perhaps appropriate to redefine MDF as process rather than a product, one that can produce an adhesive-fibre mix that can be used to produce a whole range of fibre bonded products.

Particleboard has also responded to the MDF challenge to its market area by improving performance in a number of ways. The density profile is now monitored and controlled and the importance of particle geometry in determining panel characteristics is recognised. The fixing characteristics of particleboard are not as good as for MDF but new designs of fittings have overcome much of this limitation. In another direction the possibility of using larger particles to provide a structural particleboard that reaches into the OSB market is being actively considered.

OSB was developed with a single focus, as an alternative to softwood plywood for sheathing houses, but is being used for a wide range of tasks. Many of these currently use plywood, but the cost benefit of OSB and its demonstrated performance make it an alternative that must be considered in a competitive market.

The overall picture is of the wood–based panels sector continuing to grow as products are developed that meet new customer requirements. The processes have demonstrated an adaptability to different fibre sources, both in form and in species that has allowed plants to be established in most parts of the world, primarily to meet local demand, but also with a significant export component to those areas where fibre supplies are fully committed.

CHAPTER 13

PULP AND PAPER MANUFACTURE

JOHN WALKER

1. INTRODUCTION

Pulp and paper industries are capital intensive and benefit from economies of scale. A state of the art chemical pulp mill would have a capacity of 900 000 tonnes/annum (US $1.2 billion), while a modern newsprint machine would be 300 000 tonnes/annum (US $700 million). This trend favours regions where trees grow very fast, the necessary land area is much reduced, as is the cost of trucking. Yet at the same time, new minimills (100 000 tonnes/annum) for recycled fibre compete side-by-side. These can locate close to large urban areas on the prospect of cheap fibre and low transport costs, while requiring little water and generating little waste. Yet again, one of the largest volume exports from the USA to Asia is scrap – metals, minerals and recovered paper: in 2004 some 13.4 of the 49.2 million tons of recovered paper were exported; 8.4 million tons to Asia including 5.9 million tons to China. Effectively Asia is buying energy and reduced pollution.

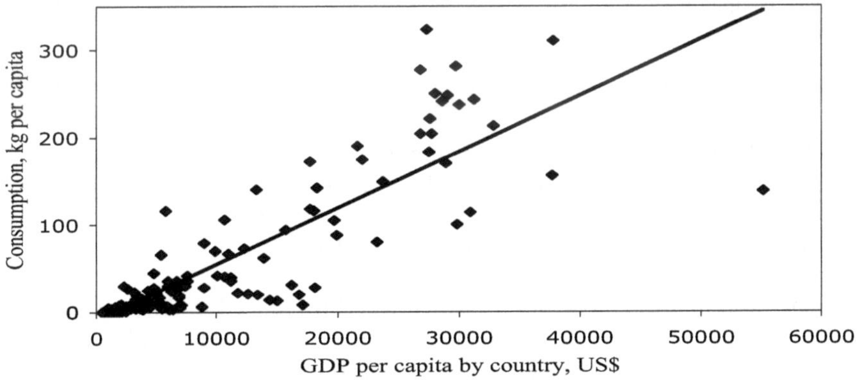

Figure 13.1. National per capita consumption of paper and paperboard in 2002.

These trends are supported by a 3.3% pa growth in demand for paper and board products, with fresh wood pulp production of 167 million tonnes and recovered fibre of 141 million tonnes in 2002 (FAO, 2004). There is a reasonable correlation between per capita consumption of paper and GDP per capita (Figure 13.1). The scatter may reflect local availability of wood, the size of the literate middle class, and the diffusion of wealth within particular countries. Increasing GDP and literacy mean that the greatest growth will be seen in Asia and South America.

Wood is the chief source of papermaking fibres, with much coming from the residues arising from the timber industry (Figure 13.2). The vast majority of paper products are made from cellulose fibres, the aggregate of fibres being known as pulps. The use of agricultural fibres remains modest, *c.* 10% of pulp production from crops like kanaf, hemp and flax (linen), or residues such as rice and wheatstraw, or cotton rags. Today greatly increased use is made of recovered paper accounting for about 42% of all paper and paperboard products. The limits to its use are not clearly defined but appear to be about 50% of consumption, a level already reached in Japan. The reuse of waste paper is limited by the ease and cost of collection and by the number of times fibres can be recycled (about 5-7 times on the basis of current technology). In practice large cities are the most viable sources of recovered paper.

Table 13.1. World production of wood and non-wood fibre pulp, millions of metric tonnes at 10% moisture content (FAO, 2004).

Year:	1969	1986	2002
Recovered paper	N/A	N/A	140.7
Mechanical pulps	23.3	31.2	34.7
Semichemical pulps	7.2	7.4	8.4
Full chemical pulps	63.0	97.5	121.8
Dissolving pulps	4.9	4.3	2.8
Non-wood fibre pulps	5.6	10.3	19.8
Total: all pulps	103.9	150.7	181.6
Chemical pulps:			
Unbleached sulphite	6.8	5.1	1.3
Bleached sulphite	5.9	4.8	4.0
Unbleached sulphate/kraft	27.0	34.7	34.0
Bleached sulphate/kraft	22.3	51.1	82.5
Dissolving, acid bisulphate and prehydrolysis kraft	4.9	4.3	2.8

World pulp production statistics are summarised in Table 13.1. There are three main categories, mechanical, semichemical and chemical pulp. The classification is based upon the process used for fibre separation. The major industrial operations are mechanical and chemical pulps, with most of the latter being manufactured by the kraft process.

Pulping describes the processes by which wood is reduced to fibre, or strictly to a mixture of fibre and fibre debris. Papermaking, although technically complex is simple in principle. Paper is made by spreading a layer of pulp fibres in suspension on the surface of a moving wire (mesh) screen so as to form a wet paper web which, after pressing to remove water and consolidate the fibre mat, is dried to form paper.

Both pulping and papermaking are energy intensive processes. Mechanical pulp production requires large amounts of electrical energy. Chemical pulping processes on the other hand use principally thermal energy and chemicals. Semichemical pulps are intermediate in their requirements.

Large quantities of water are used by the pulp industry, and also in the process of papermaking, and because of this the industry has a large impact on the environment. In modern pulp and paper mills much of the water used in process flows is continuously recycled, such that over the last 30 years water consumption in paper production has fallen from some 80 to 10 m^3 per tonne. This trend is bound to continue because of environmental pressures.

The range of paper and paperboard grades is a matter of common experience in daily life. Indeed, one of the difficulties in describing such products is their range and versatility. All paper and paperboard grades have their particular requirements. They may be coated or uncoated. Only a summary is offered.

Demand for printing and writing papers is somewhat over 100 million tonnes per year, divided equally between woodfree papers and lower value newsprint and magazine papers. Strength is essential where they are to be printed on high speed presses without frequent breaks, while high quality printing papers require a smooth surface. Papers made principally from mechanical pulp (> 50%) include short life products such as newspapers, directories and flyers as well as light weight coated papers for magazines and catalogues. Chemical pulps dominate high value coated and uncoated papers for annual reports, books, quality magazines and copy papers.

Paperboard is the term used to describe products ranging from the heavier paper grades to products such as cardboard which are made of a series of pulp layers, with the central layer generally containing the coarsest grade of pulp. The mass of paper per unit area, its grammage or basis weight (g m^{-2}), varies widely between grades and often within them. Paperboard products come in a range of grammages from 100 g m^{-2} or less for some linerboard and single-ply corrugating medium, although most products have a grammage of 150 g m^{-2} or more. This contrasts with standard newsprint paper that is around 45 g m^{-2} while high quality facial tissues may have grammages as low as 20 g m^{-2}.

Paperboard grades cover containerboard (corrugated boxes, packaging), cartonboard (cigarettes, food, pharmaceuticals) and speciality grades (wall paper, gib-board lining *etc*). Demand is somewhat over 120 million tonnes per year with containerboard being the dominant category (*c.* 80%). Paperboards are used largely for packaging and the primary requirement is strength. Containerboard has a corrugating medium sandwiched between two outer linerboards (two-ply); although double and triple-wall boards (with three fluted mediums individually separated by a sheet of linerboard: four sheets in total) are used where loadings are high. The outer ply of linerboard can be bleach or unbleached; the thicker inner ply is often of recycled fibre. Cartonboard is pigment coated for clear printing, with dense, stiff outer layers and a bulky middle ply.

Demand for tissue products is around 20 million tonnes per year. These are largely low basis weight hygiene products, e.g. high quality facial tissues may have grammages as low as 20 g m^{-2}. Quality criteria may include high wet strength for kitchen towels (more virgin softwood fibre), flexibility and stretch, softness (more hardwood fibre) and high absorbency (4-18 g/g). Sheets can be single or multiply.

The definitive overview of the industry is provided by the superb 19 book series on Papermaking Science and Technology (Finnish Paper Engineers' Association and TAPPI), to which one should refer.

2. AN OVERVIEW OF PROCESSING OPTIONS

There are a number of pulping processes, and variants of each. Most of the basic processes were developed initially to use a particular wood resource; to supply major markets where papers with specific properties were required; or to supply a market niche.

Mechanical pulps require mainly energy for their production and have high demand for electrical power. They are obtained in high yields of 85 to 96% of oven-dry wood and are basically made by two processes. Stone groundwood pulps are made by pressing roundwood billets against a rotating pulp stone in the presence of water showers. Refiner mechanical pulps on the other hand use wood chips as feed stock. All mechanical pulps have lower strength properties than chemical pulps and cannot be bleached to high brightness levels. However they have excellent printing properties and generally have high opacity. Light coloured species are preferred.

Semichemical pulps are made by treating wood chips with chemicals at high temperatures so that some lignin and hemicelluloses are removed, after which the partially softened chips are defibred, i.e. separated into their component fibres by passage between the plates of a disc refiner. The pulps are obtained in yields of 70-85%, and are generally prepared in integrated mills, going direct to the papermachine. The widest used semichemical pulping process is the neutral sulphite semichemical (NSSC) procedure. They form an important part of the packaging market, especially in products such as corrugating medium.

Chemical pulps are prepared by the digestion of wood chips with chemicals at high temperature (170-180°C) until much of the lignin has been removed. There is a concomitant loss of hemicelluloses, because of polysaccharide degradation due to chemical attack. Therefore yields are low, from 45-50%. Chemical pulping falls into two main classes, those based on pulping with sulphite liquors at various pH levels, and alkaline pulping of which the kraft process is by far the most important.

The resource requirements for such operations are very different, both in the quality and quantity of material. Long fibred outerwood is ideal for fibre cement and sack kraft where strength is sought. Certain papers, e.g. writing papers, tissues and newsprint, are superior when made from corewood. This is because the thin-walled hollow fibres flatten more readily to produce a lightweight, dense sheet with a smooth surface offering good printability, e.g. lightweight coated papers for colour magazines. Corewood generally has good tensile and burst strength, but poor tear strength. Low tear strength means corewood fibres are less suitable for purposes such as linerboard (the outer face sheets of corrugated paperboard), where the trend is to have eucalypt-pine multi-layered sheets with eucalypt on the face for good printability and increased stiffness.

More significant is the size of the resource required and the demand on utilities (Table 13.2). In this naïve 'conceptual' analysis assume initially that all the timber grown will go to pulp. In many countries perhaps only a third of the wood goes to pulp so the land area should be increased by a factor of three. Finally, it would be unrealistic to assume that all land in the vicinity of the mill would be dedicated to forest. As mills get larger wood has to be hauled from further away resulting in much higher transport costs. Doubling production with second line may be uneconomic as wood will have to be drawn from even further afield. However, economies of scale can be achieved in some countries, where cheap land is available and plantations are exceptionally productive, e.g. Brazil, Indonesia, Uruguay, Vietnam. In such countries the cost of new infrastructure – ports, roads, utilities – for the first line can be very high relative, for example, to Canada and Scandinavia where the infrastructure is largely in place. However, once in place a second line brings considerable economies of scale denied to mature economies (Figure 13.2). Building the mill inland – but with good infrastructure connections – places the resource around the mill (360°) and greatly shortens the haul distance compared to locating by a lake or large river with the resource coming from one direction (180°).

The economies of scale is illustrated by Veracel Celulose S.A., an Aracruz – Stora Enso joint venture at Eunápolis in Brazil, which it is claimed will be the mill with the lowest direct costs in the world. The first stage, opened in 2005, is a 900 000 tonne/year single-line bleached eucalypt pulp mill, with fibre for fine papers and tissue. Forest productivity (50 m^3/ha/yr) on some 90 000 ha reduces the average haul distance to around 40 km, while pulp is exported through a dedicated marine terminal 60 km away. There is provision for doubling production by putting a second line alongside. A somewhat larger joint venture between UPM and Mzetsäliito is under construction in Uruguay.

Table 13.2. Resource requirements for a greenfield bleached kraft and mechanical pulp mill, assuming here a species with a basic density of 400 kg m^{-3}.

	Bleached kraft pulp	Mechanical pulp
Electrical energy demand, kWh/tonne	900	2 600
Water requirements, m^3/tonne	100	15
Wood requirements, m^3/tonne	5.5	2.5
Minimum scale, tonnes/day	1 200	600
Annual production, tonnes	420 000	210 000
Annual wood requirement, m^3	2 310 000	525 000
Forest area (ha) at 50 m^3/ha/yr	46 000	10 500
...or at 10 m^3/ha/yr, and...	230 000	52 500
...with only 1/3rd for the pulp mill	690 000	157 500

3. PAPER

Papers are made from a variety of pulps ranging from mechanical pulps to highly bleached chemical pulps. The papermaking furnish may contain strong but lightly beaten softwood fibres, or it may contain small well-beaten even-textured hardwood fibres for use in fine writing or printing grades, or a blend from various pulp streams. The paper may be additive-free, or it may contain additives that confer wet strength, or other properties, upon it.

In the technical process of papermaking a very dilute suspension of fibres is deposited on the moving wire mesh screen of the papermachine (Figure 13.19). A layered structure made of randomly aligned and interwoven fibres is formed by draining on the wire. The wet fibre mat is then pressed to remove further water, and in the process the mat is consolidated and some fibres collapse. The wet-web is subsequently dried between heated rolls and further fibre collapse occurs. Finally the paper is wound into rolls.

It is important to appreciate that interfibre bonding in paper involves hydrogen bonding: hence the consolidation and collapse of fibres are desirable features as the area of contact between fibres is enhanced. As a consequence of hydrogen bonding, paper is very responsive to the addition of moisture, with large strength losses occurring at high humidities.

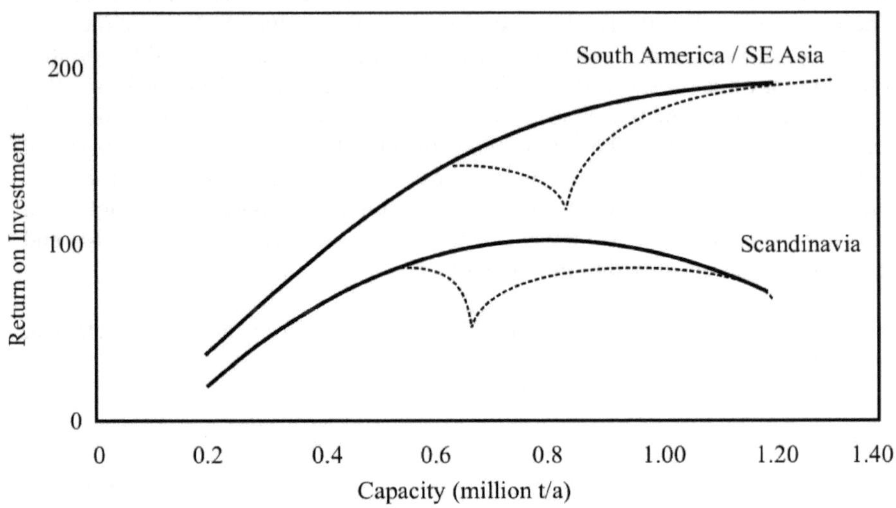

Figure 13.2. Relative return on investment as a function of capacity (*after* Diesen, 1998).

3.1. Pulp and paper tests

There are standard tests for pulp and paper products. Generally pulps are evaluated as pulp handsheets. Some of the more important tests for pulps, and those used frequently on papers, are described later. The tests and the principles behind them are illustrated in Figure 13.3.

Figure 13.3. Various tests to characterise pulp and paper (see also Appendix A). (a) The freeness test measures the drainage rate of water through a fibre mat that builds up on the screen plate (Figure 13.23). (b) Burst is a quick test that correlates with the tensile test. Burst strength is derived from the hydraulic pressure necessary to rupture the paper. (c) Tear strength is a function of the ability of paper to sustain and redistribute stresses concentrated at the root of a tear. (d) Tensile strength is the force per unit width (N m^{-1}) required to break a standard specimen. Tensile strength increases with beating. (e) Brightness is the percentage of light reflected from a thick pad of papers. Opacity is a measure of the ability of a single sheet of paper to hide colour or print on the reverse side of the sheet: it is determined by comparing the reflectance of a single sheet backed by a black body with that of a thick pad of sheets. ISO brightness is measured at a wavelength of 457 nm and TAPPI opacity is measured at 572 nm.

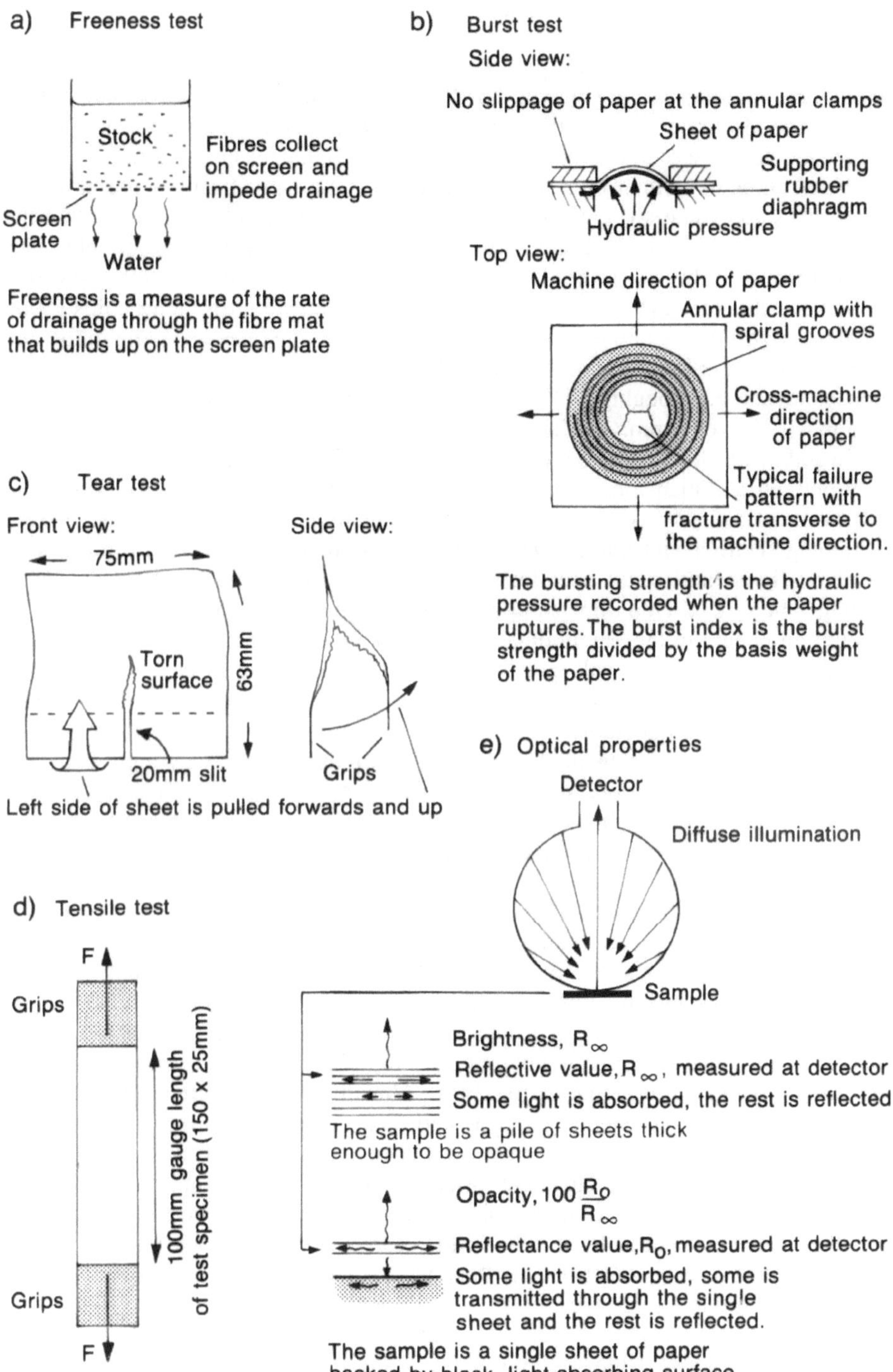
a) Freeness test
Stock
Fibres collect on screen and impede drainage
Screen plate
Water
Freeness is a measure of the rate of drainage through the fibre mat that builds up on the screen plate
b) Burst test
Side view:
No slippage of paper at the annular clamps
Sheet of paper
Supporting rubber diaphragm
Hydraulic pressure
Top view:
Machine direction of paper
Annular clamp with spiral grooves
Cross-machine direction of paper
Typical failure pattern with fracture transverse to the machine direction.
The bursting strength is the hydraulic pressure recorded when the paper ruptures. The burst index is the burst strength divided by the basis weight of the paper.
c) Tear test
Front view:
Side view:
75mm
63mm
Torn surface
20mm slit
Grips
Left side of sheet is pulled forwards and up
d) Tensile test
F
Grips
100mm gauge length of test specimen (150 x 25mm)
Grips
F
e) Optical properties
Detector
Diffuse illumination
Sample
Brightness, R∞
Reflective value, R∞, measured at detector
Some light is absorbed, the rest is reflected
The sample is a pile of sheets thick enough to be opaque
Opacity, 100 Ro/R∞
Reflectance value, Ro, measured at detector
Some light is absorbed, some is transmitted through the single sheet and the rest is reflected.
The sample is a single sheet of paper backed by black, light-absorbing surface.

4. WOOD PREPARATION

Bublitz (1980) has reviewed procedures for wood and chip storage. Wood preparation is important. In general, wood needs to be debarked. Subsequently chips can be stored in large piles for three months or more. The piles need 'turning over', by mechanical devices to prevent overheating and spontaneous combustion.

Wood chips have their own quality standards. Chip specifications are based on screen sizes and some level of allowable bark. Generally chip lengths are within the range 12 to 32 mm and about 2-10 mm thick. A 25 mm softwood chip has approximately eight 3.5 mm fibres end-to-end, two of which will be severed. Chipping techniques are compromised by the fact that in cutting long chips (good for fibre length) one is obliged to cut thick chips (retarding chemical penetration and impregnation). Further, thickness is variable depending on where the failure surface chooses to run, e.g. roughly 10% of chips are under 2 mm, 40% 2-4 mm, 40% 4-6 mm, and 10% over 6 mm. The latter accounts for the 0.5-1% of undercooked shives. Hardwoods with shorter fibres are cut into shorter, thinner chips.

Wood chip dimensions are important in chemical pulping. Mass flow and diffusion are both involved in getting the chemicals into chips in the early stages of pulping (Hartler and Stade, 1977). Diffusion is involved also in the transfer of dissolved lignin and spent chemicals out of the chips during the later stages of pulping. The crucial chip dimension in sulphite pulping is length, since most liquor penetration occurs in the longitudinal direction. However under alkaline conditions, as in kraft cooking, cell wall swelling occurs with increase in pore sizes so that diffusion of chemicals into the chips in the longitudinal, radial and tangential directions become comparable. Chip geometry means that most penetration occurs in the radial direction during alkaline cooks, so chip thickness is very important.

Reaction rates are more temperature sensitive than is diffusion. Thus non-uniform delignification is liable to occur where liquor impregnation has been poor, e.g. under conditions in which the liquor concentration is too low, chips are too thick and the cook temperature is too high. Under such circumstances most chemical is consumed in the outer parts of the chip before it can diffuse deeper inside, i.e. delignification occurs preferentially in the outer parts of the chip.

For the same reasons, the justification for high temperature washing is faster removal by diffusion of the high molecular weight organic fragments.

Kraft pulping proceeds in two phases, first impregnation (at low temperatures) then a cook phase (high temperature delignification). The initial chemical concentration and its distribution are critical. This involves pressure impregnation of chemical into degassed capillary spaces and then diffusion into the wall, but with natural variations in thickness and permeability between chips it is difficult to ensure a uniform distribution of chemical prior to cooking. Thin, uniform chips are preferred because during the final cooking phase chemicals can be consumed faster than they can be replaced by diffusion: therefore thick chips cook poorly because diffusion rates are a function of thickness squared. This imposes an upper temperature limit to aggressive cooks.

Screening before cooking seeks to eliminate oversize (undercooked) chips and fines (overcooked and liable to compromise – clog up – the recirculation of cooking liquor in the digester). Usually, this is achieved by multi-stage vibratory screens, with the oversize chips going to a re-chipper. A number of industrial systems have been developed to separate chips on the basis of thickness.

5. MECHANICAL PULPING (SUNDHOLM, 1999)

The properties of paper depend on wood species and pulping method. Mechanical pulping is used predominantly with low density softwoods, although there is increasing use of light-coloured low density hardwoods such as aspen.

In mechanical pulping, electrical energy is used to separate the fibres but during the process a high proportion of fibre fragments known collectively as 'fines' are produced. Thus mechanical pulp consists of a blend of long fibres and fines. The long fibres form the fibre matrix of the sheet within which the fines are trapped. Fines are important for two reasons (Corson, 1980). First, good quality fines aid bonding between fibres. Secondly and more important, the fines within the sheet scatter light and therefore make a very large contribution to sheet opacity. High opacity, combined with moderately high brightness, is one of the principal requirements of newsprint and other printing papers.

Mechanical pulp fibres are chemically unaltered and rich in lignin. The fibres are stiff and resist consolidation when formed into paper, so that the bonding between fibres is poor and the sheet is of low density, i.e. it is bulky and porous. The high proportion of fines is important, as indicated earlier, because of their contribution to opacity. On the other hand the bulky nature of the sheet means that paper made from mechanical pulp fibres is suited admirably to high speed printing techniques, since the sheet is resilient, can resist compression and, provided the sheet has good smoothness and reasonable brightness (75-80%), it can be used in the uncoated or coated condition. One of the disadvantages of mechanical pulps is that the lignin-rich papers yellow with age when exposed to light and there is no way of permanently bleaching such pulps.

5.1. Stone groundwood pulp (SGW)

Debarked roundwood billets, about 1.8 m long, are hydraulically pressed against the peripheral surface of a finely burred (roughened) cylindrical stone some 1.8 m in diameter that is covered with abrasive grits (Figure 13.4). The billets advance transversely at 1-2 mm s^{-1} while the stone rotates at up to 360 rpm with a peripheral velocity of up to 35 m s^{-1} using a motor delivering up to 7.5 MW. As the pulpstone rotates the abrasive grits on the surface of the stone repeatedly impact on the wood surface so that there are shear and compression/decompression cycles, as well as some associated fibre cutting. The 0.2 mm grits are at about 0.8 mm centres, such that the surface fibres are fatigued at frequencies in the range of 40-50 kHz, with the deformation pulses being dissipated within 0.5 mm of the surface (Atack, 1981). Frictional heat is generated which peaks at around 150°C about 0.1 mm from the wood surface. Salmén *et al.* (1997) emphasise that for effective fatigue the local deformation must greatly exceed the elastic limit and be accompanied by significant plastic strain. This occurs only near the locus of the deformation with much additional elastic strain energy being dissipated in the surrounding fibres. In this narrow zone lignin is softened dramatically, thus assisting the fibre separation process. After fatigue failure the fibres are 'peeled' individually from the surface.

With a freshly sharpened stone, the production rate is high and energy consumption low, but the grits cut the fibres and pulp quality is poor. As the grits wear production declines, there is less cutting, energy demand increases and pulp quality improves. The pulpmill adjusts the sharpening cycle (about 10 days) of individual stones so that mill pulp quality is approximately constant.

More is being achieved than just the physical separation of the fibres, as a separate but otherwise undamaged fibre would be far too stiff and rigid to make good paper. The internal structure of the cell wall is being delaminated. The fibre experiences both external and internal fibrillation. Also large amounts of desirable

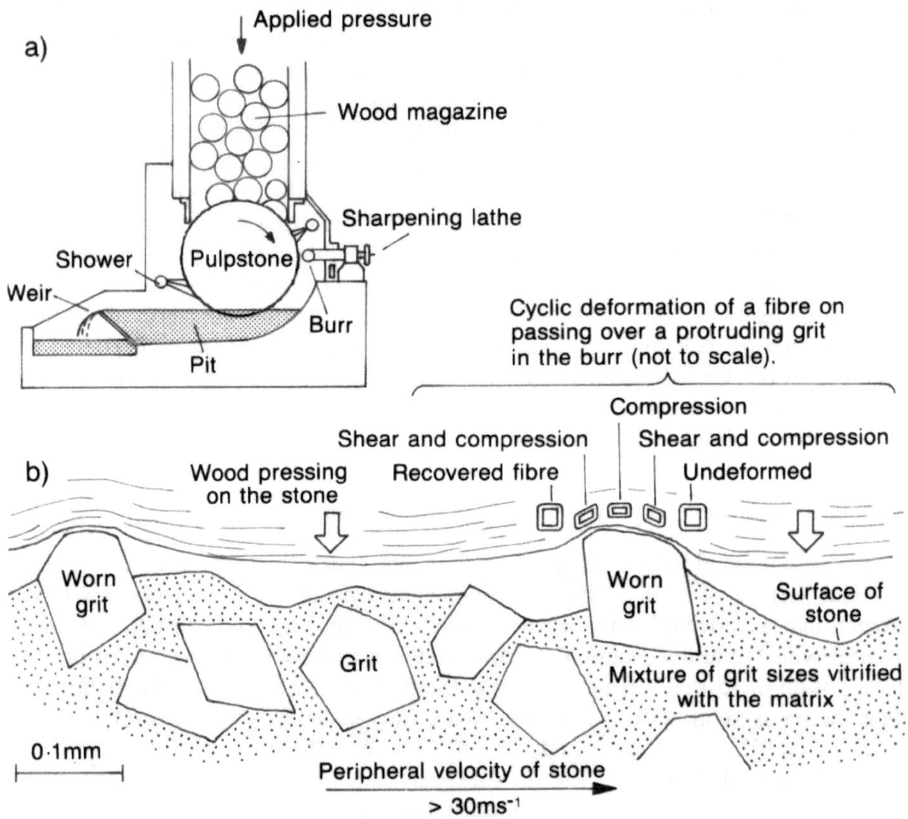

Figure 13.4. Stonegroundwood (Smook, 1982). (a) One of a number of grinder units producing SGW pulp. The stone is cooled by partial immersion in a dilute pulp suspension, and by showers, which wash off the loose fibres. (b) The deformation of fibres immediately adjacent to the stone. The exposed grinding grits become rounded as they wear away.

fines with good bonding and optical characteristics are torn from the fibres. SGW has about 50% fines on a weight/weight basis whereas for thermomechanical pulp (TMP) the fines would be around 20% on a wt/wt basis. Fibre length is largely retained.

Stone groundwood pulp uses less energy for its production than other mechanical processes and has high opacity, but its low strength means that it has to be mixed with 15-20% chemical pulp to make adequate newsprint on the high speed papermachines now available. The development of pressurised stone groundwood (PSGW) means that the temperature at the grinding zone is higher and less power is required to separate the fibres. PSGW is used in lightweight coated and super-calendered grades. The pulp has improved tear strength and has good optical properties, but has lower strength than refiner mechanical pulp.

5.2. Principles of refining

Refining is the alternative way of preparing mechanical pulps. Papermaking from wood chips in disc refiners bears some similarities to the manufacture of medium density fibreboard. The significant difference is the hugely greater energy use. MDF defibrates the wood at higher temperatures (>150°C), above the thermal softening or glass transition point of lignin (Figure 13.5). This results in ready separation of fibres in the lignin-rich middle lamella (Atack, 1972). These fibres have a smooth coating of lignin, have poor hydrogen bonding properties and are suited for MDF manufacture only with the addition of resin. In thermomechanical pulping good papermaking properties are obtained by refining at lower temperatures, between 120° and 140°C. At these temperatures the fibres are worked more, the fibres are torn open and the polysaccharide-rich regions of the primary and secondary walls are exposed. There is an appreciable proportion of fines associated with inter-lamellae failures. Both encourage interfibre hydrogen bonding that is needed to develop paper strength. Refining at a lower temperature produces a brighter pulp.

Chips enter the eye of the refiner to be broken down in the breaker bar section to fibre bundles and shives. Then within the refiner zone they undergo fibre separation and physical modification as they pass between the opposing bar patterns of the plates (Figures 12.11 and 12.12). The width and height of the bars and the frequency of dams across the intervening grooves are the main design features of the refiner segments. High bars and deep grooves provide a greater cross-section for efficient steam removal and reduced specific energy consumption, i.e. the fibres are less refined. More dams force fibre from the groves and so the pulp pad is further refined. The other feature is the minute gap between the refiner plates through which the fibres flow. There fibres suffer collisions and frictional rubbing with the refiner bars and with other fibres, so being externally defibrated and internally fibrillated: the closer the plates the greater the motor load. The energy generated is largely converted to steam, and depending on refiner conditions, some steam can flow back toward the chip feed and some can flow forwards with the fibre (Miles and May, 1990).

Refiner pulp quality is influenced by uncertain parameters. Obviously the most important are the specific energy (kWh tonne^{-1}) and intensity of refining (impact energy of each individual collision of the fibre). For a given specific energy, the severity of each impact will depend on the number of impacts. High intensity refining results in cutting while low intensity refining emphasizes fibrillating. The production rate partly determines the residence time of fibre in the refiner zone, estimated to be about 0.4-0.8 s. With the refiner operating around 1500-1800 rpm; with the fibres confined to a very narrow gap (0.2-0.4 mm) between the plates and obstructed by the bars and dams of the plates; with fibres being constantly flexed (5000 loading cycles per second); therefore an enormous amount of work is being expended that softens and fatigues the fibre with the outer layers unravelling and fragmenting and the bulk of the wall delaminating. A short residence time means less fibre in the plate gap, which in turn implies reduced clearance between the plates and increased refining intensity. Unfortunately accurate measurement of mass flow, by itself difficult, is compounded by variations in the moisture content of the infeed and the basic density of the wood. A less than anticipated throughput of fibre results in more water being converted into steam, greater steam pressure differentials within the refiner, and a faster flow of fibre through the refiner.

Typically the pulp is made in two stages: about two-thirds of the energy is applied at the first stage. In the first stage the chips are reduced to fibres or fibre bundles by high consistency refining (20-25%). In the second refiner fibre quality is improved and due to the repeated compression-decompression cycles outer lamellae are stripped from the fibres to give fibrillar fines.

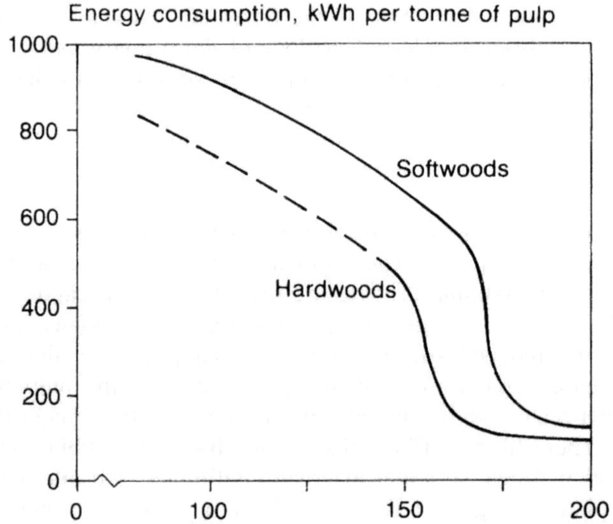

Figure 13.5. The effect of temperature on the power consumption needed to defibrate chips.

The higher energy consumption for a given freeness, for both TMP and CTMP, compared to stone groundwood is thought to be due to the greater flexibility of the fibre and fibre bundles that form more flexible interlocked fibre networks, so avoiding being cut or being beaten as severely. With groundwood the fibres remain attached to the bolt and cannot move out of the way – and so produce more fines.

5.3. Thermomechanical pulp (TMP)

In thermomechanical pulping chips are presteamed at 120-140°C for about four minutes before being defibred in a pressurised disc refiner at 150 to 500 kPa. If run at high pressure, >360 kPa at 140°C, there is the opportunity for better energy recovery since high pressure steam generated during refining can be used for drying in the papermachine, so reducing the overall demand for energy. TMP pulp fibres have a smoother appearance than stone groundwood pulp.

Production of up to 250 tonnes/day is usual for 44-68 inch (1180-1728 mm) single disc refiners operating at 1500 rpm with 2.5-15 MW motors.

Large capacity refiners (20-30 MW motors) are capable of lifting production to 450-500 tonnes/day. Here, the Sunds system performs initial refining between the faces of the discs (as in a conventional refiner), but then refines further using the conical surface on the edge/rim of the refiner – so there are two adjustable refiner gaps: between the faces; and at the edge of the disc. An alternative approach by Andritz has a rotating double-sided disc positioned between two stationary discs. Double disc refiners still need a secondary refiner to manipulate pulp quality.

5.4. Chemithermomechanical pulping (CTMP)

In chemithermomechanical pulping chips are pre-treated by steaming (to remove air) and then by soaking for about 15 minutes in 1-4% Na_2SO_3 at temperatures from 130 to 150°C, before pressurised refining. As would be expected, the greater the extent of sulphonation the lower the yield. CTMP pulps have good strength and can be bleached. This means they can substitute for more expensive kraft in some products. The main disadvantage of CTMP pulps is their low opacity. Bleached CTMP pulps from light coloured species such as aspen are recognised as a high quality product.

Alkaline peroxide mechanical pulping (APMP), i.e. sodium hydroxide and hydrogen peroxide, is attractive for low density hardwoods, in that it gives improved strength and brightness without requiring subsequent bleaching.

5.5. Latency removal

After refining the pulps go to latency tanks (2-4% consistency at about 80°C) where they are held and slightly agitated for several minutes. This improves pulp properties by slowly relieving internal stresses and removing the curled conformation of the fibres that resulted from lignin stiffening, when the pulp left the refiner.

5.6. Process energy demand

As shown in Table 13.3 mechanical pulps made from *Pinus radiata* vary widely in specific energy demand and quality. The apparent inefficiency of the process means that large amounts of energy are converted to steam. This is employed usefully in subsequent processes.

Table 13.3. Typical data for specific energy demand of mechanical pulps from radiata pine and for some of their properties (unpublished Dr. J.M. Uprichard). The pulps are of the same freeness, 100 CSF, which would be suitable for newsprint. Pulps of higher freeness require less refining and would have a lower specific energy demand. There is a general increase in strength from stone groundwood to chemithermomechanical pulp. 1 kWh = 3600 kJ.

	SGW	PGW	TMP	CTMP
Electrical energy, kWh $tonne^{-1}$	1250	1700	2400	2850
Tear, $mN.m^2g^{-1}$	3.1	4.8	10.5	10.0
Tensile, N.m g^{-1}	27.2	34.5	47.5	51.8
Burst, $kPa.m^2\ g^{-1}$	1.4	1.7	2.9	3.2
Sheet density, kg m^{-3}	383	371	370	400
Brightness, %	61.4	60.0	54.4	58.1
Scattering coefficient, $m^2\ kg^{-1}$	72.2	69.4	45.0	37.5

Chemically pretreated chips, for example with sodium sulphite in the well established CTMP process, require more energy to produce pulps of comparable freeness than corresponding TMP pulps (Table 13.3). On the other hand most softwoods CTMP pulps generally require less energy to reach a given level of strength (for example tensile index) than corresponding TMP pulps. The reasons for such differences include:

- The fibres of CTMP pulps are more supple, less prone to fibrillation and fines production, and are more easily bonded than corresponding TMP fibres produced at the same level of energy input, and
- Being highly sulphonated the chips are softened, are refined at lower temperatures, and produce less shives at a given freeness level.

Because of such behaviour the CTMP pulps are generally used at higher freeness levels than TMP pulps. Freeness although a useful index of papermaking quality is a complex parameter and is dependent on both fibre flexibility and fines content.

Interstage treatment of mechanical pulps can improve their strength while also reducing energy demand. In a process devised at the Ontario Paper Company, the so-called OPCO process, it has been shown that both post-treatment and interstage treatment of pulps improve pulp properties (Barnet and Vihmari, 1983). If pulp from the primary TMP stage is given a sulphonation treatment (using conditions similar to those used on chips) prior to secondary refining then the overall energy demand is reduced and the pulps have improved strength compared to normal CTMP pulps.

OPCO pulps have very good bonding properties, as indicated by their burst and tensile strengths, but have correspondingly low scattering coefficient and opacity.

Approaches such as these emphasize the array of techniques available to tailor the properties of mechanical pulps to particular markets. However it should be noted that the increased use of chemical in mechanical pulping will reduce pulp yield and also increase the effluent load. Hence as is usual, for each perceived benefit there are added costs.

Energy consumption is the Achilles' heel of mechanical pulping, with some processes requiring more than 3 MWh $tonne^{-1}$, e.g. for supercalendered (SC) and light weight coated papers (LWC). There is the potential to reduce this by 30% or more, by reducing the grindstone speed for SGW (more plastic deformation and a lower glass transition temperature) or with higher intensity of refining for TMP (but more fines). For a number of products either PSGW or TMP can be used so the economic decision balances higher energy costs (TMP) against higher capital costs (PSGW).

Improved fractionation methods may mean that pulp quality will no longer determined by the primary defibration and fibrillation, but by downstream fractionation and retreatments tailored to each fraction.

6. CLEANING, WASHING AND SCREENING

Regardless of the way in which pulps are prepared, all need to be freed of knots, washed and screened before bleaching or papermaking. Pulp screening and washing are critical to mill operation. Some of the procedures for both mechanical and chemical pulps are briefly described.

Hydrocyclones have long been used to separate sand, bark and other contaminants from the pulp slurry.

Washing technology is more complex, and equipment for pulp washing includes rotary vacuum washers, diffusion washers, rotary pressure washers and others. Cooked kraft pulps are washed to remove residual liquor that would otherwise contaminate subsequent processing stages, and also to recover cooking chemicals. Pulp washing is also carried out within the continuous digesters that are commonly used in the kraft pulping industry.

Some form of pulp screening is required to remove oversize particles, fibre bundles and other debris from good quality papermaking fibres. The term covers a wide range of operating modes, ranging from the removal of unwanted shives or fibre bundles in predominantly good quality pulp, to the rejection of up to 30 to 40% in mechanical pulps: the so-called reject fibres are thickened and returned to the refiners to provide in the end a better quality fibre than the accepted fibre from the first stage of screening. Coarse wood fragments are removed by bull screens, or coarse sieves, after which the pulp goes to primary and secondary screens.

Screening involves passing stock thorough a perforated screen plate to remove oversized material from the good quality fibres or fines. Screen plates are rigid metal plates in which slots have been cut or punched. There are essentially two characteristics that affect stock behaviour during screening, first the fibres and fines are flexible and secondly the stock is suspended in large volumes of water. Screens range from vibratory screens where the stock is agitated so that acceptable fibres pass through holes or slots, to modern pressure screens with oscillatory devices to allow the passage of stock at low consistency. Two screening systems in series are more effective than a single high efficiency screen. All screens have devices to keep the screen holes open, ranging from the pulses of old fashioned flat screens to the foils of modern pressure screens.

7. BLEACHING MECHANICAL PULPS

Newsprint and printing grades are generally made from light coloured softwoods (spruce and pine) or from light coloured hardwoods (aspen, poplar and some eucalypts). For example, spruce will produce unbleached pulps with a brightness of 60% to 65% (close to that of the original wood) and these can be bleached to brightness of about 80%, the level required in the higher quality printing grades.

Mechanical pulps are bleached by procedures that preserve lignin, but eliminate most coloured groups. Most mechanical pulps can be bleached to brightness levels of 75-80%, that are suitable for uncoated and coated papers. The principal bleaching procedures for mechanical pulps are reductive bleaching with sodium dithionite (hydrosulphite), and oxidative bleaching with peroxide under alkaline conditions.

Reductive diothionite bleaching. Pulp brightness can be raised by 8 to 12 brightness units using sodium dithionite ($Na_2S_2O_4$) and medium consistency bleaching. The chief limitation with dithionite is that there are diminishing benefits once the chemical charge is increased beyond 1% on pulp (wt/wt).

Oxidative peroxide bleaching. Peroxide is an effective bleaching agent at high charge levels whereas dithionite is not. In order to attain brightness levels of 80% or more it is necessary to use peroxide levels of 4 to 5% based on oven-dry pulp, to use chelating agents to remove metal contaminants, and to use high consistency bleaching (>80°C; 30 min). Modern peroxide bleaching demands capital and the efficient use of chemicals, including their recycling within the system to make use of residual bleach potential. Peroxide bleaching is carried out in alkaline conditions. The presence of alkali removes most of the extractives making the pulp more absorbent and better suited for sanitary products, e.g. disposable diapers. There is some yield loss and because of this the effluent load from the bleach plant makes a considerable addition to the effluent load from the mechanical pulp mill. Peroxide can achieve a brightness as high as 82% with spruce and 85% with poplar.

Yellowing of paper and brightness reversion. A major disadvantage of mechanical pulps is their propensity to undergo yellowing. Although lignin is colourless, under the influence of light and air some of its phenolic groups are oxidized to quinones and these and other chromophores cause paper made from mechanical pulp to yellow. Yellowing also occurs, relatively slowly, in long-time storage of paper. An economic procedure for permanently stabilizing such papers to either light or heat-induced yellowing has not been developed.

8. SEMICHEMICAL PULPING

The pulping mechanisms involved in semichemical and chemical pulping are similar. In semichemical pulping the chips are partially delignified using less chemical and shorter cooking times than in full chemical pulping. Then they are mechanically defibred with a much reduced electrical energy demand compared to mechanical pulps. Semichemical pulps generally have higher strengths than mechanical pulps. However there are few differences between low yield, fully bleached CTMP pulps and high yield semichemical pulps. Properties of semichemical and full chemical pulps are compared in Table 13.4.

Table 13.4. Typical data on the properties of semichemical and full chemical pulps for radiata pine slabwood (outerwood), compared after beating in a PFI mill (unpublished Dr. J.M.Uprichard).

	Neutral sulphite-AQ		Kraft pulp
	69% yield	53% yield	49% yield
Tear, mN.m^2 g^{-1}	14.6	16.1	21.8
Tensile, N.m g^{-1}	63.7	92.8	84.5
Burst, kPa.m^2 g^{-1}	5.5	8.2	8.1
Sheet density, kg m^{-3}	541	615	616
Brightness (unbleached), %	35.1	38.5	21.2
Scattering coefficient, m^2 kg^{-1}	17.8	18.2	17.3

The neutral sulphite semichemical (NSSC) pulping process is used on hardwood species. It involves a short pulping stage with 10-15% sodium sulphite and 4% sodium carbonate for up to an hour at high temperature (170-180°C) in a continuous digester prior to pulp defibration. Sodium carbonate is added to neutralise wood acids released by hydrolysis of the hemicelluloses and to maintain the pH between 8 and 10. Carbohydrate losses increase considerably near the end of the cook so there is a practical limit to the extent of delignification without serious loss in yield and strength. The partially delignified chips are then mechanically refined, at an energy demand of about 400 kWh tonne^{-1}. Yields are intermediate (65-75%) between mechanical and full chemical pulps. NSSC hardwood pulp is generally used for the manufacture of corrugating medium where strength and fibre length is less critical than stiffness. Corrugating medium is the unbleached fluted sheet sandwiched between the linerboards on the surfaces of corrugated cardboard boxes.

Semichemical pulps produced by the sodium bisulphite and neutral sulphite-anthraquinone pulping processes at 75% yields have properties that are rather similar to a lower yield chemical pulp shown in Table 13.4. The pulp characteristics from this process are of considerable interest, since it has been shown that chemical pulps made by the so-called ASAM process, in which the wood is delignified with a liquor containing sodium sulphite, sodium carbonate or sodium hydroxide and methanol (*c.* 40% of the liquor by volume) with a trace of anthraquinone can be delignified more rapidly and to a lower kappa number (this surrogate measure of the lignin content of pulp is explained in the appendix) than pulps made in the absence of methanol. ASAM pulps appear to have future potential, principally because pulps of very low kappa number can be produced. The significance of this is discussed in the next section.

9. CHEMICAL PULPING (GULLICHSEN AND FOGELHOLM, 2000)

Chemical pulping processes operate at high temperatures (140-180°C) and pressures (0.6-1.0 MPa), require good chemical recovery systems, and must operate on a large scale if they are to be economically competitive. The chemistry involved in the various processes will be outlined briefly. However before such discussion it is desirable to examine the digesters used by industry, since these items are common to all processes. Further some general aspects of pulping chemistry, the effects of delignification and cell wall morphology, will be described.

9.1. Digester design

Historically chemical pulping was carried out in large batch digesters (200-400 m^3). However, the trend is towards the use of continuous digesters, now accounting for about 65% of kraft production. The largest in operation have production capacities of more than 2000 tonnes of pulp a day.

Batch digesters recirculate the liquor as shown in Figure 13.6. The wood chips are fed into the digester that is then capped. The digester is filled with the required cooking liquor: it is common practice to use the residual liquor from a previous cook as make-up liquor and to add fresh chemical so that the cook has the required chemical charge and liquor-to-wood ratio, of say 3.5:1. During the cook liquor is withdrawn continuously from the digester via the strainers, to be reheated and returned to the digester. The time-temperature schedule is process and product dependent. At the end of the cook the steam pressure within the digester is decreased, the valve at the base of the digester is opened and the delignified chips go to the blow tank, where the softened and fully cooked chips are defibred to pulp. The pulp is then washed and screened.

Continuous digesters (Rydholm, 1970) offer better control of cooking and have better steam economy than batch digesters and have the added advantages of within digester washing. In the Kamyr continuous digester the chips are introduced continuously to an atmospheric chip bin (20 mins at 80°C) and then to a low pressure presteaming tube where they are steamed to remove air and non-condensibles. This is critical for uniform penetration of cooking chemicals and eliminates buoyancy of chips. From there a high pressure pocket feeder delivers a mixture of steamed chips and liquor to the top of the digester. The digester has a series of zones (Figure 13.7). Each zone is controlled separately with regard to temperature and chemical concentration. The temperatures in the various zones are controlled by forced recirculation of the cooking liquors through external heat exchangers. Radial gradients, which risk non-uniform cooking, are minimized by permitting fines to pass through the slotted screens of the strainers to be reintroduced again at the centre of the digester. The chips pass through an impregnation zone, a heating, a cooking and a diffusional washing zone to emerge at the bottom some hours later. The chips move down through the digester at a controlled rate in plug flow mode without channelling, with the digester operating at 1.0 to 1.2 MPa (145-175 psi) using hydraulic pressure to prevent the liquor boiling.

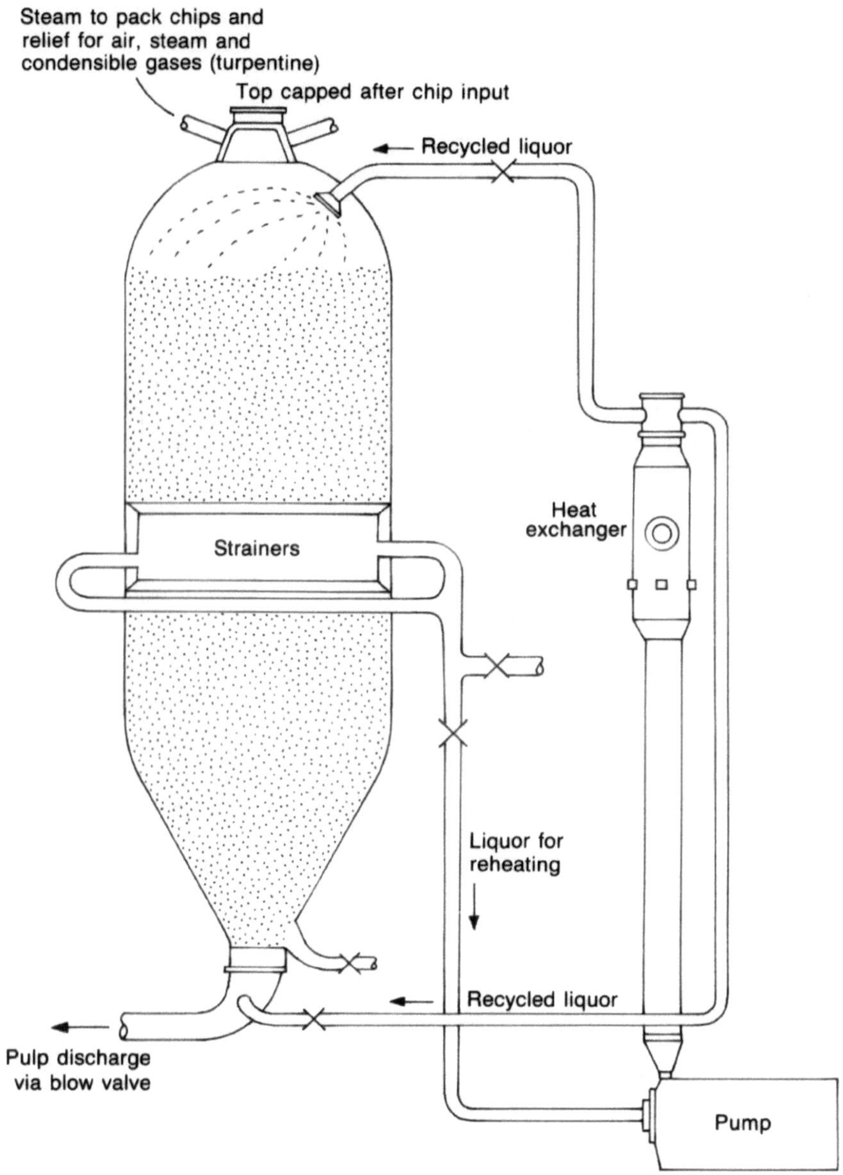

Figure 13.6. A batch digester with external heat exchange.

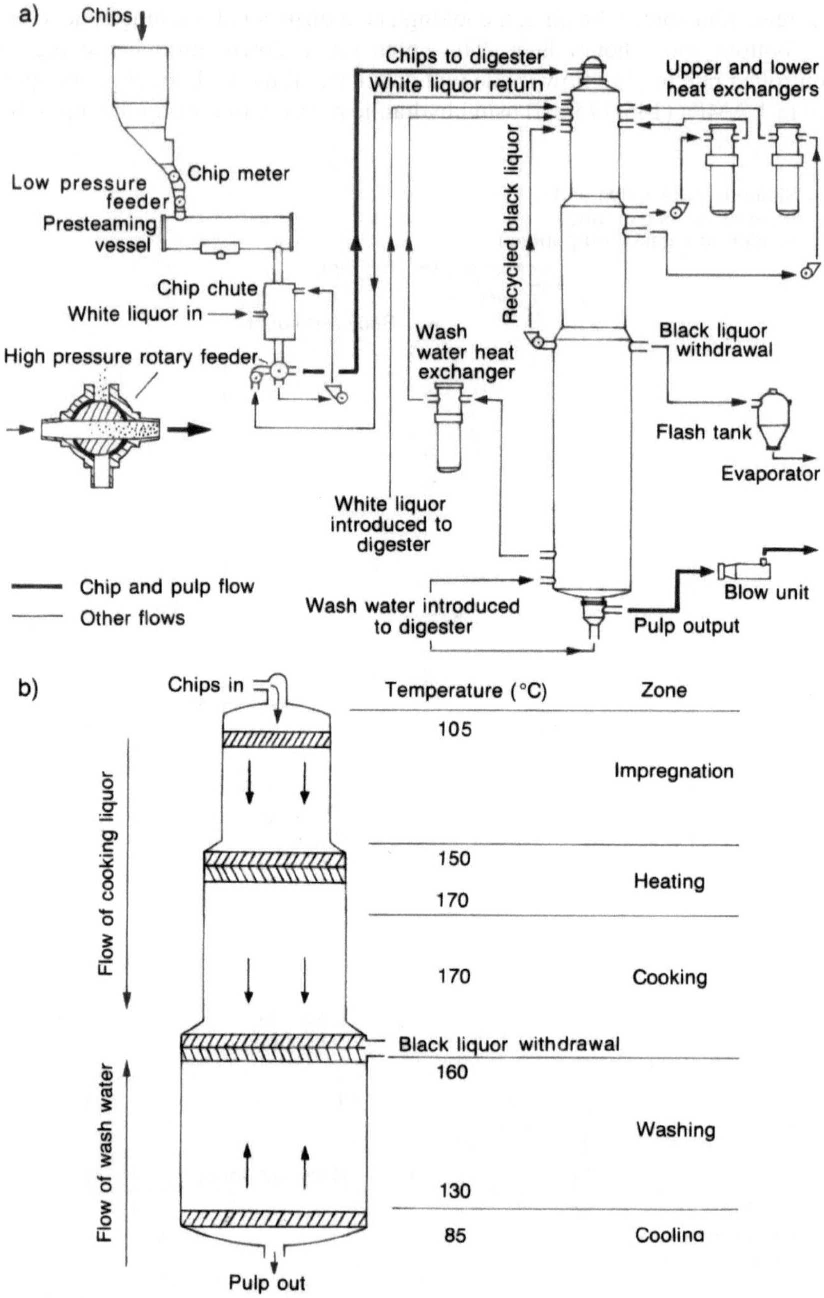

Figure 13.7. Schematic diagram of flows in a continuous digester (Kamyr of Karlstad, Sweden). The chip residence time within the digester is about 6 hr.

9.2. Delignification and wood morphology

The main aim of chemical pulping operations is to remove lignin and make the fibres available for papermaking. There is some loss of the wood polysaccharides during the delignification process, with cellulose being the least susceptible of the polysaccharides to degradation. The way in which carbohydrate loss occurs depends on the process. Technically the rate of lignin removal should be greater than that of carbohydrate loss. The term selectivity is used to describe the preferential removal of lignin, with pulping processes of high selectivity being required.

Various techniques, such as ultraviolet absorbance measurements, have been used to obtain quantitative information of lignin content within the cell wall. These show that in the early stages of both kraft and acid sulphite pulping the secondary wall is delignified at a faster rate than the corresponding compound middle lamella region (Wood and Goring, 1973). However towards the end of delignification the rates of lignin removal become similar. Kraft residual lignin is highly condensed, less reactive and more difficult to solublize than the lignin removed earlier in the cook. This means that there is less selectivity of lignin removal and correspondingly more carbohydrate loss if attempts are made to pulp below a specific kappa number, for example a kappa number of 25-30 in the case of conventional softwood pulping (Figure 13.8). Physical dissolution without mechanical aid only occurs once the yield is below 60%, but with 20-25% of all lignin remaining. At the end of a traditional cook something like 10% of the lignin remains in the pulp, as does over 50% of the hemicelluloses and about 85% of the cellulose. Kraft pulp is dark (brightness 25-30) despite the removal of much lignin because of the activation of chromophoric groups in the residual lignin. These are dissolved or modified on bleaching.

As shown in Figure 13.8 where the selectivity of acid sulphite, two stage sulphite and kraft pulping are compared, the processes differ in terms of yield, with the sulphite pulps having higher carbohydrate yields than kraft pulps over the delignification levels examined. Kraft pulping is less efficient (1-2% lower yield) than sulphite pulping because delignification is noticeably more selective in sulphite pulping. Significantly, Figure 13.8 shows that with all processes delignification takes place unselectively at both the beginning and the end of cooking.

In kraft pulping it is possible to delignify to a lower kappa number without undue carbohydrate loss by a process known as extended delignification. This involves adding alkali in two or more stages so as to maintain a fairly constant alkali concentration during the cook. The process is important as it reduces the extent of bleaching required subsequently and reduces the environmental impact of the bleach plant.

Chemical pulping and bleaching need to be considered together. In the past it was common to consider bleaching, which involves further lignin removal, merely as an extension of pulping, and the kappa number at which bleaching commenced was a purely economic decision. However, organochlorine materials are produced as by-products of pulp bleaching with molecular chlorine, and related materials include small quantities of undesirable polychlorinated compounds. A number of strategies can be employed to reduce the amount of organochlorines and dioxins. The most direct approach is to reduce the kappa number (lignin content) of the pulp before entering the bleach plant. This explains the interest in extended delignification and in procedures for producing pulps of very low kappa number that do not require chlorine bleaching (for example ASAM pulps), and why oxygen bleaching is now almost obligatory for kraft pulping. However it should be appreciated that the technique of extended delignification was examined and developed long before interest in non-chlorine bleaching methods arose. Extended delignification gives more opportunities for within-mill recycling of chemicals. Technically oxygen delignifies pulp rather than bleaches it, although it does affect some pulp brightening. However the main impact of oxygen bleaching is the reduction of pulp kappa number. It is now standard industrial practice to use extended delignification to reduce the kappa number of conventional kraft softwood pulps from 30 to 15, while oxygen delignification after extended delignification will reduce the kappa number to about 9. The benefits of both modified cooking processes and oxygen delignification become apparent when it is appreciated that the amount of chlorine required for bleaching can be reduced to a third of that used previously. The reduced environmental impact of the bleach plant is considerable.

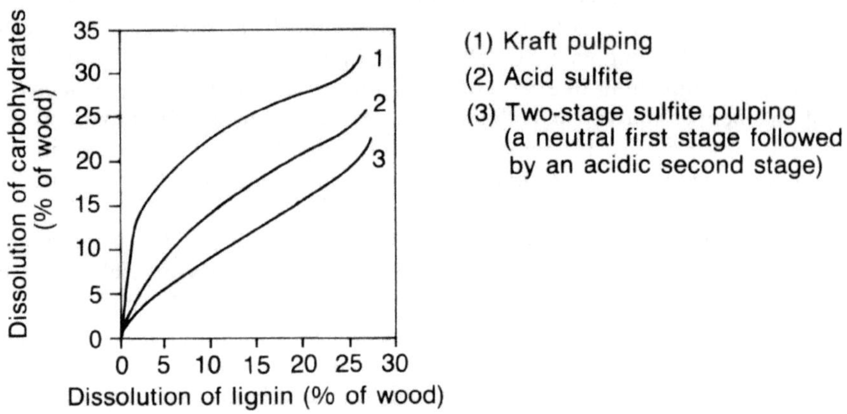

Figure 13.8. Delignification is unselective at the start of the cook (bottom left) and towards the end (top right) when carbohydrates are removed faster than lignin (Sjöström, 1981).

10. SULPHITE PULPING PROCESSES (SJÖSTRÖM, 1981)

Conventional sulphite pulping processes use aqueous solutions of sulphur dioxide at various pH levels. Sulphite solutions differ in their content of sulphur dioxide, bisulphite ions and sulphite ions, as shown in Figure 13.9. At a low pH of between 1 and 2 the sulphite liquor contains about 50% sulphurous acid and bisulphite ions respectively; at a pH of 4 to 5 it contains approximately 100% bisulphite ions; and at a pH of 8 to 10 it consists almost entirely of sulphite ions. When allowance is made for chemical charge, it is the pH and the relative amounts of bisulphite and sulphite ions that chiefly control the mode of pulping. In the original acid sulphite process

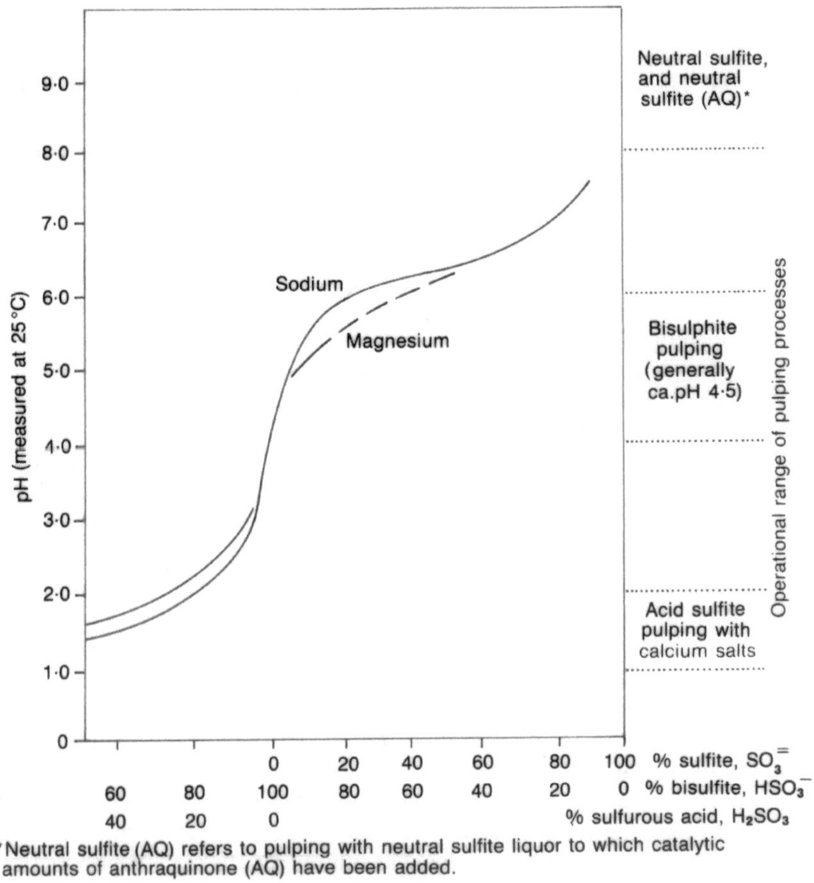

Figure 13.9. Composition of sulphite pulping liquor at various pH levels (Sjöström, 1981).

calcium was used as the base because it was cheap and spent chemicals were dumped. This is no longer acceptable. Today acid sulphite mills use sodium as base and have appropriate recovery systems

Acid sulphite and bisulphite cooks have rather similar reaction mechanisms. At about pH 4 the α-hydroxyl and α-ether groups are sulphonated by either aqueous sulphur dioxide or bisulphite ions, at an early stage of the cook. The sulphonation reaction occurs regardless of whether the phenolic group is free or etherified. Although there are only 6 to 8% α-aryl ether bonds present in the lignin their cleavage and sulphonation results in considerable fragmentation of the lignin. Once initial sulphonation is complete the temperature is raised to the final cook temperature when the hydrogen ions hydrolyse rapidly the ether linkages in the lignin. Under these conditions further sulphonation occurs and the lignin is broken into small water-soluble fragments (Sjöström, 1981, p.112).

In neutral sulphite pulping the most important reactions involve the free-phenolic lignin units. The first stage of the reaction results in the formation of a quinone methide that readily undergoes sulphonation by the attack of either a bisulphite or sulphite ion. The α-sulphonic acid group aids the displacement of the β-aryl ether substituent (in the adjacent lignin unit) and its replacement with the β-sulphonic acid group. Finally loss of the α-sulphonic substituent yields styrene β-sulphonic acid. Some loss of formaldehyde from the γ-carbon atom can occur also (Sjöström, 1981, p.112).

11. SULPHITE PULPING TECHNOLOGY

Pulps produced by sulphite processes are of lower strength than kraft but in other respects their papermaking properties are adequate. They are light in colour and easily bleached. They can be produced over a wide yield range and can be used to advantage in those products where good bonding properties rather than tear strength are of importance, e.g. for fine writing papers. Acid bisulphite pulp has the lowest tear strength while higher values are found when pulping under alkaline conditions.

Acid sulphite pulping was one of the earliest processes used for the production of chemical pulp. Sulphur dioxide is obtained by burning sulphur and absorbing the gas in the desired base, for example sodium carbonate, to give a solution of sulphur dioxide and sodium bisulphite. This is further fortified with vented sulphur dioxide from previous cooks to give a liquor which contains about 5 to 6% sulphur dioxide (H_2SO_3) and from 1.0 to 1.5% combined sulphur dioxide (Na_2SO_3).

Chip impregnation with cooking liquor is carried out at low temperatures, 110°C, for up to three hours to ensure good penetration and sulphonation of the chips before raising the temperature for the final 2 hour cook at 140°C. It is usual to relieve the digester pressure during cooking by gas relief, the emitted sulphur dioxide being recovered. Finally the cook is blown, the pulp washed, screened and freed of knots.

Under the acidic conditions of the acid sulphite cook, some depolymerization of the hemicelluloses occurs with the dissolved hemicelluloses gradually being hydrolysed to monosaccharides. Cellulose is hydrolysed too, but the loss is negligible unless a very low lignin content pulp of dissolving grade is being produced: for many years the low-yield (*c.* 35%), cellulose-rich pulp required for the dissolving pulp industry, e.g. for rayon, was prepared by this process.

The acid sulphite process is suited to non-resinous softwoods such as spruce, hemlock and fir and to some hardwoods.

Bisulphite pulping can be undertaken with a range of bases of which sodium and magnesium are the most important. Bisulphite pulping can be used successfully on hardwoods and softwoods of low extractive content provided the chips are well impregnated prior to pulping. Generally sodium bisulphite pulping is carried out at a pH of about 4.5. The commercial cooking liquor is prepared by treating sodium carbonate with sulphur dioxide. The quantity of bisulphite required for pulping depends on the yield and kappa number required. Softwood pulps of kappa number 30 require 18-22% sodium bisulphite based on oven-dry wood. Cooking is carried out at 170-175°C, at a liquor-to-wood ratio of 4:1, with a time to maximum temperature of about 1.5 hours and a time at maximum temperature of 1 to 2 hours.

Pulps from light coloured woods such as spruce and pine are light in colour and may be used unbleached in some grades of paper. Although they have good bonding properties they are restricted to products such as newsprint, tissues, fine writing papers and other products where high strength is not required.

Neutral sulphite-AQ and ASAM pulps. Neutral sulphite cooking is very effective on hardwoods but is much less effective for the delignification of softwoods, although it has been used in some pulpmills. However the addition of small amounts of anthraquinone (AQ), e.g. 0.1% based on oven-dry wood, catalyses the pulping of softwoods. The subject has been reviewed and some aspects of the chemistry and the influence of wood properties on the qualities of the pulps (from radiata pine) have been examined by Uprichard and Okayama (1984). Full chemical pulps require 30-35% sulphite charge plus added buffer on oven-dry wood. Cooking schedules are 2 to 3 hours to maximum temperature and about 2 hours at temperature (170°C).

The process gives pulps over a wide range of yield, full chemical pulp yields being 2-3% higher than corresponding kraft pulps. The pulps have excellent papermaking properties, with tear strength being only 20% or so less than that of kraft paper. A development of this process (ASAM) that uses essentially the same chemicals except for the addition of 10% vol/vol methanol to the pulping solution is especially promising since it produces both hardwood and softwood pulps of very low kappa number. These can be bleached without the use of chlorine.

12. CHEMICAL RECOVERY OF SULPHITE LIQUORS

In magnesium base sulphite cooking the liquors are burnt in a recovery furnace to give magnesium oxide and gaseous products containing sulphur dioxide. The magnesium oxide is treated with water and the resultant magnesium hydroxide reacted with sulphur dioxide to form magnesium bisulphite pulping liquor.

Both neutral sulphite and bisulphite pulping liquor can be recovered by the Tampella process (Rimpi, 1983). The residual cooking liquor is burnt in a kraft type recovery furnace and the smelt of sodium carbonate and sodium sulphide obtained as described in the section on kraft pulping. The dissolved smelt is carbonated with flue gas to form sodium hydrosulphide and sodium bicarbonate and the partially carbonated liquor stripped with steam in order to liberate hydrogen sulphide gas. The gas formed plus make up sulphur is then burnt to sulphur dioxide for the preparation of sulphite cooking liquor.

13. KRAFT PULPING

Kraft pulping, also termed the sulphate process, is the predominant process for the manufacture of chemical pulp. The kraft process offers some crucial advantages:

- An excellent recovery system;
- It can be used on any wood species and can tolerate bark in wood chips;

- Cooking times are short;
- Kraft pulp has excellent strength;
- A greenfield kraft mill can be totally self-sufficient in energy.

Kraft pulping is carried with a solution of sodium hydroxide and sodium sulphide, the solution being known as white liquor. In the terminology used by the pulp and paper industry, the chemicals are calculated as sodium equivalents and expressed as weight of NaOH or Na_2O. Sodium sulphide is hydrolysed largely to a mixture of sodium hydroxide and sodium hydrosulphide:

$$Na_2S + H_2O = NaOH + NaSH \tag{1}$$

It can be seen that one molecule of sodium sulphide releases one molecule of sodium hydroxide. Thus the effective alkali or the sodium hydroxide available for delignification is the sum of the sodium hydroxide and one-half of the sodium sulphide, when the chemicals are expressed in terms of their sodium hydroxide or sodium oxide equivalents.

$$\text{Effective alkali} = NaOH + 1/2\ Na_2S \tag{2}$$

It is general practice to express the chemical concentration used for pulping as a percentage of the effective alkali charge of sodium hydroxide (or Na_2O) based on oven-dry wood, or as the effective alkali as sodium hydroxide in g l^{-1}. This nomenclature is given here, not because it is necessary to know it in detail, but rather because it is desirable to know its background and use by industry.

The amount of sodium sulphide in the pulping liquors is expressed as the liquor sulphidity. Sulphidity is defined as:

$$\text{Sulphidity (\%)} = [\ Na_2S\ ] / \{\ [\ NaOH\ ] + [\ Na_2S\ ]\ \} \times 100\%, \tag{3}$$

where all chemicals are expressed as NaOH (or Na_2O). Sodium sulphide accelerates the rate of pulping relative to that experienced in the soda process, which uses only NaOH. The rate of delignification increases as the sulphidity of liquor increases from 0 to 24%, after which the rate of delignification continues to increase but rather more slowly.

14. PROCESS CHEMISTRY AND ALKALINE DELIGNIFICATION MECHANISMS

14.1. Carbohydrate degradation

About two-thirds of all the alkali used in alkaline pulping is consumed by the carbohydrates, the total consumption of alkali in a kraft cook being about 150 kg of sodium hydroxide per tonne of oven-dry wood (Figure 13.10). The carbohydrates are attacked early in the cooking cycle, before the maximum temperature of

165-175°C is attained. The first reaction is the hydrolysis of acetyl groups (deacetylation) on the polysaccharides, that are attached to the glucomannans of softwoods and to the xylans of hardwoods respectively. Later the polysaccharide chains are peeled from their reducing end-groups (Figure 2.2), degraded and converted to give hydroxy-carboxylic acids, generating volatiles (formic and acetic acids) and non-volatile carboxylic acids. In the case of cellulose about 50-60 glucose residues are stripped off before end-peeling is arrested by the formation of a metasaccharinic end-group with a stabilising carboxyl (Sjöström, 1981, p.134). Polysaccharides are not just subject to end-peeling, at high temperatures there is also alkaline hydrolysis of the glucosidic bonds, i.e. the polysaccharide chains are broken at points along their lengths. This means that new end-groups are formed and more end-peeling ensues. Consequently, the yield of cellulose is reduced by kraft pulping although to a lesser extent than that of the hemicelluloses that are non-crystalline and have much lower degrees of polymerization. Both peeling and stopping reactions are reviewed in Sjöström (1981).

By reducing carbohydrate losses higher yields are possible: through stabilizing the reducing end-group (increasing yield by as much as 8%) through the addition of sodium borohydride (uneconomic); by using polysulphide pulping (adding elemental sulphur to the cooking liquor to stabilize hemicelluloses at low temperatures early in the cook by oxidizing active end-groups); or by adding AQ (0.1-0.5% by weight for a 2-3% increase in yield) as a catalyst (but destroyed in the recovery furnace).

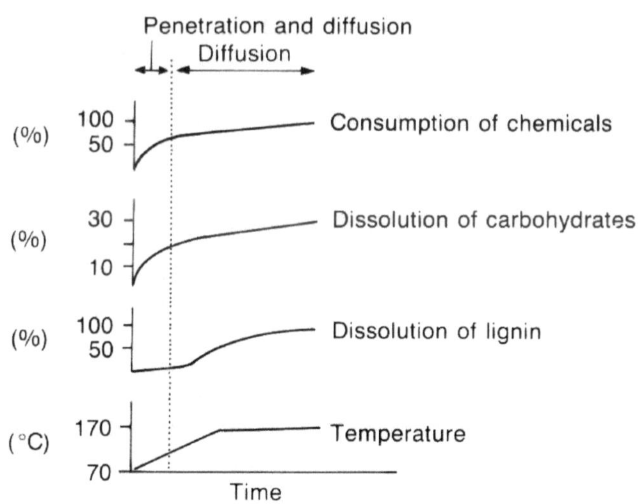

Figure 13. 10. Consumption of chemicals, and the dissolution of carbohydrates and lignin as a function of temperature and time (Hartler, 1962).

14.2. Lignin cleavage reactions

Hydrosulphide (HS^-) and hydroxyl (OH^-) ions are involved in kraft pulping, and hydroxyl ions in soda pulping. The nucleophilic hydrosulphide ions in kraft cooking greatly enhance the rate of delignification compared to soda cooking. The fragmentation of lignin depends on the cleavage of the α- and β-aryl ether linkages (C-O-C) which are the dominant linkages in both hardwood and softwood lignins. The α-aryl ether linkages are hydrolysed by the alkali present (Figure 13.11a) in both the kraft and soda processes but fragmentation of the macromolecule occurs only if the β-aryl ether linkage is absent or if it is cleaved subsequently (Figure 13.11b).

The course of the reaction of free phenolic units having β-aryl ether linkages depends on whether hydrosulphide is present (kraft pulping) or not (soda pulping). In both cases the initial step is the formation of a quinone methide from the phenolate ion. In a kraft cook the hydrosulphide ion reacts with the quinone methide, forming a thiol that is converted to a thiirane with simultaneous cleavage of the β-aryl ether bond. In the soda process the predominant reaction is the elimination of the hydroxymethyl group as formaldehyde and the formation of a styrl aryl ether structure; there is no cleavage of the β-aryl ether bond (Figure 13.11c). Etherified phenolic structures with β-aryl ether linkages are cleaved by hydroxyl ions via an oxirane intermediate, with a new phenolic group being generated by the reaction (Figure 13.11d). Other lignin reactions occur but these are a matter for the specialist.

Lignin demethylation occurs during kraft pulping, forming first methyl mercaptan which may react further to form dimethyl sulphide. These malodorous substances can give rise to air pollution problems during kraft pulping. Hardwoods being more heavily methoxylated (Figure 2.10) release a particularly strong odour when pulped by the kraft process.

15. THE KRAFT PULPING PROCESS

Kraft pulping may be carried out in either batch or continuous digesters, the use of which has been described earlier. The traditional batch operation is simpler to describe operationally.

The cooking liquor is generally a blend of white liquor and black liqour (spent cooking liquor), and is used in batch cooking at a liquor-to-wood ratio of about 3.5:1. Typical pulping conditions for softwoods are 16% effective alkali as Na_2O based on oven-dry wood, a time to maximum temperature of 75 to 90 minutes, and 60 to 90 minutes at maximum temperature (170-175°C). By the time maximum temperature is approached the carbohydrates have consumed most of the alkali, while only a small amount of lignin has been removed: the cook is at the beginning of what is termed the bulk delignification phase. During this stage the removal of lignin exceeds that of carbohydrates, and delignification is selective. Once the kappa number falls to 35 or less the rate of delignification begins to decrease again and kraft pulping enters the residual delignification phase. The cook is stopped and the

a) In an alkaline medium the cleavage of α-aryl ether bonds in phenolic arylpropane units occurs most readily

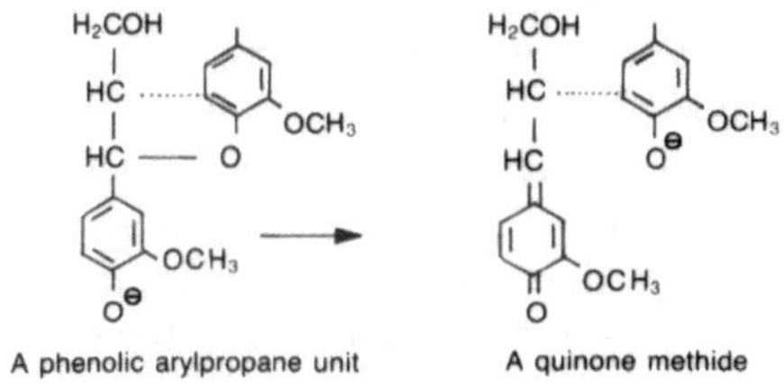

b) β-aryl ether bonds in phenolic arylpropane units are rapidly cleaved in the presence of hydrogen sulphide ions, i.e. in kraft pulping.

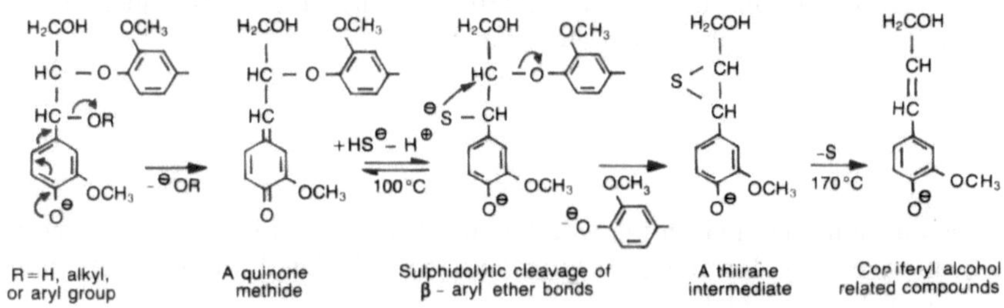

c) In the absence of sulphur, i.e. soda pulping, alkali stable structures are formed without cleavage at the β-carbon atom.

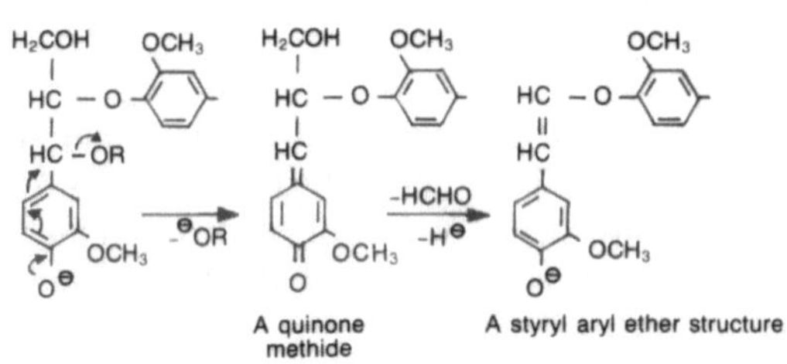

d) In both kraft and soda pulping the β-aryl ether bonds of structures with an ether link at the 4-hydroxyl position (non-phenolic units) are cleaved more slowly.

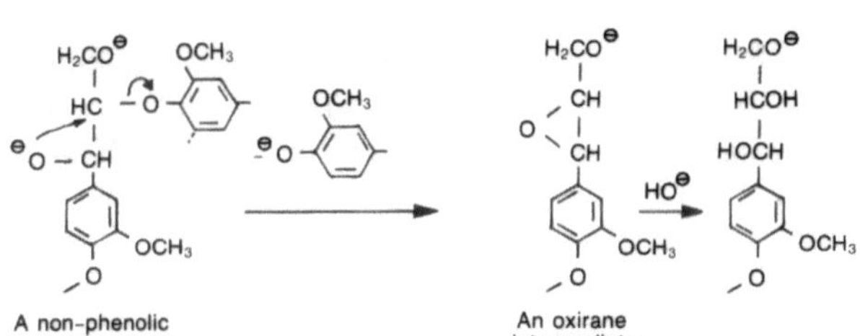

Figure 13.11. Some delignification reactions occurring in kraft and soda pulping (Gierer, 1980; Fengel and Wegener, 1984).

digester blown. The pulp is washed and screened before going to storage prior to using in the paper mill or going to the bleach plant. The progress of chips to pulp in the continuous digester follows much the same course, except that the chips have an impregnation stage and the pulp undergoes in-digester washing (Figure 13.7).

Delignification stages. An important feature of the delignification process is that it can be divided into three phases, as shown in Figure 13.12a (Kleppe, 1970). The initial phase, occurring at temperatures below 140°C, accounts for most of the carbohydrate loss by alkali but only a small proportion of the wood lignin (about 6% based on wood). The bulk phase accounts for about 16% of lignin while the residual phase a further 2% of lignin. Figure 13.12a shows delignification as a plot of residual lignin (based on wood) in pulp against the H factor.

The H factor was derived from the kinetics of kraft pulping by Vroom (1957) as follows. Under conditions of constant alkali concentration (high liquor-to-wood ratio) kraft delignification obeys first order kinetics:

$$-\frac{dL}{dt} = kL \tag{4}$$

where L is the lignin content in the wood chips at time t and k is the rate constant. The activation energy for delignification can be calculated from the Arrhenius equation, using experimental data:

$$\ln k = \ln A - \frac{E}{RT} \tag{5}$$

where T is the absolute temperature, R is the universal gas constant and A is a further constant. The activation energy, E, for bulk delignification of softwoods is about 130 to 150 kJ mol^{-1}. In order to construct a system of relative pulping rates Vroom assumed arbitrarily that the relative rate of pulping at 100°C was unity and then calculated the rates at other temperatures, obtaining:

$$\ln\,(\text{relative delignification rate}) = 43.2 - \frac{16\,113}{T} \tag{6}$$

When the values for the relative delignification rate are plotted against cooking time (in minutes), the area under the curve corresponds to the total amount of delignification that has occurred and this is expressed by a term called the H factor:

$$H\,factor = \int_0^t exp\left(43.2 - \frac{16\,113}{T}\right)dt \tag{7}$$

The H factor is important as it combines the variables of temperature and time as a single number. A normal cook to low kappa number requires an H factor of 1500 to 1800 units of which heating to temperature contributes only about 180 units.

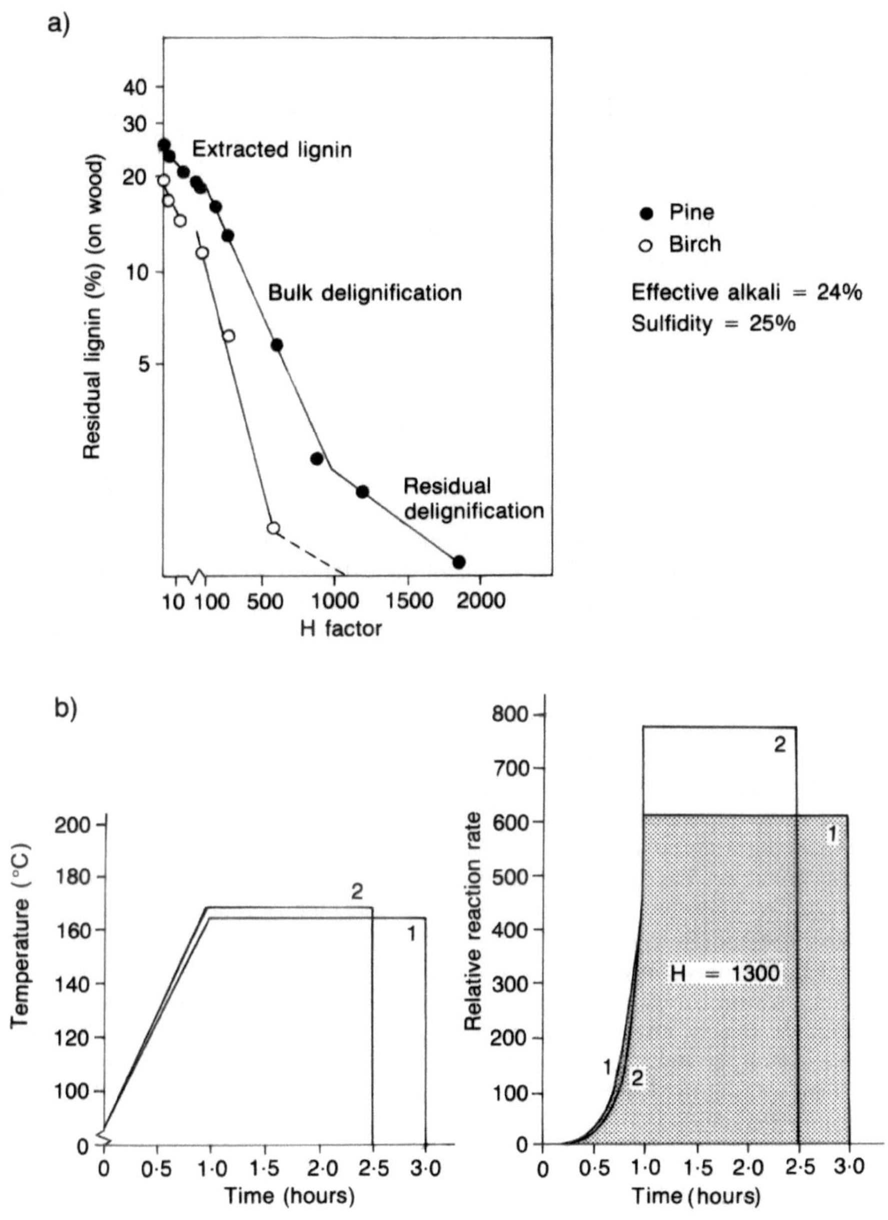

Figure 13.12. Kraft cooking. (a) Lignin removal from pine and birch as a function of the H factor (Kleppe, 1970). Little lignin is removed in the warm-up period. (b) Cooks can have the same H factor but different temperature-time profiles (Swartz *et al.*, 1969). Both cooks were to an H factor of 1300: the shaded area for the first cook corresponds to its H factor. Vroom (1957) measured relative rate values for the H factor for kraft pulping to be 1 at 100°C, 9 at 120°C, 66 at 140°C, 401 at 160°C, 610 at 165°C, 927 at 170°C and 1387 at 175°C.

The long warm up period contributes little to the H factor because reaction kinetics dictate that the rate of delignification more than doubles for every 10°C rise in temperature. The equation is useful as it allows mill operators to determine when to stop the cook even though the operating temperatures have differed from normal, for example because of fluctuations in steam supply to the digesters. Operators merely have to ensure that they cook to a constant H factor (Figure 13.12b), as pulps cooked to the same H factor have essentially the same properties, i.e. yield, lignin content (kappa number). Results can be compared provided the initial alkali and sulphidity concentrations are the same.

Kerr and Uprichard (1976) devised a predictive mathematical model of batch kraft pulping which incorporated other process variables, such as effective alkali, sulphidity and chip size, as well as the H factor.

16. EXTENDED KRAFT DELIGNIFICATION

Extended kraft delignification is a subject of increasing interest because of its potential to reduce the environmental impact of subsequent bleaching. Detailed investigations of Hartler and his collaborators (Sjoblom *et al.,* 1983, 1988; Johansson *et al.,* 1984) showed the merits of the extended delignification technique. They observed that the presence of dissolved lignin in the cooking liquor adversely affects the selectivity of delignification especially towards the end of the cook, and also that delignification occurs more efficiently under conditions of approximately constant effective alkali concentration than under conventional kraft cooking conditions where the concentration is high at the beginning but is low at the end of the cook. Figure 13.13 illustrates the difference in alkali concentrations during a modified cook and a conventional cook. A little over half the required alkali charge is added at the impregnation stage with the remainder being added partly on transfer to the main digester replacing much of the spent free liquor in the digester and partly immediately above the diffusion washing zone, where counter-current cooking with fresh liquor reduces the lignin content in the liquor. Finally counter-current washing at or near the cook temperature displaces black liquor, and accelerates diffusion of material from within the wood. The final discharge at 80-90°C at 8-9% consistency (by adding additional water) is gentle (the digester is not blown) so there is less fibre damage. The overall effect of this particular configuration, which is one of several possible, is to produce pulps with lower rejects, higher strength and – most significant – easier to bleach than conventionally cooked pulps (Dillner, 1989).

Modified cooking involves concentration profiling such that:

- The concentration of alkali should be kept low and uniform throughout the cook: too much initial alkali and too rapid a preheat give poor quality pulp; too little alkali later on favours condensation reactions so some residual alkali is essential.
- The sulphide concentration should be as high as possible early in the cook. Hydrosulphide reactions, liberating phenolic hydroxyls, render lignin more

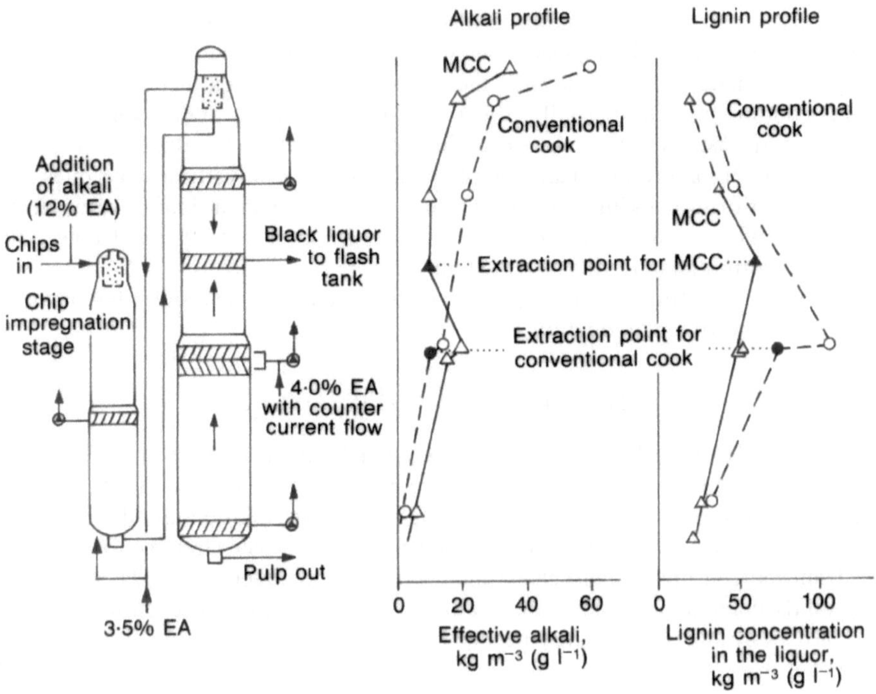

Figure 13.13. Modified continuous cooking (MCC) to obtain low kappa number pulps (Dillner, 1989). A total charge of 19.5% effective alkali (EA) is added at various points. The alkali and lignin profiles in the modified continuous cook and a conventional cook are shown. In the modified continuous cook both the alkali and lignin concentrations are evened out. The principle, of adding the alkali at stages during the cook, is applicable also to batch digesters.

susceptible to dissolution during subsequent bulk and residual delignification stages.

- The amount of dissolved lignin should be low toward the end of the cook, because the rate of kraft bulk delignification slows down as the concentration of dissolved lignin increases. Thus digesters have multiple extraction and dilution sequences.
- During the cooking stage alkali is consumed within the chip and it is critical that it is never fully exhausted, i.e. the cook needs sufficient alkali to diffuse into the chip to replace that which has been consumed. The liquor should remain above pH 11 otherwise condensation and precipitation of lignin fragments can occur. Experience suggests that depletion is too fast above 165° C so lower cook temperatures and longer cook times (1.5-2.5 hrs) are favoured. Not only is the kappa number lower, but overall there is roughly a 2% increase in yield for a given kappa.

Modified displacement cooking is also applicable to batch processing, where tank farms sit alongside the digester. The reuse of warm (95°C) and hot black liquor means less demand for steam to heat the chips. The initial treatment under mildly alkaline conditions with partially spent cooking liquor (a high hydrosulphide-to-alkali ratio) ensures the alkali-consuming reactions (the neutralization of hemi-cellulose degradation products) and the removal of both dissolved carbohydrates and lignin occur prior to cooking with fresh white liquor. The spent black liquor (from the previous white liquor final stage) is recovered and stored, by displacing with wash liquor (introducing in-digester washing).

Digesters are getting larger, in part to accommodate longer, lower temperature cooks and so improve pulp quality.

17. THE KRAFT RECOVERY CYCLE

The recovery of chemicals and energy from the black liquor is an essential feature of kraft pulping. The recovery cycle is quite elegant (Figure 13.14). In a kraft pulpmill the residual black liquor is evaporated to a highly viscous solution and then burnt in the recovery furnace. Other alkaline liquors from several waste streams, for example those from oxygen bleaching, can be included in the recovery cycle These have a lower solids content, increasing the load on the evaporators. The burning of the lignin provides most of the energy needed in the mill, and the pulping chemicals are regenerated from the molten salts that are recovered at the base of the furnace.

Black liquor from the digesters contains about 14-18% solids, with roughly 80% organics and 20% inorganic elements, e.g. sodium salts. This is concentrated in multiple-effect evaporators. Each consists of large arrays of long vertical stainless steel tubes 50 mm in diameter and 10 metres or so tall through which the liquor is passed (Figure 13.15). Typically the liquor is pumped through 5-7 such vertical evaporators while its solids content progressively rises to 50%, beyond which its viscosity increases very sharply and it becomes difficult to concentrate further. Process steam moves counter-flow and gets progressively cooler. Thus the vapour generated in one stage is the heating medium for the next stage, so that with seven effects one tonne of steam evaporates about six tonnes of water from the black liquor. New evaporators are of various falling film types – effectively the steam is inside the tubes while the viscous liquor is outside. The black liquor is recycled

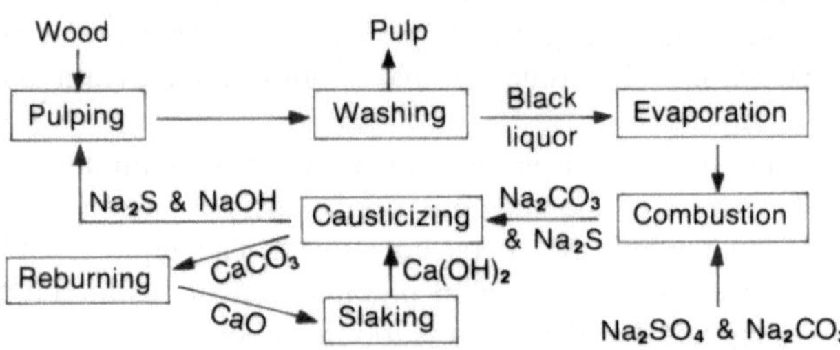

Figure 13.14. The kraft recovery cycle: it is a closed loop.

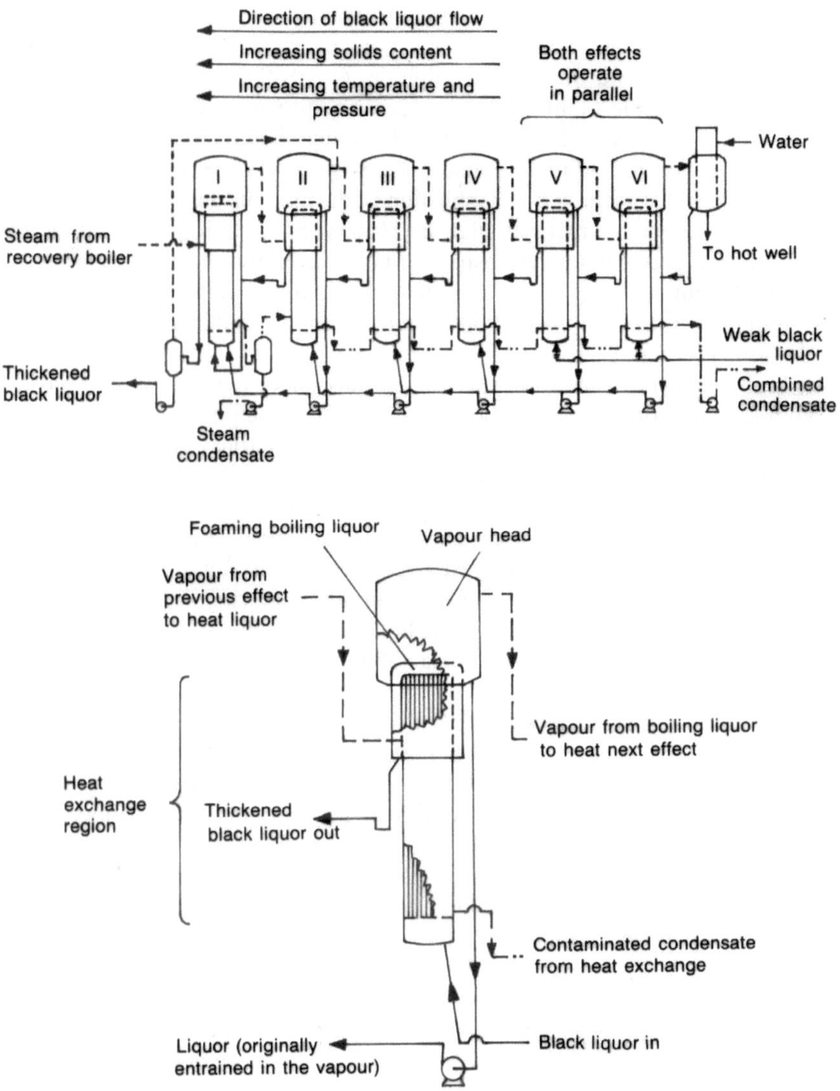

Figure 13.15. Multiple-effect evaporators.

while fresh liquor is continuously introduced (lower solids) and at the same time some is bled off to the next evaporator (higher solids). In this way successive evaporators attain progressively higher final solids content.

The black liquor needs to be concentrated to at least 65% solids before it can be burnt in the recovery furnace. Traditionally the viscous liquor was concentrated further by direct contact with the hot flue gases from the recovery furnace; for example in a cascade rotary evaporator where the liquor is picked up from a vat and passed through the flue gas stream which cools from 400 to 180°C. With this 'open' system, gases and volatile chemicals such as hydrogen sulphide, H_2S, methyl mercaptan, CH_3-S-H, and dimethyl disulphide, CH_3-S-S-CH_3, can escape unburnt from the liquor into the chimney of the recovery furnace, and so pollute the air around the mill. To reduce odour problems, the liquor should be oxidized converting the sodium sulphide to sodium sulphate. New mills have various procedures for odour elimination that include the combustion of the noxious gases in the lime kiln.

To achieve a solids content as high as 80% without having to resort to direct contact with flue gases, liquor viscosity can be lowered by a heat treatment step at 180°C for 30 minutes which involves depolymerization of the higher molecular weight lignin fragments. This occurs interstage within the multiple effect evaporator, when the solids content is around 45%.

The kraft recovery furnace (Figure 13.16) runs under reducing conditions at its base with a smelt of sodium carbonate and sodium sulphide collecting on the furnace floor. Further up the furnace secondary (and tertiary) air are introduced. Here the gases coming from the char bed are oxidized (burnt) while the droplets of black liquor are evaporated and partially burnt. Still further up high pressure steam is recovered from the furnace gases using heat-exchangers. Finally, in older mills prior to venting, flue gases are passed over the viscous liquor coming from the multiple-effect evaporators in a cyclone evaporator, capturing some of the residual heat in order to further thicken the black liquor prior injecting it into the furnace. Thermal efficiencies of recovery boilers are about 60%, in terms of the useable heat generated.

Black liquor is not an ideal fuel, partly because of its moisture and partly because of the very high 'ash' content, i.e. the inorganic elements. It has been difficult to achieve steam temperatures above 480°C, compared to 540°C with coal. The ability to generate electricity can be achieved only with higher combustion temperatures by burning at a higher solids content (80%). Newer steels and designs can limit corrosion from small amounts of chlorides and potassium. New recovery furnaces can have rated capacities of over 2500 tonnes of dry solids per day.

Losses of sodium and sulphur, e.g. in pulp wash and in flue emissions, used to be made up by the addition of sodium sulphate and sodium carbonate, hence the term the *sulphate process*. The sodium sulphate undergoes reduction to sodium sulphide in the recovery furnace. With systems moving to total closure of chemical and water cycles, the presence of small amounts of sulphur in the wood itself and in the magnesium sulphate coming to the evaporators from the oxygen delignification plant is such that the sodium intake exceeds sulphur losses. In that case the addition of Na_2SO_4 would result on excessive sulphidity. Hence only NaOH may be required.

The ideal of having only Na_2S and Na_2CO_3 in the smelt is not true; there are non-process chemicals, Na_2SO_4, $Na_2S_2O_3$ and contaminant salts, in the green liquor. The calculated causticity and sulphidity have to adjust for the presence of these salts.

The hot molten smelt at the bottom of the furnace consisting of sodium carbonate and sodium sulphide is carefully and continuously discharged through a water-cooled spout into the green liquor dissolving tank. The final stage in the recovery cycle involves causticizing the green liquor, converting the sodium carbonate to

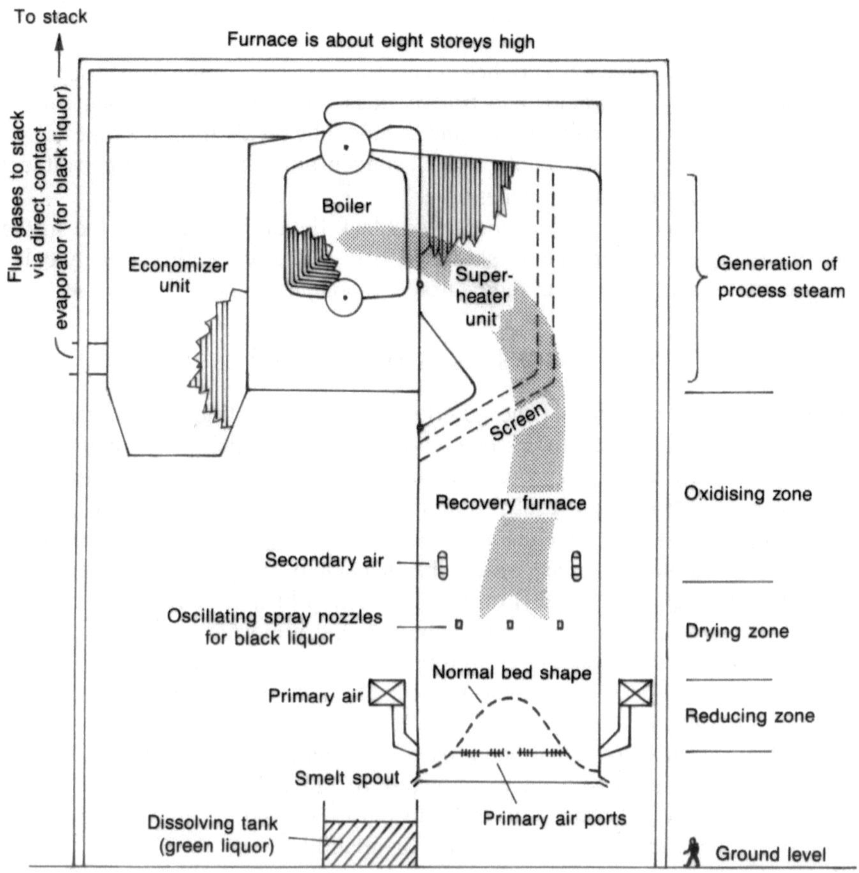

Figure 13.16. The recovery furnace.

sodium hydroxide. Treatment with calcium hydroxide precipitates out calcium carbonate while regenerating white liquor containing sodium hydroxide and sodium hydrosulphide. $Ca(OH)_2$ and $CaCO_3$ are both solids so the reaction takes place at the solid-liquid interface at just below 100°C over 1.5-2.5 hours. The insoluble calcium carbonate is filtered off, converted to quicklime in the lime kiln at 1100°C and so the cycle continues (Figure 13.14).

A bleached kraft pulp mill requires large amounts of energy (18.6 GJ tonne^{-1} in the example quoted) that is used mainly for process heat (15.2 GJ tonne^{-1}) and to a lesser extent for electrical power (3.4 GJ tonne^{-1}). The demand for thermal energy is split fairly evenly between cooking, drying, evaporation and electrical power generation (Hänninen and Ahonen, 1986). Chemical pulps require only a half to a third of the electrical energy per tonne of pulp compared to mechanical pulps. This is used mainly to circulate liquids through the digester, in the chemical recovery cycle and in the bleach plant.

18. BY-PRODUCTS OF SOFTWOOD PULPING

The heartwood of most species is richer in chemical extractives than is the sapwood. Pines contain mainly resin acids and triglycerides. Turpentine is obtained early in the pulping schedule by flashing steam and volatile chemicals from the digester, the turpentine separating from the cooled water condensate. The principal constituents of pine turpentine are α- and β-pinenes.

Tall oil is obtained from the concentrated black liquor. The soap on the surface of the black liquor concentrate is skimmed off, and after acidification yields a mixture of resin acids, fatty acids and neutral components.

19. OTHER PROCESSES AND THEIR POTENTIAL

In the 1970s it was shown that the addition of small quantities of anthraquinone (AQ) or related substances to a soda cook, i.e. one which uses only sodium hydroxide, increases the rate of delignification so that it is comparable to that in kraft pulping (Cameron *et al.*, 1982; Nomura, 1980). AQ is added in small amounts (0.05-0.25% of the oven-dry weight of wood) and although it is partially consumed it is clearly behaving as a catalyst, aiding the cleavage of the β-aryl ether linkages of the free phenolic lignin units (-OH on the C_4 position of the aromatic ring). At the same time the carbohydrates are being stabilized and protected against end-peeling by the oxidation of the reducing end-group (the terminal aldehyde, -CHO) to an alkali-stable carboxyl (-COOH). Soda-AQ pulps have strengths comparable to those of kraft pulps (*c.* 90%).

The APM mill at Burnie, Tasmania used the soda process to obtain bleachable grade pulps from eucalypts for the manufacture of fine papers, but in 1983 the mill switched to the soda-AQ process in order to pulp pine and eucalypts together, the decision being made on the grounds of wood availability.

The soda-AQ process can use the same chemical recovery system as for kraft pulping but is technically well suited to a simpler, more compact system known as DARS (direct alkali recovery system), which has been explored on the pilot plant scale. Here ferric oxide is added to the black liquor in a fluidized bed furnace. The residue of sodium ferrite particles is removed and on leaching under controlled conditions regenerates sodium hydroxide and ferric oxide:

$$Na_2CO_3 + Fe_2O_3 \rightarrow Na_2Fe_2O_4 + CO_2 \tag{8}$$

$$Na_2Fe_2O_4 + H_2O \rightarrow 2NaOH + Fe_2O_3 \tag{9}$$

Also, there is considerable interest in pulping systems that use solvents, although these have been studied for many years without commercial success. For example, a 30 tonnes per day pilot plant using the Alcell process (Pye and Lora, 1990) pulped hardwood chips at temperature in a 50:50 mixture of ethanol and water. The process relies on autohydrolysis, with acetyl groups being cleaved from the hemicelluloses to form acetic acid that proceeds to delignify the chips. Lignin fragments are more soluble and stable in the ethanol. Reasons for interest in solvent pulping systems include the desire for a simple recovery system involving no inorganic salts and the possibility of gaining value from the solvent-based fractionation of lignin and from the carbohydrate by-product. Such processes should be less dependent upon economies of scale than the kraft process, and ideally produce bleached pulp without using sulphur or chlorine.

20. BLEACHING OF CHEMICAL PULPS

The main purpose of bleaching chemical pulp is to make it whiter. Bleaching is a multistage process with the pulp being washed between stages. Traditionally molecular chlorine was used in the first stage to attack the lignin, with the bright coloured chlorinated fragments of lignin being extracted from the pulp using alkali. This section outlines some bleaching conditions, and describes some modifications in bleach technology that have been used to improve alkali extraction efficiency or reduce effluent colour.

The extent of bleaching is determined by the brightness required. This is achieved by using one of a number of bleaching sequences described as CEH, CEDED or CEHDED to name a few. To facilitate the description of the processes the industry has evolved its own notation as shown in Table 13.5. Most bleaching technology is multi-sequence and may involve up to six stages in all. Unbleached softwood pulp has a yield of 45-55% containing 65-75% cellulose, 20-30% hemicelluloses, and less than 5% residual lignin. Bleached pulp has a yield of 40-50% containing 75-80% cellulose and 20-25% hemicelluloses.

Table 13.5. Nomenclature of bleaching and oxidizing power relative to chlorine.

Nomenclature	Oxidizing power: electrons transferred/mole	Molar wt. g/mole	Equiv. wt. per electron transferred	Oxidizing power relative to chlorine
Oxygen (O)	$O_2 + 2H_2O + 4e^- \rightarrow 4OH^-$	32	8	4.43
Extraction (E)				
Chlorine (C)	$Cl_2 + 2e^- \rightarrow 2Cl^-$	71	35.5	1
Chlorine dioxide (D)	$ClO_2 + 2H_2O + 5e^- \rightarrow Cl^- + 4OH^-$	67.5	13.5	2.63
Peroxide (P)	$H_2O_2 + 2H^+ + 2e^- \rightarrow 2\ H_2O$	34	17	2.09
Ozone (Z)	$O_3 + 3H^+ + 6e^- \rightarrow 3OH^-$	48	8	4.43

Molecular chlorine, Cl_2, is an ideal chemical for delignification as it is cost-efficient and has reasonable selectivity, well capable of removing 75-90% of the residual lignin in a single stage. However its undesirable effects – waste water with chlorides that is corrosive, and a tiny by-product of chlorinated organic compounds – means that molecular chlorine has been abandoned entirely in some countries and where it is still used this is often in conjunction with chlorine dioxide (D) whose environmental footprint is some 2.6 times smaller. Today, more than 75% of pulps are bleached without any molecular chlorine. These are described as elemental chlorine-free (ECF). Only *c.* 6% of bleached pulps are total chlorine-free (TCF), using only oxygen, ozone, peroxides or peracids. The effectiveness of the various bleaching steps can be expressed in terms of oxidizing power, relative to chlorine (Table 13.5).

There is a popular wish to reduce the environmental effects of bleaching. This is being achieved by a combination of extended delignification, oxygen delignification and by greater use of chlorine dioxide relative to chlorine (less chlorine is used to affect the same degree of delignification) or by total chlorine-free bleach processes. In the bleach sequence soluble degradation products are washed out and others can be removed interstage by alkali extraction (E) with post-extraction washing. It is essential to avoid carrying over lignin fragments and other organic matter – which would consume chemical during subsequent bleaching.

Traditionally kraft softwood pulps were delignified to kappa number 25-35 (4-5% lignin in the pulp), but using extended delignification and oxygen delignification pulps can be reduced to kappa number 10 before chlorination stages. Hardwoods, because they are easier to delignify, were generally pulped to a lower kappa number of 15-20 (2.5-3.5% lignin) before bleaching. The reason for bleaching chemical pulps is to remove all but a trace of the residual lignin and any associated coloured substances to give a bright, colour-stable paper. Pulp brightness of over 90% is achievable. Cellulose degradation must be kept to a minimum during bleaching.

20.1. Oxygen delignification

Strictly the process is one of delignification rather than of brightening/bleaching. Alkaline oxygen delignification (O) is ideal for kraft operations as it takes oxidized white liquor as an alkali source and its spent liquor is reused in the kraft digester counterwash. Oxygen delignification in alkali, at about 0.5 MPa and 85-105°C, is well established (Almberg *et al.*, 1979). Oxygen is poorly soluble in alkali so needs a stable gas/liquid/solid dispersion: all reactions occur at the solid-liquid interface preferably in a high consistency (22-30%) pulp. Oxygen delignification is complex and involves free radical mechanisms, peroxy radicals being present. The treatment will remove half the residual lignin without serious cellulose degradation, but because the attack on carbohydrates increases with delignification the stage is terminated at this point. Trace amounts of transition metal elements enhances attack on carbohydrates but this is effectively countered by adding small amounts of magnesium sulphate (Mg^{++} complexes with and scavenges transition metal ions).

Oxygen is prepared by cryogenic liquefaction, or by absorbing N_2 on a molecular sieve under high pressure and venting the N_2 to the atmosphere at low pressure.

20.2. Molecular chlorine, Cl_2

Chlorination is carried out at low consistency of about 3% (because of the low solubility of chlorine in water). Chlorine (C) reacts almost instantly with all lignin groups, primarily by substitution and oxidation, with some addition of chlorine to double bonds also occurring (Sjöström, 1981). Chlorination reactions affect some lignin depolymerization to low molecular weight chlorinated products, but high molecular weight products are also formed, most of which are alkali-soluble. Sufficient hydrochloric acid is generated to keep the pH below 2.0.

$$R\text{-}H + Cl_2 \rightarrow R\text{-}Cl + 2HCl \quad (10)$$

$$R\text{-}H + Cl_2 + H_2O \rightarrow R\text{-}OH + 2HCl \quad (11)$$

Today it is common to use a chlorine dioxide/chlorine mixture (C/D) in place of chlorine (C) in the first stage. Addition of chlorine dioxide only seconds after that of the chlorine protects the cellulose at temperatures up to 70°C in the 30-60 minute treatment. Chlorine reactions with lignin are almost instantaneous but it reacts more slowly with carbohydrates, so ClO_2 can be introduced seconds later after the lignin reactions are well established and yet be in time to protect the carbohydrates. Some reaction products from acidic bleaching are only soluble in alkali. Therefore in general each oxidation stage is followed by alkali extraction and washing.

Cl_2 is prepared by electrolysis, liquified and transported to the plant, where it is stored under pressure.

20.3. Alkali extraction (NaOH) and with oxygen (E_0)

The purpose of alkali extraction, i.e. E stage (1-2% alkali, 60°C for 30-60 minutes at 10-20% consistency), is to remove alkali-soluble, coloured products from the previous delignification stage. Alkali extraction dissolves lignin fragments by forming soluble salts. Some attack of residual lignin also occurs with chloride ions being liberated and free phenolic structures being formed. The extracted liquor from the first bleach stage is responsible for much of the colour in the mill effluent.

Alkali extraction is enhanced if some oxygen is added (E_0) which aids lignin breakdown by ionizing and solublizing phenolic groups while also reducing the colour of the degraded and dissolved lignin components.

Unfortunately, alkali extraction also generates further chromophores (the pulp goes bright yellow). These must be oxidized and subsequently extracted.

20.4. Chlorine dioxide (ClO_2)

Chlorine dioxide or D stage bleaching is carried out under acidic conditions (70°C for 3-4 hours at 10-15% consistency). Chlorine dioxide is a stronger oxidizing agent than chlorine. It destroys double bonds and cleaves lignin aromatic rings. Chlorine dioxide reacts only slowly with polysaccharides and is frequently used together with chlorine (C + D stage) or added just after the chlorine (C/D stage). In these situations chlorine dioxide scavenges free radicals arising from the use of chlorine that would otherwise degrade the polysaccharides. The greater selectivity of ClO_2 as against Cl_2 preserves strength and gives brighter pulps while reducing the chlorine load in the effluent. Chlorine dioxide is much more expensive per unit of oxidizing power, and so was introduced in tandem with oxygen delignification.

The odd-electron molecule is very reactive. When warmed it explodes. ClO_2 has a bond angle of 118° and a very short bond length of 0.149 nm with six electrons on each of the oxygen atoms and seven on the chlorine. This means that there are two electrons shared between one Cl-O bond and three electrons between the other Cl-O bond. The three electron bond resonates between these two bonds. This extra electron is really a free radical that probably accounts for the instability of chlorine dioxide and its reactivity as an oxidizing agent.

For fully bleached softwood market pulps two final chlorine dioxide stages DED are common – preceded by a number of possible early sequences such as O(CD)E_oDED, ODE_oDED, or DEDED *etc.*

ClO_2 is always prepared on site by reduction of chlorate (ClO_3^-) with the gas being adsorbed in chilled water where its solubility is modest (10g/L) and stored at around 5°C. It is too unstable to transport.

20.5. Bleaching procedures

The pulp is bleached over a 8-10 hour period as it moves through the various batch treatment stages. The pulp is washed between each stage, but even with counter-current flow the fresh water demand is considerable, e.g. 18 m^3 per tonne of pulp. Thus a 500 tonne a day plant would use 9000 m^3 a day.

20.6. Bleaching strategies

One of the features of modern bleach technology is the way in which there is a combined strategy for pulping and bleaching, something which was less well defined in earlier years. The modern bleached pulp producer can plan to use modified kraft cooking and produce pulp at substantially lower kappa number and of better quality than was previously possible. This low kappa pulp (18-20) is then oxygen delignified to kappa number 9. In the bleach plant the first stage using a mixture of chlorine and chlorine dioxide, or even chlorine dioxide alone, then lowers the kappa number to about 4. In such ways the manufacturer can better control the environmental impact of pulping and bleaching.

There is increasing interest in the use of oxone (Z) for bleaching and bleach sequences such as OZEP have been examined experimentally (P stage is peroxide). It is likely that the use of oxygen and ozone in combination will be used over the next decade, with ozone replacing part or all of the chlorine. One of the disadvantages of ozone is its lack of selectivity and its tendency to attack carbohydrates as well as lignin.

21. EFFLUENT LOADS AND DISPOSAL

The main sources of pulp mill effluent are:

- Water used in debarking and wood handling.
- Digester and evaporator condensates.

- Bleach plant effluent.
- Papermachine white water (in an integrated mill).
- Unintentional fibre and liquor spills.

In part as a result of regulation, significant improvements in plant operations and in the recycling of process waters within the mill have resulted in reductions in the biological oxygen demand (BOD), the chemical oxygen demand (COD) and the total suspended solids (TSS) of mill effluent. BOD is a measure of the dissolved oxygen required for the biological oxidation of organic material in the effluent by micro-organisms. BOD_5 refers to a five-day test. Only the carbohydrate component is degraded readily over such a short time span. The COD is a measure of the total oxygen demand and relates to the oxidation of all organic material (carbohydrate, extractives and lignin fragments). The COD test is quick, based on chemical oxidation using chromic acid. To a first approximation the ratio of COD to BOD is 2:1. There is every incentive to keep the TSS loss to a minimum as this represents loss of fibre and inorganic fillers *etc.* from the paper mill, which if retained within the mill would contribute to production. Such material in the effluent is removed by screening and in the primary settlement ponds. Table 13.6 outlines some of the improvements in both effluent volumes and quality that have been achieved over a crucial 20 year period.

Both aerobic and anaerobic treatment procedures have been used to treat waste waters from mechanical pulp mills. CTMP effluent is difficult to treat anaerobically. Anaerobic procedures are generally not used for chemical effluent since sulphur compounds, e.g. hydrogen sulphide, will inhibit fermentation. The principal methods for reducing BOD of chemical pulp effluent are either lagoon treatment, in which the induced and natural bacterial populations are kept active by forced aeration in large lagoons, or by aerobic treatment in an activated sludge plant. Up to 90% of the BOD can be removed prior to discharge. However the dark colour of well treated kraft effluent still presents an aesthetic problem.

Table 13.6. Approximate changes in mill effluent per tonne of production, 1970-1990.

Mill type	Effluent volume m^3 $tonne^{-1}$		BOD_5 kg of O_2 $tonne^{-1}$		Effluent Temperature, °C	
	1970	1990	1970	1990	1970	1990
Kraft (bleached)	300	40	65	5	10-20	30-40
Sulphite (bleached)	500	60	95	5	10-20	20-30
Paper (newsprint)	60	15	10	1	30-45	45-60

22. THE MANUFACTURE OF PAPER

Mechanical pulp mills are normally on the same site and integrated with paper production, e.g. of newsprint. Chemical pulp mills can be integrated or sell market pulp that is flash dried and baled. Here, the capillary tension on drying pulls back the loose fibrillated wall elements to re-bond to the external surface of the fibre, and some of the newly created internal lamellae surfaces re-coalesce (Figure 2.11) – a process known as 'hornification' that needs reversion before making paper.

The first operation in the paper mill involves stock preparation: the slushing of baled pulp, the refining or beating of chemical pulps, the blending of pulp stocks and additives prior to forming paper.

This discussion will be confined principally to the well-established Fourdrinier papermachine and will only touch upon twin-wire machines that are now common in the industry.

Generally semichemical and chemical pulps require beating or refining before they are made into paper. Further, as the lignin content of pulps declines the pulps are easier to beat. Only the beating of chemical pulps is described. It has been shown over the years that the nature of the chemical pulp fibres used for papermaking and the way in which they have been refined or beaten largely shape the properties of the final paper.

23. BEATING OF CHEMICAL PULPS

Paper made from unbeaten chemical pulps is bulky, porous and has less tensile strength than from beaten pulps. This is because the unbeaten fibres tend to be stiffer, springy and resistant to collapse on pressing, so that there is comparatively little interfibre bonding in the sheet.

Beating is the term used to describe the process in which pulp is mechanically treated in the presence of water. Generally it is carried out at low or medium consistency by passing the pulp suspension between revolving and stationary rotors which have bars approximately aligned across the direction of stock flow. The narrow gap between opposing refiner plates is about 0.1 mm, sufficient for some 5 swollen or 20 collapsed fibres. The degree of beating is expressed either as energy input in kWh tonne^{-1}, or in the case of laboratory beaters by the number of beater revolutions, e.g. the pulp being beaten in a PFI beater for 4000 rev. The process differs from mechanical pulping as refining power is much lower, 25-150 kWh tonne^{-1}, per pass (so pulp properties are gradually developed as the pulp passes through one to three refiners in series) and it occurs in water, not steam; the refiner plates/bars on the two discs are not radially aligned as in mechanical pulping (Figure 12.12a), but are angled across the line of stock flow (Figure 13.17a) so the refining surfaces sweep along the bars; and refining is usually at medium or low consistency. Where the floc is caught between bar crossings fibres are stressed along their length: the impact, compression and dewatering of the floc at the leading edge is followed by rotation and frictional rubbing between fibres within the floc as bar faces sweep past one another (Lumiainen, 1990). After beating all but the thickest walled fibres show a degree of collapse when formed into paper. The degree of collapse increases with the amount of beating as does the extent of defibrillation.

The purpose of beating is to improve the papermaking potential of the fibre stock (Figure 13.17). The severity of beating depends on the fibre and the paper to be manufactured from it. In practice:

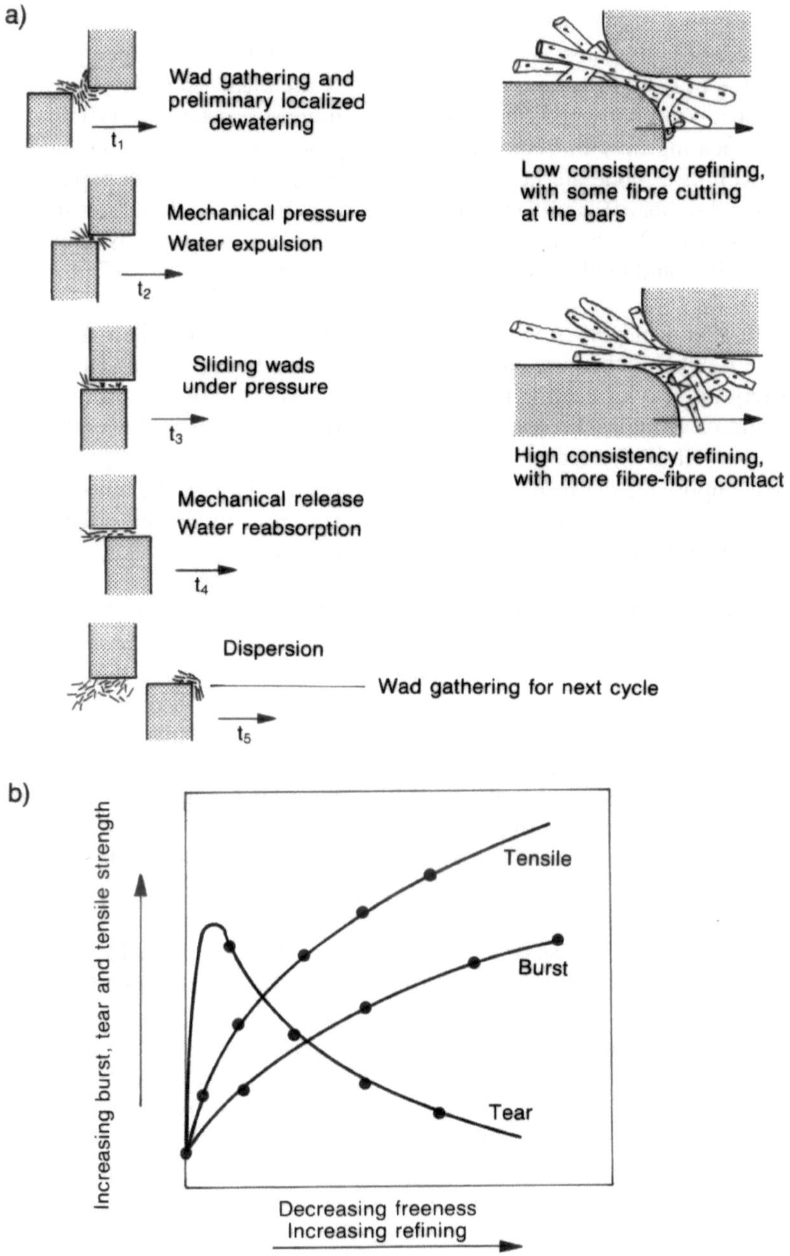

Figure 13.17. (a) Schematic representation of refining action (upper bar stationary and lower bar moving to the right (Espenmiller, 1969). (b) Typical strength development during refining (Smook, 1982).

- The response of hardwood and softwood pulps to beating is very different.
- The response of fibres to beating depends on how they have been prepared, on the degree of delignification, and whether they are bleached or unbleached. Long unbleached kraft softwood fibres display the highest resistence to refining.
- Light beating may be required for some products (filler for boxboard, tissue and towelling), moderate beating for others (bag and sack, fine papers, the top liner for boxboard) and very heavy beating for some high density papers (greaseproof). The specific energy for beating can vary by as much as 1:3 between light and moderate beating from 0.25 to 0.87 GJ tonne^{-1} (75 to 250 kWh tonne^{-1}).
- Some chemical pulp fibres respond very quickly. Thin-walled fibres rapidly become flexible on beating and may collapse while being beaten. Thicker walled cells require heavier beating to achieve the same degree of flexibility. Latewood tracheids of Douglas fir respond only slowly while their earlywood fibres are beaten very easily indeed.
- Most important, the response to beating depends largely the amount of water present when beating since there are large differences in beating effects between low (5-10%) and high consistency (20-30%) refining.

Studies by Kibblewhite (1984) and others have shown that the surfaces of chemical pulp fibres are progressively removed during the beating process. Some of the surface layers of the fibre are loosened or removed, exposing the P, S_1 and S_2 layers. The fibres become more flexible (Figure 13.18). The beaten fibres collapse on pressing, giving more intimate contact, stronger bonding and the density of the paper is increased. Better conformation plus the presence of some fibre debris arising from beating means that the modified mat drains more slowly on the wire, resulting in a more even sheet of paper.

The characteristic effects of beating (refining) on sheet properties are shown in Figure 13.17b:

- On refining the freeness of the pulp decreases while burst and tensile strength increase until they reach a plateau (from which they will decline again if beaten excessively).
- Lightly beaten softwood pulps show an initial increase in tear strength but as the fibres become better bonded they show a steady decline in tear strength.
- The opacity and porosity of the paper both decrease with refining, because despite the higher incidence of fines in the well beaten sheet, there is much better bonding between fibres (lower opacity) and fewer large pores within the sheet.
- Sheet densities increase with beating. Sheet densities in the range 500-700 kg m^{-3} are typical of beaten chemical pulps. These are higher than the 350-500 kg m^{-3} of the more bulky mechanical pulps.

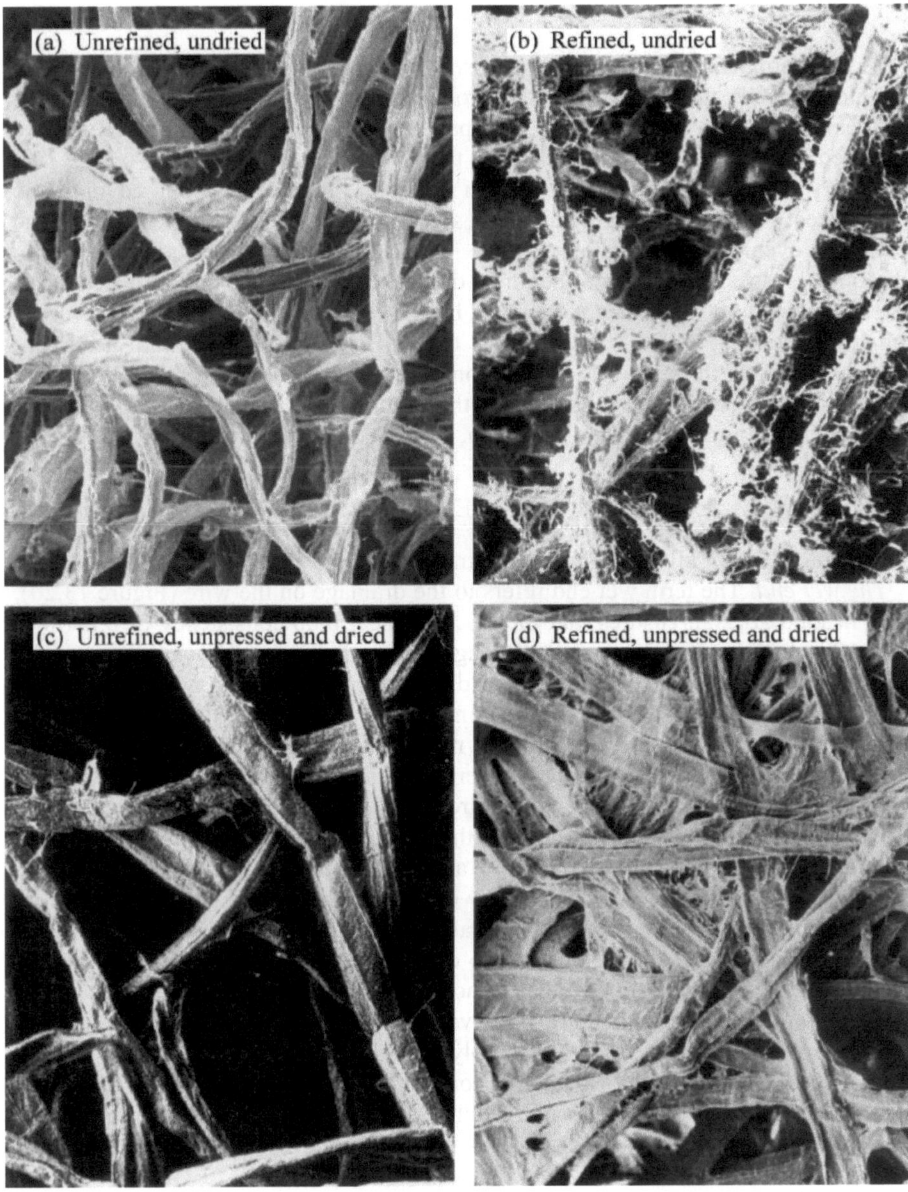

Figure 13.18. Effect of beating latewood tracheids of radiata pine, in which a web has been formed and dried without pressing (Kibblewhite, 1984). (a) Unrefined fibres after freeze drying: viewed as if in an undried state. (b) Refined fibres after freeze drying: viewed as if in an undried state. (c) Unrefined fibres after air-drying. (d) Refined fibres after air-drying.

24. PAPERMAKING

Modern papermachines are complex items. The greater the capacity of the machine (width and speed), the fewer the grades produced: newsprint machines run at speeds in excess of 30 m s^{-1}. The principles of papermaking are basically simple and are illustrated in Figure 13.19 which relates to the Fourdrinier papermachine.

A very dilute suspension of fibres in water is prepared and poured onto the forming section of a moving wire mesh screen where drainage is restricted. The fibres are evenly distributed across the wire in the forming section and only subsequently does the fibre mat begin to drain on the wire and form an even mat. The vast majority of the water is removed by the end of the wire. Them wet-web, is then lifted from the wire. More water is removed by pressing between absorbent felts while passing through rolls. The action of the rolls causes the water to leave the web and be collected in the felts. Finally the sheet is held by fabrics against heated drying cylinders and is dried under tension, before being wound on large rolls.

25. THE FOURDRINIER PAPERMACHINE

The main features of all Fourdrinier papermachines are:

The wet end. The term wet end refers to the drainage on the wire (Figure 13.20). Good headbox design ensures that the stock (typical pulp consistency of 0.2-1.0%, i.e. 2-10 g L^{-1}) is well dispersed, well agitated and distributed evenly across the width of the wire. The mix leaves the headbox through a tapering nozzle with a slit height from 8 mm (newsprint) to 25-50 mm (sack kraft papers or linerboard), i.e. the stock is being laid 8-50 mm deep on the moving wire. Obviously the higher basis weight sheets and boards require slower draining and slower Fourdrinier wire speeds. The pressure in the headbox controls the velocity of the stock through the nozzle. The tapered nozzle accelerates the mix, but it also results in machine-direction (MD) to cross-direction (CD) anisotropy: the greater the taper is, the greater the anisotropy. CD/MD anisotropy of paper, while being undesirable, is to a degree inevitable with fibres preferring to align in the machine direction.

Further, fibre orientation in the plane of the web is influenced by the relative velocities of the mix flowing through the nozzle and the fast-moving wire. Generally the velocity of the mix should correspond very closely to the velocity of the wire.

The wire is a woven, multi-layered polymer fabric that forms an endless belt up to 10 metres wide and over 70 metres long. The openings in the wire are about 0.1 mm x 0.1 mm so the wire is not capable of trapping fines and fillers. Instead, free drainage through the wire is progressively impeded by the retention of fibres that in turn trap these materials within the developing web.

Turbulence in the headbox improves the uniformity of the mix and so of the web. However on the wire turbulence is undesirable as it disrupts paper formation. Paper formation (evenness of fibre distribution) develops in the forming section and becomes fixed as water is drained or sucked through the wire using table rolls, foils and vacuum boxes (Figure 13.20). Dewatering is largely by filtration and thickening of the developing mat and as resistance increases suction is needed to maintain rapid

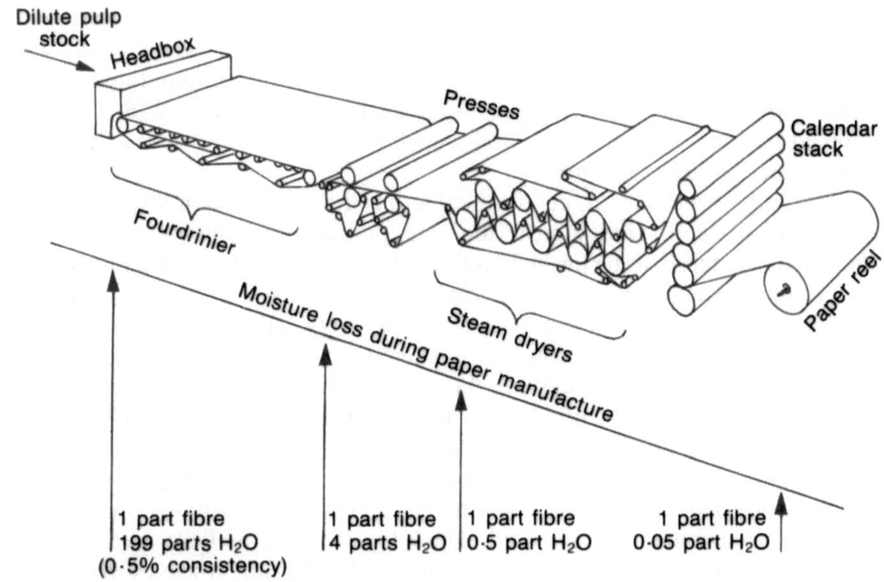

Figure 13.19. A papermachine (not to scale, the dryer section is much longer than the Fourdrinier and wet press section).

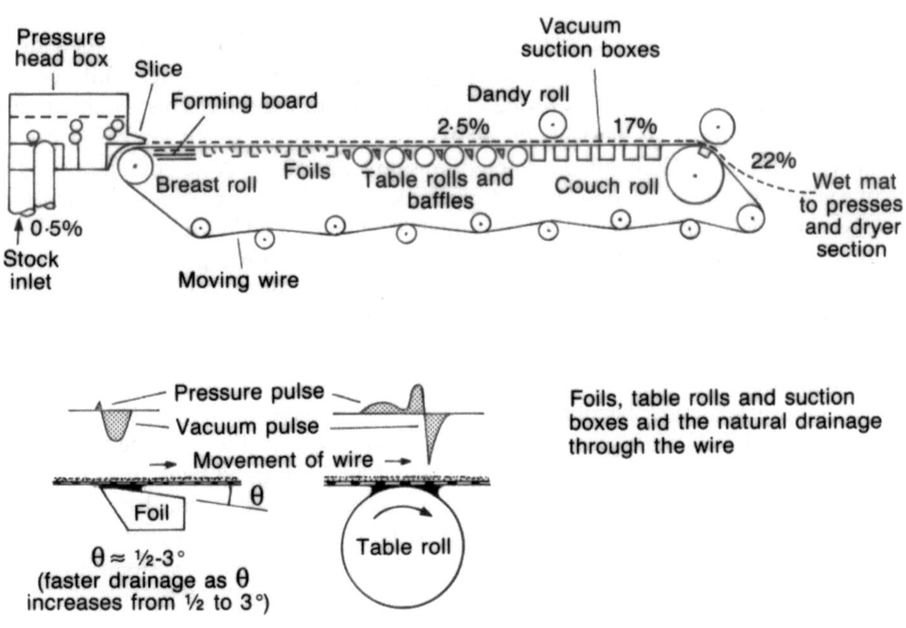

Figure 13.20. (a). Fourdrinier papermachine. (b) Drainage effects of foils and table rolls.

drainage. The rate of dewatering affects the distribution of fillers and fines within the web, and their presence in the water accounts for the term 'white water'. Paper formation is the outcome of a non-random distribution of fibres, fibre fragments, and mineral additives that form a felted (3-D entanglement) sheet. With kraft fibre reinforced papers, e.g. newsprint, the long fibre needs to form its own network in between other furnish components.

Table rolls contribute a gentle suction in the downstream expansion zone between wire and roll, while some of the water pulled through clings as a film on the underside of the fabric to be pushed up through the wire on the upstream side of the next roll: this vertical pressure pulse helps consolidate the web. Where wire speeds exceed about 25 m s^{-1} the suction increases excessively and rolls are no longer used.

Foils create a variable suction that is altered by adjusting the foil angle to the wire: the greater the angle of the foil, the greater is the suction. Water clings to the underside of the wire – to be lifted off by the leading edge of the next foil.

At the end of the Fourdrinier wire a couch roll removes more water from the web before it is transferred to the press section. At this point the moisture content of the web is about 80% water (20% consistency, 200g/L). The white water from the Fourdrinier is continuously recycled: 1000 kg of stock at 0.5% consistency is reduced to a 25 kg mat (5 kg fibre, 20 kg water) by the time it leaves the wire and 975 kg of white water is recycled to the make-up chest and thence back to the headbox. A 500 tonne a day papermachine recycles some 100 000 m^3 of water a day.

The Dandy roll on the upper surface does not dewater. It is included to improve web formation by introducing shear to the mix where it is around 2-3% consistency, when the web is very compressible.

One of the disadvantages of the Fourdrinier is that the paper is two-sided since drainage leads to the depletion of fillers and fines on the wire side, while coarse fibres and poorly bonded fillers reside on the topside of the wire. This, coupled with the opportunity to drain the web faster, has resulted in the much increased use of twin-wire formers. They all work on the same principle. A jet of stock is fed between two wires and is dewatered from both sides. Twin-wire can run four times as fast with half the water draining through each wire and through half the basis weight. Twin-wires are used for both packaging and printing grades. The general aim is to produce a sheet in which the surface density is high – for bending strength, printability and smoothness – achievable with a concentration of fines near the faces, but if this is overdone the low concentration in the centre can result in delamination.

With multi-channel headboxes the nozzle can have within it two or more compartmentalized layers so different furnish and different fillers can be supplied to these layers. These channels can deliver different pulp mixes without inter-mixing of layers, e.g. providing a layer of recycled or unbleached mechanical pulp under a layer of well-bonding bleached kraft pulp. It is a major technical challenge. More traditionally multi-ply forming for heavy board grades prepares various layers on separate formers before combining and consolidating in a single sheet.

On leaving the Fourdrinier, usually the web has to travel unsupported to the felts of the press section – a distance of a few centimetres. This unsupported transfer is called 'open draw' and its success depends on the wet-web strength with the potential for paper breaks and downtime to re-thread (tail-threading) the press and dryer. With ever increasing paper speeds a 'closed draw' is required more often, involving a suction pick-up roll to lift the web from the wire onto the press felt.

Pressing. The press section removes water and affects consolidation of the sheet. The wet-web is lifted from the wire and transferred to porous carrier felts before passing between a series of roll presses or press nips. As the wet-web is pressed, first air and then water is expressed from the web. The water is absorbed by capillary action in the pores of the carrier felts. However as the web exits the roll press it is decompressed and some of the expressed water is sucked back into the web. The effective press time is no more than 5 milliseconds. At the first nip roll the speed of flow of water from the wet-web is rate-limiting, while at the last nip roll the pressure that can be applied without crushing the sheet is rate-limiting. Traditionally moisture is squeezed out until the mat is down to 50-60% water (50-40% dry solids). Thereafter physical dewatering ceases to be efficient.

In a shoe press the web, wrapped in two felts, is squeezed between a press roll and a concave contoured shoe. The nip can be as long as 250 mm as opposed to about 50 mm in standard roll presses. Friction is minimized with a shoe belt, an endless impermeable belt/sleeve sliding over the stationary shoe on a film of hydraulic oil. Belts are grooved in the machine direction (MD) to aid water flow in the z-direction as well as the MD. Water passes from the felt to the grooves of the belt, so reducing rewetting of the paper (dry solids of 52% rather than 45%). Crucially the longer nip time permits faster running speeds. Uneven z-direction gradients in pressure mean that the sheet density (and so differences in absorption properties of the paper) varies across the thickness of the web, being denser on the side through which water is removed.

The wet-web is only a little stronger than where it left the Fourdrinier wire. It is still held together principally by van der Waals forces, surface tension, liquid cohesion and adhesion, and by frictional forces between the entangled fibres. Surface tension in particular plays an important part in sheet formation, in drawing the fibres into close contact within the fibre network. A 'closed draw' is desirable.

Drying. Drying is carried out on steam-heated rolls in a dryer section some 70 m long. The wet web is held in contact with the steam-heated dryer cylinders by means of a fabric. The steam heated rolls used for evaporative drying are 1.5-1.8 m in diameter. The temperature of the dryers increases gradually along the drying section to a maximum of 170-200°C, but in the final section the temperature is reduced again. Water is evaporated by passing the wet-web around a number of steam-filled cylinders. Each time the web loses contact with a drying cylinder moisture is flashed off before the cooling web is passed to the next heated roll (Figure 13.21a). Evaporation keeps the web temperature at around 100°C for as long as there is free water in the web. Towards the end there is danger of the web overheating and the surfaces getting scorched. To avoid this the last drying cylinders operate at lower temperatures. The temperature and moisture content of the sheet in the dryer are illustrated in Figure 13.21b. The final moisture content of the paper is 5-6%.

In a modern dryer section there are several fabric loops (with some six heated cylinders in each). This system better accommodates the changing web properties as

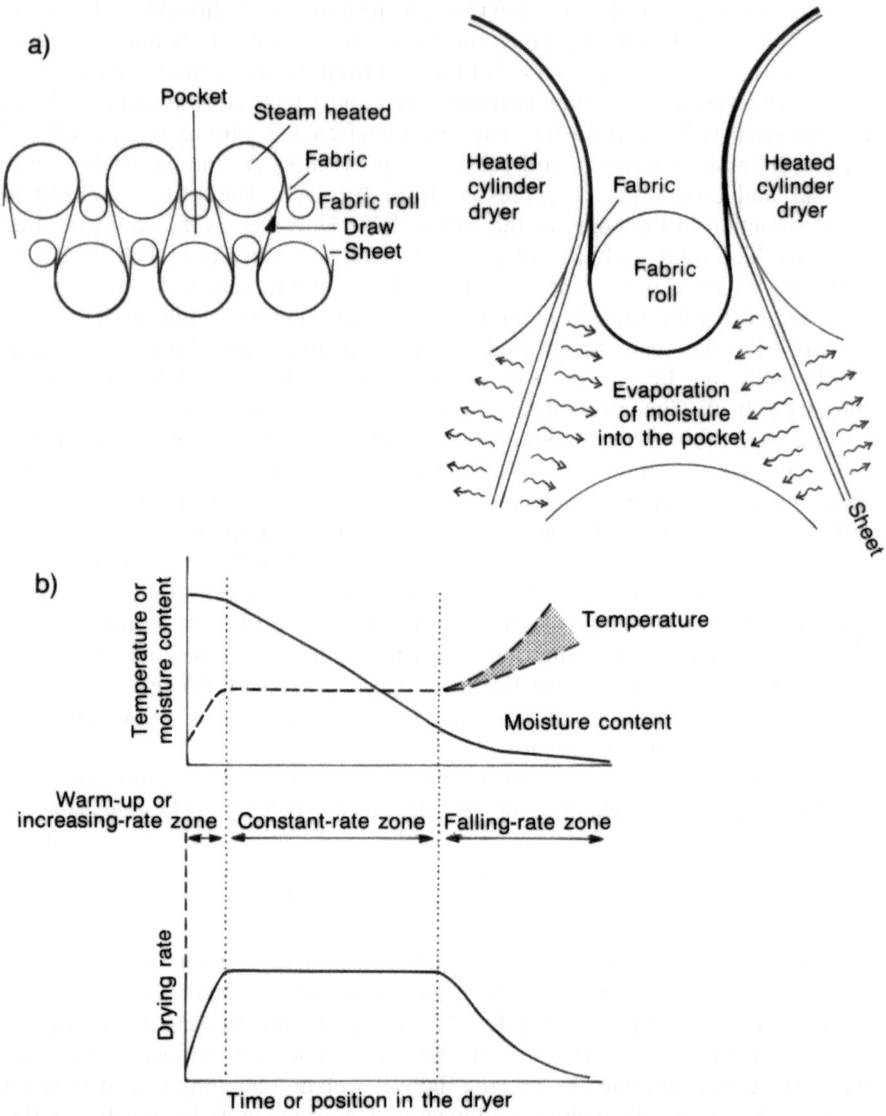

Figure 13.21. (a) A typical serpentine dryer configuration. (b) Temperature, moisture content and drying rate of the sheet as it passes through the dryer.

the paper dries and tries to shrink (MD). The paper speed decreases imperceptible between loops. Within the loops the fabric runs with the paper between the upper heated rolls and the lower cold vacuum rolls. A vacuum behind the fabric in the draws restrains CD shrinkage, i.e. the vacuum intensifies frictional resistance. This is particularly valuable in countering the tendency for the edges to dry faster (and shrink) than the centre (the greater humidity in the centre of the pocket calls for improved pocket ventilation). Uneven drying with early shrinkage on the edges creates varying tensions across the sheet, which reverses as the centre dries subsequently. Such varying tensions across the sheet at varying stages in the dryer can result in wrinkling of the paper on the winder reel.

Drying is far more energy demanding than dewatering by drainage and pressing; and energy economy and runnability are the two prerequisites, while minimizing paper breaks and the wrinkling of paper on the winder reel due to uneven shrinkage.

Calendering. Finally after the paper leaves the drying section it may be calendered, although to meet quality standards this can be done off-line on units running a slower speeds, e.g. 8 m s^{-1}. The paper is passed through a vertical stack of very smooth, friction-driven, heated rolls that are loaded against one another under pressure. With supercalendering (SC) both coated and uncoated high quality printing grades of paper (abbreviated SC papers) are passed through a stack of rolls that are alternately hard and soft, which causes rolling friction on alternate surfaces of the paper. This improves surface smoothness and gloss (Figure 13.22).

Hard rolls over densify those parts of the sheet with more fibre (arising from the natural uneven formation on the wire), but with soft nip calenders the backing roll has a soft surface that generates a lower, more even pressure within a longer nip zone. This results in a less even calliper but a surface that is more uniformly

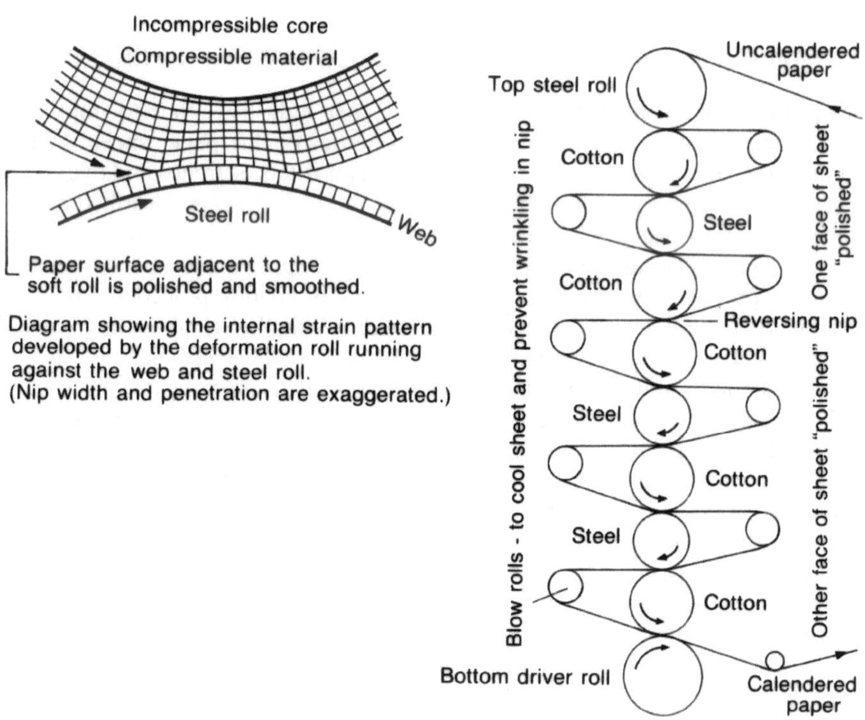

Figure 13.22. Supercalender configuration for finishing on two sides (Smook, 1982).

smooth compared to a hard nip calender. Usually one roll is soft and one roll is hot and hard, so one face is calendered at a time. Coated papers are calendered twice, a precoating to control calliper and then again after coating. In all calendering, the objective is to reduce sheet thickness, to even out caliper variation, and to impart a glazed surface finish: to reduce surface roughness and pore size for good printing.

26. CYLINDER MACHINES FOR PAPERBOARDS

Cylinder machines are used for heavy board grades, with grammages of 350 g m^{-2} or more. The products are multilayered and generally have high grade liner on their outer faces while the interior may use unbleached recycled fibre. Drainage on a Fourdrinier would be too slow resulting in uneconomic machine speeds.

The multicylinder board machine deals with the problem of slow drainage by constructing a multilayered sheet. The paper is formed on the surface of very large wire-covered cylinders, each of which rotates in its own vat containing a dilute suspension of fibres. A vacuum within the cylinders draws the white water through the wire leaving a fibre mat on the surface. The mat is pressed against a felt by a couch roll, picked up and transferred to the next cylinder. By using different stock in the various vats, the surface and interior layers can be of different grades of pulp. The paperboard is then dried in the conventional way on cylinder driers. More complex paperboard machines are available, but these are not considered here.

27. PAPER ADDITIVES, PAPER COATINGS AND WET-END CHEMISTRY

A wide range of chemicals is used by the paper industry either to improve the papermaking process or to confer special properties on the paper sheet. Additives include alum, sizing agents, clays and other mineral fillers, starches and dyes. Papers are also coated to improve printability. Some of these aspects are described briefly.

The coating of paper is a technology on its own but is mentioned because more and more paper is being coated in order to meet the higher quality demanded by the high speed printers now available. A wide variety of coating procedures is used by the industry but all involve the application of a thin layer of coating, made of pigments (clays, calcium carbonate) and binders (adhesives such as starch or polyvinyl alcohol) and water. The coating is applied to produce a smooth surface and is then dried. The end result is the better surface properties demanded by the printer and ultimately the consumer.

An important feature of papermaking is the so-called retention level, which is the amount of fibre and additives retained on the wire, generally expressed as a percentage of the amount present in the added stock. It has been shown that changes in process chemistry and the use of additives, alum or organic polymers can increase retention levels from around 50 to 70%. Papermaking has some interesting wet end chemistry that is quite complex. This is because all wood cellulose fibres have a negative charge related to their mode of production, e.g. the presence of carboxyl groups on cellulose or sulphonic acid groups on residual lignin. Fibre fines, fillers and materials used for paper sizing also carry negative charges. The fines have colloidal properties and depending on the nature of the pulps possess a zeta potential, which is generally negative and measured in millivolts. The addition of positively charged cations, for example of papermaker's alum, $Al_2(SO_4)_3$, will bridge the negatively charged colloidal fines and cause floc formation, thus improving fines retention if present in the stock. Alum is used frequently to adjust the pH of papermaking stock to acidic pH levels of between 4 and 5.

The papermaker of today uses a wide variety of organic polymers to do such things as improve fibre and fines retention, to aid filler retention, or to aid the retention of papermaking dyes. Thus in the example given above the use of a positively charged or cationic organic polymer could be used in place of alum. Anionic polymers are also used in combination with other polymers.

Traditionally papermaking is made at acid pH of about 4.5. Because of this the sizing of paper is carried out with resin acid salts in the presence of alum. Under these conditions, the resin acid anions complex with the aluminium cations and the complex formed is attracted to and deposited on the fibre surface. The purpose of sizing is to render the paper more resistant to water-based printer's ink. Today there is much interest in so-called alkaline sizing at about pH 7, which is preferred for specialist long-life papers. Here sizes such as alkyl ketene dimer replace alum. Alkaline papermaking has the further advantage that fillers such as calcium carbonate can be employed.

The inorganic fillers such as clays that are used by the industry are micron-sized particles bound together by a binder (starch) and applied with only just enough water to spread on the rough surface. Coverage per side ranges from 5-10 g m^{-2} for lightweight coated, to as much as 24-40 g m^{-2} for triple-coated. A typical coating of 5-20 g m^{-2} gives a film from 5 to 20 μm thick. In some papers coatings account for 20% of the weight of the paper. For high brightness, high value printing papers titanium oxide may be used. Fillers improve the scattering coefficient (opacity) and reduce ink absorbency. On the other hand the industry also dyes papers for aesthetic reasons, to improve the appearance of recycled paper, and so on. Overall there is much that can be done with paper.

28. FIBRE CHARACTERISTICS

Wood density is a useful indicator of potential paper properties, because it is related to cell wall thickness, and indirectly to fibre length. Traditionally low and medium density softwoods have been preferred for papermaking. Spruce pulp has long been considered as suitable material for many paper products. This is largely because its within-ring wood density variation is small, wood density is moderate-to-low and its fibres are thin-walled. Douglas fir, by contrast, displays a large within-ring density variation. The thick-walled latewood fibres of Douglas fir do not make good paper, although tear strength is high. The primary problem with thick-walled fibres is that a sheet of paper of a given grammage (g m^{-2}) made from such stock will contain fewer fibres per unit area, will have less interfibre surfaces, and even after beating these will be less-well bonded than corresponding thin-walled fibres.

From the beginning softwoods were preferred for paper manufacture because of their long tracheids, which are typically 2.5 mm long, and because they have a length to diameter ratio of around 100:1. They give strong paper and papermachines can run at high speeds using this stock. Tracheids are the predominant component of softwoods, *c.* 90-95%.

In hardwoods the structural elements are the fibres. They are quite short, *c.* 0.7-1.0 mm, with a length to diameter ratio of about 50:1. The narrowness of fibres partially compensates for their short length, giving hardwood paper sheets a more even texture and a smoother surface. Thus hardwood fibres are preferred for high grade printing papers. The vessel elements, which are short stubby wide-diameter cells, are of little use. They do not bond well in the paper sheet and can lift occasionally from the surface during printing, a phenomenon known as ‘picking’. However the use of hardwood fibre in paper and paperboards is increasing due to their low cost and availability, while the good optical properties of hardwood papers is recognised. Indeed some eucalypt pulps are now regarded as premium quality fibres for printing.

APPENDIX A: SOME DEFINITIONS AND TEST METHODS (FIGURE 13.3)

13.A.1. Analysis of pulps in suspension

Moisture content is defined in a different way in this chapter. The pulp and paper industries define moisture content as the ratio of water to total wood weight. Thus:

$$\textit{Moisture content} = \frac{\textit{Mass of water in the chips}}{\textit{Mass of water} + \textit{oven-dry wood}} \times 100\% \qquad (12)$$

Consistency is the term used to describe the percentage by weight of fibre in a mass of fibre and water. Consistency values can range from 0.3% for stock going onto the wire of a papermachine to as high as 25% in high consistency refining (beating) of pulp: in the latter case there are 25 g of oven-dry pulp in 75 g of water. As the wet-web dewaters, the preferred term becomes *solids content* (g/g), the mass of oven-dry fibre to the wet fibre mass.

Freeness, often written CSF (for Canadian Standard Freeness), is a measure of the ease with which water drains through the developing pulp mat. In the standard test a 1000 ml sample of pulp of 0.3% consistency is allowed to drain suddenly through a screen plate (wire) into a conical receiver that has only a small outlet at its base (Figure 13.23). Naturally the water backs up. Some overflows through a larger outlet further up on the side of the cone and is collected. If the pulp is free draining water passes quickly through the pad of pulp forming on the screen plate and builds up in the conical receiver before overflowing through the upper outlet where it is collected and measured. If the pulp is slow draining much of the water escapes through the lower small opening and only a small amount is collected from the upper outlet. The freeness is defined as the number of millilitres of water collected from the upper overflow. Fast draining, unbeaten pulps have high freeness numbers of 600-700 ml CSF, while slow draining pulps have low freeness numbers of 75-250 ml CSF.

The papermaker requires fast draining pulps (high freeness) when manufacturing tissues and to a lesser degree newsprint, and slow draining pulps for high quality printing papers. The freeness of the stock will be one factor determining the speed at which the papermachine can run.

The *kappa number* (K) test estimates the amount residual lignin in the pulp. The test involves treating a known mass of pulp (*c.* 1 g) with a known excess of potassium permanganate, $KMnO_4$, in acid solution (100 ml at 0.02 Molar). Under these conditions the residual double bonds are attacked and the lignin is rapidly oxidized by the permanganate while the carbohydrates react slowly and so their contribution is negligible. The kappa number is defined as the number of ml of permanganate consumed by 1 g of pulp in 10 min at 25°C. The lower the kappa number the lower the residual lignin content, e.g. for kraft softwood pulps the lignin content is approximately equal to 0.15 K, so a kappa number of 30 corresponds to a lignin content of 4.5%. It is common practice in pulping technology to refer to the kappa number rather than the lignin content as the kappa number is so quick and easy to measure. Thus in reducing the impact of bleaching on the environment it is usual to discuss the advantages of reducing the kappa number rather than to mention the corresponding lignin content.

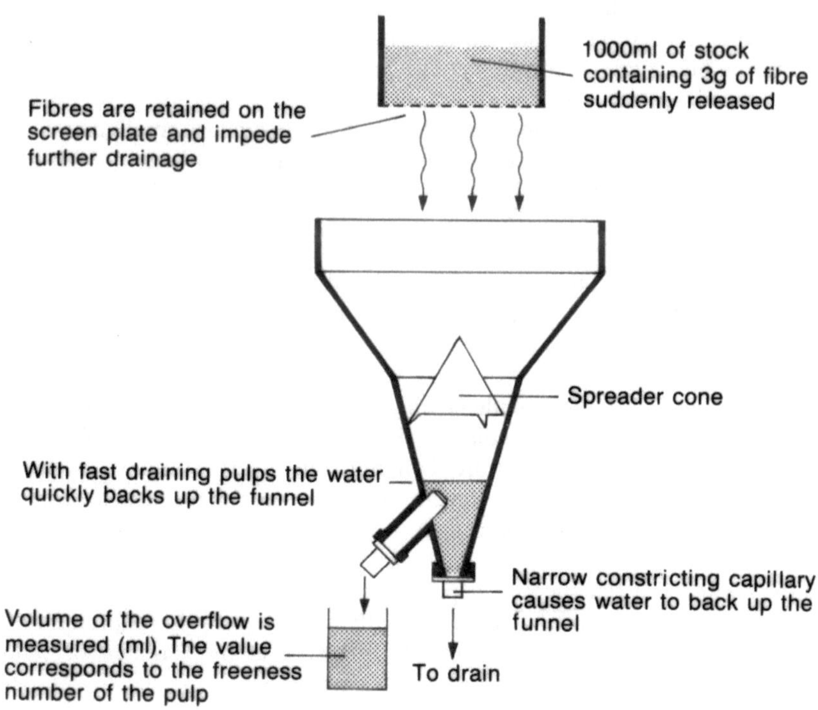

Figure 13.23. Schematic representation of the Canadian Standard Freeness Tester.

13.A.2. Pulp and paper tests

The properties of pulps are usually assessed by making handsheets and testing to determine the paper characteristics discussed below. Handsheets are prepared by draining a dilute suspension of pulp on a wire, the wet-web being removed from the wire with the aid of blotters, attached to a stainless steel plate and pressed. The sheets are generally made to a standard grammage of 60 g m^{-2} (oven-dry basis) or the corresponding air-dry sheet grammage. Handsheets are tested at 23°C and 50% relative humidity. Various standards prescribe precisely these procedures.

Brightness. The whiteness of paper is measured by the reflectance of a thick pad of sheets (R_∞) compared to the reflectance of a known standard, i.e. barium sulphate, using a standard instrument. The reflectance is normally termed brightness. The test provides a good measure of the degree of bleaching. Fully bleached kraft pulps can achieve brightness levels as high as 92%. Mechanical pulps after two-stage peroxide bleaching can reach brightness levels of 65 to 80%, depending on the brightness of the original wood.

Opacity measures the ability of a single sheet of paper to hide colour or print on the reverse side of the sheet. It is measured by determining the reflectance of a single sheet of paper backed by a perfectly black body (R_o). Opacity is the percentage ratio of this reading to the brightness of the same papers, i.e. (R_o/R_∞). Opacity is of increasing importance as the grammage of paper decreases.

Absorption and scattering coefficients are determined from R_o and R_∞ using the Kubelka-Munk theory. These are intrinsic optical properties of pulp and are of great importance in bleaching (absorption coefficient) and in printability assessment (high scattering coefficient).

The tear test measures the resistance to tearing once a small tear has been initiated. Tear strength is in essence a measure of paper brittleness. With the long fibres of softwoods there is an initial rise in tear as the pulp is slightly beaten after which tear strength declines steadily as sheet density or pulp tensile strength (measures of interfibre bonding) increase (Figure 13.17b). Some initial bonding is required, but it has been shown that the test involves fibre slippage rather than fibre rupture, in the paper rupture zone. Long fibres and thick cell walls are conducive to high tear strength in softwoods. The tear strength of hardwood pulps is about half that of softwoods.

Tensile strength increases with beating as the individual fibres become more flexible and conform better with adjacent fibres, forming a strong interfibre network. It is principally a measure of fibre bonding, but clearly fibre strength is also important. Other parameters that can be measured during the tensile test include stretch (percent elongation at rupture), elastic modulus and tensile energy (or energy to rupture).

In the *burst test* a flat sheet of paper is clamped by a circumferential ring and a small rubber diaphragm underneath is gradually inflated with fluid, forcing the sheet to bulge until it ruptures. The hydrostatic pressure at the moment of failure is measured. The virtues of the burst test are its simplicity and the speed with which it can be undertaken. The recorded hydraulic pressure offers a quantitative measure of bonding between fibres. It is linearly related to tensile strength.

CHAPTER 14

THE ENERGY SECTOR: A HIDDEN GOLIATH

CHRIS WILLIAMSON

Department of Chemical and Process Engineering, University of Canterbury, Christchurch, New Zealand

1. INTRODUCTION

Energy from wood is the most significant single use for the world's wood harvest. During the last decade more than 50% of the annual 3-3.5 billion m^3 wood production was used directly as fuel. The rest of the wood harvest provides feedstock for sawmills, pulp mills and wood-based panel manufacture. Each of these processes creates significant residues suitable for energy production, typically 20-40% of the total wood feedstock. Thus, if the wood processing industry's supply of a significant proportion of its own energy needs from residues is taken into account currently 60-70% of all wood harvested is used as fuel.

Energy from biomass (including wood) currently represents approximately 14% of the world's primary energy supply, about 40 EJ/yr – 1 EJ or 1 ExaJoule or 10^{18} J represents approximately the energy consumption of a small, developed country with a population of about 5 million people. 40 EJ/yr represents only 40% of what is easily possible: a conservative estimate of the total sustainable world biomass energy potential is about 100 EJ/yr or 30% of our current total energy consumption. Wood's share of this is estimated as 41.6 EJ/yr (Parrikka, 2004). However, projections suggest that almost 400 EJ/yr of energy could be available from biomass in 2050 (50% of the world's projected energy demand). The increase in demand for biomass (including wood) as energy depends on the availability and cost of fossil fuels. Currently oil, natural gas and coal provide cheaper, more convenient but more environmentally damaging sources of energy. The developed countries rely heavily on fossil fuels for energy, with only 2% of total energy consumption being derived from wood, although it varies from country to country. In Sweden and Finland (Ericsson *et al.*, 2004), encouraged by supportive government policy and a readily available wood supply, forest biomass contributes almost 20% of primary energy (Figure 14.1). In contrast in developing countries much of the domestic and industrial energy supply comes from wood and charcoal. Indeed in 13 developing countries, wood provides 90% or more of their total primary energy consumption and in a further 21 countries more than 70% (Trossero, 2002). In total the developing countries consume about 75% of the world's woodfuels. Unfortunately in the majority of developing countries woodfuel use is via inefficient technologies

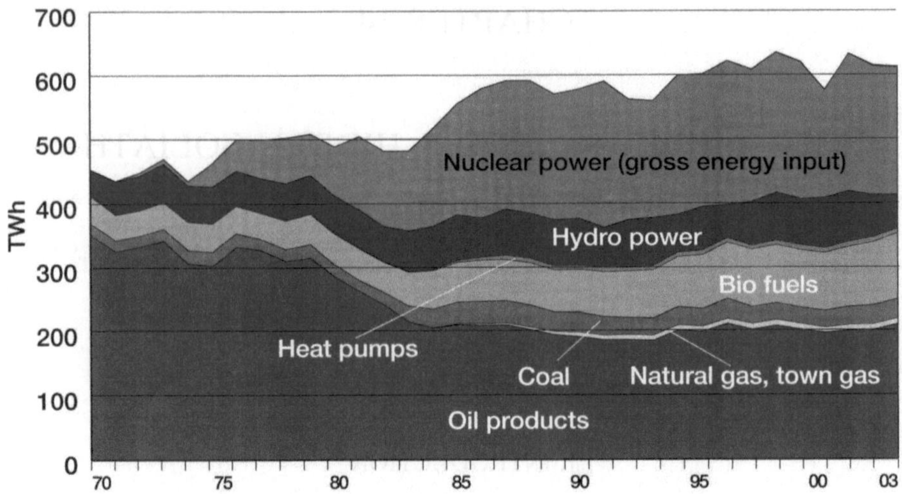

Figure 14.1. Energy supply in terawatt hours (1 TWh = 10^9 kWh) in Sweden for 1970-2003.

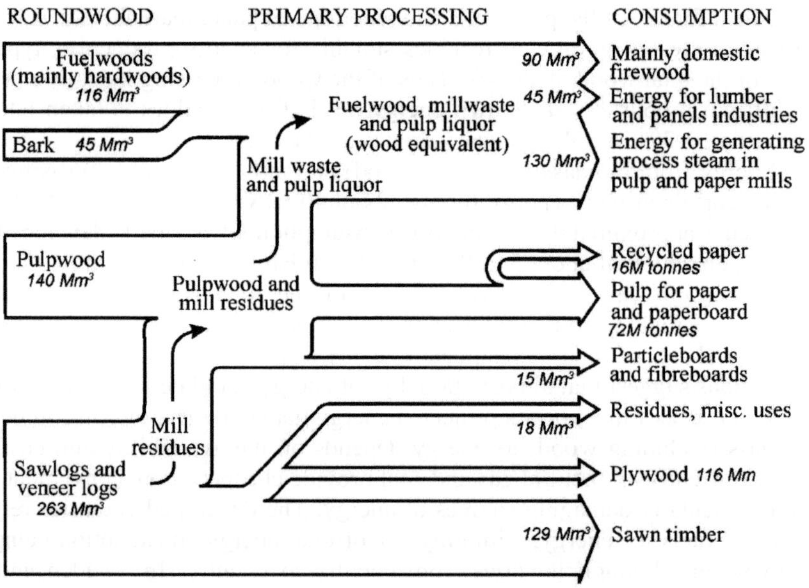

Figure 14.2. Major end uses for wood and bark in the United States, in M m^3 or M tonnes, data from Koning and Skog (1987), Waddell *et al.* (1989), Ulrich (1988) and USDA (1990).

that utilize only about 10-15% of the energy available from the supplied wood, with the majority of the energy supplied as heat alone. In contrast, the developed countries use wood wastes far more efficiently to deliver a variety of energy types (electricity, heat and liquid fuels) but this is a tiny proportion of total consumption. Figure 14.2 shows the wood flows for energy and wood products through the United States wood industry.

The challenge for energy production from wood is twofold. First to increase its share of energy supply in the developed countries; and secondly to retain and to grow its share of energy supply in the developing countries while introducing modern bioenergy production technologies. The versatility of wood as an energy source when coupled with the appropriate conversion process is shown in Figure 14.3. Choosing the appropriate pathway can deliver energy as electricity, heat, alcohols, synthetic gasoline and diesel, methane or hydrogen. In short, all of the energy products are possible that citizens of developed countries take for granted but that come largely from fossil fuels. Unfortunately, the lack of impact of this most sustainable of energy sources is poignantly summarized by noting that world fuelwood use was estimated at 1.7 billion m^3 in 1800, about the same as direct fuelwood use today (Schulz, 1993). However, with increasing urgency the world

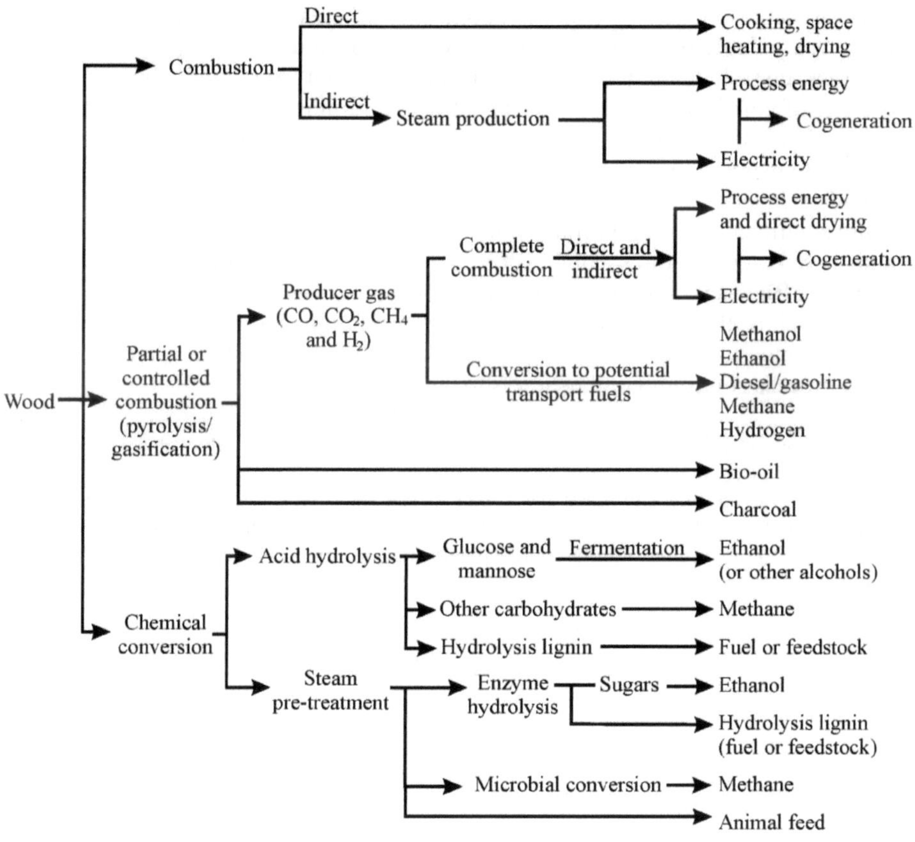

Figure 14.3. Alternative energy pathways using wood.

recognizes that climate change is inevitably associated with fossil fuel use and it is this realization that may finally awaken the sleeping giant of woody biomass energy.

2. CHARACTERISTICS OF WOOD AS A FUEL

One attractive and distinctive feature of wood is that it is a readily renewable resource. It has a very low ash content and is free of sulphur and other obviously polluting or corrosive elements. In almost all other respects the characteristics of wood are less desirable than alternative fuels:

- Wood is approximately 50% carbon, 44% oxygen and 6% hydrogen by weight. Since wood is highly oxygenated it is understandable that it has only about two-thirds of the calorific value of coal, which is pure carbon. The calorific value of oven-dry wood is about 20 MJ kg^{-1}. This varies slightly between species, with resin rich woods having higher calorific values and non-resinous species having lower calorific values. Its ash content is less than 1%, which is very low compared to coal.
- The volatile fraction of wood is about 80% whereas with coal it is only 20-30%. The volatile fraction burns first leaving the residual carbon (charcoal). The volatiles contribute to air pollution if they are not completely burnt. Clean burning of volatiles is the key to efficient wood-burning.
- Wood contains moisture, and energy must be spent unproductively in evaporating this moisture (Table 14.1). Also, the evaporating moisture forms a boundary layer on the surface of the wood, which makes ignition more difficult. More significantly, when burning wet wood the furnace temperature is depressed and at low temperatures combustion is often incomplete so further reducing thermal efficiency. While it is desirable to reduce the moisture content of the wood or chips this is not always practical or economic and some processes have to be adapted to cope with this situation. Variations in moisture content require good process control if combustion is to be efficient.

Table 14.1. The net calorific value of wood depends on its moisture content: here the values are for a typical hardwood. The reduction in available energy with moisture content is due to the need to vapourize and superheat the steam to the same temperature as the flue gases. Typically the net calorific value for a hardwood is 18.2 MJ kg^{-1}, that for a softwood is 19.2 MJ kg^{-1} (due to its higher lignin content) and that for bark is 19.7 MJ kg^{-1} (due to extractives).

Moisture content, oven-dry basis (%)	Net energy available (MJ kg^{-1})
0	18.2
15	15.4
30	13.5
45	11.9
60	10.5
100	8.0
200	4.6

- Wood is not a premium fuel. It is bulky to transport, difficult to handle unless chipped, and is of moderate calorific value. Its energy per unit volume is low: obviously denser timbers contain more woody material and therefore have a higher energy content per unit volume. Traditionally it is used for raising process steam and heating, mainly by the forest industries, and in domestic stoves.

Bark is often combined with wood residues and burnt. Bark has a fractionally higher calorific value and can be used equally in boilers and for the production of methanol following gasification. However bark can only be used in ethanol production if gasified first rather than via the conventional process based on the conversion of the polysaccharide components of wood.

When wood is heated an enormous variety of chemicals are produced: water vapour, non-condensable gases (carbon monoxide, carbon dioxide, hydrogen and methane), pyrolysis products (methanol, acetone, acetic acid, and complex hydrocarbons and volatile tars). A carbon-rich char (charcoal) remains. The relative proportions of these products vary depending on the amount of oxygen admitted, on the temperature and on the physical configuration of the furnace.

Initially heat is needed to drive off the sorbed water and raise the temperature of the wood to about 275°C. Once this temperature is reached the wood constituents begin to break down spontaneously (pyrolysis). Sufficient heat evolves for the decomposition and the release of volatile products to be self-sustaining, and the temperature rises further. Oxygen is not necessary. At even higher temperatures the tars and oxygenated hydrocarbons crack to simpler gases. These gases can be mixed with oxygen in the air supply and burnt. Combustion involves the decomposition of wood, the burning of volatile gases (degasification) and the production of a carbon-rich residue (charcoal). The furnace temperature rises further (400-600°C) which accelerates the decomposition of the wood. Eventually the volatile fractions are all driven off and the flame diminishes. The wood charcoal comes in contact with the in-coming air and becomes incandescent.

Air is passed through the fuel chamber to ensure as complete combustion as possible. In practice excess air is needed because of imperfect mixing of gases and because the additional volume of air helps prevent condensation of residual unburnt pyrolysis gases and water vapour in the flue. Condensation in the flue is more likely when the wood is wet and when the fire is starved of air. By increasing the draft the temperature at which condensation occurs in the flue is lowered, which is beneficial. However excess air means that the heating efficiency is diminished as more heat escapes up the chimney. Thermal efficiencies for cooking, hot water and spaceheating can be as high as 80% with a well designed, enclosed stove compared to 20% with an open grate or three-stone fire (Gilmour and Walker, 1995).

3. CHARCOAL

FAO (1983) estimates that some 400 Mm^3 of roundwood a year, about a quarter of all fuelwood, is first converted to charcoal before being used. In cities in developing economies charcoal is a preferred cooking fuel because it is smokeless, light and so cheaper to transport, more energy intensive than wood, and burns with a much hotter flame. In monetary terms charcoal, per unit mass, is approximately ten times more valuable than wood.

Wood for charcoal production must first be cut, split and dried because wet wood has a low heat value. In the moist tropics it is difficult to hold timber in stack for more than a couple of months without noticeable deterioration, but even in that time the moisture content can drop from 60% towards 30%. Stock holding also ensures continuity of supply.

In a traditional charcoal kiln some of the wood is burnt in order to completely dry the rest of the wood, and to raise the kiln temperature until pyrolysis reactions release sufficient heat for the process to become self sustaining (>300°C). The air vents are then sealed and the volatile gases and tars together with non-condensable gases such as carbon monoxide, carbon dioxide, hydrogen and methane are driven off. Only a proportion of these volatiles is burnt. These chemical by-products are not recovered and much of the potential heat available from pyrolysis is wasted. The temperature stabilizes around 400-450°C. When most of the volatile fraction has been driven off the chimney is also sealed and the kiln allowed to cool. The warm-up time is typically a day or so, pyrolysis takes 20-30 days to convert most of the wood to charcoal, followed by a cool-down period of equal duration before the charge is unloaded. With simple technologies the recovery of charcoal is low, about 15-20% by weight on an oven-dry basis, with at best 6 m^3 of roundwood yielding a tonne of charcoal. The charcoal has approximately 50% of the energy of the wood, while having only 20% of its original weight. With better designed kilns conversion efficiencies of 30-35% are achievable. The quality of charcoal from a traditional kiln is quite variable. It depends on the wood, its moisture content and the rate of burn. Poorly prepared charcoal can retain as much as 50% by weight of volatile material. This has the advantage of igniting easily but it burns with a smoky flame. Charcoal picks up moisture quite rapidly and when sold this can range from 5-15%, increasing with the increasing impurity of the charcoal.

An efficient kiln requires good insulation (thick walls of brick or earth) and must be well sealed so as to exclude air when desired. During the warm up-period air is drawn into the kiln through openings in the wall and these are only sealed once the colour of the smoke in the chimney changes from white to thin blue, indicating that the charge has been dried and there is no more moisture to be driven off. When pyrolysis is complete the smoke hole is sealed as well to exclude all air and the kiln left to cool. In simple kilns it is extremely hard to ensure even heating throughout the kiln. Typically part of the charcoal will be burnt while some wood elsewhere in the kiln will only be lightly charred.

Sophisticated and capital intensive kilns seek to recover the heat from the combustible gases and use this to preheat the in-coming wood, rather than relying on the partial combustion of the wood itself. Such continuous kilns operate at higher temperatures and the resultant charcoal has a much smaller volatile fraction in the charcoal, making it more suitable for many commercial purposes where purity is important, e.g. in smelting metals. This high quality charcoal is pyrolysed at higher temperatures (450-550°C) and has about 30% by weight of volatile material. Modern kilns use a variety of cheap plant residues including sawdust and bark and the residence time in the kiln is only a few hours. The fines are briquetted with the help of a starch binder.

4. FAST PYROLYSIS OF WOOD

Traditional pyrolysis of wood relies on low temperatures and long processing time to increase the charcoal yield. In contrast, modern or fast pyrolysis uses moderate temperatures (400-500°C) and very short residence times (typically only a few seconds) to maximize the production of liquids (Diebold and Bridgewater, 1997). Pyrolysis vapours are rapidly cooled and condensed after leaving the reactor to prevent cracking into smaller molecules. The main product is a 'bio-oil' obtained in yields of up to 75% by weight of the dry wood feed. Char and non-condensable gases are also produced at yields of between 10-15% of the wood feed. The bio-oil and char are commercial products while the non-condensable gases are recycled to the reactor and burnt to provide the endothermic reaction heat. Bio-oil is a dark brown, free flowing liquid mixture of several hundred different chemicals. An approximate composition is 20-25% water, 25-30% water-insoluble pyrolytic lignin, 5-12% organic acids, 5-10% non-polar hydrocarbons, 5-10% anhydrosugars and 10-25% other oxygenated compounds. It has a calorific value of 16-19 MJ/kg and is a potential substitute for fuel oil and natural gas used in boilers, gas turbines and diesel engines. Bio-oil has about 40% of the energy content of fuel oil on a mass basis and 60% on a volumetric basis because of its comparatively high density (1.2 kg/litre). The fast pyrolysis reactor has required much research but it represents only about 15% of the capital cost of the process. Feedstock for fast pyrolysis requires considerable preparation to guarantee successful processing. It must be dried to less then 10% moisture content to limit the water content of the product and ground to a small size (approximately 2 mm) to encourage rapid reaction. This adds to the final process equipment cost. Large scale bio-oil production via fast pyrolysis is nearing commercialization (Thamburaj *et al.*, 2000).

5. WOOD GASIFICATION

Wood gasification is a key stage in the production of useful energy (heat and electricity) and fuels from wood. Wood gasifies to make a 'producer gas', a mixture of hydrogen, carbon monoxide, carbon dioxide, methane and hydrocarbon/tar components. The gas has a heating value of anywhere from 3-20 MJ/Nm^3, depending on whether the gasification occurs using either air or a steam/oxygen mixture. Producer gas was used as long ago as the mid 19th century to fuel gas lights in London and, later, in the early 20th century automobiles. The carbon oxides and hydrogen can be recombined to make alcohols and liquid hydrocarbons, a far more convenient way of fuelling today's vehicles.

Gasification occurs by heating wood to a high temperature (1200-1400°C) in an oxygen deficient atmosphere, so limiting combustion. The process has four phases; drying, devolatilization, gasification and combustion. Wood fuels vary in moisture content from about 10-50% water so the first phase is heating and drying the wood.

In the second phase, of devolatilization or pyrolysis, light gases including hydrogen, carbon monoxide, carbon dioxide, water, methane, higher hydrocarbons, and tar (heavy organic and inorganic molecules) are driven off leaving a reactive char. The char gasifies by reacting with oxygen, CO_2 and water to produce more carbon oxides and hydrogen. In the gas phase, the water gas 'shift reaction' alters the relative amounts of carbon monoxide and water compared to carbon dioxide and hydrogen, while the tar and heavier hydrocarbons react with oxygen to produce lighter gases. The key to gasification success is in the controlled partial combustion to yield a gas with a high heating value.

5.1. Gasification reactions

The generalized gasification reaction can be written:

$$\underset{\text{wood}}{[CH_{1.44}O_{0.66}]} + H_2O + O_2 = H_2 + CO + CO_2 + H_2O + CH_4 \text{ (+ tar/hydrocarbons)} \quad (1)$$

The ratios of hydrogen or oxygen to carbon are almost constant in all woods and it can therefore be represented by a single molecular formula. Where air is used rather than oxygen the produced gas mixture will contain nitrogen and be less valuable. This equation is a deceptively simple summary of many reactions. The more important ones occurring, after devolatilization, can be broken into two groups. The first are char gasifying reactions;

$$C + H_2O \rightleftharpoons H_2 + CO \quad (2)$$

$$C + CO_2 \rightleftharpoons 2CO \quad (3)$$

$$C + 2H_2 \rightleftharpoons CH_4 \quad (4)$$

The second group are gas phase reactions. The hydrocarbons formed during devolatilization break down to form hydrogen and carbon monoxide:

$$C_xH_{2y} + xH_2O \rightleftharpoons xCO + (x+y)H_2 \quad (5)$$

Tar cracking reactions to produce simpler gases also occur, as long as the residence time in the reactor is sufficient. Gaseous components can rearrange via the water gas shift or methane formation reactions;

$$CO + H_2O \rightleftharpoons CO_2 + H_2 \quad (6)$$

$$CO + 3H_2 \rightleftharpoons CH_4 + H_2O \quad (7)$$

The overall result is a fuel gas mixture of carbon oxides, hydrogen, methane, water and any remaining tar and heavier hydrocarbons that escape the gasifier. Some of the reactions are exothermic and some endothermic but the overall gasification process is endothermic and requires an external energy source.

5.2. Gasifier designs

There are three variants of gasifier design: updraft, downdraft and fluidbed. The oldest form of gasifier is the updraft variation. Wood particles (chips or pellets) are fed in the top of the gasifier while gasifying agents (air or oxygen plus steam) are fed in the bottom. The wood feed dries, pyrolyses, gasifies and combusts as it moves downwards, the solids exchanging heat and reacting with the gas phase. Ash is withdrawn from the bottom and the produced gas, along with entrained wood particles and tar, leaves from the top.

In the downdraft version both the wood and gasifying agents move downwards through the gasifier, and produced gas and ash leave from the bottom. The same four stages occur but the gas produced during devolatilization must pass through the high temperature combustion section. This promotes tar breakdown and produces a cleaner gas than the updraft version. Further, when compared to the updraft version, controllability limits the operational size to smaller diameters of reaction vessel in the downdraft version

In the previous two gasifier versions the wood moves slowly through the vessel due to gravitational forces. Individual wood particles remain in contact with their neighbours. In a fluidised bed, gas is injected into the bed of wood particles at a high enough flowrate that individual particles detach and move randomly within the fluid bed volume. If gas rate is increased further a bubbling fluid bed results. Here, the wood particles continue their separate movement but bubbles of gas form rising through the bed without mixing with the solids. The solids can still be identified as mainly residing within a defined volume above the grate although there will be some small carryover of finer particles leaving with the gas stream. If the gas velocity is increased further ultimately the gas and solids flow as a single stream through the gasifier. The solids are then said to be entrained by the gas flow and the fluidised bed region is no longer identifiable. Gasifiers can operate in both the bubbling and entrained flow regimes. The main advantage of a fluidbed gasifier is that reaction conditions can be more closely controlled and similar gasification performance and gas production expected over a significant size range.

A high calorific value fuel gas mixture is produced if oxygen and steam are used. With air (20% O_2, 80% N_2) a large amount of nitrogen is included in the produced gas, inevitably reducing its calorific value. However, high purity oxygen is expensive and its extra cost makes gasification financially less attractive. Steam alone can be used but without oxygen the exothermic combustion reactions that supply energy are missing. To reduce the air supply a variant on gasification design combining an entrained and bubbling fluidised bed was developed. A fast internally circulating fluidised bed (FICFB) gasifier combusts char, from gasification, in a bed of sand fluidised in the entrained flow regime with air. The heated sand is carried out the top of the first fluid bed and sent to the gasification section, which is fed with wood chips or pellets, and the combined wood/sand bed fluidised with steam. The hot sand supplies energy and normal gasification occurs, producing a gas mixture uncontaminated with nitrogen. Figure 14.4 shows the FICFB concept (Hofbauer *et al.,* 1997).

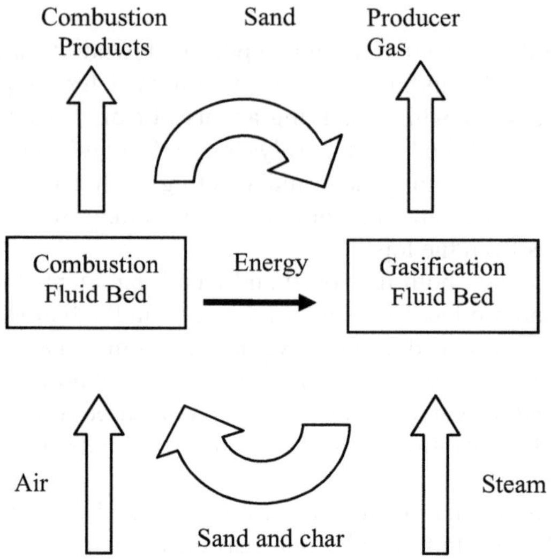

Figure 14.4. Fast, internally circulating fluidised bed gasifier.

5.3. Combined heat and power systems

Medium scale wood gasifiers (10-30 MW thermal capacity) have attracted significant interest and development investment in recent years because they can be integrated with industrial processes to supply heat and electrical power, i.e., combined heat and power (CHP) or cogeneration systems. The large size of the energy flows effectively limits these applications to fluidised bed gasifiers. The most efficient system initially burns the produced gas in a gas turbine to generate electricity. The hot exhaust gases are used in a waste heat recovery boiler to produce a lower pressure steam to run a steam turbine, generating further electricity. Any residual low grade heat in the gases from the heat recovery boiler can be used around the process plant (for example, to dry wood if the process is a sawmill) or in a district heating scheme. The total combined heat and power process is known as a biomass integrated gasification and combined cycle (BIGCC) system and is shown in Figure 14.5.

Gas turbines are sensitive to corrosion and erosion from wood particles and tar compounds that can escape from the gasifier so a necessary extra process step is gas cleaning to produce turbine quality fuel gas. Conventional cogeneration schemes consist of biomass combustion, steam generation and then electricity generation using a steam turbine. The advantages of the BIGCC system are a higher electricity generation efficiency (up to 48% *vs.* conventional 29%) and higher proportion of electricity to thermal energy production (0.9-1.1:1 *vs.* 0.29:1 for conventional systems). Its disadvantages are extra complexity and capital cost when compared to conventional cogeneration systems.

A limited number of large scale BIGCC systems have been built and operated around the world in the last 10 years. Well known examples are the ABRE scheme in the UK (8 MW electricity generation or MWe, using atmospheric biomass gasification), the Varnamo scheme in Sweden (6 MWe and 9 MW thermal or MWth capacity, using pressurized gasification) and the FICFB system at Gussing in Austria (2 MWe and 4.5 MWth, where the produced gas is burnt in a gas engine rather than fed to a combined cycle system). These demonstration projects have all been technically successful but have struggled to compete economically, mainly because of their high capital cost. However, the BIGCC technology is technically feasible and further development will see it become more successful financially.

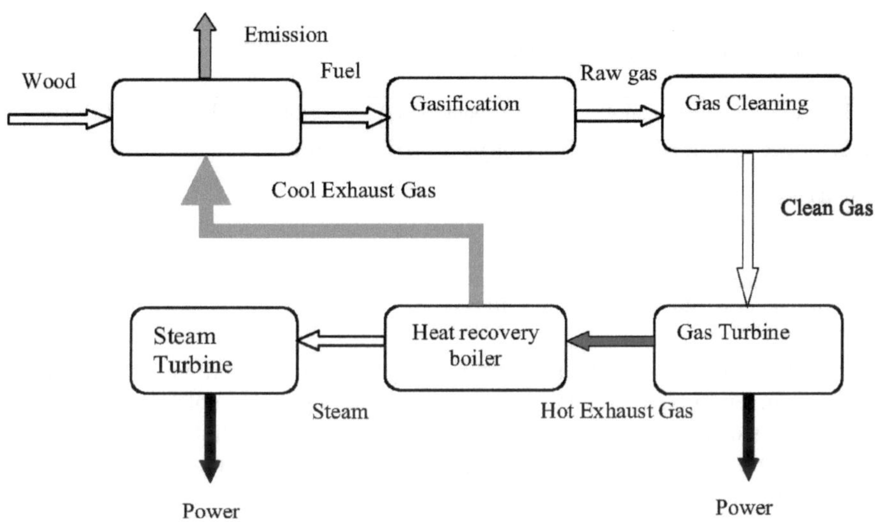

Figure 14.5. Biomass integrated gasification and combined cycle system.

6. WOOD AS A FEEDSTOCK FOR LIQUID FUELS

Wood is a mixture of carbon, hydrogen and oxygen atoms that can be rearranged into liquid fuels using appropriate processing steps just like oil, natural gas or coal. Principal liquid fuel products from wood are methanol, ethanol or Fischer-Tropsch liquids (this is a complex mixture of hydrocarbons). The alcohols can replace gasoline while Fischer-Tropsch liquids replace diesel, providing renewable rather than fossil fuels for conventional combustion engines. There are two well-developed methods for manufacturing fuels from wood. Gasification, using oxygen or a FICFB gasifier, is the key processing step in the first of these two approaches. The producer gas from gasification can be manipulated by altering the ratio of carbon oxides to hydrogen to become synthesis gas, which becomes the feedstock for producing all three liquid fuel products. Methanol and Fischer-Tropsch liquids are made via a chemical route by passing the synthesis gases over appropriate catalysts at high pressures and temperatures, while ethanol is produced biologically by fermenting the gas.

The second method is direct fermentation of wood sugars, recovered from the polysaccharides, cellulose and the hemicelluloses, by acid hydrolysis to produce ethanol. This is a proven technology whereas synthesis gas fermentation is as yet successful at only a small scale. The main advantage of synthesis gas fermentation is that all the wood components, including lignin, become a potential source of ethanol. In direct fermentation lignin must be separated out from the feed to the fermentor.

The energy efficiency of these processes (Energy in the product/Energy in the feed) is summarised in Table 14.2.

Table 14.2. Energy efficiency of wood fuel production (GJ/GJ).

Feedstock	Methanol	Ethanol (SG)	Ethanol (F)	F-T liquids
Wood	0.29-0.65	0.35*	0.3	0.16-0.43
Natural Gas	0.61			0.54-0.63
Corn			0.35	

* There is more uncertainty associated with ethanol production from syngas. SG = Synthesis Gas, F = Fermentation (Spath and Dayton, 2003).

Comparative figures for methanol and F-T liquids from natural gas are included. The higher conversion efficiencies and lower processing cost explain why natural gas is the feedstock of choice rather than wood. This is understandable because natural gas is largely methane with a low C:H ratio. Ethanol is commonly produced by direct fermentation of corn sugars in the USA at a similar efficiency to producing the alcohol from synthesis gas. However, wood is a renewable feedstock.

A second energy balance description, emphasizing renewability, is the fossil energy ratio defined as the energy in the product divided by the total fossil fuel derived energy consumed from feedstock procurement through product generation. The fossil energy ratios using wood are wood-to-methanol = 12-26, wood-to-ethanol (SG) = 16 and wood-to-F-T liquids = 6-17. Fossil energy ratios for natural gas systems are all less than 1 because the main source of energy consumption is the feedstock itself and it cannot be converted into another energy source without expenditure of energy. Life cycle analyses studying corn ethanol production suggest a fossil energy ratio of only 1.3 because of the large input of fossil fuels required for farming corn, including significant amounts of fertilizer (natural gas sourced). Clearly ethanol from wood is considerably more renewable than corn ethanol.

Wood as a source of liquid fuels is still largely untapped because of the still easy availability of cheap natural gas and the comparatively more developed and consequently efficient processing routes. The looming spectre of world climate change will accelerate adoption of wood for fuel systems in the first half of the 21st century.

7. METHANOL PRODUCTION

Synthesis gas from wood for methanol production is produced by gasifying with a mixture of steam and oxygen. The oxygen can come from either cryogenic separation of oxygen from nitrogen in liquefied air or by the electrolysis of water. The most efficient commercial gasifiers are fluidised beds that use a restricted supply of pure oxygen and operate at temperatures and pressures at or in excess of 900°C and 5 MPa. The aim is to maximise gas production with no ancillary charcoal. An alternative gasification scheme would be to use a FICFB gasifier that would produce synthesis gas of similar composition to pressurised oxygen gasification. Wood has a molar ratio of hydrogen to carbon of 1.44:1 while the corresponding ratio in methanol is 4:1. Inevitably, synthesis gas from wood contains excess carbon if used for methanol production. The proportion of hydrogen to carbon monoxide in the synthesis gas can be adjusted using the water gas shift reaction. The excess carbon monoxide is reacted exothermically with water over an iron-chrome catalyst to generate hydrogen and carbon dioxide.

$$CO + H_2O = CO_2 + H_2 \ (> 2\ \text{MPa},\ 850^{\circ}\text{C}) \qquad (8)$$

This process helps to correctly balance the gas mixture but now there will be too much carbon dioxide which must be stripped out using absorbing amine or hot bicarbonate solution.

However, if electrolysis of water were used to provide oxygen for the gasifier the hydrogen produced at the same time would correct the hydrogen/carbon monoxide ratio.

Methanol is synthesized from the carbon oxides and hydrogen using a copper based catalyst at 200-300°C and 5-10 MPa. The reaction is exothermic and the released heat can be used to generate high pressure steam:

$$CO + 2H_2 = CH_3OH \qquad (9)$$

also

$$CO_2 + 3H_2 = CH_3OH + H_2O \qquad (10)$$

Since the reaction reaches equilibrium with only a partial conversion (5%) of the reactants to methanol a system to recycle the synthesis gas is used. After cooling to condense out the methanol and water the remaining gases are recycled to the reactor in what is basically a closed loop.

A recent process development may lead to simplified plants for methanol manufacture from synthesis gas with a high carbon oxides to hydrogen ratio. Commercial methanol catalysts in fine powder form are suspended in inert mineral oil and the synthesis gas bubbled through the catalyst slurry. Heat removal by producing steam is more efficient than in a gas phase reactor and the reaction temperature is more uniform allowing higher conversion and a lower recycle rate. The water gas shift and carbon dioxide removal steps are also no longer required as synthesis gas rich in carbon oxides can be fed straight to the reactor.

The crude methanol (methanol/water and small amounts of impurities) is refined by distillation to produce fuel grade methanol. The net usable energy of the methanol represents around 30-60% of the total energy inputs to the gasification and methanol plant (Table 14.2).

Currently methanol is used largely as a feedstock for the manufacture of formaldehyde and a range of chemicals. It has considerable potential as a liquid fuel either directly or via a fuel chemical derived from methanol. For example, methyl tertiary butyl alcohol (MTBE), produced by reacting methanol with isobutene, has been a popular blending agent for gasoline because it improves octane rating. In the future, biodiesel, produced by reacting methanol with naturally occurring oils and fats, is likely to become a substitute for fossil fuel sourced diesel.

8. ETHANOL PRODUCTION FROM SYNTHESIS GAS

Ethanol is the only renewable liquid fuel made in commercial quantities and supplies about 1% of the gasoline type transport fuels used in the USA. Approximately 95% of the commercial production of ethanol in the USA is currently by direct fermentation of corn-sourced carbohydrates. However, fermentation of synthesis gas has the advantage over direct fermentation of sugars from cellulose and the hemicelluloses in that all wood components, including lignin and bark, are suitable feedstocks.

The first step in the process is gasification of wood followed by conditioning and cleaning of the synthesis gas to give a mixture rich in CO and H_2. The synthesis gas is sparged into a broth-filled tank fermentor where bacteria (*Clostridium ljungdahlii*, for example) convert CO and CO_2 to ethanol via the following two reactions:

$$6CO + 3H_2O = CH_3CH_2OH + 4CO_2 \quad (11)$$

$$2CO_2 + 6H_2 = CH_3CH_2OH + 3H_2O \quad (12)$$

The resulting fermentation broth is quite dilute, 2% ethanol or less. The ethanol can be recovered from the broth using distillation schemes developed for the existing corn to ethanol manufacturing process (Spath and Dayton, 2003).

9. FISCHER-TROPSCH LIQUIDS

Ethanol and methanol are useful gasoline substitutes. However, a diesel replacement can also be produced from wood using Fischer-Tropsch synthesis (Spath and Dayton, 2003). The first step is once again high temperature gasification using oxygen and steam to make synthesis gas, followed by cleaning of the gas to prevent damage to the catalyst and adjustment of the gas composition using the water gas shift reaction and carbon dioxide removal (Eq. 8). The required ratio of hydrogen to carbon monoxide for Fischer-Tropsch synthesis is approximately 2:1. A 'chain growth' reaction occurs over an iron or cobalt catalyst involving sequential addition of -CH_2- groups to produce an alkyl chain of increasing length:

$$nCO + 2nH_2 = C_nH_{2n+2} + nH_2O \tag{13}$$

The reactor operates either at low (200-240°C) or high (300-350°C) temperatures and between 1-4 MPa. The product mix changes from longer to shorter chain length molecules as the temperature increases. The product stream from a Fischer-Tropsch reactor is a mix of many components but by selecting the right operating conditions the mix can be adjusted so that the product stream has mostly diesel fuel properties.

10. HYDROLYSIS OF WOOD

Any treatment of wood must take account of the differing accessibilities and reactivities of the principal wood constituents. Further, any chemical or microbial method of breaking down wood has to devise conversion pathways for cellulose, the hemicelluloses and lignin, and if necessary consider ways of isolating the individual reaction products so that they can be processed separately. Hydrolysis has proved to be a most effective method of opening up the wood structure for subsequent treatments. The expression 'hydrolysis of wood' is used rather loosely. It is not technically correct since the reactions affect primarily the carbohydrate fraction of wood. Lignin is largely unaffected.

Although composed predominantly of polysaccharides wood has little value as an animal feed except as supplementary, non-nutritional roughage (Hajny, 1981). The complex lignin-carbohydrate structure of wood, the crystallinity of the cellulose, and the inaccessibility of the cell wall to large enzyme molecules makes wood resistant to the action of cellulotytic microorganisms. Softwoods are non-digestible while hardwoods are at best slightly digestible. *Populus tremuloides* is an exception having significant digestibility (40%). Presteaming hardwood chips dramatically enhances their digestibility by cellulase enzymes in a ruminant's stomach. Presteaming involves heating the wood chips in a pressure vessel at temperatures between 230 and 150°C for a period of time ranging from a few seconds to an hour or so. The mass is then explosively discharged by 'blowing' the digestor. The chips are well disintegrated. The presteaming of hardwoods is often described as autohydrolysis as the acids which catalyse hydrolysis are generated by the process itself, rather than being added as a separate ingredient as in acid hydrolysis. Presteaming of the hemicelluloses yields acetic, formic and other acids which lower the pH further so enhancing the depolymerization of the hemicelluloses and to a lesser degree the lignin, but leaving the cellulose largely unaffected. During presteaming the hardwood hemicelluloses hydrolyse readily to water-soluble, low molecular weight fragments (oligomers), the relatively abundant acetyl and carbonyl groups are cleaved, and the α-ether linkages in lignin are hydrolysed. The cleavage of many cross linkages makes the cell wall of hardwoods much more accessible to cellulase enzymes. Under optimal conditions presteaming of hardwoods is an effective pretreatment prior to a microbial or chemical route for the manufacture a variety of chemicals from the polysaccharides and lignin.

Softwoods need slightly severer conditions than those provided by steaming (Clark and Mackie, 1987). Hydrolysis proceeds much faster with the addition of sulphur dioxide which is an effective acid catalyst (Figure 14.6). The enzymatic digestibility of the cellulose in the insoluble residue increases with progressively severer cooking conditions (higher temperatures, longer cooking times and more sulphur dioxide). For example, when treating softwood chips at 215°C for 180 s their digestibility increases from around 5% to over 80% with the addition of 2.5% SO_2. The cellulose is attacked and about 25% of this is solubilized, pyrolysed or degraded. Further glucose (5%) comes from the glucomannans. The improved enzymatic digestibility appears to be related to the complete removal of hemicelluloses and a partial removal of some cellulose which allow the cell wall capillaries to enlarge. The cellulase enzymes are no longer physically excluded

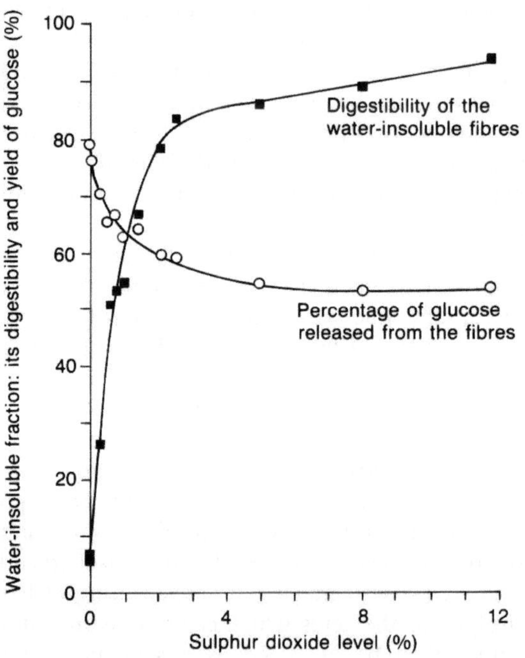

Figure 14.6. Hydrolysis of *Pinus radiata* chips (Clark *et al.,* 1989). Enzymatic digestibility and the overall carbohydrate survival are improved by the addition of SO_2 during presteaming. The water-insoluble fibre yield after washing decreases with increasing SO_2, while its digestibility (defined as the yield of glucose after 72 hr, expressed as a percentage of the theoretical yield) increases with increasing amounts of SO_2. It is clear that most benefits from using SO_2 are achieved with a SO_2 level of about 2-3%.

from the cell wall because of their size. Further, structural and chemical changes such as fibre fragmentation (frequently separating the S_2 from the rest of the cell wall) and partial depolymerization, repolymerization and coalescence of lignin into droplets within the cell wall at these high temperatures (Donaldson *et al.*, 1988) all favour subsequent enzymatic hydrolysis of the cellulosic residues. However, compared to the autohydrolysis of hardwoods, softwood lignin after acid hydrolysis is much less soluble in NaOH, presumably because of repolymerization of lignin fragments which is favoured by highly acidic conditions.

When the treated chips are blown from the digester they disintegrate into a mass of fibre and fibre fragments. Washing the exploded pulp extracts the soluble components which includes dimers and higher molecular weight fragments (oligomers). Sugar monomers are recovered after further mild hydrolysis. The solids, predominantly cellulose and lignin, can be hydrolysed under harsher conditions or treated with enzymes to extract sugars. The total sugar yield is related to polysaccharide solubilization and survival (of the water-soluble fraction), and to the enzymatic digestibility of the steam exploded fibre (the water-insoluble fraction). At near optimal conditions (180 s at 215°C with 2.5% SO_2) the total sugar yield is 57 g/100 g of oven-dry wood, consisting of 29 g of sugars from the water-soluble extract and 28 g after enzymatic digestion of the solids (Clark *et al.*, 1989). Of this approximately 40.6 g is glucose, a further 11.4 g are other hexose sugars and the balance, 5.4 g, are pentose sugars.

11. ETHANOL PRODUCTION BY ACID HYDROLYSIS AND FERMENTATION

Acid hydrolysis, with dilute sulphuric acid hydrolysis in particular, has been the traditional way of breaking down wood to recover fermentable sugars (Figure 14.7). The cellulose in the microfibrils is not readily accessible and moderately harsh conditions are necessary to make the process viable. A practical limitation in the hydrolytic decomposition of the polysaccharides to sugars is that the sugars themselves are subject to simultaneous degradation to chemicals such as furfural, so the sugars can never be recovered in full: yields can be as low as 50% of the theoretical value. The former Soviet Union developed a major commercial wood hydrolysis programme with some 40 such units. Much of this was fermented to single cell protein for cattle feed. Production was around 800 000 tonnes a year of yeast fodder (Wayman and Parekh, 1990). It is unlikely that the process would have been economic in a free market economy.

Although hardwoods have a lower lignin content and so give a higher yield of sugars, softwoods are preferred where ethanol is the desired end product. This is because hardwoods have more pentose sugars, which are not readily fermented by common yeasts. In a typical batch process sawdust and wood chips are loaded into the reactor vessel and treated with dilute sulphuric acid (0.5% concentration) at temperatures between 130 and 200°C for about three hours. Ideally, the sugars should be removed from the reaction zone before they, in turn, have time to break down.

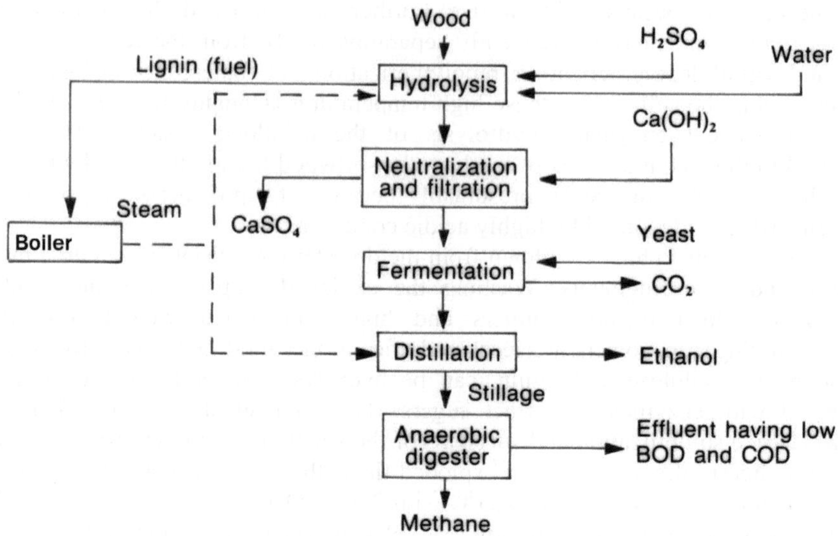

Figure 14-7. Ethanol production from wood (Burton *et al.*, 1984).

A two-stage operation is more efficient (Uprichard and Burton, 1982). Initially the temperature is 130-140°C and the dilute acid attacks the hemicelluloses, which hydrolyse very much faster than cellulose, and their sugars and other decomposition products are drawn off prior to the main acid hydrolysis stage. At this point, after about 30 minutes, the temperature is raised to 180-190°C under pressure (1.6 MPa) and cellulose hydrolysis begins. Acid is introduced continuously at the top of the vessel and the dilute sugars are drawn off continuously at the base. Acid, so necessary for hydrolysis, must be neutralised and removed prior to fermentation. Slaked lime, $Ca(OH)_2$, is added to neutralise the acid and to prevent further degradation of the sugars. The solution is filtered and flash cooled. The byproduct gypsum, $CaSO_4$, finds uses in the manufacture of wall panels and in fertilisers. The 130-140°C pretreatment not only allows recovery of the sugars from the hemicelluloses, also it increases the subsequent rate of hydrolysis of cellulose and allows the glucose to be drawn off more quickly so that simultaneous degradation of the sugars in the reaction zone is reduced. With pine the yield of sugars is about 64% of the theoretical yield. Once the carbohydrate material has been broken down into its constituent sugars the undissolved lignin in the hydrolysis tank is flushed out. It can be used for process energy or possibly as a chemical feedstock in its own right.

New developments aim to increase the yield further, by relying on fast (< 1 min), high temperature (> 240°C) and continuous hydrolysis. At higher temperatures the rate of hydrolysis of cellulose increases more rapidly than does the rate of degradation of the newly formed sugars so it should be possible to obtain a slightly better yield.

The hexose sugars, glucose and mannose, are subsequently converted to ethanol by fermentation with a yeast such as *Saccharomyces cerevisiae* within the first 12 hours at 35°C:

$$C_6H_{12}O_6 = 2C_2H_5OH + 2CO_2 \quad (14)$$

This fermentation process is the same as that used for ethanol production from cane sugar except the sugar concentration is lower, so distillation costs are greater. The ethanol is recovered and concentrated by distillation. Ethanol yields of 20% of the oven-dry weight of wood are obtainable. Purity is not a major concern as many byproducts (esters, higher alcohols etc.) are also good fuels. Galactose, the only other abundant hexose sugar, is virtually unused after 24 hours and does not contribute to the ethanol yield. It remains unconverted in the stillage. The yeast-rich stillage, containing the pentose sugars and other hydrolysis products, can be converted to methane using anaerobic bacteria. Anaerobic digestion removes much of the organic matter in the waste water system while generating a very substantial quantity of methane. The production of methane (CNG) substantially enhances the overall efficiency and process economics, while greatly relieving a major effluent problem. The overall thermal efficiency is about 50% with half arising from ethanol production (24%) and half from the surplus methane (27%) which is in excess of that needed to provide process heat (Burton *et al.*, 1984).

The process described by Burton *et al.* (1984) uses proven technology and demonstrates the point that the production of ethanol from wood is likely to be viable only when integrated as a multiproduct operation. The obvious areas for improvement are in increasing the ethanol yield and in encouraging fermentation to continue as the concentration of ethanol builds up in the solution, which would significantly reduce the cost of distillation.

12. ETHANOL VIA SIMULTANEOUS SACCHARIFICATION AND FERMENTATION

An alternative to the well established but relatively inefficient acid hydolysis of wood is enzymatic hydrolysis using the cellulase enzyme. Research at the National Renewable Energy Laboratory in the USA since the late 1980s has focused on a generation of genetically engineered cellulase enzyme systems. These are combined with yeasts to allow simultaneous saccharification (liberation of sugars from cellulose) and fermentation of those sugars to ethanol: sugars are continuously produced by the enzymes and converted by the yeasts to ethanol. Organisms capable of fermenting these sugars from cellulose and the hemicelluloses are now available, increasing the theoretical yield of ethanol in comparison to acid hydolysis processes. A recovery of about 380 litres of ethanol per tonne of wood is achievable, equivalent to 80% of the theoretical yield (Wayman and Parekh, 1990). The first step in the process is a dilute acid treatment (0.5% sulphuric acid) at high temperature (190°C). This splits up the hemicelluloses into their constituent sugars and renders the cellulose more accessible for enzymatic breakdown. The liquid hydolysate contains acetic acid and other inhibitors in addition to the hemicellulose sugars. The treated hydolysate is combined with the residual solids (cellulose and lignin) from the hydrolysis step and sent to the simultaneous saccharification and cofermentation (SSCF) tanks. The fermentors are dosed with a combined cellulase/yeast inoculum – an example fermentation microbe is *Zymomonas mobilis* which is capable of fermenting both glucose and xylose. The cellulase is produced in additional fermentors using an industrial fungus, *Trichoderma reesei* for example, that feeds on a small, diverted amount of the main fermenter feedstock. A dilute ethanol stream (approximately 2% ethanol) is then concentrated by distillation followed by molecular sieve dehydration to produce fuel grade ethanol. Waste water containing lignin and other organics is anaerobically digested to produce biogas and the residual lignin solids are burnt as fuel. SSCF is already almost cost competitive with acid hydolysis processes and it is expected that the discovery of more efficient microorganisms will make SSCF the more attractive process in the next 5-10 years. The main improvement is the anticipated development of ethanol producing microorganisms capable of working at higher ethanol concentrations (5%) and at higher temperatures (50°C). Increased ethanol concentration reduces the cost of ethanol concentration and increased temperature dramatically increases saccharification rate, reducing the cost of the required cellulose enzyme. Potentially the production cost per unit of alcohol could be halved, making lignocellulosic ethanol competitive with gasoline at current prices.

13. LIQUID FUELS

The fermentation of sugar cane to produce ethanol has been adopted by Brazil in its effort to develop a local fuel for domestic transport. Pure petrol/gasoline is no longer available and cars in that country run on either a 22-25% blend of ethanol with petrol or on pure ethanol. In 1987 ethanol consumption in transport was equivalent to 7.5 million metric tonnes of oil (Trindade and Carvalho, 1989) and it continues to be used at about that level. Brazil has the available land for sugar cane production and made the decision to develop alternative liquid fuels in 1975 despite the fact that the cost of production was very substantially greater than the cost of the equivalent oil imports. If oil prices are greater than around $US40/barrel there is an economic incentive to expand Brazilian ethanol use. However, if the oil price drops to $US30/barrel then ethanol is only just cost effective. Oil prices were less than this during the 1990s and this stalled growth in ethanol use. In 2005, because of high oil prices the Brazilian ethanol program looks set for expansion. In the United States corn-based ethanol is more expensive than compressed natural gas (CNG) and methanol but survives because of generous subsidies resulting from continued lobbying by agricultural interests and because it can be blended with petrol and used in vehicles without engine modification. Brazil and the United States account for more than 95% of all ethanol production from biomass.

Few countries apart from Brazil have developed alternative fuels to displace petrol in transport. In Europe biodiesel, a replacement for fossil diesel manufactured by reacting methanol with fats extracted from oilseeds, is starting to make an impact but consumption is still small. Elsewhere alternative fuels have made at most a limited contribution to a nation's overall liquid fuels strategy. Their use is likely to remain peripheral unless severe dislocations in energy supplies or rapidly rising fossil fuel prices make their production viable. However concerns about the uncertainty of supply and regarding previously ignored social costs due to air pollution and global warming need to be taken into account in policy. The cost to society of air pollution can only be guessed. Estimates for the United States range from only ten billion dollars to almost two hundred billion dollars annually (Sperling, 1989) without taking into account the economic impact of global warming. A major shift to alternative fuels is beginning in areas such as southern California, basically in response to appalling air quality.

Both compressed natural gas (CNG) and methanol are much less polluting than petrol. Neither is competitive with petrol on a narrow economic analysis. Of the two, methanol is probably viewed more favourably. It may appear illogical to convert natural gas into methanol rather than using it as CNG, but the ability to transform a gas (or solid if wood were to be the feedstock) to a liquid fuel outweighs the cost and energy required to effect that conversion. However, part of the support for methanol is pure inertia, emphasising the difficulties in setting up an extensive distribution system for CNG, the cost of vehicle conversion, the need for bulky fuel tanks, and the limited fuel range. Motor vehicles are designed to use liquid fuels and are burdened with redundant fuel systems when retrofitted to use compressed natural gas. By contrast modifications for methanol fuelled vehicles are much simpler.

The other major alternative fuel is ethanol. Ethanol is more polluting, although less so than petrol, but it has the advantage in that it can be blended with petrol or used by itself (with minor modifications to the engine).

The potential for manufacturing liquid fuels from wood will remain unfulfilled until the technology and economics move in its favour or the negative impact of poor air quality and global warming are considered significant enough to overturn conventional economic decision making. Few countries have the land available to dedicate to a wood fuels programme. Wayman and Parekh (1990) calculate that it would require 10% of the total land area of the United States to meet its requirements for liquid fuel – about 100 million ha of dedicated forest plantations.

14. ENERGY AND CLIMATE CHANGE

Between 1850 and 1998 atmospheric carbon dioxide levels have risen from 285 to 366 ppm, mainly because of the combustion of fossil fuels and changes in land use from forestry to agriculture. This unbalanced release of stored carbon is now generally accepted as setting us on a global climate change journey with an unknown but likely to be unpleasant destination. Combustion of fossil fuels alone contributed 6.3 Gt (gigatonnes or 10^9 tonnes) of carbon dioxide emissions annually between 1989 and 1998.

Trees have a key role to play in slowing climate change. Their growth and integration into the energy system can help in three ways:

- If wood replaces fossil fuels as an energy source it directly and permanently reduces greenhouse gas emissions. Wood energy is carbon emission friendly as trees, planted to provide fuel, grow absorbing the carbon dioxide released by transformation of wood from previous harvests (combustion, gasification or conversion into liquid fuels followed by combustion) into energy. Wood energy is almost carbon emission neutral as long as only small amounts of fossil fuel are used to grow, harvest and transport the wood fuel source.
- Even if not used for fuel forests can act as sinks, storing carbon as long as the forests are never cleared and the land subsequently used for a different purpose. The problem, especially in the period that human population has increased rapidly, is a steady trend of deafforestation to create agricultural land. It is impossible to guarantee that this will never happen to any of the world's forests so offsetting emissions is not as permanent a solution to climate change as replacing fossil fuel with woody biomass.
- Using wooden building materials can cut out emissions associated with the production of the replaced more energy intensive materials such as concrete, steel and aluminium.

Integration of wood directly into the energy supply system is the most permanent solution. Studies have shown that CO_2 emissions per kWhr in a wood fired power station are only 5-10% of the equivalent emissions from coal fired electricity generation. Ethanol from wood could potentially reduce greenhouse gas emissions by 65% on a per-mile travelled basis in comparison to gasoline if the replacement fuel is an 85% ethanol blend with petrol. If global biomass energy supply reaches 100-400 EJ yr^{-1} by 2050 then this would potentially reduce greenhouse gas emissions by between 20-55%.

Global climate change represents one of the strongest drivers for making wood to energy schemes a reality. However there will be only a slow uptake of the possible technologies as fossil fuel based systems are inevitably cheaper because conventional economics does not account for their adverse environmental effects. The Kyoto protocol is the first attempt to encourage bioenergy use by setting a target of 5% less than their 1990 greenhouse gas emissions for developed countries to meet in the first commitment period of 2008-2012. Signatories to the protocol have to find ways of meeting their targets using instruments such as carbon credits to encourage carbon friendly schemes and carbon taxes to discourage fossil fuel users. These measures slightly weight the economics of energy schemes in favour of bioenergy over fossil fuels but they are tentative first steps and more will be needed before the convenience and financial advantages of fossil fuels over biomass are removed.

REFERENCES

Acda M, Morrell J and Levien K (1997) Effect of process variables on supercritical fuild impregnation of composites with tebuconazole. *Wood and Fiber Science 29*(3) 282-90

Adair C and Camp W (2003) *Market outlook: lumber, structural panels, and engineered wood products (2003-2008)*. APA - The Engineered Wood Association, Tacoma, WA

Adair C (2004) *Regional production and market outlook for structural panels and engineered wood products (2004-2009)*. APA - The Engineered Wood Association, Tacoma, WA

Adler E (1977) Lignin chemistry - past, present and future. *Wood Science and Technology, 11*(3) 169-218

AFPA (2005a) *National design specification for wood*. American Forest and Paper Association, Washington, DC

AFPA (2005b) *Design values for wood construction: a supplement to the national design specification for wood construction*. American Forest and Paper Association, Washington, DC

AITC (1990) *Standard for preservative treatment of structural glued-laminated timber*. AITC 109. American Institute of Timber Construction, Englewood, Colorado

AITC (2001) *Standard appearance grades for structural glued-laminated timber*. AITC 110. American Institute of Timber Construction, Englewood, Colorado

AITC (2004) *Standard specification for structural glued-laminated timber of softwood species*. AITC 117-04. American Institute of Timber Construction, Englewood, Colorado

AITC (2005) *Recommended practice for protection of structural glued-laminated timber during transit, storage, and erection*. AITC 111. American Institute of Timber Construction, Englewood, Colorado

Alince B (1989) Volume contraction of cellulose-water system. In Schuerch C (ed), *Cellulose and wood: chemistry and technology*. Wiley-Interscience, New York, 379-88

Almberg L, Croon I and Jamieson A (1979) Oxygen delignification as part of future mill systems. *TAPPI, 62*(6) 33-35

Anonymous (1996) *Best Management Practices for the Use of Treated Wood in Aquatic Environments. USA version*. Western Wood Preservers' Institute, Vancouver, BC

ANSI/AITC (2002) *Structural glued laminated timber*. ANSI/AITC A190-1. American National Standards Institute and American Institute of Timber Construction, Englewood, Colorado

APA (1995) *Construction and Industrial Plywood, Voluntary Product Standard PS1*. US Department of Commerce, Washington, DC

APA (2001) *Performance Standards and qualification policy for structural-use panels*. PRP-108. APA-The Engineered Wood Association, Tacoma, Washington

APA (2004) *Regional production and market outlook for structural panels and engineered wood products (2004-2009)*. APA-The Engineered Wood Association, Tacoma, Washington

APA-EWA (2002) *Proper storage and handling of glulam beams*. EWS R540. APA-The Engineered Wood Association, Tacoma, Washington

APA-EWA (2004) *Performance rated I-joists*. EWS Z725. APA-The Engineered Wood Association, Tacoma, Washington

Araman PA and Hansen BG (1983) *Conventional processing of standard-size edge-glued blanks for furniture and cabinet parts: a feasibility study*. USDA, Forest Service, Northeastern Forest Experiment Station, Research Paper NE-524

Araman PA (1987) *Standard size rough dimension: a potential hardwood product for the European market made from secondary-quality US hardwoods.* Seminar on the valorization of secondary-quality temperate-zone hardwoods, Nancy. Economic Commission for Europe, Timber Committee

Araman PA (1987a) *New patterns of world trade in hardwood timber products.* Paper presented at the Proceedings of the 64th Agriculture Conference, Washington, DC

Aratake Y (1992) Estimation of modulus of rupture (MOR) and modulus of elasticity (MOE) of lumber using higher natural frequency of log in pile of logs - possibility of application of Sugi scaffolding board. *Mokuzai Gakkaishi, 38*(11) 995-1001

Araujo CD, MacKay AL, Hailey JRT and Whittall KP (1992) Proton magnetic resonance techniques for characterization of water in wood: application to white spruce. *Wood Science and Technology, 26* 101-13

Araujo CD, MacKay AL, Whittall KP and Hailey JRT (1993) A diffusion model for spin-spin relaxation of compartmentalized water in wood. *Journal of Magnetic Resonance Series B, 101* 248-61

Archer KJ (1985) *Bacterial modification of Douglas fir roundwood permeability.* PhD thesis, University of Canterbury, Christchurch, New Zealand

Archer RR (1986) *Growth stresses and strains in trees.* Springer-Verlag, Berlin

AS/NZ (1997) *Timber stress-graded: product requirements for mechanical stress-graded timber.* AS/NZS 1748. Australian/New Zealand Standard

Ashworth JC (1977) *The mathematical simulation of the batch-drying of softwood timber.* PhD thesis, University of Canterbury, Christchurch, New Zealand

Astley RJ, Stol KA and Harrington JJ (1998) Modelling the elastic properties of softwood. *Holz als Roh- und Werkstoff, 56*, 43-50

ASTM (2005a) *ASTM D143-94. Standard test for small clear specimens of timber.* American Society for Testing and Materials International, West Conshohocken, Pennsylvania

ASTM (2005b) *ASTM D198-05. Standard methods of static tests of lumber in structural sizes.* American Society for Testing and Materials International, West Conshohocken, Pennsylvania

ASTM (2005c) *ASTM D245-00. Standard practice for establishing structural grades and related allowable properties for visually graded lumber.* American Society for Testing and Materials International, West Conshohocken, Pennsylvania

ASTM (2005d) *ASTM D1990-00. Standard practice for establishing allowabel properties for visually-graded dimension lumber from in-grade tests of full-size specimens.* American Society for Testing and Materials International, West Conshohocken, Pennsylvania

ASTM (2005e) *ASTM D2555-98. Standard methods for establishing clearwood strength values.* American Society for Testing and Materials International, West Conshohocken, Pennsylvania

ASTM (2005f) *ASTM-2915-03. Standard practice for evaluating allowable properties for grades of structural timber.* American Society for Testing and Materials International, West Conshohocken, Pennsylvania

ASTM (2005g). *ASTM D3737-04. Standard practice for evaluating allowable properties for structural glued-laminated timber (glulam).* American Society for Testing and Materials International, West Conshohocken, Pennsylvania

ASTM (2005h). *ASTM D4761-05. Standard test methods for mechanical properties of lumber and wood-base structural material.* American Society for Testing and Materials International, West Conshohocken, Pennsylvania

ASTM (2005i). *ASTM D5055-04. Standard specification for establishing and monitoring structural capabilities of prefabricated I-joists.* American Society for Testing and Materials International, West Conshohocken, Pennsylvania

ASTM (2005j). *ASTM D5456-05. Standard specification for evaluation of structural composite lumber products.* American Society for Testing and Materials International, West Conshohocken, Pennsylvania

ASTM (2005k) *ASTM D6570-04. Standard practice for assigning allowable properties for mechanically graded lumber.* American Society for Testing and Materials International, West Conshohocken, Pennsylvania

Atack D (1972) On the characterization of pressurized refiner mechanical pulps. *Svensk Papperstidn, 73*(3) 89-94

Atack D (1981) Dynamic mechanical loss properties of wood. *Philosophical Magazine A, 43*(3) 619-25

Atalla RH (1990) The structure of cellulose. In Caulfield DF, Passaretti JD and Sobcyznski, SF (eds), *Materials interactions relevant to the pulp, paper and wood industries.* Materials Research Society, Warrendale, Pennsylvania, Vol. 197 89-98

Atalla RH (1996) Cellulose and the hemicelluloses: patterns for the assembly of lignin. In Lewis NG and Sarkanen S (eds), *Lignin and lignan biosynthesis.* American Chemical Society, Washington, DC, 172-9

Atalla RH and VanderHart DL (1999) The role of solid state C-13 NMR spectroscopy in studies of the nature of native celluloses. *Solid State Nuclear Magnetic Resonance, 15*(1) 1-19

Atalla RH (2005) *The role of the hemicelluloses in the nanobiology of wood cell walls: a systems theoretic perspective.* In Entwistle K and Walker JCF (eds), *The Hemicelluloses Workshop 2005.* Wood Technology Research Centre, University of Canterbury, Christchurch, New Zealand, 37-55

Atkins PW (1986). *Physical Chemistry* (3rd. edit). Oxford University Press.

AWPA (1997) Glue laminated timber-preservative treatment by pressure processes. Standard C 28. In *AWPA Book of Standards.* American Wood Preservers' Association, Granbury, Texas

Bailey IW and Kerr T (1935) The visible structure of the secondary wall and its significance in physical and chemical investigations of tracheary cells and fibres. *Journal of the Arnold Arboretum, 16* 273-300

Baldwin RF (1981) *Plywood manufacturing practices* (2nd Edit). Miller Freeman Publications Inc, San Francisco

Baldwin RF (1987) New developments in plywood green ends. *Forest Industries, 114*(4) 18-23

Bamber RK (1979) The origin of growth stresses. *Forpride Digest, 8*, 75-9 and 96

Bamber RK (1987) The origin of growth stresses: a rebuttal. *International Association of Wood Anatomists Bulletin, 8*(1) 80-4

Bannister MH (1962) Some variations in growth pattern of *Pinus radiata* in New Zealand. *New Zealand Journal of Science, 5*(3) 342-70

Bannister P, Carrington CG and Liu Q (1994) *Dehumidifier drying technology: new opportunities.* Proceedings 4th IUFRO International Drying Conference, Rotorua, New Zealand, 247-54

Bannister P, Carrington CG and Chen G (2002) *Heat pump dehumidifier drying technology: status, potential and prospects.* Proceedings of 7th International Energy Agency Conference on Heat Pump Technology, Beijing, Vol. 1, 219-30

Bao FC, Jiiang ZH, Jiang XM, Lu XX and Luo XQ (1998) Comparative studies on wood properties. *Scientia Silvae Sinicae, 34*(2) 63-75

Bao FC, Jiang ZH, Jiang XM, Lu XQ and Zhang SY (2001) Differences in wood properties between juvenile wood and mature wood in 10 species grown in China. *Wood Science and Technology, 35*(4) 363-75

Barber NF and Meylan BA (1964) The anisotropic shrinkage of wood. *Holzforschung, 18*(5) 146-56

Barnes D, Admiraal L, Pike RL and Mathur VNP (1976) Continuous system for the drying of lumber with microwave energy. *Forest Products Journal, 26*(5) 31-42

Barnes D (1988) Parallam: a new wood product - invention and development to the pilot scale stage. In *Timber: a material for the future*. The Marcus Wallenberg Foundation Symposium Proceedings. Falun, Sweden, *4* 5-24

Barnes HM, Williams LH and Morrell JJ (1989) *Borates as wood preserving compounds: the status of research in the United States*. International Research Group on Wood Preservation, Document No IRG/WP/3542

Barnes HM, Amburgey TL and Sanders MG (2004) Performance of zinc-based preservative systems in ground contact. *Wood Design Focus, 14*(2) 13-7

Barnes HM and Murphy RJ (1995) Wood Preservation: the classics and the new age. *Forest Products Journal, 45*(9) 16-26

Barnet AJ and Vihmari P (1983) OPCO process pilot plant. *Pulp and Paper Magazine of Canada, 84*(9) 52-55 (T215-218)

Barnett JR and Jeronimidis G (2003) Reaction wood. In Barnett JR and Jeronimidis G (eds), *Wood quality and its biological basis*. Blackwell Publishing, Oxford and CRC Press, Boca Raton, 118-36

Bascuñán A (2004) *The influence of wind on radiata pine tree shape and wood stiffness*. MForSc thesis, University of Canterbury, Christchurch, New Zealand

Bech-Andersen J and Elborne SA (1999) *The dry rot fungus (Serpula lacrymans) in nature and its history of introduction in buildings*. IRG Document IRG/WP 99-10300

Beiser von W (1933) Mikrophotographische quellungsuntersuchungen von fichten- und buchenholz an mikrotomschnitten im durchfallenden licht und an holzklötzchen im auffallenden licht. *Kolloid-Zeitschrift, 65*(2) 203-11

Berthold J, Rinaudo M and Salmén L (1996) Association of water to polar groups; estimations by an adsorption model for ligno-cellulosic materials. *Colloids and surfaces A: physiochemical and engineering aspects, 112* 117-29

Bhat KM and Indira EP (1997) *Effect of faster growth on timber quality of teak*. Kerula Forest Research Institute, Research Report 132. Peechi, Kerala, India

Bhat KM (1998) Properties of fast-grown teak wood: impact on end-user's requirements. *Journal of Tropical Forest Products, 4*(1) 1-10

Bhat KM (1999) Is fast grown teak inferior in wood quality? An appraisal from plantations of low input management. *Wood News, 8*(4), 27-31

Bhat KM, Priya P and Rugmini P (2001) Characterisation of juvenile wood in teak. *Wood Science and Technology, 34* 517-32

Bhat KM, Nair KKN, Bhat KV, Muralidharan EM and Sharma JK (2005) *Proceedings of the International Conference on Quality Timber Products of Teak from Sustainable Forest Management, December 2003*. Peechi, Kerala, India

Bilek EM and Horgan GP (1992) *The challenges of privatization: New Zealand's experience with forestry*. In IUFRO Conference, Integrated Sustainable Multiple-use Forest Management under the Market System, Pushkino Moscow Region, 119-60

Blackwell J and Kolpak FJ (1976) *Cellulose microfibrils as disordered arrays of elementary fibrils*. In Timell TE (ed) Proceedings of 8th cellulose conference. Wiley, New York, Vol. 2 751-61

Bland JD (1990) Plywood can compete, thanks to the newest spindleless lathe. *Forest Industries, 117*(3) 32-4

Bodig J (1977) *Bending properties of Douglas fir/Larch and Hem-Fir dimension lumber*. Special Report No. 6888. Department of Forestry and Wood Science, Colorado State University, Fort Collins, Colorado

Bohannan B (1966) *Effect of size on bending strength of wood members.* USDA, Forest Service, Forest Products Laboratory, Research Report, FPL 56

Booker RE (1989) *Hypothesis to explain the characteristic appearance of aspirated pits.* Proceeding 2nd Pacific Regional Wood Anatomy Conference, Oct. 1989. Forest Products Research and Development Institute, Laguna, Philippines

Booker RE (1990) Changes in transverse wood permeability during the drying of *Dacrydium cupressinum* and *Pinus radiata. New Zealand Journal of Forest Science, 20*(2) 231-44

Booker RE (2003) *Shrinkage and theories of differential shrinkage.* Conference on wood research knowledge and concepts for future demand. EMPA Wood Laboratory Research and Work Report 115/50, 29-46

Boone RS, Kozlik RJ, Bois PJ and Wengert EM (1988) *Dry kiln schedules for commercial woods: temperate and tropical.* USDA, Forest Service, Forest Products Laboratory, GTR-57

Boutelje J (1973) On the relationship between structure and the shrinkage and swelling of wood in Swedish pine (*Pinus sylvestris*) and spruce (*Picea abies*). *Svensk Papperst., 76*(2) 78-83

Boyd JD (1950) Tree growth stresses. Part 2: development of shakes and other visual failures in timber. *Australian Journal Applied Science, 1*(3) 296-312

Boyd JD (1972) Tree growth stresses. Part 5: evidence of an origin in differentiation and lignification. *Wood Science and Technology, 6*(4) 251-62

Boyd JD (1985) The key factor in growth stress generation in trees lignification or crystallisation? *International Association of Wood Anatomists Bulletin, 6*(2) 139-50

Bramhall G (1979) Sorption diffusion in wood. *Wood Science, 12*(1) 3-13

Brandão LG (1984) Presentation. In *The new eucalypt forest,* The Marcus Wallenburg Foundation Symposia Proceedings 1. Falun, Sweden, *1* 3-15

Brazier JD, Hands R and Seal DT (1985) Structural wood yields from Sitka spruce: the effect of planting spacing. *Forestry and British Timber, 14*(9) 34-7

BRE (1977) *A handbook of softwoods.* Building Research Establishment, HMSO, London

Brooks KM (2002) Characterizing the environmental response to pressure-treated wood. In *Enhancing the Durability of Lumber and Engineered Wood Products.* Forest Products Society, Madison, Wisconsin, 59-71

Brown CJ, Eaton RA, Cragg SM, Goulletquer P. Nicolaidou A. Bebianno MJ, Icely J, Daniel G, Nilsson T, Pitman AJ, and Sawyer GS (2003) Assessment of effects of chromated copper arsenate (CCA)-treated timber on nontarget epibiota by investigation of fouling community development at seven European sites. *Archives Environmental Contamination and Toxicology, 45* 37-47

Brown RM (1989) Cellulose biosynthesis and a decade of progress: a personal perspective. In C. Schuerch (ed), *Cellulose and wood: chemistry and technology.* Wiley-Interscience, New York, 639-57

Brunauer S, Emmett PH and Teller EJ (1938) Adsorption of gases in multilayers. *Journal of the American Chemical Society, 60* 309-19

Brunner-Hildebrand (1987) *Die Schnittholtrocknung [kiln-drying of sawn timber].* Brunner – Hildebrand, Hanover, Germany

Bryan EL (1977) *Low-cost profit improvement opportunities in sawmilling.* Proceeding of the 32nd Northwest Wood Products Clinic, Coeur d'Alene. Washington State University 47-59

BSI (1986) *Specification for visual strength grading of softwood, BS 4978.* British Standards Institute, London

Bublitz WJ (1980) Pulpwood. In Casey JP (ed), *Pulp and paper: chemistry and chemical technology* (3rd edit). Wiley-Interscience, New York, 113-59

Buikat ER (1971) *Operating problems with dryers and potential solutions.* In Maloney T (ed), Proceedings of 5th Washington State University Symposium on Particleboard, Pullman, Washington, 209-16

Bull DC (1998) *Chemical recovery of chromated copper arsenate wood preservative sludges.* PhD thesis, University of Canterbury, Christchurch, New Zealand

Bull DC (2001) The chemistry of chromated copper arsenate II. Preservative-wood interactions. *Wood Science and Technology, 34* 459-66

Bunn EH (1981) The nature of the resource. *New Zealand Journal of Forest Science, 26*(2) 162-99

Burdon RD (1975) Compression wood in *Pinus radiata* clones on four sites. *New Zealand Journal of Forest Science, 5* 152-64

Burdon RD, Kibblewhite RP, Walker JCF, Megraw RA, Evans R and Cown DJ (2004) Juvenile versus mature wood: a new concept, orthogonal to corewood versus outerwood, with special reference to *Pinus radiata* and *P. taeda. Forest Science, 54*(4) 399-415

Burton RJ, Horgan GP, Callandar IJ, Clark TA and Mackie KL (1984) *The co-production of ethanol and methane from wood via acid hydrolysis.* Proceedings, 6th International Alcohol Fuels Technology Symposium, Ottawa, Vol. 2 2114-20

Butterfield BG and Meylan BA (1980) *Three-dimensional structure of wood: an ultrastructural approach* (2nd. edit). Chapman and Hall, London

Butterfield BG (ed) (1998) *Proceeding of the IAWA/IUFRO Workshop on Microfibril angle in wood.* University of Canterbury, Christchurch, New Zealand

BWPA (1986) *Manual.* British Wood Preservers' Association, London

Cameron DW, Farrington A, Nelson PF, Raverty WD, Samuel EL and Vanderhoek N (1982) The effect of addition of quinonoid and hydroquinonoid derivatives in NSSC pulping. *APPITA, 35*(4) 307-15

Carroll M (1980) *We still don't boil houses.* In Maloney T (ed) Proceedings of 14th Washington State University Particleboard Symposium, Pullman, Washington, 39-54

Carson MJ (1988) *Long-internode or multinodal radiata pine: a financial analysis.* New Zealand Ministry of Forestry, Forest Research Institute Bulletin No. 115

Carson SD (1991) Genotype x environment interaction and optimal number of progeny test sites for improving *Pinus radiata* in New Zealand. *New Zealand Journal of Forest Science, 21*(1) 32-49

Carson SD (1996) Greater specialisation for improved seed-lots in New Zealand: new developments for efficient selection of parents and evaluation of performance. *New Zealand Journal of Forestry, 41*(1) 12-7

Cassens DL and Feist WC (1991) *Exterior wood in the south: selection applications and finishes.* USDA, Forest Service, Forest Products Laboratory, FPL-GTR-69

Cassens DL, Feist WC, Johnson BR and De Groot RC (1995) *Selection and use of preservative-treated wood.* Forest Products Society, Madison, Wisconsin

Cave ID (1968) The anisotropic anisotropic elasticity of the plant cell wall. *Wood Science and Technology, 2*(4) 268-78

Cave ID (1969) The longitudinal Young's modulus of *Pinus radiata. Wood Science and Technology, 3*(1) 40-8

Cave ID (1997a) Theory of X-ray measurement of microfibril angle in wood. Part 1. The condition for reflection X-ray diffraction by materials with fibre type symmetry. *Wood Science and Technology, 31*(3) 143-52

Cave ID (1997b) Theory of X-ray measurement of microfibril angle in wood. Part 2. The diffraction diagram X-ray diffraction by materials with fibre type symmetry. *Wood Science and Technology, 31*(4) 225-34

Chapman KM and Jordan PJ (2003) *Practical application of a blowline blending model.* Proceedings of 7th European Panel Products Symposium, Bangor, Wales, 102-10

Choi S, Ruddick JNR and Morris PI (2002) *The copper tolerance of mycelium vs. spores for two brown rot fungi.* IRG Document IRG/WP 02-10422

Chauhan SS (2004) *Selecting and/or processing wood according to its processing characteristics.* PhD thesis, University of Canterbury, Christchurch, New Zealand

Churchland MT (1988) Parallam - a new wood product - commercial process development. In *Timber - A material for the future.* The Marcus Wallenberg Foundation Symposium Proceedings No. 4. Falun, Sweden, *4* 25-42

Clark A and Saucier JR (1989) Influence of initial planting density, geographic location, and species on juvenile wood formation in southern pine. *Forest Products Journal, 39*(7/8) 42-8

Clark PA (1991) Panel products: past, present and future developments. *Journal of the Institute of Wood Science, 12*(4) 233-41

Clark TA and Mackie KL (1987) Steam explosion of the softwood *Pinus radiata* with sulphur dioxide addition: process optimisation. *Journal of Wood Chemistry and Technology, 7*(3) 373-403

Clark TA, Mackie KL, Dare PH and McDonald AG (1989) Steam explosion of the softwood *Pinus radiata* with sulphur dioxide addition: process characterisation. *Journal of Wood Chemistry and Technology, 9*(2) 153-66

Clemons C (2002) Wood plastic composites in the United States: the interfacing of two industries. *Forest Products Journal, 52*(6) 10-8

Clifton NC (1978) Sprinkler storage of windblow proves effective and economic. *Wood World, 19*(12) 26-7

Connell M (1999) Waste management options in the wood protection industry. *Journal of the Institute of Wood Science, 15*(1) 51-8

Cooke PW (1988) *Summary of the new European Community approach to standard development.* Publication 88-3793. National Bureau of Standards, Gaithersburg, Maryland

Cooper PA, Kazi F, Chen J and Ung T (2000) *A fixation model, based on the temperature dependence of CCA-C fixation.* International Research Group on Wood Preservation, Stockholm, Sweden

Cooper PA (2002) Minimizing preservative emissions by post treatment conditioning and fixation of CCA. In *Enhancing the Durability of Lumber and Engineered Wood Products.* Forest Products Society, Madison, Wisconsin, 197-201

Cooper PA (2003). Leaching of CCA - facts and contradictions. *Proceedings, American Wood-Preservers' Association Annual Meeting,* Boston, Maryland, *99*, 73-99

Corson SR and Lloyd JA (1978) Refiner pulp-mill effluent. Part 2. Composition of dissolved solids fraction. *Paperi Ja Puu-Paper and Timber, 60*(8) 435-9

Corson SR (1980) Fibre and fine fractions influence strength of TMP. *Pulp and Paper Magazine of Canada, 81*(5) 67-76 (T108-12)

Côté WA (1967) *Wood ultrastructure: an atlas of electron micrographs.* University of Washing Press, Seattle, Washington

Cown DJ and McConchie DL (1983) *Studies of the intrinsic properties of new-crop radiata pine: wood characteristics of 10 trees from a 24-year-old stand grown in central North Island.* New Zealand Forest Service, Forest Research Institute Bulletin No. 37

Cown DJ, Young GD and Kimberley MO (1991) Spiral grain patterns in plantation-grown *Pinus radiata. New Zealand Journal of Forest Science, 21*(2/3) 206-16

Cown DJ, McConchie DL and Kimberley MO (1992) *New Zealand radiata pine and Douglas fir: suitability for processing.* New Zealand Forest Research Institute, Bulletin No. 168

Cown DJ (1992) Corewood (juvenile wood) in *Pinus radiata* - should we be concerned? *New Zealand Journal of Forestry Science, 22*(1) 87-95

Cragg SM (2003) *Marine boring arthropods: ecology, functional anatomy, and control measures.* In Goodell B, Nicholas DD and Schultz TP (eds), Wood deterioration and preservation: advances in our changing world. ACS Symposium Series 845. American Chemical Society, Washington DC, 272-86

Crawford DM, Woodward BM and Hatfield CA (2002) *Comparison of wood preservatives in stake tests - 2000 Progress Report.* USDA, Forest Service, Forest Products Laboratory, Research Note FPL-RN-02

Creffield JW (1996) *Wood destroying insects: wood borers and termites.* (2nd edit). CSIRO, Melbourne

Cubbage F, MacDonagh P, Báez NM, Siry J, Júnior JS, Ferreira A, Hoeflich V, Ferreira G, Olmos VM, Rubilar R, Alvarez J and Donoso P (2005) *Timber Investment Returns for Plantations and Native Forests in the Americas.* Timber Investments, Forests in the Tropics Conference, Gainesville, Florida, USA. 14-15 February 2005

Dagher HJ, Shaler S, Poulin J, Abdel-Magid B, Tjoelker W and Yeh B (1998) *Ultimate strength of FRP-reinforced glulam beams made with Douglas fir and eastern Hemlock.* Proceedings ICE, January 19-22

Dagher HJ, Shaler SM and Lowry C (1999) *FRP reinforced LVL and structural I-joists.* First International Conference on Advanced Engineered Wood Composites, July 5-8. Bar Harbor, Maine

Dai C, Throughton G and Wang B (2003). Development of a new incising technology for plywood/LVL production. Part I. Incising at the lathe and its effect on veneer quality and recovery. *Forest Products Journal, 53*(3) 73-9

Daniel G (2003) *Microview of wood under degradation by bacteria and fungi.* In Goodell B, Nicholas DD and Shultz TP (eds). Wood deterioration and preservation: advances in our changing world. ACS Symposium Series 845. American Chemical Society, Washington, DC, 34-72

Davidson TC, Newman RH and Ryan MJ (2004) Variations in the fibre repeat between samples of cellulose 1 from different sources. *Carbohydrate Research, 339* 2889-93

de Villiers, AM (1973) *Observations on the timber properties of certain tropical pines grown in South Africa and their improvement by tree breeding.* In Burley J and Nikles DG (ed), Selection and tree breeding to improve some tropical conifers. Commonwealth Forestry Institute, Oxford, Vol. 2 95-115

Delmastro R, Diaz-vaz JE and Schlatte J (1982) *Variabilidad de las características tecnolóficas hereditables del Pinus radiata (D.Don).* CONAF/PNUD/FAO-CHI/76/003, Documento de trabajo No. 43. p. 89

Delmer DP (1999) Cellulose biosynthesis: exciting times for a difficult field of study. *Annual Review of Plant Physiology and Plant Molecular Biology, 50* 245-76

Denig J and Wengert EM (1982) Estimating air-drying moisture content losses for red oak and yellow-poplar lumber. *Forest Products Journal, 32*(2) 26-31

Deyev A and Keey RB (2001) Observation of the 'washboard effect' on the surface of quarter-sawn *Pinus radiata* boards under kiln-drying conditions. *New Zealand Journal of Forestry, 46*(2) 26-33

Dickinson DJ and Murphy RJ (1989) Development of boron based wood preservatives. *Record 1989 Annual Convention of the British Wood Preservers' Association*, 35-42

Dickson RL and Walker JCF (1997) Selecting wood quality characteristics for pines. In *CTIA/IUFRO International Wood Quality Workshop Timber Management toward wood quality and end-product value* Quebec City, Chapter 4 45-52

Diebold JP and Bridgewater AV (1997) Overview of fast pyrolysis of biomass for the production of liquid fuels. In Bridgewater AV and Boocock, DGB (eds), *Developments in thermochemical biomass conversion.* Blackie Academic and Professional, London, Vol. 2 5-23

Diesen M (1998) *Economics of the pulp and paper industry.* Papermaking Science and Technology Series, Vol. 1. Fapet Oy, Helsinki

Dillner B (1989) Modified continuous cooking. *Japan Pulp and Paper Journal, 26*(4) 49-55

Dinwoodie JM (1981) *Timber: its nature and behaviour.* Van Nostrand Reinhold Co., London

Dinwoodie JM (2000) *Timber: its nature and behaviour* (2nd edit). E & FN Spoon, London

Distel DL (2003) *The biology of marine wood boring bivalves and their bacterial endosymbionts.* In Goodell B, Nicholas DD and Schultz TP (eds), Wood deterioration and preservation: advances in our changing world. ACS Symposium Series 845. American Chemical Society, Washington DC, 253-71

Doe PD, Booker JD, Innes TC and Oliver AR (1996) *Optimal lumber seasoning using acoustic emission sensing and real time strain modelling.* Proceedings, 5th IUFRO International Wood Drying Conference, Quebec City, Canada, 209-12

Dohr AW (1953) Mechanical properties of Brazilian Parana pine. *Southern Lumberman, 186*(232,4) 39-42

Donaldson LA (1992) Within- and between-tree variation in microfibril angle in *Pinus radiata. New Zealand Journal of Forest Science, 22*(1) 77-86

Donaldson LA (2001) Lignification and lignin topochemistry: an ultrastructural view. *Phytochemistry, 57* 859-73

Donaldson LA, Wong KKY and Mackie KL (1988) Ultrastructure of steam-exploded wood. *Wood Science and Technology, 22*(2) 103-14

Donnelly RH, Flynn R and Shield E (2003) *The global eucalyptus wood products industry: a progress report on achieving higher value utilization.* DANA Publishing, New Zealand

Downes GM, Hudson IL, Raymond CA, Dean GH, Mitchell AJ, Schimleck LR, Evans R and Muneri A (1997) *Sampling plantation eucalpypts for wood and fibre properties.* CSIRO Publishing, Collingwood, Victoria, Australia

Doyle DV and Markwardt LJ (1966) *Properties of southern pine in relation to strength grading of dimension lumber.* USDA, Forest Service, Forest Products Laboratory, Research Paper, FPL 64

Doyle DV and Markwardt LJ (1967) *Tension parallel-to-grain properties of southern pine dimension lumber.* USDA, Forest Service, Forest Products Laboratory, Research Paper, FPL 84

Doyle DV and Markwardt LJ (1968) *Properties of No. 2 dense kiln-dried southern pine dimension lumber.* USDA, Forest Service, Forest Products Laboratory, Research Paper, FPL 96

Dueholm S. (1996) *Determination of density profiles of wood based panels; from traditional laboratory techniques to continuous, in-line, real-time monitoring.* In Maloney T (ed) Proceedings of 30th Washington State University, Pullman, Washington, 45-57

Eaton RA and Hale MD (1993) *Wood: decay, pests and protection.* Chapman and Hall, London

ECS (1995) *Structural timber: grading-requirements for visual strength grading standards. EN 518.* European Committee for Standardization

ECS (1995) *Grading of structural timber: requirements for machine strength graded timber and grading machines.* European Committee for Standardization

ECS (1995) *Design of timber structures, Eurocode 5 1-1.* European Committee for Standardization

ECS (2003) *Structural timber: strength classes, EN 338.* European Committee for Standardization

ECS (2004) *Structural timber: determination of characteristic values of mechanical properties and density, EN 384.* European Committee for Standardization

Elliott GK (1958) Spiral grain in second-growth Douglas fir and western hemlock. *Forest Products Journal, 8*(7) 205-11

Entwistle K and Walker JCF (eds) (2005) *The Hemicelluloses Workshop 2005.* Wood Technology Research Centre, University of Canterbury, Christchurch, New Zealand

Entwistle K and Walker JCF (2005b) *Workshop assessment.* The Hemicelluloses Workshop 2005, Wood Technology Research Centre, University of Canterbury, Christchurch, New Zealand, 179-190

Epmeier H, Westin M and Rapp A (2004) Differently modified wood: comparison of some selected properties. *Scandinavian Journal of Forest Research, 19 (Supplement 5)* 31-7

Erdtman H (1952) Chemistry of some heartwood constituents of conifers and their physiological and taxonomic significance. In Cook JW (ed), *Progress in Organic Chemistry.* Butterworths Scientific Publishing, London, Vol. 1 22-63

Ericsson K, Huttunen S, Nilsson LJ and Svenningsson P (2004) Bioenergy policy and market development in Finland and Sweden. *Energy Policy, 32* 1707-21

Espenmiller HP (1969) The theory and practice of refining. *Southern Pulp and Paper Manufacturers 32*(4) 50-57

Evans P (2003) Emerging technologies in wood protection. *Forest Products Journal, 53*(1) 14-22

Evans R (1999) A variance approach to the X-ray diffractometric estimation of microfibril angle in wood. *Appita Journal, 52*(4) 283-9

Evans R (2000) Measuring wood and fibre properties. In *Wood Technology Research Centre Workshop 2000*, University of Canterbury, Christchurch, New Zealand, 15-20

Evans R, Stringer S and Kibblewhite RP (2000) Variation in microfibril angle, density and fibre length in twenty-nine *Eucalyptus nitens* trees. *APPITA, 53*(5) 450-7

Evans R and Ilic J (2001) Rapid prediction of wood stiffness from microfibrill angle and density. *Forest Products Journal, 51*(3) 53-7

Fahlén J and Salmén L (2005) Pore and matrix distribution in the fibre wall revealed by atomic force microscopy and image analysis. *Biomacromolecules, 6* 433-8

FAO (1983) *Simple technologies for making charcoal.* FAO, Rome

FAO (2004) *FAO Yearbook: forest products 1998-2002.* FAO, Rome

Färber J, Lichtenegger HC, Reiterer A, Stanzl-Tschegg S and Fratzl P (2001) Cellulose microfibril angles in a spruce branch and mechanical implications. *Journal of Materials Science, 36* 5087-92

Feihl O and Godin V (1967) *Setting veneer lathes with aid of instruments.* Canadian Department for Rural Development, Forestry Branch, Publication No. 1206

Feist WC (1984) *The role of water repellents and chemicals in controlling mildew on wood exposed outdoors*: USDA, Forest Service, Forest Products Laboratory, Research Note FPL 0247

Feist WC and Hon DNS (1984) Chemistry of weathering and protection. In Rowell RM (ed), *The chemistry of solid wood,* ACS Symposium Series 207. American Chemical Society, Washington DC

Feist WC and Mraz EA (1978) *Wood finishing: water repellents and water-repellent preservatives.* USDA, Forest Service, Forest Products Laboratory, Research Note FPL-0124

Feist WC and Tarkow H (1967) A new procedure for measuring fibre saturation points. *Forest Products Journal, 17*(10) 65-8

Fengel D and Wegener G (1984) *Wood: chemistry, ultrastructure, reactions.* De Gruyter, Berlin

Fenton R (1972) Economics of radiata pine for sawlog production. *New Zealand Journal of Forest Science, 2*(3) 313-47

Fielding JM and Dadswell HE (1961) *Not sighted.* 10th Pacific Science Congress, Hawaii

Finighan R and Liversidge RM (1972a) Improving the performance of air seasoning yards: some factors affecting air-flow. *Australian Timber Journal, 38*(3) 12-27

Finighan R and Liversidge RM (1972b) Improving the performance of air seasoning yards: the effect of stack placement pattern. *Australian Timber Journal, 38*(4) 18-23

Finnish ThermoWood Association (2003) *ThermoWood Handbook.* Wood Focus Oy, Helsinki

Fisher K (1972) *Modern flaking and particle reductionizing.* In Maloney T (ed) Proceedings 6th Washington State University Symposium on Particleboard, Pullman, Washington 195-213

Fisher K (1974) *Progress in chip flaking.* In Maloney T (ed) Proceedings of 8th Washington State University Symposium on Particleboard, Washington State University, Pullman, Washington, 361-73

Floyd S (2005) *Effect of hemicellulose on longitudinal shrinkage in wood.* In Entwistle K and Walker JCF (eds), *The Hemicelluloses Workshop 2005.* Wood Technology Research Centre, University of Canterbury, Christchurch, New Zealand, 115-20

Forrer JB and Vermass HF (1987) Development of an improved moisture meter for wood. *Forest Products Journal, 37*(2) 67-71

Freeman MH, Shupe TF, Vlosky RP and Barnes HM (2003) Past, present and future of the wood preservation industry. *Forest Products Journal, 53*(10) 8-15

Froix MF and Nelson R (1975) The interaction of water with cellulose from nuclear magnetic relaxation times. *Macromolecules, 8*(6) 726-30

Fujisaki K (1985) On the relationship between the anatomical features and the wood quality of Sugi cultivars. *Bulletin, Ehime University Forest, 23* 47-58

Fujisawa Y, Ohta S and Tajima M (1993) Wood characteristics and genetic variations in sugi (*Cryptomeria japonica*). Variation in growth ring components among plus tree clones and test stands. *Mokuzai Gakkaishi, 39*(8) 875-82

Fujita M, Takabe T and Harada H (1983) *Deposition of cellulose, hemicelluloses and lignin in the differentiating tracheids.* International symposium on wood and pulping chemistry. Japanese Technical Association of the Pulp and Paper Industry *1*(2) 14-9

Fukazawa K and Imagawa H (ed) (1983) *Ultraviolet and fluorescence microscopic studies of lignin.* International symposium on wood and pulping chemistry. Japanese Technical Association of the Pulp and Paper Industry *1*(2) 14-9

Furuno T, Imamura H and Kajita H (2004) The modification of wood by treatment with low molecular weight phenol-formaldehyde resin: a properties enhancement with neutralized phenolic-resin and resin penetration into wood cell walls. *Wood Science and Technology, 37* 349-61

Gaby LI (1972) *Warping in southern pine studs.* USDA, Forest Service, Southeastern Forest Experiment Station, Research Paper SE-96

Galbraith JE and Sewell WD (1979) Thinning steep country. *Australian Forest Industries Journal, 45*(9) 20-30

Gang DR, Fujita M, Davin LB, and Lewis NG (1996) The abnormal lignins: mapping heartwood formation through the lignan biosynthetic pathway. In Lewis NG and Sarkanen S (eds), *Lignin and lignan biosynthesis* American Chemical Society, Washington, DC, 389-421

Gardner DJ, Tascioglu C and Walinder MEP (2003) Wood composite protection. In Goodell B, Nicholas DD and Schultz TP (eds) *Wood deterioration and preservation: advances in our changing world.* ACS Symposium Series 845. American Chemical Society, Washington DC, 399-419

Gardner JAF and Barton GM (1960) The distribution of dihydroquercetin in Douglas fir and western larch. *Forest Products Journal, 10*(3) 171-3

Gardner JAF and Hillis WE (1962) The influence of extractives on the pulping of wood. In Hillis WE (ed), *Wood extractives and their significance to the pulp and paper industries.* Academic Press, New York, 367-403

Gardner KH and Blackwell J (1974) Structure of Native Cellulose. *Biopolymers, 13*(10) 1975-2001

Gatenholm P and Tenkanen M (ed) (2002) *Hemicelluloses: science and technology*. ACS Symposium Series 864. American Chemical Society, Washington DC

Gaunt D (1998) If you are not winning, change the rules. *New Zealand Forest Research, Wood Processing Newletter (Sawmilling)* (23) 3

Gierer J (1980) Chemical aspects of kraft pulping. *Wood Science and Technology, 14*(4) 241-66

Gilmour IA and Walker JCF (1995) Efficiencies and emissions of woodburners. *Clean Air (Australia), 29*(1) 31-9

Girallt R, Rihouey BF, Jauneau A, Norvan C and Jarvis M (1997) Galactans and cellulose in flax fibres: putative contributions to the tensile strength. *International Journal of Biological Macromolecules, 21* 179-88

Givnish TJ (1995) Plant stems: biomechanical adaptation for energy capture and influence on species distributions. In Gartner BL (ed), *Plant stems: physiology and functional morphology*. Academic Press, San Diego 3-49

Goodell B, Nicholas DD and Schultz TP (2003) Introduction to wood deterioration and preservation. In Goodell B, Nicholas DD and Schultz TP (ed), *Wood deterioration and preservation: advances in our changing world.* ACS Symposium Series 845. American Chemical Society, Washington DC, 2-7

Gordon JE (1978) *Structures, or why things don't fall down.* Plenum Press, London

Gordon JE and Jeronimidis G (1980) Composites with high work of fracture. *Philosophical Transactions of the Royal Society of London, A 294* 545-50

Goring DAI (1965) *Thermal softening, adhesive properties and glass transitions in lignin, hemicellulose and cellulose.* Consolidation of the paper web: Transactions of the 3rd Fundamental Research Symposium, Cambridge. Technical Section of British Paper Board Makers' Association, London, 555-75

Goring DAI, Vuong R, Gancet C and Chanzy H (1979) Flatness of Lignosulfonate Macromolecules as Demonstrated by Electron-Microscopy. *Journal of Applied Polymer Science, 24*(4) 931-6

Goring DAI (1983) *Some recent topics in wood and pulping chemistry.* Japanese Technical Association of the Pulp and Paper Industry *1*(2) 3-13

Grabianowski M, Manley B and Walker J (2004) Impact of stocking and exposure on outerwood acoustic properties of *Pinus radiata* in Eyrewell Forest. *New Zealand Journal of Forestry, 49*(2) 13-7

Gran G (1982) *Blowline blending in dry process fibreboard production.* In Maloney T (ed) Proceedings of the 16th Washington State University Particleboard Symposium, Pullman, Washington, 261-7

Green DW and Evans JW (1988) *Mechanical properties of visually graded dimension lumber.* Mechanical properties of visually graded dimension lumber. National Technical Information Service, Springfield Virginia, Vols. 1–7, PB–88–159–371

Green, DW (1989) *Moisture content and the shrinkage of lumber.* USDA, Forest Service, Forest Products Laboratory, Research Paper, FPL RP 489

Green DW and Evans JW (1989) Moisture content and the mechanical properties of dimension lumber: decisions for the future. In *Proceedings of the in-grade testing of structural lumber.* Forest Products Society, Madison, Wisconsin, 44-55

Green DW, Shelley BE and Vokey HP (eds) (1989) *Proceedings of the in-grade testing of structural lumber.* Forest Products Society, Madison, Wisconsin

Green DW, Winandy JE and Kretschmann DE (1999) Mechanical properties of wood. In USDA, (1999) Agricultural Handbook No 72 *Wood handbook—Wood as an engineering material.* Chapter 4 1-45

Green DW and Evans JW (2003a) Durability of structural lumber products at high temperatures. Part I. 66°C at 75% RH and 82°C at 30% RH. *Wood and Fiber Science, 35*(4) 499-523

Green DW and Evans JW (2003b) Effect of low relative humidity on the properties of structural lumber products. *Wood and Fiber Science, 35*(2) 247-6

Guan S-Y, Mlynar J and Sarkanen S (1997) Dehyrogenative polymerization of coniferyl alcohol on macromolecular lignin templates. *Phytochemistry, 45*(5) 911-8

Guitard D, Masse HH, Yamamoto H and Okuyama T (1999) Growth stress generation: a new mechanical model of dimensional change of wood cells during maturation. *Journal of Wood Science, 45*(5) 384-91

Gullichsen J and Fogelholm C-J (2000) *Chemical pulping*. Papermaking Science and Technology, Vol. 6A. Fapet Oy, Helsinki

Guthrie RD (1974) *Guthrie and Honeyman's introduction to carbohydrate chemistry*. Clarendon Press, Oxford

Håfors B (1990) The role of the *Vasa* in the development of the polyethylene glycol preservation method. In Rowell RM and Barbour RJ (eds), *Archaeological wood: properties, chemistry and preservation,* Advances in Chemistry Series 225. American Chemical Society, Washington DC, 195-216

Hajny GJ (1981) *Biological utilization of wood for production of chemicals and foodstuffs*. USDA, Forest Service, Forest Products Laboratory, Research Paper FPL 385

Hall FD (1983) *The Kockums CanCar approach to sawmill design*. Kockums CanCar, Surrey, British Columbia

Hallock H and Lewis DW (1971) *Increasing softwood dimension yield from small logs: best opening face*. USDA, Forest Service, Forest Products Laboratory, Research Paper, FPL 166

Hallock H (1973) Best opening face for second-growth timber. In *Modern Sawmill Techniques Vol. 1*. Miller Freeman, San Francisco, 93-116

Hallock H, Stern AR and Lewis DW (1976) *Is there a 'best' sawing method?* USDA, Forest Service, Forest Products Laboratory, Research Paper FPL 280

Hanley SJ, Revol J-F, Godbout L and Gray DG (1997) Atomic force microscopy and transmission electron microscopy of cellulose from *Micrasterias dendiculata*; evidence for a chiral helical microfibril twist. *Cellulose, 4*, 209-20

Hänninen E and Ahonen A (1986) *Utilization of energy excess of pulp mills*. 22nd EUCEPA Conference: development and trends in the science and technology of pulp and paper making, Florence, Vol. 2 1-19

Hansen BG and Araman PA (1985) *Low-cost opportunity for small-scale manufacture of hardwood balks*. USDA, Forest Service, Northeastern Forest Experiment Station, Research Paper NE-559

Harada H and Côté WA (1985) Structure of wood. In Higuchi T (ed), *Biosynthesis and biodegradation of wood components*. Academic Press, London, 1-42

Harrington JJ (2002) *Hierarchical modelling of softwood hygro-elastic properties*. Ph.D thesis, University of Canterbury, Christchurch, New Zealand

Harris JF (1976) Acid hydrolysis and dehydration reactions for utilizing plant carbohydrates. In Timell TE (ed), *Proceedings of 8th cellulose conference*. Wiley, New York, Vol. 1 131-44

Harris JM (1961) *The dimensional stability, shrinkage intersection point and related properties of New Zealand timbers*. New Zealand Forest Service, Forest Research Institute, Technical Paper No. 36

Harris, JM (1969) The use of beta rays in determining wood properties. Part 2, Measuring earlywood and latewood. *New Zealand Journal of Science, 12*(2) 409-18

Harris JM (1973) *The use of beta rays to examine wood density of tropical pines in Malaysia.* In Burley J and Nikles DG (eds), Selection and tree breeding to improve some tropical conifers. Commonwealth Forestry Institute, Oxford, Vol. 2 86-94

Harris JM (1989) *Spiral Grain and Wave Phenomena in Wood Formation.* Springer-Verlag, Berlin

Harris JM (1991) Formation of wood and bark. In Kininmonth JF and Whitehouse LJ (eds), *Properties and uses of New Zealand radiata pine.* Forest Research Institute, Rotorua, New Zealand, Chapter 3 1-18

Harris JM and Cown DJ (1991) Basic wood properties. In Kininmonth JM and Whitehouse LJ (eds), *Properties and uses of New Zealand radiata pine.* Ministry of Forestry, New Zealand, Chapter 6 1-28

Harris JM (1993) Wood quality: forest management and utilization. In Walker JCF, *Primary wood processing: principles and practice* (1st edit). Chapman and Hall, London, 560-83

Harris PJ (2005) *Non-cellulosic polysaccharides in plant cell walls.* In Entwistle KM and Walker JCF (eds), The Hemicelluloses Workshop 2005, Wood Technology Research Centre, University of Canterbury, Christchurch, New Zealand, 13-35

Harris RA and Taras MA (1984) Comparison of moisture content distribution, stress distribution, and shrinkage of red oak lumber dried by a radio-frequency/vacuum drying process and a conventional kiln. *Forest Products Journal, 34*(1) 44-54

Hartler N and Stade Y (1962) [Penetration and diffusion in sulphate cooking (in Swedish)]. *Paperi Ja Puu-Paper and Timber, 44*(7) 365-74

Hartler N and Stade Y (1977) Chipper operation for improved chip quality. *Svensk Papperstidn, 80*(14) 447-53

Hartley ID and Avramidis S (1993) Analysis of the wood sorption isotherm using clustering theory. *Holzforschung, 47*(2) 163-7

Haslett AN (1988) *A guide to handling and grade-sawing plantation-grown eucalypts.* New Zealand Ministry of Forests, Forest Research Institute Bulletin 142

Hathway DE (1962) The use of hydroxystilbene compounds as taxomic tracers in the genus Eucalyptus. *Biochemical Journal, 83* 80-4

Helsen L and Van den Bulck E (2005) Review of disposal technologies for chromated copper arsenate (CCA) treated wood waste, with detailed analyses of thermochemical conversion processes. *Environmental Pollution, 134* 301-14

Hemmingway RW (1969) *Thermal instability of fats relative to surface wettability of yellow birchwood (Betula lutea).* Proceedings of a meeting of IUFRO (Section 41, For. Prod.), Melbourne. Division of Forest Products Research, CSIRO, Australia

Hieta K, Kuga S and Usuda M (1984) Electron staining of reducing ends evidences a parallel-chain structure in *Valonia* cellulose. *Biopolymers, 23*(10), 1807-10

Highley TL (1999) Biodeterioration of wood. In *Wood handbook - wood as an engineering material.* USDA, Agricultural Handbook No. 72 Chapter 13 1-16

Hill CAS and Jones D (1996) The dimensional stabilisation of Corsican pine sapwood by reaction with carboxylic acid anhydrides: the effect of chain length. *Holzforschung, 50*(5) 457-62

Hill CAS and Jones D (2006) *Wood modification: chemical, thermal and other processes.* John Wiley & Sons, New York

Hillier B and Murphy R (2000). Life cycle assessment of forest products - a good story to tell. *Journal of the Institute of Wood Science, 15*(4) 221-32

Hillis WE (ed) (1962) *Wood extractives and their significance to the pulp and paper industries.* Academic Press, New York

Hillis WE (1978) Wood quality and utilisation. In Hillis WE and Brown AG (ed), *Eucalypts for wood production.* CSIRO, Australia, 259-89

Hillis WE (1987) *Heartwood and tree exudates.* Springer-Verlag, Berlin

Hirakawa Y and Fujisawa Y (1995) The relationships between microfibril angles in the S_2 layer and latewood tracheid lengths in elite sugi trees (*Cryptomeria japonica*) clones. *Mokuzai Gakkaishi, 41*(2) 123-31

HMSO (1969) *The natural durability classification of timber*. HMSO, London

Hoadley RB (1979) *Effect of temperature, humidity and moisture content on solid wood products and end use.* In Proceedings of wood moisture content: temperature and humidity relationships symposium, Blacksburg. USDA, Forest Service, Forest Products Laboratory, North Central Experiment Station, Raleigh, North Carolina, 92-6

Hofbauer H, Veronik G, Fleck T, Rauch R, Mackinger H and Fercher E (1997) The FICFB gasification process. In Bridgewater AV and Boocock, DGB (eds), *Developments in thermochemical biomass conversion.* Blackie Academic and Professional, London, Vol. 2 1016-25

Hofbauer H, Rauch R, Loeffler G, Kaiser S, Fercher E and Tremmel H (2002) *Six years experience with the FICFB-gassification process.* 12th European Conference and Technology Exhibition on Biomass for Energy, Industry and Climate Protection, Amsterdam, The Netherlands

Höglund E (1980) Economic and technical developments in the wood-based industries: plywood. *Finnish Paper and Timber Journal* (May), 201-5

Hon DN-S and Shiraishi N (2001) *Wood and cellulosic chemistry* (2nd edit). Marcel Dekker. New York

Hon DN-S (2001) Weathering and photochemistry of wood. In Hon DN-S and Shiraishi N (eds), *Wood and cellulosic chemistry.* Marcel Dekker, New York, 513-46

Horakawa Y and Fujisawa Y (1995) The relationships between microfibril angles of the S_2 layer and latewood tracheid lengths in elite sugi tree (*Cryptomeria japonica*) clones. *Mokuzai Gakkaishi, 41*(2) 123-31

Hsu WE (1987) *A process for stabilizing OSB/waferboard.* In Maloney T (ed). Proceedings of the 21st Washington State University Particleboard Symposium, Pullman, Washington, 201-36

Huang C-L, Lindström H, Nakada R and Ralston J (2003) Cell wall structure and wood properties determined by acoustics - a selective review. *Holz als Roh- und Werkstoff, 61* 321-35

Huang L-H, Qin T-F, and Ohira Tatsuro (2004) Studies on preparations and analysis of essential oil from Chinese fir. *Journal of Forestry Research, 15*(1) 80-2

Hudson MS and Henriksson ST (1956) The oscillation pressure method of wood impregnation. *Forest Products Journal, 6*(10) 381-6

Hughes C (1982) *The natural durability of untreated timbers*. What's New in Forest Research No. 112. New Zealand Ministry of Forestry, Forest Research Institute, Rotorua

Humphrey PE and Bolton AJ (1989) The hot pressing of dry-formed wood-based composites. Part 2: A simulation model for heat and moisture transfer, and typical results. *Holzforschung, 43*(3) 199-206

Hunt GM and Garratt GA (1967) *Wood Preservation* (3rd edit). The American Forestry Series, McGraw-Hill, New York

Hutchings BF and Leicester RH (1988) *Scrimber.* Proceedings of 1988 International Conference of Timber Engineers, Seattle. Forest Products Society, Madison, Wisconsin, Vol. 2 525-33

Ibach RE (1999) Wood preservation. In *Wood handbook - wood as an engineering material.* USDA, Agricultural Handbook No. 72, Chapter 13 1-27

Imré L (1984) Aspects of solar drying. In Mujumdar AS (ed), *Drying '84.* Hemisphere, Washington, D.C, 43-50company

International Patent (2000) *Method for determining warp potential in wood.* Weyerhaeuser. World Intellectual Property Organisation, International Bureau, Geneva. WO 00/12230

Ivory EP, White DG and Upson AT (1923) *Standard grading specifications for yard lumber.* USDA, Department Circular 296, Washington, DC

Jager R (1975) *Preparation of quality particles from low quality materials.* In Maloney T (ed) Proceedings of 5th Washington State University Symposium on Particleboard, Pullman, Washington, 210-7

James WL (1988) *Electrical moisture meters for wood.* USDA, Forest Service, Forest Products Laboratory, General Technical Reportt FPL GTR-6

Jane FW (1956) *The structure of wood.* A. & C. Black, London

Jarck W (2003) *From Scrimber to TimTek.* In Maloney T (ed) Proceedings 37th International Wood Composite Materials Symposium, University of Washington, Pullman, Washington, 61-64

Jarvis MC (1998) Intercellular separation forces generated by intracellular pressure. *Plant, cell and environment, 21* 1307-10

Jarvis MC (2003) Cellulose stacks up. *Nature, 426* 611-2

Jarvis MC (2005a) *Cell adhesion and the middle lamella.* In Entwistle KM and Walker JCF (eds), The Hemicelluloses Workshop 2005, Wood Technology Research Centre, University of Canterbury, Christchurch, New Zealand, 56-67

Jarvis MC (2005b) *Cellulose structure and hemicellulose-cellulose association.* In Entwistle KM and Walker JCF (eds), The Hemicelluloses Workshop 2005, Wood Technology Research Centre, University of Canterbury, Christchurch, New Zealand, 87-102

JAS (1997) *Japanese agriculture standard for structural softwood lumber. JAS 143.* Japanese Ministry of Agriculture, Forestry and Fisheries, Tokyo

Jeffries TW (1987) Physical, chemical and biochemical considerations in the biological degradation of wood. In Kennedy JF, Phillips GO and Williams PA (eds), *Wood and cellulosics: industrial utilization, biochemistry, structure and properties.* Ellis Horwood Ltd., Chichester, England, 213-30

Johansson B, Mjoberg J, Sandstrom P and Teder A (1984) Modified continuous kraft pulping - now a reality. *Svensk Papperstidn, 87*(10) 30-5

Johnson BR and Gonzalez GE (1976) Experimental preservative treatment of three tropical hardwoods by double-diffusion processes. *Forest Products Journal, 26*(1) 39-46

Johnson JW and Kunesh RH (1975) Tensile strength of special Douglas-fir and Hem-Fir 2-inch dimension lumber. *Wood and Fiber, 6*(4) 305-18

Jozsa LA and Middleton GR (1994) *A discussion of wood quality attributes and their practical implications.* Special publication No. SP-34. Forintek Canada Corporation, Western Laboratory, Vancouver, BC

Jurasek L (1996) Experimenting with virtual lignin. In Lewis NG and Sarkanen S (eds), *Lignin and lignan biosynthesis.* American Chemical Society, Washington, DC, 276-93

Jurasek L (1998) Molecular modelling of fibre walls. *Journal of Pulp and Paper Science, 24*(7) 209-12

Kai Y (1991) Chemistry of extractives. In Hon DN-S and Shiraishi N (eds) *Wood and cellulosic chemistry* (2nd edit). Marcel Dekker, New York, 215-55

Kamke FA and Vanek M (1994) *Comparison of wood drying models.* Proceedings of the 4th IUFRO International Wood Drying Conference, Rotorua, New Zealand, 1-21

Karonen A (1985) *Scanning systems for automated lumber edging and trimming.* In First International Conference on Scanning Technology in Sawmilling, San Francisco. Industries/World Wood, Paper IX p 9

Kartal SN and Clausen CA (2001) Leachability and decay resistance of particleboard made from acid extracted and bioremediated CCA-treated wood. *International Journal of Biodeterioration and Biodegradation, 47* 18391

Kazi KMF and Cooper PA (1998) Solvent extraction of CCA-C from out of service wood. International Research Group on Wood Preservation Document 93-30024

Keckes J, Burgert I, Fruhmann K, Muller M, Kolln K, Hamilton M, Burghammer M, Roth SV, Stanzl-Tschegg S and Fratzl P (2003) Cell-wall recovery after irreversible deformation of wood. *Nature Materials, 2* 810-4

Kelly SS, Rials TG and Glasser WG (1987) Relaxation behaviour of the amorphous components of wood. *Journal of Materials Science, 22* 617-24

Kelsey KE and Clark LE (1956) The heat of sorption of water by wood. *Australian Journal of Applied Science, 7*(2) 160-75

Kerr AJ and Goring DAI (1975) The ultrastructural arrangement of the wood cell wall. *Cellulose chemistry and technology, 9* 563-73

Kerr AJ and Uprichard JM (1976) The kinetics of kraft pulping: refinement of a mathematical model. *APPITA, 30*(1) 48-54

Kho PCS, Keey RB and Walker JCF (1989) *Effects of minor board irregularities and air flows on the drying rate of softwood timber boards in kilns.* Proceedings 2nd IUFRO Wood Drying Symposium, Seattle, Washington, DC, 150-7

Kho PCS (1993) *Mass transfer from in-line slabs: application to high-temperature kiln drying of softwood timber boards.* PhD thesis, University of Canterbury, Christchurch, New Zealand

Kibblewhite RP (1984) *Web formation, consolidation and water removal and their relationships with fibre and wet web properties and behaviour.* Fundamentals of paper performance, Technical Association of Australian and New Zealand Pulp and Paper Industry, Vol. 3 1-12

Kibblewhite RP and Okayama T (1986) Some unique properties of neutral sulfite-anthraquinone pulp fibers. *Appita, 39*(2) 134-8

Kininmonth JM and Whiteside ID (1991) Log quality. In Kininmonth JM and Whitehouse LJ (eds), *Properties and uses of New Zealand radiata pine.* New Zealand Ministry of Forestry, Forest Research Institute, Chapter 5 1-21

Kleppe PJ (1970) Kraft pulping. *TAPPI, 53*(1) 35-47

Knowles RL, Hanson LW, Wedding A and Downes G (2004) Evaluation of non-destructive methods for assessing stiffness of Douglas-fir trees. *New Zealand Journal of Forestry Science, 34*(1) 87-101

Koch P (1964) *Wood Machining Processes.* Ronald Press, New York

Koning JW and Skog KE (1987) Use of wood energy in the United States: an opportunity. *Biomass, 12*, 27-36

Koshijima T and Watanabe T (2003) *Association between lignin and carbohydrates in wood and other plant tissues.* Springer-Verlag, Berlin

Kretchmann DE (2003) Velcro mechanics in wood. *Nature Materials, 2* 775-6

Kretschmann DE and Green DW (1996) Modeling moisture content-mechanical property relationships for clear Southern Pine. *Wood and Fiber Science, 28*(3) 320-37

Kubler H (1987) Growth stresses in trees and related wood properties. *Forestry Abstracts, 48*(3) 131-89

Kumar S, Ball R and Cown D (2001) Analysis of wood density and spiral grain at seven contrasting sites. *New Zealand Radiata Pine Breeding Cooperative,* Unpublished report

Kunesh RH and Johnson JW (1972) Effect of single knots on strength of 2 by 8-inch Douglas-fir dimension lumber. *Forest Products Journal, 22*(1) 32-6

Kunesh RH and Johnson JW (1974) Effect of size on tensile strength of clear Douglas-fir and Hem-Fir dimension lumber. *Forest Products Journal, 24*(8) 32-6

Kuo ML and Arganbright DC (1980) Cellular distribution of extractives in redwood and incense cedar: Part 1, radial variation in cell wall extractive content. *Holzforschung, 34*(1) 17-22

Laks PE (2004) *Protection of wood-based composites.* Proceeding of the American Wood Preservers' Association Annual Meeting, *100* 78-82

Lande S, Eikenes M and Westin M (2004a) Chemistry and ecotoxicology of furfurylated wood. *Scandinavian Journal of Forest Research, 19 (Supplement 5)* 14-21

Lande S, Westin M and Schneider M (2004b) Properties of furfurylated wood. *Scandinavian Journal of Forest Research, 19 (Supplement 5)* 22-30

Langmuir I (1918) The adsorption of gases on plane surfaces of glass, mica and platinum. *Journal of the American Chemical Society, 40* 1361-403

Langrish TAG, Keey RB, Kho PCS and Walker JCF (1993) Time-dependent flow in arrays of timber boards: flow visualisation, mass-transfer coefficients and numerical simulation. *Chemical Engineering Science, 48*(12) 2211-23

Langrish TAG, Brooke AS, Davis CL, Musch HE and Barton GW (1997) An improved drying schedule for Australian ironbark timber: optimisation and experimental validation. *Drying Technology, 15*(1) 47-70

Larsson PT (2003) Interaction between cellulose 1 and hemicelluloses studied by spectral fitting of CP/MAS ^{13}C-NMR spectra. In Gatenholm P and Tenkanen M (eds), *Hemicelluloses: science and technology*, Advances in Chemistry Series 864. American Chemical Society, Washington, DC, 254-68

Lasserre J-P, Mason E and Watt M (2004) The influence of initial stocking on corewood stiffness in a clonal experiment of 11 yr-old *Pinus radiata* D.Don. *New Zealand Journal of Forestry, 49*(2) 18-23

Laufenberg TL, Rowlands RE and Krueger GP (1984) Economic feasibility of synthetic fibre reinforced laminated veneer lumber. *Forest Products Journal, 34*(4) 15-22

Lavers GM (1969). *The strength properties of timbers.* Princes Risborough Laboratory Bulletin 50, HMSO, London

Lavers GM (1974). The strength properties of timbers. In *The strength properties of timbers.* Building Research Establishment, MTP Construction, 3-66

Lebow ST, Brooks KM and Simonsen J (2002) Environmental impact of treated wood in service. In *Enhancing the durability of lumber and engineered wood products.* Forest Products Society, Madison, Wisconsin, 205-15

Lebow ST and Morrell JJ (1995) Interactions of ammoniacal copper zinc arsenate (ACZA) with Douglas fir. *Wood and Fiber Science, 27*(2) 105-18

Lebow ST and Tippie M (2001) *Guide for minimizing the effect of preservative-treated wood on sensitive environments.* USDA, Forest Service, Forest Products Laboratory, General Technical Report FPL-GTR-122

Lee HL, Chen GC and Rowell RM (2004) Fungal decay resistance of wood reacted with phosphorus pentoxide-amine system. *Holzforschung, 58* 311-5

Lee HS (1990) *Flow visualization on high-temperature wood drying.* B.E. (Chemical and Process) Report, University of Canterbury, Christchurch, New Zealand

Lehman WF (1977) *Durability of composition board products.* In Maloney T (ed) Proceedings of the 11th Washington State University Particleboard Symposium, Pullman, Washington, 351-68

Lehman WF (1986) *Outline of a fast durability test for UF and PF adhesives in composite materials.* In Maloney T (ed) Proceedings of the 20th Washington State University Particleboard Symposium, Pullman, Washington, 105-22

Leicester RH (1988) *Timber engineering standards for tropical countries.* Proceedings of 1988 International Conference of Timber Engineers, Seattle. Forest Products Society, Madison, Wisconsin, Vol. 1 177-85

Lenz M (2002) Termite problem species and management of termite problems in Australia. *Sociobiology 40*(1), 11-2

Lenz M and Runko S (1993) *Protection of buildings, other structures and materials in ground contact from attack by subterranean termites with a physical barrier – a fine mesh of high grade stainless steel.* IRG Document IRG/WP 93-10014

Lewin M and Roldan LG (1975) Oxidation and alkaline-degradation of mercerized cotton - morphological study. *Textile Research Journal, 45*(4) 308-14

Lewis NG and Davin LB (1996) The biochemical control of monolignol coupling and structure during lignan and lignin biosynthesis. In Lewis NG and Sarkanen S (eds), *Lignin and lignan biosynthesis*. American Chemical Society, Washington, DC, 334-61

Li J and Pang S (2005) *Evaluation of BIGCC technologies developed oversea.* Confidential Report, University of Canterbury, Christchurch, New Zealand

Li T-Q, Henriksson U and Odberg L (1995) Determination of pore volume in cellulose fibres by the pulsed gradient spin-echo NMR technique. *Journal of Colloid and Interface Science, 169* 376-9

Lichtenegger H, Reiterer A, Stanzl-Tschegg SE and Fratzl P (1999) Variation of cellulose microfibril angles in softwoods and hardwoods: a possible strategy of mechanical optimization. *Journal of Structural Biology, 128* 257-69

Liese W (1984) Wet storage of windblown conifers in Germany. *New Zealand Journal of Forestry, 29*(1) 119-36

Lima JT, Breese MC and Cahalan CM (2004) Variation in microfibril angle in eucalyptus clones. *Holzforschung, 58* 160-6

Lindström H, Harris P, Sorensson CT and Evans R (2004) Stiffness and wood variation in 3-year old *Pinus radiata* clones. *Wood Science and Technology, 38* 579-97

Lindström T (1986) The concept and measurement of fiber swelling. In Bristow AJ and Kolsseth P (eds), *Paper: structure and properties.* Marcel Dekker, New York, 75-97

Littleford TW (1978) *Flexural properties of dimension lumber from western Canada. VP-X-179.* Environment Canada, Forestry Directorate and Western Forest Products Laboratory, Vancouver, BC

Littleford TW and Abbott RA (1978) *Parallel-to-grain compressive properties of dimension lumber from western Canada. VP-X-180.* Environment Canada, Forestry Directorate and Western Forest Products Laboratory, Vancouver, BC

Lloyd JA (1978) Distribution of extractives in *Pinus radiata* earlywood and latewood. *New Zealand Journal of Forestry Science, 8*(2) 288-94

Loehnertz SP (1988) A continuous press dryer for veneer. *Forest Products Journal, 38*(9), 61-3

Lu Xi-Xian, Wang De-Long and Zhou Ming (1987) Influence of the extractives of Chinese fir wood upon their natural resistance to fungus and termite damage (In Chinese). *Scientia Silvae Sinicae, 22*(4) 454-62

Lumiainen JJ (1990) New theory can improve practice. *Pulp and Paper International, 32*(7) 46

Lunstrum S (1982) What have we learned from the sawmill improvement program after nine years? *Southern Lumberman, 234*(12) 42-4

Lutz JF (1971) *Wood and log characteristics affecting veneer production.* USDA, Forest Service, Forest Products Laboratory, Research Paper FPL 150

Lutz JF (1974) *Techniques for peeling, slicing, and drying veneer.* USDA, Forest Service, Forest Products Laboratory, Research Paper FPL 228

Macarthur RS (1952) *The influence of initial stand density on growth, increment and timber quality of Pinus radiata.* Internal report of the Forest Research Institute, Rotorua, New Zealand

Macdonald E and Hubert J (2002) A review of the effects of silviculture on timber quality of Sitka spruce. *Forestry, 75*(2) 107-38

MacLean JD (1952) Effect of temperature on the dimensions of green wood. *Proceeding of the American Wood Preservers' Association, 48*, 136-57

Madsen B (1975) *Moisture content-strength relationship for lumber subjected to bending.* Structural Research Service Report No. 11. Department of Civil Engineering, University of British Columbia, Vancouver BC

Madsen B (1976) *In-grade testing: size investigation on lumber subjected to bending.* Structural Research Service Report No. 15. Dept. of Civil Engineering, University of British Columbia, Vancouver, BC

Madsen B (1978) In-grade testing: problem analysis. *Forest Products Journal, 28*(4) 42-50

Madsen B and Buchanan AH (1986) Size effect in timber explained by a modified weakest link theory. *Canadian Journal of Civil Engineering, 2* 218-32

Maeglin RR and Boone RS (1983) Manufacture of quality yellow poplar studs using the saw-dry-rip (S-D-R) concept. *Forest Products Journal, 33*(3) 10-8

Maeglin RR and Boone RS (1988) *Saw-dry-rip improves quality of random-length yellow-poplar 2 by 4's.* USDA, Forest Service, Forest Products Laboratory, Research Paper FPL RP-490

Mandal AK and Chawan PH (2003) *Investigations on inheritance of growth and wood properties and their interrelationships in teak.* International conference on quality timber products of teak from sustainable forest management. Kerala Forest Research Institute, Peechi, Kerala, India

Marchessault RH and Sundararajan PR (eds) (1983). *Cellulose* Vol. 2. Academic Press, New York

Markstrom DC and Gjovik LR (1999) *Service life of fence posts treated by double-diffusion methods.* USDA, Forest Service, Rocky Mountain Research Station, Research Paper RMRS-RP-17

Marszalek PE, Oberhauser AF, Pang YP and Fernandez JM (1998) Polysaccharide elasticity governed by chair-boat transitions of the glucopyranose ring. *Nature, 396* 661-4

Marx-Figini M (1969) *On the biosynthesis of cellulose in higher and lower plants.* In Marchessault RH (ed), Proceedings of 6th cellulose conference. Wiley, New York

Mason DC (ed) (1975) *Sawmill Techniques for Southeast Asia.* Miller Freeman, San Francisco

Mason WH (1926) US Patent 1578609

Maxwell JW Gran G and Waters GD (1984) *Experiments with blowline blending for MDF.* In Maloney T (ed), Proceedings of the 18th Washington State University Particleboard Symposium, Pullman, Washington, 117-43

McCarthy ET and Smart WA (1979) Computerized veneer drying. *Modern Plywoood Techniques, 7* 62-9

McDaniels DK (1984) *The sun, our future energy source* (2nd. edit). J. Wiley, New York

McKimmy MD (1986) The genetic potential for improving wood quality. In Oliver CD, Hanley DP and Johnson JA (eds), *Douglas fir: stand management for the future.* College of Forest Resources, University of Washington, 118-22

McMillen JM (1955) Drying stresses in red oak. *Forest Products Journal, 5*(2) 71-6

Megraw RA (1985) *Wood quality factors in loblolly pine.* TAPPI, Atlanta

Megraw RA (1986) Douglas fir wood properties. In Oliver CD, Hanley DP and Johnson JA (eds), *Douglas fir: stand management for the future.* College of Forest Resources, University of Washington, 81-96

Megraw RA, Leaf G and Bremer D (1998) Longitudinal shrinkage and microfibril angle in loblolly pine. In Butterfield BG (ed), *Proceeding of the IAWA/IUFRO Workshop on Microfibril angle in wood.* University of Canterbury, Christchurch, New Zealand, 27-61

Meier H (1985) Localization of polysaccharides in wood cell walls. In Higuchi T (ed), *Biosynthesis and biodegradation of wood components.* Academic Press, Orlando, 43-50

Merchant MV (1957) A study of water-swollen cellulose fibres which have been liquid-exchanged and dried from hydrocarbons. *TAPPI, 33*(8) 771-81

Methol R (2001) *Comparisons of approaches to modelling tree taper, stand structure and stand dynamics in forest plantations.* PhD thesis, University of Canterbury, Christchurch, New Zealand

Methol R (2002) *INIA DSS growth and yield simulation model for Uruguay, [using] industry forestry regimes and industry cost/price estimates.* Unpublished report

Meyer JA (1984) Wood-polymer materials. In Rowell RM (ed), *The Chemistry of solid wood.* Advances in Chemistry Series 207. American Chemical Society, Washington DC, 257-89

Meylan BA (1968) Cause of high longitudinal shrinkage in wood. *Forest Products Journal, 18*(4) 75-8

Miles KB and May WD (1990) The flow of pulp in chip refiners. *Journal of Pulp and Paper Science, 16*(2) J60-72

Miller DG (1971) Combining radio-frequency heating with kiln-drying to provide fast drying without degrade. *Forest Products Journal, 21*(12) 17-21

Miltz H (2002) Heat treatment technologies in Europe: scientific background and technological state of the art. In *Enhancing the durability of lumber and engineered wood products.* Forest Products Society, Madison, Wisconsin, 239-49

Minor JL (1983) *Chemical linkage of polysaccharides to residual lignin in pine kraft pulps.* International symposium on wood and pulping chemistry. Japanese Technical Association of the Pulp and Paper Industry, Vol. 1 153-8

Morrell JJ, Love CS and Freitag CM (1996) Integrated remedial protection of wood in bridges. In Ritter MA, Duwadi SR and Lee PDH (eds), *National conference on wood transportation structures.* USDA, Forest Service, Forest Products Laboratory, General Technical Report FPL- GTR-94

Morrell JJ, Freitag CM and Chen H (2005) Sequential treatments with fluoride and copper: effects of solution concentration and dipping time on treatment. *Forest Products Journal, 55*(7/8) 57-62

Morrell JJ, Freitag CM and Chen H (1995) Use category: a proposal for a new approach to treating standards. *Proceedings of the American Wood Preservers' Association, Annual Meeting, 91* 63-81

Morris S, Hanna S and Miles MJ (2004) The self-assembly of plant wall components by single force spectroscopy and Monte Carlo modelling. *Nanotechnology, 15*, 1296-301

Moslemi A (ed) (1991) *Inorganic bonded wood and fiber composite materials.* Forest Products Society, Madison, Wisconsin

Muhlethaler K (1969) *Fine structure of natural polysaccharide systems.* In Marchessault RH (ed), Proceedings of 6th cellulose conference. Wiley, New York, 57-67

Murphy RJ, Barnes HM and Dickinson DJ (2002) Vapor boron technology. In *Enhancing the durability of lumber and engineered wood products.* Forest Products Society, Madison, Wisconsin, 251-6

Murton K (1998) *The effect of preheating and refining conditions on radiata pine MDF fibre quality.* Proceedings of the 52nd Appita Annual General Meeting, Vol. 2 493-500

Nanassy AJ (1974) Water sorption in green and remoistened wood studied by the broad-line component of the wide-line NMR spectra. *Wood Science, 7*(1) 61-8

Natus G (1996) *12 foot wide OSB lines: a successful step forward.* In Maloney T (ed) Proceedings of the 18th Washington State University Particleboard Symposium, Pullman, Washington, 177-82

Neely JE and Bertone TJ (2003) *Practical metallurgy and materials of industry* (6th edit). Prentice Hall, New York

Newman RH (1997). How stiff is an individual cellulose microfibril? In B. G. Butterfield (ed), *Proceeding of the IAWA/IUFRO Workshop on Microfibril angle in wood.* University of Canterbury, Christchurch, New Zealand, 81-93

Newman RH (2005) *Solid-state NMR as a tool for studying dancing molecules.* In Entwistle KM and Walker JCF (eds), The Hemicelluloses Workshop 2005, Wood Technology Research Centre, University of Canterbury, Christchurch, New Zealand, 77-86

NHLA (1991) *An introduction to grading of hardwood lumber.* National Hardwood Lumber Association, Memphis, Tennessee

Nicholas DD (1973a) *Wood deterioration and its prevention by preservative treatments. Vol. I. Degradation and protection of wood.* Syracuse University Press, Syracuse, NY

Nicholas DD (1973b) *Wood deterioration and its prevention by preservative treatments, Vol. 2. Preservatives and preservative systems.* Syracuse University Press, Syracuse, NY

Nicholls JMP and Dadswell HE (1965) *Assessment of wood qualities for tree breeding.* CSIRO, Division of Forest Products Technological Paper No. 37

Nijdam JJ (1998) *Reducing moisture-content variations in kiln-dried timber.* PhD thesis, University of Canterbury, Christchurch, New Zealand

Nijdam JJ and Keey RB (1996) Influence of local variations of air velocity and flow direction reversals on the drying of stacked timber boards in a kiln. *Transactions of the Institution of Chemical Engineers, 74A* 882-92

Nijdam JJ and Keey RB (2002) New timber-kiln designs for promoting uniform airflows within a timber stack. *Chemical Engineering Research and Design, 80*(7) 739-44

Nishiyama Y, Chanzy H and Langan P (2002) Crystal structure and hydrogen-bonding system in cellulose 1b from synchrotron x-ray and neutron fiber diffraction. *Journal of the American Chemical Society, 124*(31) 9074-82

Nishiyama Y, Sugiyama J, Chanzy H and Langan P (2003) Crystal structure and hydrogen-bonding system in cellulose Ia from synchrotron x-ray and neutron fiber diffraction. *Journal of the American Chemical Society, 125*(47) 14300-6

Nishiyama Y, Kim U-J, Kim D-Y, Katsumata KS, May RP and Langan P (2003) Periodic disorder along ramie cellulose microfibrils. *Biomacromolecules, 4* 1013-7

NIST (1995) *Voluntary product standard PS 1-95. Construction and industrial plywood.* US Department of Commerce, National Institute of Standards and Technology, Gaithersburg, Maryland

NIST (2004) *Performance Standard for Wood-Based Structural-Used Panels, Voluntary Product Standard. PS 2-04.* US Department of Commerce, Washington, DC

Nomura Y (1980) Quinone additive cooking. *Japan TAPPI, 34*(1) 50-5

Nurmi AJ and Lindroos L (1994) *Recycling of treated timber by copper smelter.* International Research Group on Wood Preservation, Document No. 94-500-30

Nylinder P (1958) Aspects of quality production: Part 2 (in Swedish). *Skogen, 45*(23) 714 and 717-8

O'Carroll C (2001) Trends in Surfacing. *Wood Based Panels International, 21*(3) 50-1

O'Dea DJ (1983) *Forest industries economies of scale.* Ministry of Works, Wellington, New Zealand. Central North Island Forest and Transport Planning Study, Technical Paper 6

Okuma M and Lee JJ (1985) The effective use of low grade domestic plantation timber: the properties of the new structural sheet material, LVB, made from small sized veneers. *Mokuzai Gakkaishi, 31*(12) 1015-20

Olsson A-M and Salmén L (2004) The association of water to cellulose and hemicellulose in paper examined by FTIR spectroscopy. *Carbohydrate Research, 339* 813-8

Orman RH (1955) *The response of N.Z. timbers to fluctuations in atmospheric moisture condition.* New Zealand Forest Service, Forest Research Institute, Technical Paper No. 8

Ormarsson S (1999) *Numerical analysis of moisture-related distortion in sawn timber.* PhD thesis, Chalmers University of Technology, Göteborg, Sweden

Pang S, Keey RB and Walker JCF (1994) *Modelling of the high-temperature drying of mixed sap and heartwood boards.* Proceedings of 4th International Wood Drying Conference, Rotorua, New Zealand, 430-9

Pang S (1994) *High-temperature drying of Pinus radiata boards in a batch kiln.* PhD thesis, University of Canterbury, Christchurch, New Zealand

Panshin AJ and de Zeeuw C (1980) *Textbook of wood technology* (4th. edit). McGraw-Hill Book Co. New York

Parrikka M (2004) Global biomass fuel resources. *Biomass and Bioenergy, 27* 613-20

Pasek EA (2003) Minimizing preservative losses: fixation. A report of the P4 migration/ fixation/depletion task force. *Proceedings of the American Wood Preservers' Association Annual Meeting, 99*, 100-24

Pease D (1980) Platen veneer drying systems cut shrinkage and energy costs. *Forest Industries, 107*(7) 29-31

Pellerin RF and Ross RJ (2002) *Nondestructive evaluation of wood.* Forest Products Society, Madison, Wisconsin

Pentoney RE (1953) Mechanism affecting tangential vs radial shrinkage. *Forest Products Journal, 3*(2) 27-33

Peter RK (1967) Influence of sawing methods on lumber grade yield from yellow poplar. *Forest Products Journal, 17*(11) 19-24

Petty JA and Puritch GS (1970) The effects of drying on the structure and permeability of the wood of *Abies grandis. Wood Science and Technology, 4*(2) 140-54

Petty JA (1972) The aspiration of bordered pits in conifer wood. *Proceedings of the Royal Society, London, B 181*(1065) 395-406

Petty JA (1981) Fluid flow through the vessels and intervascular pits of sycamore wood. *Holzforschung, 35*(5) 213-6

Pizzi A (1989) *Wood adhesives: chemistry and technology.* Marcel Dekker, New York

Plumtre RA (1979) Simple solar heated lumber dryers: design, performance and commercial viability. *Commonwealth Forestry Review, 46*(4) 298-309

Plumtre RA (1984) *Pinus Caribaea, Volume 2: Wood Properties.* Tropical Forestry Paper 17. Commonwealth Forestry Institute, University of Oxford

Pnevmaticos SM and Bousquet DW (1972) Sawing pattern effect on the yield of lumber and furniture components from medium and low-grade maple logs. *Forest Products Journal, 22*(3) 34-41

Poliquin L (1998) *Wood-based panel products special report: a technology roadmap.* Forintek Canada Corp, Vancouver, BC

Pordage LJ and Langrish TAG (2000) Optimisation of hardwood drying schedules allowing for biological variability. *Drying Technology, 18*(8) 1797-815

Pound J (1973) *Radio frequency heating in the timber industry.* Spon, London

Preston AF (2000) Wood preservation: trends of today that will influence the industry tomorrow. *Forest Products Journal, 50*(9) 12-9

Preston RD (1974) *The physical biology of plant walls.* Chapman and Hall, London

PRL (1972) *A handbook of hardwoods.* Princes Risborough Laboratory, Building Research Establishment, HMSO, London

Pye EK and Lora JH (1991) The Alcell process: proven alternative to kraft pulping. *TAPPI, 74*(3), 113-8

Ramiah MV and Goring DAI (1965) The thermal expansion of cellulose, hemicellulose and lignin. *Journal of Polymer Science C, 11* 27-43

Rast ED, Sonderman DL and Gammon GL (1979) *A guide to hardwood log grading (revised).* USDA, Forest Service, General Technical Report NE-1

Raymond CA and Anderson DW (2005) Prior land-use influences wood properties of *Pinus radiata* in New South Wales. *New Zealand Journal of Forest Science, 35*(1) 72-90

Rayner ADM and Boddy L (1988) *Fungal decomposition of wood: its biology and ecology.* John Wiley and Sons, New York

Reiterer A, Lichtenegger H, Tschegg S and Fratzl P (1999) Experimental evidence for a mechanical function of the cellulose microfibril angle in the wood cell walls. *Philosopical Magazine, A 79*(9) 2173-84

Reiterer A, Lichtenegger H, Fratzl P and Stanzl-Tschegg SE (2001) Deformation and energy absorption of wood cell walls with different nanostructure under tensile loading. *Journal of Materials Science, 36* 4681-6

Reynolds HW and Gatchell CJ (1982) *New technology for low-grade hardwood utilisation: system 6.* USDA, Forest Service, Research Paper NE-504

Reynolds HW, Gatchell CJ, Araman PA and Hansen BG (1983) *System 6: used to make kitchen C2F blanks from small-diameter, low-grade red oak.* USDA, Forest Service, Research Paper NE-525

Richards DB, Adkins WK, Hallock H and Bulgrin EH (1980) *Lumber values from computerised simulation of hardwood log sawing.* USDA, Forest Service, Research Paper FPL 359

Richardson BA (1993) *Wood Preservation* (2nd edit). E & FN SPON, London

Richardson SD (1978) *Appropriate operational scale in forest industries.* Proceedings 8th World Forestry Conference, FAO, Rome Document FID II/21-18

Rietz RC and Page RH (1971) *Air drying of lumber: a guide to industry practice.* USDA, Forest Service, Agriculture Handbook No. 402

Rietz RC (1972) *A calendar for air-drying lumber in the upper midwest.* USDA, Forest Service, Forest Products Laboratory, Research Note FPL-0224

Riggin MT, Sharp AR, Kaiser R and Schneider MH 1979 Transverse NMR relaxation of water in wood. *Journal of Applied Polymer Science, 23* 3147-54

Rimpi PK (1983) *Chemical recovery in the first SAP-market pulp producing mill.* Proceedings 1983 Pulping Conference, Houston. TAPPI Press, Atlanta, Georgia, Vol. 1 59-61

Robards AW (1969) The effect of gravity on the formation of wood. *Science Progress, 57* 513-32

Roffael E (1993) *Formaldehyde release from particleboard and other wood-based panels.* Report 37. Malaysian Forest Research Institute, Kuala Lumpur

Rosen HN (1979) Psychrometric relationships and equilibrium moisture content of wood at temperatures above 212°F. *Symposium on wood moisture content temperature and humidity relationships.* USDA, Forest Service, Forest Products Laboratory, North Central Forest Experiment Station

Ross RJ, McDonald KA, Green DW and Schad KC (1997) Relationship between log and lumber modulus of elasticity. *Forest Products Journal, 47*(2) 89-92

Rowell RM and Banks WB (1985) *Water repellency and dimensional stability of wood.* USDA, Forest Service, Forest Products Laboratory, FPL GTR-50

Rowell RM and Konkol P (1987) *Treatments that enhance physical properties of wood.* USDA, Forest Service, Forest Products Laboratory, FPL GTR-55

Rowell RM, Imamura Y, Kowai S and Norimoto M (1989) Dimensional stability, decay resistance and mechanical properties of veneer faced low-density particleboards made from acetylated wood. *Wood and Fiber Science, 21*(1) 67-79

Rowell RM, Kawai S and Inoue M (1995) Dimensionally stabilized, very low density fibreboard. *Wood and Fiber Science, 27*(4) 428-36

Rozsa AN (1994) *Dielectric vacuum drying of hardwood.* Proceedings 4th IUFRO International Drying Conference, Rotorua, New Zealand, 271-8

Ruddick JNR (1987) *Proceeding of the incising workshop*, Richmond, British Colombia, 1986. Special Publication 28. Forinteck Canada Corp., Vancouver, BC

Ruel K, Barnoud F and Goring DAI (1978) Lamellation in the S_2 layer of softwood tracheids as demonstrated by scanning transmission electron microscopy. *Wood Science and Technology, 12* 287-91

Ruel K and Joseleau J-P (2005) *Deposition of hemicelluloses and lignins during secondary wood cell wall assembly.* In Entwistle KM and Walker JCF (eds), The Hemicelluloses Workshop 2005, Wood Technology Research Centre, University of Canterbury, Christchurch, New Zealand, 103-13

Russell WR, Forrester AR, Chesson A and Burkitt MJ (1996) Oxidative coupling during lignin polymerisation is determined by unpaired electron delocalization within parent phenylpropenoid radicals. *Archives of Biochemistry and Biophysics, 332*(2) 357-66

Russell WR, Forrester AR and Chesson A (2000) Predicting the macromolecular structure and properties of lignin and comparison with synthetically produced polymers. *Holzforschung, 54*(5) 505-10

Rydholm SA (1970) *Continuous pulping processes.* TAPPI, New York

SAA (1986a) *Timber: Softwood visually stress-graded for structural purposes.* AS2858. Standards Association of Australia, Sydney

SAA (1986b) *Timber: Classification into strength groups.* AS2878. Standards Association of Australia, Sydney

Saka S and Thomas RJ (1982) A study of lignification in loblolly-pine tracheids by the SEM-EDAX technique. *Wood Science and Technology, 16*(3) 67-79

Saka S and Goring DAI (1985) Localization of lignins in wood cell walls. In Higuchi T (ed), *Biosynthesis and biodegradation of wood components.* Academic Press, Orlando, 51-62

Saka S (2001) Wood-inorganic composites as prepared by the sol-gel process. In Hon DN-S and Shiraishi N (eds), *Wood and cellulosic chemistry.* Marcel Dekker, New York, 781-94

Sakamoto S (1987) *Veneer lathe for small diameter logs. Longitudinal veneer slicer.* FAO Wood-based Panels: Proceedings Expert Consultation, FAO, Rome, 180-2

Salin J-G (1996) *Prediction of heat and mass transfer coefficients for individual boards and board surfaces.* Proceedings 5th IUFRO Wood Drying Symposium, Quebec, Canada, 49-58

Salmén L, Dumail JF and Uhmeier A (1997) Compression behaviour of wood in relation to mechanical pulping. In *Proceedings International Mechanical Pulping Conference, SPCI* Stockholm, 207

SANS (1997) *Sawn softwood timber.* SANS1783. Standards South Africa, Pretoria

SANZ (1988) *Visual grading rules.* NZS3631. Standards Association of New Zealand, Wellington

Sarko A (1986) Recent X-ray crystallographic studies of cellulose. In Young RA and Rowell RM (eds), *Cellulose structure, modification and hydrolysis.* Interscience, New York, 29-49

Savage RA (2003) Tree growth and wood quality. In Barnett JR and Jeronimidis G (eds), *Wood quality and its biological basis.* Blackwell Publishing, Oxford and CRC Press, Boca Raton, 1-29

Scallan AM (1971) Quantitative Picture of Fringed Micellar Model of Cellulose. *Textile Research Journal, 41*(8) 647-53

Schaffer EL, Jokerst RW, Moody RC, Peters CC, Tschernitz IL and Zahn JJ (1972) *Feasibility of producing a high-yield laminated structural product: general summary.* USDA, Forest Service, Forest Products Laboratory, Research Paper FPL 175

Schaffer EL, Jokerst RW, Moody RC, Peters CC, Tschernitz IL and Zahn JJ (1977) *Press-lam: progress in technical development of laminated veneer structural products.* USDA, Forest Service, Forest Products Laboratory, Research Paper FPL 279

Schaffner RD (1991) Drying costs: a brief introduction. In *Australian Timber Seasoning Manual.* Australian Furniture Research and Development Institute, Launceston, Tasmania

Scheffer TC and Morrell JJ (1998) *Natural durability of wood: A worldwide checklist of species.* Forest Research Laboratory, Oregon State University. Research contribution 22
Schiffmann RF (1995) Microwave and dielectric drying. In Mujumdar AS (ed), *Handbook of industrial drying* (2nd. Edit). Marcel Dekker, New York, Vol. 1 345-72
Schneider MH and Witt AE (2004) History of wood polymer composite commercialisation. *Forest Products Journal, 54*(4) 19-24
Schultz TP Nicholas DD (2003) *A brief overview of non-arsenical wood preservative systems.* preservation. In Goodell B, Nicholas DD and Schultz TP (ed), *Wood deterioration and preservation: advances in our changing world.* ACS Symposium Series 845. American Chemical Society, Washington DC, 420-32
Schulz H (1993) The development of wood utilization in the 19th, 20th and 21st centuries. *The Forestry Chronicle, 69*(4) 413-8
Seale D (2004) *TimTek technology.* Conference on Wood Utilization Solutions to Hazardous Fuels. December 14-15, Spokane, Washington
Sellers T (1985) *Plywood and Adhesive Technology.* Marcel Dekker, New York
Shelbourne CJA (1997) *Genetics of adding value to the end-products of radiata pine.* In Burdon RD and Moore JM (ed), IUFRO '97 Genetics of Radiata Pine. New Zealand, Forest Research Institute Bulletin No. 203, 129-41
Shen KC (1988) *Binderless composite products.* In Burdon RD (ed) Proceedings of the Composite Wood Products Symposium. New Zealand Forest Research Institute, Bulletin Bulletin 153 105-7
Shi SQ (2001) Chapter 11: Fibre reinforced polymer (FRP) - wood hybrid composites. In Williamson TE (ed), *APA Handbook on Engineered Wood.* McGraw-Hill, New York
Shukla NK, Rajput SS, Lal M, and Khanduri AK (1994) Studies on the variation of density and strength properties from pith to periphery in *Populus deltoides. Journal of Indian Academy of Wood Science, 2*(2) 1-6
Siau JF (1984) *Transport processes in wood.* Springer-Verlag, Berlin
Sigglekow IL (1974) *A comparision of tension wood in some New Zealand trees.* MSc thesis, University of Canterbury, Christchurch, New Zealand
Simperingham P (1997) *A structural engineer's approach to forestry production.* Wood Technology Research Centre, Workshop 1997. University of Canterbury, Christchurch, New Zealand, 5p
Simpson W (1980) Sorption theories for wood. *Wood Fiber, 12*(3) 183-95
Sinclair RM and Vincent TA (1964) Yellowing of radiata pine timber. *New Zealand Journal of Science, 7*(2) 196-206
Sjoblom K, Hartler N, Mjoberg J and Sjoden L (1983) A new technique for pulping to low kappa numbers in batch pulping: results of mill trials. *TAPPI, 66*(9) 97-102
Sjoblom K, Mjoberg J, Soderqvist-Lindblad M and Hartler N (1988) Extended delignification in kraft cooking through improved selectivity. *Paperi Ja Puu-Paper and Timber, 70*(5), 452-460
Sjöström E (1981) *Wood chemistry: fundamentals and applications.* Academic Press, Orlando
Skaar C (1988) *Wood-water relations.* Springer-Verlag, Berlin
Smith D (1998) *North American MDF blending technology.* In Bradfield J (ed), Proceedings 1998 Resin and Blending Seminar. Composite Panel Association, Gaithersburg, Maryland, 71-8
Smith R, Alderman D and Araman P (2002) What's stopping the recycling of recovered CCA-treated lumber. In *Enhancing the durability of lumber and engineered wood products.* Forest Products Society, Madison, Wisconsin, 47-50
Smith WB, Smith A and Neauhauser EF (1994) *Radio-frequency/vacuum drying of red oak: energy quality value.* Proceedings 4th IUFRO International Drying Conference, Rotorua, New Zealand, 263-70

Smook GA (1982) *Handbook for pulp and paper technologists.* Tappi, Atlanta, Georgia & Canadian Pulp and Paper Association, Montreal, Quebec

So C-L, Via BK, Groom LH, Schimleck LR, Shupe TF, Kelley SS and Rials TG (2004) Near infrared spectroscopy in the forest products industry. *Forest Products Journal, 54*(3) 6-16

Solo-Gabriele H and Townsend T (1999) Disposal practices and management alternatives for CCA-treated wood waste. *Waste Management & Research, 17* 378-89

Solo-Gabriele H, Townsend T, Messick B and Calitu V (2002) Characteristics of chromated copper arsenate-treated wood ash. *Journal of Hazardous Materials, B89* 213-32

Sorenson J (1985a) Spindleless lathe enables plywood comeback. *Canadian Forest Industries, 105*(8) 33-4

Sorenson J (1985b) Rotary veneer clipper ups production, reduces noise. *Canadian Forest Industries, 105*(6) 12-6

Sorensson C, Nepveu G and Kimberley M (2004) *Intra- and inter-tree modelling of wood stiffness of clears of age-10 radiata pine clones, and simulated response of stiffness to increased genetic selection intensity.* Connection between silviculture and wood quality through modelling, Harrison Hot Springs, British Colombia (2002). IUFRO, 104-17

Spalt HA (1958) The fundamentals of water vapour sorption by wood. *Forest Products Journal, 8*(10) 288-95

Spath P and Dayton DC (2003) *Preliminary screening: technical and economic assessment of synthesis gas to fuels and chemicals with emphasis on the potential for biomass-derived syngas.* National Renewable Energy Laboratory Report Number NREL/TP-510-34929

Spelter H (1988) New panel technologies and their potential impact. In Hamel MP (ed), *Structural Wood Composites: New Technologies for Expanding Markets.* Forest Products Society, Madison, Wisconsin, 136-41

Spelter H and Sleet G (1989) Potential reductions in plywood manufacturing costs resulting from improved technology. *Forest Products Journal, 39*(1) 8-15

Springate NC and Roubicek TT (1979) Improvements in lathe and clipper production. *Modern Plywood Techniques, 7* 51-61

Stamm AJ (1959) Bound-water diffusion into wood in the fibre direction. *Forest Products Journal, 9*(1) 27-32

Stamm AJ (1964) *Wood and cellulose science.* The Ronald Press Co, New York

Stamm AJ (1967a) The movement of fluids in wood - Part 1: Flow of fluids in wood. *Wood Science and Technology, 1*(2) 122-41

Stamm AJ (1967b) Movement of fluids in wood - Part 2: Diffusion. *Wood Science and Technology, 1*(3) 205-30

Statton WO (1967) *The meaning of crystallinity when judged by X-rays.* In Price FP (ed), The meaning of crystallinity in polymers. Journal of Polymer Science Part C 33-50

Stöckmann VE (1972). Developing a hypothesis: native cellulose elementary fibrils are formed with metastable structure. *Biopolymers, 11*, 251-270

Stone JE and Scallan AM (1968) The effect of component removal upon the porous structure of the cell wall of wood. *Pulp and Paper Magazine of Canada, 69*(6) T288-93

Su N-Y and Scheffrahn RH (1991) *Population suppression of subterranean termites by slow-acting toxicants.* In Haverty MI and Wilcox WW (eds), Proceeding of the symposium on current research on wood-destroying organisms and future prospects for protecting wood in use, Bend, Oregon, 1989. USDA, Forest Service, General Technical Report PSW-128

Suchsland O and Woodson GE (1986) *Fiberboard manufacturing practices in the United States*: USDA, Forest Service, Agriculture Handbook No. 640

Suckling ID, Pasco MF and Gifford JS (1993) *Origin of effluent toxicity from a simulated BCTMP process.* Tappi Environmental Conference, Boston, Vol. 1 449

Sugiyama J and Imai T (1999) Aspects of native cellulose microfibrils at molecular resolution. *Trends in Glycoscience and Glycotechnolgy, 57*(11) 23-31

Sundholm J (1999) *Mechanical pulping.* Papermaking Science and Technology, Vol. 5. Fapet Oy, Helsinki

Sunley JG (1968) *Grade stresses for structural timbers.* Forest Products Research Bulletin, No. 47. HMSO, London

Sunley JG (1979) Timber engineering: the international scene. *Journal of the Institute of Wood Science, 8*(3) 121-6

Sutton WRJ (1970) *Effect of initial spacing on branch size.* Symposium No. 12. Pruning and thinning practice. New Zealand Forest Service, Forest Research Institute, Vol. 2 106-7

Sutton WRJ (1984) *New Zealand experience with radiata pine.* HR MacMillan Lecture, February 1984, University of British Colombia, Vancouver, BC

Swartz JN, Muhonen JM, LaMarche LJ, Hambaugh PC and Richter FH (1969) Alkaline pulping. In Macdonald RG and Franklin JN (eds), *Pulp and paper manufacture. Vol. 1 The pulping of wood.* McGraw-Hill Book Co., New York, 439-575

Tabor D (1971) *Gases, liquids and solids.* Penguin Books, Harmondsworth, England

Tamblyn NE (1978) Preservation and preserved wood. In Hillis WE and Brown AG (eds), *Eucalypts for wood production.* CSIRO, Australia, 343-52

Tarkow H and Krueger J (1961) Distribution of hot-water soluble material in cell wall and cavities of redwood. *Forest Products Journal, 11*(5) 228-9

Tarkow H (1971) On a reinterpretation of anomalous moisture adsorption isobars below 0°C. *TAPPI, 54*(4) 593

Thamburaj R and the Dynamic Energy Systems Corporation (2000) *Fast pyrolysis of biomass for green-power generation.* First World Conference and Exhibition on Biomass for Energy and Industry

Thomas RJ and Kringstad KP (1971) The role of hydrogen bonding in pit aspiration. *Holzforschung, 25*(5), 143-9

Thorbjornsson S (1985) *Effects on mat density and springback in multi-nip precompressors.* In Maloney T (ed), Proceedings of the 19th Washington State University Particleboard Symposium, Pullman, Washington, 451-64

Timell TE (1967) Recent progress in the chemistry of wood hemicelluloses. *Wood Science and Technology, 1*(1) 45-70

Timell TE (1986) *Compression wood in Gymnosperms: Vols 1-3.* Springer-Verlag, Berlin

Tjeerdsma BF, Boonstra M, Pizzi A, Tekely P and Militz H (1998) Characterisation of thermally modified wood: molecular reasons for wood performance improvement. *Holz als Roh- und Werkstoff, 56*, 149-53

Todd KL (1982) *An evaluation of potential markets for New Zealand softwood plywood exports.* MForSc, University of Canterbury, Christchurch, New Zealand

Tokoh C, Takabe K, Fujita M and Saiki H (1998) Cellulose synthesized by *Acetobacter xylinun* in the presence of acetyl glucomannan. *Cellulose, 5* 249-61

TRADA (1974) *Visual stress grading of timber.* Timber Research and Development Association, High Wycombe, UK

Treybal RE (1980) *Mass-transfer operations*, p. 23. McGraw-Hill, New York

Trindade SC and Carvalho AV de (1989) Transportation fuels policy issues and options: the case of ethanol fuels in Brazil. In Spirling D (ed), *Alternative transportation fuels: an environmental and energy solution.* Quorum Books, New York, 163-85

Turner GPA (1967) *Introduction to paint chemistry.* Chapman and Hall, London

Turner IW and Ferguson WJ (1995) A study of the power density distribution generated during the combined microwave and convective drying of softwood. *Drying Technology, 13* 1411-30

Tyree MT and Dixon MA (1983) Cavitation events in *Thuja occidentalis* L.? Ultrasonic acoustic emissions from the sapwood can be measured. *Plant Physiology, 72*(4) 1094-9

Ulrich AH (1988) *U.S. Timber production, trade, consumption, and price statistics 1950-86.* USDA, Forest Service, Miscellaneous Publication 1460

Ung TY and Cooper PA (2005) Copper stabilization in ACQ-D treated wood: retention, temperature and species effects. *Holz als Roh- und Werkstoff, 63* 186-91

Uprichard JM (1962) The extractives content of New Zealand grown larch species (*Larix decidua* and *Larix leptolepis*). *Holzforschung, 17*(5) 129-34

Uprichard, JM (1971) Cellulose and lignin content in *Pinus radiata* within-tree variation in chemical composition, density, and tracheid length. *Holzforschung, 25*(4) 97-105

Uprichard JM and Lloyd JA (1980) Influence of tree age on the chemical composition of radiata pine. *New Zealand Journal of Forestry Science, 10*(3) 551-7

Uprichard JM and Burton RJ (1982) *Ethanol from wood.* Proceedings, 5th International Alcohol Fuel Technology Symposium, Auckland, Vol. 1 317-24

Uprichard JM and Okayama T (1984) Neutral sulphite-anthraquinone pulping of corewood and slabwood samples of new crop *Pinus radiata. TAPPI, 37*(7), 560-75

Uprichard JM (2002) *Pulp and paper from radiata pine.* Appita and Forest Research.

US Patent (2000) Log cutting optimisation system (Weyerhaeuser Inc.). No. 6026689

USDA (1965) *Western wood density survey: report No. 1.* USDA, Forest Service, Forest Products Laboratory, Research Paper FPL 27

USDA (1987) *Wood handbook: wood as an engineering material.* USDA, Agricultural Handbook No. 72. US Government Printing Office, Washington DC

USDA (1990) *Analysis of the timber situation in the United States: 1989-2040.* USDA, Forest Service, Rocky Mountain Forest and Range Experimental Station, General Technical Report RM-199

USDA (1999) *Wood Handbook: Wood as an engineering material.* USDA, Agricultural Handbook No. 72, US Government Printing Office, Washington DC

USDC (2005) *American softwood lumber standard.* Product Standard PS20-05. US Department of Commerce, Washington, DC

Uzonovic A, Byrne A, Dian-Qing Yang and Morris PI (2003) *Review of mold issues in North America and mold research at Forintek.* IRG Document, IRG/WP 03-10458

Vajda P (1976) A comparative evaluation of the economics of particleboard and fibreboard manufacture. In *Non-wood plant fibre pulping progress report No. 7.* Tappi, Atlanta, Georgia, pp. 93-111

Valadez L (1990) Avoid saws' critical speeds to improve cutting accuracy. *Forest Industries, 117*(9) 21-2

Viëtor RJ, Newman RH, Ha M-A, Apperley DC and Jarvis MC (2002). Conformational features of crystal-surface cellulose from higher plants. *The Plant Journal, 30*(6) 721-31

Vinden P, Burton R, Bergervoet A, Nasheri K and Page D (1990) *Vapour boron treatment.* What's New in Forest Research, No. 200. New Zealand Ministry of Forestry, Forest Research Institute

Vinden P and Torgnikov G (2002) *Microwave conditioning of wood for combined drying and preservative treatment.* Timber Preservation 2002: technology and product opportunities to improve performance. Forest Industry Engineering Association, Rotorua, New Zealand, 7 p

Vinden P and Drysdale J (1990). *Thickened boron "diffuse" - a new approach to a traditional treatment.* What's New in Forest Research No. 193. New Zealand Ministry of Forestry, Forest Research Institute

Von Wedel KW, Zobel BJ and Shelbourne CJA (1968) Prevalence and effect of knots in young loblolly pine. *Forest Products Journal, 18*(9) 97-103

Vroom KE (1957) The H Factor: the means of expressing cooking times and temperatures as a single variable. *Pulp and Paper Magazine of Canada, 58*(3) 228-31

Waddell KL, Oswald DD and Powell DS (1989). *Forest statistics of the United States, 1987.* USDA, Forest Service, Pacific Northwest Research Station, PNW RB-168

Wadsworth J (2005) New investment returns to the western MDF industry. *Wood Based Panels International, 3* 13-8

Walker JCF (1984) New forests: new but predetermined opportunities. *Commonwealth Forestry Review, 63*(4) 255-62

Walker JCF (1993) *Primary wood processing: principles and practice.* Chapman and Hall, London

Walker JCF and Nakada R (1999) Understanding corewood in some softwoods: a selective review on stiffness and acoustics. *International Forestry Review, 1*(4) 251-9

Walker JCF (2004) Good wood and blind prejudice. *New Zealand Journal of Forestry, 49*(2) 36-7

Walser DC (1978) New developments in veneer peeling. *Modern Plywood Techniques, 6* 6-18

Wang X, Ross RJ, Mattson JA, Erickson JR, Forman JW, Geske EA and Wehr MA (2001) *Several non-destructive evaluation techniques for assessing stiffness and MOE of small diameter logs.* USDA, Forest Service, Forest Products Laboratory, Research Paper RP-600

Wang X, Ross RJ, Bradshaw BK, Punches JRJ, Erickson JR, Forman JW and Pellerin RF (2004) Diameter effect on stress-wave evaluation on modulus of logs. *Wood and Fiber Science, 36*(3) 368-77

Wardrop AB and Dadswell HE (1950) The swelling behaviour of conifer tracheids and the concept of a skin substance. *Proceedings APPITA, 4* 198-226

Watson AJ and Dadswell HE (1962) Influence of fibre morphology on paper properties. Part 2, Early wood and late wood. *Appita, 15*(6) 116-28

Wayman M and Parekh SR (1990) *Biotechnology of biomass conversion.* Prentice Hall, Englewood Cliffs, New Jersey

Weatherwax RC and Tarkow H (1968) Density of wood substance: importance of penetration and adsorption compression of the displacement fluid. *Forest Products Journal, 18*(7) 44-6

Wellons JD, Krahmer RL, Sandoe MD and Jokerst RW (1983) Thickness loss in hot-pressed plywood. *Forest Products Journal, 33*(1) 27-34

Welzbacher CR and Rapp AO (2002) *Comparison of thermally modified wood originating from four industrial scale processes - durability.* International Research Group on Wood Preservation. Document No. IRG/WP 02-20262

Wengert EM (1979) Lumber predrying: a timely examination. *Timber Process Industries, 4*(9) 20-24

Weyman M and Parekh SR (1986) *Biotechnology of biomass conversion.* Prentice-Hall, Englewood Cliff, NJ

Wilkie KCB (1970) Polysaccharides in normal wood tissue. *Supplement to Forestry* (The wood we grow), 36-45

Williams DH and Kininmonth JA (1984) *High-temperature kiln drying of radiata pine sawn timber.* New Zealand Forest Service, Forest Research Institute Bulletin 73

Williams LH (1990) *Potential benefits of diffusible preservatives for wood protection: an analysis with emphasis on building timbers.* First International Conference on Wood Protection with Diffusible Preservatives. Forest Products Research Society, 29-34

Williamson TG (ed) (2002) *APA Engineered Wood Handbook.* McGraw-Hill, New York, NY

Williston EM (1981) *Small log sawmills: profitable product selection, process design and operation.* Miller Freeman, San Francisco

Williston EM (1988) *Lumber manufacturing: the design and operation of sawmills and planar mills.* Miller Freeman, San Francisco

Williston EM (1989) *Saws: design, selection, operation, maintenance* (2nd Edit.). Miller Freeman, San Francisco

Wilson BF and Archer RR (1979) Tree design: some biological solutions to mechanical problems. *Bioscience, 29*(5) 293-7

Wilson K, and White DJB (1986) *The anatomy of wood: its diversity and variability.* Stobart and Son Ltd., London

Winandy JE (1995a) *Effects of waterborne preservative treatment on mechanical properties: A review.* Proceedings, 91st annual meeting of American Wood Preservers' Association, New York, *91* 17-33

Winandy JE (1995b) The influence of time-to-failure on the strength of CCA-treated lumber. *Forest Products Journal, 45*(2) 82-5

Winandy JW (1995c) Effects of moisture content on strength of CCA-treated lumber. *Wood and Fiber Science, 27*(2) 168-77

Winandy JE and Morrell JJ (1998) Effects of incising on lumber strength and stiffness: relationships between incision density and depth, species, and MSR grade. *Wood and Fiber Science, 32*(2) 185-87

Wingate-Hill R and MacArthur IJ (1991) Debarking small-diameter eucalypts. In Kerruish CM and Rawlins WHM (eds), *The Young Eucalypt Report.* CSIRO, Melbourne, Australia, 107-51

Wolcott MP and Englund K (1996) *A technical review of wood-plastic composites.* In Maloney T (ed) Proceedings of the 33rd Washington State University Particleboard Symposium, Pullman, Washington, 117-43

Wolcott MP (2004) *Engineered wood-plastic composites.* Conference on Wood Utilization Solutions to Hazardous Fuels. December 14-15. Spokane, WA

Wood JR and Goring DAI (1973) The distribution of lignin in fibres produced by kraft and acid sulphite pulping of spruce wood. *Pulp and Paper Magazine of Canada, 74*(9), 117-21 (T309-13)

Wood LW (1951) *Relation of strength of wood to duration of load.* USDA, Forest Service, Forest Products Laboratory, Report No. 1916

Woodfin RO (1973) Wood losses in plywood production - four species. *Forest Products Journal, 23*(9) 98-106

Woods B and Calnan CD (1976) Toxic woods. *British Journal of Dermatology, 95* (suppl 13)

Wu J, Fukazawa K and Ohtani J (1992) Distribution of syringyl and guaiacyl lignins in hardwoods in relation to habitat and porosity form in wood. *Holzforschung, 46* 181-5

Wu Q (1989) *An investigation of some problems in drying of Tasmanian eucalypt timbers.* MEngSci thesis, University of Tasmania, Hobart

Wyk van JL (1990) *An overview of value added strategies* FIE Conference: technology towards 2000, Auckland. Forest Industries Engineering Association, New Zealand

Xu P (2000) *The mechanical properties and stability of radiata pine structural timber.* PhD thesis, University of Canterbury, Christchurch, New Zealand

Xu P and Walker JCF (2004) Stiffness gradients in radiata pine trees. *Wood Science and Technology, 38*(1) 1-9

Yamamoto H (1998) Generation mechanism of growth stresses in wood cell walls: roles of lignin and cellulose microfibril during cell wall maturation. *Wood Science and Technology, 32* 171-82

Yamamoto H, Kojima Y, Okuyama T, Abasolo WP and Gril J (2002) Origin of the biomechanical properties of wood related to the fine structure of the mulit-layered cell wall. *Journal of Biomechanical Engineering, 124*(4) 432-40

Yang JL and Evans R (2003) Prediction of MOE of eucalypt wood from microfibril angle and density. *Holz als Roh- und Werkstoff, 61* 440-52

Ying L, Kretschmann DE and Bendtsen BA (1994) Longitudinal shrinkage in fast-grown loblolly pine plantation wood. *Forest Products Journal, 44*(1) 58-62

Yokota T and Tarkow H (1962) Changes in dimension on heating green wood. *Forest Products Journal, 12*(1) 43-5

Young RA (1986) Structure, swelling and bonding of cellulose fibers. In Young RA and Rowell RM (ed), *Cellulose: structure, modification and hydrolysis.* Wiley-Interscience, New York, 91-128

Young S (2005) *Emissions from wood-based products: a world update on regulations and trends.* Proceedings of the Asia-Pacific Wood-Based Panels Conference, Shanghai, Paper 15

Zabel RA and Morrell JJ (1992) *Wood microbiology: decay and its prevention.* Academic Press, Inc., San Diego, California

Zelinka SL and Rammer DR (2005) *Review of test methods used to determine the corrosion rate of metals in contact with treated wood.* USDA, Forest Service, Forest Products Laboratory, General Technical Report FPL-GTR-156

Zobel BJ (1963) *Not sighted.* World Consultation on Forest Genetics and Tree Improvement, Section 7, Stockholm

Zobel BJ (1975) Using the juvenile wood concept in the southern pines. *South Pulp and Paper Manufacturers, 38*(9) 14-6

Zobel BJ and van Buijtenen JP (1989) *Wood variation: its causes and control.* Springer-Verlag, Berlin

Zobel BJ and Jett JB (1995) *Genetics of wood production.* Springer-Verlag, Berlin

INDEX

MIX
Papier aus verantwortungsvollen Quellen
Paper from responsible sources
FSC® C105338

If you have any concerns about our products,
you can contact us on
ProductSafety@springernature.com

In case Publisher is established outside the EU,
the EU authorized representative is:
Springer Nature Customer Service Center GmbH
Europaplatz 3, 69115 Heidelberg, Germany

Printed by Libri Plureos GmbH
in Hamburg, Germany